Eiichi Bannai, Etsuko Bannai, Tatsuro Ito, and Rie Tanaka
Algebraic Combinatorics

De Gruyter Series in Discrete Mathematics and Applications

Volume 5

Eiichi Bannai, Etsuko Bannai, Tatsuro Ito, and
Rie Tanaka

Algebraic Combinatorics

DE GRUYTER

Authors
Prof. Dr. Eiichi Bannai
Professor Emeritus of Kyushu University
Home address: Asagaya-minami 3-2-33,
Suginami-ku
166-0004 Tokyo
Japan
bannai@math.kyushu-u.ac.jp

Prof. Dr. Etsuko Bannai
Retired from Kyushu University
Home address: Asagaya-minami 3-2-33,
Suginami-ku
166-0004 Tokyo
Japan
et-ban@rc4.so-net.ne.jp

Prof. Dr. Tatsuro Ito
Anhui University
111 Jiulong Road
230601 Hefei
People's Republic of China
tito@staff.kanazawa-u.ac.jp

Dr. Rie Tanaka
Part-time lecturer
hrie196560@gmail.com

ISBN 978-3-11-062763-3
e-ISBN (PDF) 978-3-11-063025-1
e-ISBN (EPUB) 978-3-11-062773-2
ISSN 2195-5557

Library of Congress Control Number: 2020948103

Bibliographic information published by the Deutsche Nationalbibliothek
The Deutsche Nationalbibliothek lists this publication in the Deutsche Nationalbibliografie; detailed bibliographic data are available on the Internet at http://dnb.dnb.de.

Typesetting: VTeX UAB, Lithuania
Printing and Binding: LSC Communications, United States

www.degruyter.com

Preface

This is the translation of the book *Introduction to Algebraic Combinatorics*, published by Kyoritsu Shuppan (Kyoritsu Publisher), Tokyo, Japan in 2016, by Eiichi Bannai, Etsuko Bannai, and Tatsuro Ito, written in Japanese. In making the English version, there were two important changes: The title was changed to *Algebraic Combinatorics*, dropping the "Introduction" from the original, because this will perhaps describe the nature of the book more accurately. Another was that Rie Tanaka joined the group of authors, and she took most of the job of translation from Japanese.

It is explained in the preface to the original Japanese version why and how this book was written, and so we will not repeat the details. But let us mention that this book is not exactly the *Algebraic Combinatorics II* that is a sequel to *Algebraic Combinatorics I* by Bannai and Ito. It was written as a kind of preparation for Part II, at least in our minds. The authors are getting old, and Part II may not be completed in the desired way. But we will endeavour, keeping optimistic that younger generations will follow us and develop "Algebraic Combinatorics" beyond our personal dreams. We would be very happy if this book could serve the purpose.

Here, we would like to add some personal comments by each of the authors of this book.

Eiichi Bannai

I dreamed to be a mathematician since the time of junior high school. So I was very lucky to have been able to do mathematics all the way up to now, even after retirement. I was first interested in discovering a new finite simple group, but it turned out that I came too late. Also, the classification problem of finite simple groups was too difficult and too complicated for me, although I felt strongly that this is the most important problem. So, I was interested in studying combinatorial objects from the perspective of and with the underlying philosophy of finite group theory. With Tatsuro Ito, we wrote the book *Algebraic Combinatorics I* (1984) to emphasize this philosophy. Actually, there was a very strong flow of thought in this direction by many researchers in many parts of the world somehow independently and also somehow related to each other. I feel I was also lucky that we could take part of this flow of "Algebraic Combinatorics." I also feel that I am lucky that I still can basically believe in this flow, including the contents of this new book. Finally, I would like to thank all the individuals who helped and supported me to do mathematics, in various capacities: parents, teachers, friends, fellow researchers and coauthors, many former colleagues and students, and all my family members. I regret that many of these people have already passed away or we have lost contact before I could express my personal gratitude to them.

Etsuko Bannai

I think I have come a long way. When I was a middle school child, I liked solving problems in elementary geometry. At the same time I read about Marie Curie and I wanted

https://doi.org/10.1515/9783110630251-201

to be a chemist like her. I entered University of Tokyo hoping to have such kind of career, and majored in chemistry. Then I met Eiichi. He was studying mathematics. We read the book *Group theory* (by Asano and Nagao, written in Japanese) together with some other fellow students. That is how I became interested in mathematics. I entered graduate school at Tokyo Metropolitan University to study mathematics. Around that time Eiichi took a job at Ohio State University and we went to Columbus, Ohio (we were already married). After a while, I entered the Graduate School of Ohio State University, and got my Ph. D. under John Hsia. My Ph. D. thesis was on quadratic forms. At that time, Eiichi and Tatsuro were writing *Algebraic Combinatorics I*. I also knew that they had a strong wish to write *Algebraic Combinatorics II*. After we returned to Japan, Eiichi and I wrote *Algebraic Combinatorics on Spheres* (in Japanese). Then Eiichi, Tatsuro, and I wrote the Japanese version of this book. I knew that many people wanted to have an English version of this book. I am very happy that Rie joined us and did great work to publish this English version. I have known her for many years. I feel I have come a long way. I would like to thank all the people around us who helped us to materialize this book.

Tatsuro Ito

I stayed at Ohio State University for a year twice, invited by Eiichi. It was fruitful visits. I discussed mathematics with Eiichi every day at university and played Go with him every weekend at his home after Etsuko prepared dinner. The first visit (1980–1981) led to the publication in 1984 of *Algebraic Combinatorics I*. Part II ought to have shortly followed the second (1986–1987): we prepared manuscripts that roughly covered Chapters 3 and 4 of this book, but they were not completed. Then came the 1990s, and we were occupied with other projects separately: in 1989, Eiichi moved to Kyushu University to start a new life in Fukuoka, and I went to Osaka Kyoiku University and then in 1997 to Kanazawa University for a new life in Osaka and then in Kanazawa. I believe that this book is the best we can offer as what should follow Part I at present, and I am glad to see the English version, owing to the enormous efforts of Rie. A Japanese writer wrote that there are two kinds of adventurers: the historian type and the geographer type. The historian type of people, for example, visit the same restaurant repeatedly, and the geographer type as many restaurants as possible. If we make a wild application of his theory to mathematics and mathematicians, there are two kinds of them: the converging type and the diverging type. Algebraic combinatorics includes both: it is a young branch of mathematics and it tends to diverge or converge as we do. Readers will find typical diverging features in Chapter 5 and converging ones in Chapter 6.

Rie Tanaka

This is my first experience to translate a book. (I did not translate the whole book; Eiichi Bannai and Etsuko Bannai translated Chapter 5 and Tatsuro Ito translated the last subsection of Chapter 6.) When I read the original Japanese version, my curiosity was stimulated so much that I became eager to challenge some of the open problems introduced in the book. I hope that this English version also stimulates many readers to study algebraic combinatorics. I would like to thank Hajime Tanaka for his valuable linguistic advice and for pointing out some mathematical details in this book. He also undertook the laborious task of converting the figures, which were embedded in the .tex file of the original Japanese version, into .eps files. I would also like to thank the authors of the original Japanese version to give me this valuable opportunity.

To conclude, we would like to thank those who helped us to prepare the English version. First, we would like to thank Kyoritsu Shuppan for giving us official permission to use the original Japanese version freely. We thank Hajime Tanaka, who greatly helped Rie Tanaka to prepare the English version. We thank Da Zhao, Yan Zhu, and Wei-Hsuan Yu for checking the manuscript of the English version. We would like to thank Yaokun Wu and Tullio Ceccherini-Silberstein, who helped us to find the publisher of the English version.

September 2020
Eiichi Bannai
Etsuko Bannai
Tatsuro Ito
Rie Tanaka

Preface to the Japanese version

The purpose of this book is to give an introduction to algebraic combinatorics.

There is no explicit definition of algebraic combinatorics. In *Algebraic Combinatorics I* (1984), written by Eiichi Bannai and Tatsuro Ito, algebraic combinatorics is described as "a group theory without groups" or "a character theoretical study of combinatorial objects." Specifically speaking, we pursue the study of combinatorics as an extension or a generalization of the study of finite permutation groups. In this book, we keep this direction in our mind. This is also the approach, initiated by Philippe Delsarte, which enables us to look at a wide range of combinatorial objects such as graphs, codes, and designs from a unified viewpoint. Based on these thoughts, we explain Delsarte's theory and its various extensions.

When we finished writing *Algebraic Combinatorics I*, we wanted to write the sequel *Algebraic Combinatorics II* soon. We regret we did not write it up at that time, but due to various reasons, we could not. We would not say that it has nothing to do with our laziness, but if we could use an excuse, it seems that the developments in the field are not enough to complete a book, and the range of mathematical objects that we are interested in has expanded too widely to handle. Therefore, we thought we could and should write on algebraic combinatorics on spheres, and in 1999, Eiichi Bannai and Etsuko Bannai published *Algebraic Combinatorics on Spheres*, written in Japanese. We did not prepare an English edition because we planned to write *Algebraic Combinatorics II* in the future, and we regard the above book as a preparation for it. In the present book, we want to present another subject, Delsarte's theory in association schemes. Besides, we would like to start writing on the classification of P- and Q-polynomial association schemes, which is the most important problem in *Algebraic Combinatorics I*, through the study of Terwilliger algebras by Terwilliger and Ito. In this sense, writing up this book enabled us to work seriously on *Algebraic Combinatorics II*. Before we know it, we get older. Counting the remaining time in our lives, we would like to write one more work.

We give an overview on the contents of this book in the following. Chapter 1 is the introduction to classical combinatorics. The contents are suitable for undergraduate courses. In Japan, there are not so many universities offering lectures on combinatorics for undergraduates. Therefore, we selected basic subjects which beginners in combinatorics should learn. The contents are slightly long for a one-semester course. Chapter 2 is the introduction to association schemes. Some contents overlap with *Algebraic Combinatorics I* (and with *Algebraic Combinatorics on Spheres*). We start with the basics, so the chapter is comprehensive for readers who are not familiar with this area. Chapter 3 is the introduction to Delsarte's theory, which is the theory of codes and designs in association schemes. The description is faithful to the original paper by Delsarte (1973). The contents up to this chapter are understandable to undergraduate students. Chapter 4 is the extension of Delsarte's theory. Our aim is to introduce

https://doi.org/10.1515/9783110630251-202

several papers by Delsarte. The contents are not so difficult compared to the previous chapter, but there remain several parts where we wonder if we make the best use of our "ingredients." In Chapter 5, away from association schemes, we study finite sets (and algebraic combinatorics) on spheres or on other spaces. As for the case of spheres, Delsarte–Goethals–Seidel (1977) is the starting point of the study in this direction, and the main content of this chapter is to introduce this theory. Of course, some contents of this chapter overlap with *Algebraic Combinatorics on Spheres*. In some parts, we refer to the above book to avoid duplication. Chapter 5 also contains a survey of the recent works by Eiichi Bannai and Etsuko Bannai and the topics they are recently interested in. The reason to include this subject is to show the similarity of the progress in the study of algebraic combinatorics on association schemes with that on spheres, which is what we think is the essence of algebraic combinatorics.

In Chapter 6, we return to association schemes. As is seen in *Algebraic Combinatorics I* and also in the previous chapters, Bose–Mesner algebras play an important role in the study of association schemes. We explain the importance of Terwilliger algebras, which are the refined concept of Bose–Mesner algebras, and the role they play in the study of association schemes in the past and the future. This chapter is written by Tatsuro Ito himself on years of research by Terwilliger and him, and the contents are very original. The main theme of Chapter 6 is the classification of P- and Q-polynomial schemes, which overlaps with Chapter 3 of *Algebraic Combinatorics I* in many parts. We revised the representation theory of Bose–Mesner algebras, which was discussed in the above book, by using principal representations of Terwilliger algebras. Therefore, Leonard's theorem appearing in *Algebraic Combinatorics I*, which claims that dual systems of orthogonal polynomials are Askey–Wilson polynomials, is reformulated into the theorem on the classification of Leonard pairs (L-pairs). While the proof of the theorem requires many tools and becomes longer than the direct proof in *Algebraic Combinatorics I*, the logic used here is much clearer. The concept of L-pairs is obtained by generalizing principal representations of Terwilliger algebras, and a much wider concept called tridiagonal pairs (TD-pairs) arises from general irreducible representations of Terwilliger algebras. In fact, the classification of TD-pairs is completed as they become some sort of tensor products of L-pairs. (Some of the results are still unpublished.) In this book, however, we do not cover this topic because it requires preparations for the representation theory of the quantum affine algebra $U_q(\widehat{\mathfrak{sl}}_2)$ and the contents would be too long. If the reader is interested in this topic, we recommend the original paper [260]. We introduce basic facts with proofs on TD-pairs which will be needed to read the original paper. In Section 6.4 of Chapter 6, we give the list of known P- and Q-polynomial schemes, where we add Hemmeter schemes, Ustimenko schemes, and twisted Grassmann schemes as new families. In the last section, the author (T. Ito) states his personal opinion on the present situation and the prospect toward the classification of P- and Q-polynomial schemes.

As we mentioned above, the writing styles of Chapter 6 and the other chapters are quite different. The authors are sure that algebraic combinatorics is located at the junction of the two streams. In this sense, with confidence, we put them together in this book.

Finally, we would like to thank many people who supported us during the preparation of this book. Actually, it was about 20 years ago when we were requested to write this book. While we thought we could write anytime, two decades have passed. We apologize to the publisher and editors (especially Kei Akagi) for the inconvenience. We could have written this book several years ago, and we recognize that we should have completed the book earlier. It may sound like an excuse, but the environments around the authors have drastically changed during this period. In 2008 Etsuko Bannai and in 2009 Eiichi Bannai retired from Kyushu University. In 2011, Eiichi Bannai got a position at Shanghai Jiao Tong University and moved to Shanghai with Etsuko Bannai. Tatsuro Ito retired from Kanazawa University in 2014, and got a position at Anhui University in China. We are trying to develop algebraic combinatorics in new environments. We would like to thank many people who gave us the opportunity to study in China, especially the following people: Zhexian Wan, Hao Shen, Yaokun Wu, Yangxian Wang, Suogang Gao, Michel Deza, Paul Terwilliger, Xiaodong Zhang, Yuehui Zhang, and Yizheng Fan.

It has been a long time since we started writing this book. During this period, in 2011 Nagayoshi Iwahori and in 2015 Noboru Ito passed away. We are sorry that we cannot show them this book. We do not know for how long we can continue, but we would like to study mathematics for as long as possible.

Finally, we would like to thank the people who read the draft and made various comments. Especially, we are grateful to Ryuzaburo Noda, Masaaki Harada, Makoto Tagami, Hajime Tanaka, and Tsuyoshi Miezaki. We would also like to thank Takamichi Ookoshi, the editor, for helping with proofreading.

April 2016
Eiichi Bannai
Etsuko Bannai
Tatsuro Ito

Contents

1 Classical design theory and classical coding theory

The purpose of this chapter is to give a brief introduction on theory of designs and codes, which is a starting point for algebraic combinatorics. The contents of this chapter are standard and appropriate for beginners at combinatorics, and do not contain any new approach. This chapter is based on the lecture titled *Introduction to Combinatorics* given by one of the authors for undergraduate students as a half-year course. The reader who is familiar with this topic can read through this chapter quickly.

1.1 Introduction to graph theory

First, we introduce graph theory before studying design theory.

Graphs

A *graph* is a pair (V, E) of a set V and a set E, where V is called the *vertex* set and E the *edge* set. We give examples of graphs below. The shapes of the plane figures does not matter. What matters is whether vertices are joined or not. A graph is said to have a *loop* if there is an edge joining $x \in V$ to itself. A graph is said to have *multiple edges* if there exist vertices $x, y \in V$ joined by more than one edge. A graph is called a *directed graph* if each edge has a direction from one vertex to another. See Figure 1.1, Figure 1.2 and Figure 1.3.

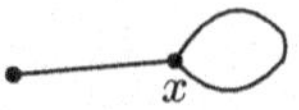

Figure 1.1: Loop.

Figure 1.2: Multiple edge.

Figure 1.3: Directed graph.

A graph is called a *simple graph* if it does not have loops or multiple edges and is not directed. We consider simple graphs in what follows. A graph $\Gamma = (V, E)$ is *regular* if the number of edges coming from a vertex does not depend on the choice of a vertex. The number of edges coming from each vertex of a regular graph is called the *degree*. All of the graphs in the following figure consist of 6 vertices of degree 3.

By renumbering the vertices, we can verify that the graph in Figure 1.4 is the same as the graph in Figure 1.6, and that the graph in Figure 1.5 is the same as the graph in Figure 1.7.

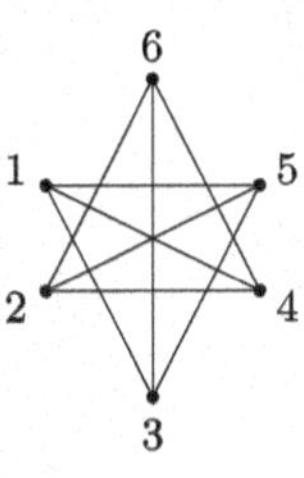

Figure 1.4

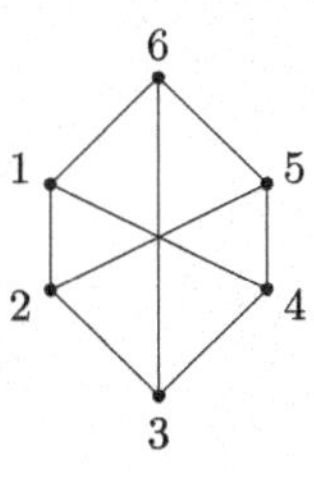

Figure 1.5

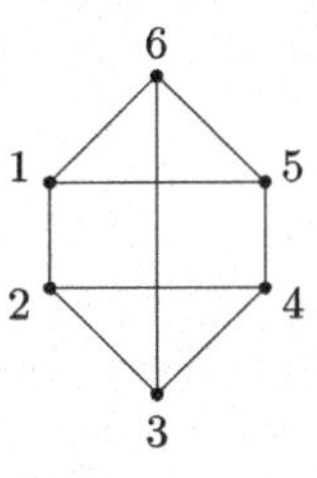

Figure 1.6

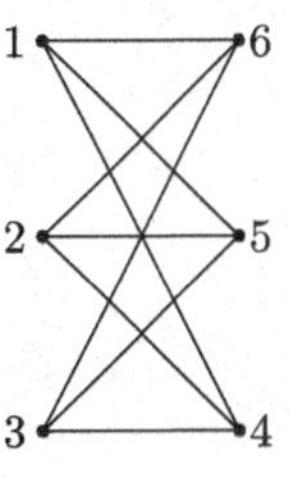

Figure 1.7

https://doi.org/10.1515/9783110630251-001

There does not always exist a regular graph with given number of vertices and degree. For example, there is no regular graph with 7 vertices of degree 3. Since 3 edges come from each vertex, $3 \cdot 7$ must be twice the number of edges, which is impossible.

What kinds of simple graphs can exist? What kinds of simple graphs are interesting in terms of combinatorial aspects? We consider these things toward the classification of simple graphs. As in the examples, we have observed that two graphs which look different are identical by a suitable transformation. We need to define when two graphs are identical mathematically. We have been vague about the definition of graphs. Here we give a slightly more precise definition of them. If there is an edge $e \in E$ joining 2 vertices $x_1, x_2 \in V$, then x_1, x_2 are called the *endpoints* of e. We say e is a loop when $x_1 = x_2$. When we consider a simple graph, we can think of the edge set E as a subset of the set consisting of 2-element subsets of V.

Definition 1.1. Two graphs (V_1, E_1) and (V_2, E_2) are defined to be isomorphic and denoted as $(V_1, E_1) \cong (V_2, E_2)$ if they satisfy the following conditions (1) and (2).
(1) There exist bijections $\varphi_V : V_1 \longrightarrow V_2$ and $\varphi_E : E_1 \longrightarrow E_2$.
(2) For $e_1 \in E_1$, let x_1, x_2 be the endpoints of e_1. Then $\varphi_V(x_1), \varphi_V(x_2)$ are the endpoints of $\varphi_E(e_1)$.

In what follows, consider the edge set E of a simple graph as a subset of $\{\{x, y\} \mid x, y \in V, x \neq y\}$.

Adjacency matrices of graphs

When we study a graph $\Gamma = (V, E)$, the use of algebra is effective. Defining the following matrix enables algebraic analysis of graphs. Let v be the number of vertices and write $V = \{1, 2, \dots, v\}$. The square matrix $A = (a_{i,j})$ of size v which is defined by the following equation is called the *adjacency matrix* of $\Gamma = (V, E)$:

$$a_{i,j} = \begin{cases} 1, & \text{if } \{i, j\} \in E, \\ 0, & \text{otherwise.} \end{cases}$$

The adjacency matrix A of a simple graph is a real symmetric matrix. When we study combinatorics, our interest is to find a set of finite points with a good structure. So, when we study graphs we would like to consider which type of graph is good or beautiful. To this end, we use the adjacency matrices to study the properties of graphs with good structures. The adjacency matrices of the graphs in Figure 1.4, Figure 1.5, Figure 1.6, and Figure 1.7 are the following in this order:

$$A_1 = \begin{bmatrix} 0 & 0 & 1 & 1 & 1 & 0 \\ 0 & 0 & 0 & 1 & 1 & 1 \\ 1 & 0 & 0 & 0 & 1 & 1 \\ 1 & 1 & 0 & 0 & 0 & 1 \\ 1 & 1 & 1 & 0 & 0 & 0 \\ 0 & 1 & 1 & 1 & 0 & 0 \end{bmatrix}, \quad A_2 = \begin{bmatrix} 0 & 1 & 0 & 1 & 0 & 1 \\ 1 & 0 & 1 & 0 & 1 & 0 \\ 0 & 1 & 0 & 1 & 0 & 1 \\ 1 & 0 & 1 & 0 & 1 & 0 \\ 0 & 1 & 0 & 1 & 0 & 1 \\ 1 & 0 & 1 & 0 & 1 & 0 \end{bmatrix},$$

$$A_3 = \begin{bmatrix} 0 & 1 & 0 & 0 & 1 & 1 \\ 1 & 0 & 1 & 1 & 0 & 0 \\ 0 & 1 & 0 & 1 & 0 & 1 \\ 0 & 1 & 1 & 0 & 1 & 0 \\ 1 & 0 & 0 & 1 & 0 & 1 \\ 1 & 0 & 1 & 0 & 1 & 0 \end{bmatrix}, \quad A_4 = \begin{bmatrix} 0 & 0 & 0 & 1 & 1 & 1 \\ 0 & 0 & 0 & 1 & 1 & 1 \\ 0 & 0 & 0 & 1 & 1 & 1 \\ 1 & 1 & 1 & 0 & 0 & 0 \\ 1 & 1 & 1 & 0 & 0 & 0 \\ 1 & 1 & 1 & 0 & 0 & 0 \end{bmatrix}.$$

Let P_1 and P_2 be the permutation matrices of size 6 corresponding to the permutation $(1,5)(3,6)$ and the transposition $(2,5)$, respectively. Then $A_3 = P_1^{-1}A_1P_1$ and $A_4 = P_2^{-1}A_2P_2$ hold. In other words, the graphs in Figure 1.4 and Figure 1.6 are isomorphic as graphs and so are the graphs in Figure 1.5 and Figure 1.7, despite the names of their vertices being different. Moreover, we can verify that the characteristic polynomials of A_1 and A_2 are $t^2(t-1)(t-3)(t+2)^2$ and $t^4(t-3)(t+3)$. In this way, by using information of adjacency matrices, we conclude that the graphs in Figure 1.4 and Figure 1.5 are not isomorphic.

Problem 1.2. Find the adjacency matrix and the eigenvalues of the graph $\Gamma = (V,E)$, where V is the vertex set and E is the edge set of the regular n-gon.

Problem 1.3. Let $\Gamma = (V,E)$ be the graph defined by the following figure. Find the adjacency matrix and the eigenvalues of Γ.

Connected graphs

For a pair of vertices $x,y \in V$ of a graph $\Gamma = (V,E)$, x and y are joined by a *path* if there exists a sequence of vertices $x_0 = x, x_1, \ldots, x_\ell = y$, where $\{x_0,x_1\}, \{x_1,x_2\}, \ldots, \{x_{\ell-1}, x_\ell\} \in E$ and $x_i \neq x_j$ whenever $i \neq j$. We call ℓ the *length* of the path $x_0 = x, x_1, \ldots, x_\ell = y$. A graph Γ is *connected* if each pair of vertices in Γ is joined by a path.

For a connected graph $\Gamma = (V,E)$, we define the *distance* between vertices x and y by the minimum length of the paths joining x and y, and denote it by $d(x,y)$. The distance $d(x,y)$ satisfies the axioms of distance such as the triangle inequality. The maximum distance between 2 vertices in Γ, i. e., $d = \max\{d(x,y) \mid x,y \in V\}$, is called the *diameter* of Γ.

Next, we consider the properties of the eigenvalues of the adjacency matrix of $\Gamma = (V,E)$.

Theorem 1.4. *Suppose $\Gamma = (V,E)$ is a regular graph with v vertices of degree k. Let A be the adjacency matrix of Γ and θ an eigenvalue of A. Then θ is real and $|\theta| \leq k$ holds. Moreover, k is an eigenvalue of A.*

Proof. Since A is real and symmetric, every eigenvalue θ is real. There are k 1's and $v-k$ 0's in each row of A. Thus, $A\,{}^t(1,1,\ldots,1) = k\,{}^t(1,1,\ldots,1)$. Hence k is an eigenvalue of A and ${}^t(1,1,\dots,1)$ is an eigenvector corresponding to k. Let $\mathbf{x} = {}^t(x_1,x_2,\dots,x_v)$ be an eigenvector corresponding to an eigenvalue θ and let $i_0 \in \{1,2,\dots,v\}$ be an integer satisfying $|x_{i_0}| \geq |x_j|$ $(1 \leq j \leq v)$. As $\mathbf{x} \neq \mathbf{0}$, we have $|x_{i_0}| > 0$. Since Γ has degree k, for the i_0-th row vector $(a_{i_0,1},a_{i_0,2},\dots,a_{i_0,v})$ of A, there exist $j_1,j_2,\dots,j_k \in \{1,2,\dots,v\}$ such that $a_{i_0,j_1} = a_{i_0,j_2} = \cdots = a_{i_0,j_k} = 1$ and $a_{i_0,j} = 0$ for $j \notin \{j_1,j_2,\dots,j_k\}$ $(1 \leq j \leq v)$. By calculating the i_0-th entry of both sides of $A\mathbf{x} = \theta\mathbf{x}$, we obtain $x_{j_1} + x_{j_2} + \cdots + x_{j_k} = \theta x_{i_0}$. Then we have $k|x_{i_0}| \geq |x_{j_1} + x_{j_2} + \cdots + x_{j_k}| = |\theta||x_{i_0}|$, which implies $k \geq |\theta|$. □

Theorem 1.5. *Suppose $\Gamma = (V,E)$ is a regular graph with v vertices of degree k. The multiplicity of the eigenvalue k of the adjacency matrix A of Γ is equal to the number of connected components of Γ.*

Proof. Let $V = \{1,2,\dots,v\}$. Let $\mathbf{x} = {}^t(x_1,x_2,\dots,x_v)$ be an eigenvector of A corresponding to the eigenvalue k and let i_0 be an integer satisfying $x_{i_0} \geq x_i$ $(1 \leq i \leq v)$. By comparing the i_0-th entry of both sides of $A\mathbf{x} = k\mathbf{x}$, we show that there exist $j_1,j_2,\dots,j_k$ satisfying $x_{j_1}+x_{j_2}+\cdots+x_{j_k} = kx_{i_0}$. Note that $\{j \mid \{i_0,j\} \in E\} = \{j_1,j_2,\dots,j_k\}$. As $x_{j_1}+x_{j_2}+\cdots+x_{j_k} = kx_{i_0} \geq x_{j_1} + x_{j_2} + \cdots + x_{j_k}$, we have $x_{j_1} = x_{j_2} = \cdots = x_{j_k} = x_{i_0}$. Since $a_{i_0,j_1} = a_{i_0,j_2} = \cdots = a_{i_0,j_k} = 1$, we have $\{i_0,j_\ell\} \in E$ $(1 \leq \ell \leq k)$. In other words, $\{j_1,j_2,\dots,j_k\}$ is the set of vertices adjacent to i_0. For $j \in V$, let $i_0 = \ell_0,\ell_1,\dots,\ell_r = j$ be a path joining i_0 and j. Then, as $\{i_0,\ell_1\} \in E$, we have $\ell_1 \in \{j_1,j_2,\dots,j_k\}$, and thus $x_{\ell_1} = x_{i_0} \geq x_\ell$ $(1 \leq \ell \leq v)$. If we repeat the same argument for l_1 as we did for i_0, we can show that $x_{\ell_1} = x_{\ell_2}$. By repeating this argument, we obtain $x_{i_0} = x_{\ell_1} = x_{\ell_2} = \cdots = x_{\ell_r} = x_j$. So if Γ is connected, the eigenspace for the eigenvalue k is the 1-dimensional subspace spanned by ${}^t(1,1,\dots,1)$. When Γ is not connected, let $V = V_1 \cup V_2 \cup \cdots \cup V_r$ be the decomposition of V into the connected components and define $E_i = E \cap (V_i \times V_i)$ $(1 \leq i \leq r)$. Then (V_i,E_i) is a connected regular graph of degree k. Let A_i be the adjacency matrix of (V_i,E_i) and express A in the order of the above decomposition of V. The following holds:

$$A = \begin{bmatrix} A_1 & 0 & \dots & \dots & 0 \\ 0 & A_2 & 0\dots & \dots & 0 \\ \vdots & \ddots & \ddots & \ddots & \vdots \\ \vdots & \ddots & \ddots & \ddots & 0 \\ 0 & \dots & \dots & 0 & A_r \end{bmatrix}.$$

Here the eigenspace of A_i for the eigenvalue k is 1-dimensional. Therefore the dimension of the eigenspace of A for the eigenvalue k, i. e., the multiplicity of k, is exactly equal to the number r of the connected components. □

Complementary graphs

For a simple graph $\Gamma = (V,E)$, the graph $\overline{\Gamma} = (V,\overline{E})$ having the vertex set V and the edge set $\overline{E} = \{\{x,y\} \subset V \mid \{x,y\} \notin E, x \neq y\}$ is called the *complementary graph* of Γ.

1.2 Strongly regular graphs and Moore graphs

Among regular graphs, strongly regular graphs are known for their particularly good properties.

Definition 1.6. A graph $\Gamma = (V, E)$ is called a *strongly regular graph* with *parameters* (v, k, λ, μ) if the following hold:

(1) $|V| = v$;

(2) Γ is a regular graph of degree k;

(3) if $\{x, y\} \in E$, then $|\{z \in V \mid \{x, z\}, \{z, y\} \in E\}|$ is equal to λ;

(4) if $\{x, y\} \notin E$, then $|\{z \in V \mid \{x, z\}, \{z, y\} \in E\}|$ is equal to μ.

Example 1.7.

$(v, k, \lambda, \mu) = (4, 2, 0, 2)$

$(v, k, \lambda, \mu) = (5, 2, 0, 1)$

$(v, k) = (6, 2)$, not strongly regular

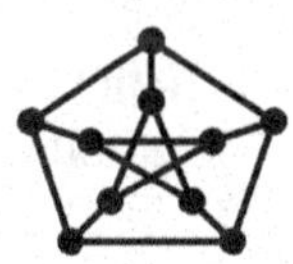

$(v, k, \lambda, \mu) = (10, 3, 0, 1)$

$(v, k, \lambda, \mu) = (4, 1, 0, 0)$

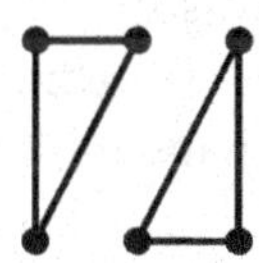

$(v, k, \lambda, \mu) = (6, 2, 1, 0)$

Proposition 1.8. *A connected strongly regular graph $\Gamma = (V, E)$ with parameters (v, k, λ, μ) satisfies the following equation:*

$$k(k - \lambda - 1) = \mu(v - k - 1). \tag{1.1}$$

Proof. For $x \in V$, we count the number of $\{y, z\}$ contained in the set $\{\{y, z\} \mid \{x, y\} \in E, \{x, z\} \notin E, z \neq x\}$.

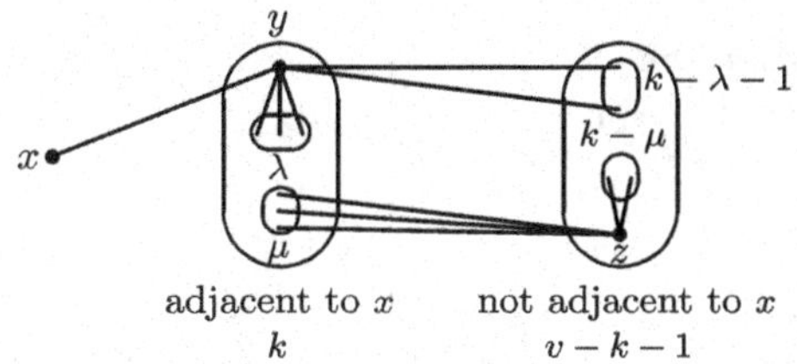

If we fix y, the number of choices of z is $k - \lambda - 1$. If we fix z, the number of choices of y is μ. Therefore $k(k - \lambda - 1) = (v - k - 1)\mu$. □

It is an interesting problem to determine parameters (v, k, λ, μ) of which strongly regular graphs exist.

Proposition 1.9. *The complementary graph of a strongly regular graph is a strongly regular graph.*

Proof. Let $\Gamma = (V, E)$ be a strongly regular graph with parameters (v, k, λ, μ) and let $\overline{\Gamma} = (V, \overline{E})$ be the complementary graph of Γ. Since the number of vertices adjacent to a vertex x of V in $\overline{\Gamma}$ is $v-k-1$, $\overline{\Gamma}$ is a regular graph of degree $\overline{k} = v-k-1$. In the following figure, the relations among vertices in Γ are described for the case where x and y are adjacent and the case where x and y are not adjacent.

$\overline{\lambda} = v - (2k - \mu) - 2 = v - 2k + \mu - 2,$ $\overline{\mu} = v - (2k - \lambda - 2) - 2 = v - 2k + \lambda$

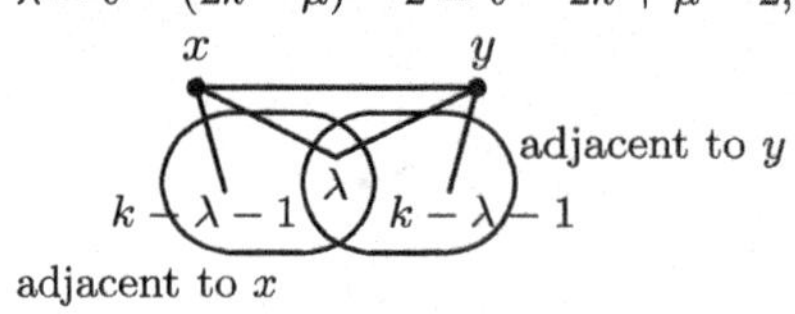

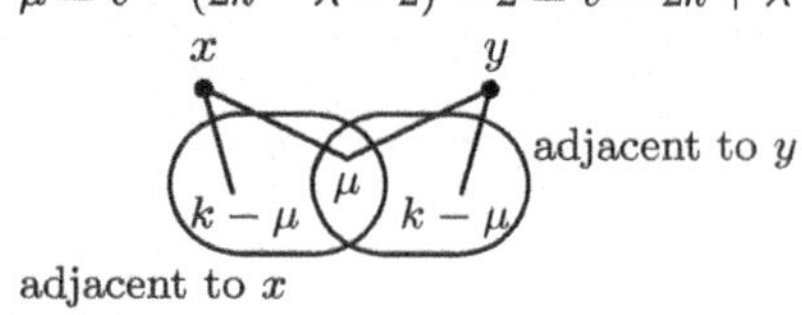

□

Example 1.10 ($T(m)$ graph). Let m be an integer of at least 3 and $V = \{\{i, j\} \mid 1 \leq i \neq j \leq m\}$. Two elements $x = \{i, j\}$, $y = \{k, l\}$ of V are joined by an edge if they satisfy $|x \cap y| = 1$. Namely, if we define $E = \{(x, y) \in V \times V \mid |x \cap y| = 1\}$, then (V, E) is a strongly regular graph with parameters $(\frac{m(m-1)}{2}, 2(m-2), m-2, 4)$. The graph is called the ***T*(*m*)** graph.

Proof. The number of vertices is $|V| = \binom{m}{2} = \frac{m(m-1)}{2}$. Since the set of vertices adjacent to $\{i_0, j_0\} \in V$ is $\{\{i_0, \ell\}, \{j_0, \ell\} \mid \ell \neq i_0, j_0\}$, the degree is $k = 2(m-2)$. Since the set of vertices adjacent to both $\{i_0, \ell_0\}$ and $\{j_0, \ell_0\}$, where $i_0 \neq j_0$, $i_0 \neq \ell_0$, $j_0 \neq \ell_0$, is $\{\{\ell_0, \ell\} \mid \ell \neq \ell_0, i_0, j_0\} \cup \{\{i_0, j_0\}\}$, we have $\lambda = m-2$. Finally, since the number of vertices adjacent to both $\{i_1, j_1\}, \{i_2, j_2\}$ for distinct i_1, i_2, j_1, j_2 is exactly 4, $\{i_1, i_2\}$, $\{i_1, j_2\}$, $\{j_1, i_2\}$, $\{j_1, j_2\}$, we have $\mu = 4$. □

The $T(4)$ graph and the $T(4)$ graph embedded in $\mathbb{R}^3$ are given in the following figures.

The $T(4)$ graph

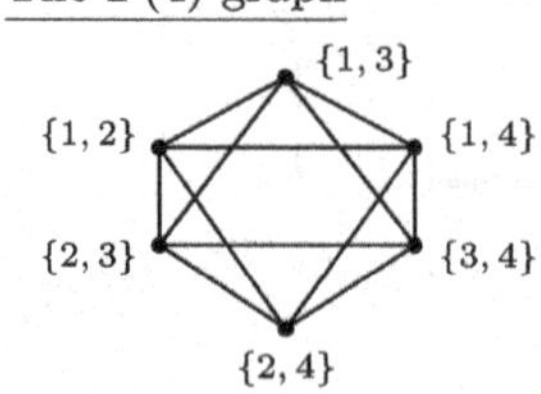

The $T(4)$ graph in $\mathbb{R}^3$

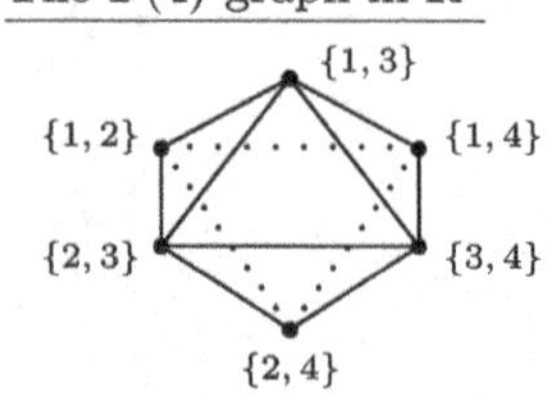

The complementary graph of the $T(5)$ graph is the *Petersen graph*.

The $T(5)$ graph

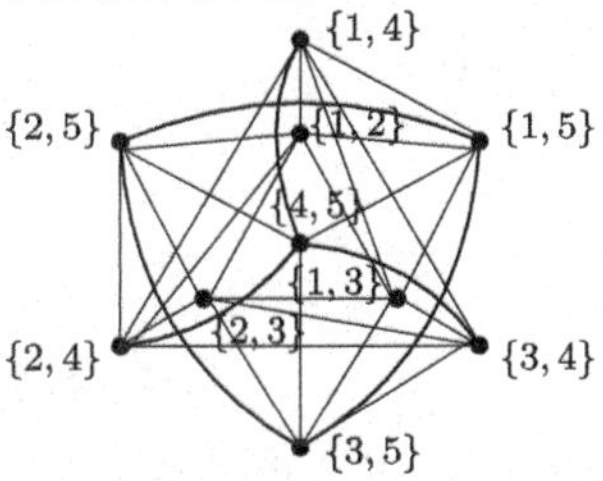

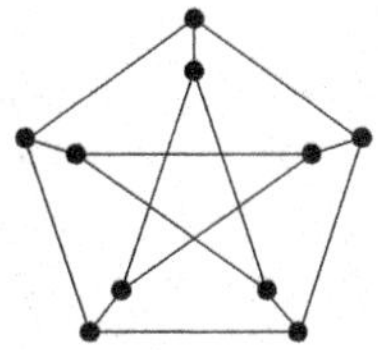

The Petersen graph

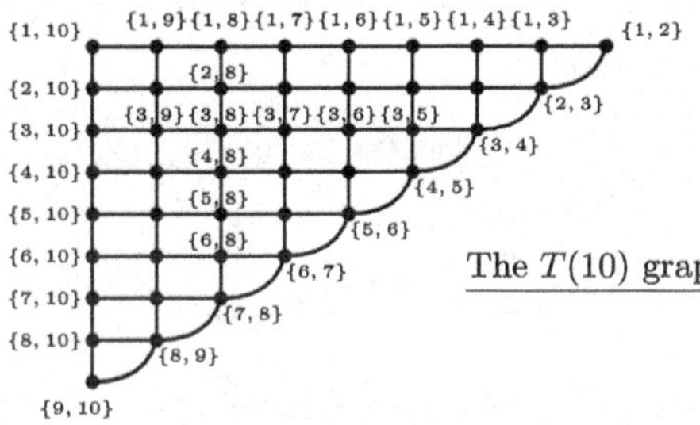

The $T(10)$ graph

For example, for the case of $T(10)$ depicted in the above figure, the vertices adjacent to the vertex $\{3,8\}$ are the sequence of vertices from $\{1,3\}$ to $\{3,10\}$, that is, $\{1,3\},\{2,3\},\{3,4\},\{3,5\},\{3,6\},\{3,7\},\{3,9\},\{3,10\}$, and the sequence of vertices from $\{1,8\}$ to $\{8,10\}$, that is, $\{1,8\},\{2,8\},\{4,8\},\{5,8\},\{6,8\},\{7,8\},\{8,9\},\{8,10\}$, and thus there are 16 vertices in total.

Example 1.11 (Complete graph). A graph is called a *complete graph* if any 2 distinct vertices are joined by an edge. We denote the complete graph with v vertices as K_v. Namely, $K_v = (V,E)$, where $E = \{\{x,y\} \subset V \mid x,y \in V, x \neq y\}$.

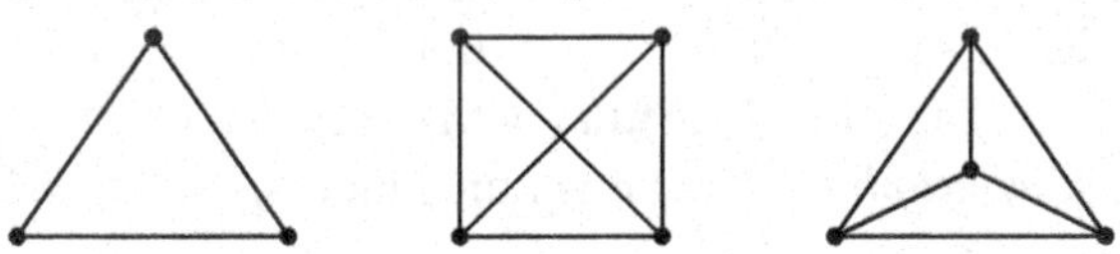

K_3 K_4 K_4

The adjacency matrix of K_v is $A = J - I$, a square matrix of degree v, where I denotes the identity matrix and J denotes the all 1's matrix.

Example 1.12 (Discrete graph). A graph is called a *discrete graph* if there is no edge. Namely, a discrete graph is the complementary graph of a complete graph.

Complete graphs and discrete graphs are strongly regular graphs.

Example 1.13 (Complete bipartite graph). A graph is called the *complete bipartite graph* if it is a strongly regular graph with $2v$ vertices and parameters $(2v, v, 0, v)$. The graph is denoted by $K_{v,v}$.

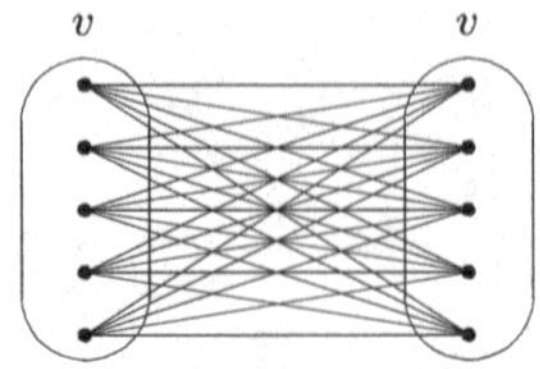

Example 1.14 (Lattice graph). Let $X = \{1, 2, \ldots, m\}$, $V = X \times X$, and $L(m) = (V, E)$. For $(x_1, y_1), (x_2, y_2) \in V$, we define $\{(x_1, y_1), (x_2, y_2)\} \in E$ if $x_1 = x_2$ and $y_1 \neq y_2$, or $y_1 = y_2$ and $x_1 \neq x_2$; $L(m)$ is a strongly regular graph with parameters $(m^2, 2(m-1), m-2, 2)$. The graph is called the *lattice graph.*

The $L(3)$ graph

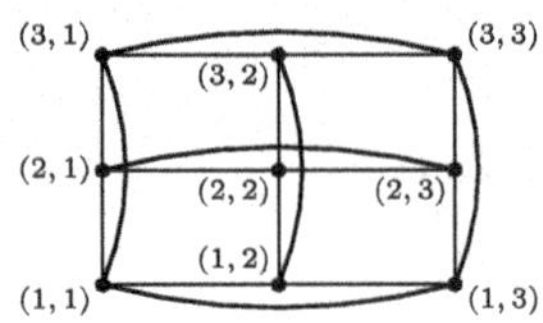

Proof. Since the set of vertices adjacent to the vertex (x_1, y_1) is $\{(x_1, y) \mid y \neq y_1\} \cup \{(x, y_1) \mid x \neq x_1\}$, we have $k = 2(m-1)$. Since the number of vertices adjacent to both (x_1, y_1) and (x_2, y_1) where $x_1 \neq x_2$ is $|\{(x, y_1) \mid x \neq x_1, x_2\}| = m - 2$ and the number of vertices adjacent to both (y_1, x_1) and (y_1, x_2) is $|\{(y_1, x) \mid x \neq x_1, x_2\}| = m - 2$, we have $\lambda = m - 2$. When $x_1 \neq x_2$, $y_1 \neq y_2$, the number of vertices adjacent to both (x_1, y_1) and (x_2, y_2) is $|\{(x_1, y_2), (x_2, y_1)\}|$, and so we have $\mu = 2$. □

Example 1.15 (Paley graph). Let q be a prime power satisfying $q \equiv 1 \pmod 4$. Let F_q be the finite field of order q. Let $V = F_q$. We define $E = \{\{x, y\} \subset V \mid x - y \text{ is a non-zero square}\}$ as the edge set. It is well known that -1 is a square if $q \equiv 1 \pmod 4$ (the reader is referred to books on number theory). So we obtain an undirected graph $\Gamma = (V, E)$; Γ is strongly regular with parameters $(q, \frac{q-1}{2}, \frac{1}{4}(q-5), \frac{1}{4}(q-1))$, and is called the *Paley graph* ($F_q^\times$ contains non-squares). For $q = 5, 17$, we have the following tables.

$q = 5$

x	0	1	2	3	4
x^2	0	1	4	4	1

The Paley graph ($q = 5$)

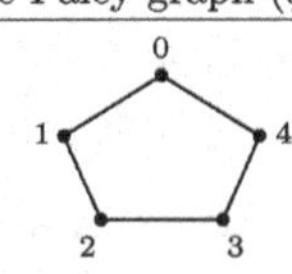

$q = 17$

x	0	1	2	3	4	5	6	7	8
x^2	0	1	4	9	16	8	2	15	13

x	9	10	11	12	13	14	15	16	
x^2	13	15	2	8	16	9	4	1	

We define $f : F_q^\times \longrightarrow F_q^\times$ by $f(x) = x^2$. Then f is a homomorphism and $1, -1$ are contained in $\mathrm{Ker} f$. Hence there exists a non-square $\eta \in F_q^\times$. Let ξ be a generator of $F_q^\times$. Let $H = \langle \xi^2 \rangle$. Then H is a cyclic subgroup of $F_q^\times$ of order $\frac{q-1}{2}$ and of index 2. Moreover, every element of H is a square and every element of ηH is a non-square. Therefore, for any vertex $x \in F_q$, a vertex y defined by $y = x - z$ $(z \in H)$ satisfies $x - y \in H$, and so it is adjacent to x. Thus $k = \frac{q-1}{2}$. The explicit values for λ and μ can be found in Section 1.3.

Strongly regular graphs with at most 36 vertices are classified ([200, page 227]).[1]

There are lots of open problems on strongly regular graphs. For instance, for which parameters do strongly regular graphs exist? Here we discuss strongly regular graphs in terms of linear algebra. Let $\Gamma = (V, E)$ be a connected strongly regular graph with parameters (v, k, λ, μ) and let A be the adjacency matrix. Let I be the identity matrix of size v and let J be the square matrix of size v in which every entry is 1. Since Γ is a regular graph of degree k, we have $AJ = kJ$, $JA = kJ$. Moreover the following holds:

$$A^2 = kI + \lambda A + \mu(J - A - I). \tag{1.2}$$

This is because, if we denote the (x, y)-entry of A by $A(x, y)$, we have

$$(A^2)(x,y) = \sum_{z\in V} A(x,z)A(z,y) = \big|\{z \mid \{x,z\},\ \{z,y\} \in E\}\big|,$$

and then we have

$$(A^2)(x,y) = \begin{cases} k, & \text{if } x = y, \\ \lambda, & \text{if } \{x,y\} \in E, \\ \mu, & \text{if } \{x,y\} \notin E,\ x \neq y. \end{cases}$$

Because A and J commute under matrix multiplication, they are simultaneously diagonalizable by an orthogonal matrix. The rank of J is 1 and the eigenvalues are v and 0. The multiplicity of the eigenvalue v is 1. Let $\boldsymbol{j}$ be the column vector in which every entry is 1. Then we have $J\boldsymbol{j} = v\boldsymbol{j}$. By Theorem 1.5, k is an eigenvalue of A with multiplicity 1. We also have $A\boldsymbol{j} = k\boldsymbol{j}$. By equation (1.2), we have $k^2\boldsymbol{j} = A^2\boldsymbol{j} = k\boldsymbol{j} + \lambda A\boldsymbol{j} + \mu(J - A - I)\boldsymbol{j} = (k + \lambda k + \mu(v - k - 1))\boldsymbol{j}$. Hence $k^2 = k + \lambda k + \mu(v - k - 1)$ holds. This gives another proof for the formula in Proposition 1.8:

$$k(k - \lambda - 1) = \mu(v - k - 1).$$

Next we observe the eigenvalues other than k. Since the diagonal entries of A are all 0, A has an eigenvalue other than k (because the trace of A is 0). Suppose A has exactly 2

1 It is known that there are 15 strongly regular graphs with parameters (25, 8, 3, 2), although it is written that there are 10 such graphs in [200].

eigenvalues k and ρ ($\neq k$). By Theorem 1.5, we have $k+(v-1)\rho = 0$. Thus ρ is an algebraic integer and a rational number, which means $\rho = -\frac{k}{v-1}$ is a rational integer. This implies that a graph with exactly 2 eigenvalues must be the complete graph (with eigenvalues $k = v - 1$ and -1). Therefore a graph which is not a complete graph must have at least 3 eigenvalues.[2] Next we assume that Γ is not a complete graph and observe its eigenvalues. Let $\boldsymbol{u}$ be an eigenvector of A with eigenvalue ρ. Since $\boldsymbol{u}$ is orthogonal to $\boldsymbol{j}$, $J\boldsymbol{u} = \boldsymbol{0}$. Thus by equation (1.2), we have $\rho^2\boldsymbol{u} = A^2\boldsymbol{u} = k\boldsymbol{u} + \lambda A\boldsymbol{u} - \mu(A+I)\boldsymbol{u} = (k + \lambda\rho - \mu\rho - \mu)\boldsymbol{u}$ and $\boldsymbol{u} \neq \boldsymbol{0}$, and therefore we have

$$\rho^2 = k - \mu + (\lambda - \mu)\rho. \tag{1.3}$$

Since every eigenvalue ρ of A satisfies equation (1.3), equation (1.3) does not have multiple roots. Hence the eigenvalues of A are k, r, and s:

$$r = \frac{1}{2}(\lambda - \mu + \sqrt{(\lambda - \mu)^2 + 4(k - \mu)}), \tag{1.4}$$

$$s = \frac{1}{2}(\lambda - \mu - \sqrt{(\lambda - \mu)^2 + 4(k - \mu)}). \tag{1.5}$$

Next let f and g be the multiplicities of r and s. Then we have

$$v = f + g + 1. \tag{1.6}$$

Also if we calculate the trace of A, we have

$$k + fr + gs = 0. \tag{1.7}$$

Therefore, by equations (1.4), (1.5), (1.6), and (1.7), we obtain the following:

$$f = \frac{1}{2}\left(v - 1 + \frac{(\mu - \lambda)(v-1) - 2k}{\sqrt{(\mu - \lambda)^2 + 4(k - \mu)}}\right), \tag{1.8}$$

$$g = \frac{1}{2}\left(v - 1 - \frac{(\mu - \lambda)(v-1) - 2k}{\sqrt{(\mu - \lambda)^2 + 4(k - \mu)}}\right). \tag{1.9}$$

Theorem 1.16. *Let $\Gamma = (V, E)$ be a connected strongly regular graph with parameters (v, k, λ, μ). Suppose Γ is not a complete graph. Then one of the following holds:*
(1) *$k = v - k - 1$, $\mu = \lambda + 1 = \frac{k}{2}$, $f = g = k$;*
(2) *$D = (\lambda - \mu)^2 + 4(k - \mu)$ is a square. Moreover,*
   (i) *if v is even, $\sqrt{D}$ divides $2k + (\lambda - \mu)(v - 1)$;*
   (ii) *if v is odd, $2\sqrt{D}$ divides $2k + (\lambda - \mu)(v - 1)$.*

2 In general, it is known that a graph of diameter d has at least $d + 1$ distinct eigenvalues ([200, Lemma 8.12.1 on page 186]).

Proof. (1) Note that f, g are natural numbers. If

$$(\mu - \lambda)(v - 1) - 2k = 0,$$

then $f = g = \frac{v-1}{2}$. If we put $\ell = v - k - 1$, then $\ell \geq 0$. If we substitute $v = \ell + k + 1$ in the above equation, then we obtain $(\mu - \lambda)(k + \ell) - 2k = 0$, and hence $\mu - \lambda > 0$. We also have $0 \leq (\mu - \lambda)\ell = (2 - (\mu - \lambda))k$. Namely, $\mu - \lambda = 1$ or 2. If $\mu - \lambda = 2$, then $\ell = 0$ and so $v = k + 1$, which implies Γ is the complete graph. Therefore by the assumption, we may set $\mu - \lambda = 1$. Then we have $k = \ell = v - k - 1$, and by $k(k - \lambda - 1) = \mu(v - k - 1) = (\lambda + 1)k$, we get $k = 2(\lambda + 1)$.

(2) If

$$(\mu - \lambda)(v - 1) - 2k \neq 0$$

holds, D must be a square because f, g are natural numbers. Since $2g = v - 1 + \frac{2k+(\lambda-\mu)(v-1)}{\sqrt{D}}$ is an integer, $\sqrt{D}$ divides $2k + (\lambda - \mu)(v - 1)$. Moreover, if v is odd, $\frac{2k+(\lambda-\mu)(v-1)}{2\sqrt{D}}$ must be an integer. □

Theorem 1.16 is a powerful tool for classification of strongly regular graphs.

Let $\Gamma = (V, E)$ be a regular connected simple graph. It is known that the number of vertices of a connected graph of degree k and diameter d is at most $1 + k + k(k - 1) + \cdots + k(k - 1)^{d-1}$, as shown in the following figure.

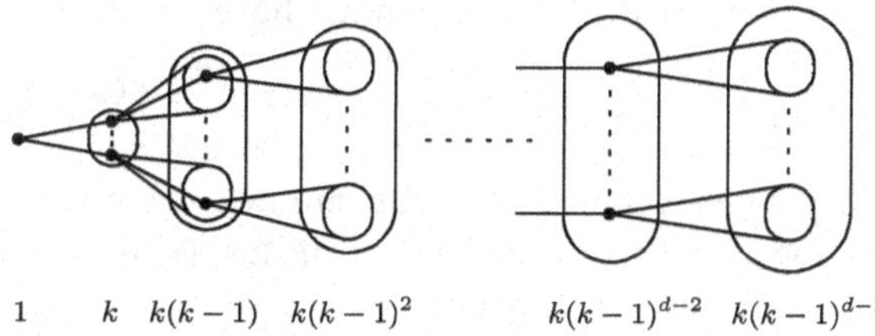

Definition 1.17 (Moore graph). A regular simple graph of degree k and diameter d whose number of vertices is exactly $1 + k + k(k - 1) + \cdots + k(k - 1)^{d-1}$ is called the *Moore graph* of *type* (d, k).

The next theorem immediately follows from Definition 1.17.

Theorem 1.18. *The graph consisting of* $(2d + 1)$ *vertices and* $(2d + 1)$ *edges of the* $(2d + 1)$*-gon is a Moore graph of degree* 2*. Conversely, a Moore graph of degree* 2 *is the graph defined by the vertices and edges of the* $(2d + 1)$*-gon.*

Theorem 1.19. *A Moore graph* $\Gamma = (V, E)$ *of diameter* 2 *is a strongly regular graph with parameters* (v, k, λ, μ)*, where* $\lambda = 0$, $\mu = 1$, $v = k^2 + 1$*, and* $k = 2, 3, 7, 57$.

Remark 1.20. If $k = 2$, Γ is isomorphic to the graph defined by the vertices and the edges of the pentagon (Theorem 1.18). If $k = 3$, we can easily show that Γ is isomorphic to the Petersen graph. If $k = 7$, Hoffman and Singleton [232] proved that Γ is isomorphic to the Hoffman–Singleton graph. The existence of a Moore graph of diameter 2 and degree 57 is an open problem.

Proof of Theorem 1.19. By the assumption, $v = |V| = 1 + k + k(k-1) = k^2 + 1$ holds. We also have $\lambda = 0$, $\mu = 1$. Firstly, we consider the case where Theorem 1.16 (1) occurs. In this case, $k = v - k - 1 = k^2 - k$ holds. So we have $k = 2$, $v = 5$, and Γ is isomorphic to the graph with the vertices and edges of the pentagon. Next we consider the case of Theorem 1.16 (2): $D = (\lambda-\mu)^2 + 4(k-\mu) = 4k-3$ is a square; $\sqrt{D}$ divides $2k + (\lambda-\mu)(v-1) = 2k - k^2$. Namely, $\frac{k(k-2)}{\sqrt{4k-3}}$ must be an integer. Therefore $\frac{k^2(k-2)^2}{4k-3}$ is an integer. Then $\frac{256k^2(k-2)^2}{4k-3} = 64k^3 - 208k^2 + 100k + 75 + \frac{3^2 5^2}{4k-3}$ is also an integer. Thus k is 2, 3, 7, or 57. If $k = 2$, then $D = 5$ and D is not a square. So Theorem 1.16 (2) occurs only if $k = 3, 7$, or 57. □

Theorem 1.21 (Bannai–Ito [59], Damerell [156]). *A Moore graph of diameter $d \geq 3$ and degree $k \geq 3$ does not exist.*

Remark 1.22. The non-existence of a Moore graph of $d = 3$ was proved by Hoffman and Singleton [232] in 1960. The non-existence of a Moore graph of $d \geq 4$ was proved by Eiichi Bannai and Tatsuro Ito, and by Damerell independently in 1973.

We close this section by giving the sketch of the proof of Theorem 1.21.

We assume that there exists a Moore graph $\Gamma = (V, E)$ of type (d, k), where $d, k \geq 3$, and derive a contradiction. By the assumption, we have

$$v = |V| = 1 + k + k(k-1) + \cdots + k(k-1)^{d-1} = 1 + \frac{k((k-1)^d - 1)}{k-2}.$$

Let $A = (a_{i,j})$ be the adjacency matrix of Γ. The minimal polynomial of A is written as a polynomial of degree $d + 1$ with the form $(x - k)F_d(x)$. Here the polynomial $F_d(x)$ of degree d is given by the following recurrence formula:

$$F_0(x) = 1, \quad F_1(x) = x + 1,$$
$$F_i(x) = xF_{i-1}(x) - (k-1)F_{i-2}(x), \quad 2 \leq i \leq d.$$

In general, it is known that such a polynomial has d distinct zeros. So the adjacency matrix A has $d + 1$ distinct eigenvalues denoted by $\theta_0 (= k), \theta_1, \theta_2, \ldots, \theta_d$. As we have seen before, k is the largest eigenvalue. Let θ_d be the smallest eigenvalue and assume $\theta_i < \theta_{i-1}$ for $1 \leq i \leq d$. Since $\operatorname{tr}(A) = 0$, we have $\theta_d < 0$. We observe that $\theta_1, \ldots, \theta_d$ are distributed on the real line as in the following figure.

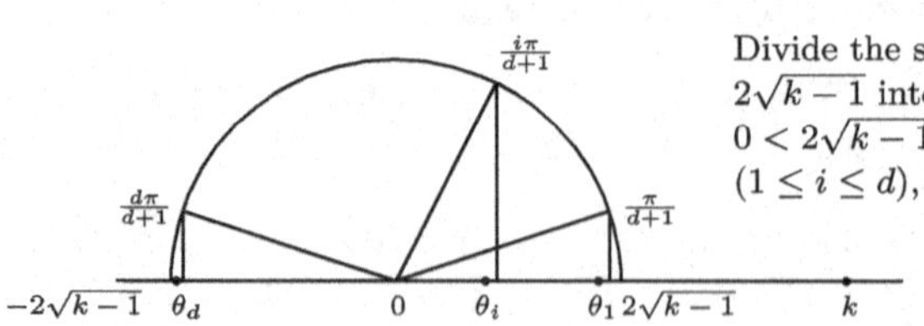

Divide the semicircle of radius $2\sqrt{k-1}$ into $d+1$ equal parts. $0 < 2\sqrt{k-1}\cos(\frac{i\pi}{d+1}) - \theta_i < \varepsilon$ $(1 \leq i \leq d)$, ε is very small.

Moreover, if we let $m(\theta_i)$ be the multiplicity of θ_i, then we observe that

$$m(\theta_i) = \frac{vk(k-1)^{d-1}}{(k-\theta_i)F'_d(\theta_i)F_{d-1}(\theta_i)} \quad \text{for } 1 \le i \le d.$$

By using them, we can roughly estimate $m(\theta_i)$.

Since $\theta_1, \dots, \theta_d$ are eigenvalues of A whose entries are 0 or 1, they are roots of a monic polynomial, i. e., a polynomial whose leading coefficient is 1. Namely, they are algebraic integers. If θ_i and θ_j are algebraically conjugate over the rational field $\mathbb{Q}$ (i. e., if θ_i and θ_j are contained in an orbit of the Galois group over $\mathbb{Q}$), $m(\theta_i) = m(\theta_j)$ holds. Moreover, the following hold (except for several cases with small d and k):

$$\begin{aligned} &m(\theta_1) < m(\theta_2), \dots, m(\theta_{d-1}), \\ &m(\theta_d) < m(\theta_2), \dots, m(\theta_{d-1}), \\ &-1 < \theta_1 + \theta_d < 0. \end{aligned}$$

Therefore it is shown that neither θ_1 nor θ_d is algebraically conjugate with $\theta_2, \dots, \theta_{d-1}$ over $\mathbb{Q}$. Hence $\theta_1 + \theta_d$ must be invariant under the action of the Galois group. In other words, $\theta_1 + \theta_d$ must be a rational integer (an element of $\mathbb{Z}$). By using this fact, we derive a contradiction and prove the non-existence theorem of Moore graphs. (For more detail, see [59, 156, 60]. This is also related to the theory of association schemes in Chapter 2.)

As a generalization of strongly regular graphs, we introduce *distance-regular graphs*. Distance-regular graphs have very good properties, which we will discuss in the next chapter. Certainly Moore graphs are examples of distance-regular graphs. A regular graph $\Gamma = (V, E)$ of degree k and diameter d is called a distance-regular graph if it satisfies the following condition: When we fix $x \in V$ and $y \in V$ with $d(x,y) = i$, the numbers

$$\begin{aligned} a_i &= |\{z \in V \mid d(x,z) = i,\ d(y,z) = 1\}|, \\ b_i &= |\{z \in V \mid d(x,z) = i+1,\ d(y,z) = 1\}|, \\ c_i &= |\{z \in V \mid d(x,z) = i-1,\ d(y,z) = 1\}| \end{aligned}$$

are constants which do not depend on the choice of x, y, and depend only on the choice of i.

A graph satisfying this condition is called a distance-regular graph of type

$$\begin{bmatrix} * & c_1 & c_2 & \cdots & c_{d-1} & c_d \\ a_0 & a_1 & a_2 & \cdots & a_{d-1} & a_d \\ b_0 & b_1 & b_2 & \cdots & b_{d-1} & * \end{bmatrix}. \tag{1.10}$$

A Moore graph is of type

$$\begin{bmatrix} * & 1 & 1 & \cdots & 1 & 1 \\ 0 & 0 & 0 & \cdots & 0 & k-1 \\ k & k-1 & k-1 & \cdots & k-1 & * \end{bmatrix}.$$

1.3 Classical *t*-designs: definitions and basic properties

The essence of *design theory* is to find a good subset which approximates the whole set. In later chapters, various designs will appear. In this section, we describe standard designs in combinatorics, especially *t-designs*. They are important among various kinds of concepts of designs and have a long history. (For the history of designs, see Remark 1.70 in Section 1.4.)

Definition 1.23 (t-(v,k,λ) **Design).** Let t, k, v, λ be natural numbers (= positive integers), and assume that $t \le k \le v$. We consider a finite subset V consisting of v points and the set $V^{(k)}$ consisting of the k-element subsets of V.[3] A pair $(V,\mathcal{B})$ of V and a subset $\mathcal{B}$ of $V^{(k)}$ is called a *t-(v,k,λ) design* (or simply a *t-design*) if there exists a natural number λ such that for any $T \in V^{(t)}$, the following holds:

$$\left|\{B \in \mathcal{B} \mid T \subset B\}\right| = \lambda.$$

For a t-design $(V,\mathcal{B})$, an element of V is called a *point*, and an element of $\mathcal{B}$ is called a *block*. A design is also called a *block design*.

Example 1.24 (The 2-(7,3,1) **design).** As shown in the following figure, V is a 7-element set. There are 7 blocks consisting of 3 edges and 3 medians of the triangle, and 1 circle inscribed in the triangle.

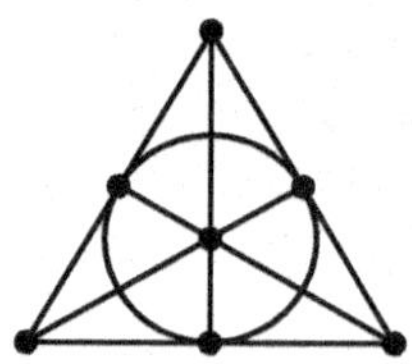

Problem 1.25 (Trivial t-**design).** Let $\mathcal{B} = V^{(k)}$. Prove that $(V,\mathcal{B})$ is a t-design. Find the value of λ. Such a t-(v,k,λ) design is called a *trivial t-design*. Usually we consider nontrivial t-designs.

Remark 1.26 (Block design with repeated blocks). Usually for a t-design, a set $\mathcal{B}$ of blocks is assumed to be a subset of $V^{(k)}$. There is another way of thinking of t-designs which allows the case where a subset appears repeatedly as elements of $\mathcal{B}$. From the viewpoint of pure mathematics such as group theory and combinatorics, it is more natural to not allow repeated blocks. From the viewpoint of statistics, however, there seems to be no problem with repeated blocks. Such a design is called a *design with repeated blocks*. If repeated blocks are not allowed, a design is also called a *simple design*. In what follows, we consider simple designs unless otherwise stated.

3 For $1 \le \ell \le v$, let $V^{(\ell)}$ be the set of ℓ-element subsets of V.

Definition 1.27 (Isomorphism of block designs). Two t-designs $(V, \mathcal{B})$ and $(V', \mathcal{B}')$ are said to be isomorphic if there exists a bijection from V to V', and the bijection induces a bijection from $\mathcal{B}$ to $\mathcal{B}'$, and moreover $p \in B$ in $(V, \mathcal{B})$ implies $p^\sigma \in B^\sigma$ in $(V', \mathcal{B}')$, where σ denotes the bijection.

Remark 1.28. To be precise, the above definition is relevant for simple designs. For designs with repeated blocks, we should define $(V, \mathcal{B})$ and $(V', \mathcal{B}')$ to be isomorphic if there exists a bijection σ from V to V' and a bijection ρ from $\mathcal{B}$ to $\mathcal{B}'$ such that $p \in B$ in $(V, \mathcal{B})$ implies $p^\sigma \in B^\rho$ in $(V', \mathcal{B}')$. However we may mostly ignore this remark because we mainly consider simple designs.

Remark 1.29. The set of isomorphisms of designs from $(V, \mathcal{B})$ to itself forms a group. This group is called the *automorphism group* of the design $(V, \mathcal{B})$ and denoted by $\mathrm{Aut}(V, \mathcal{B})$. The automorphism group $\mathrm{Aut}(V, \mathcal{B})$ can be regarded as a *permutation group* on a set V of points and also regarded as a permutation group on a set $\mathcal{B}$ of blocks. Actions of these permutation groups are not necessarily *transitive*.

The next fact is easy but important.

Proposition 1.30. *Let $(V, \mathcal{B})$ be a t-(v, k, λ) design. For any integer s with $0 \le s \le t$, $(V, \mathcal{B})$ is an s-design. Namely, if we let*

$$\lambda_s = \frac{\binom{v-s}{t-s}}{\binom{k-s}{t-s}} \lambda, \tag{1.11}$$

then λ_s is a natural number and $(V, \mathcal{B})$ is an s-(v, k, λ_s) design. (The concept of a 0-design has no special meaning but λ_0 can be regarded as the number $|\mathcal{B}|$ of blocks.)

Proof. For $S \in V^{(s)}$, let $\lambda(S) = |\{B \in \mathcal{B} | S \subseteq B\}|$. We prove that $\lambda(S)$ is independent of the choice of S as follows. By counting the number of pairs (T, B) of $T \in V^{(t)}$ and B such that $S \subseteq T \subseteq B (\in \mathcal{B})$ in two ways, we obtain

$$\lambda(S)\binom{k-s}{t-s} = \binom{v-s}{t-s}\lambda.$$

(If we choose T first, we will get the right-hand side, and if we choose B first, we will get the left-hand side.) Therefore $\lambda(S) = \frac{\binom{v-s}{t-s}}{\binom{k-s}{t-s}}\lambda = \lambda_s$, and $\lambda(S)$ is independent of the choice of S. Hence λ_s is a natural number and $(V, \mathcal{B})$ is an s-(v, k, λ_s) design. □

From the above proof, we obtain the necessary and sufficient condition for the existence of t-designs.

Proposition 1.31. *A t-(v, k, λ) design exists only if λ_s is a natural number for any integer s $(0 \le s \le t)$.*

Remark 1.32. If we apply the above proposition to the case of $t = 2$, the following holds.

Let $(V, \mathcal{B})$ be a 2-(v, k, λ) design and let r be the number of blocks containing a point and b the number of blocks $(= |\mathcal{B}|)$. Then $r = \lambda_1 = \frac{v-1}{k-1}\lambda$, $b = \lambda_0 = \frac{v(v-1)}{k(k-1)}\lambda$ hold. Namely, we have

$$r(k-1) = (v-1)\lambda, \tag{1.12}$$

$$bk = vr. \tag{1.13}$$

These two equations are important. A non-trivial 2-(v, k, λ) design is also called a *balanced incomplete block design (BIBD)*, and a t-$(v, k, 1)$ design with $\lambda = 1$ is called a *Steiner system*.

Definition 1.33 (Incidence matrix of a design $(V, \mathcal{B})$). For a design $(V, \mathcal{B})$, we define a matrix M whose rows are indexed by $\mathcal{B}$ and whose columns are indexed by V as follows. For $B \in \mathcal{B}$, $P \in V$, define the (B, P)-entry of M as

$$M(B,P) = \begin{cases} 1 & (\text{if } P \in B), \\ 0 & (\text{if } P \notin B). \end{cases}$$

The matrix M is called the *incidence matrix* of a design $(V, \mathcal{B})$.

The incidence matrix of the 2-(7, 3, 1) design in Example 1.24 is given as follows:

$$M = \begin{bmatrix} 0 & 1 & 1 & 0 & 1 & 0 & 0 \\ 0 & 0 & 1 & 1 & 0 & 1 & 0 \\ 0 & 0 & 0 & 1 & 1 & 0 & 1 \\ 1 & 0 & 0 & 0 & 1 & 1 & 0 \\ 0 & 1 & 0 & 0 & 0 & 1 & 1 \\ 1 & 0 & 1 & 0 & 0 & 0 & 1 \\ 1 & 1 & 0 & 1 & 0 & 0 & 0 \end{bmatrix}.$$

Here we give a basic definition.

Definition 1.34 (Complementary design). For a t-(v, k, λ) design $(V, \mathcal{B})$, define $\mathcal{B}' = \{V \backslash B \mid B \in \mathcal{B}\}$. Then $(V, \mathcal{B}')$ becomes a block design. Let t-(v, k', λ') be its parameters. Then $k' = v - k$, $\lambda' = \frac{\binom{v-k}{t}}{\binom{k}{t}}\lambda$ holds. The design $(V, \mathcal{B}')$ is called the *complementary design* of $(V, \mathcal{B})$. For details, we refer the reader to [319, corollary on page 192] or [366].

Let M be the incidence matrix of a t-(v, k, λ) design $(V, \mathcal{B})$. In general, each row of M has exactly k 1's, and each column of M has exactly r 1's. Let J be the square matrix whose entries are all 1. Namely, we have $MJ = kJ$ and $JM = rJ$. (Note that J has different sizes for different equations. To avoid making things complicated, we often use J and the identity matrix I with their sizes unspecified.) In particular, if $(V, \mathcal{B})$ is a 2-design, for any two different columns of M, there are exactly λ rows in which both columns have 1's. So we have the following:[4]

$$ {}^{t}MM = \begin{bmatrix} r & & & \lambda \\ & r & & \\ & & \ddots & \\ \lambda & & & r \end{bmatrix}. \tag{1.14} $$

Problem 1.35. Prove that for the incidence matrix M of a 2-design, the determinant of ${}^{t}MM$ is given as follows:

$$ \det({}^{t}MM) = (r + (v-1)\lambda)(r-\lambda)^{v-1}. $$

Theorem 1.36 (Fisher type inequality). *For a 2-(v, k, λ) design, assume $v > k$. Then the inequality $b \geq v$ holds.*

Proof. By the assumption $k < v$ and by (1.12), we have $r > \lambda$. Then the determinant $\det({}^{t}MM)$ is non-zero (Problem 1.35). Namely, ${}^{t}MM$ is a non-singular matrix of size v, and this implies $b \geq v$. (In general, for matrices A and B, the rank of AB does not exceed the rank of A or B.) □

Definition 1.37 (Symmetric design). A 2-(v, k, λ) design with $b = v$ is called a *symmetric (v, k, λ) design.*

Remark 1.38. By (1.13), for a symmetric (v, k, λ) design, $r = k$ holds.

Now, let M be the incidence matrix of a symmetric (v, k, λ) design. Then M is a square matrix and $MJ = JM = kJ$ holds. Besides, (1.14) can be expressed as

$$ {}^{t}MM = (r-\lambda)I + \lambda J. $$

Moreover, by the proof of Theorem 1.36, M is non-singular. Therefore we have

$$ {}^{t}M = \{(r-\lambda)I + \lambda J\}M^{-1}. $$

If we consider $M\,{}^{t}M$, we have

$$ \begin{aligned} M\,{}^{t}M &= M\{(r-\lambda)I + \lambda J\}M^{-1} = \{(r-\lambda)I + \lambda J\}MM^{-1} \\ &= \{(r-\lambda)I + \lambda J\} = \begin{bmatrix} r & & & \lambda \\ & r & & \\ & & \ddots & \\ \lambda & & & r \end{bmatrix}. \end{aligned} \tag{1.15} $$

This means that two blocks B_i and B_j contain exactly λ common points. With this in mind, we state the following proposition.

4 ${}^{t}M$ denotes the transpose of M.

Proposition 1.39. *Assume that a 2-(v,k,λ) design satisfies $v > k$. Then the following four conditions are equivalent:*
(1) $b = v$;
(2) $r = k$;
(3) *any two blocks have exactly λ common points;*
(4) *any two blocks have exactly m common points for a constant m.*

Proof. Since $bk = vr$, we have (1) $\Leftrightarrow$ (2); (2) $\Rightarrow$ (3) follows from the above discussion; and (3) $\Rightarrow$ (4) is obvious.

We show (4) $\Rightarrow$ (1). We exchange the roles of points for those of blocks. Namely, the matrix ${}^tM$ has exactly r 1's in each row, and ${}^tM$ has exactly k 1's in each column. Besides, any two columns of ${}^tM$ have exactly m 1's in common. Therefore ${}^tM$ is the incidence matrix of a 2-(b,r,m) design. Each block of this design contains r points and each point is contained in exactly k blocks. Applying (1.13) to the given design, we have $bk = vr$. Moreover, since $v > k$, we have $b > r$. Next if we apply (1.12) to the 2-(b,r,m) design, we have $k(r-1) = (b-1)m$, and thus $k > m$. By Problem 1.35, we have $\det(M\,{}^tM) = (k+(b-1)m)(k-m)^{b-1} > 0$. Hence $M\,{}^tM$ is a non-singular matrix of size b, and $b \le v$ holds. On the other hand, by applying the Fisher type inequality to the given design $(V,\mathcal{B})$, we have $b \ge v$, and thus $b = v$. □

Remark 1.40. As we considered in the proof of the above proposition, if condition (4) of Proposition 1.39 holds for a 2-(v,k,λ) design $(V,\mathcal{B})$, the existence of a 2-(b,r,m) design is easily shown by exchanging the roles of points for those of blocks. The 2-(b,r,m) design is called the *dual structure* or the *dual design* of $(V,\mathcal{B})$.

Remark 1.41. To sum up, the incidence matrix M of a symmetric 2-(v,k,λ) design satisfies

$$MJ = JM = kJ(= rJ),\ {}^tMM = M\,{}^tM = (r-\lambda)I + \lambda J.$$

Remark 1.42. As was stated in M. Hall, Jr. [210, Section 10.2], for a symmetric 2-(v,k,λ) design $(V,\mathcal{B})$, $(V,\mathcal{B})$ has the same parameters as the dual design $(\mathcal{B},V)$, but they are not necessarily isomorphic.

Important theorems for designs

As we stated before, Proposition 1.31 together with equation (1.11) is a necessary condition for the existence of a t-(v,k,λ) design. How strong is it? How close is it to a sufficient condition? It is not clear in general, but for the case of $t = 2$, we know it is very strong.

For example, a 2-$(v,3,1)$ design, which is called a *Steiner triple system*, has been studied a lot. For a 2-$(v,3,1)$ design, Proposition 1.31 with equation (1.11) is shown to be a sufficient condition for the existence of a design. And it is immediately proved that they are equivalent to

$$v \equiv 1,3 \pmod 6.$$

(For details, see [319, Example 19.11, Example 19.15].) Of course, it is not true for t-(v,k,λ) designs in general. Many examples such that it fails to be a sufficient condition are known. The following Wilson theorem is a definitive result in this direction of study. For details, we refer the reader to Wilson [515, 516, 518] and Ray-Chaudhuri and Wilson [399].

Theorem 1.43 (Wilson). *Suppose that k,λ are given. There exists a number v_0 determined by k,λ such that there is a 2-(v,k,λ) design with $r = \lambda(v-1)/(k-1)$ and $b = \lambda v(v-1)/k(k-1)$ whenever $v \geq v_0$, $\lambda(v-1) \equiv 0 \pmod{k-1}$, $\lambda v(v-1) \equiv 0 \pmod{k(k-1)}$.*

Remark 1.44. It would be desirable if we could show a similar result for $t \geq 3$; however, it is an open problem. If we allow repeated blocks, this necessary condition is known to be very close to a sufficient condition. (For details, see Graver and Jurkat [204], Wilson [517], or Khosrovshahi and Tayfeh-Rezaie [276].[5])

Regarding the necessary condition for the existence of a symmetric 2-(v,k,λ) design, the next theorem is known (see [326, 116]).

Theorem 1.45 (The Bruck–Ryser–Chowla theorem). *For a symmetric 2-(v,k,λ) design, if we let $n = k - \lambda$, the following hold:*
(1) *If v is even, then n is a square.*
(2) *If v is odd, then $z^2 = nx^2 + (-1)^{(v-1)/2}\lambda y^2$ has a solution in integers x,y,z, not all of which are 0.*

Proof. (1) The left-hand side of $\det({}^tMM) = (r+(v-1)\lambda)(r-\lambda)^{v-1}$ is a square. By symmetry of the design, we have $r = k$. Moreover, by (1.12), we have $r + (v-1)\lambda = k^2$. Therefore $(r-\lambda)^{v-1}$ is a square, and if we note that $v-1$ is odd, $n = k - \lambda$ must be a square.

The proof of (2) is very interesting but quite complicated. So we refer the reader to other books. (Several kinds of proofs are known. See M. Hall, Jr. [210], K. Yamamoto [523], and E. S. Lander [303].) □

1.4 Examples of designs

Finite projective planes

Let K be a field and consider the $(n+1)$-dimensional vector space K^{n+1} over K. Two non-zero vectors in K^{n+1} are defined to be equivalent if one is a scalar multiple of the

5 Note that the study of the extension of the theorem to the case $t \geq 3$ has made progress in recent years [184, 293].

other, and the quotient space $KP^n = (K^{n+1}\backslash\{0\})/\sim$ with this equivalence relation $\sim$ is called the n-dimensional *projective space*. We call an element of KP^n a point. Namely, a point of the projective space KP^n corresponds to a 1-dimensional subspace of K^{n+1}. A subset of KP^n corresponding to a 2-dimensional subspace of K^{n+1} is often called a line. Moreover a subset of KP^n corresponding to a 3-dimensional subspace of K^{n+1} is often called a plane. In particular, the case $n = 2$ is the *projective plane* over K, which consists of points and lines. In particular, if K is the finite field F_q of order q, we obtain the projective space or projective plane (over the finite field) which consists of finitely many points. They are, in a sense, standard projective spaces or projective planes, but with regard to projective planes it is natural and common to see them as an axiomatization of an incidence relation of points and lines defined by inclusion or non-inclusion, like the definition given in the following. The system of such points and lines or the structure consisting of them is called finite geometry. Finite geometry is not necessarily geometry over a finite field, which is a much wider concept.

Definition 1.46 (Projective plane). Let P be a set of points and let L be a set of lines. An incidence relation $\in$ of inclusion is defined between points and lines. The triple $(P, L, \in)$ is called a projective plane if it satisfies the following three axioms.
Axiom (1): For given 2 points, there exists a unique line connecting them.
Axiom (2): Every 2 lines meet at exactly 1 point.
Axiom (3): There exists a subset of 4 points, any 3 of which are not collinear.

Remark 1.47. As for the above definition, we allow the case where P is infinite. Moreover, in many cases, the definition of an incidence relation $\in$ is unspecified as an implicit assumption. So we usually denote a projective plane by $\pi = (P, L)$ in what follows. Due to Axiom (3), we can eliminate cases such as the following figure, which is, in a sense, trivial and degenerate.

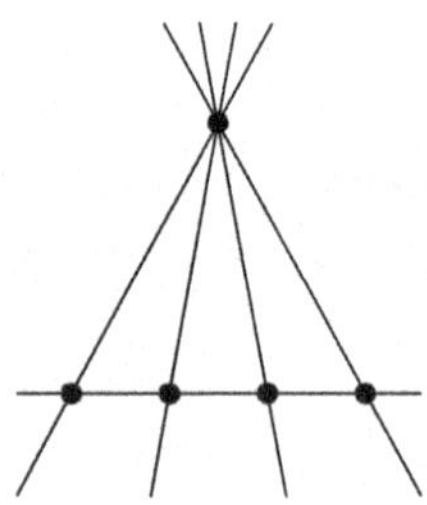

Remark 1.48. In particular, if P is a finite set, we call $\pi = (P, L)$ a *finite projective plane*. In this case, it is known that there exists a natural number n greater than 1 such that the number of points and that of lines of π are $n^2 + n + 1$, a line contains $n + 1$ points, and a point is contained in $n + 1$ lines (we will discuss details later). We call the number n the *order* of a finite projective plane. The projective plane over the field F_q is a projective plane of order q and called a *Desarguesian plane*. Other than projective planes over finite fields, several non-standard projective planes are known. They are

called *non-Desarguesian planes*. Up to the present, the orders of known finite projective planes are prime powers. It is an important *open problem* whether the order of a finite projective plane is always a prime power. Specialists are divided over the above issue. In search of algebraic structures, weaker structures than fields, which give coordinate systems of non-Desarguesian planes, many studies have been conducted for a long time. Also many attempts have been made to classify non-Desarguesian planes. They are very difficult problems and only few things are known. Considering only finite projective planes, you will find many interesting open problems. Of course, if n is given, there are "finitely many" projective planes of order n up to isomorphism. However, as is often the case, the condition "finitely many" is not so powerful. You might think that computers can solve all the problems on finite things, but it is not true at all. If the number of points is very small, computers may be useful more or less, but if the number of points are increasing, computers become useless quickly.

In fact, the concept of finite projective planes can be characterized in terms of designs. We state this fact as the following problem. (Readers would find it interesting if they try to prove it.)

Problem 1.49. Prove that a finite projective plane of order n exists if and only if a 2-$(n^2+n+1, n+1, 1)$ design exists.

We also give the following facts as problems.

Problem 1.50.

(1) There exists a unique finite field with q elements for each prime power $q = p^f$ up to isomorphism. Check this fact by textbooks on algebra. We denote the finite field by F_q. (There is a way to denote it by $GF(q)$ instead.)

(2) Verify that for each finite field $K = F_q$, KP^2 is a finite projective plane of order q. In other words, consider the 3-dimensional vector space F_q^3 over the finite field F_q. Let V be the set of 1-dimensional subspaces of F_q^3, and let $\mathcal{B}$ be the set of 2-dimensional subspaces of F_q^3. (Note that a block, i. e., an element of $\mathcal{B}$, consists of $q+1$ points of V.) Then show that $(V, \mathcal{B})$ is a 2-$(q^2+q+1, q+1, 1)$ design, in other words, a projective plane of order q.

Theorem 1.51. *For a projective plane $\pi = (P, L)$, the following are equivalent:*

(1) *A line contains exactly $n+1$ points.*

(2) *A point is contained in exactly $n+1$ lines.*

(3) *Every line contains exactly $n+1$ points.*

(4) *Every point is contained in exactly $n+1$ lines.*

(5) *There exist n^2+n+1 points in total.*

(6) *There exist n^2+n+1 lines in total.*

For the proof of the above theorem, we refer the reader to other books. (It is not so difficult if you proceed step by step. For details, see [210]. When the author (Eiichi

Bannai) lectured on this topic, he gave a proof only for the beginning part and the rest was left for the students as an assignment.)

When considering the proof of the theorem, note that if three axioms (1), (2), (3) of a projective plane in Definition 1.46 hold, then the following Axiom (3′) holds.

Axiom (3′): There exists a set of 4 lines, any 3 of which do not meet at 1 point.

Note that Axiom (3′) is the dual condition of Axiom (3), and guarantees that the dual structure $\pi' = (L, P)$ obtained by exchanging the roles of points and lines of a projective plane $\pi = (P, L)$ becomes a projective plane.

Here we make some comments in regard to projective planes and projective geometries. If we apply the Bruck–Ryser–Chowla theorem to a projective plane, i. e., a 2-(v, k, λ) design, we obtain the following.

Theorem 1.52. *Let $n \equiv 1, 2 \pmod 4$. If there exists a projective plane of order n, then n can be expressed as $n = x^2 + y^2$, a sum of 2 squares.*

Corollary 1.53. *There exist no projective planes of order $n = 6, 14, 21, 22, \ldots$.*

Remark 1.54. How many projective planes of order n do exist for a given n? Many studies have been conducted on this problem. Still there are a lot of things left unsolved. It is known that for $n = 2, 3, 4, 5, 7, 8$, the projective plane of order n is unique up to isomorphism. For the case of order 9, the classification was completed by concluding that there are 4 non-isomorphic projective planes [301]. For the case of order 10, there is a history of efforts by so many people. At one time there was a legend that people who attacked this problem lost their mind because of the huge amount and complexity of calculations. Through the early works of M. Hall, Jr., Parker, MacWilliams, Sloane, and Thompson, at last this problem was solved in 1989; Lam, Thiel, and Swiercz [302] proved the non-existence (see also [300]). The proof is essentially based on a brute force search, and it is said that they run the fastest large computer at that time more than 1000 hours. In that sense, this solution made a great impact in the world of mathematics with regard to the role of computers like the proof of the four-color theorem and that of the Kepler conjecture (the conjecture on dense sphere packings in 3-dimensional Euclidean space) did. Furthermore, following the case of order 10, several people seriously study the case of order 12 (or 15). As stated before, whether the order n of a projective plane is a prime-power is the most important problem in this area. In particular, if n is a prime p, whether a projective plane of order p is unique up to isomorphism is also an interesting open problem. (No counter-example is known yet.) In addition, it is known that depending on n, so many non-isomorphic projective planes of order n exist.

Affine planes

Next we discuss the concept of affine planes, which are closely related to and as important as projective planes.

Definition 1.55 (Affine plane). If a set P of points and a set L of lines satisfy the following conditions, (P, L) is said to be an *affine plane*.
(1) There exists a unique line passing through 2 points.
(2) For a line and a point not on that line, there exists a unique line which passes through the point and does not intersect that line.
(3) There exists a set of 3 points which are not collinear.

Remark 1.56. For an affine plane, it is known that the following are equivalent. (The proof is an easy exercise for the reader.)
(1) A line contains exactly n points.
(2) A point is contained in exactly $n + 1$ lines.
(3) There exist n^2 points in total.
(4) There exist $n^2 + n$ lines in total.

An affine plane satisfying the above is called an affine plane of *order n*.

Problem 1.57. Prove that there exists an affine plane of order n if and only if there exists a 2-$(n^2, n, 1)$ design $(n \geq 2)$.

Problem 1.58. The existence of an affine plane of order n and that of a projective plane of order n are equivalent. Prove that a 2-$(n^2, n, 1)$ design exists if and only if a 2-$(n^2 + n + 1, n + 1, 1)$ design exists.

Remark 1.59. The construction of a projective plane of order n from an affine plane of order n is essentially the same as the construction of the (real) projective plane by adding the line at infinity to the (real) Euclidean plane. (For details, see textbooks on projective geometry [492], [210], [167], [243].) Conversely, by removing a line and $n + 1$ points on that line from a projective plane of order n, we obtain an affine plane of order n. In general, we can uniquely construct a projective plane of order n from an affine plane of order n. On the other hand, since the construction of an affine plane of order n from a projective plane of order n depends on the choice of a line, there is a possibility that non-isomorphic affine planes of order n are obtained. (In fact, several examples are known. The smallest example is the case of order 9.)

Multiply (t-)transitive groups and t-designs

Definition 1.60 (t-Transitive groups). Let G be a permutation group on a set X. (Unless otherwise stated, we only consider the case where X is finite, i. e., G is a finite permutation group.) Then G is called a transitive permutation group (or simply, transitive group) on X if there exists an element $g \in G$ such that $\alpha^g = \beta$ for any elements α, β of X, where α^g denotes the element of X obtained by the action of g on α. Moreover, G is called a t-transitive permutation group (or t-transitive group) on X if there exists an element $g \in G$ such that for any two ordered sets $(\alpha_1, \alpha_2, \ldots, \alpha_t)$, $(\beta_1, \beta_2, \ldots, \beta_t)$ of distinct t elements of X, $\alpha_i^g = \beta_i$ $(1 \leq i \leq t)$ holds. A t-transitive permutation group for

$t \geq 2$ is called a multiply transitive group. As is well known, the symmetric group S_n is an n-transitive group if it acts naturally on a set of n letters and the alternating group A_n is an $(n-2)$-transitive group. The symmetric group S_n and the alternating group A_n are called *trivial multiply transitive groups*.

Remark 1.61 (t-Designs obtained from multiply transitive groups). The concept of t-designs is deeply related to the concept of t-transitive permutation groups in group theory. Particularly, in a historical view, we can say that the concept of t-designs arises from the concept of t-transitive groups. The following proposition is very important in this sense.

Proposition 1.62. *Let G be a t-transitive group acting on a finite set $X = \{1, 2, \ldots, n\}$. Suppose B is any subset of X and $|B| \geq t$. Then $(X, \mathcal{B})$ is a t-design, where $\mathcal{B} = \{B^\sigma | \sigma \in G\}$. (Note that it possibly becomes the trivial design.)*

Proof. The proof seems almost obvious. If $T,\ T' \in X^{(t)}$, then there exists $g \in G$ such that $T^g = T'$. Since $\lambda(T) = |\{B \in \mathcal{B} | T \subset B\}|$, we have $\lambda(T') = |\{B^g \in \mathcal{B} | T \subset B\}|$, and thus $\lambda(T) = \lambda(T')$ holds. □

Next we will introduce the concept of the *t-homogeneous group*, which is similar to but a little weaker than a t-transitive group.

Definition 1.63 (t-Homogeneous group). A permutation group G on X is called a t-homogeneous group if it satisfies the following condition:

For any subsets $\{\alpha_1, \alpha_2, \ldots, \alpha_t\}$, $\{\beta_1, \beta_2, \ldots, \beta_t\}$ of (unordered) t elements of X, there exists an element $g \in G$ such that $\{\alpha_1, \alpha_2, \ldots, \alpha_t\}^g = \{\beta_1, \beta_2, \ldots, \beta_t\}$.

By the definition, it is obvious that a t-transitive permutation group is a t-homogeneous group. Also a t-homogeneous group is an $(n-t)$-homogeneous group, where $n = |X|$. So we usually assume $t \leq n/2$ for a t-homogeneous group.

Remark 1.64. Proposition 1.62 holds for t-homogeneous groups exactly in the same way. The proof seems to be obvious. Conversely, it is clear from the definition that t-homogeneous groups can be characterized as groups satisfying this property. (Consider the case $|B| = t$.) The following theorem is a very interesting theorem which claims that a t-homogeneous group becomes a t-transitive group for $t \geq 5$. Here it is important that this holds for finite permutation groups. (In general, it is not true for infinite groups.)

Theorem 1.65 (The Livingstone–Wagner theorem [320]). *A t-homogeneous group becomes a t-transitive group for $t \geq 5$. (Here we assume $t \leq n/2$.)*

Remark 1.66. As will be stated below, non-existence of non-trivial t-homogeneous groups for $t \geq 6$ (and non-trivial t-transitive groups for $t \geq 6$) can be proved by using the classification of the finite simple groups, which was completed in the early 1980s. In that sense, the theorem can be viewed as the theorem with very few examples. However, the proof of the theorem by Livingstone and Wagner (1965) is quite elementary and very beautiful. This is my (Eiichi Bannai) favorite theorem among many theorems on permutation groups. For $2 \leq t \leq 4$, the classification of the finite t-homogeneous groups which are not t-transitive groups was completed (Kantor [271]).

Remark 1.67 (Multiply transitive groups, esp. $PGL(n,q)$, $PSL(n,q)$). We will reserve the detailed explanation for other books dealing with group theory, but the most important fact is that the actions of the general linear group $GL(n,q)$ and the special linear group $\mathrm{SL}(n,q)$ on the set of points of the projective space are 2-transitive. See Proposition 1.62. Note that $\mathrm{GL}(n,q)$ is the group consisting of invertible $n \times n$ matrices over a finite field F_q and $\mathrm{SL}(n,q)$ is its subgroup of index $q-1$ consisting of the matrices with determinant 1. If we view $\mathrm{GL}(n,q)$, $\mathrm{SL}(n,q)$ as permutation groups on the points of the projective space, we obtain the projective general linear group and the projective special linear group, and denote them by $\mathrm{PGL}(n,q)$, $\mathrm{PSL}(n,q)$, respectively. If $n = 3$, $\mathrm{PGL}(3,q), \mathrm{PSL}(3,q)$ act doubly transitive on the set of q^2+q+1 points of a Desarguesian finite projective plane of order q. Note that the order of the general linear group $\mathrm{GL}(n,q)$ is $|\mathrm{GL}(n,q)| = q^{n(n-1)/2}(q^n-1)(q^{n-1}-1)\cdots(q-1)$, and $|\mathrm{PGL}(n,q)| = |\mathrm{SL}(n,q)| = |\mathrm{GL}(n,q)|/(q-1)$, $|\mathrm{PGL}(n,q) : \mathrm{PSL}(n,q)| = (n,q-1)$, where $(n,q-1)$ denotes the greatest common divisor of n and $q-1$. Besides, $\mathrm{PGL}(2,q)$ is a triply transitive permutation group on the $q+1$ points of a projective line over F_q. There are many other interesting examples of 2- or 3-transitive groups. For details, we refer the reader to books on group theory.

Definition 1.68 (Extension of a permutation group). If G is a t-transitive group on a set X, for an element α of X, the stabilizer G_α becomes a $(t-1)$-transitive group on the set $X\backslash\{\alpha\}$ obtained from X by deleting α. Conversely, if G is a t-transitive group on a set X, and if for a transitive group $\hat{G}$ on the union $X \cup \{\omega\}$ of X and a new point ω, the stabilizer $\hat{G}_\omega$ is isomorphic to G (as a permutation group on X), the permutation group $\hat{G}$ is called a transitive extension of the permutation group G. (In this case, $\hat{G}$ is always a $(t+1)$-transitive group.)

In general, it is difficult to determine which permutation group admits to a transitive extension and many studies have been conducted on this problem. The group $\mathrm{PSL}(3,4)$ is a 2-transitive group on 21 points of a projective plane over F_4, and it is extended 3 times repeatedly to provide a 3-transitive group M_{22}, a 4-transitive group M_{23}, and a 5-transitive group M_{24}. The orders are $|M_{24}| = 24|M_{23}|$, $|M_{23}| = 23|M_{22}|$, $|M_{22}| = 22|\mathrm{PSL}(3,4)|$, $|\mathrm{PSL}(3,4)| = 20160$. Besides, it is known that by extending a 3-transitive group on 10 letters repeatedly we obtain other Mathieu groups, a 4-transitive group M_{11} and a 5-transitive group M_{12}. The orders are $|M_{12}| = 12|M_{11}|$, $|M_{11}| = 7920 = 2^4 \cdot 3^2 \cdot 5 \cdot 11$. These Mathieu groups were found 100 years ago, but no other t-transitive group for $t \geq 4$ was found despite various attempts to find them.

Remark 1.69 (Finite simple groups and multiply transitive groups). As we stated before, by using the classification of finite simple groups, all multiply transitive permutation groups can be classified. For $t = 2, 3$, there are several infinite families and several exceptional cases. On the other hand, for $t \geq 6$, there is no non-trivial one; for $t = 5$, there are two Mathieu groups M_{12}, M_{24} only; and for $t = 4$, there are exactly 4 Mathieu groups $M_{11}, M_{12}, M_{23}, M_{24}$. Since there is no non-trivial t-transitive group for $t \geq 6$, it is desirable to show this fact directly using the strong condition of being a multiply transitive group (not using the very difficult classification of finite simple groups), but there is no idea how to solve it at present. We sincerely hope that some readers find a breakthrough to this problem.

Remark 1.70. It seems that, historically, the concept of t-designs appeared in connection with t-transitive permutation groups, but the relation between them is not so simple. As we discussed before, it was proved that there is no non-trivial finite t-transitive permutation group for $t \geq 6$. However, the situation is quite different for t-designs. The problem of the existence of non-trivial t-(v, k, λ) designs has a long history. Various examples were known for small t. Especially up to $t = 5$ several examples were known, but for $t \geq 4$ known examples were limited to t-designs constructed from t-transitive permutation groups as of the early 1960s. In subsequent years, for $t = 4, 5$ several examples were found, which are not directly associated with multiply transitive groups but even in the early 1970s, there was no known example of a non-trivial t-design for $t \geq 6$. Afterwards, a number of non-trivial t-designs for $t = 6$ were found but it had not been known that for any (large) t, there exists a non-trivial design until it was first shown in the paper by Teirlinck [457]. However, λ is very large for the case of Teirlinck. It is still an open problem whether a t-design exists for any small λ. See also the following remark.

Remark 1.71. A t-(v, k, λ) design with $\lambda = 1$ is called a Steiner system or a Witt system and has been studied for a long time compared to general t-(v, k, λ) designs in the area of geometry. It is an interesting open problem whether there exists a t-design with $\lambda = 1$ for any (large) t. (To be precise, we should say it "was" an open problem. It seems that Keevash recently solved the problem [273].)

Problem 1.72. Let V be the set of 1-dimensional vectors of the m-dimensional vector space F_q^m over the finite field F_q, and let $\mathcal{B}$ be the set of ℓ-dimensional subspaces of F_q^m, where $\ell \geq 2$. Prove that $(V, \mathcal{B})$ becomes a 2-design, and find the parameters v, k, λ of the design.

(Answer:

$$v = (q^m - 1)/(q - 1), \quad k = (q^\ell - 1)/(q - 1),$$

$$\lambda = \frac{(q^{m-2} - 1)(q^{m-3} - 1) \cdots (q^{m-k+1} - 1)}{(q^{\ell-2} - 1)(q^{\ell-3} - 1) \cdots (q - 1)}.$$

We recommend the reader to verify these values.)

Note that we can prove it directly without using groups, but we can obtain the result immediately by the fact that $\mathrm{PGL}(m,q)$ is a 2-transitive group on V, which follows from Proposition 1.62.

Remark 1.73. As we stated before, a 2-(v,k,λ) design with $b = v$ is called a *symmetric 2-design*. There are so many symmetric 2-designs. A projective space, i. e., a 2-$(n^2+n+1,n,1)$ design, is an example of a symmetric 2-design. Note that if we set $\ell = m-1$ in the previous problem, the design becomes a symmetric 2-design.

Remark 1.74. For a non-trivial t-(v,k,λ) design, a stronger and generalized *Fisher type inequality* holds: $b \geq \binom{v}{[t/2]}$. Moreover, if equality holds, it is called a tight t-design. (To be precise, when t is odd, a much stronger inequality holds (Chapter 3, Section 3.6), and it is called a *tight t-design*. Therefore we should think that the above definition of a tight t-design is valid only for the case where t is even.) A tight 2-design is a symmetric 2-design. We refer the reader to Chapter 3 for the generalized Fisher type inequality and its various extensions.

Latin squares and orthogonal arrays

Definition 1.75 (Latin square). Let $S = \{1,2,\ldots,n\}$. An $n \times n$ matrix $A = (a_{ij})$ $(a_{ij} \in S)$ is called a *Latin square* of order n if each number appears exactly once in each row and each column.

Definition 1.76 (Orthogonal Latin squares). Two Latin squares $A = (a_{ij})$ and $B = (b_{ij})$ of the same order n are said to be *orthogonal* if each element (x,y) in $S \times S$ appears exactly once in n^2 pairs $\{(a_{ij},b_{ij})\}$ of elements. Moreover, r Latin squares $A_1,A_2,\ldots,A_r$ of order n are said to be mutually orthogonal if any two of them are orthogonal.

Definition 1.77 (Orthogonal square). An $n\times n$ matrix $A = (a_{ij})$ with entries in S is called a *square* of order n. In the same way as Latin squares, two squares $A = (a_{ij}), B = (b_{ij})$ of order n are said to be *orthogonal* if any element (x,y) in $S \times S$ appears exactly once in n^2 pairs $\{(a_{ij},b_{ij})\}$ of elements. Moreover, r squares $A_1,A_2,\ldots,A_r$ of order n are said to be mutually orthogonal if any two of them are orthogonal.

More generally, we define the following concept.

Definition 1.78 (Orthogonal array). Let $S = \{1,2,\ldots,n\}$. An $n^2 \times k$ matrix $A = (a_{ij})$ with entries in S is called an orthogonal array $\mathrm{OA}(n,k)$ if for any two columns of A, say, the p-th and q-th columns, each element (x,y) in $S \times S$ appears exactly once in $\{(a_{ip},a_{iq})\}$ $(1 \leq i \leq n^2)$.

Theorem 1.79. *The following three conditions are equivalent:*

(1) *There exists* $\mathrm{OA}(n,k)$.
(2) *There exist* $k-2$ *mutually orthogonal Latin squares of order* n.
(3) *There exist* k *mutually orthogonal squares of order* n.

Proof. The proof is left as an exercise for the reader. □

The following theorem holds. The proof is also left as an exercise for the reader.

Theorem 1.80. *The following five conditions are equivalent:*
(1) *There exists* $\mathrm{OA}(n, n+1)$.
(2) *There exist* $n+1$ *mutually orthogonal squares of order* n.
(3) *There exist* $n-1$ *mutually orthogonal Latin squares of order* n.
(4) *There exists an affine plane of order* n *(i. e.,* $2\text{-}(n^2, n, 1)$ *design).*
(5) *There exists a projective plane of order* n *(i. e.,* $2\text{-}(n^2+n+1, n+1, 1)$ *design).*

Remark 1.81. A $2\text{-}(v, k, \lambda)$ design (BIBD) or a Steiner system $2\text{-}(v, k, 1)$ are special cases of t-designs. Likewise, $\mathrm{OA}(n, k)$ can be considered as a special case of a more general orthogonal array $t\text{-}\mathrm{OA}(n, k; \lambda)$.

Next we define a general orthogonal array $t\text{-}\mathrm{OA}(n, k; \lambda)$.

Definition 1.82 (Orthogonal array $t\text{-}OA(n, k; \lambda)$). A $\lambda n^t \times k$ matrix $A = (a_{ij})$ with entries in S is called an orthogonal array $t\text{-}\mathrm{OA}(n, k; \lambda)$ if the following holds. A matrix constructed from any t columns of A contains any element $(x_1, \ldots, x_t)$ of a direct product S^t of S exactly λ times. (Clearly, if $t = 2$, $\lambda = 1$, we obtain an orthogonal array $\mathrm{OA}(n, k)$ which we first defined.) Note that this general orthogonal array $t\text{-}\mathrm{OA}(n, k; \lambda)$ will appear as a t-design in the Hamming association scheme and will play an important role in the following chapters (the last part of Section 3.3 in Chapter 3, and Chapter 4).

Disproof of Euler's conjecture

It is clear that there exists a Latin square of order n for any natural number n. For instance, the matrix with the first row $(1, 2, \ldots, n)$ whose second and subsequent rows are obtained by permuting the first row cyclically gives an example of a Latin square. Then, for which n do two mutually orthogonal Latin squares exist? It is relatively easy to show the existence if n is odd or a multiple of 4. Also clearly, there is no such example for $n = 2$. When $n = 6$, Euler proposed and studied the 36 officers problem: Is it possible to arrange 36 officers from 6 regiments and 6 ranks in the square so that each row and column contains all regiments and ranks? He had a feeling that it is impossible. Moreover, in 1782 he made a conjecture that it is impossible for $n = 4k+2$. In 1900, the French mathematician Tarry first proved that it is impossible for $n = 6$ [456]. The proof was made by a brute force method and there was a dispute over the rigor of the proof, but now it can be easily verified by using computers. After that, Euler's conjecture was believed to be true for a long time, and papers of false proofs were published. For example, in the first edition of a Japanese book written by Teiji Takagi [446], he introduced a proof that Euler's conjecture is true, which is, of course, wrong. In 1960, Bose, Shrikhande, and Parker [102] gave a construction of two mutually orthogonal Latin squares for all $n = 4k + 2$ except for $n = 2, 6$, and disproved Euler's conjecture. It was a great surprise at that time.

Now Euler's conjecture was disproved, but it does not mean the study of this area ended. For example, the problem of the maximum number $N(n)$ of mutually orthogonal Latin squares of order n is interesting and has been studied a lot. In general, $2 \leq N(n) \leq n-1$ holds except for $n = 2, 6$, and $N(n) = n-1$ if and only if there exists a projective plane (or an affine plane) of order n. If $n = 10$, for instance, we can conclude that $2 \leq N(10) \leq 8$, but the problem of determining the value of $N(10)$, or determining whether $N(10) \geq 3$ or not is still open. On the other hand, there are some cases where $N(n)$ is reasonably well estimated (for example, see [210]).

Hadamard matrices

Consider an $n \times n$ matrix $A = (a_{i,j})$. Here $a_{i,j}$ is a real number such that $|a_{i,j}| \leq 1$ for each i, j. Then for the determinant $\det A$ of $A = (a_{i,j})$, we have $|\det A| \leq n^{n/2}$. (Why? Answer: Denote the row vectors of the matrix A by $\boldsymbol{a}_1, \boldsymbol{a}_2, \ldots, \boldsymbol{a}_n$. Note that $|\det A|$ is an absolute value of the volume of the n-parallelotope spanned by the vectors $\boldsymbol{a}_1, \boldsymbol{a}_2, \ldots, \boldsymbol{a}_n$. Now the length of each vector $\boldsymbol{a}_i$ is $\sqrt{a_{i1}^2 + a_{i2}^2 + \cdots + a_{in}^2}$, which is bounded from above by $\sqrt{n}$ since the absolute value of each a_{ij} is bounded from above by 1. Therefore the volume of the n-parallelotope takes the maximum absolute value if and only if the length of $\boldsymbol{a}_i$ equals $\sqrt{n}$ for all i, i. e., every $a_{i,j}$ is 1 or -1, and the vectors $\boldsymbol{a}_1, \boldsymbol{a}_2, \ldots, \boldsymbol{a}_n$ are mutually orthogonal. Thus, $|\det A|$ is bounded from above by $n^{n/2}$.)

We consider the case the equality $|\det A| = n^{n/2}$ holds. If equality holds, the matrix A is called a *Hadamard matrix*. Namely, we have the following.

Definition 1.83 (Hadamard matrix). An $n \times n$ matrix H is said to be a Hadamard matrix of order n if each entry of H is either 1 or -1 and $|\det H| = n^{n/2}$.

Immediately, we obtain the following proposition.

Proposition 1.84. *For an $n \times n$ matrix H with each entry 1 or -1, the following three conditions are equivalent:*
(1) $|\det H| = n^{n/2}$ *(i. e., H is a Hadamard matrix);*
(2) $H^tH = nI$ *(the orthogonality relation for the rows);*
(3) ${}^tHH = nI$ *(the orthogonality relation for the columns).*

Usually we define Hadamard matrices by using the orthogonality condition such as (2) or (3) in the above proposition. The major problems of this area are determining n such that a Hadamard matrix of order n exists, and classifying Hadamard matrices of order n for each n, or if the complete classification is impossible, classifying them in some form.

Example 1.85. The following matrices are Hadamard matrices:
(1) $H = [1]$;
(2) $H = \left[\begin{smallmatrix} 1 & 1 \\ 1 & -1 \end{smallmatrix}\right]$;
(3) $H = \begin{bmatrix} 1 & 1 & 1 & 1 \\ 1 & 1 & -1 & -1 \\ 1 & -1 & 1 & -1 \\ 1 & -1 & -1 & 1 \end{bmatrix}, H' = \begin{bmatrix} 1 & -1 & 1 & 1 \\ 1 & -1 & -1 & -1 \\ -1 & -1 & -1 & 1 \\ 1 & 1 & -1 & 1 \end{bmatrix}$.

The following theorem is fundamental.

Theorem 1.86. *If a Hadamard matrix of order n exists, n is* 1, 2, *or a multiple of* 4.

Remark 1.87. The existence problem of a Hadamard matrix of order $4k$ for any natural number k is a famous open problem. The conjecture that there exists a Hadamard matrix of order $4k$ for any natural number k is called the *Hadamard conjecture*. We have an impression that the number of specialists who believe the conjecture is true is bigger than the number of those who do not, but we are not sure. At the time we started writing this book, 428 was the smallest order of a Hadamard matrix whose existence was unknown, but later the existence was proved by Kharaghani and Tayfeh-Rezaie in 2005 [274]. As of December 2015, the smallest order of a Hadamard matrix whose existence is unknown is possibly $668 = 4 \cdot 167$ [274].

Proof of Theorem 1.86. A Hadamard matrix remains a Hadamard matrix after permutations or negating of rows and columns. (We say two Hadamard matrices are *equivalent*, if one Hadamard matrix is obtained from the other by these transformations. For example, it can be verified that H and H' in the above Example 1.85 (3) are equivalent. Note also that this defines an equivalence relation.) From now on, we assume $n \geq 3$. By using the above transformations, we can set the entries of the first row of H as all 1's. Consider the second and third rows. Let x, y, z, w be the numbers of columns whose pair of the entries of the second and the third rows becomes $(1, 1), (1, -1), (-1, 1), (-1, -1)$, respectively. If we calculate the inner products of the first row and itself, the first row and the second row, the first row and the third row, and the second row and the third row, by the orthogonality of rows, we obtain $x+y+z+w = n$, $x + y - z - w = 0$, $x - y + z - w = 0$, $x - y - z + w = 0$. Thus $x = y = z = w = n/4$ holds, and n must be a multiple of 4. □

Hadamard designs

Let H be a Hadamard matrix of order n. Then, by multiplying some rows and columns by -1 if necessary, we can normalize H so that the first row and the first column are all 1's, so we may write

$$H = \begin{bmatrix} 1 & 1 & \cdots & 1 \\ 1 & & & \\ \vdots & & H' & \\ 1 & & & \end{bmatrix},$$

where H' denotes an $(n-1) \times (n-1)$ matrix. The $(0, 1)$-matrix obtained from H' by substituting -1 with 0 is denoted by A. Let $n = 4t$. Then the number of 1's which appear in each row (or each column) of A is $2t - 1$, and the number of common 1's which appear in two distinct columns of A is $t - 1$. Namely, we have $AJ = (2t-1)J$ and ${}^tAA = tI + (t-1)J$. Hence A is an incidence matrix of a 2-$(4t-1, 2t-1, t-1)$ design. To sum up, we can obtain a 2-$(4t-1, 2t-1, t-1)$ design from a Hadamard matrix of order $n = 4t$.

A design with these parameters, i. e., a 2-$(4t-1, 2t-1, t-1)$ design, is called a *Hadamard 2-design*. Conversely, it is clear that a Hadamard matrix of order $4n$ can be constructed from a 2-$(4t-1, 2t-1, t-1)$ design. It is also clear that Hadamard matrices of order $4t$ constructed from two isomorphic 2-$(4t-1, 2t-1, t-1)$ designs are equivalent. Note that, however, given two equivalent Hadamard matrices of order $n = 4t$, we can normalize the first rows and columns of them as all 1's in various ways, so there is a possibility that we may obtain several non-isomorphic 2-$(4t-1, 2t-1, t-1)$ designs. (This in fact occurs.)

Remark 1.88. Among symmetric 2-(v, k, λ) designs, a projective plane, i. e., a 2-$(n^2 + n+1, n+1, 1)$ design, and a Hadamard design, i. e., a 2-$(4t-1, 2t-1, t-1)$ design, can be viewed as two extremes and other symmetric 2-(v, k, λ) designs take a position in the middle of them in the following sense. For symmetric 2-(v, k, λ) designs, we set $n = k-\lambda$ and compare the numbers v of the points of designs as functions of the variable n. For a projective plane, v is $n^2 + n + 1$, and for a Hadamard design, v is $4n - 1$. For other symmetric 2-(v, k, λ) designs, it is known that $4n - 1 \le v \le n^2 + n + 1$. (For details, see [523, page 112].) If we regard the number of points as a function of $k-\lambda = n$, a projective plane takes the largest value and a Hadamard design takes the smallest. Note also that a symmetric 2-(v, k, λ) design and the complementary design, i. e., a 2-$(v, v-k, v-2k+\lambda)$ design, take the same values of $n = k - \lambda$.

Kronecker products of matrices

The Kronecker product of an $n \times n$ matrix $A = (a_{ij})$ and an $m \times m$ matrix $B = (b_{kl})$ is an $nm \times nm$ matrix defined as follows:

$$A \otimes B = \begin{bmatrix} a_{11}B & a_{12}B & \cdots & a_{1n}B \\ a_{21}B & a_{22}B & \cdots & a_{2n}B \\ \vdots & \vdots & \ddots & \vdots \\ a_{n1}B & a_{n2}B & \cdots & a_{nn}B \end{bmatrix}.$$

Then the following equation hold:
(1) ${}^t(A \otimes B) = {}^tA \otimes {}^tB$;
(2) if A, C are $n \times n$ matrices and B, D are $m \times m$ matrices, we have

$$(A \otimes B)(C \otimes D) = (AC) \otimes (BD).$$

From these equations, if H_n is a Hadamard matrix of order n and H_m is a Hadamard matrix of order m, then we can verify that $H_{mn} = H_n \otimes H_m$ is a Hadamard matrix of order nm. (We recommend the reader to verify it.)

Immediately this implies that there exists a Hadamard matrix of order a power of 2, say, 2^r. The reader who has knowledge of group theory will understand that this Hadamard matrix coincides with the character table of the elementary abelian group of order 2^r. Therefore the problem of existence of a Hadamard matrix of general order is reduced to the problem of existence of a Hadamard matrix of order 4×(odd number). Next, we discuss the existence of Hadamard matrices of special orders.

Let F_q^* be the multiplicative group of a finite field F_q with q elements, where $q = p^r$ for a prime p. Also note that F_q^* is a cyclic group of order $q-1$. (Please go over a textbook on algebra.) Consider the case q is odd. Define a map χ from F_q^* to $\{1, -1\}$ by $\chi(x) = 1$ if x is a square and $\chi(x) = -1$ otherwise. Then χ becomes a homomorphism. Moreover the kernel of χ coincides with $\{a^2 \mid a \in \mathbb{F}_q^*\}$ (the set of *quadratic residues*) and becomes a subgroup of F_q^* of index 2. Sometimes we extend the domain of the map χ to the whole F_q by setting $\chi(0) = 0$. For example, for the prime field F_7 of order 7, we have the following table.

a	0	1	2	3	4	5	6
$\chi(a)$	0	1	1	−1	1	−1	−1

Also for the case of F_{23}, the set of a with $\chi(a) = 1$ (i. e., quadratic residues) consists of $1, 2, 3, 4, 6, 8, 9, 12, 13, 16, 18$, and the value of χ is -1 for other non-zero elements $5, 7, 10, 11, 14, 15, 17, 19, 20, 21, 22$.

Remark 1.89. We should pay attention to the difference between a finite field F_q and a finite ring $\mathbb{Z}/q\mathbb{Z}$. Of course, they coincide if q is a prime, but they do not otherwise. We also note that if q is odd, F_q^* becomes a cyclic group of even order, which implies $\chi(-1) = 1$ if $q \equiv 1 \pmod 4$ and $\chi(-1) = -1$ if $q \equiv 3 \pmod 4$.

Lemma 1.90. *For the function χ over a finite field F_q (q is odd) defined above,*

$$\sum_{b \in F_q} \chi(b)\chi(b + c) = -1$$

holds for any $c \in F_q^*$.

Proof. By $\chi(0) = 0$, we have $\chi(0)\chi(0 + c) = 0$. If $b \neq 0$, there is a unique $z \in F_q$ such that $b + c = bz$. Since $c \in F_q^*$, we have $z \neq 1$. We put $\tau(b) = z$ for this correspondence. We have $\tau(-c) = 0$. Therefore τ becomes a bijection from F_q^* to $F_q \setminus \{1\}$. Since $\chi(0) = 0$, we have

$$\begin{aligned}\sum_{b \in F_q} \chi(b)\chi(b + c) &= \sum_{b \in F_q^*} \chi(b)\chi(b\tau(b)) = \sum_{b \in F_q^*} \chi(b^2)\chi(\tau(b)) \\ &= \sum_{b \in F_q^*} \chi(\tau(b)) = \sum_{z \in F_q, z \neq 1} \chi(z) \\ &= \sum_{z \in F_q} \chi(z) - \chi(1) = -1.\end{aligned}$$ □

Remark 1.91. The reader who is familiar with group theory will soon notice that the equation in the above lemma can be obtained immediately from the second orthogonality relation of the character table of a cyclic group (or a finite abelian group).

Next, we arrange the order of the elements in F_q as $a_1, a_2, \ldots, a_q$ so that $a_{q+1-i} = -a_i$ $(i = 1, 2, \ldots, q)$ holds. Put $q_{ij} = \chi(a_i - a_j)$. Define a matrix Q as follows:

$$Q = (q_{ij})_{1\le i\le q, 1\le j\le q}.$$

Note that since $q_{ij} = \chi(a_i - a_j) = \chi(-1)\chi(a_j - a_i)$, we have $q_{ij} = -q_{ji}$ if $q \equiv 3 \pmod 4$ and $q_{ij} = q_{ji}$ if $q \equiv 1 \pmod 4$.

Lemma 1.92. *For the matrix Q defined above, $Q\,{}^tQ = qI - J$, $QJ = JQ = 0$ hold.*

Proof. Set $Q\,{}^tQ = B = (b_{ij})$. We have $b_{ij} = \sum_{t=1}^{q} \chi(a_i - a_t)\chi(a_j - a_t)$. If $i = j$, we have $b_{ij} = q - 1$. If $i \ne j$, by applying Lemma 1.90 to the case $b = a_i - a_t$, $c = a_j - a_i$, we obtain $b_{ij} = -1$. Therefore we have $Q\,{}^tQ = qI - J$; $QJ = JQ = 0$ follows from $\sum_{z\in F_q} \chi(z) = 0$. □

Theorem 1.93. *If $q = p^r \equiv 3 \pmod 4$, there exists a Hadamard matrix of order $n = p^r + 1$.*

Proof. Assume $q = p^r \equiv 3 \pmod 4$. Then note that the matrix Q defined above is a *skew symmetric matrix*. Now we put

$$S^{(-)}_{q+1} = \begin{bmatrix} 0 & 1 & \cdots & 1 \\ -1 & & & \\ \vdots & & Q & \\ -1 & & & \end{bmatrix}.$$

By Lemma 1.92, ${}^tS^{(-)}_{q+1} = -S^{(-)}_{q+1}$, $S^{(-)}_{q+1}\,{}^tS^{(-)}_{q+1} = qI_{q+1}$ hold. Put $H_{q+1} = I_{q+1} + S^{(-)}_{q+1}$. Then we have

$$\begin{aligned} H_{q+1}\,{}^tH_{q+1} &= (I_{q+1} + S^{(-)}_{q+1})\,{}^t(I_{q+1} + S^{(-)}_{q+1}) = (I_{q+1} + S^{(-)}_{q+1})(I_{q+1} + {}^tS^{(-)}_{q+1}) \\ &= I_{q+1} + S^{(-)}_{q+1} + {}^tS^{(-)}_{q+1} + S^{(-)}_{q+1}\,{}^tS^{(-)}_{q+1} = (q+1)I. \end{aligned} \tag{1.16}$$

So H_{q+1} is a Hadamard matrix of order $n = q + 1$. □

The Hadamard matrix constructed in the above theorem is called a *Paley type Hadamard matrix*.

Remark 1.94. When $q = p^r \equiv 1 \pmod 4$, consider

$$S^{(+)}_{q+1} = \begin{bmatrix} 0 & 1 & \cdots & 1 \\ 1 & & & \\ \vdots & & Q & \\ 1 & & & \end{bmatrix}.$$

Put

$$H_{2(q+1)} = S^{(+)}_{q+1} \otimes \begin{bmatrix} 1 & 1 \\ 1 & -1 \end{bmatrix} + I_{q+1} \otimes \begin{bmatrix} 1 & -1 \\ -1 & -1 \end{bmatrix}.$$

Then $H_{2(q+1)}$ becomes a Hadamard matrix of order $2(q + 1)$. This Hadamard matrix is also called a (Type II) Paley type Hadamard matrix.

The conjecture of Noboru Ito and Hadamard groups

Various constructions of Hadamard matrices are known, but the scope of each application is not so wide and the Hadamard conjecture seems far from completion. However, there might be a possibility that this problem can be solved very easily. We introduce a conjecture which may become a clue for the construction.

Let $G(n)$ be a group defined by $G(n) = \langle a, b \mid a^{4n} = e, a^{2n} = b^2, b^{-1}ab = a^{-1}\rangle$; $G(n)$ has order $8n$ and has a unique element $e^* = a^{2n} = b^2$ of order 2. The element e^* is central in $G(n)$. A Sylow 2-subgroup of $G(n)$ is a generalized quaternion group. The quotient group of $G(n)$ by the normal subgroup $\langle e^*\rangle$ of order 2 is a cyclic group of order $4n$. Next, let D be a $4n$-element subset of $G(n)$ such that $G(n) = D \cup De^*$ and $D \cap De^* = \emptyset$. Such D is called a *transversal* of $G(n)$ with respect to $\langle e^*\rangle$.

Conjecture (Noboru Ito ([248], page 370)). *For any n, there exists a transversal D of a group $G(n)$ such that $|D \cap Dg| = 2n$ holds for any $g \in G(n)$ $(g \neq e, e^*)$.*

A group $G(n)$ having such D is called a *Hadamard group*. It is known that if $G(n)$ is a Hadamard group, there exists a Hadamard matrix of order $4n$. In addition, the automorphism group of the Hadamard matrix obtained above contains $G(n)$ (as a part of it) ([247], [248]). In other words, this conjecture says that "a group $G(n)$ is a Hadamard group."

Remark 1.95. The Hadamard matrices of order $n \leq 28$ are completely classified. It is known that the number of equivalence classes is 1 if $n = 1, 2, 4, 8, 12$; is 5 if $n = 16$; is 3 if $n = 20$ (M. Hall, Jr. [208, 209]); is 60 if $n = 24$ (Ito, Leon, and Longyear [249], Kimura [277]); and is 487 if $n = 28$ (Kimura [278]). For $n \geq 32$, the classification of Hadamard matrices of order n is open for any multiple n of 4, but recently it was reportedly solved for $n = 32$ [275]. It seems that there is no good estimation of the number of equivalent Hadamard matrices. We are not sure how important it is, but a formula which (in principle) gives the number of equivalence classes of Hadamard matrices of order n was given by Eliahou [178] in 1994. In practice, we cannot calculate this value, and so far we do not understand what the point of this formula is. Anyway, the challenge of solving the Hadamard conjecture may attract amateur mathematicians as well.

1.5 Introduction to classical coding theory

Codes

The essence of design theory is, roughly speaking, "to find a subset which globally approximates the whole set very well." Likewise the aim of coding theory is "to find a subset whose points are locally scattered as far as possible in the whole set." Here we can consider anything as "the whole set." For example, we consider the n-dimensional vector space F_q^n over a finite field F_q. In this case, a subset C of F_q^n is called a code, and

n is called the length of the code C. In particular, if C is a subspace, it is called a linear code.

The essence of coding theory is to find C with $|C| = N$ for a fixed N such that the *minimum distance*

$$d(C) = \min_{x,y \in C, x \neq y} \partial(\boldsymbol{x}, \boldsymbol{y})$$

is as large as possible, or to find C with $d(C) = d$ for a fixed d such that $|C|$ is as large as possible. Here, $\partial(\boldsymbol{x}, \boldsymbol{y})$ usually denotes the *Hamming distance*. Namely, for $\boldsymbol{x} = (x_1, x_2, \ldots, x_n)$, $\boldsymbol{y} = (y_1, y_2, \ldots, y_n) \in F_q^n$, define $\partial(\boldsymbol{x}, \boldsymbol{y}) = |\{j \mid 1 \leq j \leq n,\ x_j \neq y_j\}|$. Of course, there are various arrangements and extensions, such as considering other distances or other spaces, which will appear in later chapters. Here we give some notation and symbols. The ***e**-neighbor* of an element $\boldsymbol{x} \in F_q^n$ is defined as $\Sigma_e(\boldsymbol{x}) = \{\boldsymbol{y} \in F_q^n \mid \partial(\boldsymbol{x}, \boldsymbol{y}) \leq e\}$.

Firstly, we take a look at how to use codes for *communications*.

A *binary channel* is a channel where one can transmit one of two symbols $\{0, 1\}$, and it is said to be *symmetric* if the probabilities that 0 is changed to 1 and that 1 is changed to 0 due to noise are both equal to p $(0 \leq p \leq 1/2)$. Therefore if a *message* $\boldsymbol{u} = u_1 u_2 \cdots u_k$ of length k where $u_i \in \{0, 1\}$ is transmitted, the probability that it is sent correctly is $(1-p)^k$. Of course, the probability that one u_i is sent correctly is $1-p$. Suppose that, when $k = 1$, the message 000 is sent instead of 0 and the message 111 is sent instead of 1. Even if the received message is not necessarily 000 or 111, we regard it as 0 if the received message contains more than two 0's, and we regard it as 1 if the received message contains more than two 1's. Then the probability that a message of length 3 is sent correctly is $(1-p)^3 + 3p(1-p)^2$. This is larger than $1-p$ if p is small. In other words, one u_i is sent more correctly. In this way, by sending a longer message $\boldsymbol{x} = x_1 x_2 \cdots x_n$ of length n instead of a message $\boldsymbol{u} = u_1 u_2 \cdots u_k$ of length k, we obtain more correct transmission. This is the idea of codes. The above example is the case where $k = 1$ and $n = 3$. However, it is thought that if the ratio k/n of k to n is larger (or it is closer to 1), the transmission is more efficient. On the other hand, a code with higher probability of a correct transmission is better. Here the process of constructing $\boldsymbol{x} = x_1 x_2 \cdots x_n$ from $\boldsymbol{u} = u_1 u_2 \cdots u_k$ is called a *coding*, and the process of inferring the original message $\boldsymbol{u} = u_1 u_2 \cdots u_k$ from the received message $\hat{\boldsymbol{x}} = \hat{x}_1 \hat{x}_2 \cdots \hat{x}_n$ (where $\hat{x}_i$ is 0 or 1) which is obtained by the transmission of the coding message $\boldsymbol{x} = x_1 x_2 \cdots x_n$ is called a *decoding*. In practice, a code with easy coding and decoding is good. Here, however, we ignore this aspect and consider that a code with larger ratio k/n and a higher probability of correct transmission is better.

We said that the higher the probability of the correct transmission, the better the code is, but it is quite complicated to find the probability actually. Therefore instead of finding the probability precisely, we consider an *e-error correcting code* (or *e-code*) as the first approximation. Namely, let F_2^k be the set of messages of length k and we embed C into a k-dimensional subspace of F_2^n. In other words, we replace a message

$\boldsymbol{u} = u_1 u_2 \cdots u_k$ by an element $\boldsymbol{x} = x_1 x_2 \cdots x_n$ of C ($\subset F_2^n$) (coding). Then C is called an *e-error correcting code* (or *e-code*) if for any 2 distinct elements $\boldsymbol{x}, \boldsymbol{y}$ of C, $\Sigma_e(\boldsymbol{x}) \cap \Sigma_e(\boldsymbol{y}) = \emptyset$ holds. Here if the received message $\hat{\boldsymbol{x}} = \hat{x}_1 \hat{x}_2 \cdots \hat{x}_n$ belongs to one of $\Sigma_e(\boldsymbol{z})$ ($\boldsymbol{z} \in C$), we can infer the element $\boldsymbol{x}$ of C which must have been sent, and decode the original message $\boldsymbol{u}$ (an element of F_2^k). If e is larger, an e-code C admits the higher probability of the correct transmission of a message. Hence the problem is to find a code C satisfying the following two requirements as much as possible:

(1) Let k/n be as large as possible.
(2) Find C which becomes an e-code where e is as large as possible.

These two requirements are incompatible. So the purpose of coding theory is to find an equilibrium such that both requirements are satisfied as much as possible.

Proposition 1.96. *A code C with minimum distance d is a $[\frac{d-1}{2}]$-code.*

Proof. This follows immediately from the fact that the Hamming distance $\partial(\boldsymbol{x}, \boldsymbol{y})$ satisfies the triangle inequality. □

Remark 1.97. Therefore the above requirement (2) to let e be as large as possible can be replaced by the following condition:
(2′) Let the minimum distance d of a code C be as large as possible.

Remark 1.98. So far, when we discuss a code, we consider a *binary* channel, but theoretically, it is possible to consider a q-ary channel. Again the purpose is to find $C \subset \mathbb{F}_q^n$ satisfying requirements (1), (2) as much as possible. Here C is called a *q-ary code*. The above Proposition 1.96 remains true for a q-ary code. The distance $\partial(\boldsymbol{x}, \boldsymbol{y})$ is a Hamming distance again. Moreover, in the above explanation, C is a k-dimensional subspace, but we may consider a subset which is not a subspace. In this case, we should use $\log_q(|C|)/n$ instead of k/n.

Remark 1.99. In terms of the above requirements (1), (2), a code with the following condition would be an ideal e-code:

(P) For a code C, $\{\Sigma_e(\boldsymbol{x}) \mid \boldsymbol{x} \in C\}$ becomes a partition of $C \subset F_q^n$. Namely, $\{\Sigma_e(\boldsymbol{x}) \mid \boldsymbol{x} \in C\}$ covers $C \subset \mathbb{F}_q^n$ without overlappings.

An e-code satisfying such ideal condition (P) is called a *perfect code*. In what follows, we make various comments on such codes. In general, this condition is too strong. So, perfect codes rarely exist, but if they exist, they are very good codes. Therefore the problem on the existence of such codes is quite interesting and problems on the existence and the classification of codes which are similar to perfect codes are also interesting. We will discuss such problems in Chapter 4.

Definition 1.100 (Parameters of a code). If a code C is a subset of F_q^n, it is called a code of *length n*. Moreover, if C is a k-dimensional subspace of F_q^n with minimum distance d, C is called an $[n, k, d]$-code. If the minimum distance d of C is not specified, we simply call it an $[n,k]$-code. In addition, if C is a subset of F_q^n with minimum distance d, C is called an $(n,|C|,d)$-code. (For the above notation, we do not specify that C is a code over a field of order q, but if we want to specify that, we write it as an $[n,k,d]_q$-code. For $q=2$, we usually omit the index 2 and write $[n,k,d]$-code.)

Example 1.101 (Hamming $[7,4,3]$-code). We consider a system of linear equations:

$$\begin{array}{cccccccl} x_1 & & +x_3 & +x_4 & +x_5 & & & =0, \\ x_1 & +x_2 & & +x_4 & & +x_6 & & =0, \\ & x_2 & +x_3 & +x_4 & & & +x_7 & =0. \end{array} \tag{1.17}$$

If we write it in matrix form, we have

$$\begin{bmatrix} 1 & 0 & 1 & 1 & 1 & 0 & 0 \\ 1 & 1 & 0 & 1 & 0 & 1 & 0 \\ 0 & 1 & 1 & 1 & 0 & 0 & 1 \end{bmatrix} \begin{bmatrix} x_1 \\ x_2 \\ \vdots \\ x_7 \end{bmatrix} = \begin{bmatrix} 0 \\ 0 \\ 0 \end{bmatrix},$$

where the coefficients and variables of the above system of equations are elements of F_2. Let C be the subspace consisting of the solutions. Each of x_5, x_6, x_7 is determined by the first, the second, and the third equation, respectively, so we can choose x_1, x_2, x_3, x_4 freely for the solutions. In other words, C forms a 4-dimensional subspace. Moreover, we can easily verify that the minimum distance d of C is 3. Therefore, this code C is a $[7,4,3]$-code. We can also easily verify that this code is a perfect 1-code. This code is called a Hamming $[7,4,3]$-code.

Definition 1.102 (Isomorphisms of codes). Two codes $C, C' \subset F_2^n$ are isomorphic if there exists a permutation on the n coordinates which maps C to C'. The set of permutations (on the n coordinates) which map C to itself forms a group. We call it the automorphism group of C and denote it by $\mathrm{Aut}(C)$.

Remark 1.103 (Isomorphisms of codes over non-binary finite fields.). We should be careful in defining isomorphisms of codes over non-binary finite fields. There are several different definitions. We call isomorphisms defined in Definition 1.102 *permutation isomorphisms*. If C is mapped to C' by multiplying a non-zero element of the field to each coordinate of C or by permuting n coordinates of C, we say $C, C' \subset F_q^n$ are *monomially isomorphic*. The automorphism group of C in this sense is called the monomial automorphism. Moreover, we sometimes consider actions of the Galois group, i. e., the automorphism group of the field, over the prime field of the field besides monomial isomorphisms, and in this case the automorphism group of C is called the full automorphism group. The permutation automorphism, the monomial automorphism, and the full automorphism coincide for a code over the binary field.

Definition 1.104 (Hamming codes and perfect 1-codes). Consider an $r \times (2^r - 1)$-matrix H whose columns consist of the base-2 expansions of the natural numbers from 1 to 2^r.

Here each entry of H is an element of the binary field F_2. In particular, if $r = 3$, we have

$$H = \begin{bmatrix} 0 & 0 & 0 & 1 & 1 & 1 & 1 \\ 0 & 1 & 1 & 0 & 0 & 1 & 1 \\ 1 & 0 & 1 & 0 & 1 & 0 & 1 \end{bmatrix}.$$

(This matrix can be obtained by a suitable permutation of columns of the matrix H for the $[7,4,3]$-code, which was introduced in Example 1.101 without explaining where the matrix comes from. Therefore if we consider the subspace generated by the rows of the 3×7-matrix as a code over F_2^7, these two $[7,4,3]$-codes are isomorphic.)

For general r, let C be the set of row vectors $\mathbf{x} = (x_1, x_2, \ldots, x_{2^r-1}) \in F_2^{2^r-1}$ such that for an $r \times (2^r - 1)$-matrix H, the following holds:

$$H\,{}^t\mathbf{x} = \mathbf{0}.$$

We can easily verify that C is a $[2^r - 1, 2^r - r - 1]$-code. Since any 2 columns of H are distinct, the weight (the number of non-zero coordinates) of a solution $\mathbf{x}$ of $H\,{}^t\mathbf{x} = \mathbf{0}$ cannot be less than or equal to 2. Therefore C becomes a $[2^r - 1, 2^r - r - 1, 3]$-code. We can also verify that this code C is a perfect 1-code. This code is called the Hamming $[2^r - 1, 2^r - r - 1, 3]$-code. In general, for a linear code C, the matrix H having such properties is called the *parity check matrix* of C.

The Hamming [8, 4, 4]-code and extension of a code

Consider the 3×7-matrix H given in Definition 1.104. Clearly, the solution space of $H\,{}^t\mathbf{x} = \mathbf{0}$ is a 4-dimensional subspace over F_2 whose basis consists of the three row vectors of H and $(1,1,1,1,1,1,1)$. Therefore the row vectors of the following matrix G form a basis of C:

$$G = \begin{bmatrix} 0 & 0 & 0 & 1 & 1 & 1 & 1 \\ 0 & 1 & 1 & 0 & 0 & 1 & 1 \\ 1 & 0 & 1 & 0 & 1 & 0 & 1 \\ 1 & 1 & 1 & 1 & 1 & 1 & 1 \end{bmatrix}.$$

In general, the matrix G whose row vectors generate a linear code C is called the *generator matrix* of C. We can define a linear code by using the parity check matrix as we stated before and by using the generator matrix as well. If the parity check matrix of a linear code C is the following $(n-k) \times n$-matrix which has the form

$$H = (A, I_{n-k}),$$

then the $k \times n$-matrix G defined by

$$G = (I_k, -{}^tA)$$

becomes a generator matrix of C. The converse is also true. Note that we add the negative sign before ${}^tA$, which is not needed for the case of the binary field, so that it holds for a general case of a finite field F_q.

Now we go back to the Hamming $[7,4,3]$-code, and consider the generator matrix G. Consider a 4×8-matrix obtained from G by adding a column (in this case, the first column) so that each row sum becomes 0:

$$\tilde{G} = \begin{bmatrix} 0 & 0 & 0 & 0 & 1 & 1 & 1 & 1 \\ 0 & 0 & 1 & 1 & 0 & 0 & 1 & 1 \\ 0 & 1 & 0 & 1 & 0 & 1 & 0 & 1 \\ 1 & 1 & 1 & 1 & 1 & 1 & 1 & 1 \end{bmatrix}.$$

We can easily verify that the code $\tilde{C}$ defined by the generator matrix $\tilde{G}$ is an $[8,4,4]$-code. This code $\tilde{C}$ is called the *extended Hamming* $[8,4,4]$*-code*. Note that a code defined by a generator matrix which is obtained from the original generator matrix of a code by adding a column so that each row sum equals 0 is called the *extended code* of an original code.

1.6 Examples and existence problems of codes

Weight enumerators of codes and dual codes

In what follows, we consider linear codes of F_2^n unless otherwise stated. We use the Hamming distance $\partial(-,-)$ defined in Section 1.5. We define the *weight* $w(\boldsymbol{x})$ of an element $\boldsymbol{x} = (x_1, x_2, \ldots, x_n)$ by

$$w(\boldsymbol{x}) = \big|\{i \mid x_i \neq 0\ (1 \le i \le n)\}\big| = \partial(\boldsymbol{x}, \boldsymbol{0}).$$

We also define a polynomial W_C with 2 variables x, y as follows:

$$W_C(x,y) = \sum_{\boldsymbol{c}\in C} x^{n-w(\boldsymbol{c})} y^{w(\boldsymbol{c})}. \tag{1.18}$$

Let A_i be the number of elements of C with weight i. Then we have

$$W_C(x,y) = \sum_{i=0}^{n} A_i x^{n-i} y^i. \tag{1.19}$$

This homogeneous polynomial of degree n is called the *weight enumerator* of a code. The weight enumerator of the Hamming $[7,4,3]$-code is

$$x^7 + 7x^4y^3 + 7x^3y^4 + y^7,$$

and the weight enumerator of the Hamming $[8,4,4]$-code is

$$x^8 + 14x^4y^4 + y^8.$$

We define the inner product of 2 elements $\boldsymbol{x} = (x_1, x_2, \dots, x_n)$, $\boldsymbol{y} = (y_1, y_2, \dots, y_n)$ of F_2^n by

$$\boldsymbol{x} \cdot \boldsymbol{y} = x_1y_1 + x_2y_2 + \cdots + x_ny_n \in F_2.$$

There are various choices of inner product for F_2^n; however, at present almost all the works of coding theory are limited to the case of the above inner product. It seems that it does not work for other inner products. Is there no possibility of using other inner products?[6]

Definition 1.105 (Dual codes). For a code C, we define the dual code $C^\perp$ as follows:

$$C^\perp = \{\boldsymbol{x} \in F_2^n \mid \boldsymbol{x} \cdot \boldsymbol{y} = 0,\ \forall \boldsymbol{y} \in C\}.$$

In particular, if $C^\perp = C$ holds, we call C the *self-dual code*.

Remark 1.106. It is obvious from linear algebra that if C is an $[n, k]$-code, $C^\perp$ is an $[n, n-k]$-code. We also note that if G and H are the generator matrix and the parity check matrix of a linear code C, respectively, then H and G are the generator matrix and the parity check matrix of the dual code $C^\perp$, respectively.

If the minimum distance of a code C is given, that of the dual code $C^\perp$ cannot be determined immediately in general; however, if the weight enumerator $W_C(x, y)$ of C is given, the weight enumerator $W_{C^\perp}(x, y)$ of the dual code $C^\perp$ can be computed in principle. The following *MacWilliams identity* supports this claim.

Theorem 1.107 (The MacWilliams identity). *For a linear code over the binary field F_2, the following holds:*

$$W_{C^\perp}(x, y) = \frac{1}{|C|} W_C(x + y, x - y).$$

In other words, if we denote the number of elements of $C^\perp$ of weight i by A_i', we have

$$\sum_{k=0}^{n} A_k' x^{n-k} y^k = \frac{1}{|C|} \sum_{l=0}^{n} A_l (x + y)^{n-l} (x - y)^l.$$

Note that for a code over F_q, we have

$$W_{C^\perp}(x, y) = \frac{1}{|C|} W_C(x + (q-1)y, x - y).$$

We omit the proof of this theorem, which is given in many books. As you will see in later chapters, this theorem is very important in coding theory.

---

6 Da Zhao pointed out that there are some comments on other bilinear forms in [373].

Binary Golay codes

Next, we define a [23, 12, 7]-code over the binary field and the extended [24, 12, 8]-code. They are called *binary Golay codes*, which are important in various senses. First, we define the Golay [24, 12, 8]-code as follows. Consider a 12×24-matrix $[I, G]$. We regard the entries of the matrix as elements of the binary field F_2. Here, I denotes the identity matrix of order 12, and G denotes the following square matrix of order 12:

$$\begin{bmatrix} 0&1&1&1&1&1&1&1&1&1&1&1\\ 1&1&1&0&1&1&1&0&0&0&1&0\\ 1&0&1&1&0&1&1&1&0&0&0&1\\ 1&1&0&1&1&0&1&1&1&0&0&0\\ 1&0&1&0&1&1&0&1&1&1&0&0\\ 1&0&0&1&0&1&1&0&1&1&1&0\\ 1&0&0&0&1&0&1&1&0&1&1&1\\ 1&1&0&0&0&1&0&1&1&0&1&1\\ 1&1&1&0&0&0&1&0&1&1&0&1\\ 1&1&1&1&0&0&0&1&0&1&1&0\\ 1&0&1&1&1&0&0&0&1&0&1&1\\ 1&1&0&1&1&1&0&0&0&1&0&1 \end{bmatrix}.$$

The 11×11-submatrix obtained by deleting the first row and column is the matrix whose rows are given by permuting $(1, 1, 0, 1, 1, 1, 0, 0, 0, 1, 0)$ cyclically. The vector $(1, 1, 0, 1, 1, 1, 0, 0, 0, 1, 0)$ is related to quadratic residues modulo 11. Namely, the first entry of the vector is 1 (regarding 0 as a quadratic residue), and for i with $2 \le i \le 11$, the i-th entry is 1 if $i - 1$ is a quadratic residue modulo 11, and 0 otherwise. For example, the 4-th entry is 1 since $3 \equiv 5^2 \pmod{11}$.

Let C be the linear code with the generator matrix $[I, G]$. Immediately it follows that C has dimension 12, and any two rows of the generator matrix are orthogonal with respect to the inner product over F_2^{24}. Moreover, we can show that the weight of each element of the code is a multiple of 4, and the minimum distance is 8. For the determination of the minimum distance, we need a detailed discussion, so we refer the reader to other books. It is, by using the term which will be stated later, a *doubly even* self-dual [24, 12, 8]-code. This code is called the Golay [24, 12, 8]-code. It is proved that a doubly even self-dual code with these parameters is unique, and equivalently, it is isomorphic to the Golay code. We omit the proof. The weight enumerator of this code is as follows:

$$W_C(x, y) = x^{24} + 759x^8y^{16} + 2576x^{12}y^{12} + 759x^{16}y^8 + y^{24}. \tag{1.20}$$

We can verify that the automorphism group of the Golay [24, 12, 8]-code is the Mathieu group M_{24}, a 5-transitive group on 24 letters. Moreover, a (truncated) code C' obtained from the Golay [24, 12, 8]-code by deleting one coordinate position becomes a

$[23, 12, 7]$-code. The weight enumerator of C' can be computed as follows, for which we also need some discussion (the code C is the extended code of C'):

$$W_{C'}(x,y) = x^{23} + 253x^7y^{16} + 506x^8y^{15} + 1288x^{11}y^{12} \\ + 1288x^{12}y^{11} + 506x^{15}y^8 + 253x^{16}y^7 + y^{23}.$$

It is very interesting that the $[23, 12, 7]$-code is a perfect 3-code. This follows immediately from the facts that the number of elements contained in the 3-neighbor of $\mathbf{c} \in C'$ is given by $|\Sigma_3(\mathbf{c})| = 1 + \binom{23}{1} + \binom{23}{2} + \binom{23}{3} = 2^{11}$ and that this code is a 3-code (since the minimum distance is 7). As will be stated in Chapter 4, the $[23, 12, 7]$-code is the only non-trivial perfect e-code over the binary field with $e \geq 2$. For the case of the ternary code, the following perfect 2-code is known.

Ternary Golay code

We consider the following 6×11-matrix:[7]

$$\begin{bmatrix} 1 & 0 & 0 & 0 & 0 & 0 & 1 & 1 & 1 & 1 & 1 \\ 0 & 1 & 0 & 0 & 0 & 0 & 0 & 1 & -1 & -1 & 1 \\ 0 & 0 & 1 & 0 & 0 & 0 & 1 & 0 & 1 & -1 & -1 \\ 0 & 0 & 0 & 1 & 0 & 0 & -1 & 1 & 0 & 1 & -1 \\ 0 & 0 & 0 & 0 & 1 & 0 & -1 & -1 & 1 & 0 & 1 \\ 0 & 0 & 0 & 0 & 0 & 1 & 1 & -1 & -1 & 1 & 0 \end{bmatrix},$$

where the entries are elements of the ternary field $F_3 = \{0, 1, -1 = 2\}$. (The 5×5-submatrix in the lower right part is obtained by rotating the row vector $(0, 1, -1, -1, 1)$ cyclically.) The subspace C spanned by the 6 row vectors of the matrix has dimension 6 and minimum distance 5. (To prove this fact, we need some discussion.) Therefore the code is an $[11, 6, 5]_3$-code, and the weight enumerator is

$$W_C(x,y) = x^{11} + 132x^6y^5 + 132x^5y^6 + 330x^3y^8 + 110x^2y^9 + 24y^{11}.$$

The code is a perfect 2-code. (This is because $|\Sigma_2(\mathbf{c})| = 1+2\binom{11}{1}+2^2\binom{11}{2} = 3^5$ holds and the code has minimum distance 5.) If we consider the extended code, in other words, the subspace $\tilde{C}$ spanned by the 6 row vectors of the matrix which is obtained by adding a column to the above 6×11-matrix so that each row sum equals 0, then $\tilde{C}$ is a self-dual $[12, 6, 6]_3$-code. The weight enumerator of $\tilde{C}$ is given as follows:

$$W_{\tilde{C}}(x,y) = x^{12} + 264x^6y^6 + 440x^9y^3 + 24y^{12}.$$

7 You may think the matrix appears "out of the blue." Japanese mathematicians often use the Japanese word "amakudari," which means a descent from heaven, if something appears out of the blue like the above matrix.

(Hence, the weight of each element of $\tilde{C}$ is a multiple of 3.) These two codes $C, \tilde{C}$ are called ternary Golay codes. The (permutation) automorphisms of these codes are Mathieu groups ($\mathrm{Aut}(C) = M_{11}, \mathrm{Aut}(\tilde{C}) = M_{12}$), which are a 4-transitive group on 11 letters and a 5-transitive group on 12 letters, respectively.

Some other important topics are the general theory of cyclic codes, quadratic residue codes, BCH codes, and so on. For details, we refer the reader to other books [326, 319].

The relation between Golay codes and Witt designs

Golay codes and *Steiner systems* 5-(24, 8, 1) and 4-(23, 7, 1), which we call the Witt designs, are closely related.

The automorphism group of the Golay [24, 12, 8]-code is the Mathieu group M_{24}, which is 5-transitive. On the other hand, if we let the set V of points be the set of 24 coordinates, and we let the set $\mathcal{B}$ of blocks be the set of 8-element sets of supports of the 759 elements of the Golay [24, 12, 8]-code with weight 8, then $(V, \mathcal{B})$ is a 5-(24, 8, 1) design. Conversely, if we start from the 5-(24, 8, 1) design $(V, \mathcal{B})$ and view the incidence matrix (of size 759×24) as the matrix over the binary field F_2 and consider the subspace of F_2^{24} spanned by the row vectors, then it forms a 12-dimensional subspace C and C is the Golay [24, 12, 8]-code. (There is a similar relation between the Golay code C' and the Steiner system 4-(23, 7, 1). There is a similar relation among the ternary Golay $[12, 6, 6]_3$-code, the 5-transitive group Mathieu M_{12}, and the Steiner system 5-(12, 6, 1) design as well. For details, the reader is referred to other books.)

Self-dual codes and Gleason's theorem

Let C be a self-dual linear code over F_q of length n (i. e., $C \subset F_q^n$). Consider the codes satisfying the following:

"The weight of each element of the code is a multiple of c_0 for a natural number $c_0 > 0$."

The codes with the following 4 types are interesting:
(1) the case where $q = 2$, $c_0 = 2$ (Type I code);
(2) the case where $q = 2$, $c_0 = 4$ (Type II code);
(3) the case where $q = 3$, $c_0 = 3$ (Type III code);
(4) the case where $q = 4$, $c_0 = 2$ (Type IV code).

Any self-dual code over F_2 is of Type I. The codes satisfying the above condition in "..." are essentially classified into these 4 types, which is known as the Gleason–Pierce theorem, where we omit the proof. It is also known that for codes of Types I, II, III, and IV, the lengths of codes are 2, 8, 4, and multiples of 2, respectively. In addition, the minimum distance d of each code satisfies the following:

$$d \leq \begin{cases} 2[\frac{n}{8}] + 2, & \text{Type I,} \\ 4[\frac{n}{24}] + 4, & \text{Type II,} \\ 3[\frac{n}{12}] + 2, & \text{Type III,} \\ 2[\frac{n}{6}] + 3, & \text{Type IV.} \end{cases}$$

A code which attains the bound in each of these inequalities is called an *extremal code* of each type. We can explain why these inequalities hold in terms of weight enumerators. For a code C of length n, in (1.19), we define the weight enumerator by $W_C(x,y) = \sum_{i=0}^{n} A_i x^{n-i} y^i$. Consider $W_C(x,y)$ over the space $\mathbb{C}[x,y]$ consisting of bivariate polynomials with complex coefficients. The group $G = \langle \sigma_1, \sigma_2 \rangle$ generated by the following two matrices naturally acts on $\mathbb{C}[x,y]$:

$$\sigma_1 = \begin{bmatrix} 1 & 0 \\ 0 & \sqrt{-1} \end{bmatrix}, \quad \sigma_2 = \frac{1}{\sqrt{2}} \begin{bmatrix} 1 & 1 \\ 1 & -1, \end{bmatrix}$$

$$x \longmapsto x, \qquad x \longmapsto \tfrac{1}{\sqrt{2}}(x+y),$$

$$y \longmapsto \sqrt{-1}y, \qquad y \longmapsto \tfrac{1}{\sqrt{2}}(x-y).$$

The weight enumerator $W_C(x,y)$ of a Type II code C is invariant under this action. It is invariant under the action of σ_1 since the code is doubly even (i. e., the weight of each element is a multiple of 4), and the invariance under the action of σ_2 follows from the self-duality of C and from the fact that

$$W_C(x,y) = W_{C^\perp}(x,y) = W_C\left(\frac{x+y}{\sqrt{2}}, \frac{x-y}{\sqrt{2}}\right)$$

holds by the MacWiliams transform. Therefore $W_C(x,y)$ is invariant under the action of the subgroup $G = \langle \sigma_1, \sigma_2 \rangle$ of $\mathrm{GL}(2,\mathbb{C})$. This is nothing but the well-known fact of group theory that G is a *unitary reflection group* of order 192. This is the group called No. 9 in the classification of *finite complex reflection groups* by Shephard and Todd in 1954 [424]. Moreover it is known that the *invariant subspace* $\mathbb{C}[x,y]^G = \{f \in \mathbb{C}[x,y] \mid \sigma f = f, {}^\forall \sigma \in G\}$ of $\mathbb{C}[x,y]$ by the subgroup G is generated by the following polynomials f and g as a polynomial ring:

$$f(x,y) = x^8 + 14x^4y^4 + y^8,$$
$$g(x,y) = x^4y^4(x^4 - y^4)^4.$$

This is known as *Gleason's theorem* (1970). The proof uses *Molien's theorem* (1897) [351] effectively ([434, 373, 326]). Therefore we can show that for a Type II code, the length n of the code is a multiple of 8 and moreover the minimum distance d satisfies $d \leq 4[\frac{n}{24}]+4$ ([326]). The code is called an extremal code if equality holds in this inequality.

As will be stated in the *Assmus–Mattson theorem* later, if we let C be a Type II code of length n, the set of elements of C of weight k forms a t-(v, k, λ) design. Moreover, for an extremal code, it is known that the value of t is given as follows:

$$t = \begin{cases} 5, & \text{if } n = 24m, \\ 3, & \text{if } n = 24m + 8, \\ 1, & \text{if } n = 24m + 16. \end{cases} \tag{1.21}$$

In particular, if $n = 24m$ and $t = 5$, we have $v = n = 24m$, $k = 4m + 4$, and $\lambda = \binom{5m-2}{m-1}$.

Classification of extremal Type II codes is an extremely interesting open problem, and a lot of studies have been conducted. If an extremal Type II code exists, n must be a multiple of 8. For $n = 8, 16, 24$, the classification is completed. For $n = 8$, there exists only the Hamming $[8, 4, 4]$-code e_8; for $n = 16$, there exist only two, $e_8 \oplus e_8$ and d_{16}^{+}; and for $n = 24$, there exists only the Golay $[24, 12, 8]$ code g_{24}. Moreover, it was proved that for $n = 32$, there exist only 5 such codes (Conway and Pless (1980) [145], Conway, Pless, and Sloane (1992) [146], and Koch (1989) [282]). For $n = 40, 48, 56, 64, 80, 88, 104, 112, 136$, the existence of such codes is known. But the classification is not completed for those n except 40. The situations are different depending on n. For $n = 40$, King [279] showed that there exist quite a few such codes, and recently Betsumiya, Harada, and Munemasa [89] announced that there are exactly 16470 such codes. For $n = 48$, there is only one known example. Moreover, it is known that if such a code exists, n must be at most about $4,000$ (see Mallows, Odlyzko, and Sloane [332] and Zhang [527], and the revised version of the former). It seems that there is no such n for which the non-existence is proved. We refer the reader to Harada [211] for the latest results as of March 2011. For earlier results, we refer the reader to Conway and Sloane [147] and Nebe, Rains, and Sloane [373]. The existence of an extremal Type II code of $n = 72$ is a famous open problem. Since Nebe found an *extremal even unimodular lattice* of dimension 72 [368], it seems that many researchers think an extremal Type II code of $n = 72$ may exist as well. (For even unimodular lattices, see Section 5.1.3 of Chapter 5.)

Existence theorem for codes

Finally, we give a brief introduction of Shannon's theory, which forms part of the basics of coding theory. For details, we refer the reader to books on coding theory or information theory.

For p $(0 \le p \le 1/2)$, we define the *capacity* by the following:

$$C(p) = 1 + p \log_2 p + (1 - p) \log_2(1 - p).$$

Theorem 1.108 (Shannon's theorem). *For any $\varepsilon > 0$ and any $R \le C(p)$, if n is sufficiently large, there exists a (binary) $[n, k]$-code such that $k/n \ge R$ and the probability of error at the receiver is at most ε.*

Remark 1.109. This theorem, which was proved more than half a century ago, is so basic and important that it has become a starting point of coding theory and information theory. This theorem proves the existence of such codes and does not find or construct explicit examples. Construction of examples is a very hard problem. In that sense, it is interesting. In combinatorics, it is often the case that the existence is proved but examples are unknown.[8]

So many conditions on parameters or necessary conditions for the existence of $[n, k, d]$-codes are known. For example, the following condition is called the *sphere packing bound* or the *Hamming bound.*

Proposition 1.110. *If a (binary) $[n, k, d]$-code C exists, the following inequality holds:*

$$|C| \leq 2^n / \left(1 + n + \binom{n}{2} + \cdots + \binom{n}{e}\right),$$

where $e = [(d-1)/2]$.

The proof of this proposition is clear. (The denominator of the right-hand side is the number of elements of F_2^n contained in $|\Sigma_e(\boldsymbol{x})|$, or the e-neighbor of $\boldsymbol{x} \in C$.) There are several tens, or more, of bounds named "… bound" for necessary conditions on the existence of an $[n, k, d]$-code. Besides the Hamming bound (sphere packing bound), there are famous bounds such as the Singleton bound, the Plotkin bound, the Griesmer bound, the Elias bound, and so on. For details, we refer the reader to textbooks on coding theory ([326]). One of the most powerful methods to find necessary conditions is the linear programming bound due to Delsarte [159]. We will discuss details in later chapters. A strengthening of the linear programming bound based on semidefinite programming has also been proposed (for example, see [414], [194]).

Here we gave Shannon's theorem and related theorems for (binary) codes. Note that they can be formulated for codes over other finite fields.

For a sufficient condition on the existence of codes, the following theorem is known. (There are many reference books. For example, see [326].)

Theorem 1.111 (Gilbert–Varshamov bound). *If the inequality*

$$1 + \binom{n-1}{1} + \cdots + \binom{n-1}{d-2} < 2^r$$

holds, there exists a (binary) $[n, k, d]$-code such that $k \geq n - r$.

Remark 1.112. The Gilbert–Varshamov theorem is also an existence theorem and does not give any construction of specific codes. Similar formulations are possible for codes over other finite fields.

8 Recently, progress has been made in the construction of codes close to the Shannon limit.

Remark 1.113. In general, to measure how well codes can perform, we consider the following problem. Let δ be a real number satisfying $0 \le \delta \le 1$. Let $A(n, n\delta)$ be the maximum size of (not necessarily linear) codes of length n with minimum distance at least $n\delta$. Then put

$$\alpha(\delta) = \limsup_{n\to\infty} \frac{1}{n} \log_2 A(n, \delta n),$$

and consider the behavior of the function $\alpha(\delta)$ $(0 \le \delta \le 1)$.

(At present, the existence of $\lim_{n\to\infty} \frac{1}{n} \log_2 A(n, \delta n)$ is unknown. See [333, 334].)

Upper and lower bounds of the function $\alpha(\delta)$ $(0 \le \delta \le 1)$ are known. In particular, it is known that if $1/2 \le \delta \le 1$, $\alpha(\delta) = 0$.

A lower bound of $\alpha(\delta)$ can be obtained from the Gilbert–Varshamov bound, which has the following form:

$$\alpha(\delta) \ge 1 - H(\delta),$$

where $0 \le \delta \le 1/2$ and $H(x)$ is the entropy function

$$H(x) = -x \log_2 x - (1 - x) \log_2(1 - x).$$

For upper bounds of $\alpha(\delta)$, there are studies from various viewpoints. In general, the most powerful bound is the McEliece–Rodemich–Rumsey–Welch bound, which is obtained as an application of Delsarte's *linear programming* method [339]. It is interesting to determine to which bound $\alpha(\delta)$ is actually closer, and this is an open problem.

Remark 1.114 (Algebraic geometry code). For a code over a general finite field F_q, if we consider

$$\alpha(\delta) = \limsup_{n\to\infty} \frac{1}{n} \log_q A_q(n, \delta n),$$

then the value of $\alpha(\delta)$ in the interval $[(q-1)/q, 1]$ is 0, and in the interval $[0, (q-1)/q]$, an upper bound and a lower bound are given. A lower bound is the Gilbert–Varshamov bound over a finite field F_q (this is a little different from the one given before, but leads to essentially the same estimation of a lower bound of $\alpha(\delta)$), that is,

$$A_q(n, \delta) \ge q^n / V_q(n, \delta - 1),$$

where $V_q(a, b) = \sum_{i=0}^{b} \binom{a}{i} (q-1)^i$, and so we obtain

$$\alpha(\delta) \ge 1 - H_q(\delta).$$

Note that

$$\begin{aligned} &H_q(0) = 0, \\ &H_q(x) = x \log_q(q-1) - x \log_q x - (1-x) \log_q(1-x), \quad (0 < \delta \le (q-1)/q). \end{aligned}$$

For upper bounds, there is an F_q version of the McEliece–Rodemich–Rumsey bound.

For this case, it was also an open problem which bound the actual value of $\alpha(\delta)$ is closer to. Since there is room for improvement of the upper bound, many researchers conjectured that the lower bound, i. e., the Gilbert–Varshamov bound, would give the actual value. In the paper [486] of 1982, Tsfasman–Vlădut–Zink showed that if p is a prime number of at least 7 and $q = p^2$, there is a case that $\alpha(\delta)$ exceeds the Gilbert–Varshamov bound, by considering a family of *algebraic geometry code* called the *Goppa codes*. This made a great impact on coding theory, and since then a lot of studies on algebraic geometry codes have been conducted. So far, the condition that $q \geq 49$ has not been improved. For the case of the binary filed F_2, it is still an open problem whether $\alpha(\delta)$ exceeds the Gilbert–Varshamov bound ([485]).

2 Association schemes

In this chapter, we discuss association schemes. Association schemes can be regarded as an extension of the concept of groups. Indeed, if a group acts transitively on a finite set, the concept of an association scheme naturally arises. Strongly regular graphs which are discussed in Chapter 1 become association schemes of class 2. For general reference books on algebraic combinatorics, please see Biggs [94], Bannai and Ito [60], Brouwer, Cohen, and Neumaier [113], Godsil [197], Godsil and Royle [200]. For basic references, see Delsarte [159, 158, 161], Delsarte, Goethals, and Seidel [163], and for general survey papers, see Bannai [24, 29, 30].

2.1 The definition of association schemes

Definition 2.1 (Association schemes). If a pair $\mathfrak{X} = (X, \{R_i\}_{0 \le i \le d})$ of a finite set X and a set $\{R_0, R_1, \ldots, R_d\}$ of subsets of the direct product $X \times X$ satisfies the following conditions (1), (2), (3), and (4), then $\mathfrak{X} = (X, \{R_i\}_{0 \le i \le d})$ is called an *association scheme* of *class d*. In what follows, R_i is called the (i-th) *relation*.

(1) $R_0 = \{(x, x) \mid x \in X\}$.
(2) $X \times X = R_0 \cup R_1 \cup \cdots \cup R_d$, and $R_i \cap R_j = \emptyset$ if $i \neq j$. In other words, $\{R_0, R_1, \ldots, R_d\}$ gives a *partition* of $X \times X$.
(3) Define ${}^tR_i = \{(x, y) \mid (y, x) \in R_i\}$ for R_i $(0 \le i \le d)$. Then there exists $i' \in \{0, 1, \ldots, d\}$ such that ${}^tR_i = R_{i'}$.
(4) Fix $i, j, k \in \{0, 1, \ldots, d\}$. Then $p_{i,j}(x, y) = |\{z \in X \mid (x, z) \in R_i, (z, y) \in R_j\}|$ is constant for any $(x, y) \in R_k$. In other words, the number is independent of the choice of (x, y) in R_k, and depends only on i, j, k. The number is denoted by $p_{i,j}^k$ and called the *intersection number*.

Moreover, if the following condition holds, $\mathfrak{X} = (X, \{R_i\}_{0 \le i \le d})$ is called a *commutative association scheme*.

(5) (Commutativity) For any $i, j, k \in \{0, 1, \ldots, d\}$, $p_{i,j}^k = p_{j,i}^k$.

Also, if the following condition holds, $\mathfrak{X} = (X, \{R_i\}_{0 \le i \le d})$ is called a *symmetric association scheme*.

(6) (Symmetry) For all $i \in \{0, 1, \ldots, d\}$, ${}^tR_i = R_i$, (i. e., $i' = i$).

Problem 2.2. Prove that if an association scheme $\mathfrak{X}$ is symmetric, $\mathfrak{X}$ is commutative.

Example 2.3 (The association scheme coming from the action of a group). Suppose a finite group G acts transitively on a finite set Ω. We define the action of G on $\Omega \times \Omega$ by $(x, y)^g = (x^g, y^g)$ for $x, y \in \Omega$ and $g \in G$. Since G is transitive on Ω, $R_0 = \{(x, x) \mid x \in \Omega\}$ forms an orbit. Let $R_0 \cup R_1 \cup \cdots \cup R_d$ be the *G-orbit decomposition* of $\Omega \times \Omega$. Then $(\Omega, \{R_i\}_{0 \le i \le d})$ becomes an association scheme.

https://doi.org/10.1515/9783110630251-002

Proof. Clearly, (1)–(3) hold. We prove (4). Let $(x_1, y_1), (x_2, y_2) \in R_k$. Since R_k is a G-orbit, there exists $g \in G$ such that $(x_2, y_2) = (x_1, y_1)^g$. The mapping which maps $x \in \Omega$ to x^g is a bijection from Ω to itself. Moreover, $z_1 \in \{z \in \Omega \mid (x_1, z) \in R_i, (z, y_1) \in R_j\}$ if and only if $z_1{}^g \in \{z \in \Omega \mid (x_2, z) \in R_i, (z, y_2) \in R_j\}$. Therefore, $|\{z \in \Omega \mid (x, z) \in R_i, (z, y) \in R_j\}|$ is independent of the choice of $(x, y) \in R_k$ and depends only on i, j, k. □

Remark 2.4. Association schemes defined in Example 2.3 are not commutative in general.

Problem 2.5. Let $\Omega = \{1, 2, 3, 4, 5\}$ be the vertex set of the regular pentagon and let D_{10} be the *dihedral group* of order 10. The dihedral group acts transitively on the regular pentagon. We regard D_{10} as the subgroup of the symmetric group S_5 generated by (12345) and (14)(23). Determine the class d of the association scheme on Ω. Also determine the relations $R_0, R_1, \ldots, R_d$ explicitly. Moreover, find all intersection numbers $p_{i,j}^k$.

Example 2.6 (The Hamming association scheme $H(d, q)$). Let F be a q-element set and let X be the direct product F^d of d copies of F. We also define the *Hamming distance* between elements (we say "points" for the remainder) of X. Namely, for $x = (x_1, x_2, \ldots, x_d)$, $y = (y_1, y_2, \ldots, y_d) \in X$, let $\partial(x, y) = |\{j \mid x_j \neq y_j, 1 \leq j \leq d\}|$. Next, for each i $(0 \leq i \leq d)$, define the relation $R_i \subset X \times X$ as $R_i = \{(x, y) \mid \partial(x, y) = i\}$. Then $(X, \{R_i\}_{0 \leq i \leq d})$ becomes a symmetric association scheme of class d. This association scheme is called the *Hamming association scheme*, or simply the *Hamming scheme*, and is denoted by $H(d, q)$.

Proof. Let S_q be the symmetric group on the set F and S_d the symmetric group on the index set $\{1, 2, \ldots, d\}$ of F^d. Let $S = S_q \times S_q \times \cdots \times S_q$ be the group of the direct product of d copies of S_q. For an element $\sigma = (\sigma_1, \sigma_2, \ldots, \sigma_d)$ of S and $x = (x_1, x_2, \ldots, x_d)$, define $x^\sigma = (x_1{}^{\sigma_1}, x_2{}^{\sigma_2}, \ldots, x_d{}^{\sigma_d})$. Then S can be identified with a subgroup of the symmetric group S_X on X. Also by defining $x^\tau = (x_{1^\tau}, x_{2^\tau}, \ldots, x_{d^\tau})$ for $\tau \in S_d$ and $x \in X$, S_d can be identified with a subgroup of the symmetric group on X. Then we can verify that $x^{\tau^{-1}\sigma\tau} = (x_1{}^{\sigma_{1^\tau}}, x_2{}^{\sigma_{2^\tau}}, \ldots, x_d{}^{\sigma_{d^\tau}})$ for $\tau \in S_d$ and $\sigma = (\sigma_1, \sigma_2, \ldots, \sigma_d) \in S$, so we have $\tau^{-1}\sigma\tau = (\sigma_{1^\tau}, \sigma_{2^\tau}, \ldots, \sigma_{d^\tau}) \in S$. Therefore S_d normalizes S, and SS_d becomes a subgroup of the symmetric group S_X on X; SS_d is denoted by $S_q \wr S_d$ and called the *wreath product* of S_q and S_d. The order of $S_q \wr S_d$ is $d!(q!)^d$. Clearly the group S acts transitively on X, and so does $S_q \wr S_d$. Moreover, it is not difficult to show that each R_i becomes an orbit of the action of $S_q \wr S_d$ on $X \times X$. Clearly, ${}^tR_i = R_i$. Hence $(X, \{R_i\}_{0 \leq i \leq d})$ is a symmetric association scheme. □

Example 2.7 (The Johnson association scheme $J(v, d)$). Let V be the set of v points and let $X = \{x \mid x \subset V, |x| = d\}$, where $d \leq \frac{v}{2}$. Define the relation R_i $(0 \leq i \leq d)$ as $R_i = \{(x, y) \in X \times X \mid |x \cap y| = d - i\}$. Then $(X, \{R_i\}_{0 \leq i \leq d})$ becomes a symmetric association scheme of class d. This association scheme is called the *Johnson association scheme* or simply the *Johnson scheme*, and is denoted by $J(v, d)$.

Proof. The symmetric group S_v on the set V acts transitively on X. Then $X \times X = R_0 \cup R_1 \cup \cdots \cup R_d$ gives the orbit decomposition. Moreover, by the definition of each R_i, it is symmetric. So $(X, \{R_i\}_{0 \le i \le d})$ is a symmetric association scheme. □

Example 2.8 (Conjugacy classes of a group G). Let G be a finite group and let $\{C_0 = \{1\}, C_1, \ldots, C_d\}$ be the set of conjugacy classes of G, where 1 denotes the identity element of G. Define the relation R_i $(0 \le i \le d)$ by $R_i = \{(x, y) \in G \times G \mid y^{-1}x \in C_i\}$. Then $\mathfrak{X}(G) = (G, \{R_i\}_{0 \le i \le d})$ becomes a commutative association scheme of class d. It is not necessarily symmetric.

Proof. Define the action of the direct product $G \times G$ of the group G as follows. For $x \in G$ and $(g, h) \in G \times G$, define $x^{(g,h)} = g^{-1}xh$. The action is transitive on G. By definition, each R_i is invariant as a set under the action of $G \times G$. Also, for each $i \ne 0$, if we let $(x_l, y_l) \in R_i$, i. e., $y_l^{-1}x_l \in C_i$, where $l = 1, 2$, there exists $h \in G$ such that $h^{-1}(y_1^{-1}x_1)h = y_2^{-1}x_2$. Then by putting $g = x_1hx_2^{-1}$, we have $(x_1, y_1)^{(g,h)} = ((x_1hx_2^{-1})^{-1}x_1h, (x_1hx_2^{-1})^{-1}y_1h) = (x_2, y_2)$. Thus R_i is a $G \times G$-orbit and $G \times G = R_0 \cup R_1 \cup \cdots \cup R_d$ gives the orbit decomposition. Moreover, for each i, j, k, $p_{i,j}^k$ is determined by the following. Fix $(x, y) \in R_k$. Then we have $\{z \mid (x, z) \in R_i, (z, y) \in R_j\} = \{z \mid z^{-1}x \in C_i, y^{-1}z \in C_j\} = xC_i^{-1} \cap yC_j = y(y^{-1}xC_i^{-1} \cap C_j)$. So by putting $y^{-1}x = a \in C_k$, $p_{i,j}^k = |aC_i^{-1} \cap C_j|$. Since C_j is a conjugacy class, $aC_ja^{-1} = C_j$. Then we have $p_{j,i}^k = |aC_j^{-1} \cap C_i| = |C_ja^{-1} \cap C_i^{-1}| = |aC_ja^{-1} \cap aC_i^{-1}| = |aC_i^{-1} \cap C_j| = p_{i,j}^k$. Therefore $\mathfrak{X}(G)$ is a commutative association scheme. □

Problem 2.9. Let $\mathbb{Z}_n$ be the cyclic group of order n. Let $\mathfrak{X}(\mathbb{Z}_n)$ be the association scheme arising from the conjugacy classes of $\mathbb{Z}_n$ as defined in Example 2.8. Prove $\mathfrak{X}(\mathbb{Z}_n)$ is not symmetric if $n \ge 3$.

We have seen various examples of association schemes coming from the actions of groups as in Example 2.3. It is known that the following association scheme cannot be obtained by the construction as in Example 2.3.

Example 2.10 (Shrikhande graph). Let Γ be the graph with the vertex set X consisting of 16 vertices as in the following figure. Define the relations R_0, R_1, R_2 on X as follows:

$$
\begin{aligned}
R_0 &= \{(x, x) \mid x \in X\}, \\
R_1 &= \{(x, y) \mid x \text{ and } y \text{ are adjacent}\}, \\
R_2 &= \{(x, y) \mid x \text{ and } y \text{ are not adjacent}\}.
\end{aligned}
$$

Then $(X, \{R_i\}_{0 \le i \le 2})$ becomes a symmetric association scheme of class 2. The intersection numbers $p_{i,j}^k$ of this association scheme coincide with those of the Hamming scheme $H(2, 4)$. However, $H(2, 4)$ is defined by the action of the transitive group whereas the association scheme of the Shrikhande graph cannot be obtained as the orbit decomposition of $X \times X$ by the action of a transitive group.

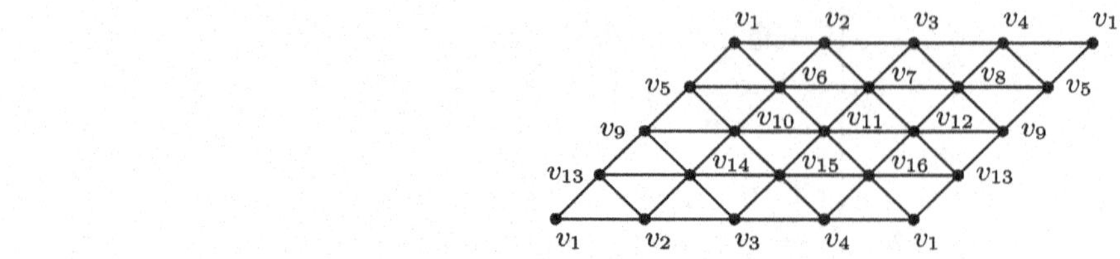

The Shrikhande graph

In the above figure of the Shrikhande graph, we identify the vertices having the same name. (The first row and the fifth row, and the first column and the fifth column are identified.)

Proof. We derive a contradiction by assuming that this association scheme can be defined by using the action of G which is transitive on $X = \{v_i \mid 1 \le i \le 16\}$. Assume that $X \times X = R_0 \cup R_1 \cup R_2$ is the orbit decomposition of the action of G. Then each element of G is an automorphism of the graph Γ. For a vertex a of X, define $\Gamma_i(a) = \{x \in X \mid (x, a) \in R_i\}$ for $i = 1, 2$. Then there exists $g \in G$ such that $(x, a)^g = (y, a)$ for $x, y \in \Gamma_i(a)$. Then g is contained in the stabilizer G_a of a. Therefore G_a acts transitively on $\Gamma_1(a)$ and $\Gamma_2(a)$. Say, $a = v_6$. We have $\Gamma_1(v_6) = \{v_1, v_2, v_5, v_7, v_{10}, v_{11}\}$, and $\Gamma_2(v_6) = \{v_3, v_4, v_8, v_9, v_{12}, v_{13}, v_{14}, v_{15}, v_{16}\}$. Let $\mathrm{Aut}(\Gamma)$ be the automorphism group of the Shrikhande graph. Since $G_{v_6} \subset \mathrm{Aut}(\Gamma)$, the stabilizer subgroup $\mathrm{Aut}(\Gamma)_{v_6}$ of $\mathrm{Aut}(\Gamma)$ must be transitive on $\Gamma_2(v_6)$; however, we can show that it is impossible as follows. Assume there exists $g \in \mathrm{Aut}(\Gamma)_{v_6}$ such that $v_3{}^g = v_{16}$. Then $\Gamma_1(v_6)^g = \Gamma_1(v_6)$ and $\Gamma_1(v_3)^g = \Gamma_1(v_{16})$. Since $\Gamma_1(v_6) \cap \Gamma_1(v_3) = \{v_2, v_7\}$ and $\Gamma_1(v_6) \cap \Gamma_1(v_{16}) = \{v_{11}, v_1\}$, we have $\{v_2, v_7\}^g = \{v_{11}, v_1\}$. However, v_7 and v_2 are joined but v_{11} and v_1 are not. Therefore, it contradicts $g \in \mathrm{Aut}(\Gamma)_{v_6}$. That is, the relation R_2 of the Shrikhande graph is not a G-orbit. □

Problem 2.11. Prove that the intersection numbers of the association scheme constructed from the Shrikhande graph are the same as those of the Hamming scheme $H(2, 4)$.

Exercise 2.12 (Cyclotomic schemes). Let q be a prime power and $\mathrm{GF}(q)$ the *Galois field* (a finite field of q elements). Let d be a divisor of $q-1$ and let $r = \frac{q-1}{d}$. Let ω be a generator of the multiplicative group $\mathrm{GF}(q)^*$ of $\mathrm{GF}(q)$, and let $C_i = \omega^{i-1}H_r$ $(1 \le i \le d)$ be the coset of $\mathrm{GF}(q)^*$ by the subgroup $H_r = \langle \omega^d \rangle$ of order r. Therefore the coset decomposition is $\mathrm{GF}(q)^* = \bigcup_{i=1}^{d} C_i$. On the other hand, define $C_0 = \{0\}$. Moreover, define the relations on $\mathrm{GF}(q)$ by $R_i = \{(x, y) \mid x - y \in C_i\}$ $(1 \le i \le d)$. Then $(\mathrm{GF}(q), \{R_i\}_{0 \le i \le d})$ becomes a commutative association scheme of class d.

Proof. Let G be the set of maps σ from $\mathrm{GF}(q)$ to itself such that $x^\sigma = ax + b$, for $a \in H_r$, $b \in \mathrm{GF}(q)$; G is a subgroup of the symmetric group of $\mathrm{GF}(q)$ and acts transitively on $\mathrm{GF}(q)$. We can also verify that the orbit decomposition of $\mathrm{GF}(q) \times \mathrm{GF}(q)$ is given by $\bigcup_{i=0}^{d} R_i$. By using the facts that for $y - x \in C_k$, $p_{i,j}^k = |\{z \mid x - z \in C_i, z - y \in C_j\}| = |(-C_i + x) \cap (C_j + y)|$ and that for any subset Y and any $b \in \mathrm{GF}(q)$, $|Y| = |-Y| = |Y + b|$, we can show that $p_{i,j}^k = p_{j,i}^k$. □

2.2 Bose–Mesner algebras

We start with the notation. For a finite set X, we consider $|X| \times |X|$-matrices whose rows and columns are indexed by the elements of X. Let $M_X(\mathbb{C})$ be the full matrix algebra of such matrices over the complex field. For $x, y \in X$, $M(x, y)$ denotes the (x, y)-entry of a matrix $M \in M_X(\mathbb{C})$. Let I be the identity matrix of $M_X(\mathbb{C})$, and let J be the matrix in $M_X(\mathbb{C})$ whose entries are all 1. For a matrix $M \in M_X(\mathbb{C})$, ${}^tM$ denotes the transpose of M. For any matrices M_1, M_2 in $M_X(\mathbb{C})$, we define the *Hadamard product* $M_1 \circ M_2$ by

$$(M_1 \circ M_2)(x, y) = M_1(x, y)M_2(x, y), \quad (x, y) \in X \times X.$$

Namely, the Hadamard product is the entry-wise product of matrices. (In elementary linear algebra, this product is forbidden.)

Let $\mathfrak{X} = (X, \{R_i\}_{0 \le i \le d})$ be an association scheme. For each relation R_i $(0 \le i \le d)$, we define the matrix $A_i \in M_X(\mathbb{C})$ as follows:

$$A_i(x, y) = \begin{cases} 1, & \text{if } (x, y) \in R_i, \\ 0, & \text{if } (x, y) \notin R_i; \end{cases} \tag{2.1}$$

A_i is called the *adjacency matrix* of the relation R_i. Then by conditions (1), (2), (3), and (4) in Definition 2.1, we obtain the following conditions (1′), (2′), (3′), and (4′):

(1′) $A_0 = I$;

(2′) $A_0 + A_1 + \cdots + A_d = J$;

(3′) for each i $(0 \le i \le d)$, there exists $i' \in \{0, 1, \ldots, d\}$ such that ${}^tA_i = A_{i'}$;

(4′) for each i, j $(0 \le i, j \le d)$, there exist non-negative integers $p_{i,j}^k$ $(0 \le k \le d)$ such that

$$A_iA_j = \sum_{k=0}^{d} p_{i,j}^k A_k.$$

Moreover, if $\mathfrak{X}$ is commutative,

(5′) for any i, j $(0 \le i, j \le d)$, we have $A_iA_j = A_jA_i$.

Also, if $\mathfrak{X}$ is symmetric,

(6′) for any i $(0 \le i \le d)$, A_i is a symmetric matrix (i. e., $A_i = A_{i'} = {}^tA_i$).

Conversely, suppose that for a finite set X, there exist $(0, 1)$-matrices $A_0, A_1, \ldots, A_d \in M_X(\mathbb{C})$ satisfying the above conditions (1′), (2′), (3′), and (4′). For each i $(0 \le i \le d)$, we define a subset of $X \times X$ as

$$R_i = \{(x, y) \in X \times X \mid A_i(x, y) = 1\}.$$

Then $\mathfrak{X} = (X, \{R_i\}_{0 \le i \le d})$ becomes an association scheme. If (5′) holds, $\mathfrak{X}$ is commutative, and if (6′) holds, $\mathfrak{X}$ is symmetric.

Problem 2.13. Prove the above fact.

Let $\mathfrak{X} = (X, \{R_i\}_{0\le i\le d})$ be an association scheme of class d. Let $\mathfrak{A}$ be the linear subspace spanned by the adjacency matrices $A_0, A_1, \dots, A_d$ of $\mathfrak{X}$. By (4′), $\mathfrak{A}$ is closed under the ordinary matrix product. Namely, $\mathfrak{A}$ is a $(d+1)$-dimensional subalgebra of $M_X(\mathbb{C})$. Moreover, by (2′), for any $i, j \in \{0, 1, \dots, d\}$, we have

$$A_i \circ A_j = \delta_{i,j} A_i.$$

This implies $\mathfrak{A}$ is closed under the Hadamard product $\circ$ as well, and $\mathfrak{A}$ has two algebraic structures; $\mathfrak{A}$ is not commutative with the ordinary matrix product in general, but it is commutative with the Hadamard product. In general, $\mathfrak{A}$ is called the *adjacency algebra*. In particular, if $\mathfrak{A}$ is commutative with the ordinary matrix product, it is called the *Bose–Mesner algebra*.

Lemma 2.14. *Let $\mathfrak{M}$ be a vector subspace of $M_X(\mathbb{C})$. Assume that $\mathfrak{M}$ is closed under the Hadamard product. Then there exists a basis $\{A_0, A_1, \dots, A_d\}$ of $\mathfrak{M}$ such that $A_i \circ A_j = \delta_{i,j} A_i$ for $0 \le i, j \le d$.*

Proof. Let M be a matrix in $\mathfrak{M}$ and let $\{\beta_1, \beta_2, \dots, \beta_r\}$ be the set of distinct non-zero entries of M. For each β_i, define the matrix $M^{(i)}$ as follows:

$$M^{(i)}(x,y) = \begin{cases} 1, & \text{if } M(x,y) = \beta_i, \\ 0, & \text{otherwise.} \end{cases}$$

Let $M^{\circ j} = M \circ \dots \circ M$ be the j-th power of M with respect to the Hadamard product. Then $M^{\circ j} = \sum_{i=1}^{r} \beta_i^j M^{(i)}$ and since the determinant of the coefficient matrix satisfies

$$\begin{vmatrix} \beta_1 & \beta_2 & \cdots & \beta_r \\ \beta_1^2 & \beta_2^2 & \cdots & \beta_r^2 \\ \vdots & \vdots & & \vdots \\ \beta_1^r & \beta_2^r & \cdots & \beta_r^r \end{vmatrix} \neq 0,$$

each $M^{(i)}$ is a linear combination of $M, M^{\circ 2}, \dots, M^{\circ r}$. In particular, $M^{(i)}$ belongs to $\mathfrak{M}$. Next, let $B_0, B_1, \dots, B_d$ form an arbitrary basis of $\mathfrak{M}$, and for each B_j, we construct $(0,1)$-matrices $B_j^{(1)}, B_j^{(2)}, \dots, B_j^{(r_j)}$ by using the above method; $B_j^{(i)}$ is an element of $\mathfrak{M}$, and B_j is a linear combination of $B_j^{(1)}, B_j^{(2)}, \dots, B_j^{(r_j)}$. Therefore $\{B_j^{(i)} \mid 1 \le i \le r_j,\ 0 \le j \le d\}$ spans $\mathfrak{M}$. Among them, we can choose a basis $\{A_0, A_1, \dots, A_d\}$ of $\mathfrak{M}$. Next assume $A_i \circ A_j \neq 0$ for $i \neq j$. Let $C_1 = A_i \circ A_j$, $C_2 = A_i - C_1$. If necessary, we can exchange i and j so that $C_1, C_2 \neq 0$. Since $C_1, C_2 \in \mathfrak{M}$, $\{A_0, A_1, \dots, A_{i-1}, C_1, C_2, A_{i+1}, \dots, A_d\}$ spans $\mathfrak{M}$. We obtain a basis of $\mathfrak{M}$ by removing one element from this generating set. By this operation, the total number of 1's appearing in the basis is decreasing. By repeating this operation, we obtain the desired basis. □

Proposition 2.15 (Theorem 2.6.1 in [113]). *Let $\mathfrak{M}$ be a vector subspace of $M_X(\mathbb{C})$. Suppose $\mathfrak{M}$ is closed under the Hadamard product and the ordinary matrix product. Moreover suppose the following conditions are satisfied:*

(1) *if* $M \in \mathfrak{M}$, *then* ${}^tM \in \mathfrak{M}$;
(2) *for each* $M \in \mathfrak{M}$, *there exists a constant* $\alpha(M,I)$ *such that* $M \circ I = \alpha(M,I)I$;
(3) $I, J \in \mathfrak{M}$.

Then there exist $(0,1)$-*matrices* $A_0(=I), A_1, \dots, A_d$ *which satisfy the previous conditions* $(1'), (2'), (3'), (4')$, *and form a basis of* $\mathfrak{M}$.

Proof. Applying the method used in the proof of Lemma 2.14, there exists a basis of $\mathfrak{M}$ consisting of $(0,1)$-matrices $A_0, A_1, \dots, A_d$. Also $A_i \circ A_j = \delta_{i,j}A_i$ holds. Since $I \in \mathfrak{M}$, we may write $I = \sum_{i=0}^d a_iA_i$. Since $I \circ I = I$, $a_i = 0$ or 1. Then for j such that $a_j = 1$, by (2), we have $\alpha(A_j, I)I = A_j \circ I = A_j$. Therefore there exists j such that $A_j = I$. We set $A_0 = I$. Next, put $J = \sum_{i=0}^d b_iA_i$. Then $A_j = A_j \circ J = \sum_{i=0}^d b_i(A_j \circ A_i) = b_jA_j$. Hence $b_0 = b_1 = \cdots = b_d = 1$ and $(2')$ holds. Since ${}^tJ = J$, we have $A_0 + A_1 + \cdots + A_d = {}^tA_0 + {}^tA_1 + \cdots + {}^tA_d$. Moreover, each ${}^tA_i$ is a sum of some of $A_0, A_1, \dots, A_d$, and ${}^tA_i \circ {}^tA_j = {}^t(A_i \circ A_j) = \delta_{i,j}{}^tA_i$. This implies there is exactly one A_l which appears in the sum ${}^tA_i$. Thus $(3')$ holds. Since $\mathfrak{M}$ is closed under the ordinary matrix product, $(4')$ follows. □

2.3 Commutative association schemes

In this section, we assume $\mathfrak{X} = (X, \{R_i\}_{0 \le i \le d})$ is a commutative association scheme. Consider the property of the intersection numbers $p_{i,j}^k$ of $\mathfrak{X}$. By the assumption that $\mathfrak{X}$ is commutative, we have $p_{i,j}^k = p_{j,i}^k$. Here we introduce some notation. Fix $x \in X$ and define $\Gamma_i(x) = \{z \in X \mid (x,z) \in R_i\}$. By the definition, $\Gamma_{i'}(x) = \{z \in X \mid (z,x) \in R_i\}$ $(0 \le i \le d)$ holds. Then for any $(x,y) \in R_k$, we have $|\Gamma_i(x) \cap \Gamma_{j'}(y)| = p_{i,j}^k$. We define the following:

$$k_i = p_{i,i'}^0. \tag{2.2}$$

Since $(x,z) \in R_i$ if and only if $(z,x) \in R_{i'}$, we have $k_i = |\Gamma_i(x)|$ for any $x \in X$. We call k_i the *valency* of the relation R_i.

By the definition of the intersection numbers, the following holds.

Proposition 2.16. *We have:*
(1) $k_0 = 1$;
(2) $k_i = k_{i'}$;
(3) $|X| = k_0 + k_1 + \cdots + k_d$.

Moreover, the following holds.

Proposition 2.17. *We have:*
(1) $p_{i,0}^k = \delta_{i,k}$;
(2) $p_{0,j}^k = \delta_{j,k}$;
(3) $p_{i,j}^0 = k_i\delta_{i,j'}$;

(4) $p^k_{i,j} = p^{k'}_{i',j'}$;
(5) $\sum_{j=0}^d p^k_{i,j} = k_i$;
(6) $k_\ell p^\ell_{i,j} = k_j p^j_{i',\ell} = k_i p^i_{\ell,j'}$;
(7) $\sum_{\alpha=0}^d p^\alpha_{i,j} p^\ell_{k,\alpha} = \sum_{\alpha=0}^d p^\alpha_{k,i} p^\ell_{\alpha,j}$.

Proof. By the definition, (1), (2), (3), and (4) follow immediately.
(5) Fix a pair $(x,y) \in R_k$. Then $\Gamma_i(x) = \bigcup_{j=0}^d (\Gamma_i(x) \cap \Gamma_{j'}(y))$. Since the right-hand side is a disjoint union, we obtain (5).
(6) The following holds:

$$\begin{aligned}&\{(x,y,z) \mid (x,y) \in R_\ell, (x,z) \in R_i, (z,y) \in R_j\}\\ &\quad = \bigcup_{x\in X} \bigcup_{y\in\Gamma_\ell(x)} \{(x,y,z) \mid (x,z) \in R_i, (z,y) \in R_j\}\\ &\quad = \bigcup_{z\in X} \bigcup_{y\in\Gamma_j(z)} \{(x,y,z) \mid (z,x) \in R_{i'}, (x,y) \in R_\ell\}\\ &\quad = \bigcup_{x\in X} \bigcup_{z\in\Gamma_i(x)} \{(x,y,z) \mid (x,y) \in R_\ell, (y,z) \in R_{j'}\}.\end{aligned}$$

Therefore we obtain (6).
(7) Fix a pair $(x,y) \in R_\ell$. Then the following holds:

$$\begin{aligned}&\{(z,w) \mid (z,w) \in R_i, (x,z) \in R_k, (w,y) \in R_j\}\\ &\quad = \bigcup_{\alpha=0}^d \bigcup_{z\in\Gamma_k(x)\cap\Gamma_{\alpha'}(y)} \{(z,w) \mid (z,w) \in R_i, (w,y) \in R_j\}\\ &\quad = \bigcup_{\alpha=0}^d \bigcup_{w\in\Gamma_\alpha(x)\cap\Gamma_{j'}(y)} \{(z,w) \mid (x,z) \in R_k, (z,w) \in R_i\}.\end{aligned}$$

Therefore we obtain (7). □

Next, we consider the Bose–Mesner algebra $\mathfrak{A}$ of $\mathfrak{X}$. We proceed the discussion in a more general setting. For an algebra which is commutative under the ordinary matrix product, by the method used in the proofs of Lemma 2.14 and Proposition 2.15, we can prove the following lemma.

Lemma 2.18. *Let $\mathfrak{M}$ be a vector subspace of $M_X(\mathbb{C})$. Assume $\mathfrak{M}$ is closed under the ordinary matrix product and is commutative. Moreover, assume that $\mathfrak{M}$ satisfies the following conditions:*
(1) *if $M \in \mathfrak{M}$, then ${}^tM, \overline{M} \in \mathfrak{M}$;*
(2) *for any $M \in \mathfrak{M}$, there exists a constant $\alpha(M,J)$ such that $JM = \alpha(M,J)J$;*
(3) *$I, J \in \mathfrak{M}$.*

Then there exists a basis $E_0(= \frac{1}{|X|}J), E_1, \ldots, E_d$ of $\mathfrak{M}$ satisfying the following:

(1″) $|X|E_0 = J$;
(2″) $E_0 + E_1 + \cdots + E_d = I$, $E_iE_j = \delta_{i,j}E_i$;
(3″) *for each* i, *there exists* $\hat{i} \in \{0, 1, \ldots, d\}$ *such that* ${}^tE_i = E_{\hat{i}}$.

Proof. For any $M \in \mathfrak{M}$, $M{}^t\overline{M} = {}^t\overline{M}M$ holds, i. e., M is a normal matrix. Therefore $\mathfrak{M}$ has a basis of mutually commuting normal matrices. Since the elements of $\mathfrak{M}$ commute with each other, they are simultaneously diagonalizable by a unitary matrix U. Namely, every matrix contained in ${}^t\overline{U}\mathfrak{M}U$ is a diagonal matrix. For a set of diagonal matrices, the ordinary matrix product and the Hadamard product are identical. Since ${}^t\overline{U}\mathfrak{M}U$ is closed under the Hadamard product, by Lemma 2.14, there exists a basis $\Lambda_0, \Lambda_1, \ldots, \Lambda_d$ of ${}^t\overline{U}\mathfrak{M}U$ such that $\Lambda_i\Lambda_j = \Lambda_i \circ \Lambda_j = \delta_{i,j}\Lambda_i$. If we set $E_i = U\Lambda_i{}^t\overline{U}$ $(0 \le i \le d)$, then $\{E_0, E_1, \ldots, E_d\}$ is a basis of $\mathfrak{M}$, and $E_iE_j = \delta_{i,j}E_j$ holds. Next set $J = |X| \sum_{j=0}^{d} a_jE_j$. Since $J^2 = |X|J$, each a_j is 0 or 1. In the same way as the proof of Proposition 2.15, by (2), there exists j such that $|X|E_j = J$. By renumbering the indices, we set $E_0 = \frac{1}{|X|}J$. We also set $I = \sum_{j=0}^{d} b_jE_j$. Then $I = I^2 = \sum_{j=0}^{d} b_j^2E_j$, and we obtain $b_0 = b_1 = \cdots = b_d = 1$. In this way, we complete the proof of (1″) and (2″). The decomposition of I into mutually orthogonal idempotents as in (2″) is easily shown to be unique, and hence (3″) follows. □

Regarding Lemma 2.18, for the case of symmetric matrices, see [113], and for the non-commutative case, see [219].

Now we go back to the Bose–Mesner algebra $\mathfrak{A}$ of a commutative association scheme $\mathfrak{X}$. There exists a basis for $\mathfrak{A}$ consisting of adjacency matrices $A_0, A_1, \ldots, A_d$. Moreover, by Lemma 2.18, there exists a basis $E_0, E_1, \ldots, E_d$ of primitive idempotents. Since $\mathfrak{A}$ is closed under the Hadamard product, the following equations hold:
(4″) $E_i \circ E_j = \frac{1}{|X|} \sum_{k=0}^{d} q_{i,j}^kE_k$;
(5″) $E_i \circ E_j = E_j \circ E_i$.

The structure constants $q_{i,j}^k$ of $\mathfrak{A}$ with respect to the Hadamard product, which appeared in (4″), are called the *Krein numbers*. They are not necessarily integers, but it is known that they are non-negative real numbers. For details, we refer the reader to [159, 217, 95, 417].

Problem 2.19. Let $E_0, E_1, \ldots, E_d$ form the basis of primitive idempotents for $\mathfrak{A}$. Prove $\overline{E_i} = {}^tE_i$ for any i, where $\overline{E_i}$ is the complex conjugate of E_i.

2.4 Character tables of association schemes

Let $\mathfrak{X} = (X, \{R\}_{0 \le i \le d})$ be a commutative association scheme. Let V be an $|X|$-dimensional vector space over the complex field indexed by the elements of X. As was seen in the previous section, $\mathfrak{A}$ has two kinds of bases $A_0, A_1, \ldots, A_d$ and $E_0, E_1, \ldots, E_d$. For each i, define $V_i = E_i V$, which is a subspace of V. Then by (2″) and by $\overline{E_i} = {}^tE_i$ (Problem 2.19), we have

$$V = V_0 \perp V_1 \perp \cdots \perp V_d \quad \text{(orthogonal direct sum)}, \tag{2.3}$$

with respect to the complex inner product. In particular, since $E_0 = \frac{1}{|X|}J$, V_0 is a 1-dimensional vector space spanned by the all 1's vector $\mathbf{1}$. By (2") in Lemma 2.18, for any matrix M of $\mathfrak{A}$ and any E_i, there exists a constant $\alpha(M, E_i)$ such that $ME_i = \alpha(M, E_i)E_i$. So the subspace V_i appearing in the orthogonal direct sum of V is a common eigenspace of all matrices of $\mathfrak{A}$. In particular, we denote the eigenvalue of each adjacency matrix A_i on V_j by $P_i(j)$. Then the following holds:

$$A_i = \sum_{j=0}^{d} P_i(j)E_j, \quad 0 \le i \le d. \tag{2.4}$$

Conversely, since $A_0, A_1, \ldots, A_d$ is a basis of $\mathfrak{A}$, we have

$$E_i = \frac{1}{|X|}\sum_{j=0}^{d} Q_i(j)A_j, \quad 0 \le i \le d. \tag{2.5}$$

These base change matrices

$$P = (P_j(i))_{\substack{0 \le i \le d,\\ 0 \le j \le d}}, \quad Q = (Q_j(i))_{\substack{0 \le i \le d,\\ 0 \le j \le d}},$$

which are the matrices of size $d + 1$ whose (i, j)-entry is $P_j(i)$, $Q_j(i)$, respectively, are called the *first eigenmatrix* (or the *character table*) and the *second eigenmatrix* (or the *second character table*) of the association scheme $\mathfrak{X}$, respectively. Note that since each $P_j(i)$ is an eigenvalue of A_j, the first and second eigenmatrices of a symmetric association scheme are real matrices.

Moreover, by the definition, we can easily verify that

$$PQ = QP = |X|I. \tag{2.6}$$

Proposition 2.20. *Let m_i be the dimension of the direct summand V_i in equation (2.3) (i. e., the rank of E_i). Then the following hold.*

(1) $\operatorname{tr}(A_i) = \begin{cases} |X|, & \text{if } i=0, \\ 0, & \text{if } i\neq 0; \end{cases}$

(2) $JA_i = A_iJ = k_iJ$;

(3) $\operatorname{tr}(E_i) = m_i$.

Proof. (1) Since $A_0 = I$, $\operatorname{tr}(A_0) = |X|$. If $i \neq 0$, $\operatorname{tr}(A_i) = 0$ since $A_i(x, x) = 0$.
(2) We have $(JA_i)(x, y) = \sum_{z\in X} J(x, z)A_i(z, y) = \sum_{z\in X} A_i(z, y) = |\{z \mid (z, y) \in R_i\}| = k_i$.

(3) Since $E_i^2 = E_i$, the eigenvalues of E_i are 0 or 1. Since E_i is diagonalizable, $\mathrm{tr}(E_i) = m_i$. □

We call m_i the *multiplicity* of $\mathfrak{X}$.

Proposition 2.21. *For any $i \in \{0,1,\dots,d\}$, the following hold:*
(1) $P_0(i) = 1$;
(2) $P_i(0) = k_i$;
(3) $Q_0(i) = 1$;
(4) $Q_i(0) = m_i$.

Proof. (1) Since $A_0 = I = E_0 + E_1 + \cdots + E_d$, $P_0(i) = 1$ holds for any i.
(2) By Proposition 2.20, $E_0A_i = \frac{1}{|X|}JA_i = \frac{1}{|X|}k_iJ = k_iE_0$ holds. On the other hand, by equation (2.4), $E_0A_i = E_0\sum_{j=0}^d P_i(j)E_j = P_i(0)E_0$ holds. Therefore $P_i(0) = k_i$.
(3) We have $E_0 = \frac{1}{|X|}J = \frac{1}{|X|}(A_0 + A_1 + \cdots + A_d)$. By equation (2.4), $Q_0(i) = 1$ for any i.
(4) By Proposition 2.20, we have $m_i = \mathrm{tr}(E_i)$. By equation (2.4),

$$\mathrm{tr}(E_i) = \frac{1}{|X|}\sum_{j=0}^d Q_i(j)\mathrm{tr}(A_j) = Q_i(0).$$

Therefore $Q_i(0) = m_i$ for any i. □

Theorem 2.22. *For the character tables of a commutative association scheme, the following hold:*
(1) $P_{i'}(j) = \overline{P_i(j)}$;
(2) $Q_{\hat{i}}(j) = \overline{Q_i(j)}$;
(3) $\frac{Q_j(i)}{m_j} = \frac{\overline{P_i(j)}}{k_i}$;
(4) $\sum_{\nu=0}^d \frac{1}{k_\nu}P_\nu(i)\overline{P_\nu(j)} = \delta_{i,j}\frac{|X|}{m_i}$ *(the first orthogonality relation);*
(5) $\sum_{\nu=0}^d m_\nu P_i(\nu)\overline{P_j(\nu)} = \delta_{i,j}|X|k_i$ *(the second orthogonality relation);*
(6) $P_i(\ell)P_j(\ell) = \sum_{k=0}^d p_{i,j}^k P_k(\ell)$;
(7) $Q_i(\ell)Q_j(\ell) = \sum_{k=0}^d q_{i,j}^k Q_k(\ell)$;
(8) $P_i(j)Q_j(\ell) = \sum_{k=0}^d p_{i,k}^\ell Q_j(k)$.

Proof. (1) By equation (2.4) and Problem 2.19, we have

$$A_{i'} = {}^tA_i = \overline{{}^tA_i} = \sum_{j=0}^d \overline{P_i(j)}\,\overline{{}^tE_j} = \sum_{j=0}^d \overline{P_i(j)}E_j.$$

Therefore $P_{i'}(j) = \overline{P_i(j)}$ holds.
(2) Since $E_{\hat{i}} = {}^tE_i = \overline{E_i}$, we can show (2) by using equation (2.5).

(3) By equation (2.5), $A_i \circ E_j = \frac{1}{|X|} Q_j(i) A_i$. Consider the sum of the entries in the matrix of the left-hand side. Then we have

$$\sum_{x \in X} \sum_{y \in X} (A_i \circ E_j)(x,y) = \sum_{x \in X} \sum_{y \in X} E_j(x,y) A_i(x,y) = \sum_{x \in X} (E_j{}^t A_i)(x,x)$$
$$= \operatorname{tr}(E_j A_{i'}) = \operatorname{tr}(P_{i'}(j) E_j) = P_{i'}(j) m_j = m_j \overline{P_i(j)}.$$

On the other hand, the sum of the entries of the right-hand side is

$$\frac{1}{|X|} Q_j(i) \sum_{x \in X} \sum_{y \in X} A_i(x,y) = k_i Q_j(i).$$

Hence we obtain (3).
(4) Compute the entries of both sides of $PQ = |X|I$ and use (3).
(5) Compute the entries of both sides of $QP = |X|I$ and use (3).
(6) On the one hand, we have

$$A_i A_j = \sum_{k=0}^{d} p_{i,j}^k A_k = \sum_{k=0}^{d} p_{i,j}^k \sum_{\ell=0}^{d} P_k(\ell) E_\ell,$$

and on the other hand,

$$A_i A_j = \sum_{\ell=0}^{d} P_i(\ell) E_\ell \sum_{k=0}^{d} P_j(k) E_k = \sum_{\ell=0}^{d} P_i(\ell) P_j(\ell) E_\ell,$$

so we obtain the desired result.
(7) By (4″),

$$E_i \circ E_j = \frac{1}{|X|} \sum_{k=0}^{d} q_{i,j}^k E_k = \frac{1}{|X|^2} \sum_{k=0}^{d} q_{i,j}^k \sum_{\ell=0}^{d} Q_k(\ell) A_\ell.$$

On the other hand,

$$E_i \circ E_j = \frac{1}{|X|^2} \left(\sum_{\ell=0}^{d} Q_i(\ell) A_\ell \right) \circ \left(\sum_{v=0}^{d} Q_j(v) A_v \right) = \frac{1}{|X|^2} \sum_{\ell=0}^{d} Q_i(\ell) Q_j(\ell) A_\ell.$$

(8) We have

$$A_i E_j = \frac{1}{|X|} \sum_{k=0}^{d} Q_j(k) A_i A_k = \frac{1}{|X|} \sum_{k=0}^{d} Q_j(k) \sum_{\ell=0}^{d} p_{i,k}^\ell A_\ell = \frac{1}{|X|} \sum_{\ell=0}^{d} \left(\sum_{k=0}^{d} p_{i,k}^\ell Q_j(k) \right) A_\ell.$$

On the other hand,

$$A_i E_j = P_i(j) E_j = \frac{1}{|X|} P_i(j) \sum_{\ell=0}^{d} Q_j(\ell) A_\ell = \frac{1}{|X|} \sum_{\ell=0}^{d} P_i(j) Q_j(\ell) A_\ell.$$

□

Theorem 2.23. *The intersection numbers and the Krein numbers of a commutative association scheme are determined by the character tables as follows:*
(1) $q_{i,j}^{\ell} = \frac{m_i m_j}{|X|} \sum_{\nu=0}^{d} \frac{1}{k_\nu{}^2} P_\nu(i) P_\nu(j) \overline{P_\nu(\ell)}$;
(2) $p_{i,j}^{\ell} = \frac{k_i k_j}{|X|} \sum_{\nu=0}^{d} \frac{1}{m_\nu{}^2} Q_\nu(i) Q_\nu(j) \overline{Q_\nu(\ell)}$.

Proof. (1) Calculate the traces of both sides of the following equation:

$$\frac{1}{|X|} q_{i,j}^{\ell} E_\ell = (E_i \circ E_j) E_\ell.$$

The trace of the left-hand side is $\frac{1}{|X|} q_{i,j}^{\ell} m_\ell$. As for the right-hand side, we obtain the following:

$$\begin{aligned}
\operatorname{tr}(E_i \circ E_j)E_\ell &= \sum_{x\in X} ((E_i \circ E_j)E_\ell)(x,x) \\
&= \sum_{x\in X} \sum_{y\in X} E_i(x,y) E_j(x,y) E_\ell(y,x) \\
&= \sum_{x\in X} \sum_{y\in X} (E_i \circ E_j \circ {}^t E_\ell)(x,y) \\
&= \sum_{x\in X} \sum_{y\in X} \left(\frac{1}{|X|^3} \sum_{\nu=0}^{d} Q_i(\nu) Q_j(\nu) Q_{\hat{\ell}}(\nu) A_\nu(x,y) \right) \\
&= \frac{1}{|X|^2} \sum_{\nu=0}^{d} Q_i(\nu) Q_j(\nu) \overline{Q_\ell(\nu)} k_\nu.
\end{aligned}$$

Moreover, by using Theorem 2.22 (3) and Theorem 2.22 (1), we obtain (1).
(2) Compute the sum of the entries of the matrix on both sides of the following equation:

$$p_{i,j}^{\ell} A_\ell = (A_i A_j) \circ A_\ell.$$

As for the left-hand side, we obtain $|X| p_{i,j}^{\ell} k_\ell$. As for the right-hand side, we obtain the following:

$$\begin{aligned}
\sum_{x\in X} \sum_{y\in X} ((A_i A_j) \circ A_\ell)(x,y) &= \sum_{x\in X} \sum_{y\in X} (A_i A_j)(x,y)\, {}^t A_\ell(y,x) \\
&= \operatorname{tr}(A_i A_j A_{\ell'}) = \operatorname{tr}\left(\sum_{\nu=0}^{d} P_i(\nu) P_j(\nu) P_{\ell'}(\nu) E_\nu \right) \\
&= \sum_{\nu=0}^{d} P_i(\nu) P_j(\nu) P_{\ell'}(\nu) m_\nu \\
&= \frac{k_i k_j k_\ell}{m_\nu{}^2} \sum_{\nu=0}^{d} \overline{Q_\nu(i) Q_\nu(j) Q_\nu(\ell)}.
\end{aligned}$$

Since the left-hand side is real, so is the right-hand side. Therefore, we obtain the desired equation. □

Since $k_{\nu'} = k_\nu$, the following holds:

$$\begin{aligned}\sum_{\nu=0}^{d} \frac{1}{k_\nu{}^2} P_\nu(i)P_\nu(j)\overline{P_\nu(\ell)} &= \sum_{\nu'=0}^{d} \frac{1}{k_{\nu'}{}^2} P_{\nu'}(i)P_{\nu'}(j)\overline{P_{\nu'}(\ell)} \\ &= \sum_{\nu=0}^{d} \frac{1}{k_\nu{}^2} \overline{P_\nu(i)P_\nu(j)}P_\nu(\ell) = \sum_{\nu=0}^{d} \frac{1}{k_\nu{}^2} \overline{P_\nu(i)P_\nu(j)\overline{P_\nu(\ell)}}.\end{aligned}$$

Therefore, the right-hand side of the equation obtained in Theorem 2.23 (1) is real. Hence, as mentioned before, the Krein numbers $q^\ell_{i,j}$ are real. We need more information to show that they are non-negative. The Krein numbers satisfy the following equations, which are similar to the equations for $p^\ell_{i,j}$. We omit the proof except for (6). (For details, see [60], [32].)

Proposition 2.24. *The Krein numbers satisfy the following equations:*

(1) $q^k_{0,j} = \delta_{j,k}$;
(2) $q^k_{i,0} = \delta_{i,k}$;
(3) $q^0_{i,j} = \delta_{i,\hat j} m_i$;
(4) $q^k_{i,j} = q^{\hat k}_{\hat i,\hat j}$;
(5) $\sum_{j=0}^d q^k_{i,j} = m_i$;
(6) $m_k q^k_{i,j} = m_j q^j_{\hat i,k} = m_i q^i_{k,\hat j}$;
(7) $\sum_{\alpha=0}^d q^\alpha_{i,j} q^\ell_{k,\alpha} = \sum_{\alpha=0}^d q^\alpha_{k,i} q^\ell_{\alpha,j}$.

Proof. We prove (6) only. Since $m_{\hat i} = m_i$, by Theorem 2.23 (1), the following holds:

$$\begin{aligned} m_j q^j_{\hat i,k} &= m_j \frac{m_i m_k}{|X|} \sum_{\nu=0}^{d} \frac{1}{k_\nu^2} P_\nu(\hat i) P_\nu(k) \overline{P_\nu(j)} \\ &= m_j \frac{m_i m_k}{|X|} \sum_{\nu=0}^{d} \frac{1}{k_\nu^2} \overline{P_\nu(i)} P_\nu(k) \overline{P_\nu(j)} = m_j \frac{m_i m_k}{|X|} \sum_{\nu=0}^{d} \frac{1}{k_\nu^2} P_{\nu'}(i) P_\nu(k) P_{\nu'}(j) \\ &= m_k \frac{m_i m_j}{|X|} \sum_{\nu'=0}^{d} \frac{1}{k_{\nu'}^2} P_{\nu'}(i) P_{\nu'}(j) \overline{P_{\nu'}(k)} = m_k q^k_{i,j}. \end{aligned}$$

Therefore we obtain $m_i q^i_{k,\hat j} = m_i q^i_{\hat j,k} = m_k q^k_{j,i} = m_k q^k_{i,j}$. □

Problem 2.25. Prove Proposition 2.24.

At the end of this section, we give the following theorem.

Theorem 2.26 (The Krein condition). *The Krein numbers of a commutative association scheme are non-negative real numbers.*

Proof. In short, the proof is as follows. Since E_i and E_j are non-negative Hermitian matrices, $E_i \otimes E_j$ is a non-negative Hermitian matrix, too. Since $E_i \circ E_j$ is a principal minor of $E_i \otimes E_j$, it is a non-negative Hermitian matrix, too. Therefore the eigenvalues $q^k_{i,j}$ of $E_i \circ E_j$ are non-negative real numbers (Biggs [95]).

We give an alternative proof in the following ([159]). In general, for a complex matrix, we define $\|M\|^2 = \sum_{x\in X}\sum_{y\in X}|M(x,y)|^2$. Then the following holds:

$$\mathrm{tr}(M\,{}^t\overline{M}) = \|M\|^2 \geq 0. \tag{2.7}$$

Let U be a unitary matrix which diagonalize all matrices in the Bose–Mesner algebra $\mathfrak{A}$. Let X' be the index set for the columns of U. Then $|X'| = |X|$. Note that $E_0, E_1, \ldots, E_d$ are mutually orthogonal (condition (2″)). Then X' can be partitioned into $d+1$ subsets $X'_0, X'_1, \ldots, X'_d$ satisfying the following:

$${}^t\overline{U}E_iU = T_i, \quad T_i(x,y) = \begin{cases} 1, & \text{if } x,y \in X'_i,\ x = y, \\ 0, & \text{otherwise.} \end{cases}$$

Namely, T_i is a matrix whose diagonal entries consist of m_i 1's and 0's. Let $U = [U_0, U_1, \ldots, U_d]$ be the partition of the unitary matrix U associated with the partition of X'. Each U_i is indexed by $X \times X'_i$. Then $E_i = UT_i{}^t\overline{U} = U_i{}^t\overline{U_i}$ holds. So we fix an eigenvector $\boldsymbol{v}$ of E_k with eigenvalue 1, and define the diagonal matrix Δ as follows:

$$\Delta(x,y) = \delta_{x,y}\boldsymbol{v}(x), \quad x,y \in X.$$

Let $M = {}^t\overline{U_i}\Delta U_j$. Compute $\mathrm{tr}(M\,{}^t\overline{M})$. Then we have the following:

$$\begin{aligned}
\mathrm{tr}(M\,{}^t\overline{M}) &= \mathrm{tr}({}^t\overline{U_i}\Delta U_j{}^t\overline{U_j}\,\overline{\Delta}U_i) = \mathrm{tr}({}^t\overline{U_i}\Delta E_j\overline{\Delta}U_i) \\
&= \sum_{x\in X'_i}\sum_{t_1\in X}\sum_{t_2\in X}{}^t\overline{U_i}(x,t_1)\Delta(t_1,t_1)E_j(t_1,t_2)\overline{\Delta}(t_2,t_2)U_i(t_2,x) \\
&= \sum_{t_1\in X}\sum_{t_2\in X}\boldsymbol{v}(t_1)E_j(t_1,t_2)\overline{\boldsymbol{v}(t_2)}\sum_{x\in X'_i}U_i(t_2,x){}^t\overline{U_i}(x,t_1) \\
&= \sum_{t_1\in X}\sum_{t_2\in X}\boldsymbol{v}(t_1)E_j(t_1,t_2)\overline{\boldsymbol{v}(t_2)}E_i(t_2,t_1) \\
&= \sum_{t_1\in X}\sum_{t_2\in X}(E_i \circ E_{\hat{j}})(t_2,t_1)\boldsymbol{v}(t_1)\overline{\boldsymbol{v}(t_2)} \\
&= \sum_{\ell=0}^{d} q^{\ell}_{i,\hat{j}}\sum_{t_1\in X}\sum_{t_2\in X}E_\ell(t_2,t_1)\boldsymbol{v}(t_1)\overline{\boldsymbol{v}(t_2)} = \sum_{\ell=0}^{d} q^{\ell}_{i,\hat{j}}\sum_{t_2\in X}\overline{\boldsymbol{v}(t_2)}(E_\ell\boldsymbol{v})(t_2) \\
&= q^k_{i,\hat{j}}\sum_{t_2\in X}\overline{\boldsymbol{v}(t_2)}\boldsymbol{v}(t_2) = q^k_{i,\hat{j}}\sum_{t_2\in X}|\boldsymbol{v}(t_2)|^2.
\end{aligned} \tag{2.8}$$

Then equation (2.7) is followed by $q^k_{i,\hat{j}} \geq 0$. □

Problem 2.27. Prove the following inequalities hold:

$$|P_i(j)| \leq k_i, \quad |Q_i(j)| \leq m_i.$$

(Hint: $P_i(j)$ is an eigenvalue of A_i. Consider an eigenvector $\boldsymbol{v}$ such that $A_i\boldsymbol{v} = P_i(j)\boldsymbol{v}$.)

2.5 Intersection matrices and Bose–Mesner algebras

Let $\mathfrak{X} = (X, \{R_i\}_{0\leq i\leq d})$ be a commutative association scheme of class d and let $\mathfrak{A}$ be its Bose–Mesner algebra; $\mathfrak{A}$ becomes an algebra with respect to the ordinary matrix product and the Hadamard product. Moreover, we denote $\mathfrak{A}^\circ$ instead of $\mathfrak{A}$ if we view it as an algebra with respect to the Hadamard product.

Intersection matrices and regular representations of Bose–Mesner algebras

We consider the following square matrix B_i of size $d+1$ whose entries are the intersection numbers $p_{i,j}^k$ of $\mathfrak{X}$. Namely, the (j,k)-entry of B_i is defined as follows:

$$B_i(j,k) = p_{i,j}^k,\ 0 \leq j,k \leq d.$$

The matrices $B_0, B_1, \dots, B_d$ are called the *intersection matrices*. Let $\mathfrak{B}$ be the subalgebra of the full matrix algebra $M_{d+1}(\mathbb{C})$ over the complex field generated by the intersection matrices with respect to the ordinary matrix product. Then the following holds.

Theorem 2.28. *The mapping which maps A_i to B_i for each i ($0 \leq i \leq d$) gives an algebra isomorphism between the Bose–Mesner algebra $\mathfrak{A}$ and $\mathfrak{B}$.*

Proof. For each $M \in \mathfrak{A}$, define a linear map $L_M \in \mathrm{End}(\mathfrak{A})$ where $\mathfrak{A}$ is viewed as an $\mathfrak{A}$-module by $L_M(N) = MN$ ($N \in \mathfrak{A}$). Let $\rho(M)$ be the matrix representation of L_M with respect to the basis $\{A_0, A_1, \dots, A_d\}$ of $\mathfrak{A}$. Then we obtain an algebra homomorphism $\rho : \mathfrak{A} \longrightarrow M_{d+1}(\mathbb{C})$. (The representation ρ is called the *left regular representation* of $\mathfrak{A}$.) If $\rho(M) = 0$, then $L_M(N) = 0$ holds for any $N \in \mathfrak{A}$. In particular, $M = L_M(I) = 0$, which means ρ is a faithful representation (i. e., injection). Since $A_iA_j = \sum_{k=0}^d p_{i,j}^k A_k$, by the definition, we have $\rho(A_i) = {}^tB_i$. Since $\mathfrak{A}$ is commutative, the mapping $A_i \mapsto B_i = {}^t(\rho(A_i))$ gives an algebra isomorphism between $\mathfrak{A}$ and $\mathfrak{B}$. □

By Theorem 2.28, we obtain the following.

Corollary 2.29. *The adjacency matrix A_i and the intersection matrix B_i have the same minimal polynomial.*

Dual intersection matrices and regular representations of Bose–Mesner algebras

Consider the following matrix B_i^* of size $d+1$ by using the Krein numbers $q_{i,j}^k$ of $\mathfrak{X}$. Namely, the (j,k)-entry of B_i^* is defined as follows:

$$B_i^*(j,k) = q_{i,j}^k,\ 0 \leq j,k \leq d.$$

The matrices $B_0^*, B_1^*, \dots, B_d^*$ are called the *dual intersection matrices*. Let $\mathfrak{B}^*$ be the subalgebra of the full matrix algebra $M_{d+1}(\mathbb{C})$ over the complex field generated by the dual intersection matrices with respect to the ordinary matrix product. Then the following holds.

Theorem 2.30. *The mapping which maps $|X|E_i$ to B_i^* for each i $(0 \le i \le d)$ gives an algebra isomorphism between the algebra $\mathfrak{A}^\circ$ with respect to the Hadamard multiplication and $\mathfrak{B}^*$ (the algebra with respect to the ordinary matrix product).*

Proof. Similarly to the proof of Theorem 2.28, for each $M \in \mathfrak{A} = \mathfrak{A}^\circ$, define a linear transformation $L_M^* \in \mathrm{End}(\mathfrak{A}^\circ)$ by $L_M^*(N) = M \circ N$ $(N \in \mathfrak{A})$. Let $\rho^*(M)$ be the matrix representation of $L^*(M)$ with respect to the basis $|X|E_0, |X|E_1, \ldots, |X|E_d$ of $\mathfrak{A}^\circ$. Then we obtain an algebra homomorphism $\rho^* : \mathfrak{A}^\circ \longrightarrow M_{d+1}(\mathbb{C})$. This representation ρ^* is faithful. (The representation ρ^* is called the left regular representation of $\mathfrak{A}^\circ$.) By condition (4″) on primitive idempotents in Section 2.3, $\rho^*(|X|E_i) = {}^tB_i^*$ follows from $|X|E_i \circ (|X|E_j) = \sum_{k=0}^{d} q_{i,j}^k |X|E_k$. Since $\mathfrak{A}^\circ$ is commutative, the mapping $|X|E_i \mapsto B_i^* = {}^t(\rho^*(|X|E_i))$ gives an isomorphism between the algebras $\mathfrak{A}^\circ$ and $\mathfrak{B}^*$. □

Proposition 2.31. *We have:*
(1) $P\,{}^tB_iP^{-1} = \mathrm{diag}(P_i(0), P_i(1), \ldots, P_i(d))$;
(2) $Q\,{}^tB_i^*Q^{-1} = \mathrm{diag}(Q_i(0), Q_i(1), \ldots, Q_i(d))$.

Here $\mathrm{diag}(x_1, \ldots, x_m)$ *denotes the diagonal matrix whose i-th diagonal entry is x_i.*

Proof. We prove (1) only. The proof of (2) is similar. Compute the (ν, j)-entry of the matrix $\mathrm{diag}(P_i(0), P_i(1), \ldots, P_i(d))P$. By Theorem 2.22 (6), we have

$$(\mathrm{diag}(P_i(0), P_i(1), \ldots, P_i(d))P)(\nu, j)$$
$$= P_i(\nu)P_j(\nu) = \sum_{k=0}^{d} p_{i,j}^k P_k(\nu) = (P\,{}^tB_i)(\nu, j),$$

and we obtain $\mathrm{diag}(P_i(0), P_i(1), \ldots, P_i(d))P = P\,{}^tB_i$. □

2.6 Dual Bose–Mesner algebras and Terwilliger algebras

In this section, we introduce the *Terwilliger algebra* T. Let $\mathfrak{X} = (X, \{R_i\}_{0 \le i \le d})$ be a commutative association scheme. As we mentioned at the beginning of the previous section, the Bose–Mesner algebra $\mathfrak{A}$ of $\mathfrak{X}$ becomes an algebra with respect to the ordinary matrix product and the Hadamard product. The algebra $\mathfrak{A}^\circ$ with respect to the Hadamard product is in a sense the dual of $\mathfrak{A}$. (For details, see [60, Theorem 5.9].) The Terwilliger algebra is introduced to be an algebra which contains both $\mathfrak{A}$ and $\mathfrak{A}^\circ$. More specifically, we make a copy of $\mathfrak{A}^\circ$ in the full matrix algebra $M_X(\mathbb{C})$, which is called the *dual Bose–Mesner algebra* denoted by $\mathfrak{A}^*$, and define the Terwilliger algebra to be an algebra containing both $\mathfrak{A}$ and $\mathfrak{A}^*$.

Firstly, we define the dual Bose–Mesner algebra. Let P, Q be the first and second eigenmatrices of $\mathfrak{X}$. Fix a point x_0 of X. We identify the standard basis of the complex vector space $V = \mathbb{C}^{|X|}$ with the set X of points. Namely, $x \in X$ also means a unit vector of V. For i $(0 \le i \le d)$, let $\Gamma_i(x_0) = \{x \mid (x_0, x) \in R_i\}$. Let V_i^* be the subspace of V spanned by $\Gamma_i(x_0)$. Therefore $\dim(V_i^*) = k_i$ and we have the following orthogonal decomposition:

$$V = V_0^* \perp V_1^* \perp \cdots \perp V_d^*.$$

Let E_i^* be the matrix representation of the orthogonal projection from V to V_i^*. By definition, E_i^* is the diagonal matrix satisfying

$$E_i^*(x,y) = \begin{cases} 1, & \text{if } x = y \text{ and } (x_0, x) \in R_i, \\ 0, & \text{otherwise.} \end{cases}$$

Therefore, we obtain $E_i^* E_j^* = \delta_{i,j} E_i^*$ and $I = E_0^* + E_1^* + \cdots + E_d^*$. Next, for i $(0 \le i \le d)$, we define

$$A_i^* = \sum_{\alpha=0}^{d} Q_i(\alpha) E_\alpha^*. \tag{2.9}$$

By definition, A_i^* is a diagonal matrix, and if we let $(x_0, x) \in R_\alpha$, we have

$$A_i^*(x,x) = \sum_{\nu=0}^{d} Q_i(\nu) E_\nu^*(x,x) = Q_i(\alpha) = |X| E_i(x_0, x). \tag{2.10}$$

Then the following proposition holds.

Proposition 2.32. *We have:*
(1) $E_i^* = \frac{1}{|X|} \sum_{j=0}^{d} P_i(j) A_j^*$;
(2) $A_i^* A_j^* = \sum_{k=0}^{d} q_{i,j}^k A_k^*$, *where* $q_{i,j}^k$ *is the Krein number;*
(3) $\{Q_i(0), Q_i(1), \ldots, Q_i(d)\}$ *coincides with the set of eigenvalues of* A_i^*.

Proof. (1) It is clear since $PQ = |X|I$.
(2) Since A_i^*, A_j^* are diagonal matrices, $A_i^* A_j^*$ is also diagonal, and for $x \in X$, by (2.10), we have

$$\begin{aligned}(A_i^* A_j^*)(x,x) &= A_i^*(x,x) A_j^*(x,x) = |X| E_i(x_0,x) |X| E_j(x_0,x) \\ &= ((|X|E_i) \circ (|X|E_j))(x_0,x) = \sum_{k=0}^{d} q_{i,j}^k |X| E_k(x_0,x) = \sum_{k=0}^{d} q_{i,j}^k A_k^*(x,x).\end{aligned}$$

(3) By definition, it is clear. □

Definition 2.33. The subalgebra $\mathfrak{A}^* = \mathfrak{A}^*(x_0) = \langle A_0^*, A_1^*, \ldots, A_d^* \rangle$ of the full matrix algebra $M_X(\mathbb{C})$ generated by $\{A_0^*, A_1^*, \ldots, A_d^*\}$ is called the *dual Bose–Mesner algebra* with respect to x_0.

(In the above definition, we omit the base point x_0 from the notation for the basis of the dual Bose–Mesner algebra. In later chapters, there are cases where we need the information of the base point. In such cases, we write $\mathfrak{A}^*(x_0)$, $E_0^*(x_0), E_1^*(x_0), \ldots, E_d^*(x_0)$, or $A_0^*(x_0), A_1^*(x_0), \ldots, A_d^*(x_0)$ without omitting x_0.)
We have $|X|E_i = \sum_{j=0}^{d} Q_i(j)A_j$, where $A_0, A_1, \ldots, A_d$ are the primitive idempotents of $\mathfrak{A}^\circ$. On the other hand, by definition, $E_0^*, E_1^*, \ldots, E_d^*$ are primitive idempotents of $\mathfrak{A}^*$ and we have $A_i^* = \sum_{j=0}^{d} Q_i(j)E_j^*$. So by Proposition 2.32 (2), the mapping $|X|E_i \mapsto A_i^*$ becomes an isomorphism between $\mathfrak{A}^\circ$ and $\mathfrak{A}^*$. To sum up, we have the following proposition.

Proposition 2.34. *The dual Bose–Mesner algebra $\mathfrak{A}^* = \mathfrak{A}^*(x_0)$ has the following properties:*
1. $\mathfrak{A}^* = \langle E_0^*, E_1^*, \ldots, E_d^* \rangle$;
2. *by the mapping $|X|E_i \mapsto A_i^*$, the Bose–Mesner algebra $\mathfrak{A}$ is isomorphic to $\mathfrak{A}^*$ as an algebra with respect to the Hadamard product (note that since all matrices in $\mathfrak{A}^*$ are diagonal matrices, the Hadamard product and the ordinary matrix product are the same in $\mathfrak{A}^*$);*
3. *by the mapping $A_i^* \mapsto B_i^*$, $\mathfrak{A}^*$ is isomorphic to the algebra $\mathfrak{B}^*$ generated by the dual intersection matrices with respect to the ordinary matrix product.*

Proof. (1) It follows from the fact that P and Q are non-singular matrices.
(2) As was stated before, by Proposition 2.32, the mapping $|X|E_i \mapsto A_i^*$ becomes an isomorphism from $\mathfrak{A}^\circ$ to $\mathfrak{A}^*$.
(3) It is clear from (2) and Theorem 2.30. □

Definition 2.35 (Terwilliger algebras [468, 469, 470]). Let $\langle \mathfrak{A}, \mathfrak{A}^* \rangle$ be the subalgebra of the full matrix algebra $M_X(\mathbb{C})$ generated by the Bose–Mesner algebra $\mathfrak{A}$ and the dual Bose–Mesner algebra $\mathfrak{A}^* = \mathfrak{A}^*(x_0)$. Then $\langle \mathfrak{A}, \mathfrak{A}^* \rangle$ is called the *Terwilliger algebra* with respect to x_0 and denoted by $T = T(x_0)$. The point x_0 is called the *base point*.

Firstly, we mention some basic facts on subalgebras $\mathfrak{A}, \mathfrak{A}^*$ generating the Terwilliger algebra. Here we introduce a symbol. For matrices $A, B \in M_X(\mathbb{C})$, define the inner product (A, B) as follows:

$$(A, B) = \tau(A \circ \overline{B}) = \mathrm{tr}(A^t \overline{B}), \tag{2.11}$$

where $\tau(M) = \sum_{(x,y) \in X \times X} M(x, y)$ for $M \in M_X(\mathbb{C})$.

Lemma 2.36. *For any non-negative integers $\alpha, \beta, \gamma, i, j, k$ at most d, the following hold:*
1. $(E_\alpha A_\beta^* E_\gamma, E_i A_j^* E_k) = \delta_{\alpha,i}\delta_{\beta,j}\delta_{\gamma,k} q_{\alpha,\beta}^{\gamma} m_\gamma$, *where* $m_\gamma = Q_\gamma(0) = \mathrm{tr}(E_\gamma) = \dim(V_\gamma)$;
2. $(E_\alpha^* A_\beta E_\gamma^*, E_i^* A_j E_k^*) = \delta_{\alpha,i}\delta_{\beta,j}\delta_{\gamma,k} p_{\alpha,\beta}^{\gamma} k_\gamma$, *where* $k_\gamma = P_\gamma(0) = \mathrm{tr}(E_\gamma^*) = \dim(V_\gamma^*)$.

The above lemma gives an alternative proof of the non-negative property of the Krein numbers. Namely, the following corollary holds.

Corollary 2.37. *For any non-negative integers α, β, γ at most d, the following hold:*

(1) $q^{\gamma}_{\alpha,\beta} \geq 0$;
(2) $E_\alpha A^*_\beta E_\gamma = 0$ *if and only if* $q^{\gamma}_{\alpha,\beta} = 0$;
(3) $E^*_\alpha A_\beta E^*_\gamma = 0$ *if and only if* $p^{\gamma}_{\alpha,\beta} = 0$.

Remark 2.38. In general, if an association scheme is symmetric, for any non-negative integers α, β, γ at most d, $p^{\gamma}_{\alpha,\beta} k_\gamma$ and $q^{\gamma}_{\alpha,\beta} m_\gamma$ are symmetric with respect to α, β, γ (Proposition 2.17 (6) and Proposition 2.24 (6)). This property is important when we consider P-polynomial schemes and Q-polynomial schemes, which will appear in later sections.

The proof of Lemma 2.36. We transform the left-hand side of the equality in (1). Note that ${}^t\overline{E}_k = E_k$, and $\mathrm{tr}(AB) = \mathrm{tr}(BA)$ holds for square matrices A, B. By (2.11), we have the following:

$$\begin{aligned}
(E_\alpha A^*_\beta E_\gamma, E_i A^*_j E_k) &= \delta_{\alpha,i}\delta_{\gamma,k} \,\mathrm{tr}(E_\alpha A^*_\beta E_\gamma \overline{A^*_j}) \\
&= \delta_{\alpha,i}\delta_{\gamma,k} \sum_{x,y\in X} E_\alpha(x,y) A^*_\beta(y,y) E_\gamma(y,x) \overline{A^*}_j(x,x) \\
&= \delta_{\alpha,i}\delta_{\gamma,k} \sum_{x,y\in X} E_\alpha(x,y) |X| E_\beta(x_0,y) E_\gamma(y,x) |X| \overline{E_j}(x_0,x) \\
&= \delta_{\alpha,i}\delta_{\gamma,k} |X|^2 \sum_{x,y\in X} E_\beta(x_0,y)(E_\gamma \circ {}^tE_\alpha)(y,x) E_j(x,x_0) \\
&= \delta_{\alpha,i}\delta_{\gamma,k} |X|^2 (E_\beta(E_\gamma \circ {}^tE_\alpha)E_j)(x_0,x_0) \\
&= \delta_{\alpha,i}\delta_{\gamma,k} |X| \sum_{\nu=0}^{d} q^{\nu}_{\hat{\alpha},\gamma} (E_\beta E_\nu E_j)(x_0,x_0) \\
&= \delta_{\alpha,i}\delta_{\beta,j}\delta_{\gamma,k} |X| q^{\beta}_{\hat{\alpha},\gamma} E_\beta(x_0,x_0) = \delta_{\alpha,i}\delta_{\beta,j}\delta_{\gamma,k} q^{\beta}_{\hat{\alpha},\gamma} Q_\beta(0) \\
&= \delta_{\alpha,i}\delta_{\beta,j}\delta_{\gamma,k} q^{\beta}_{\hat{\alpha},\gamma} m_\beta.
\end{aligned}$$

Therefore by Proposition 2.24 (6), we obtain (1). Next, we transform the left-hand side of the equality in (2). Similarly to (1), we have

$$\begin{aligned}
(E^*_\alpha A_\beta E^*_\gamma, E^*_i A_j E^*_k) &= \delta_{\gamma,k} \,\mathrm{tr}(E^*_\alpha A_\beta E^*_\gamma \,{}^tA_j E^*_i) \\
&= \delta_{\alpha,i}\delta_{\gamma,k} \,\mathrm{tr}(E^*_\alpha A_\beta E^*_\gamma \,{}^tA_j) \\
&= \delta_{\alpha,i}\delta_{\gamma,k} \sum_{x,y\in X} E^*_\alpha(x,x) A_\beta(x,y) E^*_\gamma(y,y) \,{}^tA_j(y,x) \\
&= \delta_{\alpha,i}\delta_{\gamma,k} \sum_{\substack{x\in\Gamma_\alpha(x_0),\\ y\in\Gamma_\gamma(x_0)}} A_\beta(x,y) A_j(x,y) \\
&= \delta_{\alpha,i}\delta_{\beta,j}\delta_{\gamma,k} \sum_{y\in\Gamma_\gamma(x_0)} \sum_{x\in\Gamma_\alpha(x_0)} A_\beta(x,y) = \delta_{\alpha,i}\delta_{\beta,j}\delta_{\gamma,k} k_\gamma p^{\gamma}_{\alpha,\beta}.
\end{aligned}$$

□

Next, we consider the action of the Terwilliger algebra on the complex vector space $V = \mathbb{C}^{|X|}$. Similarly to Section 2.6, we identify the standard basis of $V = \mathbb{C}^{|X|}$ with the set X. Namely, $x \in X$ also means a unit vector of V.

In what follows, we introduce some basic properties on T-invariant subspaces of V, i. e., T-modules; V itself is called the *standard module* of T.

Proposition 2.39. *If a subspace W of V is T-invariant, the orthogonal complement $W^{\perp}$ is also T-invariant.*

Proof. Since $T = \langle \mathfrak{A}, \mathfrak{A}^* \rangle$, T is closed under conjugate transposition. Therefore if W is T-invariant, so is $W^{\perp}$. □

Let $\mathbf{1}$ denote the all 1's vector in V. Then $Jx_0 = \mathbf{1} \in V_0 = E_0V$.

Lemma 2.40. *We have:*
(1) $E_i^*\mathbf{1} = A_i x_0$;
(2) $E_i x_0 = \frac{1}{|X|} A_i^* \mathbf{1}$;
(3) $\mathfrak{A}x_0 = \mathfrak{A}^*\mathbf{1}$ *and* $\dim(\mathfrak{A}x_0) = \dim(\mathfrak{A}^*\mathbf{1}) = d+1$.

Proof. (1) We have $E_i^*\mathbf{1} = \sum_{x \in \Gamma_i(x_0)} x = A_i x_0$.
(2) By (1) and equation (2.9) in the definition of A_i^*, we have

$$E_i x_0 = \frac{1}{|X|} \sum_{\alpha=0}^{d} Q_i(\alpha) A_\alpha x_0 = \frac{1}{|X|} \sum_{\alpha=0}^{d} Q_i(\alpha) E_\alpha^* \mathbf{1} = \frac{1}{|X|} A_i^* \mathbf{1}.$$

(3) This is clear from the definition and (1), (2). □

Definition 2.41 (Principal T-modules, primary T-modules). The subspace $\mathfrak{A}x_0 = \mathfrak{A}^*\mathbf{1}$ appearing in the above lemma is T-invariant; $\mathfrak{A}x_0$ is called the *principal* ***T*-*module*** or the *primary* ***T*-*module***.

Definition 2.42. For the principal T-module $\mathfrak{A}x_0$, we define the following.
(1) Let $v_i = E_i^*\mathbf{1}$ $(i = 0, 1, \ldots, d)$. We call $v_0, v_1, \ldots, v_d$ the *standard basis* of $\mathfrak{A}x_0$. Then

$$v_i = E_i^*\mathbf{1} = A_i x_0 \in V_i^* \tag{2.12}$$

and

$$v_0 + v_1 + \cdots + v_d = \mathbf{1} \in V_0 \tag{2.13}$$

follow from Lemma 2.40.
(2) Let $v_i^* = E_i x_0$ $(i = 0, 1, \ldots, d)$. We call $v_0^*, v_1^*, \ldots, v_d^*$ the *dual standard basis* of $\mathfrak{A}x_0$. Then

$$v_i^* = E_i x_0 = \frac{1}{|X|} A_i^* \mathbf{1} \in V_i \tag{2.14}$$

and

$$v_0^* + v_1^* + \cdots + v_d^* = x_0 \in V_0^* \tag{2.15}$$

follow from Lemma 2.40.

We can easily verify the following proposition.

Proposition 2.43. *For the principal T-module $\mathfrak{A}x_0$, the following hold:*
(1) $\dim(E_i\mathfrak{A}x_0) = 1$ *for* $i = 0, 1, \ldots, d$;
(2) $\dim(E_i^*\mathfrak{A}x_0) = 1$ *for* $i = 0, 1, \ldots, d$;
(3) $A_j v_i = \sum_{k=0}^{d} p_{j,i}^k v_k$; *namely, with regard to the standard basis, the matrix of the action of A_j on $\mathfrak{A}x_0$ is the transpose of the intersection matrix $B_j = (p_{j,i}^k)$ (Section 2.5);*
(4) $A_j^* v_i^* = \sum_{k=0}^{d} q_{j,i}^k v_k^*$; *namely, with regard to the dual standard basis, the matrix of the action of A_j^* on $\mathfrak{A}x_0$ is the transpose of the dual intersection matrix $B_j^* = (q_{j,i}^k)$ (Section 2.5).*

Proposition 2.43 implies that the principal T-module $\mathfrak{A}x_0$ has all the information about the intersection numbers and the Krein numbers. Moreover, since we have

$$v_j = A_j x_0 = \sum_{i=0}^{d} P_j(i) E_i x_0 = \sum_{i=0}^{d} P_j(i) v_i^*, \tag{2.16}$$

$$v_j^* = \frac{1}{|X|} A_j^* \mathbf{1} = \frac{1}{|X|} \sum_{i=0}^{d} Q_j(i) E_i^* \mathbf{1} = \frac{1}{|X|} \sum_{i=0}^{d} Q_j(i) v_i, \tag{2.17}$$

the first eigenmatrix $P = (P_j(i))_{\substack{0 \le i \le d \\ 0 \le j \le d}}$ and the second eigenmatrix $Q = (Q_j(i))_{\substack{0 \le i \le d \\ 0 \le j \le d}}$ are the transition matrices for the standard basis $v_0, v_1, \ldots, v_d$ and the dual standard basis $v_0^*, v_1^*, \ldots, v_d^*$.

2.7 Various concepts on association schemes

When we proceed the study of association schemes, it is essential to consider which association schemes are important. Of course, there are various viewpoints and various definitions of importance. In this section, we introduce several concepts coming from the study of association schemes which are important from the various viewpoints.

2.7.1 Duality of association schemes

Let $\mathfrak{X} = (X, \{R_i\}_{0 \le i \le d})$ be a commutative association scheme of class d. Let $\mathfrak{A}$ be the Bose–Mesner algebra of $\mathfrak{X}$.

Definition 2.44. A linear isomorphism Ψ of $\mathfrak{A}$ is called a *duality* of $\mathfrak{A}$ if the following hold:
(1) $\Psi(MN) = \Psi(M) \circ \Psi(N)$ for any $M, N \in \mathfrak{A}$;
(2) $\Psi^2(M) = |X|^t M$ for any $M \in \mathfrak{A}$.

We follow the definition by Jaeger ([47]).

Definition 2.45. If $\mathfrak{A}$ has a duality, we say $\mathfrak{A}$ and the association scheme $\mathfrak{X}$ are *self-dual.*

Problem 2.46. Let Ψ be a duality of $\mathfrak{A}$. Then prove the following holds:

$$^t\Psi(M) = \Psi(^tM), \qquad \Psi(M \circ N) = \frac{1}{|X|}\Psi(M)\Psi(N).$$

Proposition 2.47. *For a commutative association scheme, the following (1) and (2) are equivalent.*
(1) *$\mathfrak{X}$ is self-dual;*
(2) *we can rearrange the order of the relations $R_0, R_1, \dots, R_d$ of $\mathfrak{A}$ so that the character tables P, Q satisfy $P = \overline{Q}$.*

Proof. (1)$\Longrightarrow$(2) Let Ψ be a duality of $\mathfrak{A}$. By Problem 2.46, for any $M \in \mathfrak{A}$, we have $\Psi(M) = \Psi(J \circ M) = \frac{1}{|X|}\Psi(J)\Psi(M) = \Psi(E_0)\Psi(M)$. Therefore $\Psi(E_0) = I = A_0$. Moreover, since $\Psi(E_i) = \Psi(E_i{}^2) = \Psi(E_i) \circ \Psi(E_i)$, the entries of $\Psi(E_i)$ are 0 or 1. On the other hand, since Ψ is an isomorphism, $\Psi(E_0), \Psi(E_1), \dots, \Psi(E_d)$ form a basis of $\mathfrak{A}$. Hence $\{\Psi(E_0), \Psi(E_1), \dots, \Psi(E_d)\} = \{A_0, A_1, \dots, A_d\}$ as sets. Thus by rearranging the order of the relations $R_0, R_1, \dots, R_d$, we obtain $\Psi(E_i) = A_i$. Then the following holds:

$$A_i = \Psi(E_i) = \frac{1}{|X|}\sum_{j=0}^{d} Q_i(j)\Psi(A_j) = \frac{1}{|X|}\sum_{j=0}^{d} Q_i(j)\Psi^2(E_j)$$
$$= \sum_{j=0}^{d} Q_i(j)\, E_{\hat{j}} = \sum_{j=0}^{d} Q_i(j)\, \overline{E_j}.$$

Therefore we have

$$A_i = \overline{A_i} = \sum_{j=0}^{d} \overline{Q_i(j)}E_j,$$

and we obtain $P_i(j) = \overline{Q_i(j)}$.
(2)$\Longrightarrow$(1) Rearrange the order of the relations $R_0, R_1, \dots, R_d$ so that $P = \overline{Q_i(j)}$ holds. Immediately it follows that $\Psi(E_i) = A_i$ $(0 \le i \le d)$ becomes a duality of $\mathfrak{A}$. □

It is known that the Hamming association scheme $H(d, q)$ and the association scheme $\mathfrak{X}(G)$ constructed from the conjugacy classes of a finite abelian group G are self-dual. We will give proofs for these examples in Section 2.10. We will look at many other examples in Chapter 6.

2.7.2 Fusion schemes of association schemes

In this subsection, we discuss *fusion schemes* of association schemes.

Let $\mathfrak{X} = (X, \{R_i\}_{0 \le i \le d})$ be a commutative association scheme. For a partition of the index set $\{0, 1, \dots, d\} = \{0\} \cup \Lambda_1 \cup \cdots \cup \Lambda_{\tilde{d}}$, define $\tilde{R}_0 = R_0$, $\tilde{R}_1 = \bigcup_{j \in \Lambda_1} R_j, \dots, \tilde{R}_{\tilde{d}} = \bigcup_{j \in \Lambda_{\tilde{d}}} R_j$.

If $\tilde{\mathfrak{X}} = (X, \{\tilde{R}_i\}_{0\le i\le \tilde{d}})$ becomes an association scheme, $\tilde{\mathfrak{X}}$ is called a fusion scheme of $\mathfrak{X}$. (In some references, it is called a subscheme instead of a fusion scheme. In this book, we will use the term "subscheme" in Section 2.7.4, so we use the term "fusion scheme.")

There are interesting problems on fusion schemes. For which association schemes do fusion schemes exist? Which association schemes appear as fusion schemes? For group association schemes, we refer the reader to Bannai [28], Iwakata [261], etc. The following lemma, which is called the Bannai–Muzychuk criterion, gives a necessary and sufficient condition for the existence of a fusion scheme of a commutative association scheme and it is very useful.

Lemma 2.48 (Bannai–Muzychuk criterion [28, 365]). *Let $\mathfrak{X} = (X, \{R_i\}_{0\le i\le d})$ be a commutative association scheme. With reference to the notation above, $\tilde{\mathfrak{X}} = (X, \{\tilde{R}_i\}_{0\le i\le \tilde{d}})$ is a fusion scheme of $\mathfrak{X}$ if and only if the following conditions (1), (2), and (3) hold:*

(1) *For each i, ${}^t\tilde{R}_i \in \{\tilde{R}_0, \tilde{R}_1 \dots, \tilde{R}_{\tilde{d}}\}$.*
(2) *There exists another partition of the index set $\{0, 1, \dots, d\} = \{0\} \cup F_1 \cup \dots \cup F_{\tilde{d}}$ such that ${}^t\tilde{E}_i \in \{\tilde{E}_0, \tilde{E}_1 \dots, \tilde{E}_{\tilde{d}}\}$, where $\tilde{E}_0 = E_0$, $\tilde{E}_1 = \sum_{j\in F_1} E_j, \dots, \tilde{E}_{\tilde{d}} = \sum_{j\in F_{\tilde{d}}} E_j$.*
(3) *Let $P|_{F_k\times\Lambda_\ell}$ be the submatrix of the character table P of $\mathfrak{X}$ obtained by restricting rows and columns of P on F_k and Λ_ℓ, respectively. Then $P|_{F_k\times\Lambda_\ell}$ has a constant row sum. Namely, for $i \in F_k$, $\sum_{j\in\Lambda_\ell} P_j(i)$ is constant and independent of the choice of i.*

Proof. For each ℓ $(0 \le \ell \le \tilde{d})$, define $\tilde{A}_\ell = \sum_{i\in\Lambda_\ell} A_i$.
(Necessary condition) Assume $\tilde{\mathfrak{X}}$ is a fusion scheme. Then $\tilde{A}_0 = A_0, \tilde{A}_1, \dots, \tilde{A}_{\tilde{d}}$ are the adjacency matrices which form a basis of the Bose–Mesner algebra $\tilde{\mathfrak{A}}$ of $\tilde{\mathfrak{X}}$. Clearly, (1) holds. Also let $\tilde{E}_0 = E_0, \tilde{E}_1, \dots, \tilde{E}_{\tilde{d}}$ be the basis of primitive idempotents of $\tilde{\mathfrak{A}}$. Then each $\tilde{E}_k$ is an orthogonal idempotent of $\mathfrak{A}$, so (2) holds. Next, if we let $\tilde{P}$ be the first eigenmatrix of $\tilde{\mathfrak{X}}$, then $\tilde{A}_\ell\tilde{E}_k = \tilde{P}_\ell(k)\tilde{E}_k$. Therefore we have

$$\sum_{j\in\Lambda_\ell} A_j \sum_{i\in F_k} E_i = \sum_{i\in F_k}\sum_{j\in\Lambda_\ell} A_jE_i = \sum_{i\in F_k}\sum_{j\in\Lambda_\ell} P_j(i)E_i = \tilde{P}_\ell(k) \sum_{i\in F_k} E_i,$$

and we obtain $\sum_{j\in\Lambda_\ell} P_j(i) = \tilde{P}_\ell(k)$ for any $i \in F_k$. Therefore (3) holds.
(Sufficient condition) Assume (1), (2), and (3) hold. It is obvious that conditions (1′), (2′), and (3′) for an association scheme appearing in Section 2.2 hold. Moreover, $\{\tilde{A}_0 = A_0, \tilde{A}_1, \dots, \tilde{A}_{\tilde{d}}\}$ and $\{\tilde{E}_0 = E_0, \tilde{E}_1, \dots, \tilde{E}_{\tilde{d}}\}$ are linearly independent. Furthermore, $\{\tilde{E}_0 = E_0, \tilde{E}_1, \dots, \tilde{E}_{\tilde{d}}\}$ is a set of orthogonal idempotents. Then by (3), the following holds:

$$\begin{aligned}\tilde{A}_\ell &= \sum_{j\in\Lambda_\ell} A_j = \sum_{j\in\Lambda_\ell}\sum_{v=0}^{d} P_j(v)E_v = \sum_{k=0}^{\tilde{d}}\sum_{v\in F_k}\sum_{j\in\Lambda_\ell} P_j(v)E_v \\ &= \sum_{k=0}^{\tilde{d}}\sum_{v\in F_k} \tilde{P}_\ell(k)E_v = \sum_{k=0}^{\tilde{d}} \tilde{P}_\ell(k)\tilde{E}_k. \end{aligned} \tag{2.18}$$

Therefore the subspace spanned by $\{\tilde{A}_0, \tilde{A}_1, \ldots, \tilde{A}_{\tilde{d}}\}$ coincides with the subspace spanned by $\{\tilde{E}_0, \tilde{E}_1, \ldots, \tilde{E}_{\tilde{d}}\}$. Since the subspace spanned by $\{\tilde{E}_0, \tilde{E}_1, \ldots, \tilde{E}_{\tilde{d}}\}$ is closed under the ordinary matrix product, $\tilde{A}_k \tilde{A}_\ell \in \langle \tilde{E}_0, \tilde{E}_1, \ldots, \tilde{E}_{\tilde{d}} \rangle = \langle \tilde{A}_0, \tilde{A}_1, \ldots, \tilde{A}_{\tilde{d}} \rangle$ holds for any integers k, ℓ with $0 \le k,\ \ell \le \tilde{d}$. Namely, Condition (4′) of Section 2.2 holds. □

2.7.3 Primitive association schemes, distribution graphs, representation graphs

Primitive association schemes, which we will define below, correspond to simple groups in group theory. As simple groups are building blocks of finite groups, primitive association schemes play an important role in the construction of association schemes.

Definition 2.49. Let $\mathfrak{X} = (X, \{R_i\}_{0 \le i \le d})$ be a commutative (not necessarily symmetric) association scheme of class d; $\mathfrak{X}$ is said to be *primitive* if for any i with $1 \le i \le d$, the (directed) graph (X, R_i) is connected. $\mathfrak{X}$ is said to be *imprimitive* if $\mathfrak{X}$ is not primitive.

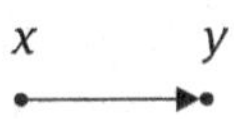

$$(x, y) \in R_i \quad \Updownarrow \quad y \in \Gamma_i(x)$$

Remark 2.50. In the above definition, for 2 vertices $x, y \in X$ of the directed graph (X, R_i), a sequence $x = u_0, u_1, \ldots, u_r = y$ of vertices satisfying $(u_{\nu-1}, u_\nu) \in R_i$ $(0 \le \nu \le r)$ is called a *path* from x to y. The graph (X, R_i) is said to be connected if for every 2 vertices $x, y \in X$, there exists a path from x to y. We call r the length of a path. We consider that for each vertex x, there exists a path from x to x of length 0.

Proposition 2.51. *For the graph (X, R_i) of a commutative association scheme $\mathfrak{X}$, which is not necessarily symmetric, the following (1) and (2) hold:*

(1) *Let $\Gamma^{(i)}(x) = \{y \in X \mid$ There exists a path from x to y in $(X, R_i)\}$. Then $|\Gamma^{(i)}(x)|$ is a constant which is independent of x.*
(2) *For $x, y \in X$, if there exists a path from x to y, then there exists a path from y to x.*

Proof. (1) Let $\Lambda_0 = \{0\}$, $\Lambda_1 = \{i\}$. For $2 \le \ell \le d$, define $\Lambda_\ell = \{\nu \mid p^{\nu}_{\mu,i} > 0, \mu \in \Lambda_{\ell-1}, \nu \notin \Lambda_j (0 \le j \le \ell - 1)\}$. Then $|\Gamma^{(i)}(x)| = \sum_{\ell=0}^{d} \sum_{\nu \in \Lambda_\ell} k_\nu$.

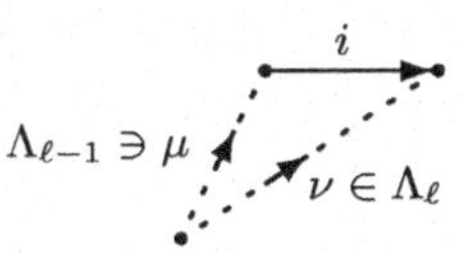

(2) If there exists a path from x to y, then $y \in \Gamma^{(i)}(x)$. Next we assume there exists a path from x to y, and show that there exists a path from y to x. For $z \in \Gamma^{(i)}(y)$, there exists a path from x to z via y. Hence $z \in \Gamma^{(i)}(x)$. Namely, $\Gamma^{(i)}(y) \subset \Gamma^{(i)}(x)$. It turns out that $\Gamma^{(i)}(y) = \Gamma^{(i)}(x)$, which implies $x \in \Gamma^{(i)}(y)$. □

By Theorem 2.28, Corollary 2.29, or Proposition 2.31, the eigenvalues of the intersection matrix B_i coincide with $P_i(0), P_i(1), \dots, P_i(d)$. The (j,k)-entry of B_i is the intersection number $p_{i,j}^k \geq 0$. By Proposition 2.17 (5), we have $\sum_{j=0}^d p_{i,j}^k = k_i = P_i(0)$. Therefore $P_i(0)$ is the *Perron–Frobenius eigenvalue* of B_i. Now consider the graph Δ_{A_i}, whose 2 distinct vertices j, k of the index set $\{0, 1, \dots, d\}$ are adjacent if $p_{i,j}^k > 0$. If $\mathfrak{X}$ is symmetric, by Proposition 2.17 (6), $p_{i,j}^k > 0$ if and only if $p_{i,k}^j > 0$, and so we obtain an undirected graph.

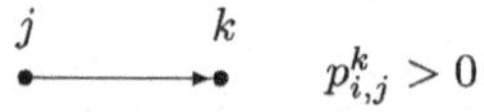

Definition 2.52 (Distribution graphs). For a commutative association scheme $\mathfrak{X}$, which is not necessarily symmetric, the graph Δ_{A_i} defined above is called the *distribution graph* of $\mathfrak{X}$ for the adjacency matrix A_i.

Lemma 2.53. *For $i, \nu, \mu \in \{0, 1, \dots, d\}$, the following are equivalent:*
(1) *For any $(x_0, y) \in R_\mu$, there exist $x \in \Gamma_\nu(x_0)$ and a path from x to y in (X, R_i).*
(2) *For some $(x_0, y) \in R_\mu$, there exist $x \in \Gamma_\nu(x_0)$ and a path from x to y in (X, R_i).*
(3) *There exists a path from ν to μ in Δ_{A_i}.*

Proof. (1) $\Rightarrow$ (2): This is clear.
(2) $\Rightarrow$ (3): By the assumption, there exists a path $x = u_0, u_1, \dots, u_r = y$ from $x \in \Gamma_\nu(x_0)$ to $y \in \Gamma_\mu(x_0)$ in (X, R_i). Let $\nu_0 = \nu$, $\nu_r = \mu$, and $(x_0, u_j) \in R_{\nu_j}$ for $j = 1, 2, \dots, r$. Then $p_{i,\nu}^{\nu_1} = p_{\nu,i}^{\nu_1} > 0$, $p_{i,\nu_1}^{\nu_2} > 0, \dots, p_{i,\nu_{r-1}}^{\nu_r} = p_{i,\nu_{r-1}}^{\mu} > 0$. Therefore there exists a path $\nu = \nu_0, \nu_1, \dots, \nu_r = \mu$ from ν to μ in the distribution graph Δ_{A_i}.

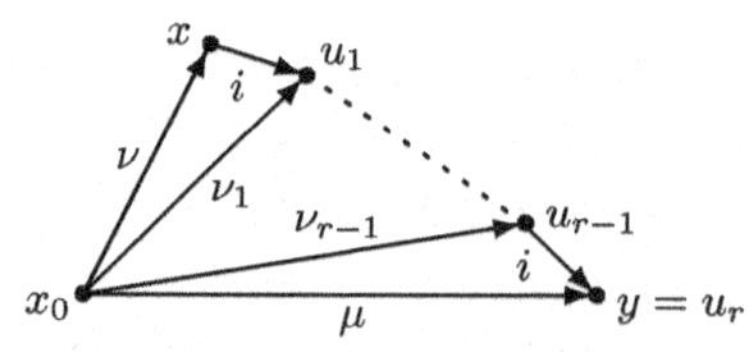

(3)$\Rightarrow$(1): By the assumption, there exists a sequence $p_{i,\nu}^{\nu_1} = p_{\nu,i}^{\nu_1}, p_{i,\nu_1}^{\nu_2}, \dots, p_{i,\nu_{r-1}}^{\nu_r} = p_{i,\nu_{r-1}}^{\mu}$ of positive intersection numbers. Therefore, for any $(x_0, y) \in R_\mu$, we can take a sequence $u_r = y, u_{r-1}, \dots, u_1, u_0 = x$ such that $(x_0, u_{j-1}) \in R_{\nu_{j-1}}$ and $(u_{j-1}, u_j) \in R_i$. Then $x \in \Gamma_\nu(x_0)$ and $x = u_0, u_1, \dots, u_r = y$ is a path from x to y in (X, R_i). □

Corollary 2.54. *For the distribution graph Δ_{A_i}, if there exists a path from ν to μ, then there exists a path from μ to ν.*

Proof. Since condition (3) of Lemma 2.53 holds, by Lemma 2.53 (1), for $(x_0, y) \in R_\mu$, there exist $x \in \Gamma_\nu(x_0)$ and a path from x to y in (X, R_i). Thus by Proposition 2.51 (2), there exists a path from y to x in (X, R_i). This means Lemma 2.53 (2) holds. Therefore there exists a path from μ to ν in Δ_{A_i}. □

For $\nu, \mu \in \{0, 1, \ldots, d\}$, if there exists a path from ν to μ in Δ_{A_i}, we write $\nu \sim_{\Delta_{A_i}} \mu$. By Corollary 2.54, $\sim_{\Delta_{A_i}}$ becomes an equivalence relation on the set $\{0, 1, \ldots, d\}$. An equivalence class of $\sim_{\Delta_{A_i}}$ is called a connected component of Δ_{A_i}.

Proposition 2.55. *Let Ω be the connected component of the distribution graph Δ_{A_i} containing 0. Let $R_\Omega = \bigcup_{\alpha \in \Omega} R_\alpha$. Then the following (1) and (2) hold:*
(1) *for $x, y \in X$, $(x, y) \in R_\Omega$ if and only if there exists a path from x to y in (X, R_i);*
(2) *R_Ω is an equivalence relation on X.*

Proof. (1) Let $(x, y) \in R_\mu$ and assume there exists a path from x to y in (X, R_i). By Lemma 2.53, by setting $\nu = 0$, $x_0 = x$, there exists a path from 0 to μ in Δ_{A_i}. Thus we obtain $\mu \in \Omega$. Conversely, suppose $(x, y) \in R_\Omega$. Then there exists $\mu \in \Omega$ such that $(x, y) \in R_\mu$. Therefore, there exists a path from 0 to μ in Δ_{A_i}. By Lemma 2.53, by setting $\nu = 0$, $x_0 = x$, there exists a path from x to y in (X, R_i).
(2) We define a relation on X as $x \sim_{R_i} y$ if there exists a path from x to y in (X, R_i). Then by Proposition 2.51, $\sim_{R_i}$ is an equivalence relation on X. The above (1) claims that $x \sim_{R_i} y$ if and only if $(x, y) \in R_\Omega$. Therefore R_Ω is an equivalence relation. □

Corollary 2.56. *The graph (X, R_i) is connected if and only if Δ_{A_i} is connected.*

Proof. The graph (X, R_i) is connected if and only if $x \sim_{R_i} y$ for any $x, y \in X$. Also $x \sim_{R_i} y$ for $(x, y) \in R_\mu$ if and only if $\mu \in \Omega$. Therefore (X, R_i) is connected if and only if $\Omega = \{0, 1, \ldots, d\}$ if and only if Δ_{A_i} is connected. □

Proposition 2.57. *For a commutative association scheme $\mathfrak{X}$, the following (1) and (2) are equivalent:*
(1) *$\mathfrak{X}$ is imprimitive;*
(2) *there exists Ω satisfying $\{0\} \subsetneq \Omega \subsetneq \{0, 1, \ldots, d\}$ such that $R_\Omega = \bigcup_{\alpha \in \Omega} R_\alpha$ is an equivalence relation on X.*

Proof. (1)⇒(2): By the assumption, there exists $i \in \{1, \ldots, d\}$ such that (X, R_i) is not connected. Therefore, by Corollary 2.56, Δ_{A_i} is not connected. Let Ω be the connected component of Δ_{A_i} containing 0. Then by Proposition 2.55, R_Ω satisfies condition (2).
(2)⇒(1): Let $i \in \Omega$ and $i \neq 0$. If $x \sim_{R_i} y$, x and y are contained in the same equivalence class with regard to the equivalence relation R_Ω. Therefore (X, R_i) cannot be connected. □

By Proposition 2.31 (2), the eigenvalues of the dual intersection matrix B_i^* are $Q_i(0), Q_i(1), \ldots, Q_i(d)$. Moreover, each entry of B_i^* is a non-negative real number, and if $\mathfrak{X}$ is symmetric, $Q_i(0), Q_i(1), \ldots, Q_i(d)$ are real. By Proposition 2.24 (5), we have $\sum_{j=0}^{d} q_{i,j}^k = m_i = Q_i(0)$. Therefore $Q_i(0)$ is the Perron–Frobenius eigenvalue of B_i^*. Now consider the graph Δ_{E_i} with vertex set $\{0, 1, \ldots, d\}$, whose 2 distinct vertices j, k are adjacent if $q_{i,j}^k > 0$. If $\mathfrak{X}$ is symmetric, by Proposition 2.24 (6), $q_{i,k}^j > 0$ if and only if $q_{i,j}^k > 0$, and so we obtain an undirected graph.

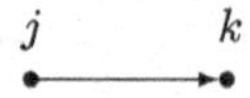

$q^k_{i,j} > 0$

Definition 2.58 (Representation graphs). For a commutative association scheme $\mathfrak{X}$, which is not necessarily symmetric, the graph Δ_{E_i} defined above is called the *representation graph* of $\mathfrak{X}$ for the primitive idempotent E_i.

The following proposition holds. (This is known as a special case of the Perron–Frobenius theorem. See Theorem 1.4 and Theorem 1.5 for the adjacency matrices of a graph.)

Proposition 2.59. *Let $\mathfrak{X}$ be a commutative association scheme, which is not necessarily symmetric. Then the following hold:*

(1) *For every eigenvalue $P_j(i)$ of B_j, we have $|P_j(i)| \le P_j(0) = k_j$.*
(2) *The eigenvalue $P_j(0)$ of B_j has multiplicity 1 if and only if the distribution graph Δ_{A_j} is connected.*
(3) *For every eigenvalue $Q_j(i)$ of B_j^*, we have $|Q_j(i)| \le Q_j(0) = m_j$.*
(4) *The eigenvalue $Q_j(0)$ of B_j^* has multiplicity 1 if and only if the representation graph Δ_{E_j} is connected.*

Proof. Since the proofs of (1) and (3) are similar, we prove (1) only. Let $(x_0, x_1, \dots, x_d)$ be an eigenvector of B_j for the eigenvalue $P_j(i)$. Let $j_0 \in \{0, 1, \dots, d\}$ be an index satisfying $|x_{j_0}| \ge |x_j|$ $(0 \le j \le d)$. Then if we compare the j_0-entry of both sides of $(x_0, x_1, \dots, x_d)B_j = P_j(i)(x_0, x_1, \dots, x_d)$, we have

$$\sum_{\ell=0}^{d} B_j(\ell, j_0)x_\ell = P_j(i)x_{j_0},$$

and since $B_j(\ell, j_0) = p^{j_0}_{j,\ell}$ is a non-negative integer, we obtain

$$|P_j(i)x_{j_0}| \le \sum_{\ell=0}^{d} |x_\ell| \cdot B_j(\ell, j_0) \le |x_{j_0}| \sum_{\ell=0}^{d} p^{j_0}_{j,\ell} = |x_{j_0}|k_j$$

(Proposition 2.17 (5)). Therefore we obtain $|P_j(i)| \le k_j$. Similarly, since $B_j^*(\ell, j_0) = q^{j_0}_{j,\ell}$ is a non-negative real number and $\sum_{\ell=0}^{d} q^{j_0}_{j,\ell} = m_j$ holds, we obtain (3) (Proposition 2.24 (5) and Problem 2.27).

Since the proofs of (2) and (4) are similar, we prove (4) only. Without loss of generality, we may set $j = 1$. Firstly, assume the representation graph Δ_{E_1} is connected and prove that the m_1-eigenspace of B_1^* has dimension 1. Let $\mathbf{x} = (x_0, x_1, \dots, x_d)$ be the eigenvector of B_1^* for the eigenvalue m_1. Let $i_0 \in \{0, 1, \dots, d\}$ be an index satisfying $x_{i_0} \ge x_\ell$ $(0 \le \ell \le d)$. Here we may assume $x_{i_0} > 0$. (Otherwise we may use $-\mathbf{x}$.) Let $\{v_1, v_2, \dots, v_s\} = \{k \mid q^{i_0}_{1,k} > 0\}$. By the choice of the index, we have $\sum_{k=1}^{s} q^{i_0}_{1,v_k} = m_1$.

Considering the i_0-entry of the equation $\mathbf{x}B_1^* = m_1\mathbf{x}$, we have

$$m_1 x_{i_0} = \sum_{l=0}^{d} x_l q_{1,l}^{i_0} = \sum_{k=1}^{s} x_{v_k} q_{1,v_k}^{i_0} \le x_{i_0} \sum_{k=1}^{s} q_{1,v_k}^{i_0} = x_{i_0} m_1.$$

Therefore $x_{v_1} = \cdots = x_{v_s} = x_{i_0}$. If $s = d+1$, then $\mathbf{x} = x_{i_0}(1,1,\ldots,1)$ and it turns out that the eigenspace has dimension 1. If $s < d+1$, let $j \notin \{v_1, v_2, \ldots, v_s\}$. Let $j = j_r, j_{r-1}, \ldots, j_2, j_1, j_0 = i_0$ be a path from j to i_0 in Δ_{E_1}. Since j_1 and i_0 are adjacent, we have $q_{1,j_1}^{i_0} > 0$, which implies $j_1 \in \{v_1, v_2, \ldots, v_s\}$. Therefore $x_{j_1} = x_{i_0}$. By repeating the above discussion using j_1 instead of i_0, if we note $q_{1,j_2}^{j_1} > 0$, we obtain $x_{i_0} = x_{j_1} = x_{j_2}$. In this way, we obtain $x_{i_0} = x_{j_2} = \cdots = x_{j_r} = x_j$. Finally, we obtain $x_0 = x_1 = \cdots = x_d = x_{i_0}$. It turns out that the m_1-eigenspace of B_1^* has dimension 1.

Next, we consider the case where Δ_{E_1} is disconnected. We partition the index set $\{0, 1, \ldots, d\}$ of the rows and columns of B_1^* into the connected components of the representation graph Δ_{E_1}. Then B_1^* can be written as follows:

$$B_1^* = \begin{bmatrix} M_1 & 0 & \cdots & 0 \\ 0 & \ddots & \cdots & \vdots \\ \vdots & \ddots & \ddots & 0 \\ 0 & \cdots & 0 & M_r \end{bmatrix}.$$

Namely, the indices of $M_1, \ldots, M_r$ correspond to the connected components of the graph. Since each of $M_1, \ldots, M_r$ has column sum $m_1 = Q_1(0)$, they have the eigenvalue m_1 with multiplicity 1. Therefore the dimension of the eigenspace of the matrix B_1^* for m_1 agree with the number $r > 1$ of connected components. □

Proposition 2.60. *The following are equivalent:*
(1) *There exists $j \neq 0$ such that Δ_{A_j} is disconnected.*
(2) *There exists $i \neq 0$ such that Δ_{E_i} is disconnected.*

Proof. (2)⇒(1): By Proposition 2.59 (4), the multiplicity of the eigenvalue $Q_i(0)$ of B_i^* is greater than 1. Therefore there exists $j \neq 0$ such that $Q_i(j) = Q_i(0)$. By Proposition 2.21 and Theorem 2.22 (3), we have $1 = \frac{Q_i(j)}{Q_i(0)} = \frac{\overline{P_j(i)}}{P_j(0)}$. Therefore $P_j(i) = P_j(0) = k_j$ holds and by Proposition 2.59 (2), the graph Δ_{A_j} is disconnected. The proof of (1)⇒(2) is similar. □

To summarize, we obtain the following proposition.

Proposition 2.61. *Let $\mathfrak{X}$ be a commutative association scheme, which is not necessarily symmetric. Then the following are equivalent:*
(1) *$\mathfrak{X}$ is primitive;*
(2) *for any $j \neq 0$, Δ_{A_j} is connected;*
(3) *for any $i \neq 0$, Δ_{E_i} is connected.*

Finally, we introduce new symbols. Let Ω be a subset of the index set $\{0,1,\ldots,d\}$ of $\mathfrak{X}$ and let $A_\Omega = \sum_{\alpha\in\Omega} A_\alpha$ and $R_\Omega = \bigcup_{\alpha\in\Omega} R_\alpha$. Moreover, for two subsets $\Omega,\Omega' \subset \{0,1,\ldots,d\}$, we define the product $\Omega\Omega' \subset \{0,1,\ldots,d\}$ as follows:

$$\Omega\Omega' = \{k \mid \text{There exist } i \in \Omega \text{ and } j \in \Omega' \text{ such that } p_{i,j}^k \neq 0\}. \tag{2.19}$$

Proposition 2.62. *We have*
(1) $\Omega\Omega' = \{k \mid (A_\Omega A_{\Omega'}) \circ A_k \neq 0\}$;
(2) $(\Omega\Omega')\Omega'' = \Omega(\Omega'\Omega'') = \{k \mid (A_\Omega A_{\Omega'} A_{\Omega''}) \circ A_k \neq 0\}$.

Proof. The proof is left to the reader. □

Lemma 2.63. *For a subset Ω of $\{0,1,\ldots,d\}$, the following are equivalent:*
(1) $\Omega^2 = \Omega$;
(2) R_Ω *is an equivalence relation on* X.

Proof. (1)$\Longrightarrow$(2): Since $\Omega^2 = \Omega$, the transitive law is clear. Let $\alpha \in \Omega$ and consider the graph (X, R_α). Let $x_0, x_1, \ldots, x_l$ be a path in (X, R_α). Since X is finite, we may assume l is large enough so that there exist vertices x_r, x_s in the path such that $x_r = x_s$ $(r < s)$. Since $\Omega = \Omega^2$, we can verify that for any i,j with $i < j$, $(x_i, x_j) \in R_\Omega$. In particular, since $(x_r, x_r) = (x_r, x_s) \in R_\Omega$, we obtain $0 \in \Omega$, which means the reflexive law holds. Since $(x_s, x_{s-1}) = (x_r, x_{s-1}) \in R_\Omega$, we obtain $\alpha' \in \Omega$ (if $r = s-1$, we use $0 \in \Omega$ shown above). Hence the symmetric law holds.

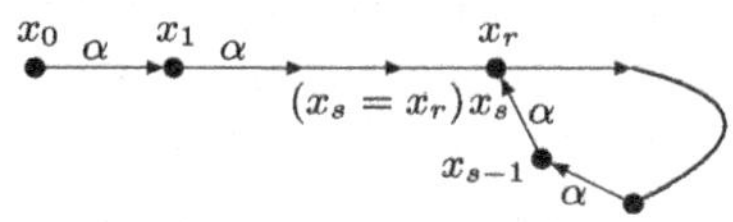

(2)$\Rightarrow$(1): We display the adjacency matrix so that each equivalence class of R_Ω forms a block. Then we obtain the following:

$$A_\Omega = \begin{bmatrix} J & & & \\ & J & & \\ & & \ddots & \\ & & & J \end{bmatrix}.$$

Here J is the all 1's square matrix of degree k_Ω, where $k_\Omega = \sum_{\alpha\in\Omega} k_\alpha$. Then $(A_\Omega)^2 = k_\Omega A_\Omega$, and then by Proposition 2.62, we have $\Omega^2 = \Omega$. □

2.7.4 Subschemes and quotient schemes

In what follows, we assume $\mathfrak{X} = (X, \{R_i\}_{0\le i\le d})$ is imprimitive. Then by Proposition 2.57 and Lemma 2.63, there exists a set Ω satisfying $\{0\} \subsetneq \Omega \subsetneq \{0,1,2,\ldots,d\}$ such that

$\Omega^2 = \Omega$ and R_Ω becomes an equivalence relation on X. Let Ω be one of such sets. Let $X = X_1 \cup X_2 \cup \cdots \cup X_r$ be the partition of equivalence classes of R_Ω and let $\Sigma = \{X_1, X_2, \ldots, X_r\}$. We call Σ the *system of imprimitivity* of $\mathfrak{X}$. For any R_α $(\alpha \in \Omega)$, the relation graph (X, R_α) is disconnected and the sets X_i $(1 \le i \le r)$ of system of imprimitivity give a partition of the graph (X, R_α). As we have seen in Theorem 1.4, Theorem 1.5, and Proposition 2.59, if we display the adjacency matrices A_α $(\alpha \in \Omega)$ as the block matrices whose submatrices are indexed by X_i $(1 \le i \le r)$, then they become block diagonal. Since we consider adjacency matrices of the commutative association scheme, strong conditions hold compared to the case of general graphs. Indeed, if we let $k_\Omega = \sum_{\alpha\in\Omega} k_\alpha$, we have $|X_1| = |X_2| = \cdots = |X_r| = k_\Omega$. Each block $A_\alpha|_{X_i\times X_i}$ is a $k_\Omega \times k_\Omega$ square matrix and becomes the adjacency matrix over X_i of size k_Ω (denoted by $\star$ in the following figure). Let $A_\Omega = \sum_{\alpha\in\Omega} A_\alpha$. Then $A_\Omega|_{X_i\times X_i}$ is the all 1's matrix. Therefore $k_\Omega = \frac{|X|}{r}$. We have

$$A_\alpha = \begin{array}{c} \\ X_1 \\ X_2 \\ \vdots \\ \vdots \\ X_r \end{array}\!\!\begin{array}{c} \begin{array}{cccccc} X_1 & X_2 & \cdots & \cdots & X_r \end{array} \\ \begin{pmatrix} \star & 0 & \cdots & \cdots & 0 \\ 0 & \star & 0 & \ddots & \vdots \\ \vdots & \ddots & \ddots & \ddots & \vdots \\ \vdots & \ddots & 0 & \star & 0 \\ 0 & \cdots & \cdots & 0 & \star \end{pmatrix} \end{array}, \quad A_\Omega = \begin{array}{c} \\ X_1 \\ X_2 \\ \vdots \\ \vdots \\ X_r \end{array}\!\!\begin{array}{c} \begin{array}{cccccc} X_1 & X_2 & \cdots & \cdots & X_r \end{array} \\ \begin{pmatrix} J & 0 & \cdots & \cdots & 0 \\ 0 & J & 0 & \ddots & \vdots \\ \vdots & \ddots & \ddots & \ddots & \vdots \\ \vdots & \ddots & 0 & J & 0 \\ 0 & \cdots & \cdots & 0 & J \end{pmatrix} \end{array},$$

$|X_i| = k_\Omega$.

Let $Y = X_i$ be an equivalence class of the system of imprimitivity and consider $\langle A_\alpha|_{Y\times Y} \mid \alpha \in \Omega\rangle$. For each $\alpha \in \Omega$, $A_\alpha|_{Y\times Y}$ becomes the adjacency matrix of the graph (Y, R_α) with valency k_α. Also, by Proposition 2.62 and $\Omega^2 = \Omega$, the vector space spanned by $\{A_\alpha \mid \alpha \in \Omega\}$ is closed under the ordinary matrix product. This implies that if we let $\mathfrak{A}_\Omega = \langle A_\alpha \mid \alpha \in \Omega\rangle$, $\mathfrak{A}_\Omega$ is a subalgebra of the Bose–Mesner algebra $\mathfrak{A}$ of $\mathfrak{X}$. Since $0 \in \Omega$, the subalgebra $\mathfrak{A}_\Omega$ contains $A_0 = I$. Moreover, $\langle A_\alpha|_{Y\times Y} \mid \alpha \in \Omega\rangle$ has dimension $|\Omega|$, which equals the dimension of $\mathfrak{A}_\Omega$. Hence $\langle A_\alpha|_{Y\times Y} \mid \alpha \in \Omega\rangle$ is isomorphic to $\mathfrak{A}_\Omega$ as an algebra. From the above discussion, we conclude that $\langle A_\alpha|_{Y\times Y} \mid \alpha \in \Omega\rangle$ contains the identity matrix $I = A_0|_{Y\times Y}$ and the all 1's matrix $J = A_\Omega|_{Y\times Y}$, and satisfies conditions (1′)–(5′) of the Bose–Mesner algebra of an association scheme appearing in Section 2.2. Namely, $(Y, \{R_\alpha|_{Y\times Y}\}_{\alpha\in\Omega})$ becomes a commutative association scheme and $\langle A_\alpha|_{Y\times Y} \mid \alpha \in \Omega\rangle$ is its Bose–Mesner algebra.

Definition 2.64 (Subschemes). As seen above, the association scheme $\mathfrak{Y} = (Y, \{R_\alpha|_{Y\times Y}\}_{\alpha\in\Omega})$ constructed on each connected component Y of the system of imprimitivity of an imprimitive association scheme $\mathfrak{X}$ is called a *subscheme*.

Let $|\Omega| = s + 1$. Then $s + 1$ primitive idempotents of $\mathfrak{A}_\Omega$ are uniquely determined up to ordering. Each primitive idempotent is a sum of some $E_0, E_1, \ldots, E_d$. Namely, there exists a partition $\{0,1,\dots,d\} = \Lambda_0 \cup \Lambda_1 \cup \cdots \cup \Lambda_s$ such that $\{E_{\Lambda_i} = \sum_{\alpha\in\Lambda_i} E_\alpha \mid 0 \le i \le s\}$ forms a set of primitive idempotents of the subalgebra $\mathfrak{A}_\Omega$. Therefore the basis of primitive idempotents of the Bose–Mesner algebra $\langle A_\alpha|_{Y\times Y} \mid \alpha \in \Omega\rangle$ is $\{E_{\Lambda_i}|_{Y\times Y} \mid 0 \le i \le s\}$. On the other hand, since $A_\Omega|_{Y\times Y} = J \in \langle A_\alpha|_{Y\times Y} \mid \alpha \in \Omega\rangle$, there exists $E_{\Lambda_i}|_{Y\times Y}$, which coincides with $\frac{1}{k_\Omega}A_\Omega|_{Y\times Y} = \frac{1}{k_\Omega}J$. We renumber $\{\Lambda_i \mid 0 \le i \le s\}$ so that $E_{\Lambda_0}|_{Y\times Y} = \frac{1}{k_\Omega}J$ holds. Now we let $\Lambda = \Lambda_0$.

Proposition 2.65. *With the above notation, regarding a subscheme $\mathfrak{Y}$ of a commutative association scheme $\mathfrak{X}$, the following hold:*

(1) $0 \in \Lambda$;

(2) *let* $\Lambda \circ \Lambda = \{k \mid$ *There exist* $i, j \in \Lambda$ *such that* $q_{i,j}^k \ne 0\}$; *then* $\Lambda \circ \Lambda = \Lambda$;

(3) *let* $(P_j(\Lambda_i))_{\substack{0\le i\le s\\ j\in\Omega}}$ *be the first eigenmatrix of* $\mathfrak{Y}$; *then* $P_j(\Lambda_i) = P_j(\alpha)$ *for any* $\alpha \in \Lambda_i$ *and* $j \in \Omega$;

(4) *let* $(Q_{\Lambda_j}(i))_{\substack{i\in\Omega\\ 0\le j\le s}}$ *be the second eigenmatrix of* $\mathfrak{Y}$; *then* $Q_{\Lambda_j}(i) = \frac{|Y|}{|X|}\sum_{\alpha\in\Lambda_j} Q_\alpha(i)$ *for* $i \in \Omega$. *We also have* $\sum_{\alpha\in\Lambda_j} Q_\alpha(i) = 0$ *if* $i \notin \Omega$.

Proof. (1) Since $E_0E_\Lambda = \frac{1}{|X|k_\Omega}JA_\Omega = \frac{1}{|X|}J = E_0$, we have $0 \in \Lambda$.
(2) It follows that $\Lambda \circ \Lambda \subset \Lambda$ from $k_\Omega E_\Lambda = A_\Omega$, $A_\Omega \circ A_\Omega = A_\Omega$, and

$$E_\Lambda \circ E_\Lambda = \frac{1}{|X|}\sum_{i,j\in\Lambda}\sum_{k=0}^{d} q_{i,j}^k E_k.$$

The other inclusion is immediate from (1).
(3) Since the mapping $A_\alpha \mapsto A_\alpha|_{Y\times Y}$ becomes an isomorphism from $\mathfrak{A}_\Omega$ to $\langle A_\alpha|_{Y\times Y} \mid \alpha \in \Omega\rangle$, we have $A_j = \sum_{i=0}^{s} P_j(\Lambda_i)E_{\Lambda_i}$ for $j \in \Omega$. On the other hand, we have $A_j = \sum_{\alpha=0}^{d} P_j(\alpha)E_\alpha$. So we obtain (3).
(4) By the discussion in (3), for $j \in \{0,1,\dots,s\}$, we have $E_{\Lambda_j} = \frac{1}{|Y|}\sum_{i\in\Omega} Q_{\Lambda_j}(i)A_i$, where $|Y| = k_\Omega$. On the other hand, we have $E_\alpha = \frac{1}{|X|}\sum_{i=0}^{d} Q_\alpha(i)A_i$ and $E_{\Lambda_j} = \sum_{\alpha\in\Lambda_j} E_\alpha$. This implies (4). □

The character tables P and Q of $\mathfrak{X}$ are displayed as follows ($\mathbf{1}_i$ denotes the all 1's vector of size $|\Lambda_i|$ for $0 \le i \le s$):

$$P = \begin{array}{c} \\ \Lambda_0 \\ \vdots \\ \Lambda_i \\ \vdots \\ \Lambda_s \end{array}\!\!\begin{array}{c} \Omega \qquad \overset{j}{\vee} \\ \left(\begin{array}{ccc} & P_j(\Lambda_0)\mathbf{1}_0 & \cdots \\ \vdots & \vdots & \cdots \\ & P_j(\Lambda_i)\mathbf{1}_i & \cdots \\ \vdots & \vdots & \cdots \\ & P_j(\Lambda_s)\mathbf{1}_s & \cdots \end{array}\right) \end{array}, \qquad \begin{array}{l} |\Omega| = s+1, \\ \text{if } \alpha,\beta \in \Lambda_i,\ j \in \Omega, \\ P_j(\alpha) = P_j(\beta) = P_j(\Lambda_i), \end{array}$$

$$Q = \begin{array}{c} \\ \\ \Omega\, i) \\ \\ \\ \end{array}\!\!\begin{array}{c} \begin{array}{ccccc} \Lambda_0 & \cdots & \Lambda_j & \cdots & \Lambda_s \end{array} \\ \begin{pmatrix} - & \cdots & - & \cdots & - \\ \vdots & \vdots & \vdots & \vdots & \vdots \\ - & \cdots & - & \cdots & - \\ \vdots & \vdots & \vdots & \vdots & \vdots \\ - & \cdots & - & \cdots & - \end{pmatrix} \end{array}, \qquad \text{the row sum of } \Lambda_j: \quad \sum_{\alpha\in\Lambda_j} Q_\alpha(i) = \begin{cases} \frac{|X|}{|Y|} Q_{\Lambda_j}(i), & \text{if } i \in \Omega, \\ 0, & \text{if } i \notin \Omega. \end{cases}$$

Next we consider the subalgebra $\mathfrak{A}_\Lambda = \langle E_i \mid i \in \Lambda\rangle$ of $\mathfrak{A}$. Let $|\Lambda| = t + 1$. Then $\dim(\mathfrak{A}_\Lambda) = t+1$. Since $\Lambda = \Lambda \circ \Lambda$, $\mathfrak{A}_\Lambda$ is closed under the Hadamard product. Moreover since $E_\Lambda = \frac{1}{k_\Omega} A_\Omega$ and ${}^tA_\Omega = A_\Omega$, for $i \in \Lambda$, there exists $j \in \Lambda$ such that ${}^tE_i = E_j$. Namely, the subalgebra is closed under the transposition. Since $0 \in \Lambda$, we have $|X|E_0 = J \in \mathfrak{A}_\Lambda$. In other words, $\mathfrak{A}_\Lambda$ contains the unit element J with respect to the Hadamard product. Furthermore, $\mathfrak{A}_\Lambda$ is commutative under the ordinary matrix product. In particular, for any $j \in \Lambda$, $E_\Lambda E_j = E_j E_\Lambda = E_j$ holds. So E_Λ is the identity element of the subalgebra $\mathfrak{A}_\Lambda$ with respect to the ordinary matrix product. By the discussion similar to the proof of Proposition 2.15, $\mathfrak{A}_\Lambda$ has the basis of primitive idempotents (matrices with entries 1 or 0) with respect to the Hadamard product. Since idempotents with respect to the Hadamard product contained in the Bose–Mesner algebra $\mathfrak{A}$ must be a sum of some adjacency matrices, there exists a partition of the index set $\{0, 1, \ldots, d\} = \Omega_0 \cup \cdots \cup \Omega_t$ such that the primitive idempotents of $\mathfrak{A}_\Lambda$ with respect to the Hadamard product can be expressed as $A_{\Omega_i} = \sum_{\alpha\in\Omega_i} A_\alpha$ $(0 \le i \le t)$. And we have $\mathfrak{A}_\Lambda = \langle A_{\Omega_i} \mid 0 \le i \le t\rangle$. In particular, let Ω_0 satisfy $0 \in \Omega_i$. Then we have $J = |X|E_0 = A_{\Omega_0} + A_{\Omega_1} + \cdots + A_{\Omega_t}$.

Proposition 2.66. *With the above notation, the following (1)–(3) hold:*

(1) *For any* i $(0 \le i \le t)$, $A_{\Omega_i} A_\Omega = A_\Omega A_{\Omega_i} = k_\Omega A_{\Omega_i}$.

(2) *For the system of imprimitivity* $X_1, \ldots, X_r$, *the* $(X_\alpha \times X_\beta)$*-block* $A_{\Omega_i}|_{X_\alpha\times X_\beta}$ *of* A_{Ω_i} *is the zero matrix or the all 1's matrix* J.

(3) *For each* i $(0 \le i \le t)$, *define the adjacency matrix* D_i *on* $\Sigma = \{X_1, X_2, \ldots, X_r\}$ *as follows:*

$$D_i(\alpha, \beta) = \begin{cases} 1, & \text{if } A_{\Omega_i}|_{X_\alpha\times X_\beta} = J, \\ 0, & \text{if } A_{\Omega_i}|_{X_\alpha\times X_\beta} = 0. \end{cases}$$

Then $A_{\Omega_i} = D_i \otimes J$.

Proof. (1) We have $\mathfrak{A}_\Lambda = \langle E_j \mid j \in \Lambda\rangle = \langle A_{\Omega_i} \mid 0 \le i \le t\rangle$, and it is obvious that $E_\Lambda = \frac{1}{k_\Omega} A_\Omega$ is the identity element of $\mathfrak{A}_\Lambda$.

(2) Compare the $(X_\alpha \times X_\beta)$-block of both sides of $A_\Omega A_{\Omega_i} = k_\Omega A_{\Omega_i}$. Since A_Ω is block diagonal and each diagonal block is J, if we let $(A_{\Omega_i})_{\alpha,\beta}$ be the $(X_\alpha \times X_\beta)$-block of A_{Ω_i}, we obtain $J(A_{\Omega_i})_{\alpha,\beta} = k_\Omega (A_{\Omega_i})_{\alpha,\beta}$. Here both J and $(A_{\Omega_i})_{\alpha,\beta}$ are square matrices of size k_Ω, and any entry of $(A_{\Omega_i})_{\alpha,\beta}$ is 0 or 1. Hence equality holds if and only if the entries of $(A_{\Omega_i})_{\alpha,\beta}$ are all 1, or all 0.

(3) It is immediate from (2). □

Since each $E_j \in \mathfrak{A}_\Lambda$ is a linear combination of $\{A_{\Omega_\ell} \mid 0 \le \ell \le t\}$, by Proposition 2.66 (3), there exists a constant $\tilde{Q}_j(\ell)$ such that $E_j = \frac{1}{|X|}\sum_{\ell=0}^{t} \tilde{Q}_j(\ell)A_{\Omega_\ell} = (\frac{1}{r}\sum_{\ell=0}^{t}\tilde{Q}_j(\ell)D_\ell)\otimes(\frac{1}{k_\Omega}J)$. Now we define

$$F_j = \frac{1}{r}\sum_{\ell=0}^{t}\tilde{Q}_j(\ell)D_\ell,\ j = 0,1,\dots,t. \tag{2.20}$$

Note that $rk_\Omega = |X|$.

Proposition 2.67.

(1) *Define a mapping φ from $\langle E_i \mid i \in \Lambda\rangle$ to $\langle D_i \mid 0 \le i \le t\rangle$ by $\varphi(\frac{1}{k_\Omega}A_{\Omega_i}) = D_i$. Then φ becomes an algebra isomorphism.*

(2) *For each i $(0 \le i \le t)$, there exists $j \in \{0,1,\dots,t\}$ such that ${}^tD_i = D_j$.*

(3) *We have $A_\Omega = A_{\Omega_0}$. Therefore $D_0 = I$, $\Omega_0 = \Omega$.*

(4) *The algebra $\langle D_i \mid 0 \le i \le t\rangle$ is the Bose–Mesner algebra of the commutative association scheme on the system of imprimitivity Σ of $\mathfrak{X}$.*

(5) *The matrices $F_0, F_1, \dots, F_t$ defined in equation (2.20) form the basis of primitive idempotents of the Bose–Mesner algebra $\langle D_i \mid 0 \le i \le t\rangle$. Note that for $j \in \Lambda$, $\varphi(E_j) = F_j$.*

Proof. (1) Since $(\frac{1}{k_\Omega}J)^2 = \frac{1}{k_\Omega}J$, by the property of tensor products (Kronecker products) of matrices, the mapping $\varphi : D_i \otimes \frac{1}{k_\Omega}J \mapsto D_i$ becomes an isomorphism from $\mathfrak{A}_\Lambda = \langle A_{\Omega_i|0\le i\le t}\rangle$ to $\langle D_i \mid 0 \le i \le t\rangle \subset M_r(\mathbb{C})$.

(2) Let $\Omega_i' = \{\alpha' \mid \alpha \in \Omega_i\}$; ${}^tA_{\Omega_i} = A_{\Omega_i'}$ holds for $i = 0,1,\dots,t$. Since the subalgebra $\mathfrak{A}_\Lambda$ is closed under transposition, we have $A_{\Omega_i'} \in \mathfrak{A}_\Lambda$. Moreover, since $\{A_{\Omega_0}, A_{\Omega_1}, \dots, A_{\Omega_t}\}$ are primitive idempotents of $\mathfrak{A}_\Lambda$ with respect to the Hadamard product, we have $A_{\Omega_i'} \in \{A_{\Omega_0}, A_{\Omega_1}, \dots, A_{\Omega_t}\}$. Namely, there exists j such that $\Omega_i' = \Omega_j$. Therefore for each i, there exists $j \in \{0,1,\dots,t\}$ such that ${}^tD_i = D_j$.

(3) Since $A_\Omega = k_\Omega E_\Lambda \in \mathfrak{A}_\Lambda$, A_Ω is a sum of some matrices in $\{A_{\Omega_0}, A_{\Omega_1}, \dots, A_{\Omega_t}\}$. On the other hand, we have $A_\Omega = I \otimes J$ and $A_{\Omega_0} = \sum_{\ell\in\Omega_0} A_\ell = A_0 + \cdots$. Hence we have $A_\Omega = A_{\Omega_0}$. Thus we obtain $\Omega_0 = \Omega$, $D_0 = I$.

(4) Since $J = A_{\Omega_0} + A_{\Omega_1} + \cdots + A_{\Omega_t}$, we have $D_0 + D_1 + \cdots + D_t = J$. Therefore all conditions $(1'), (2'), \dots, (5')$ of the Bose–Mesner algebra of an association scheme in Section 2.2 are satisfied.

(5) This is clear from (1). □

Definition 2.68 (Quotient schemes). The commutative association scheme on the system of imprimitivity $\Sigma = \{X_1, X_2, \dots, X_r\}$ of $\mathfrak{X}$ which has the Bose–Mesner algebra $\langle D_i \mid 0 \le i \le t\rangle$ of Proposition 2.67 (4) is called the *quotient scheme* of $\mathfrak{X}$.

Proposition 2.69. *Let $\tilde{P}$ and $\tilde{Q}$ be the first and second eigenmatrices of the quotient scheme of $\mathfrak{X}$. Then the following hold:*

(1) $\tilde{P}_j(i) = \frac{1}{k_\Omega}\sum_{\alpha\in\Omega_j} P_\alpha(i)$ $(i \in \Lambda,\ 0 \le j \le t)$;
if $i \notin \Lambda$ and $0 \le j \le t$, we have $\sum_{\alpha\in\Omega_j} P_\alpha(i) = 0$;

(2) $\tilde{Q}_j(i) = Q_j(\alpha)$ $(\alpha \in \Omega_i,\ j \in \Lambda)$.

Proof. For the quotient scheme, we have the following:

$$
\begin{aligned}
D_j &= \sum_{i\in\Lambda} \tilde{P}_j(i)F_i,\ 0 \le j \le t,\\
F_j &= \frac{1}{r}\sum_{i=0}^{t} \tilde{Q}_j(i)D_i,\ j \in \Lambda.
\end{aligned}
\tag{2.21}
$$

Considering the inverses $\varphi^{-1}(D_j)$ and $\varphi^{-1}(F_j)$, where φ is defined in Proposition 2.67 (1), we have the following equations for the bases of the subalgebra $\mathfrak{A}_\Lambda = \langle E_i \mid i \in \Lambda\rangle = \langle A_{\Omega_i} \mid 0 \le i \le t\rangle$:

$$
\begin{aligned}
\frac{1}{k_\Omega}A_{\Omega_j} &= \sum_{i\in\Lambda} \tilde{P}_j(i)E_i,\ 0 \le j \le t,\\
E_j &= \frac{1}{|X|}\sum_{i=0}^{t} \tilde{Q}_j(i)A_{\Omega_i},\ j \in \Lambda.
\end{aligned}
\tag{2.22}
$$

On the other hand, we have

$$
\begin{aligned}
A_{\Omega_j} &= \sum_{\alpha\in\Omega_j} A_\alpha = \sum_{i=0}^{d}\Big(\sum_{\alpha\in\Omega_j} P_\alpha(i)\Big)E_i,\\
E_j &= \frac{1}{|X|}\sum_{\alpha=0}^{d} Q_j(\alpha)A_\alpha.
\end{aligned}
\tag{2.23}
$$

Thus we obtain (1) and (2). ◻

The character tables P and Q can be displayed as follows (for i $(0 \le i \le t)$, $\mathbf{1}_i$ denotes the all 1's vector of size $|\Omega_i|$):

$$
P = \begin{array}{c} \\ \\ \Lambda\ i) \\ \\ \\ \\ \end{array}
\begin{array}{c}
\begin{array}{ccccc} \Omega_0 & \cdots & \Omega_j & \cdots & \Omega_t \end{array}\\
\left(\begin{array}{ccccc}
 & \cdots & & \cdots & \\
- & \cdots & - & \cdots & -\\
 & & & & \\
 & & & & \\
- & \cdots & - & \cdots & -
\end{array}\right)
\end{array},
\quad
\begin{array}{l}
\text{the row sum of } \Omega_j\text{: } \sum_{\alpha\in\Omega_j} P_\alpha(i)\\
= \begin{cases} k_\Omega \tilde{P}_j(i), & \text{if } i \in \Lambda,\\ 0, & \text{if } i \notin \Lambda,\end{cases}
\end{array}
$$

$$
Q = \begin{array}{c} \\ \Omega_0 \\ \vdots \\ \Omega_i \\ \vdots \\ \Omega_t \end{array}
\begin{array}{c}
\begin{array}{ccc} \Lambda & \overset{j}{\vee} & \end{array}\\
\left(\begin{array}{ccc}
 & \tilde{Q}_j(0)\mathbf{1}_0 & \cdots\\
\vdots & \vdots & \vdots\\
 & \tilde{Q}_j(i)\mathbf{1}_i & \cdots\\
\vdots & \vdots & \vdots\\
 & \tilde{Q}_j(t)\mathbf{1}_t & \cdots
\end{array}\right)
\end{array},
\quad
\begin{array}{l}
|\Lambda| = t + 1,\\
\text{if } \alpha,\beta \in \Omega_i,\ j \in \Lambda,\\
Q_j(\alpha) = Q_j(\beta) = \tilde{Q}_j(i).
\end{array}
$$

2.8 Distance-regular graphs and P-polynomial association schemes

In Chapter 1, Section 1.2, we defined distance-regular graphs, which are undirected connected regular graphs with parameters given in (1.10). Let $\Gamma = (X, E)$ be a distance-regular graph of diameter d. Let $\partial(x, y)$ be the distance between 2 vertices $x, y \in X$. For $x \in V$, define $\Gamma_i(x) = \{y \in X \mid \partial(x, y) = i\}$. With this notation, the definition of distance-regular graphs is reformulated as follows. For any integer i with $0 \le i \le d$ and for any $x \in X$ and $y \in \Gamma_i(x)$, $|\Gamma_{i+1}(x) \cap \Gamma_1(y)| = b_i$, $|\Gamma_{i-1}(x) \cap \Gamma_1(y)| = c_i$ are constants which do not depend on the choice of x and y, where we define $c_0 = b_d = 0$. In fact, if we let k be the valency of a distance-regular graph, then $b_0 = k$, and also if we let $z \in \Gamma_1(y)$, then the following hold:

$$\partial(x,z) + \partial(z,y) \ge \partial(x,y) = i,$$
$$\partial(x,y) + \partial(y,z) \ge \partial(x,z).$$

So we have $i - 1 \le \partial(x, z) \le i + 1$, which means $z \in \Gamma_{i-1}(x) \cup \Gamma_i(x) \cup \Gamma_{i+1}(x)$. Therefore by setting $a_i = |\Gamma_i(x) \cap \Gamma_1(y)|$, we have $a_i + b_i + c_i = |\Gamma_1(y)| = k$. That is to say, a_i is a constant which does not depend on the choice of x and y. We also have $c_1 = |\Gamma_0(x) \cap \Gamma_1(y)| = |\{x\}| = 1$. Moreover, since the diameter is d, we have $c_i > 0$ for $1 \le i \le d$.

Problem 2.70. Let $k_i = |\Gamma_i(x)|$ for $0 \le i \le d$. Prove that $k_0 = 1$, $k_1 = k$, $k_{i+1} = \frac{k_i b_i}{c_{i+1}}$ hold for $0 \le i \le d-1$. (Hint: Count the number of edges contained in $\{\{z, y\} \mid \partial(x, y) = i, \partial(x, z) = i + 1\}$.)

Proposition 2.71. *Let Γ be a distance-regular graph. Define the subset R_i of $X \times X$ as $R_i = \{(x, y) \mid \partial(x, y) = i\}$. Then $\mathfrak{X} = (X, \{R_i\}_{0 \le i \le d})$ is a symmetric association scheme.*

Proof. Define the matrix A_i as in equation (2.1). Then the previous conditions (1′), (2′), (3′) and (6′) on the adjacency matrices of an association scheme hold. Hence it suffices to show that (4′) holds. We have

$$(A_j A_1)(x,y) = \sum_{z \in X} A_j(x,z) A_1(z,y) = |\Gamma_j(x) \cap \Gamma_1(y)|$$
$$= \begin{cases} c_{j+1}, & \text{if } \partial(x,y) = j+1, \\ a_j, & \text{if } \partial(x,y) = j, \\ b_{j-1}, & \text{if } \partial(x,y) = j-1, \\ 0, & \text{if } |\partial(x,y) - j| \ge 2. \end{cases}$$

So for $1 \le j \le d - 1$, we obtain

$$A_j A_1 = c_{j+1} A_{j+1} + a_j A_j + b_{j-1} A_{j-1}. \tag{2.24}$$

Therefore we can verify that A_j is a polynomial in A_1 of degree j inductively. Moreover since the graph Γ has degree $k = b_0$, we have

$$(A_1 - kI)J = A_1 J - kJ = 0. \tag{2.25}$$

Thus we obtain $(A_1 - kI)(A_0 + A_1 + \cdots + A_d) = 0$. Since $A_0, A_1, \ldots, A_d$ are linearly independent, the minimal polynomial of A_1 is the polynomial in A_1 of degree $d+1$ given by the left-hand side of equation (2.25). Therefore the vector space spanned by the matrices $A_0, A_1, \ldots, A_d$ is closed under the ordinary matrix product. Namely, there exists a non-negative integer $p_{i,j}^{\ell}$ such that $A_iA_j = \sum_{\ell=0}^{d} p_{i,j}^{\ell} A_\ell$. Thus we obtain (4′), proving that $\mathfrak{X} = (X, \{R_i\}_{0\le i\le d})$ is a symmetric association scheme. □

By Proposition 2.71, if $\partial(x,y) = \ell$, then

$$|\Gamma_i(x) \cap \Gamma_j(y)| = p_{i,j}^{\ell}$$

holds and the left-hand side is independent of the choice of x and y. In this way, the following concept arises from distance-regular graphs.

Definition 2.72 (P-polynomial scheme). Let $\mathfrak{X} = (X, \{R_i\}_{0\le i\le d})$ be a symmetric association scheme; $\mathfrak{X}$ is called a *P-polynomial association scheme* or a *P-polynomial scheme* with respect to the ordering $R_0, R_1, \ldots, R_d$ if the following condition holds: For each i $(0 \le i \le d)$, there exists a polynomial $v_i(x)$ of degree i in the variable x such that $A_i = v_i(A_1)$ holds by a suitable rearrangement of the ordering of the adjacency matrices of $\mathfrak{X}$.

By definition, the Bose–Mesner algebra of a P-polynomial scheme becomes an algebra generated by the matrix A_1.

The following theorem holds.

Theorem 2.73.
(1) *Let $\mathfrak{X} = (X, \{R_i\}_{0\le i\le d})$ be a P-polynomial scheme with respect to the ordering $R_0, R_1, \ldots, R_d$. The graph $\Gamma = (X, E)$ with vertex set X and edge set $E = \{\{x,y\} \mid (x,y) \in R_1\}$ is a distance-regular graph.*
(2) *Let $\Gamma = (X, E)$ be a distance-regular graph of diameter d. If we define $R_i \subset X \times X$ as $R_i = \{(x,y) \mid \partial(x,y) = i\}$ for each i $(0 \le i \le d)$, then $\mathfrak{X} = (X, \{R_i\}_{0\le i\le d})$ is a P-polynomial scheme of class d.*

Problem 2.74. Prove Theorem 2.73. (Note that (2) is clear from Proposition 2.71.)

Definition 2.75. Let $M = (m_{i,j})_{0\le i,j\le d}$ be a square matrix of size $d+1$. If $m_{i,j} = 0$ for any i,j with $|i-j| \ge 2$ $(0 \le i,j \le d)$, then M is called a *tridiagonal matrix*. We denote M as follows:

$$M = \begin{bmatrix} * & m_{0,1} & m_{1,2} & \cdots & m_{d-2,d-1} & m_{d-1,d} \\ m_{0,0} & m_{1,1} & m_{2,2} & \cdots & m_{d-1,d-1} & m_{d,d} \\ m_{1,0} & m_{2,1} & m_{3,2} & \cdots & m_{d,d-1} & * \end{bmatrix}.$$

Let $\mathfrak{X} = (X, \{R_i\}_{0\le i\le d})$ be a P-polynomial scheme with respect to the ordering $R_0, R_1, \ldots, R_d$. Then by Theorem 2.73, it turns out that the intersection matrix B_1 becomes a tridiagonal matrix.

Theorem 2.76. *Let $\mathfrak{X} = (X, \{R_i\}_{0\le i\le d})$ be a symmetric association scheme. Then the following are equivalent:*
(1) *$\mathfrak{X}$ is a P-polynomial scheme with respect to the ordering $R_0, R_1, \ldots, R_d$;*
(2) *the intersection matrix B_1 is a tridiagonal matrix with the following form:*

$$B_1 = \begin{bmatrix} * & c_1 & c_2 & \cdots & c_{d-1} & c_d \\ 0 & a_1 & a_2 & \cdots & a_{d-1} & a_d \\ b_0 & b_1 & b_2 & \cdots & b_{d-1} & * \end{bmatrix},$$

where $b_i \neq 0$ $(0 \le i \le d-1)$, $c_i \neq 0$ $(1 \le i \le d)$, $b_0 = p^0_{1,1} = k_1$, and $c_1 = 1$;
(3) *if we let P be the first eigenmatrix of $\mathfrak{X}$ and we let $\theta_j = P_1(j)$ for $j = 0, 1, \ldots, d$, then for each i $(0 \le i \le d)$, there exists a polynomial $v_i(x)$ of degree i in the variable x such that $P_i(j) = v_i(\theta_j)$ holds for $j = 0, 1, \ldots, d$.*

(Remark: In (3), $\theta_0, \theta_1, \ldots, \theta_d$ are the eigenvalues of A_1. Therefore for a P-polynomial scheme, any eigenvalue of A_j $(2 \le j \le d)$ is determined by the eigenvalues of A_1.)

Proof. First we prove (1)⇒(2). Firstly, we have $p^0_{1,0} = 0$, $b_0 = p^0_{1,1} = k_1$, $c_1 = p^1_{1,0} = 1$. By the assumption of (1), for each i, there exists a polynomial $v_i(x)$ of degree i such that $A_i = v_i(A_1)$. Since $xv_i(x)$ is a polynomial of degree $i+1$, it can be expressed as a linear combination of $v_0(x), v_1(x), \ldots, v_{i+1}(x)$. So we may write $xv_i(x) = \sum_{\ell=0}^{i+1} c_\ell v_\ell(x)$. In particular, $c_{i+1} \neq 0$. If we substitute A_1 in the equation, we have $A_1 v_i(A_1) = \sum_{\ell=0}^{i+1} c_\ell v_\ell(A_1)$. Therefore we obtain $A_1 A_i = \sum_{\ell=0}^{i+1} c_\ell A_\ell$. Since the left-hand side of the equation is $\sum_{\ell=0}^{d} p^\ell_{1,i} A_\ell$, we have $p^\ell_{1,i} = 0$ for any ℓ with $\ell \ge i+2$. We also have $p^{i+1}_{1,i} = c_{i+1} \neq 0$. Since $\mathfrak{X}$ is symmetric, by Proposition 2.17 (6), we have $k_\ell p^\ell_{1,i} = k_i p^i_{1,\ell}$. Therefore $p^i_{1,\ell} = 0$ holds for any ℓ with $\ell \ge i+2$ and $p^i_{1,i+1} \neq 0$ holds as well.
Next we prove (2)⇒(1). Since $a_0 = 0$, $a_i = p^i_{1,i}$ $(1 \le i \le d)$, $b_i = p^i_{1,i+1}$ $(0 \le i \le d-1)$, and $c_i = p^i_{1,i-1}$ $(1 \le i \le d)$, $p^j_{1,i} = 0$ $(|i-j| \ge 2)$, we have $A_1A_i = p^{i-1}_{1,i}A_{i-1} + p^i_{1,i}A_i + p^{i+1}_{1,i}A_{i+1} = b_{i-1}A_{i-1} + a_iA_i + c_{i+1}A_{i+1}$. Define a polynomial $v_i(x)$ of degree i as $v_0(x) \equiv 1$, $v_1(x) = x$ and define a polynomial $v_{i+1}(x)$ of degree $i+1$ by using a three-term recurrence $xv_i(x) = b_{i-1}v_{i-1}(x) + a_iv_i(x) + c_{i+1}v_{i+1}(x)$. Then we have $A_{i+1} = v_{i+1}(A_1)$.
Finally, we prove (1)⇔(3). Let E_i $(0 \le i \le d)$ be a primitive idempotent of $\mathfrak{X}$. We have $A_1 = \sum_{j=0}^{d} P_1(j)E_j = \sum_{j=0}^{d} \theta_j E_j$. Then for any polynomial $v_i(x) = \sum_{\ell=0}^{i} \lambda_\ell x^\ell$ of degree i, we obtain

$$v_i(A_1) = \sum_{\ell=0}^{i} \lambda_\ell A_1{}^\ell = \sum_{\ell=0}^{i} \lambda_\ell \sum_{j=0}^{d} \theta_j{}^\ell E_j = \sum_{j=0}^{d}\left(\sum_{\ell=0}^{i} \lambda_\ell \theta_j{}^\ell\right)E_j = \sum_{j=0}^{d} v_i(\theta_j)E_j.$$

Therefore (1) and (3) are equivalent. □

Problem 2.77.
(1) Prove that the Hamming scheme $H(d, q)$, which is defined in Example 2.6 in Section 2.1, is a P-polynomial scheme.

(2) Prove that the polynomials $K_i(x)$, which give the first character table P of $H(d,q)$ (i. e., the polynomials of degree i which satisfy $P_i(j) = K_i(j)$), satisfy the following *three-term recurrence relation*:

$$(i+1)K_{i+1}(x) = \big(i + (q-1)(d-i) - qx\big)K_i(x) - (q-1)(d-i+1)K_{i-1}(x).$$

Problem 2.78. Prove that the Johnson scheme $J(v,d)$ defined in Example 2.7 in Section 2.1 is a P-polynomial scheme.

Proposition 2.79. *Let $\mathfrak{X}$ be a P-polynomial scheme. Then the eigenvalues $\theta_0, \theta_1, \ldots, \theta_d$ of A_1, which are defined in Theorem 2.76 (3), are mutually distinct real numbers.*

Proof. Suppose that there exist integers $0 \le j \ne \ell \le d$ such that $\theta_j = \theta_\ell$. Then for any integer $i = 0, 1, \ldots, d$, we have $P_i(j) = v_i(\theta_j) = v_i(\theta_\ell) = P_i(\ell)$, where v_i $(0 \le i \le d)$ is the polynomial of degree i defined in Theorem 2.76 (3). Then the j-th row and the ℓ-th row of the first eigenmatrix P are equal, which contradicts the fact that P is nonsingular. □

Here we make a remark on the origin of the concepts and the names of distance-regular graphs and distance-transitive graphs. Higman (1967) [215] studied distance-transitive graphs, which he called permutation groups of maximal diameter. It was in the late 1960s when the relation between permutation groups of rank 3 and strongly regular graphs was known to the group theorists, and so Higman undoubtedly knew the concept of distance-regular graphs at that time. Historically, Tutte [490] (1947) is the old original paper that started the study in this direction in a sense. Also, we should mention that there was a tradition of Russian school started around the end of 1960's, by Lehman, Weisfeiler, Faradjev, A. A. Ivanov, A. V. Ivanov, Shpectrov, Klin and others. Cf. [262], [182] and many other papers. Biggs (1971) [92] introduced the term and the concept of distance-transitive graphs, and a little later, he introduced the term of distance-regular graphs. It is said that the concept itself was clearly understood by Biggs around 1970 ([91] and [113, page 128]). Delsarte [159] independently figured out the concept of distance-regular graphs, which he called metrically regular graphs or P-polynomial association schemes. Delsarte also made an important contribution by introducing Q-polynomial schemes, which will be discussed in the following section.

2.9 Q-polynomial association schemes

In this section, we consider a symmetric association scheme $\mathfrak{X}$. The Bose–Mesner algebra of $\mathfrak{X}$ has a structure of a commutative algebra which is closed under the ordinary matrix product and the Hadamard product. For the algebraic structure with respect to the Hadamard product, we can define a concept similar to P-polynomial schemes.

Definition 2.80. Let $\mathfrak{X} = (X, \{R_i\}_{0 \le i \le d})$ be a symmetric association scheme; $\mathfrak{X}$ is called a *Q-polynomial association scheme* or a *Q-polynomial scheme* with respect to the ordering $E_0, E_1, \ldots, E_d$ if the following condition holds: For each i $(0 \le i \le d)$, there exists a polynomial $v_i^*(x)$ of degree i in the variable x such that $|X|E_i = v_i^*(|X|E_1)$ holds by a suitable rearrangement of the ordering of the primitive idempotents of $\mathfrak{X}$.

When we substitute a matrix into the polynomial $v_i^*(x)$, we use the Hadamard product for products of matrices.

Distance-regular graphs give a combinatorial meaning of P-polynomial schemes. Unlike P-polynomial schemes, Q-polynomial schemes do not have such combinatorial objects. However, the following proposition holds.

Proposition 2.81. *Let $\mathfrak{X} = (X, \{R_i\}_{0 \le i \le d})$ be a symmetric association scheme. Then the following are equivalent:*

(1) *$\mathfrak{X}$ is a Q-polynomial scheme with respect to the ordering $E_0, E_1, \ldots, E_d$;*
(2) *the dual intersection matrix B_1^* is a tridiagonal matrix with the following form:*

$$B_1^* = \begin{bmatrix} * & c_1^* & c_2^* & \cdots & c_{d-1}^* & c_d^* \\ 0 & a_1^* & a_2^* & \cdots & a_{d-1}^* & a_d^* \\ b_0^* & b_1^* & b_2^* & \cdots & b_{d-1}^* & * \end{bmatrix},$$

where $b_i^* \ne 0$ $(0 \le i \le d-1)$, $c_i^* \ne 0$ $(1 \le i \le d)$, $b_0^* = q_{1,1}^0 = m_1$, and $c_1^* = 1$;
(3) *if we let Q be the second eigenmatrix of $\mathfrak{X}$ and we let $\theta_j^* = Q_1(j)$ for $0 \le j \le d$, then for each i $(0 \le i \le d)$, there exists a polynomial $v_i^*(x)$ of degree i in the variable x such that $Q_i(j) = v_i^*(\theta_j^*)$ for any j $(0 \le j \le d)$.*

Proof. The proof is similar to the proof of Theorem 2.76 for P-polynomial schemes. □

Problem 2.82. Prove Proposition 2.81.

The property similar to Proposition 2.79 holds for Q-polynomial schemes.

Proposition 2.83. *Let $\mathfrak{X}$ be a Q-polynomial scheme. Then the following hold:*

(1) *for $1 \le i \le d-1$, we have*

$$A_i^* A_1^* = c_{i+1}^* A_{i+1}^* + a_i^* A_i^* + b_{i-1}^* A_{i-1}^*;$$

in particular, $q_{i,j}^k = 0$ holds for any integers $0 \le i, j, k \le d$ with $i + j < k$;
(2) *$\theta_0^*, \theta_1^*, \ldots, \theta_d^*$ defined in Proposition 2.81 (3) are the eigenvalues of A_1^* and they are mutually distinct real numbers.*

Proof. (1) This is clear from Proposition 2.32 (2) and Proposition 2.81 (2).
(2) By equation (2.9) of the definition in Section 2.2, we have $A_1^* E_k^* = \sum_{\alpha=0}^{d} Q_1(\alpha) E_\alpha^* E_k^* = Q_1(k) E_k^* = \theta_k^* E_k^*$. Therefore $\theta_0^*, \theta_1^*, \ldots, \theta_d^*$ are the eigenvalues of A_1^*. The proof that they are mutually distinct is similar to that of Proposition 2.79. □

The polynomial systems arising from P-polynomial schemes and Q-polynomial schemes form systems of orthogonal polynomials satisfying the so-called three-term recurrence relations. Many association schemes which are important in terms of applications to combinatorics and other areas have the properties of both P-polynomial and Q-polynomial schemes. The association schemes having these properties are called the *P- and Q-polynomial schemes*, and the corresponding orthogonal polynomials have been extensively studied in the theory of orthogonal polynomials. The Hamming schemes and the Johnson schemes defined in Examples 2.6 and 2.7 in Section 2.1 of this chapter provide important examples of P- and Q-polynomial schemes. In the next section, we will explicitly compute their character tables and show that they are indeed P- and Q-polynomial schemes. Besides these examples, the q-Johnson schemes and the association schemes of bilinear forms also have the structure of P- and Q-polynomial schemes. These will be discussed in Chapter 6.

2.10 Character tables of various association schemes

2.10.1 Association schemes on finite abelian groups

Let G be a finite abelian group. Every conjugacy class of G consists of a single element. Here we consider the association scheme $\mathfrak{X}(G)$ arising from the conjugacy classes of G, which is defined in Example 2.8. In what follows, the operation on G is written in additive notation and the identity element is denoted by 0. If we define $R_g = \{(x,y) \in G \times G \mid x - y = g\}$ for each $g \in G$, we can write $\mathfrak{X}(G) = (G, \{R_g\}_{g\in G})$. The adjacency matrix A_g corresponding to R_g of $\mathfrak{X}(G)$ is indexed by $G \times G$ and its (x,y)-entry can be written as $A_g(x,y) = \delta_{x-y,g}$. Then for $g, h \in G$, the (x,y)-entry of A_gA_h can be calculated as follows:

$$\begin{aligned}(A_gA_h)(x,y) &= \sum_{z\in G} A_g(x,z)A_h(z,y) = \sum_{z\in G} \delta_{x-z,g}\delta_{z-y,h}\\ &= \begin{cases} 1, & \text{if } x-y = g+h,\\ 0, & \text{if } x-y \neq g+h.\end{cases}\end{aligned}$$

So it turns out that $A_gA_h = A_hA_g = A_{g+h}$. Let P be the character table of $\mathfrak{X}(G)$ and E_x $(x \in G)$ the primitive idempotents of $\mathfrak{X}(G)$. Since we have

$$A_x = \sum_{g\in G} P_x(g)E_g,$$

the following holds:

$$\begin{aligned}\sum_{g\in G} P_{x+y}(g)E_g = A_{x+y} = A_xA_y &= \Big(\sum_{g\in G} P_x(g)E_g\Big)\Big(\sum_{h\in G} P_y(h)E_h\Big)\\ &= \sum_{g\in G} P_x(g)P_y(g)E_g.\end{aligned} \tag{2.26}$$

Therefore we obtain $P_{x+y}(g) = P_x(g)P_y(g)$ for each $g, x, y \in G$. If we define $\psi_g(x) = P_x(g)$, then ψ_g becomes an element of the character group $\hat{G}$ of G. Since P is nonsingular, $\{\psi_g \mid g \in G\}$ is identical to $\hat{G}$, and P is identical to the character table of the irreducible characters of G. We can rearrange the ordering of R_x $(x \in G)$ so that P is symmetric. Since any finite abelian group is isomorphic to a direct product of finite cyclic groups by the fundamental theorem of finite abelian groups, we explain the case of cyclic groups only. Let $G = \mathbb{Z}/m\mathbb{Z} = \{0, 1, \dots, m-1\}$ be the cyclic group of order m. Let ζ be a primitive m-th root of unity. If we define $\psi_y(x) = P_x(y) = \zeta^{xy}$, ψ_y is a linear character of $G = \mathbb{Z}/m\mathbb{Z}$. Obviously, $P_x(y) = P_y(x)$. Thus P becomes symmetric. Then by Theorem 2.22 (3), it turns out that $P = \overline{Q}$, and by Proposition 2.47, $\mathfrak{X}(G)$ is self-dual.

2.10.2 The character table of the Hamming scheme $H(d, q)$

At the end of Section 2.8 in this chapter, the proof of the fact that the Hamming scheme $H(d, q)$ becomes a P-polynomial scheme is left as an exercise for the reader. Here, by calculating the eigenmatrices P, Q of $H(d, q)$, we want to show that $H(d, q)$ is a self-dual P- and Q-polynomial scheme. When we defined the Hamming scheme $H(d, q)$ in Example 2.6 in Section 2.1, F was defined to be a q-element set which does not have any algebraic structure. Here we regard F as an abelian group, and write $F = F_q$. Namely, $F = F_q$ is an additive group of order q. In fact, this does not make any difference to the definition of the Hamming scheme. For $\mathbf{x} \in X$, we define the *weight* as $w(\mathbf{x}) = |\{i \mid 1 \le i \le d, x_i \ne 0\}|$. Then the Hamming distance can be expressed as $\partial(\mathbf{x}, \mathbf{y}) = w(\mathbf{x} - \mathbf{y})$. Thus we have $R_i = \{(\mathbf{x}, \mathbf{y}) \mid w(\mathbf{x} - \mathbf{y}) = i\}$. We define $X_i = \{\mathbf{x} \in X \mid w(\mathbf{x}) = i\}$. Then $X_0, X_1, \dots, X_d$ give a partition of X. As is mentioned in Section 2.10 of this chapter, let $\hat{F}_q$ be the character group consisting of the linear characters of the finite abelian group F_q; $\hat{F}_q$ can be indexed by F_q. If we put $\hat{F}_q = \{\psi_x \mid x \in F_q\}$, we can rearrange the ordering of the elements of $\hat{F}_q$ so that $\psi_x(y) = \psi_y(x)$ holds; $X = F_q^d$ is also a finite abelian group. For $\mathbf{x} = (x_1, x_2, \dots, x_d)$, $\mathbf{y} = (y_1, y_2, \dots, y_d)$, if we define $\Psi_{\mathbf{x}}(\mathbf{y}) = \prod_{i=1}^d \psi_{x_i}(y_i)$, then $\Psi_{\mathbf{x}}$ becomes a linear character of X. It turns out that the character group of X is $\hat{X} = \{\Psi_{\mathbf{x}} \mid \mathbf{x} = (x_1, x_2, \dots, x_d) \in F_q^d\}$.

Proposition 2.84. *Let $D = \{0, 1, \dots, d\}$. For given $j, k \in D$ and $\mathbf{u} \in X_j$, we have*

$$\sum_{\mathbf{x} \in X_k} \Psi_{\mathbf{u}}(\mathbf{x}) = K_k(j) = \sum_{i=0}^{k} (-1)^i (q-1)^{k-i} \binom{d-j}{k-i} \binom{j}{i}.$$

Remark 2.85. The polynomial $K_k(u)$, which appears in Proposition 2.84, becomes a polynomial of order k in u. In general, the polynomials of the following form are called

Krawtchouk polynomials and form a system of orthogonal polynomials:

$$K_k(u) = \sum_{i=0}^{k}(-1)^i(q-1)^{k-i}\binom{d-u}{k-i}\binom{u}{i}. \tag{2.27}$$

The following transformation is also known:

$$K_k(u) = \sum_{i=0}^{k}(-q)^i(q-1)^{k-i}\binom{d-i}{k-i}\binom{u}{i}. \tag{2.28}$$

The *generating function* of the Krawtchouk polynomial is $(1+(q-1)z)^{d-u}(1-z)^u$. Namely, $K_k(u)$ appears as the coefficient of the term z^k in the generating function. Consider the following transformation of the generating function:

$$(1+(q-1)z)^{d-u}(1-z)^u = (1+(q-1)z)^d\left(\frac{1-z}{1+(q-1)z}\right)^u.$$

Then we can prove equation (2.28). We omit the proof ([445]). Here we extend the definition of the binomial coefficients as $\binom{i}{j} = 0$ if $i < 0$ or $j < 0$, or $i < j$.

Proof of Proposition 2.84. Let $D^\times = D\backslash\{0\}$. For each k-element subset M of $D^\times$, define $X_M \subset X_k$ by

$$X_M = \{\boldsymbol{x} \in X_k \mid x_\nu \neq 0,\ \nu \in M\}.$$

For $M_1, M_2 \subset D^\times$ with $|M_1| = |M_2| = k$ and $M_1 \neq M_2$, $X_{M_1} \cap X_{M_2} = \emptyset$ holds. Therefore X_k is partitioned as follows:

$$X_k = \bigcup_{\substack{M\subset D^\times,\\ |M|=k}} X_M.$$

Let $\boldsymbol{x} = (x_1, x_2, \ldots, x_d) \in X_M$. Then $x_\nu = 0$ for $\nu \notin M$, and $x_\nu \neq 0$ for $\nu \in M$. So we have $\Psi_{\boldsymbol{u}}(\boldsymbol{x}) = \prod_{\nu=1}^{d}\psi_{u_\nu}(x_\nu) = \prod_{\nu\in M}\psi_{u_\nu}(x_\nu)$. Therefore we have

$$\sum_{\boldsymbol{x}\in X_M}\Psi_{\boldsymbol{u}}(\boldsymbol{x}) = \sum_{\boldsymbol{x}\in X_M}\left(\prod_{\nu\in M}\psi_{u_\nu}(x_\nu)\right) = \prod_{\nu\in M}\sum_{x_\nu\in F_q^\times}\psi_{u_\nu}(x_\nu).$$

Since ψ_{u_ν} is a linear character of F_q, we have $\sum_{x_\nu\in F_q}\psi_{u_\nu}(x_\nu) = \delta_{0,u_\nu}q$, and thus we have

$$\sum_{x_\nu\in F_q^\times}\psi_{u_\nu}(x_\nu) = \begin{cases} q-1, & \text{if } u_\nu = 0,\\ -1, & \text{if } u_\nu \neq 0.\end{cases}$$

Next we define $i = |\{\nu \in M \mid u_\nu \neq 0\}|$. Then

$$\prod_{\nu\in M}\sum_{x_\nu\in F_q^\times}\psi_{u_\nu}(x_\nu) = (-1)^i(q-1)^{k-i}.$$

On the other hand, since $\boldsymbol{u} \in X_j$, for each i $(0 \le i \le d)$, the number of $M \subset D^\times$ satisfying $i = |\{\nu \in M \mid u_\nu \ne 0\}|$ and $|M| = k$ is exactly equal to $\binom{j}{i}\binom{d-j}{k-i}$. Therefore we obtain

$$\sum_{\boldsymbol{x}\in X_k} \Psi_{\boldsymbol{u}}(\boldsymbol{x}) = \sum_{\substack{M\subset D^\times \\ |M|=k}} \sum_{\boldsymbol{x}\in M} \Psi_{\boldsymbol{u}}(\boldsymbol{x}) = \sum_{i=0}^{d} \binom{j}{i}\binom{d-j}{k-i}(-1)^i(q-1)^{k-i}. \qquad \square$$

Theorem 2.86. *Let P, Q be the first and second eigenmatrices of the Hamming scheme $H(d,q)$. Then the following holds:*

$$P_k(i) = Q_k(i) = K_k(i).$$

In particular, $H(d,q)$ is self-dual.

Proof. Define a matrix H_k, whose rows are indexed by X and columns by X_k, as follows. For $(\boldsymbol{x},\boldsymbol{y}) \in X \times X_k$,

$$H_k(\boldsymbol{x},\boldsymbol{y}) = \Psi_{\boldsymbol{x}}(\boldsymbol{y}).$$

Then for any $(\boldsymbol{x},\boldsymbol{y}) \in R_i$,

$$\begin{aligned}(H_k H_k^*)(\boldsymbol{x},\boldsymbol{y}) &= \sum_{\boldsymbol{z}\in X_k} \Psi_{\boldsymbol{x}}(\boldsymbol{z})\overline{\Psi_{\boldsymbol{z}}(\boldsymbol{y})} = \sum_{\boldsymbol{z}\in X_k} \Psi_{\boldsymbol{x}}(\boldsymbol{z})\Psi_{-\boldsymbol{y}}(\boldsymbol{z}) \\ &= \sum_{\boldsymbol{z}\in X_k} \Psi_{\boldsymbol{x}-\boldsymbol{y}}(\boldsymbol{z}) = K_k(i).\end{aligned}$$

So we have

$$\frac{1}{|X|} H_k H_k^* = \frac{1}{|X|} \sum_{i=0}^{d} K_k(i) A_i.$$

Moreover, for $(\boldsymbol{x},\boldsymbol{y}) \in X_k \times X_j$, we have

$$(H_k^* H_j)(\boldsymbol{x},\boldsymbol{y}) = \sum_{\boldsymbol{z}\in X} \overline{\Psi_{\boldsymbol{z}}(\boldsymbol{x})}\Psi_{\boldsymbol{z}}(\boldsymbol{y}) = \sum_{\boldsymbol{z}\in X} \Psi_{\boldsymbol{z}}(\boldsymbol{y}-\boldsymbol{x}) = \delta_{\boldsymbol{x},\boldsymbol{y}}|X|.$$

So we obtain

$$H_k^* H_j = \delta_{k,j}|X| I.$$

Therefore, if we define $E_j = \frac{1}{|X|} H_j H_j^*$ for $j \in D$, then $E_0, E_1, \dots, E_d$ are mutually orthogonal idempotents and satisfy

$$E_j = \frac{1}{|X|} \sum_{i=0}^{d} K_j(i) A_i.$$

Therefore we have

$$Q_j(i) = K_j(i).$$

Next, note that $m_i = Q_i(0) = K_i(0) = (q-1)^i\binom{d}{i}$ and $k_j = |X_j| = (q-1)^j\binom{d}{j}$. By Theorem 2.22 (3), we obtain

$$\begin{aligned}
P_j(i) &= \frac{k_j}{m_i}Q_i(j) = \frac{(q-1)^j\binom{d}{j}}{(q-1)^i\binom{d}{i}}K_i(j)\\
&= \frac{(q-1)^{j-i}i!(d-i)!}{j!(d-j)!}\sum_{\ell=0}^{j}(-1)^\ell(q-1)^{i-\ell}\binom{d-j}{i-\ell}\binom{j}{\ell}\\
&= \sum_{\ell=0}^{j}(-1)^\ell(q-1)^{j-\ell}\frac{i!(d-i)!}{j!(d-j)!}\frac{(d-j)!}{(i-\ell)!(d-j-i+\ell)}\frac{j!}{\ell!(j-\ell)!}\\
&= \sum_{\ell=0}^{j}(-1)^\ell(q-1)^{j-\ell}\frac{i!(d-i)!}{(i-\ell)!\ell!(d-i-j+\ell)(j-\ell)!}\\
&= K_j(i).
\end{aligned}$$

□

Thus we have shown that $H(d,q)$ is self-dual. In Problem 2.78, $H(d,q)$ was proved to be a P-polynomial scheme. Theorem 2.86 shows that $H(d,q)$ is a P- and Q-polynomial scheme.

2.10.3 The character table of the Johnson scheme $J(v,d)$

Let V be a v-element set and X the set $V^{(d)}$ of d-element subsets of V. Assume $0 < d \le \frac{v}{2}$. The Johnson scheme $J(v,d)$ is defined on X and its relations $R_0, R_1, \ldots, R_d$ are defined by $R_i = \{(x,y) \in X\times X \mid |x\cap y| = d-i\}$. Let $A_0, A_1, \ldots, A_d$ be the adjacency matrices and $\mathfrak{A}$ the Bose–Mesner algebra. Let $D = \{0,1,\ldots,d\}$.

Proposition 2.87. *The valency $k_i (i \in D)$ of $J(v,d)$ is given by the following equation:*

$$k_i = \binom{d}{i}\binom{v-d}{i}.$$

Proof. Fix $x \in X$. Then $k_i = |\{y \in X \mid (x,y) \in R_i\}| = |\{y \in X \mid |x\cap y| = d-i\}|$. Thus by choosing $d-i$ elements from x and choosing i elements from $V\backslash x$, we obtain $y \in X$ satisfying the condition. Therefore, $k_i = \binom{d}{d-i}\binom{v-d}{i} = \binom{d}{i}\binom{v-d}{i}$. □

Next, we define a matrix $C_i \in \mathfrak{A}$ $(i \in D)$ as follows:

$$C_i = \sum_{\ell=i}^{d}\binom{\ell}{i}A_{d-\ell}. \tag{2.29}$$

Proposition 2.88. *We have:*

(1) *$C_0, C_1, \ldots, C_d$ form a basis of $\mathfrak{A}$;*

(2) *if we let $(x,y) \in R_{d-k}$, then $C_i(x,y)$ is the number of i-element subsets contained in $x \cap y$.*

Proof. (1) By the definition, $C_{d-i} = A_i + (d-i+1)A_{i-1} + \cdots$ for $i = 0,1,\ldots,d$. Hence $\langle C_d, C_{d-1},\ldots,C_1,C_0\rangle = \langle A_0,A_1,\ldots,A_{d-1},A_d\rangle = \mathfrak{A}$.
(2) It is clear because $C_i(x,y) = \binom{k}{i}$ and $|x\cap y| = k$. □

Proposition 2.89. *We have*

$$\binom{d-j}{r-i}\binom{j}{i} = \sum_{\ell=i}^{r}(-1)^{\ell-i}\binom{\ell}{i}\binom{d-\ell}{r-\ell}\binom{j}{\ell}.$$

Proof. We use induction on j. For any d,r, when $j=0$, both the left-hand side and the right-hand side become $\binom{d}{r}$ if $i=0$ and they become 0 if $i>0$. For any d,r,i, we assume the equation holds for $0\le j\le k$, and will show that it holds for $j=k+1$. We have

$$\begin{aligned}
&\sum_{\ell=i}^{r}(-1)^{\ell-i}\binom{\ell}{i}\binom{d-\ell}{r-\ell}\binom{k+1}{\ell}\\
&\quad=\sum_{\ell=i}^{r}(-1)^{\ell-i}\binom{\ell}{i}\binom{d-\ell}{r-\ell}\frac{(k+1)!}{\ell!(k+1-\ell)!}\\
&\quad=(k+1)\sum_{\ell=i}^{r}(-1)^{\ell-i}\binom{\ell}{i}\binom{d-\ell}{r-\ell}\frac{k!}{\ell!(k-(\ell-1))!}\\
&\quad=(k+1)\sum_{\nu=i-1}^{r-1}(-1)^{\nu+1-i}\binom{\nu+1}{i}\binom{d-1-\nu}{r-1-\nu}\frac{k!}{(\nu+1)!(k-\nu)!}\\
&\quad=\frac{k+1}{i}\sum_{\nu=i-1}^{r-1}(-1)^{\nu-(i-1)}\frac{\nu!}{(i-1)!(\nu-(i-1))!}\binom{d-1-\nu}{r-1-\nu}\frac{k!}{\nu!(k-\nu)!}\\
&\quad=\frac{k+1}{i}\sum_{\nu=i-1}^{r-1}(-1)^{\nu-(i-1)}\binom{\nu}{i-1}\binom{d-1-\nu}{r-1-\nu}\binom{k}{\nu}.
\end{aligned}$$

By the induction hypothesis, for $d-1, i-1, r-1$, the equation holds for $0\le j\le k$. Therefore we have

$$\begin{aligned}
&\sum_{\ell=i}^{r}(-1)^{\ell-i}\binom{\ell}{i}\binom{d-\ell}{r-\ell}\binom{k+1}{\ell}\\
&\quad=\frac{k+1}{i}\binom{d-1-k}{r-1-(i-1)}\binom{k}{i-1}=\binom{d-(k+1)}{r-i}\binom{k+1}{i}.
\end{aligned}$$ □

Proposition 2.90. *We have*

$$C_rC_s=\sum_{\ell=0}^{\min\{r,s\}}\binom{d-\ell}{r-\ell}\binom{d-\ell}{s-\ell}\binom{v-r-s}{v-d-\ell}C_\ell.$$

Proof. For $k\in D$, let $V^{(k)}$ be the set of k-element subsets of V ($X = V^{(d)}$). We want to compute the (x,y)-entry of the matrix C_rC_s. Let $(x,y)\in R_{d-u}$, i. e., $x,y\in X=V^{(d)}$, $|x\cap y| = u$. Then we have

$$(C_rC_s)(x,y)=\sum_{z\in V^{(d)}}C_r(x,z)C_s(z,y).$$

By Proposition 2.88 (2), $C_r(x,z) = |\{\xi \subset x \cap z \mid |\xi| = r\}|$ and $C_s(z,y) = |\{\eta \subset z \cap y \mid |\eta| = s\}|$. So we have

$$
\begin{aligned}
(C_rC_s)(x,y) &= \sum_{z\in X} |\{\xi \subset x \cap z \mid |\xi| = r\}|\,|\{\eta \subset z \cap y \mid |\eta| = s\}| \\
&= |\{(\xi,\eta,z) \in V^{(r)} \times V^{(s)} \times V^{(d)} \mid \xi \subset x, \eta \subset y, \xi \cup \eta \subset z\}|.
\end{aligned} \tag{2.30}
$$

For each $i \in D$, we have

$$|\{\xi \subset x \mid \xi \in V^{(r)}, |\xi \cap y| = i\}| = \binom{d-u}{r-i}\binom{u}{i}.$$

If we fix $\xi \in V^{(r)}$ satisfying $\xi \subset x$ and $|\xi \cap y| = i$, and $j \in D$, we obtain

$$|\{\eta \subset y \mid \eta \in V^{(s)}, |\eta \cap \xi| = j\}| = \binom{d-i}{s-j}\binom{i}{j}.$$

Moreover, if we fix $\xi \in V^{(r)}$ and $\eta \in V^{(s)}$ satisfying $\xi \subset x$, $\eta \subset y$, $|\xi \cap y| = i$ and $|\xi \cap \eta| = j$, we obtain

$$|\{z \in V^{(d)} \mid \xi \cup \eta \subset z\}| = \binom{v-r-s+j}{v-d}.$$

Therefore we obtain

$$(C_rC_s)(x,y) = \sum_{i,j\in D} \binom{d-u}{r-i}\binom{u}{i}\binom{d-i}{s-j}\binom{i}{j}\binom{v-r-s+j}{v-d}. \tag{2.31}$$

Now we apply Proposition 2.89 to $\binom{d-u}{r-i}\binom{u}{i}$. We obtain

$$
\begin{aligned}
&(C_rC_s)(x,y) \\
&= \sum_{i,j\in D}\sum_{\ell=i}^{r} (-1)^{\ell-i}\binom{\ell}{i}\binom{d-\ell}{r-\ell}\binom{u}{\ell}\binom{d-i}{s-j}\binom{i}{j}\binom{v-r-s+j}{v-d} \\
&= \sum_{\ell=0}^{r}\sum_{i=0}^{\ell}\sum_{j=0}^{s} (-1)^{\ell-i}\binom{\ell}{i}\binom{d-\ell}{r-\ell}\binom{d-i}{s-j}\binom{i}{j}\binom{v-r-s+j}{v-d}\binom{u}{\ell} \\
&= \sum_{\ell=0}^{r}\sum_{i=0}^{\ell}\sum_{j=0}^{s} (-1)^{\ell-i}\binom{\ell}{i}\binom{d-\ell}{r-\ell}\binom{d-i}{s-j}\binom{i}{j}\binom{v-r-s+j}{v-d} C_\ell(x,y).
\end{aligned}
$$

We simplify the coefficient of $C_\ell(x,y)$ by using Proposition 2.89. We have

$$
\begin{aligned}
&\sum_{i=0}^{\ell}\sum_{j=0}^{s} (-1)^{\ell-i}\binom{\ell}{i}\binom{d-\ell}{r-\ell}\binom{d-i}{s-j}\binom{i}{j}\binom{v-r-s+j}{v-d} \\
&= \binom{d-\ell}{r-\ell}\sum_{j=0}^{s}(-1)^{\ell-j}\binom{v-r-s+j}{v-d}
\end{aligned}
$$

$$\times \sum_{i=j}^{\ell}(-1)^{j-i}\binom{i}{j}\binom{d-i}{d-s+j-i}\binom{\ell}{i}$$
$$=\binom{d-\ell}{r-\ell}\sum_{j=0}^{s}(-1)^{\ell-j}\binom{v-r-s+j}{v-d}\binom{d-\ell}{d-s}\binom{\ell}{j}$$
$$=\binom{d-\ell}{r-\ell}\binom{d-\ell}{d-s}\sum_{j=0}^{\ell}(-1)^{\ell-j}\binom{v-r-s+j}{v-d}\binom{\ell}{j}.$$

Here we put $\nu = \ell - j$. Again by Proposition 2.89, we obtain the following:

$$\sum_{j=0}^{\ell}(-1)^{\ell-j}\binom{v-r-s+j}{v-d}\binom{\ell}{j} = \sum_{\nu=0}^{\ell}(-1)^{\nu}\binom{v-r-s+\ell-\nu}{d-r-s+\ell-\nu}\binom{\ell}{\nu}$$
$$=\binom{v-r-s+\ell-\ell}{d-r-s+\ell} = \binom{v-r-s}{v-d-\ell}. \tag{2.32}$$

By combining the above calculation, the proof is complete. □

Next, we consider the relation between the basis $E_0, E_1, \ldots, E_d$ of primitive idempotents for the Bose–Mesner algebra $\mathfrak{A}$ and $C_0, C_1, \ldots, C_d$.

Proposition 2.91. *By a suitable rearrangement of the ordering of $E_0, E_1, \ldots, E_d$, the following (1), (2) hold:*

(1) *for any $r = 0, 1, \ldots, d$, $C_r = \sum_{i=0}^{r}\binom{d-i}{r-i}\binom{v-r-i}{d-r}E_i$;*

(2) $m_i = \mathrm{rank}(E_i) = \mathrm{rank}(C_i) - \mathrm{rank}(C_{i-1})$.

Proof. (1) By the definition of C_i $(0 \le i \le d)$, we have $\mathfrak{A} = \langle A_i \mid 0 \le i \le d\rangle = \langle C_i \mid 0 \le i \le d\rangle$. If we define $\mathfrak{A}_r = \langle C_i \mid 0 \le i \le r\rangle$, then the following holds:

$$\mathfrak{A}_0 \subset \mathfrak{A}_1 \subset \cdots \subset \mathfrak{A}_d.$$

Each $\mathfrak{A}_r$ is an ideal of $\mathfrak{A}$, and as a vector space, its dimension is $\dim(\mathfrak{A}_r) = r+1$. Therefore by a suitable rearrangement of the ordering of $E_0, E_1, \ldots, E_d$, we have $\mathfrak{A}_r = \langle C_i \mid 0 \le i \le r\rangle = \langle E_i \mid 0 \le i \le r\rangle$. Then for each $r = 0, 1, \ldots, d$, there exists a real number $\rho_{r,i}$ $(0 \le i \le r)$ satisfying the following:

$$C_r = \sum_{i=0}^{r}\rho_{r,i}E_i.$$

Now suppose $0 \le s \le r \le d$, and we compute C_rC_s. Then we obtain the following:

$$C_rC_s = \sum_{i=0}^{r}\rho_{r,i}E_i\sum_{j=0}^{s}\rho_{s,j}E_j = \sum_{j=0}^{s}\rho_{r,j}\rho_{s,j}E_j$$
$$= \rho_{r,s}\rho_{s,s}E_s + \sum_{j=0}^{s-1}\rho_{r,j}\rho_{s,j}E_j$$

$$= \rho_{r,s}\left(C_s - \sum_{j=0}^{s-1} \rho_{s,j}E_j\right) + \sum_{j=0}^{s-1} \rho_{r,j}\rho_{s,j}E_j$$

$$= \rho_{r,s}C_s + \sum_{j=0}^{s-1} \rho_{s,j}(\rho_{r,j} - \rho_{r,s})E_j.$$

Since $\sum_{j=0}^{s-1} \rho_{s,j}(\rho_{r,j} - \rho_{r,s})E_j \in \mathfrak{A}_{s-1}$, by Proposition 2.90, we obtain

$$\rho_{r,s} = \binom{d-s}{r-s}\binom{v-r-s}{v-d-s} = \binom{d-s}{r-s}\binom{v-r-s}{d-r}.$$

This completes the proof of (1).

(2) Since $E_0, E_1, \ldots, E_d$ are mutually orthogonal primitive idempotents, we have $\operatorname{rank}(C_r) = \sum_{i=0}^{r} \operatorname{rank}(E_i) = \sum_{i=0}^{r} m_i$. By this, we can prove (2). □

Next, in order to find the rank of C_i, we define the following matrix. Let $0 \le i \le d$. We define the matrix M_i, whose rows are indexed by $X = V^{(d)}$ and whose columns are indexed by $V^{(i)}$, as follows. For $(x, \xi) \in X \times V^{(i)}$, the (x, ξ)-entry of M_i is defined by

$$M_i(x, \xi) = \begin{cases} 1, & \text{if } \xi \subset x, \\ 0, & \text{if } \xi \not\subset x. \end{cases} \tag{2.33}$$

Then the following holds.

Proposition 2.92. *We have*

$$C_i = M_i\, {}^t M_i, \quad 0 \le i \le d.$$

Proof. Let $(x, y) \in R_{d-\ell}$. Then by the definition, $C_i(x, y) = \binom{\ell}{i}$. On the other hand, since $|x \cap y| = \ell$, we have

$$\sum_{\xi \in V^{(i)}} M_i(x, \xi) M_i(y, \xi) = |\{\xi \in V^{(i)} \mid \xi \subset x \cap y\}| = \binom{\ell}{i}.$$ □

The next lemma was given by Kantor [272].[1]

Lemma 2.93 (Kantor 1972). *With the above definitions and notations, the following holds:*

$$\operatorname{rank}(M_i) = |V^{(i)}| = \binom{v}{i}.$$

1 Tsuyosi Miezaki pointed out that this result is introduced as Gottlieb's theorem [202] in the book: Jiří Matoušec, *Thirty-three Miniatures: Mathematical and Algorithmic Applications of Linear Algebra*, AMS, 2010.

Proof. Suppose the column vectors of M_i are not linearly independent. Then there exists $\xi^* \in V^{(i)}$ such that

$$M_i(x, \xi^*) = \sum_{\xi \in V^{(i)}, \xi \neq \xi^*} \alpha_\xi M_i(x, \xi), \quad \alpha_\xi \in \mathbb{R},\ x \in X.$$

Let S_i be the symmetric group on ξ^* and S_{v-i} the symmetric group on $V \backslash \xi^*$. Let $G = S_i \times S_{v-i}$ act on V. More precisely, for $g = (\sigma, \tau) \in G$ and $a \in V$, we have

$$a^g = \begin{cases} a^\sigma, & \text{if } a \in \xi^*, \\ a^\tau, & \text{if } a \notin \xi^*. \end{cases}$$

Then for any $g \in G$ and for subsets $\xi \in V^{(i)}, x \in X = V^{(d)}$ of V satisfying $\xi \subset x$, we have $\xi^g \in V^{(i)}$, $x^g \in X$, $\xi^g \subset x^g$. We also have $(\xi^*)^g = \xi^*$. Therefore for $g \in G$, $(x, \xi) \in X \times V^{(i)}$, we have $M_i(x^g, \xi^g) = M_i(x, \xi)$ and the following holds:

$$\begin{aligned} M_i(x, \xi^*) = 7M_i(x^g, (\xi^*)^g) = M_i(x^g, \xi^*) &= \sum_{\xi \in V^{(i)}, \xi \neq \xi^*} \alpha_\xi M_i(x^g, \xi) \\ = \sum_{\substack{\xi \in V^{(i)}, \\ \xi \neq \xi^*}} \alpha_\xi M_i(x, \xi^{g^{-1}}) &= \sum_{\substack{\xi \in V^{(i)}, \\ \xi \neq \xi^*}} \alpha_{\xi^g} M_i(x, \xi). \end{aligned} \tag{2.34}$$

If we take the summation over $g \in G$ for both sides of equation (2.34), we obtain

$$|G| M_i(x, \xi^*) = \sum_{g \in G} \sum_{\substack{\xi \in V^{(i)}, \\ \xi \neq \xi^*}} \alpha_{\xi^g} M_i(x, \xi) = \sum_{\substack{\xi \in V^{(i)}, \\ \xi \neq \xi^*}} M_i(x, \xi) \Big(\sum_{g \in G} \alpha_{\xi^g} \Big). \tag{2.35}$$

Now consider the action of G on $V^{(i)}$. If we define $V_j^{(i)} = \{\xi \in V^{(i)} \mid |\xi \cap \xi^*| = j\}$, then $V_j^{(i)}$ becomes an orbit and $V^{(i)} = \bigcup_{j=0}^{i} V_j^{(i)}$ is the orbit decomposition of $V^{(i)}$ by G. In particular, $V_i^{(i)} = \{\xi^*\}$. Moreover, for $\xi_1, \xi_2 \in V_j^{(i)}$, both of which belong to the same orbit of $V^{(i)}$, we have $\sum_{g \in G} \alpha_{\xi_1^g} = \sum_{g \in G} \alpha_{\xi_2^g}$. Here we define

$$b_j = \sum_{g \in G} \alpha_{\xi^g}, \quad \xi \in V_j^{(i)},\ j = 0, \dots, i-1.$$

Then we obtain the following:

$$M_i(x, \xi^*) = \sum_{j=0}^{i-1} b_j \sum_{\xi \in V_j^{(i)}} M_i(x, \xi), \quad x \in X. \tag{2.36}$$

Next, for each $j = 0, 1, \dots, i-1$, fix $x^{(j)} \in X$ satisfying $|x^{(j)} \cap \xi^*| = j$. Then the following holds:

$$M_i(x^{(j)}, \xi^*) = 0, \quad 0 \le j \le i-1\ (\xi^* \not\subset x^{(j)}), \tag{2.37}$$

$$\sum_{\xi \in V_\ell^{(i)}} M_i(x^{(j)}, \xi) \neq 0, \quad 0 \leq \ell \leq j \leq i-1 \text{ (there exist } \xi \subset x^{(j)},\ \xi \in V_\ell^{(i)}), \tag{2.38}$$

$$\sum_{\xi \in V_\ell^{(i)}} M_i(x^{(j)}, \xi) = 0, \quad 0 \leq j < \ell \leq i-1 \ (\xi \not\subset x^{(j)}). \tag{2.39}$$

If we substitute x in (2.36) by $x^{(j)}$ $(0 \leq j \leq i-1)$, we obtain exactly i linear equations in b_ℓ $(0 \leq \ell \leq i-1)$ as follows:

$$\sum_{\ell=0}^{i-1} b_\ell \sum_{\xi \in V_\ell^{(i)}} M_i(x^{(j)}, \xi) = 0, \quad 0 \leq j \leq i-1. \tag{2.40}$$

By (2.37)–(2.39), it turns out that the coefficient matrix of (2.40) is a triangular matrix with diagonal entries non-zero. Thus $b_0 = b_1 = \cdots = b_{i-1} = 0$. By (2.36), $M_i(x, \xi^*) = 0$ holds for all $x \in X$. However, $x \in X$ satisfying $\xi^* \subseteq x$ always exists, which is a contradiction. Therefore, the column vectors of M_i are linearly independent, which means $\operatorname{rank}(M_i) = |V^{(i)}| = \binom{v}{i}$. □

By Proposition 2.91, Proposition 2.92, and Lemma 2.93, we obtain the following.

Proposition 2.94. *The multiplicity of the Johnson scheme is given by the following:*

$$m_i = \operatorname{rank}(E_i) = \binom{v}{i} - \binom{v}{i-1}, \quad 0 \leq i \leq d.$$

Theorem 2.95. *Let P, Q be the first and second eigenmatrices of the Johnson scheme $J(v,d)$, respectively. Then the following hold:*

$$P_j(i) = \sum_{\ell=0}^{j} (-1)^{j-\ell} \binom{d-\ell}{d-j} \binom{d-i}{\ell} \binom{v-d+\ell-i}{\ell}, \tag{2.41}$$

$$Q_j(i) = m_j k_i^{-1} P_i(j). \tag{2.42}$$

Proof. By (2.29), $A_0 = C_d$. In general, we can verify that

$$A_{d-k} = \sum_{r=k}^{d} (-1)^{r-k} \binom{r}{k} C_r. \tag{2.43}$$

Hence by Proposition 2.91(1), we have

$$A_{d-k} = \sum_{r=k}^{d} (-1)^{r-k} \binom{r}{k} \sum_{\ell=0}^{r} \binom{d-\ell}{r-\ell} \binom{v-r-\ell}{d-r} E_\ell. \tag{2.44}$$

If we set $j = d-r$, $u = d-k$ in (2.44), then we obtain

$$P_u(i) = \sum_{j=0}^{u} (-1)^{u-j} \binom{d-j}{d-u} \binom{d-i}{j} \binom{v-d+j-i}{j}. \tag{2.45}$$

□

Thus, we have determined the eigenmatrices of the Johnson scheme $J(v,d)$ specifically. In Problem 2.78, the Johnson scheme $J(v,d)$ was shown to be P-polynomial. If we express as $\boldsymbol{E}_j(x) = \sum_{\ell=0}^{j}(-1)^{j-\ell}\binom{d-\ell}{d-j}\binom{d-x}{\ell}\binom{v-d+\ell-x}{\ell}$, then $\boldsymbol{E}_j(x)$ is a polynomial of degree $2j$ in x; $P_j(i)$ is a polynomial of degree j in $P_1(i)$, where $P_1(i) = (d-i)(v-d-i+1)-d$. This polynomial $\boldsymbol{E}_j(x)$ is an orthogonal polynomial, which has long been known. In Delsarte [159], $\boldsymbol{E}_j(x)$ is called the *Eberlein polynomial* (this name seems not very common nowadays). For details, see Chapter 6. Also, by using the formulas in Theorem 2.95, the second eigenmatrix Q can be expressed in terms of an orthogonal polynomial. We omit the proof because the calculations are too complicated. In this way, we can verify that the Johnson scheme $J(v,d)$ is a P- and Q-polynomial scheme.

2.11 Embeddings into spheres

In this section, we consider a symmetric association scheme $\mathfrak{X} = (X, \{R_i\}_{0\le i\le d})$. Since the eigenvalues of a symmetric matrix are real, the eigenmatrices P, Q of $\mathfrak{X}$ are real matrices and the basis of primitive idempotents consists of real matrices. Therefore, in this section, V denotes the vector space spanned by X over the real field, i. e., $V = \mathbb{R}^{|X|} = \mathbb{R}^n$, where $n = |X|$. Here we identify X with the canonical basis of $\mathbb{R}^n$, that is, $x \in X$ is also the basis vector in $\mathbb{R}^n$ whose x-entry is 1. The *canonical inner product* of $\mathbb{R}^n$ is denoted by $\langle\ ,\ \rangle$. Fix $E = E_1$ for a primitive idempotent E_1. For each i $(0 \le i \le d)$, let $\theta_i^* = Q_1(i)$.

Definition 2.96 (Spherical embeddings). Define a mapping $\rho = \rho_E$ from X to $V = \mathbb{R}^n$ by $\rho(x) = \sqrt{n}Ex$. The mapping ρ is called the *spherical representation* of $\mathfrak{X}$.

The next lemma holds.

Lemma 2.97.
(1) *If $(x,y) \in R_i$, then $\langle\rho(x),\rho(y)\rangle = \theta_i^*$. Note that $\theta_i^* = Q_1(i)$.*
(2) *There exists a constant $\alpha \in \mathbb{R}$ such that $\sum_{y\in\Gamma_i(x)}\rho(y) = \alpha\rho(x)$ holds, where $\Gamma_i(x) = \{y \in X \mid (x,y) \in R_i\}$. In this case, $\alpha = P_i(1)$.*

Proof. (1) By definition, we have the following:

$$\langle\rho(x),\rho(y)\rangle = n\langle Ex, Ey\rangle = n^t x^t E_1 E_1 y = n^t x E_1 y$$
$$= {}^t x\left(\sum_{l=0}^{d} Q_1(l)A_l\right)y = Q_1(i).$$

(2) By definition, we have the following:

$$\sum_{y\in\Gamma_i(x)}\rho(y) = \sqrt{n}E_1\sum_{y\in\Gamma_i(x)} y = \sqrt{n}E_1A_i x$$
$$= \sqrt{n}P_i(1)E_1 x = P_i(1)\rho(x).$$

Hence we have $\alpha = P_i(1)$. □

By definition, the set $\{\rho(x) \mid x \in X\}$ is a subset of $V_1 = E_1 V \subset V = \mathbb{R}^{|X|}$; V_1 is an m_1-dimensional subspace of V. Moreover, by Lemma 2.97 (1), we have $\langle \rho(x), \rho(x) \rangle = \theta_0^* = Q_1(0) = m_1$. So each element of $\{\rho(x) \mid x \in X\}$ is on the sphere of radius $\sqrt{m_1}$. The reason why we call $\rho = \rho_E$ the spherical representation is based on this fact.

Definition 2.98. The spherical representation $\rho = \rho_E$ is said to be *non-degenerate* (resp. *weakly non-degenerate*) if $\theta_0^*, \theta_1^*, \ldots, \theta_d^*$ are distinct (resp. if $\theta_0^* \neq \theta_i^*, 1 \leq i \leq d$).

Remark 2.99. If $\mathfrak{X}$ is Q-polynomial with respect to $E = E_1$, then $\rho = \rho_E$ is non-degenerate.

Proof. It follows from Proposition 2.83. □

In what follows, we consider a weakly non-degenerate spherical embedding $\rho = \rho_E$.

Definition 2.100 (Terwilliger [464, 472]). For any $x, y \in X$ and $i, j \in \{0, 1, \ldots, d\}$, the spherical representation $\rho = \rho_E$ is said to be *balanced* if the following holds: There exists a real number $\alpha \in \mathbb{R}$ such that

$$\sum_{z \in \Gamma_i(x) \cap \Gamma_j(y)} \rho(z) - \sum_{z \in \Gamma_j(x) \cap \Gamma_i(y)} \rho(z) = \alpha(\rho(x) - \rho(y)). \tag{2.46}$$

In this case, if we let $(x, y) \in R_k$ and define

$$\gamma_{i,j}^k = p_{i,j}^k \frac{\theta_i^* - \theta_j^*}{\theta_0^* - \theta_k^*}, \tag{2.47}$$

then $\alpha = \gamma_{i,j}^k$. Moreover, if (2.46) holds for specific $i, j \in \{0, 1, \ldots, d\}$, ρ is said to be *balanced with respect to* $\{i, j\}$ (or $\{i, j\}$*-balanced*). Equation (2.46) is known as the *balanced set condition* ([464, 472]).

Proof. If we compute the inner product of $\rho(x)$ and the left-hand side of (2.46), then we obtain

$$p_{i,j}^k \theta_i^* - p_{j,i}^k \theta_j^* = p_{i,j}^k (\theta_i^* - \theta_j^*). \tag{2.48}$$

If we compute the inner product of $\rho(x)$ and the right-hand side of (2.46), then we obtain $\alpha(\theta_0^* - \theta_k^*)$. □

Next we consider the representation graph Δ_E, which was introduced in Definition 2.58 in Section 2.7.3 of this chapter.

Proposition 2.101. *Let Δ_E be the representation graph for $E = E_1$. Then the following are equivalent:*

(1) *The spherical representation $\rho = \rho_E$ of $\mathfrak{X}$ is weakly non-degenerate.*
(2) *The representation graph $\Delta = \Delta_E$ of $\mathfrak{X}$ for E is connected.*

Proof. By Proposition 2.59, Δ is connected if and only if the $Q_1(0)$-eigenspace of B_1^* is 1-dimensional if and only if $Q_1(j) \neq Q_1(0)$ for any $j \neq 0$. □

An alternative proof of Proposition 2.101*: (1)⟹(2).* We use the dual Bose–Mesner algebra $\mathfrak{A}^* = \mathfrak{A}^*(x_0)$ defined in Section 2.6 of this chapter (Proposition 2.32 and Definition 2.33). Here x_0 is a fixed point of X. Let Δ' be a connected component of Δ. Let $E_{\Delta'} = \sum_{i\in\Delta'} E_i$. By a suitable rearrangement of the ordering of $A_1^*, \ldots, A_d^*$, we set $A^* = A_1^*$. Firstly, we show that

$$E_{\Delta'}A^* = E_{\Delta'}A^*E_{\Delta'} = A^*E_{\Delta'}. \tag{2.49}$$

By Corollary 2.37 (2), we have $E_iA^*E_j = 0$ if and only if $q_{1,i}^j = 0$. Therefore we obtain the following:

$$\begin{aligned} E_{\Delta'}A^*I &= \Big(\sum_{i\in\Delta'} E_i\Big)A^*\Big(\sum_{j=0}^{d} E_j\Big) = \sum_{\substack{i\in\Delta'\\ 0\le j\le d}} E_iA^*E_j \\ &= \sum_{i,j\in\Delta'} E_iA^*E_j = E_{\Delta'}A^*E_{\Delta'}. \end{aligned} \tag{2.50}$$

Similarly, we can prove $IA^*E_{\Delta'} = E_{\Delta'}A^*E_{\Delta'}$. Hence (2.49) holds. Let $E_{\Delta'} = \sum_{i=0}^{d} \alpha_iA_i$. By (2.49), we obtain

$$0 = E_{\Delta'}A^* - A^*E_{\Delta'} = \sum_{i=1}^{d} \alpha_i(A_iA^* - A^*A_i). \tag{2.51}$$

$$\begin{aligned} (A_iA^* - A^*A_i)(x_0, y) &= A_i(x_0, y)A^*(y, y) - A^*(x_0, x_0)A_i(x_0, y) \\ &= \begin{cases} \theta_i^* - \theta_0^*, & \text{if } (x_0, y) \in R_i, \\ 0, & \text{if } (x_0, y) \notin R_i. \end{cases} \end{aligned} \tag{2.52}$$

Since the spherical representation ρ is weakly non-degenerate, $\theta_i^* \neq \theta_0^*$ for any $i \neq 0$. Therefore $\alpha_1 = \alpha_2 = \cdots = \alpha_d = 0$, and $E_{\Delta'} = \alpha_0 I$. Since $E_{\Delta'}$ is an idempotent, $\alpha_0^2 = \alpha_0$, which implies $\alpha_0 = 0$ or 1. If $\alpha_0 = 0$, Δ' is an empty set. If $\alpha_0 = 1$, then $\Delta' = \{0, 1, \ldots, d\}$. Thus Δ is connected. □

Theorem 2.102 (Terwilliger [464, 472]). *Let E be a primitive idempotent of a symmetric association scheme $\mathfrak{X}$ and let $\rho = \rho_E$ and $\Delta = \Delta_E$ be the weakly non-degenerate spherical representation and the representation graph, respectively. Then the following are equivalent:*

(1) *$\rho = \rho_E$ is balanced;*

(2) *$\Delta = \Delta_E$ is a tree.*

The next corollary immediately follows from Proposition 2.81 (2) and Theorem 2.102.

Corollary 2.103 (Terwilliger [464, 472]). *If $\mathfrak{X}$ is Q-polynomial with respect to $E = E_1$, then the spherical representation $\rho = \rho_E$ is balanced.*

Proof of Theorem 2.102. We use the dual Bose–Mesner algebra $\mathfrak{A}^* = \mathfrak{A}^*(x_0)$ again. Let $A^* = A_1^* \in \mathfrak{A}^*$. We define a linear subspace $\mathcal{L}$ of the Terwilliger algebra T as follows:

$$\mathcal{L} = \mathcal{L}(x_0) = \mathrm{Span}\{MA^*N - NA^*M \mid M, N \in \mathfrak{A}\}. \tag{2.53}$$

Firstly, we prove the next lemma and corollary.

Lemma 2.104.
(1) *The set* $\{E_iA^*E_j - E_jA^*E_i \mid 0 \le i, j \le d,\ i \ne j,\ q_{1,i}^j \ne 0\}$ *is a basis of* $\mathcal{L}$.
(2) *The subset* $\{A^*A_k - A_kA^* \mid 1 \le k \le d\}$ *of* $\mathcal{L}$ *is linearly independent.*

Proof. (1) Let $M = \sum_{i=0}^d \alpha_i E_i$ and $N = \sum_{j=0}^d \beta_j E_j$. By Corollary 2.37 (2), $E_iA^*E_j = 0$ if and only if $q_{1,i}^j = 0$. So the following holds:

$$MA^*N - NA^*M = \sum_{\substack{i,j \\ q_{1,i}^j \ne 0}} \alpha_i\beta_j(E_iA^*E_j - E_jA^*E_i).$$

Therefore we obtain

$$\mathcal{L} = \mathrm{Span}\{E_iA^*E_j - E_jA^*E_i \mid 0 \le i, j \le d, i \ne j,\ q_{1,i}^j \ne 0\}.$$

Next, we show the above generating set is linearly independent. Assume

$$\sum_{\substack{i,j,i\ne j \\ q_{1,i}^j \ne 0}} \alpha_{i,j}(E_iA^*E_j - E_jA^*E_i) = 0, \tag{2.54}$$

and multiply both sides of (2.54) by E_i from the left and by E_j from the right. Then the left-hand side becomes $\alpha_{i,j}E_iA^*E_j$, and hence $\alpha_{i,j}E_iA^*E_j = 0$. Since $q_{1,i}^j \ne 0$, we have $E_iA^*E_j \ne 0$. Thus $\alpha_{i,j} = 0$.
(2) Similarly to (2.52) in the alternative proof of Proposition 2.101, the following holds:

$$(A^*A_k - A_kA^*)(x_0, y) = \begin{cases} \theta_0^* - \theta_k^*, & \text{if } (x_0, y) \in R_k, \\ 0, & \text{if } (x_0, y) \notin R_k. \end{cases}$$

Assume

$$\sum_{k=1}^d \alpha_k(A^*A_k - A_kA^*) = 0.$$

For all $1 \le k \le d$, compute the (x_0, y)-entry of both sides where $(x_0, y) \in R_k$. Then we obtain $\alpha_k(\theta_0^* - \theta_k^*) = 0$. Since ρ is weakly non-degenerate, $\theta_0^* - \theta_k^* \ne 0$, and hence $\alpha_k = 0$. □

Corollary 2.105.
(1) *The dimension* $\dim(\mathcal{L})$ *of* $\mathcal{L}$ *equals the number of edges of the representation graph* $\Delta = \Delta_E$.
(2) *We have* $d \le \dim(\mathcal{L})$. *Equality holds if and only if the representation graph* $\Delta = \Delta_E$ *is a tree.*

Proof. (1) This is clear from Lemma 2.104 (1).
(2) By Lemma 2.104 (2), we obtain $d \le \dim(\mathcal{L})$. Note that $d+1$ is the number of vertices of the representation graph Δ and $\dim(\mathcal{L})$ is the number of edges of Δ. Moreover, since ρ is weakly non-degenerate, Δ is connected by Proposition 2.101. Therefore $\dim(\mathcal{L}) = d$ if and only if Δ is a tree. □

Here we return to the proof of Theorem 2.102. As we mentioned above, Δ is a tree (Theorem 2.102 (2)) if and only if $d = \dim(\mathcal{L})$ if and only if $\mathcal{L} = \mathrm{Span}\{A^*A_k - A_kA^* \mid 1 \le k \le d\}$ if and only if for any $i,j \in \{0,1,\dots,d\}$, there exists a real number $y_{i,j}^k$ such that

$$A_iA^*A_j - A_jA^*A_i = \sum_{k=1}^{d} y_{i,j}^k (A^*A_k - A_kA^*). \tag{2.55}$$

If (2.55) holds, we have

$$y_{i,j}^k = p_{i,j}^k \frac{\theta_i^* - \theta_j^*}{\theta_0^* - \theta_k^*}. \tag{2.56}$$

This can be proved as follows. For $(x_0,y) \in R_k$, the (x_0,y)-entry of the right-hand side of (2.55) is $y_{i,j}^k(\theta_0^* - \theta_k^*)$, and the (x_0,y)-entry of the left-hand side is

$$\begin{aligned}\sum_{z\in X} A_i(x_0,z)A^*(z,z)A_j(z,y) - \sum_{z\in X} A_j(x_0,z)A^*(z,z)A_i(z,y)\\ = p_{i,j}^k(\theta_i^* - \theta_j^*).\end{aligned} \tag{2.57}$$

Therefore (2.56) holds. Next, we show the following lemma.

Lemma 2.106. *Let $\rho = \rho_E$ be the spherical representation and let $A^* = A_1^*(x_0) \in \mathfrak{A}^*(x_0)$. Then the following hold:*
(1) *for any $z \in X$, we have $A^*(z,z) = \langle \rho(x_0), \rho(z)\rangle$;*
(2)

$$(A^*A_k - A_kA^*)(x,y) = \begin{cases} \langle \rho(x_0), \rho(x) - \rho(y)\rangle, & \text{if } (x,y) \in R_k,\\ 0, & \text{if } (x,y) \notin R_k.\end{cases}$$

Proof. (1) If $(x_0,z) \in R_k$, by definition, $A^*(z,z) = Q_1(k) = \theta_k^*$ ((2.9) and (2.10)). Hence by Lemma 2.97 (1), $A^*(z,z) = \langle \rho(x_0), \rho(z)\rangle$.
(2) Similarly to (2.52) in the alternative proof of Proposition 2.101, we obtain

$$\begin{aligned}(A^*A_k - A_kA^*)(x,y) &= A^*(x,x)A_k(x,y) - A_k(x,y)A^*(y,y)\\ &= \begin{cases} A^*(x,x) - A^*(y,y), & \text{if } (x,y) \in R_k,\\ 0, & \text{if } (x,y) \notin R_k.\end{cases}\end{aligned} \tag{2.58}$$

Hence by (1), we obtain (2). □

By Lemma 2.106, the (x,y)-entry of the left-hand side of (2.55) can be calculated as follows:

$$\begin{aligned}(A_iA^*A_j - A_jA^*A_i)(x,y) &= \sum_{z\in X} A_i(x,z)A^*(z,z)A_j(z,y) - \sum_{z\in X} A_j(x,z)A^*(z,z)A_i(z,y)\\ &= \sum_{z\in\Gamma_i(x)\cap\Gamma_j(y)} A^*(z,z) - \sum_{z\in\Gamma_j(x)\cap\Gamma_i(y)} A^*(z,z)\\ &= \left\langle \rho(x_0), \sum_{z\in\Gamma_i(x)\cap\Gamma_j(y)} \rho(z) - \sum_{z\in\Gamma_j(x)\cap\Gamma_i(y)} \rho(z)\right\rangle. \end{aligned} \tag{2.59}$$

For $(x,y) \in R_k$, the (x,y)-entry of the right-hand side of (2.55) is $\gamma_{i,j}^k\langle\rho(x_0),\rho(x)-\rho(y)\rangle$. Since x_0 is arbitrary, for any $(x,y)\in R_k$, (2.55) and (2.46) are equivalent. From the above discussion, we have shown that the spherical representation ρ is balanced if and only if Δ is a tree. This completes the proof of Theorem 2.102. □

We close this section by proving the next theorem.

Theorem 2.107 (Terwilliger [472]). *Let $\mathfrak{X} = (X,\{R_i\}_{0\le i\le d})$ be a P-polynomial scheme of class $d \ge 3$. Let $\rho = \rho_E$ be the spherical representation of $\mathfrak{X}$ for a primitive idempotent $E = E_1$. Suppose ρ is weakly non-degenerate and balanced with respect to $\{1,2\}$ ($\{1,2\}$-balanced). Then $\mathfrak{X}$ is a Q-polynomial scheme with respect to E_1.*

Proof. The proof consists of four parts: **I**, **II**, **III**, and **IV**.
I. Let $(x,y)\in R_k$. Then the constant α of the balanced set condition (2.46) in Definition 2.100 is given by $\alpha = \gamma_{1,2}^k = p_{1,2}^k\frac{\theta_1^*-\theta_2^*}{\theta_0^*-\theta_k^*}$.
II. For $A^* = A_1^*(x_0)$, the following holds:

$$A_1A^*A_2 - A_2A^*A_1 = \sum_{k=1}^d \gamma_{1,2}^k(A^*A_k - A_kA^*). \tag{2.60}$$

Proof. Let $(x,y) \in R_k$. Compute the (x,y)-entry of both sides of (2.60). By Lemma 2.106 (1), we have

$$\begin{aligned}(A_1A^*A_2 - A_2A^*A_1)(x,y) &= \sum_{z\in X} A_1(x,z)A_1^*(z,z)A_2(z,y) - \sum_{z\in X} A_2(x,z)A_1^*(z,z)A_1(z,y)\\ &= \sum_{z\in\Gamma_1(x)\cap\Gamma_2(y)} \langle\rho(x_0),\rho(z)\rangle - \sum_{z\in\Gamma_2(x)\cap\Gamma_1(y)} \langle\rho(x_0),\rho(z)\rangle. \end{aligned} \tag{2.61}$$

On the other hand, we have

$$\begin{aligned}&\left(\sum_{l=1}^d \gamma_{1,2}^l(A^*A_l - A_lA^*)\right)(x,y)\\ &= \gamma_{1,2}^k(A^*(x,x) - A^*(y,y)) = \gamma_{1,2}^k\langle\rho(x_0),\rho(x)-\rho(y)\rangle. \end{aligned} \tag{2.62}$$

Hence by **I** and the $\{1,2\}$-balanced set conditions (2.61), (2.62), we have **II**. □

III. Let $A = A_1$. Then there exist $\beta, \gamma, \delta \in \mathbb{R}$ such that

$$A^3A^* - A^*A^3 - (\beta+1)(A^2A^*A - AA^*A^2) - \gamma(A^2A^* - A^*A^2) \\ - \delta(AA^* - A^*A) = 0. \tag{2.63}$$

Proof. Since $\mathfrak{X}$ is a P-polynomial scheme, for any integer k with $4 \le k \le d$, we have $p_{1,2}^k = 0$. Hence, by **I**, we have $\gamma_{1,2}^k = 0$. Then by **II** (2.60),

$$A_1A^*A_2 - A_2A^*A_1 \\ = \gamma_{1,2}^3(A^*A_3 - A_3A^*) + \gamma_{1,2}^2(A^*A_2 - A_2A^*) + \gamma_{1,2}^1(A^*A_1 - A_1A^*). \tag{2.64}$$

The case where $\gamma_{1,2}^3 \neq 0$.
Since $\mathfrak{X}$ is a P-polynomial scheme, for each $i = 0, 1, \dots, d$, there exists a polynomial $v_i(x)$ of degree i such that $A_i = v_i(A_1) = v_i(A)$. Let $v_3(x) = a_3x^3 + a_2x^2 + a_1x + a_0$, where $a_3 \neq 0$. Then we have

$$\begin{aligned} A^*A_3 - A_3A^* &= A^*v_3(A) - v_3(A)A^* \\ &= A^*(a_3A^3 + a_2A^2 + a_1A + a_0I) - (a_3A^3 + a_2A^2 + a_1A + a_0I)A^* \\ &= a_3(A^*A^3 - A^3A^*) + a_2(A^*A^2 - A^2A^*) + a_1(A^*A - AA^*). \end{aligned} \tag{2.65}$$

By (2.64) and (2.65), $A^3A^* - A^*A^3$ is a linear combination of $A^*A - AA^*, A^*A^2 - A^2A^*, AA^*A^2 - A^2A^*A$.

The case where $\gamma_{1,2}^3 = 0$.
We will prove this case does not occur. Suppose $\gamma_{1,2}^3 = 0$. Since $\mathfrak{X}$ is a P-polynomial scheme, $p_{1,2}^3 \neq 0$ holds (Theorem 2.76). Therefore by **I**, we have $\theta_1^* = \theta_2^*$. Hence by **I**, $\gamma_{1,2}^k = 0$ holds for $k = 1, 2, \dots, d$. Thus by **II**, we obtain

$$A_1A^*A_2 - A_2A^*A_1 = 0. \tag{2.66}$$

Multiplying both sides of (2.66) by E_i from the left and by E_j from the right, we have

$$0 = E_i(A_1A^*A_2 - A_2A^*A_1)E_j = (P_1(i)P_2(j) - P_2(i)P_1(j))E_iA^*E_j. \tag{2.67}$$

If we let $\theta_k = P_1(k)$, then $P_2(k) = v_2(\theta_k)$ (Theorem 2.76), and we have

$$(\theta_i v_2(\theta_j) - v_2(\theta_i)\theta_j)E_iA^*E_j = 0.$$

Here we consider i, j with $i \neq j$ and $q_{1,i}^j \neq 0$. By Corollary 2.37, we have $E_iA^*E_j \neq 0$. Set $v_2(x) = \frac{1}{c_2}(x^2 - a_1x - k)$, where c_2 and a_1 are parameters of the distance-regular graph (Theorem 2.76). Then we have

$$\theta_i v_2(\theta_j) - v_2(\theta_i)\theta_j = \frac{1}{c_2}(\theta_j - \theta_i)(\theta_i\theta_j + k).$$

By Proposition 2.79, $\theta_0, \theta_1, \dots, \theta_d$ are distinct real numbers, and hence for i, j with $q_{1,i}^j \neq 0$, we have $\theta_i\theta_j = -k \neq 0$. Since ρ is weakly non-degenerate, by Proposition 2.59, the representation graph Δ of $\mathfrak{X}$ is connected. Since $q_{1,i}^j \neq 0$, i, j are adjacent in Δ. Therefore there exist three vertices i, j, j' in Δ such that i is adjacent to both j and j'.

Therefore, there exists j' such that $q^{j'}_{1,i} \neq 0$, and we have $\theta_i\theta_j = \theta_i\theta_{j'} = -k \neq 0$. This does not occur because $\theta_j \neq \theta_{j'}$, which means $\gamma^3_{1,2} = 0$ does not occur. □

IV. The graph Δ is a path. This means $\mathfrak{X}$ is a Q-polynomial scheme with respect to E.

Proof. Since ρ is weakly non-degenerate, Δ is connected. It suffices to show that the valencies of the vertices of Δ are at most 2. Let i be any vertex of Δ. Suppose j and i are adjacent, which means $q^j_{1,i} \neq 0$. As mentioned above, $E_iA^*E_j \neq 0$ holds. By multiplying both sides of **III** (2.63) by E_i from the left and by E_j from the right, we obtain the following:

$$\begin{aligned}0 &= E_i(A^3A^* - A^*A^3 - (\beta+1)(A^2A^*A - AA^*A^2) - \gamma(A^2A^* - A^*A^2)\\&\quad - \delta(AA^* - A^*A))E_j\\&= (\theta_i^3 - \theta_j^3 - (\beta+1)(\theta_i^2\theta_j - \theta_i\theta_j^2) - \gamma(\theta_i^2 - \theta_j^2) - \delta(\theta_i - \theta_j))E_iA^*E_j\\&= (\theta_i - \theta_j)(\theta_i^2 - \beta\theta_i\theta_j + \theta_j^2 - \gamma(\theta_i + \theta_j) - \delta)E_iA^*E_j.\end{aligned}$$

Therefore, if we fix θ_i, then θ_j is a solution of a quadratic equation (again, we use the fact that $\theta_0, \theta_1, \ldots, \theta_d$ are distinct real numbers). Hence the number of vertices which are adjacent to i is at most 2. □

This completes the proof of Theorem 2.107. □

3 Codes and designs in association schemes (Delsarte theory on association schemes)

In this chapter, we consider subsets, namely, codes and designs, in association schemes. The theory of codes and that of designs are unified in the framework of association schemes. This is a basic idea of Delsarte's theory, which was introduced in his remarkable Ph. D. thesis. In this chapter, we explain the basic part of Delsarte's theory following his thesis.

3.1 Introducing linear programming

In this section, we consider the application of *linear programming* to association schemes. We adopt the notation by Delsarte [159], which is suitable for the application to the character tables of association schemes. (For linear programming, see [210], etc.)

Let d be a positive integer. Let $D = \{0, 1, 2, \ldots, d\}$, $M \subset D$, $0 \in M$, $D^{\times} = D \backslash \{0\}$, and $M^{\times} = M \backslash \{0\}$. Let C be a real square matrix of size $d + 1$ indexed by D. Let $C_j(i)$ denote the (i, j)-entry of C. Let $C_0(i) = 1$ for any $i \in D$. Now we consider the following two problems.

Problem (C, M):

Maximize $g = \sum\limits_{i \in M} a_i C_0(i) = \sum\limits_{i \in M} a_i$ subject to

Constraint 1: $\sum\limits_{i \in M} a_i C_j(i) \geq 0,\ j \in D^{\times}, a_i \geq 0\ (i \in M^{\times})$.

Problem $(C, M)'$:

Minimize $\gamma = \sum\limits_{j \in D} \alpha_j C_j(0)$ subject to

Constraint 2: $\sum\limits_{j \in D} \alpha_j C_j(i) \leq 0, i \in M^{\times}, \alpha_j \geq 0\ (j \in D^{\times})$.

Let $|M| = m + 1$. Let $\boldsymbol{a} \in \mathbb{R}^{m+1}$ be a vector indexed by M whose i-th entry is a_i $(i \in M)$. The vector $\boldsymbol{a}$ is called a program of (C, M) if it satisfies $a_0 = 1$ and Constraint 1. A program is called a *maximal program* if it gives the maximum value of g. Let $\boldsymbol{\alpha} \in \mathbb{R}^{d+1}$ be a vector indexed by D whose i-th entry is α_i $(i \in D)$. The vector $\boldsymbol{\alpha}$ is called a program of $(C, M)'$ if it satisfies $\alpha_0 = 1$ and Constraint 2. A program is called a *minimal program* if it gives the minimum value of γ. Problem (C, M) and Problem $(C, M)'$ are said to be *feasible* if there exist $\boldsymbol{a}$ and $\boldsymbol{\alpha}$ satisfying Constraint 1 and Constraint 2, respectively.

Proposition 3.1. *Let $\boldsymbol{a}$ and $\boldsymbol{\alpha}$ be programs of (C, M) and $(C, M)'$, respectively. Then $g \leq \gamma$.*

https://doi.org/10.1515/9783110630251-003

Proof. By Constraint 1, we have

$$a_0 C_j(0) + \sum_{i \in M^\times} a_i C_j(i) \geq 0,$$

$$\sum_{i \in M^\times} a_i C_j(i) \geq -a_0 C_j(0) = -C_j(0) \quad (j \in D^\times).$$

Therefore, we have

$$\sum_{j \in D^\times} \alpha_j \Big(\sum_{i \in M^\times} a_i C_j(i) \Big) \geq - \sum_{j \in D^\times} \alpha_j C_j(0) = 1 - \sum_{j \in D} \alpha_j C_j(0) = 1 - \gamma.$$

By Constraint 2, we have

$$\sum_{j \in D^\times} \alpha_j C_j(i) \leq -\alpha_0 C_0(i) = -1 \quad (i \in M^\times),$$

$$\sum_{i \in M^\times} a_i \Big(\sum_{j \in D^\times} \alpha_j C_j(i) \Big) \leq - \sum_{i \in M^\times} a_i = 1 - g.$$

Consequently, we have $1 - \gamma \leq 1 - g$, from which we obtain $g \leq \gamma$. □

The following is the duality theorem for linear programming.

Theorem 3.2. *If Problem (C, M) and Problem $(C, M)'$ are feasible, then there exist a maximal program and a minimal program, and $g_0 = \gamma_0$ holds, where g_0 is the maximum value of g and γ_0 is the minimum value of γ.*

As a preliminary for the proof of the theorem, we state some basic facts.

A closed subset Ω of $\mathbb{R}^n$ is called a *closed convex cone* if it satisfies the following conditions:

(1) For any $\boldsymbol{x}, \boldsymbol{y} \in \Omega$ and any real number λ with $0 \leq \lambda \leq 1$, we have $\lambda \boldsymbol{x} + (1 - \lambda)\boldsymbol{y} \in \Omega$.
(2) For any $\boldsymbol{x} \in \Omega$ and any non-negative real number $\alpha \geq 0$, we have $\alpha \boldsymbol{x} \in \Omega$.

For a closed convex cone Ω of $\mathbb{R}^n$, define $\Omega^* = \{\boldsymbol{u} \in \mathbb{R}^n \mid \boldsymbol{u} \cdot \boldsymbol{x} \geq 0, \boldsymbol{x} \in \Omega\}$. Then Ω^* is also a closed convex cone, and it is well known that $(\Omega^*)^* = \Omega$. (For proofs, see [210, Chapter 8].) In general, we define $\boldsymbol{x} \geq 0$ if $\boldsymbol{x} = {}^t(x_1, \ldots, x_n) \in \mathbb{R}^n$ satisfies $x_i \geq 0$ for all $1 \leq i \leq n$.

Theorem 3.3. *Let A be an $m \times n$-matrix and $\boldsymbol{c} \in \mathbb{R}^n$. For any $\boldsymbol{u} \in \mathbb{R}^n$ satisfying $A\boldsymbol{u} \geq 0$, if $\boldsymbol{c} \cdot \boldsymbol{u} \geq 0$ holds, then there exists $\boldsymbol{x} \in \mathbb{R}^m$ such that $\boldsymbol{c} = {}^t A \boldsymbol{x}$ and $\boldsymbol{x} \geq 0$.*

Proof. Let $\Omega = \{{}^t A \boldsymbol{v} \mid \boldsymbol{v} \in \mathbb{R}^m, \boldsymbol{v} \geq 0\}$. Namely, if we let $\boldsymbol{a}_i$ be the i-th row of A, we have $\Omega = \{\sum_{i=1}^m v_i {}^t\boldsymbol{a}_i \mid v_1, v_2, \ldots, v_m \geq 0\}$. Then Ω is a closed convex cone of $\mathbb{R}^n$. By the assumption, for any $\boldsymbol{u}$ satisfying $A\boldsymbol{u} \geq 0$, $\boldsymbol{c}$ satisfies $\boldsymbol{c} \cdot \boldsymbol{u} \geq 0$. If $A\boldsymbol{u} \geq 0$ holds, then ${}^t\boldsymbol{a}_i \cdot \boldsymbol{u} \geq 0$ $(1 \leq i \leq m)$. Therefore, for any $\boldsymbol{z} \in \Omega$, we have $\boldsymbol{z} \cdot \boldsymbol{u} \geq 0$. Namely, $\boldsymbol{u} \in \Omega^*$. It turns out that $\boldsymbol{c} \in (\Omega^*)^* = \Omega$. Thus, there exist $x_1, \ldots, x_m \geq 0$ such that $\boldsymbol{c} = \sum_{i=1}^m x_i {}^t\boldsymbol{a}_i$. That is to say, there exists $\boldsymbol{x}$ such that $\boldsymbol{c} = {}^t A \boldsymbol{x}$ and $\boldsymbol{x} \geq 0$. □

Lemma 3.4. *Let A be an $m \times n$ matrix and let $\boldsymbol{b} \in \mathbb{R}^m$ and $\kappa \in \mathbb{R}$. Assume that there exists $\boldsymbol{u} \in \mathbb{R}^n$ such that $A\boldsymbol{u} \geq \boldsymbol{b}$. Then the following two conditions on $\boldsymbol{c} \in \mathbb{R}^n$ are equivalent:*
(1) *For any $\boldsymbol{v} \in \mathbb{R}^n$ satisfying $A\boldsymbol{v} \geq \boldsymbol{b}$, we have $\boldsymbol{v} \cdot \boldsymbol{c} \geq \kappa$.*
(2) *There exists $\boldsymbol{x} \in \mathbb{R}^m$ such that $\boldsymbol{c} = {}^tA\boldsymbol{x}$, $\boldsymbol{x} \cdot \boldsymbol{b} \geq \kappa$ and $\boldsymbol{x} \geq 0$.*

Proof. First we show (2)⇒(1). By the assumption, there exists $\boldsymbol{x} \in \mathbb{R}^m$ which satisfies the condition in (2). Then if we let $A\boldsymbol{v} \geq \boldsymbol{b}$, we obtain $\boldsymbol{v} \cdot \boldsymbol{c} = \boldsymbol{v} \cdot {}^tA\boldsymbol{x} = {}^t\boldsymbol{x}A\boldsymbol{v} \geq \boldsymbol{x} \cdot \boldsymbol{b} \geq \kappa$. Next we assume (1) and define the matrix A' and the vector $\boldsymbol{c}'$ as follows:

$$A' = \begin{bmatrix} A & -\boldsymbol{b} \\ 0 & 1 \end{bmatrix}, \quad \boldsymbol{c}' = \begin{bmatrix} \boldsymbol{c} \\ -\kappa \end{bmatrix}.$$

We show A' and $\boldsymbol{c}'$ satisfy the assumption of Theorem 3.3. To be precise, we show the following:

$$\text{For any } \boldsymbol{v}' \in \mathbb{R}^{n+1} \text{ satisfying } A'\boldsymbol{v}' \geq 0, \text{ we have } \boldsymbol{c}' \cdot \boldsymbol{v}' \geq 0. \tag{3.1}$$

Let $\boldsymbol{v}' = \left[\begin{smallmatrix} \boldsymbol{v} \\ v_{n+1} \end{smallmatrix}\right]$. By the assumption, $v_{n+1} \geq 0$.
(i) The case $v_{n+1} > 0$:
Since $0 \leq A'\boldsymbol{v}'$, we have $A\boldsymbol{v} - v_{n+1}\boldsymbol{b} \geq 0$. Hence we have $A(\frac{1}{v_{n+1}}\boldsymbol{v}) \geq \boldsymbol{b}$. Therefore by (1), we have $\frac{1}{v_{n+1}}\boldsymbol{v} \cdot \boldsymbol{c} \geq \kappa$. Then we obtain $\boldsymbol{c}' \cdot \boldsymbol{v}' = \boldsymbol{c} \cdot \boldsymbol{v} - v_{n+1}\kappa \geq 0$. Thus (3.1) holds.
(ii) The case $v_{n+1} = 0$:
By the assumption of the proposition, there exists $\boldsymbol{u} \in \mathbb{R}^n$ such that $A\boldsymbol{u} \geq \boldsymbol{b}$. Since we assume (1), we have $\boldsymbol{u} \cdot \boldsymbol{c} \geq \kappa$. Let $\boldsymbol{u}' = \left[\begin{smallmatrix} \boldsymbol{u} \\ 1 \end{smallmatrix}\right] \in \mathbb{R}^{n+1}$ and $\boldsymbol{v}'' = \boldsymbol{v}' + \varepsilon\boldsymbol{u}'$, where ε is a positive real number. Then we have $A'\boldsymbol{v}'' = A'\boldsymbol{v}' + \varepsilon A'\boldsymbol{u}' = A'\boldsymbol{v}' + \varepsilon\left[\begin{smallmatrix} A\boldsymbol{u}-\boldsymbol{b} \\ 1 \end{smallmatrix}\right] \geq 0$. Since the $(n+1)$-entry of $\boldsymbol{v}''$ is ε (> 0), we can apply (i). Namely, we have $\boldsymbol{c}' \cdot \boldsymbol{v}'' \geq 0$. Therefore, for any $\varepsilon > 0$, we have $0 \leq \boldsymbol{c}' \cdot \boldsymbol{v}' + \varepsilon\boldsymbol{c}' \cdot \boldsymbol{u}' = \boldsymbol{c}' \cdot \boldsymbol{v}' + \varepsilon(\boldsymbol{c} \cdot \boldsymbol{u} - \kappa)$. Since $\boldsymbol{c} \cdot \boldsymbol{u} - \kappa \geq 0$, we have $\boldsymbol{c}' \cdot \boldsymbol{v}' \geq 0$.

By the above discussion, we showed (3.1). So we can apply Theorem 3.3 to A' and $\boldsymbol{c}'$. Therefore there exists $\boldsymbol{x}' \in \mathbb{R}^{m+1}$ such that $\boldsymbol{c}' = {}^tA'\boldsymbol{x}'$ and $\boldsymbol{x}' \geq 0$. If we let $\boldsymbol{x}' = \left[\begin{smallmatrix} \boldsymbol{x} \\ x_{m+1} \end{smallmatrix}\right]$, we have $\boldsymbol{x} \in \mathbb{R}^m$, $\boldsymbol{x} \geq 0$, $x_{m+1} \geq 0$, and $\left[\begin{smallmatrix} \boldsymbol{c} \\ -\kappa \end{smallmatrix}\right] = \boldsymbol{c}' = {}^tA'\boldsymbol{x}' = \left[\begin{smallmatrix} {}^tA & 0 \\ -{}^t\boldsymbol{b} & 1 \end{smallmatrix}\right]\left[\begin{smallmatrix} \boldsymbol{x} \\ x_{m+1} \end{smallmatrix}\right]$. Namely, we have $\boldsymbol{c} = {}^tA\boldsymbol{x}$ and $-\kappa = -\boldsymbol{b} \cdot \boldsymbol{x} + x_{m+1}$, and thus $\boldsymbol{b} \cdot \boldsymbol{x} = \kappa + x_{m+1} \geq \kappa$. □

Proof of Theorem 3.2. Suppose Problem (C, M) and Problem $(C, M)'$ are feasible. Then there exist a program $\boldsymbol{a} \in \mathbb{R}^{m+1}$ of Problem (C, M) and a program $\boldsymbol{\alpha} \in \mathbb{R}^{d+1}$ of Problem $(C, M)'$. Let $g_{\boldsymbol{a}} = \sum_{i \in M} a_i C_0(i) = \sum_{i \in M} a_i$ and $\gamma_{\boldsymbol{\alpha}} = \sum_{j \in D} \alpha_j C_j(0)$. Let $\boldsymbol{x} \in \mathbb{R}^{m+1}$ and $\boldsymbol{y} \in \mathbb{R}^{d+1}$ be programs of Problem (C, M) and Problem $(C, M)'$, respectively, and let $g_{\boldsymbol{x}} = \sum_{i \in M} x_i C_0(i) = \sum_{i \in M} x_i$ and $\gamma_{\boldsymbol{y}} = \sum_{j \in D} y_j C_j(0)$. Then by Proposition 3.1, we have $g_{\boldsymbol{x}} \leq \gamma_{\boldsymbol{\alpha}}$ and $g_{\boldsymbol{a}} \leq \gamma_{\boldsymbol{y}}$. Therefore, subject to

Condition 1: $\sum_{i \in M} x_i C_j(i) \geq 0$, $x_i \geq 0$ $(j \in D^\times,\ i \in M^\times)$,

$g_{\boldsymbol{x}}$ is bounded from above. Namely, there exists a program $\boldsymbol{x}_0 \in \mathbb{R}^{m+1}$ which gives the maximum value g_0 of $g_{\boldsymbol{x}}$. Similarly, subject to

Condition 2: $\sum_{j\in D} y_j C_j(i) \le 0$, $y_j \ge 0$ $(i \in M^\times,\ j \in D^\times)$,

$y_{\boldsymbol{y}}$ is bounded from below and there exists a program $\boldsymbol{y}_0 \in \mathbb{R}^{d+1}$ which gives the minimum value y_0. For any program $\boldsymbol{y}$ of Problem $(C,M)'$, we have $y_{\boldsymbol{y}} \ge g_0$. Therefore, $y_0 \ge g_0$. In order to use Lemma 3.4, we state Problem $(C,M)'$ in another way. Let $\boldsymbol{y} = {}^t(y_0, y_1, \dots, y_d) \in \mathbb{R}^d$ be a program of Problem $(C,M)'$ (hence, $\boldsymbol{y} \ge 0$, $y_0 = 1$). Let $\tilde{\boldsymbol{y}} = {}^t(y_1, \dots, y_d) \in \mathbb{R}^d$. Let A be the $m \times d$-matrix whose (i,j)-entry is defined by the following:

$$A(i,j) = -C_j(i) \quad (1 \le i \le m,\ 1 \le j \le d).$$

Since $\boldsymbol{y}$ is a program of Problem $(C,M)'$, we have $1+\sum_{j=1}^{d} y_j C_j(i) \le 0$ $(1 \le i \le m, 1 \le j \le d)$. Therefore, if we let $\tilde{\boldsymbol{b}} \in \mathbb{R}^m$ be the all 1's vector, we have $A\tilde{\boldsymbol{y}} \ge \tilde{\boldsymbol{b}}$. Define $\tilde{\boldsymbol{c}} = {}^t(c_1, \dots, c_d) \in \mathbb{R}^d$ by $c_j = C_j(0)$ $(1 \le j \le d)$. By using the above notation, Problem $(C,M)'$ is equivalent to the problem to minimize $\tilde{\boldsymbol{c}} \cdot \tilde{\boldsymbol{y}} = y_{\boldsymbol{y}} - 1$ subject to $A\tilde{\boldsymbol{y}} \ge \tilde{\boldsymbol{b}}$. Since there exists a vector to make Problem $(C,M)'$ feasible and there exists the minimum value y_0, if we let $\kappa = y_0 - 1$ in Lemma 3.4, then Lemma 3.4 (1) holds for $\tilde{\boldsymbol{c}}$. Therefore, Lemma 3.4 (2) holds. Namely, there exists $\tilde{\boldsymbol{x}} = {}^t(x_1, \dots, x_m) \in \mathbb{R}^m$ such that $\tilde{\boldsymbol{c}} = {}^tA\tilde{\boldsymbol{x}}$, $\tilde{\boldsymbol{x}} \cdot \tilde{\boldsymbol{b}} \ge y_0 - 1$, and $\tilde{\boldsymbol{x}} \ge 0$. Let $\boldsymbol{x} = {}^t(1, {}^t\tilde{\boldsymbol{x}})$. Since $\tilde{\boldsymbol{c}} = {}^tA\tilde{\boldsymbol{x}}$, we have

$$x_0 C_j(0) = c_j = \sum_{i=1}^{m} A(i,j)x_i = -\sum_{i=1}^{m} C_j(i)x_i,$$

and hence we have

$$\sum_{i=0}^{m} C_j(i)x_i = 0.$$

Namely, $\boldsymbol{x}$ is a program of Problem (C,M). Therefore $1 + \sum_{i=1}^{m} x_i = g_{\boldsymbol{x}} \le g_0$. On the other hand, since $\tilde{\boldsymbol{x}} \cdot \tilde{\boldsymbol{b}} \ge y_0 - 1$, we have $1 + \sum_{i=1}^{m} x_i \ge y_0 \ge g_0$. Therefore, $y_0 = g_0$ and $\boldsymbol{x}$ is a maximal program of Problem (C,M). □

Lemma 3.5. *Let $\boldsymbol{a}$ be a maximal program of (C,M) and $\boldsymbol{\alpha}$ a minimal program of $(C,M)'$. Then we have*

$$\alpha_j\Big(\sum_{i\in M} a_i C_j(i)\Big) = 0 \quad (j \in D^\times), \tag{3.2}$$

$$a_i\Big(\sum_{j\in D} \alpha_j C_j(i)\Big) = 0 \quad (i \in M^\times). \tag{3.3}$$

Conversely, if (3.2) and (3.3) hold for programs $\boldsymbol{a}$ and $\boldsymbol{\alpha}$, then $\boldsymbol{a}$ is a maximal program of (C,M) and $\boldsymbol{\alpha}$ is a minimal program of $(C,M)'$.

Proof. Let g_0 be the maximum value of g and y_0 the minimum value of y. Then we have

$$g_0 = y_0 = \sum_{i\in M} a_i = \sum_{j\in D} \alpha_j C_j(0).$$

Since $a_i \geq 0$, by Constraint 2, we have

$$\begin{aligned}
0 &\geq \sum_{i\in M^\times} a_i \sum_{j\in D} \alpha_j C_j(i) = \sum_{j\in D} \alpha_j \sum_{i\in M^\times} a_i C_j(i) \\
&= \sum_{j\in D} \alpha_j \Big(\sum_{i\in M} a_i C_j(i) - a_0 C_j(0) \Big) \\
&= \sum_{j\in D^\times} \alpha_j \sum_{i\in M} a_i C_j(i) + \sum_{i\in M} \alpha_0 a_i C_0(i) - \sum_{j\in D} \alpha_j a_0 C_j(0) \\
&= \sum_{j\in D^\times} \alpha_j \sum_{i\in M} a_i C_j(i) + g_0 - \gamma_0 = \sum_{j\in D^\times} \alpha_j \sum_{i\in M} a_i C_j(i) \geq 0.
\end{aligned} \tag{3.4}$$

Therefore we obtain

$$\sum_{i\in M^\times} a_i \sum_{j\in D} \alpha_j C_j(i) = \sum_{j\in D^\times} \alpha_j \sum_{i\in M} a_i C_j(i) = 0.$$

By the assumption, we have $a_i \geq 0$ and $\sum_{j\in D} \alpha_j C_j(i) \leq 0$ for any $i \in M^\times$, and we have $\alpha_j \geq 0$ and $\sum_{i\in M} a_i C_j(i) \geq 0$ for any $j \in D^\times$. So (3.2) and (3.3) hold. Conversely, if (3.2) and (3.3) hold, by the process of the proof, it turns out that $\boldsymbol{a} = (a_0, a_1, \ldots, a_m)$ is a maximal program of (C, M) and $\boldsymbol{\alpha} = (\alpha_0, \alpha_1, \ldots, \alpha_m)$ is a minimal program of $(C, M)'$. □

3.2 Subsets of association schemes

3.2.1 Subsets of association schemes

Let $\mathfrak{X} = (X, \{R_i\}_{0\leq i\leq d})$ be a symmetric association scheme. Let $D = \{0, 1, \ldots, d\}$. Let P, Q be the first and second eigenmatrices of $\mathfrak{X}$, respectively. We follow the notation used in the previous chapter for adjacency matrices, primitive idempotents, and so on. We introduce the following concepts for a non-empty subset $Y \subset X$. Let $\boldsymbol{a}_Y = (a_0, a_1, \ldots, a_d)$ be the $(d+1)$-dimensional row vector whose entries are defined as follows:

$$a_i = \frac{1}{|Y|} |R_i \cap (Y \times Y)| \quad (i \in D).$$

The vector $\boldsymbol{a}_Y$ is called the *inner distribution* of Y. We also define $\boldsymbol{a}_Y^* = \boldsymbol{a}_Y Q$ and call it the *dual distribution* of Y. We denote $\boldsymbol{a}_Y^* = (a_0^*, a_1^*, \ldots, a_d^*)$. Let B_Y be the matrix indexed by $X \times D$ whose entries are defined as follows. For $(x, i) \in X \times D$,

$$B_Y(x, i) = \big|R_i \cap (\{x\} \times Y)\big| = \big|Y \cap \Gamma_i(x)\big|.$$

The matrix B_Y is called the *outer distribution* of Y. Define the column vector ψ_Y indexed by X as follows:

$$\psi_Y(x) = \begin{cases} 1, & \text{if } x \in Y, \\ 0, & \text{if } x \notin Y. \end{cases}$$

We call ψ_Y the *characteristic vector* of Y. In particular, if $Y = \{x\}$, we write $\psi_{\{x\}} = \psi_x$. For any vector $\boldsymbol{u} = (u_1, u_2, \ldots, u_r)$, let $\Delta_{\boldsymbol{u}}$ denote the diagonal matrix whose diagonal entries are $u_1, \ldots, u_r$.

Proposition 3.6. *For the inner distribution $\boldsymbol{a}_Y = (a_0, a_1, \ldots, a_d)$ of $Y \subset X$, the following hold:*

(1) $a_0 = 1$;

(2) $a_0^* = \sum_{i=0}^{d} a_i = |Y|$.

Proof. By the definition, it is clear. □

Proposition 3.7. *We have:*

(1) $a_i = \frac{1}{|Y|} {}^t\psi_Y A_i \psi_Y$;

(2) $B_Y = [A_0\psi_Y, A_1\psi_Y, \ldots, A_d\psi_Y]$;

(3) $\boldsymbol{a}_Y = \frac{1}{|Y|} {}^t\psi_Y B_Y$.

Proof. (1) We have $\frac{1}{|Y|} {}^t\psi_Y A_i \psi_Y = \frac{1}{|Y|} \sum_{(x,y)\in Y\times Y} A_i(x,y) = a_i$.
(2) For $(x, i) \in X \times D$, we have

$$(A_i\psi_Y)(x) = \sum_{z\in Y} A_i(x,z)\psi_Y(z) = |Y \cap \Gamma_i(x)| = B_Y(x,i).$$

This means $A_i\psi_Y$ is the i-th column vector of B_Y.
(3) We have

$$\begin{aligned}\frac{1}{|Y|} {}^t\psi_Y B_Y &= \left(\frac{1}{|Y|} {}^t\psi_Y A_0\psi_Y, \frac{1}{|Y|} {}^t\psi_Y A_1\psi_Y, \ldots, \frac{1}{|Y|} {}^t\psi_Y A_d\psi_Y\right)\\ &= (a_0, a_1, \ldots, a_d) = \boldsymbol{a}_Y.\end{aligned}$$
□

Theorem 3.8. *For $Y \subset X$, we have*

$${}^tB_Y B_Y = \frac{|Y|}{|X|} {}^tP\Delta_{\boldsymbol{a}_Y^*}P.$$

Proof. By Proposition 3.7 (2), we have

$$\begin{aligned}({}^tB_Y B_Y)(i,j) &= {}^t(A_i\psi_Y)(A_j\psi_Y) = {}^t\psi_Y A_i A_j \psi_Y\\ &= \sum_{k=0}^{d} p_{i,j}^k {}^t\psi_Y A_k \psi_Y = |Y| \sum_{k=0}^{d} p_{i,j}^k a_k.\end{aligned} \tag{3.5}$$

On the other hand, since $\boldsymbol{a}_Y = \frac{1}{|X|}\boldsymbol{a}_Y^* P$, we have $a_k = \frac{1}{|X|}\sum_{\ell=0}^{d} a_\ell^* P_k(\ell)$, and hence

$$\begin{aligned}({}^tB_Y B_Y)(i,j) &= \frac{|Y|}{|X|} \sum_{k=0}^{d} p_{i,j}^k \sum_{\ell=0}^{d} a_\ell^* P_k(\ell)\\ &= \frac{|Y|}{|X|} \sum_{\ell=0}^{d} a_\ell^* \sum_{k=0}^{d} p_{i,j}^k P_k(\ell).\end{aligned} \tag{3.6}$$

Therefore, by Theorem 2.22 (6), we obtain

$$\begin{aligned}({}^tB_YB_Y)(i,j) &= \frac{|Y|}{|X|}\sum_{\ell=0}^{d} a_\ell^* P_i(\ell)P_j(\ell)\\ &= \frac{|Y|}{|X|}({}^tP\Delta_{\boldsymbol{a}_Y^*}P)(i,j). \end{aligned} \tag{3.7}$$

□

Corollary 3.9. *The rank of the matrix B_Y is equal to the number of non-zero entries of the dual distribution $\boldsymbol{a}_Y^*$ of Y.*

Theorem 3.10. *For a subset $Y \subset X$ of a symmetric association scheme $\mathfrak{X}$, each entry a_k^* of the dual distribution $\boldsymbol{a}_Y^*$ is a non-negative real number. Moreover, the following are equivalent:*
(1) $a_k^* = 0$;
(2) $B_YQ_k = \mathbf{0}$;
(3) $E_k\psi_Y = \mathbf{0}$.

Here Q_k denotes the k-th column vector of the second eigenmatrix Q of the association scheme $\mathfrak{X}$.

Proof. By Theorem 3.8, we have

$${}^tQ{}^tB_YB_YQ = \frac{|Y|}{|X|}{}^tQ({}^tP\Delta_{\boldsymbol{a}_Y^*}P)Q = |X||Y|\Delta_{\boldsymbol{a}_Y^*}.$$

If we compute the diagonal entries of both sides, we obtain

$$\sum_{x\in X}((B_YQ)(x,k))^2 = |X||Y|a_k^*,\ k = 0,1,\ldots,d.$$

Thus, we have $a_k^* \geq 0$, and $(B_YQ)(x,k) = 0$ holds for any $x \in X$ if and only if $a_k^* = 0$. Next, by Proposition 3.7 (1), we have $a_i = \frac{1}{|Y|}{}^t\psi_YA_i\psi_Y$. Therefore, we have

$$\begin{aligned} a_k^* &= \sum_{i=0}^{d} a_iQ_k(i) = \sum_{i=0}^{d}\frac{1}{|Y|}{}^t\psi_YA_i\psi_YQ_k(i) = \frac{|X|}{|Y|}{}^t\psi_YE_k\psi_Y\\ &= \frac{|X|}{|Y|}{}^t\psi_YE_kE_k\psi_Y = \frac{|X|}{|Y|}{}^t(E_k\psi_Y)(E_k\psi_Y). \end{aligned} \tag{3.8}$$

Hence $a_k^* = 0$ if and only if $E_k\psi_Y = \mathbf{0}$. □

We are interested in which a_k^* of the dual distribution becomes 0. The property of the dual distribution $\boldsymbol{a}_Y^*$ of a subset Y in a symmetric association scheme is important when we define designs in Q-polynomial schemes.

Definition 3.11. Let $\boldsymbol{a}_Y = (a_0, a_1, \ldots, a_d)$ and $\boldsymbol{a}_Y^* = (a_0^*, a_1^*, \ldots, a_d^*)$ be the inner distribution and the dual distribution of Y, respectively.

(1) Define

$$\delta = \min\{i \mid a_i \neq 0, i \geq 1\}, \qquad \delta^* = \min\{i \mid a_i^* \neq 0, i \geq 1\}.$$

We call δ the *minimum distance* of Y, and δ^* the *dual minimum distance* of Y.

(2) Define

$$s = \left|\{i \mid a_i \neq 0, i \geq 1\}\right|, \quad s^* = \left|\{i \mid a_i^* \neq 0, i \geq 1\}\right|.$$

We call s the *degree* of Y, and s^* the *dual degree* of Y.

(3) Define

$$t = \max\{i \mid a_1^* = a_2^* = \cdots = a_i^* = 0,\ i \geq 1\} = \delta^* - 1.$$

We call t the *strength* of Y.

The above definitions depend on the ordering of the adjacency matrices and the primitive idempotents, which becomes important if we consider the structures of P-polynomial schemes and Q-polynomial schemes.

Next we introduce the *MacWilliams inequality*, which plays an important role in the theory of codes and designs in association schemes. For a vector $\boldsymbol{u} = (u_0, u_1, \ldots, u_d)$ of $\mathbb{R}^{d+1}$, define

$$t(\boldsymbol{u}) = \max\{i \mid u_1 = u_2 = \cdots = u_i = 0\}.$$

If $u_1 \neq 0$, we define $t(\boldsymbol{u}) = 0$. We also define

$$s(\boldsymbol{u}) = \left|\{i \mid u_i \neq 0,\ 1 \leq i \leq d\}\right|.$$

Theorem 3.12 (MacWilliams inequality). *Let $\mathfrak{X}$ be a symmetric association scheme and let P and Q be the first and second eigenmatrices of $\mathfrak{X}$, respectively. Let $\boldsymbol{u} = (u_0, u_1, \ldots, u_d)$ be a vector in $\mathbb{R}^{d+1}$ and let $u_0 \neq 0$.*

(1) *If $\mathfrak{X}$ is a P-polynomial scheme, we have*

$$s(\boldsymbol{u}Q) \geq \left[\frac{t(\boldsymbol{u})}{2}\right].$$

(2) *If $\mathfrak{X}$ is a Q-polynomial scheme, we have*

$$s(\boldsymbol{u}P) \geq \left[\frac{t(\boldsymbol{u})}{2}\right].$$

Before the proof of Theorem 3.12, we give a corollary.

Corollary 3.13. *Let $\mathfrak{X}$ be a Q-polynomial scheme with the ordering of the primitive idempotents $E_0, E_1, \ldots, E_d$. Let Y be a subset of X in $\mathfrak{X}$ with degree s and strength t. Then $s \geq [\frac{t}{2}]$.*

Proof. We use Theorem 3.12 (2). Let $\boldsymbol{u} = \boldsymbol{a}_Y^* = \boldsymbol{a}_Y Q$. Then by Proposition 3.6, we have $u_0 = a_0^* > 0$. Therefore, $s = s(\boldsymbol{a}_Y) = s(\boldsymbol{u}P) \geq [\frac{t(\boldsymbol{u})}{2}] = [\frac{t(\boldsymbol{a}_Y^*)}{2}] = [\frac{t}{2}]$. □

Proof of Theorem 3.12. The proof of (2) is similar to that of (1), so we only prove (1). Let $\theta_i = P_1(i)$ $(0 \leq i \leq d)$. Since $\mathfrak{X}$ is a P-polynomial scheme, for each k $(0 \leq k \leq d)$, there exists a polynomial $v_k(z)$ of degree k such that $P_k(i) = v_k(\theta_i)$. By definition, $u_1 = u_2 = \cdots = u_{t(\boldsymbol{u})} = 0$, $u_{t(\boldsymbol{u})+1} \neq 0$. We assume $s(\boldsymbol{u}Q) < [\frac{t(\boldsymbol{u})}{2}]$ for the sake of contradiction. Let $s = s(\boldsymbol{u}Q)$ and $\{v_1, v_2, \ldots, v_s\} = \{i \mid \boldsymbol{u}Q_i \neq 0, i \neq 0\}$. Let $h(z)$ be a polynomial of degree $[\frac{t(\boldsymbol{u})}{2}] - s - 1 (\geq 0)$ satisfying $h(\theta_i) \neq 0$ for any $i = 0, 1, \ldots, d$. Let

$$f(z) = h(z) \prod_{j=0}^{s} (z - \theta_{v_j}),$$

where $\theta_{v_0} = \theta_0$. Then the following holds:

$$f(\theta_i) = 0 \quad \text{if and only if} \quad i \in \{0, v_1, v_2, \ldots, v_s\}.$$

Since $\deg(f(z)^2) = 2[\frac{t(\boldsymbol{u})}{2}]$, we have

$$f(z)^2 = b_0 v_0(z) + b_1 v_1(z) + \cdots + b_{2[\frac{t(\boldsymbol{u})}{2}]} v_{2[\frac{t(\boldsymbol{u})}{2}]}(z).$$

Define $b_i = 0$ for $2[\frac{t(\boldsymbol{u})}{2}] + 1 \leq i \leq d$, and write $\boldsymbol{b} = (b_0, b_1, \ldots, b_d)$. Since $2[\frac{t(\boldsymbol{u})}{2}] + 1 = t(\boldsymbol{u})$ or $t(\boldsymbol{u}) + 1$, we obtain

$$\boldsymbol{u}Q\,{}^t(\boldsymbol{b}^t P) = \boldsymbol{u}QP\,{}^t\boldsymbol{b} = |X|\boldsymbol{u}^t\boldsymbol{b} = |X| \sum_{i=0}^{d} u_i b_i = |X| u_0 b_0.$$

Moreover, since

$$(\boldsymbol{b}\,{}^t P)(v_j) = \sum_{k=0}^{d} b_k P_k(v_j) = \sum_{k=0}^{d} b_k v_k(\theta_{v_j}) = (f(\theta_{v_j}))^2 = 0, \quad 0 \leq j \leq s,$$

we have

$$\boldsymbol{u}Q\,{}^t(\boldsymbol{b}^t P) = 0.$$

Since $u_0 \neq 0$, we have $b_0 = 0$. Therefore we obtain

$$\sum_{i=0}^{d} f(\theta_i)^2 Q_i(0) = \sum_{i=0}^{d} \sum_{j=0}^{d} b_j P_j(i) Q_i(0) = |X| b_0 = 0.$$

This is impossible as $f(\theta_i)^2 > 0$ and $Q_i(0) = m_i > 0$ for $i \notin \{0, v_1, \ldots, v_s\}$. □

3.2.2 Codes in P-polynomial schemes

In Chapter 1, Section 1.5, we consider codes, namely, subsets or subspaces, of the n-dimensional vector space over F_q equipped with the Hamming distance. For a subset Y of X in a P-polynomial scheme $\mathfrak{X}$, we adopt the distance coming from the distance-regular graph (X, R_1) and consider e-error correcting codes. The following properties on P-polynomial schemes are well known. Every eigenvalue $\theta_i = P_1(i)$ $(0 \le i \le d)$ of A_1 is real, and $k_1 = \theta_0 \ge \theta_i$ for $0 \le i \le d$. Moreover, $\theta_i = \theta_j$ if and only if $i = j$. We consider the inner distribution $\boldsymbol{a}_Y$ of Y. Now the graph (X, R_1) is a distance-regular graph. We denote the distance between $x, y \in X$ by $\partial(x, y)$. Then $(x, y) \in R_i$ if and only if $\partial(x, y) = i$. The minimum distance δ of Y is expressed as $\delta = \min\{\partial(x, y) \mid x, y \in Y, x \ne y\}$. In Chapter 1, we denote the minimum distance by d. Here we use δ instead of d since d denotes the class of the association scheme. By definition, we have $\delta = t(\boldsymbol{a}_Y) + 1$, and hence by the MacWilliams inequality, the dual degree $s^* = s(\boldsymbol{a}_Y^*)$ of Y satisfies $s^* \ge [\frac{\delta-1}{2}]$. A code Y is called a *perfect e-code* if X is partitioned as

$$X = \bigcup_{y \in Y} \{x \in X \mid \partial(x, y) \le e\}.$$

Since $|\{x \in X \mid \partial(x, y) \le e\}| = \sum_{i=0}^{e} k_i$, if Y is a perfect code, then $|Y|(k_0 + k_1 + \cdots + k_e) = |X|$. In general, if the minimum distance δ of Y satisfies $\delta \ge 2e + 1$, then $\{x \in X \mid \partial(x, y_1) \le e\} \cap \{x \in X \mid \partial(x, y_2) \le e\} = \emptyset$ for distinct $y_1, y_2 \in Y$, and hence $|Y|(k_0 + k_1 + \cdots + k_e) \le |X|$.

Theorem 3.14 (Lloyd type theorem). *Let Y be a code of minimum distance δ in a P-polynomial scheme $\mathfrak{X}$. Let $e = [\frac{\delta-1}{2}]$. Then the following (1) and (2) hold:*

(1) *We have*

$$|Y|(k_0 + k_1 + \cdots + k_e) \le |X|.$$

(2) *If equality holds in (1), the dual degree $s^* = s(\boldsymbol{a}_Y^*)$ of Y is equal to e. Moreover, the e distinct eigenvalues θ_k such that $a_k^* \ne 0$ are the zeros of the polynomial*

$$\Psi_e(z) = v_0(z) + v_1(z) + \cdots + v_e(z),$$

where $v_i(z)$ is the polynomial of degree i such that $A_i = v_i(A_1)$ (and hence $P_i(j) = v_i(\theta_j)$).

The polynomial $\Psi_e(z)$ is called the Lloyd polynomial.

Proof. Although (1) was already proved, we give an alternative proof based on linear programming. Let $M = \{0, \delta, \delta + 1, \ldots, d\}$. We apply linear programming to M and the second eigenmatrix Q. We have

$$0 \le \boldsymbol{a}_Y Q_k = \sum_{i=0}^{d} a_i Q_k(i) = \sum_{i \in M} a_i Q_k(i),$$

$$|Y| = \sum_{i \in M} a_i.$$

We also have $a_0 = 1$, and $a_i \geq 0$ for any $i \in M^\times$. Therefore $\boldsymbol{a}_Y$ is a program of Problem (Q, M). Next, we define a vector $\boldsymbol{\alpha} = (\alpha_0, \alpha_1, \dots, \alpha_d)$ in $\mathbb{R}^{d+1}$ by

$$\alpha_j = \left(\frac{\Psi_e(\theta_j)}{K_e}\right)^2, \ 0 \leq j \leq d,$$

where $K_e = \Psi_e(\theta_0)$ $(= \sum_{j=0}^{e} P_j(0) = \sum_{j=0}^{e} k_j)$. We have $\alpha_0 = (\frac{\Psi_e(\theta_0)}{K_e})^2 = 1$, and $\alpha_j \geq 0$ for any $j \in D^\times$. Moreover, the eigenmatrices P, Q are real matrices and satisfy

$$Q_j(i) = \frac{m_j P_i(j)}{k_i} = \frac{m_j v_i(\theta_j)}{k_i}.$$

Therefore we have

$$\begin{aligned}(\boldsymbol{\alpha}^t Q)(i) &= \sum_{j=0}^{d} \alpha_j Q_j(i) = \sum_{j=0}^{d} \left(\frac{\Psi_e(\theta_j)}{K_e}\right)^2 \frac{m_j v_i(\theta_j)}{k_i} \\ &= \frac{1}{K_e^2 k_i} \sum_{j=0}^{d} \Psi_e(\theta_j)^2 m_j v_i(\theta_j).\end{aligned} \tag{3.9}$$

Since $e = [\frac{\delta-1}{2}]$, the degree of $\Psi_e(z)^2$ is $2e$, and hence we have

$$\Psi_e(z)^2 = \sum_{\nu=0}^{2e} c_\nu v_\nu(z)$$

for some $c_\nu \in \mathbb{R}$. Note that $2e = \delta - 1$ if δ is odd, and $2e = \delta - 2$ if δ is even (hence $2e < \delta$). By Theorem 2.22 (5), we have

$$\delta_{\nu,i}|X|k_\nu = \sum_{j=0}^{d} m_j P_\nu(j) P_i(j) = \sum_{j=0}^{d} m_j v_\nu(\theta_j) v_i(\theta_j).$$

Thus, for $i \in M^\times$, we have

$$\begin{aligned}\sum_{j \in D} \alpha_j Q_j(i) &= \frac{1}{K_e^2 k_i} \sum_{j=0}^{d} \Psi_e(\theta_j)^2 m_j v_i(\theta_j) \\ &= \frac{1}{K_e^2 k_i} \sum_{j=0}^{d} \sum_{\ell=0}^{2e} c_\ell v_\ell(\theta_j) m_j v_i(\theta_j) = 0.\end{aligned}$$

Therefore, $\boldsymbol{\alpha} = (\alpha_0, \alpha_1, \dots, \alpha_d)$ is a program of Problem $(Q, M)'$. Next, we will show

$$c_0 = K_e. \tag{3.10}$$

Note that P is a real matrix. By Theorem 2.22 (5), we have

$$\sum_{\nu=0}^{d} m_\nu (\Psi_e(\theta_\nu))^2 = \sum_{\nu=0}^{d} m_\nu \sum_{i=0}^{e} \sum_{j=0}^{e} P_i(\nu) P_j(\nu) = \sum_{i=0}^{e} |X| k_i = |X| K_e.$$

On the other hand, we have

$$\sum_{v=0}^{d} m_v(\Psi_e(\theta_v))^2 = \sum_{v=0}^{d} Q_v(0) \sum_{i=0}^{2e} c_i P_i(v) = |X|c_0.$$

Therefore we obtain (3.10). Next, by (3.9), we have

$$\begin{aligned} y = \sum_{j\in D} \alpha_j Q_j(0) &= \frac{1}{K_e^2 k_0} \sum_{j=0}^{d} \Psi_e(\theta_j)^2 m_j v_0(\theta_j) \\ &= \frac{1}{K_e^2} \sum_{j=0}^{d} \sum_{v=0}^{2e} c_v v_v(\theta_j) m_j v_0(\theta_j) = \frac{1}{K_e^2} \sum_{v=0}^{2e} c_v \sum_{j=0}^{d} m_j P_v(j) P_0(j) \\ &= \frac{1}{K_e^2} c_0 |X|. \end{aligned} \tag{3.11}$$

By (3.10), (3.11), we have $y = \frac{|X|}{K_e}$ and

$$|Y| = \sum_{i\in D} a_i = \sum_{i\in M} a_i = g \le y = \frac{|X|}{K_e}.$$

This proves (1). Here, if equality holds, $\boldsymbol{a}$ is a maximal program of (Q, M) and $\boldsymbol{\alpha}$ is a minimal program of $(Q, M)'$. Therefore, by Lemma 3.5 (equation (3.2)), we have

$$\alpha_j\Big(\sum_{i\in M} a_i Q_j(i)\Big) = \alpha_j \boldsymbol{a}_j^* = 0, \ j \in D^\times.$$

Therefore, for any $j \in D^\times$ satisfying $\boldsymbol{a}_j^* \neq 0$, we have $\alpha_j = 0$, and hence $\Psi_e(\theta_j) = 0$; $\Psi_e(z)$ is a polynomial of degree e, and $\theta_0, \theta_1, \ldots, \theta_d$ are distinct real numbers. Consequently, the dual degree $s^* = s(\boldsymbol{a}_Y^*)$ of Y is at most e, which implies $s^* \le e$. Furthermore, by the MacWilliams inequality, we have $s^* = s(\boldsymbol{a}_Y Q) \ge [\frac{t(\boldsymbol{a}_Y)}{2}] = [\frac{\delta-1}{2}] = e$, and hence $s^* = e$. □

3.2.3 Designs in Q-polynomial schemes

In Chapter 1, Section 1.3, we discuss combinatorial designs. We will extend the theory of designs in the framework of Q-polynomial schemes. In Section 3.2.1 of this chapter, we investigated the inner distribution $\boldsymbol{a}_Y$ and the dual distribution $\boldsymbol{a}_Y^* = \boldsymbol{a}_Y Q$ of a subset Y of X in an association scheme $\mathfrak{X}$, and proved that a_i^* is a non-negative real number for each $i \in D$. We first considered the case of linear codes in F_q^n, where F_q is the finite field of order q. As we introduced in Theorem 1.107 for the case $q = 2$, in general, the weight enumerators of a linear code $C \subset F_q^n$ and its dual code $C^\perp$ are related to each other in terms of the *MacWilliams identity*,

$$W_{C^\perp}(x, y) = \frac{1}{|C|} W_C(x + (q-1)y, x - y). \tag{3.12}$$

We omit the proof because there are many books which discuss this identity. Let $W_C(x,y) = \sum_{i=0}^{n} a_i x^{n-i} y^i$ and $\boldsymbol{a} = (a_0, a_1, \ldots, a_n)$. Also let $W_{C^\perp}(x,y) = \sum_{i=0}^{n} a_i' x^{n-i} y^i$ and $\boldsymbol{a}' = (a_0', a_1', \ldots, a_n')$. Then (3.12) can be reformulated in terms of the Hamming scheme $H(n,q)$ as follows:

$$\begin{aligned}
W_{C^\perp}(x,y) &= \sum_{i=0}^{n} a_i' x^{n-i} y^i \\
&= \frac{1}{|C|} W_C(x + (q-1)y, x - y) = \frac{1}{|C|} \sum_{\nu=0}^{n} a_\nu (x + (q-1)y)^{n-\nu} (x-y)^\nu \\
&= \frac{1}{|C|} \sum_{\nu=0}^{n} a_\nu \sum_{\ell=0}^{n-\nu} \sum_{k=0}^{\nu} \binom{n-\nu}{\ell} \binom{\nu}{k} (-1)^{\nu-k} (q-1)^{n-\nu-\ell} x^{\ell+k} y^{n-\ell-k} \\
&= \frac{1}{|C|} \sum_{\mu=0}^{n} \sum_{\nu=0}^{n} a_\nu \sum_{\ell=0}^{\mu} \binom{n-\nu}{\ell} \binom{\nu}{\mu-\ell} (-1)^{\nu-\mu+\ell} (q-1)^{n-\nu-\ell} x^{\mu} y^{n-\mu} \\
&= \frac{1}{|C|} \sum_{\mu=0}^{n} \sum_{\nu=0}^{n} a_\nu \sum_{\ell=0}^{n-\mu} \binom{n-\nu}{\ell} \binom{\nu}{n-\mu-\ell} (-1)^{n-\nu-\mu-\ell} (q-1)^{n-\nu-\ell} x^{n-\mu} y^{\mu} \\
&= \frac{1}{|C|} \sum_{\mu=0}^{n} \sum_{\nu=0}^{n} a_\nu \sum_{\ell=0}^{\mu} (-1)^{\ell} (q-1)^{\mu-\ell} \binom{n-\nu}{\mu-\ell} \binom{\nu}{\ell} x^{n-\mu} y^{\mu} \\
&= \frac{1}{|C|} \sum_{\mu=0}^{n} \sum_{\nu=0}^{n} a_\nu Q_\mu(\nu) x^{n-\mu} y^{\mu}.
\end{aligned} \tag{3.13}$$

Thus, we obtain

$$\boldsymbol{a}' = \frac{1}{|C|} \boldsymbol{a} Q, \tag{3.14}$$

where Q is the second eigenmatrix of $H(n,q)$ (Theorem 2.86). Since the vector $\boldsymbol{a}$ is the inner distribution of C, the dual distribution $\boldsymbol{a}^*$ of C satisfies $\boldsymbol{a}^* = |C|\boldsymbol{a}'$. Therefore, the minimum distance δ^* of the dual code $C^\perp$ coincides with the dual minimum distance which is defined in Definition 3.11 for a Q-polynomial scheme. This is why the maximum index t satisfying $a_1^* = a_2^* = \cdots = a_t^* = 0$ is important for the dual distribution $\boldsymbol{a}^* = \boldsymbol{a}Q = (a_0^*, a_1^*, \ldots, a_n^*)$ of a linear code C in F_q^n, which is a key to the definition of t-designs in a Q-polynomial scheme $\mathfrak{X}$. By generalizing this definition, we can define t-designs for codes, which are not necessarily linear, in a Hamming scheme over a finite set, which is not necessarily a finite field. Similarly, we can define t-designs in symmetric association schemes, in particular, in Q-polynomial schemes. This is the starting point of the successful work of Delsarte. For a subset T of the index set $D^\times = D\backslash\{0\}$, Delsarte defines T-designs in general symmetric association schemes, which are not necessarily Q-polynomial. In what follows, we consider the case where $T = \{1, 2, \ldots, t\}$ for a Q-polynomial scheme.

The definition of a t-design in a Q-polynomial scheme $\mathfrak{X}$ is given as follows.

Definition 3.15 (*t*-Designs in Q-polynomial schemes). Let Y be a non-empty subset of X; Y is called a *t-design* if there exists a natural number t such that $\boldsymbol{a}_i^* = 0$ for any i with $1 \le i \le t$.[1]

In Definition 3.11, we defined the degree s and the strength t for a non-empty subset Y of X. In Corollary 3.13 of Theorem 3.12, we proved that $s \ge [\frac{t}{2}]$ holds for a Q-polynomial scheme. For a t-design Y in a Q-polynomial scheme $\mathfrak{X}$, the following theorem holds.

Theorem 3.16. *Let Y be a t-design in a Q-polynomial scheme $\mathfrak{X}$ and let $e = [\frac{t}{2}]$. Then the following (1) and (2) hold:*

(1) *We have*

$$|Y| \ge m_0 + m_1 + \cdots + m_e.$$

(2) *If equality holds in (1), the degree $s = s(\boldsymbol{a}_Y)$ of Y is equal to e. Moreover, the e distinct eigenvalues θ_k^* such that $a_k \neq 0$ are the zeros of the polynomial*

$$\Psi_e^*(z) = v_0^*(z) + v_1^*(z) + \cdots + v_e^*(z),$$

where $v_i^*(z)$ is the polynomial of degree i such that $|X|E_i = v_i^*(|X|E_1)$ (and hence $Q_i(j) = v_i^*(\theta_j^*)$). The polynomial $v_i^*(z)$ is called the Wilson polynomial.

Before the proof of the theorem, we give the following definition.

Definition 3.17 (Tight 2*e*-design). Let Y be a $2e$-design in a Q-polynomial scheme $\mathfrak{X}$. If equality holds in (1) of Theorem 3.16, Y is called a *tight 2e-design* in $\mathfrak{X}$.

Remark 3.18. In Section 3.6, we will discuss tight t-designs for odd t.

Proof of Theorem 3.16. The proof is based on linear programming. Let Y be a t-design. Let $\boldsymbol{b} = \frac{1}{|Y|}\boldsymbol{a}_Y^*$ and write $\boldsymbol{b} = (b_0, b_1, \ldots, b_d)$. Then $b_0 = \frac{1}{|Y|}a_0^* = 1$ and $b_i = \frac{1}{|Y|}a_i^* \ge 0$ $(1 \le i \le d)$. Let $M = \{0, t+1, \ldots, d\}$. Since Y is a t-design, $b_i = \frac{1}{|Y|}a_i^* = 0$ for $1 \le i \le t$. Therefore, for any $k \in D$, we have

$$\sum_{i \in M} b_i P_k(i) = \sum_{i=0}^{d} b_i P_k(i) = \sum_{i=0}^{d} \frac{1}{|Y|} \sum_{\ell=0}^{d} a_\ell Q_i(\ell) P_k(i) = \frac{|X|}{|Y|} a_k \ge 0,$$

and hence $\boldsymbol{b}$ is a program of Problem (P, M), i. e., the problem to maximize $g = \sum_{i \in M} b_i$. Then $g = \frac{|X|}{|Y|}$. Next, let

$$\Psi_e^*(z) = v_0^*(z) + v_1^*(z) + \cdots + v_e^*(z),$$

1 By Theorem 3.10, $\boldsymbol{a}_i^* = 0$ if and only if $E_i \psi_Y = 0$.

and define

$$\beta_j = \left(\frac{\Psi_e^*(\theta_j^*)}{K_e^*}\right)^2, \quad 0 \le j \le d.$$

Since $\Psi_e^*(\theta_0^*) = v_0^*(\theta_0^*) + v_1^*(\theta_0^*) + \cdots + v_e^*(\theta_0^*) = Q_0(0) + Q_1(0) + \cdots + Q_e(0) = m_0 + m_1 + \cdots + m_e = K_e^*$, we have $\beta_0 = (\frac{\Psi_e^*(\theta_0^*)}{K_e^*})^2 = 1$. On the other hand, since $(\Psi_e^*(z))^2$ has degree $2e$, we may write $(\Psi_e^*(z))^2 = \sum_{l=0}^{2e} c_l^* v_l^*(z)$ for some real numbers c_l^* $(0 \le l \le 2e)$. Then for $i \in M^\times$, we have

$$\begin{aligned}\sum_{j\in D} \beta_j P_j(i) &= \frac{1}{K_e^{*2}} \sum_{j\in D} \Psi_e^*(\theta_j^*)^2 P_j(i) = \frac{1}{K_e^{*2}} \sum_{j\in D} \sum_{\ell=0}^{2e} c_\ell^* Q_\ell(j) P_j(i) \\ &= \frac{1}{K_e^{*2}} \sum_{\ell=0}^{2e} c_\ell^* \sum_{j\in D} Q_\ell(j) P_j(i) = \frac{1}{K_e^{*2}} \sum_{\ell=0}^{2e} c_\ell^* |X| \delta_{i,\ell}. \end{aligned} \tag{3.15}$$

Since $e = [\frac{t}{2}]$, if $i \in M^\times$, then $i \ge t+1 > 2e$, and by (3.15), we have

$$\sum_{j\in D} \beta_j P_j(i) = 0. \tag{3.16}$$

Therefore $\boldsymbol{\beta} = (\beta_0, \beta_1, \ldots, \beta_d)$ is a program of Problem $(P, M)'$, i. e., the problem to minimize

$$\gamma = \sum_{j\in D} \beta_j P_j(0).$$

Next, we will show

$$c_0^* = K_e^*. \tag{3.17}$$

We have

$$\begin{aligned}\sum_{\ell=0}^{d} k_\ell (\Psi_e^*(\theta_\ell^*))^2 &= \sum_{\ell=0}^{d} k_\ell \sum_{i=0}^{e} \sum_{j=0}^{e} Q_i(\ell) Q_j(\ell) \\ &= \sum_{i=0}^{e} \sum_{j=0}^{e} \sum_{\ell=0}^{d} m_i P_\ell(i) Q_j(\ell) \\ &= |X| \sum_{i=0}^{e} m_i = |X| K_e^*. \end{aligned} \tag{3.18}$$

Similarly, we have

$$\begin{aligned}\sum_{\ell=0}^{d} k_\ell (\Psi_e^*(\theta_\ell^*))^2 &= \sum_{\ell=0}^{d} k_\ell \sum_{j=0}^{2e} c_j^* Q_j(\ell) \\ &= \sum_{j=0}^{2e} c_j^* \sum_{\ell=0}^{d} P_\ell(0) Q_j(\ell) = |X| c_0^*. \end{aligned} \tag{3.19}$$

So we get $c_0^* = K_e^*$. Then we have

$$\gamma = \sum_{j\in D} \beta_j P_j(0) = \frac{|X|c_0^*}{K_e^{*2}} = \frac{|X|}{K_e^*}. \tag{3.20}$$

Therefore $\frac{|X|}{|Y|} = g \le \gamma = \frac{|X|}{K_e^*}$, and thus $|Y| \ge K_e^* = m_0 + m_1 + \cdots + m_e$. If equality holds, $\boldsymbol{b}$ is a maximal program of Problem (P, M) and $\boldsymbol{\beta}$ is a minimal program of Problem $(P, M)'$. By (3.2) in Lemma 3.5, we obtain

$$\beta_j\left(\sum_{i\in M} b_i P_j(i)\right) = 0,\ j \in D^\times.$$

Note that $\sum_{i\in M} b_i P_j(i) = \frac{|X|}{|Y|} a_j$. Then, for any $j \in D^\times$ with $a_j \neq 0$, we have $\beta_j = 0$, i. e., $\Psi_e^*(\theta_j^*) = 0$. Since $\Psi_e^*(z)$ is a polynomial of degree e and $\theta_0^*, \theta_1^*, \ldots, \theta_d^*$ are distinct real numbers, we obtain $s \le e$. Therefore we have $s = e = [\frac{t}{2}]$. □

3.2.4 On the strength and the degree of a Q-polynomial scheme

In Definition 3.11, we introduced the strength t and the degree s of a subset in a symmetric association scheme. These concepts depend on the orderings of the adjacency matrices and the primitive idempotents of the Bose–Mesner algebra. In particular, for a Q-polynomial scheme, the strength and the degree, which are defined with respect to the ordering of the primitive idempotents associated with the Q-polynomial structure, have nice properties. In this section, we give the following theorem on the strength and the degree of a subset Y in a Q-polynomial scheme $\mathfrak{X}$.

Theorem 3.19. *We retain the above notation. Let Y be a subset of a Q-polynomial scheme $\mathfrak{X}$ with strength t and degree s. Let $e = [\frac{t}{2}]$. Then the following (1) and (2) hold:*

(1) *We have $|Y| \le m_0 + m_1 + \cdots + m_s$.*
(2) *If equality holds in (1), we have $s = e$. Therefore, equality holds in 3.16 of Theorem 3.16.*

Proof. We follow the Ph. D. thesis of Delsarte. As was seen in the proof of Theorem 2.26, all matrices in the Bose–Mesner algebra are simultaneously diagonalizable by a unitary matrix U. Since $\mathfrak{X}$ is a symmetric association scheme, we may assume U is an orthogonal matrix. The rows of U are indexed by X. Let X' be the index set of the columns of U. Then $|X'| = |X|$. For each $i \in D = \{0, 1, \ldots, d\}$, $T_i = {}^tUE_iU$ is a diagonal matrix whose diagonal entries are m_i 1's and $|X| - m_i$ 0's. Define $X_i' = \{x \in X' \mid T_i(x, x) = 1\}$. Then $|X_i'| = m_i$ and $X' = X_0' \cup X_1' \cup \cdots \cup X_d'$ is a partition of X'. For each i, define the $|Y| \times m_i$-matrix H_i indexed by $Y \times X_i'$ as follows:

$$H_i(x, y) = \sqrt{|X|}U(x, y), \quad (x, y) \in Y \times X_i'. \tag{3.21}$$

Then H_0 is the Y-dimensional vector whose entries are all 1's. We have

$$\begin{aligned}(H_i{}^tH_i)(x,y) &= \sum_{z\in X_i'} H_i(x,z)H_i(y,z) = |X| \sum_{z\in X_i'} U(x,z)U(y,z)\\ &= |X| \sum_{z\in X} U(x,z)T_i(z,z)U(y,z) = |X|E_i(x,y)\\ &= \sum_{\ell=0}^{d} Q_i(\ell)A_\ell(x,y) = \sum_{\ell=0}^{d} v_i^*(\theta_\ell^*)A_\ell(x,y),\end{aligned} \tag{3.22}$$

where $\theta_\ell^* = Q_1(\ell)$ and v_i^* is the polynomial of degree i associated with the Q-polynomial structure, i. e., $Q_i(\ell) = v_i^*(\theta_\ell^*)$. Since Y has degree s, for the inner distribution $\boldsymbol{a}_Y = (a_0, a_1, \ldots, a_d)$ of Y, we may choose $\{\ell_1, \ell_2, \ldots, \ell_s\} \subset D^\times$ so that $\{\ell_1, \ell_2, \ldots, \ell_s\} = \{\ell \mid a_\ell \neq 0,\ \ell \in D^\times\}$. Let $M = \{\ell_0, \ell_1, \ell_2, \ldots, \ell_s\}$, where $\ell_0 = 0$. Next define the polynomial $F(z)$ as follows:

$$F(z) = \frac{|Y|}{\prod_{i=1}^{s}(\theta_0^* - \theta_{\ell_i}^*)} \prod_{i=1}^{s}(z - \theta_{\ell_i}^*). \tag{3.23}$$

Then we have

$$F(\theta_0^*) = F(\theta_{\ell_0}^*) = |Y|. \tag{3.24}$$

Moreover, F is a polynomial of degree s whose zeros are $\theta_{\ell_1}^*, \theta_{\ell_2}^*, \ldots, \theta_{\ell_s}^*$. If we write

$$F(z) = \sum_{j=0}^{s} f_j v_j^*(z), \quad f_j \in \mathbb{R},\ j = 0, 1, \ldots, s, \tag{3.25}$$

we obtain the following equation:

$$|Y| = F(\theta_0^*) = \sum_{j=0}^{s} f_j v_j^*(\theta_0^*) = f_0 m_0 + f_1 m_1 + \cdots + f_s m_s. \tag{3.26}$$

Now we consider the sum $\sum_{j=0}^{s} f_j H_j{}^t H_j$. By (3.22), we obtain

$$\begin{aligned}\sum_{j=0}^{s} f_j(H_j{}^tH_j)(x,y) &= \sum_{j=0}^{s} f_j \sum_{\ell=0}^{d} v_j^*(\theta_\ell^*)A_\ell(x,y)\\ &= \sum_{\ell=0}^{d}\sum_{j=0}^{s} f_j v_j^*(\theta_\ell^*)A_\ell(x,y)\\ &= \sum_{\ell=0}^{d} F(\theta_\ell^*)A_\ell(x,y).\end{aligned} \tag{3.27}$$

Since $(x,y) \in Y \times Y$, the ℓ-th term of the sum $\sum_{\ell=0}^{d} F(\theta_\ell^*)A_\ell(x,y)$ is non-zero if (x,y) belongs to $R_\ell \cap (Y \times Y)$ and $R_\ell \cap (Y \times Y) \neq \emptyset$. Therefore we obtain

$$\begin{aligned}\sum_{j=0}^{s} f_j(H_j{}^tH_j)(x,y) &= \sum_{i=0}^{s} F(\theta_{\ell_i}^*)A_{\ell_i}(x,y)\\ &= F(\theta_{\ell_0}^*)A_{\ell_0}(x,y) = \delta_{x,y}|Y|.\end{aligned} \tag{3.28}$$

Define the $|Y| \times (m_0 + m_1 + \cdots + m_s)$-matrix H by $H = [H_0, H_1, \ldots, H_s]$. Moreover, we define the diagonal matrix Λ indexed by $(\bigcup_{i=0}^{s} X_i) \times (\bigcup_{i=0}^{s} X_i)$ as $\Lambda(x,x) = f_i$ for $x \in X_i$. Then (3.28) can be written as follows:

$$H\Lambda^t H = |Y|I. \tag{3.29}$$

Since the right-hand side is a non-singular matrix of size $|Y|$, we have $f_i \neq 0$ $(0 \leq i \leq s)$ and $H\Lambda^t H$ has rank $|Y|$. Therefore, H has rank $|Y|$, which implies $|Y| \leq m_0+m_1+\cdots+m_s$.
(2) Suppose equality holds in (1), i. e., $|Y| = m_0 + m_1 + \cdots + m_s$. By (3.26), we have

$$m_0 + m_1 + \cdots + m_s = f_0 m_0 + f_1 m_1 + \cdots + f_s m_s. \tag{3.30}$$

By the assumption, H is non-singular. Therefore we obtain

$$\Lambda^{-1} = \frac{1}{|Y|}{}^t HH. \tag{3.31}$$

Since ${}^t HH$ is a positive definite symmetric matrix, we have $f_i > 0$ for $i = 0, \ldots, s$. In particular, $f_0 = 1$. Next we show $0 < f_i \leq 1$ for any $i = 0, 1, \ldots, s$. By (3.23), $F(\theta_\ell^*) = 0$ for $\ell \in M^\times$. Therefore, if we substitute $v_j^*(\theta_\ell^*) = Q_j(\ell)$ in (3.25), we get

$$\sum_{j\in D} f_j Q_j(\ell) = 0, \tag{3.32}$$

where we define $f_j = 0$ $(s+1 \leq j \leq d)$. Next we compute the following sum in two ways:

$$\sum_{\ell\in M} a_\ell Q_i(\ell) \sum_{j\in D} f_j Q_j(\ell) \quad (i \in D). \tag{3.33}$$

Note that $\ell \in D^\times$ and $a_\ell = 0$ for $\ell \notin M$. By (3.32) and (3.24), we have

$$\begin{aligned}
&\sum_{\ell\in M} a_\ell \sum_{j\in D} f_j Q_j(\ell) Q_i(\ell) \\
&\quad = a_0 \sum_{j\in D} f_j Q_j(0) Q_i(0) + \sum_{\ell\in M^\times} a_\ell Q_i(\ell) \sum_{j\in D} f_j Q_j(\ell) \\
&\quad = a_0 Q_i(0) \sum_{j\in D} f_j Q_j(0) = |Y| m_i.
\end{aligned} \tag{3.34}$$

Next, if we compute the sum (3.33) by using Theorem 2.22 (7), Proposition 2.24 (3), and Proposition 3.6, we obtain the following:

$$\begin{aligned}
&\sum_{\ell\in M} a_\ell Q_i(\ell) \sum_{j\in D} f_j Q_j(\ell) \\
&\quad = \sum_{\ell\in D} a_\ell \sum_{j\in D} f_j \sum_{k\in D} q_{i,j}^k Q_k(\ell) = \sum_{k\in D} \sum_{j\in D} f_j q_{i,j}^k \sum_{\ell\in D} a_\ell Q_k(\ell)
\end{aligned}$$

$$
\begin{aligned}
&= \sum_{j\in D} f_j q_{i,j}^0 \sum_{\ell\in D} a_\ell Q_0(\ell) + \sum_{k\in D^\times}\sum_{j\in D} f_j q_{i,j}^k \sum_{\ell\in D} a_\ell Q_k(\ell) \\
&= f_i m_i |Y| + \sum_{k\in D^\times}\sum_{j\in D} f_j q_{i,j}^k a_k^*. \qquad (3.35)
\end{aligned}
$$

By (3.34) and (3.35), we have

$$
|Y| m_i (1 - f_i) = \sum_{k\in D^\times}\sum_{j\in D} f_j q_{i,j}^k a_k^*. \qquad (3.36)
$$

Since each term of the right-hand side of (3.36) is non-negative real, we have $0 \le f_i \le 1$. Therefore by (3.30), it turns out that $f_0 = f_1 = \cdots = f_s = 1$. Consequently, the matrix Λ in (3.29) is the identity matrix, and so ${}^tHH = H{}^tH = |Y|I$. Hence for any integers i, j with $0 \le i, j \le s$, we obtain

$$
{}^tH_iH_j = \begin{cases} |Y|I, & \text{if } i = j, \\ \mathbf{0}, & \text{if } i \ne j. \end{cases} \qquad (3.37)
$$

If we compute the sum $\|{}^tH_iH_j\|^2$ of the squares of each entry of the matrix ${}^tH_iH_j$, we have

$$
\begin{aligned}
\|{}^tH_iH_j\|^2 &= \sum_{\substack{x\in X_i', \\ y\in X_j'}} (({}^tH_iH_j)(x,y))^2 = \sum_{\substack{x\in X_i', \\ y\in X_j'}} \Big(\sum_{z\in Y} H_i(z,x)H_j(z,y)\Big)^2 \\
&= \sum_{\substack{x\in X_i', \\ y\in X_j'}} \Big(\sum_{z\in Y} H_i(z,x)H_j(z,y)\Big)\Big(\sum_{w\in Y} H_i(w,x)H_j(w,y)\Big) \\
&= \sum_{z,w\in Y} \Big(\sum_{x\in X_i'} H_i(z,x)H_i(w,x)\Big)\Big(\sum_{y\in X_j'} H_j(z,y)H_j(w,y)\Big) \\
&= \sum_{z,w\in Y} |X|^2 E_i(z,w)E_j(z,w) = |X| \sum_{z,w\in Y} |X|(E_i \circ E_j)(z,w) \\
&= |X| \sum_{z,w\in Y}\sum_{\ell=0}^{d} q_{i,j}^\ell E_\ell(z,w) = |X| \sum_{\ell=0}^{d} q_{i,j}^\ell\, {}^t\psi_Y E_\ell \psi_Y. \qquad (3.38)
\end{aligned}
$$

Hence for any $0 \le i \ne j \le s$, we have

$$
\sum_{\ell=0}^{d} q_{i,j}^\ell\, {}^t\psi_Y E_\ell \psi_Y = 0. \qquad (3.39)
$$

Since $q_{i,j}^\ell$ is a non-negative real number and ${}^t\psi_Y E_\ell \psi_Y = {}^t(E_\ell\psi_Y)E_\ell\psi_Y \ge 0$, we have

$$
q_{i,j}^\ell\, {}^t\psi_Y E_\ell \psi_Y = 0 \qquad (3.40)
$$

for $0 \le i \ne j \le s$ and $\ell \in D$. If we let $1 \le \ell \le 2s-1$, there exist $0 \le i \ne j \le s$ with $i+j=\ell$. Since $q_{i,j}^{i+j} > 0$, we have

$$ {}^t\psi_Y E_\ell \psi_Y = 0, \quad \ell = 1, 2, \dots, 2s-1. \tag{3.41} $$

Therefore we get $a_\ell^* = 0$ for $\ell = 1, 2, \dots, 2s-1$. So, Y is a $(2s-1)$-design. Moreover, if we let $i = j = s$, by (3.37), (3.38), and (3.41), we obtain

$$ \begin{aligned} m_s|Y|^2 &= \|{}^tH_sH_s\|^2 \\ &= |X|q_{s,s}^0\,{}^t\psi_Y E_0 \psi_Y + |X|q_{s,s}^{2s}({}^t\psi_Y E_{2s}\psi_Y) \\ &= m_s|Y|^2 + |X|q_{s,s}^{2s}({}^t\psi_Y E_{2s}\psi_Y). \end{aligned} \tag{3.42} $$

Since $\mathfrak{X}$ is Q-polynomial, we have $q_{s,s}^{2s} \ne 0$. It follows that ${}^t\psi_Y E_{2s}\psi_Y = 0$. Hence we have $E_\ell\psi_Y = \mathbf{0}$ for any $1 \le \ell \le 2s$. So, Y is a $2s$-design. Therefore we have $2s \le t$. By Corollary 3.13 of Theorem 3.12, we obtain $s = [\frac{t}{2}] = e$. □

3.3 Combinatorial designs and designs in Johnson schemes

In Chapter 2, Section 2.10.3, we defined Johnson schemes and showed that they are P- and Q-polynomial schemes. In this section, we consider the connection between combinatorial t-(v,k,λ) designs, which were introduced in Chapter 1, Section 1.3, and t-designs in Johnson schemes. In what follows, let V be a v-element set and let $X = \{x \mid x \subset V, |x| = k\}$ be the set of k-element subsets of V. Let $J(v,k) = (X, \{R_i\}_{0\le i\le k})$ be the Johnson scheme, where $R_i = \{(x,y) \in X \times X \mid |x \cap y| = k-i\}$. We will explain that the concept of combinatorial designs is identical to that of designs in $J(v,k)$. To be precise, we will prove the following theorem. Note that the set of blocks of a t-(v,k,λ) design is a subset of X, which we will denote by Y rather than $\mathcal{B}$ in this context.

Theorem 3.20. *Let Y be a subset of the Johnson scheme $J(v,k) = (X, \{R_i\}_{0\le i\le k})$. Then Y is a t-design in $J(v,k)$ if and only if Y is a t-(v,k,λ) design for some integer $\lambda > 0$.*

Proof. In this proof, we use many propositions appearing in Chapter 2, Section 2.10.3. First, suppose Y is a t-(v,k,λ) design. Note that $Y \subset V^{(k)} = X$. For $i \le t$, let $V^{(i)}$ be the set of i-element subsets of V. We define the matrix M_i indexed by $X \times V^{(i)}$ as follows ((2.33) in Chapter 2, Section 2.10.3). For $(x,\xi) \in X \times V^{(i)}$, let

$$ M_i(x,\xi) = \begin{cases} 1, & \text{if } \xi \subset x, \\ 0, & \text{if } \xi \not\subset x. \end{cases} \tag{3.43} $$

By Proposition 1.30 in Chapter 1, Section 3.3, for any integer i with $0 \le i \le t$, Y is an i-(v,k,λ_i) design. We also have

$$ \lambda_i = \frac{\binom{v-i}{t-i}}{\binom{k-i}{t-i}}\lambda = |Y|\frac{\binom{v-i}{k-i}}{\binom{v}{k}} = \frac{|Y|}{|X|}\binom{v-i}{k-i}. \tag{3.44} $$

In particular, $|Y| = \lambda_0 = \frac{\binom{v}{t}}{\binom{k}{t}}\lambda$. Let ψ_X and ψ_Y be the characteristic vectors of X and Y, respectively. Note that ψ_X is the all 1's vector.

Fix $\xi \in V^{(i)}$. Then we have

$$\begin{aligned}({}^tM_i\psi_Y)(\xi) &= \sum_{x\in X} M_i(x,\xi)\psi_Y(x) = \sum_{y\in Y} M_i(y,\xi)\\ &= \big|\{y \in Y \mid \xi \subset y\}\big|. \end{aligned} \tag{3.45}$$

Namely, $({}^tM_i\psi_Y)(\xi)$ is the number of points (or blocks) in Y each of which contains the i points in ξ. Since Y is an i-(v,k,λ_i) design, we have $({}^tM_i\psi_Y)(\xi) = \lambda_i$. On the other hand, for ψ_X, we have the following:

$$({}^tM_i\psi_X)(\xi) = \sum_{x\in X} M_i(x,\xi) = \big|\{x \in X \mid \xi \subset x\}\big| = \binom{v-i}{k-i}. \tag{3.46}$$

By (3.44), (3.45), and (3.46), we obtain

$${}^tM_i\psi_Y = \frac{|Y|}{|X|}{}^tM_i\psi_X. \tag{3.47}$$

Next, let C_i be the matrix defined by (2.29). In Proposition 2.92, we proved $C_i = M_i{}^tM_i$ for $0 \le i \le k$. Therefore, by (3.47), for any i with $0 \le i \le t$, we have $C_i\psi_Y = \frac{|Y|}{|X|}C_i\psi_X$. Moreover, by Proposition 2.91, for any $r \in D = \{0,1,\dots,k\}$, we have $\langle C_0, C_1, \dots, C_r\rangle = \langle E_0, E_1, \dots, E_r\rangle$. Note that the ordering of E_i $(i \in D)$ is associated with the Q-polynomial structure of $J(v,k)$. From the above facts, we have $E_i\psi_Y = \frac{|Y|}{|X|}E_i\psi_X$ for any $i = 0,1,\dots,t$. Since ψ_X is the all 1's vector, we have $E_i\psi_X = \mathbf{0}$ for $i > 0$. It follows that $E_i\psi_Y = \mathbf{0}$ for $i = 1,2\dots,t$. By Theorem 3.10, Y is a t-design in $J(v,k)$.

Next, suppose Y is a t-design in $J(v,k)$ and show that Y is a combinatorial t-(v,k,λ) design. By the assumption, $E_i\psi_Y = \frac{|Y|}{|X|}E_i\psi_X$ for any i with $0 \le i \le t$, because if $i \ne 0$, both sides equal $\mathbf{0}$ and if $i = 0$, we have $E_0\psi_Y = \frac{|Y|}{|X|}\psi_X = \frac{|Y|}{|X|}E_0\psi_X$ since $E_0 = \frac{1}{|X|}J$. By following the above discussion in reverse order, we obtain $M_i{}^tM_i\psi_Y = \frac{|Y|}{|X|}M_i{}^tM_i\psi_X$, and this implies that

$$(M_i{}^tM_i)\left(\psi_Y - \frac{|Y|}{|X|}\psi_X\right) = \mathbf{0}.$$

Hence we have ${}^tM_i(\psi_Y - \frac{|Y|}{|X|}\psi_X) = \mathbf{0}$, that is to say,

$${}^tM_i\psi_Y = \frac{|Y|}{|X|}{}^tM_i\psi_X,$$

for $i = 0,1,\dots,t$. By (3.45), for any $i \in D$ and $\xi \in V^{(i)}$, we have $({}^tM_i\psi_Y)(\xi) = |\{y \in Y \mid \xi \subset y\}|$ and $({}^tM_i\psi_X)(\xi) = |\{x \in X \mid \xi \subset x\}| = \binom{v-i}{k-i}$. So we get

$$({}^tM_i\psi_Y)(\xi) = \frac{|Y|}{|X|}\binom{v-i}{k-i} = |Y|\frac{\binom{v-i}{k-i}}{\binom{v}{k}}. \tag{3.48}$$

If we let $\lambda_i = |Y|\frac{\binom{v-i}{k-i}}{\binom{v}{k}}$, Y is an i-(v,k,λ_i) design for any i with $0 \le i \le t$. □

The concept of t-designs in the Hamming scheme $H(d,q)$ is equivalent to that of so-called *orthogonal arrays*. Let $X = F \times \cdots \times F$ be the vertex set of $H(d,q)$. A subset Y of X is called an *orthogonal array* if it satisfies the following condition. If we fix $z = (z_1, z_2, \ldots, z_d) \in X$ and a subset $L = \{\ell_1, \ell_2, \ldots, \ell_t\}$ of $D^\times$ which consists of t distinct integers, the cardinality of the set $\{y = (y_1, y_2, \ldots, y_d) \in Y \mid y_\ell = z_\ell,\ \ell \in L\}$ equals a constant λ which is independent of the choices of z and L. The constant λ is called the *index* of the orthogonal array. If t is the largest integer which satisfies this condition, t is called the *strength* of the orthogonal array Y. Usually, an orthogonal array Y is displayed as a $|Y| \times d$-matrix, where each element of Y is a row vector. By the discussion similar to the case of Johnson schemes, we can show that Y is a t-design in the Q-polynomial scheme $H(d,q)$ if and only if the strength of the orthogonal array Y is at least t. Design theories of this sort will be generalized further in Chapter 4, Sections 4.1 and 4.2. The proof of the above fact will be found there as a special case of the general theorem. The detailed discussions are left as an exercise for the reader.

3.4 Codes in Hamming schemes

In Chapter 1, Section 1.5, we mainly consider linear codes over a finite field F_q. In this chapter, we consider codes on the Hamming scheme $H(n,q)$. We use the notation introduced in Example 2.6 of Chapter 2. Let F be a q-element set, where we do not mind the algebraic structure of F. Let q be a natural number of at least 2, which is not necessarily a prime power. Let $H(n,q)$ be the Hamming scheme with vertex set $X = F^n$. Let $\partial(-,-)$ be the Hamming distance on $X = F^n$. For $c \in X$ and a natural number e, we call $\Sigma_e(c) = \{x \in X \mid \partial(x,c) \le e\}$ the e-neighborhood of c. A subset C of $X = F^n$ is called a *perfect e-code* of X if $\{\Sigma_e(c) \mid c \in C\}$ gives a partition of X. The existence problem of perfect e-codes in the Hamming scheme $H(n,q)$ has a long history in coding theory ([318], [482, 483]). The complete classification of the perfect e-codes in $H(n,2)$ with $e \ge 2$ was given by Tietäväinen and Perko (1971) [484]. For the case where q is a prime power (and hence X is the n-dimensional vector space over the finite field $\mathrm{GF}(q)$), the existence of the perfect e-codes with $e \ge 2$ was solved by Tietäväinen et al. (1973) [482]. For $H(n,q)$, the following examples of perfect e-codes Y ($e \ge 2$) are known:

(1) $e \ge n$, that is, $|Y| = 1$ (Y is called trivial);
(2) e is any natural number, $n = 2e + 1$, $q = 2$, $|Y| = 2$ (Y is called almost trivial);
(3) $e = 3$, $n = 23$, $q = 2$, $|Y| = 2^{12}$ (Y is called the binary Golay code, which is related to the Mathieu group M_{23});
(4) $e = 2$, $n = 11$, $q = 3$, $|Y| = 3^6$ (Y is called the ternary Golay code, which is related to the Mathieu group M_{11}).

Remark 3.21. We refer the reader to [317, 119]. Moreover, for the relation between group theory and codes, the work of Ward [508] is interesting.

It is conjectured that if $e \geq 2$, there exists no other example of a perfect e-code in $H(n, q)$.

Theorem 3.22 (Tietäväinen (1973) [482], some parts are proved by van Lint et al.). *If $e \geq 2$ and q is a prime power, then a perfect e-code in $H(n, q)$ is one of the above (1)–(4).*

Remark 3.23. In the proof by Tietäväinen and van Lint, we use the Lloyd type theorem and a trivial necessary condition called the *sphere packing condition*:

$$(1 + k_1 + \cdots + k_e) \mid |X|, \qquad \text{where } k_i = (q-1)^i \binom{n}{i},\ |X| = q^n. \tag{3.49}$$

The condition that q is a prime power is effective when we use the above condition.

The general case where q is not a prime power was considered to be difficult. For $e = 3$ and several special e's, There are some works on the existence and the non-existence of perfect e-codes in $H(n, q)$, where q is not necessarily a prime power, such as the Ph. D. thesis of Reuvers (1977) [402]. Eiichi Bannai proved that for a given $e \geq 3$, there are only finitely many perfect e-codes in $H(n, q)$ for any (n, q) (1977). The main purpose of this section is to explain this result, or the following theorem.

Theorem 3.24 (Bannai (1977) [21]). *For each $e \geq 3$, there are only finitely many non-trivial perfect e-codes in the Hamming scheme $H(n, q)$. (Both n and q are bounded above by a function of e.)*

Sketch of the proof of Theorem 3.24. (For details, see [21].) We use the Lloyd type theorem only. We prove this theorem by contradiction. Precisely, we will show that the zeros $x_1, x_2, \ldots, x_e$ of the polynomial

$$\Psi_e(x) = \sum_{j=0}^{e} (-q)^j (q-1)^{e-j} \binom{n-1-j}{e-j} \binom{x-1}{j}$$

can not be all integers for infinitely many values of n and q. (Remark: Note that the eigenvalues of $H(n, q)$ are given by $\theta_i = n(q-1) - qi$ $(0 \leq i \leq n)$. The polynomial $\Psi_e(x)$ above is obtained by the polynomial $\Psi_e(z)$ in Theorem 3.14 by the affine transformation $z = n(q-1) - qx$.) Let $\alpha = (x_1 + x_2 + \cdots + x_e)/e$, $x = \alpha + m$, $\Psi_e(x) := \Psi_e(\alpha + m) := (-1)^e F_e(m)$. We regard m as a variable. Then $F_e(m)$ is expanded as follows:

$$F_e(m) = \sum_{\substack{0 \leq b \leq e \\ 0 \leq c \leq e}} \beta_{b,c} (n-e)^c m^b \quad (\beta_{b,c} \in Q[q]).$$

By using the generating function of $\Psi_e(m)$, we can show that:

(1) $b + 2c > e \Longrightarrow \beta_{b,c} = 0$;

(2) $b + 2c = e \Longrightarrow \beta_{b,c} = [(-1)^c \binom{e}{2c} \cdot (2c-1)!!] \frac{1}{e!} q^b (q-1)^c$;

(3) $b + 2c = e - 1 \Longrightarrow \beta_{b,c} = [(-1)^{c-1} \binom{e-3}{2(c-1)} \cdot (2(c-1)-1)!!] \cdot \frac{e(e-1)(e-2)}{6} \cdot \frac{1}{e!} \cdot q^b (q-1)^c \cdot (q-2)$;

where we define $(2r-1)!! = 1\cdot 3\cdot 5\cdots(2r-1), (-1)!! = 1$. (The proof of this part is very difficult.) We define the Hermite polynomial $H_n(x)$ by

$$H_n(x) := (-1)^n e^{-x^2/2}\cdot\frac{d}{dx}(e^{-x^2/2}) = \sum_{r=0}^{[n/2]}\binom{n}{r}(2r-1)!!x^{n-2r}.$$

(Different books have slightly different definitions of the Hermite polynomials.) The parts in the brackets of the right-hand sides of $\beta_{b,c}$ of the above (2) and (3) denote the coefficient of the term of degree $e-2c$ of the Hermite polynomial of degree e and that of the term of degree $e-3-2(c-1)$ of the Hermite polynomial of degree $e-3$, respectively. Therefore, if we let

$$\varphi_{e,i}(m) = \sum_{b+2c=i}\beta_{b,c}(n-e)^c m^b,$$

then we obtain

$$\begin{aligned} F_e(m) &= \varphi_{e,e}(m) + \varphi_{e,e-1}(m) + (\text{the rest terms}),\\ \varphi_{e,e}(m) &= \frac{1}{e!}\{(n-e)(q-1)\}^{e/2}\cdot H_e\left(\frac{q}{\sqrt{(n-e)(q-1)}}m\right),\\ \varphi_{e,e-1}(m) &= \frac{1}{e!}\cdot\frac{e(e-1)(e-2)}{6}\{(n-e)(q-1)\}^{(e-3)/2}\\ &\quad\cdot H_{e-3}\left(\frac{q}{\sqrt{(n-e)(q-1)}}m\right)(n-e)(q-1)(q-2). \end{aligned} \tag{3.50}$$

Let $\beta = \sqrt{(n-e)(q-1)}/q$ and consider the situation when $\beta\to+\infty$. Let

$$x_{(-[e/2])} < \cdots < x_{(-1)} < (x_{(0)}) < x_{(1)} < \cdots < x_{([e/2])}$$

be the zeros of $\Psi_e(x)$, where $x_{(0)}$ appears only if e is odd. We also let

$$\xi_{-[e/2]} < \cdots < \xi_{-1} < (\xi_0) < \xi_1 < \cdots < \xi_{[e/2]}$$

be the zeros of $H_e(x)$, where ξ_0 appears only if e is odd. Then it turns out that if $\beta\to+\infty$,

$$x_{(i)} \longrightarrow \alpha + \beta\xi_i + \lambda_i.$$

(Hence the difference between both sides tends to 0.) Here we have

$$\lambda_i = \frac{(q-2)}{q}\left(\frac{e-1}{6} - \frac{\xi_i^2}{6}\right).$$

When e is large, the location of $x_{(1)}, x_{(2)}, x_{(-1)}, x_{(-2)}$ is described as follows.

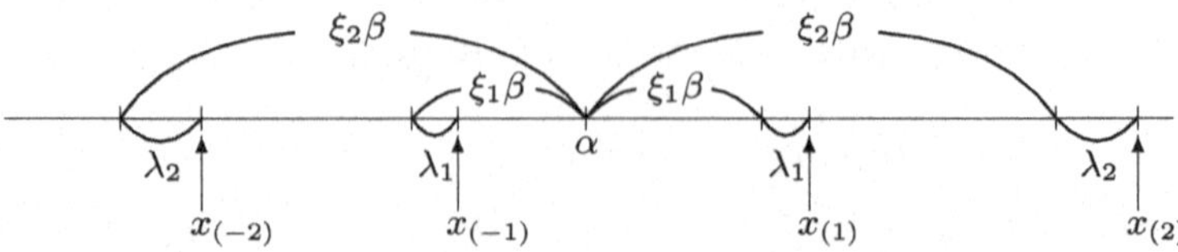

Then we have $x_{(1)} + x_{(-1)} - x_{(2)} - x_{(-2)} \longrightarrow [(q-2)/q] \cdot (2/6) \cdot (\xi_2^2 - \xi_1^2)$. Since the case $q = 2$ was solved by Tietäväinen and Perko [484] (see also [318, 432]), we consider the case $q > 2$ only. Approximate values of the zeros of Hermite polynomials are well known. In particular, if e is a little large, we have

$$0 < \frac{q-2}{q} \cdot \frac{2}{6}(\xi_2^2 - \xi_1^2) < 1.$$

If $\beta \longrightarrow +\infty$, this contradicts the Lloyd type theorem since $x_{(1)} + x_{(-1)} - x_{(2)} - x_{(-2)}$ is always an integer. This implies that β is bounded above by a function of e. It is relatively easy to show that both n and q are bounded above by a function of e only. If e is small (but $e \geq 3$), we need a special discussion but can prove it similarly. This completes the proof of Theorem 3.24. (If $q = 2$, $x_{(i)}$'s are symmetric with respect to α, and the technique that evaluates the deviation is not available. However, we can use another technique.) □

Remark 3.25. It is possible to find a function of e which bounds n and q, although it was not found in Bannai (1977) [21].

In the Ph. D. thesis of Best (1982) [88], he refined the technique used in Theorem 3.24 and showed the non-existence except for the cases $e = 6$ and $e = 8$. Hong eliminated these cases and completed the proof of the non-existence of a perfect e-code in $H(n, q)$ with $e \geq 3$ except for the examples (1)–(4) listed at the beginning of this section [237, 238] (1984). If $q = 2$ or q is a prime power, the sphere packing condition (3.49) and the Lloyd type theorem are used in the proof. On the other hand, if q is not necessarily a prime power, as was seen in the sketch of the proof of Theorem 3.24, only the Lloyd type theorem was used in the method of Bannai, Best, and Hong. To be precise, if $q = 2$, the zeros of the Lloyd polynomial are located symmetrically with respect to the average, and hence the technique that evaluates the deviation is not available. Since Hamming schemes are self-dual, i. e., $P = Q$, the Lloyd polynomial of a perfect e-code and the Wilson polynomial of a tight $2e$-design are identical. Hence the result of Bannai, Best, and Hong almost solved the classification of tight $2e$-designs in $H(n, q)$ with $q \neq 2$ and $e \geq 3$ at the same time.

Remark 3.26 (Some remarks).
(1) If $q = 2$, the classification of tight $2e$-designs in $H(n, q)$ is not complete (Ph. D. thesis of Delsarte). It is conjectured that there does not exist a non-trivial case.
(2) If t is small, the exceptional examples of tight t-designs in Hamming schemes are considered in the paper by Ryuzaburo Noda [379, 380]. For other interesting papers of Noda, see [381, 382]. The cases $t = 3, 4, 5$ seem unsolved.
(3) Quite a few examples of perfect 1-codes in $H(n, q)$ are known, and the classification seems almost impossible. There are a lot of reference books. We refer the reader to [326, page 180].
(4) The classification of perfect 2-codes in $H(n, q)$ is not complete. When q is a prime power, as was stated in Theorem 3.22, it is complete (Tietäväinen, 1973). The case

where q has exactly two different prime factors ($q = p_1^{r_1} p_2^{r_2}$) and the case where q has exactly three prime factors ($q = p_1^{r_1} p_2^{r_2} p_3^{r_3}$) are solved in [483] and [299], respectively. They are non-existence theorems, and it is conjectured that there are no perfect 2-codes except for examples (1)–(4) listed at the beginning of this section.

3.5 Tight designs in Johnson schemes

In Section 3.2, we discussed the general theory of tight t-designs in Q-polynomial schemes (especially for the case t is even), which is an important part of Delsarte's theory. Now we focus on the case of the Johnson scheme $J(v, d)$.

3.5.1 Existence and non-existence of tight designs

Let $V = \{1, 2, \dots, v\}$, and let $X = \binom{V}{d}$ be the set of d-element subsets of V. We assume $1 \le t \le d \le \frac{v}{2}$. Let $(X, \{R_i\}_{0 \le i \le d})$ be the Johnson scheme $J(v, d)$. Recall that a $2e$-design $Y \subset X$ is tight if $|Y| = m_0 + m_1 + \cdots + m_e = \sum_{i=0}^{e} (\binom{v}{i} - \binom{v}{i-1}) = \binom{v}{e}$ holds. In general, a $2e$-design Y satisfies the Fisher type inequality $|Y| \ge \binom{v}{e}$. This inequality was first obtained by Petrenjuk [395] for the case $e = 2$. Afterwards, Ray-Chaudhuri and Wilson announced that it holds for any e [514]. For the proof, see Ray-Chaudhuri and Wilson [400]. Delsarte (1974) also gave a proof, and as was written in the book of Hiroshi Nagao [366], Noda and Bannai obtained the same result independently in 1972. If there exists a tight $2e$-design, by using Theorem 3.16, we can show the following theorem.

Theorem 3.27. *If there exists a tight $2e$-design in the Johnson scheme $J(v, d)$, then all e zeros of the polynomial*

$$\Psi_e(x) = \sum_{i=0}^{e} (-1)^{e-i} \frac{\binom{v-e}{i}\binom{k-i}{e-i}\binom{k-1-i}{e-i}}{\binom{e}{i}} \binom{x}{i}$$

are positive integers. This polynomial Ψ_e is called the Wilson polynomial or the Ray-Chaudhuri–Wilson polynomial.

When $e = 1$, Y is a tight 2-design if and only if $b = v$ ($b = |Y|$) if and only if Y is a symmetric 2-design (Chapter 1, Section 1.3, Definition 1.37). There exist quite a few symmetric 2-designs and their classification seems to be almost impossible.

When $e = 2$, the non-trivial tight 4-designs are the *Witt design* 4-(23, 7, 1) and its complementary design 4-(23, 16, 52) only. (In the latter case, $d \le \frac{v}{2}$ does not hold.) The classification was started by Noboru Ito [245, 246] and was almost completed by Enomoto, Ito, and Noda [179]. To be precise, the classification was completely solved by Bremner [111] and Stroeker [440] by determining a rational integral solution of the Diophantine equation $3x^4 - 4y^4 - 2x^2 + 12y^2 - 9 = 0$, which is related to an elliptic function. In the next subsection, we present the detailed proof by Noda.

When $e = 3$, the non-existence of a non-trivial tight 6-design was shown by Peterson [394]. The proof is relatively easy.

For the case $e \geq 4$, we have the following theorem.

Theorem 3.28 (Bannai (1977) [22]). *For each $e \geq 4$, there exist only finitely many non-trivial tight 2e-designs.*

Sketch of the proof of Theorem 3.28. We want to show that the e zeros of the polynomial $\Psi_e(x)$ are not all integers. The method is quite similar to the proof of Theorem 3.24 for the case of perfect e-codes in $H(n, q)$. In fact, Hermite polynomials play an essential role in evaluating the first-order and the second-order approximations of zeros. In this case, the polynomial $\Psi_e(x)$ is complicated, which makes the calculation very hard. If $v = 2k \pm 1$, unlike the case $v \neq 2k \pm 1$ (similar to the case $q = 2$ in the proof of Theorem 3.24), the locations of zeros are completely symmetric with respect to the average, and the technique used in Theorem 3.24 that evaluates the deviation of the location of zeros is not available. However, we can use a result on the distribution of prime numbers by Schur to show the non-existence. For details, see [22]. (In [22], the proof was given only for the case $e \geq 5$.) The proof for the case $e = 4$ was almost finished by Bannai and Ito (unpublished), where a few possible cases still remain. (The difficult part is about an integral solution of a Diophantine equation.) There was no progress for a long time about this problem, but recently, Dukes and Short-Gershman [173] showed the non-existence for the case $5 \leq e \leq 9$. They refined the discussion of [22]. The non-existence for the case $e = 4$ was proved by Xiang [522]. As of May 2020, the case $e \geq 10$ is open. □

Remark 3.29. The existence of perfect e-codes in the Johnson scheme $J(v, d)$ is an open problem. In this case, zeros of the polynomial $\Psi_e(x)$ have a certain symmetry, so we cannot apply the technique that evaluates the deviation. The discussion of Diophantine equations is also difficult since it becomes 3-variate. For a partial result, see [23]. Later, some partial results were obtained (for example, see [181, 405, 426]). Any results on extensions like Theorem 3.24 are unknown. We expect the reader to try it.

Remark 3.30. The existence of tight t-designs whose automorphism groups are t-transitive is interesting from the viewpoint of multiply transitive groups. For a $2s$-design whose automorphism group is $2s$-transitive, a generalized Fisher type inequality can be easily proved by using the representation theory of symmetric groups. This, together with Ryuzaburo Noda's influence, motivated Eiichi Bannai to study this area around 1972.

Let G be a $2s$-transitive group on $\Omega = \{1, 2, \ldots, n\}$ and let H be the stabilizer subgroup of $\Delta = \{1, 2, \ldots, s\}$ as a set, i. e., $H = \{g \in G \mid \Delta^g = \Delta\}$. If there exists a subgroup K of G which satisfies $(1_H)^G = (1_K)^G$ ($(1_H)^G$ denotes the induced character to G of the identity character 1_H of the subgroup H) and is not conjugate to H, then it induces a tight $2s$-design. Conversely, if there exists a non-trivial $2s$-design whose automorphism group is $2s$-transitive, then the stabilizer subgroup K of a block as a set satisfies the above condition. The case $s = 1$ is well known. If we use the classification of multiply transitive groups, we can show that there does not exist a $2e$-transitive group ($e \geq 2$) from which a tight $2e$-design arises except for M_{23}, although we do not want to use it because it means the use of the classification of the finite simple groups.

Remark 3.31. It is a subtle question how to define a tight t-design for $t = 2e + 1$. When t is odd, Ray-Chaudhuri and Wilson [400] proved that $b \geq \frac{v}{d}\binom{v-1}{[\frac{t}{2}]}$ by applying a Fisher type inequality to a $(t-1)$-design which fixes a point. So it seems natural to define that Y is a tight t-design if equality holds in the above inequality. Since we may assume $d \leq \frac{v}{2}$, we have $b \geq 2\binom{v-1}{[\frac{t}{2}]}$, in general. This definition seems a bit too strong (Section 3.6).

3.5.2 Classification of tight 4-designs in Johnson schemes

Let $(V, \mathcal{B})$ be a tight 4-design in the Johnson scheme $J(v, k)$. Suppose $(V, \mathcal{B})$ is non-trivial. Assume $k \leq \frac{v}{2}$. Note that if $k > \frac{v}{2}$, the complementary design is also a tight 4-design. In this section, we prove the following theorem.

Theorem 3.32 (Enomoto–Ito–Noda [179]). *A non-trivial tight 4-design in the Johnson scheme is the* 4-(23, 7, 1) *design or its complementary design* 4-(23, 16, 52) *only.*

Remark 3.33. Noboru Ito [245, 246] started the proof of this theorem, and Enomoto, Ito, and Noda (1979) [179] almost completed the proof by correcting errors. In a part of the proof, a number theoretic result on the solution of a Diophantine equation was used (Bremner [111], Stroeker [440]). The proof given here is based on an unpublished note by Ryuzaburo Noda, which was written soon after [179]. We are grateful to Professor Noda, who permits us to use the contents of his note. It is similar to [179] that the problem is transformed into the Diophantine equation. However, compared to the proof combining three papers [245, 246, 179], this proof is clearer and easier to read.

The proof of Theorem 3.32 consists of steps (A)–(K).

(A) Let i, j be the cardinalities of intersections of two distinct blocks. Let $i < j$. Then i, j are the roots of the following quadratic equation:

$$X^2 - \left(\frac{2(k-1)(k-2)}{v-3} + 1\right)X + \frac{k(k-1)^2(k-2)}{(v-2)(v-3)} = 0. \tag{3.51}$$

(B) We have

$$(v-2)(v-3) | 2k(k-1)(k-2). \tag{3.52}$$

Proof. Since $b = \lambda_0 = \binom{v}{2}$, $\lambda_4 = \frac{k(k-1)(k-2)(k-3)}{2(v-2)(v-3)}$ is an integer. Therefore, $2\lambda_4 = \frac{k(k-1)(k-2)(k-3)}{(v-2)(v-3)}$ is an integer. On the other hand, by (A), $\frac{k(k-1)^2(k-2)}{(v-2)(v-3)}$ is an integer, and hence $\frac{k(k-1)^2(k-2)}{(v-2)(v-3)} - \frac{k(k-1)(k-2)(k-3)}{(v-2)(v-3)} = \frac{2k(k-1)(k-2)}{(v-2)(v-3)}$ is also an integer. □

(C) We have $(j-i)|(k-j)$.

Proof. Let A be the incidence matrix of points and blocks, and let N be the adjacency matrix between blocks. Namely, A is a $v \times b$-matrix whose entries are defined as follows:

$$A(p,B) = \begin{cases} 1, & \text{if } p \in B, \\ 0, & \text{if } p \notin B; \end{cases}$$

N is a $b \times b$-matrix whose entries are defined as follows:

$$N(B,C) = \begin{cases} 1, & \text{if } |B \cap C| = i, \\ 0, & \text{if } |B \cap C| \neq i. \end{cases}$$

Then the following equation holds (E is the identity matrix and J is the all 1's matrix):

$${}^tA\,A = kE + iN + j(J - E - N) = (k-j)E + (i-j)N + jJ.$$

Since ${}^tA\,A$ has 0 as an eigenvalue, we have $(k-j)+(i-j)\alpha = 0$, where α is an eigenvalue of N. Since the eigenvalues of N are integers, $\frac{k-j}{j-i}$ is an integer. □

(D) Define integers e, a by $e = \frac{k-j}{j-i} (\geq 0)$, $a = j - i (> 0)$, respectively. Then we have $e \geq a$.

Proof. Assume $e < a$ for the sake of contradiction. Since $e \leq a-1$, we have $\frac{k-j}{a} \leq a-1$. Then $k - \frac{i+j+a}{2} \leq a(a-1)$. Hence $2k-(i+j) \leq 2a(a-1)+a$. Moreover, $2(k-1)-(i+j-1)+1 \leq 2a^2 - a$. Therefore, we obtain

$$2(k-1) - (i+j-1) \leq 2a^2 - a - 1 \leq 2(a^2-1). \tag{3.53}$$

On the other hand, by (A), we have

$$i + j - 1 = \frac{2(k-1)(k-2)}{v-3}, \tag{3.54}$$

$$ij = \frac{k(k-1)^2(k-2)}{(v-2)(v-3)}. \tag{3.55}$$

Thus by $a^2 = (j-i)^2 = (i+j)^2 - 4ij$, we get

$$a^2 - 1 = \frac{4(k-1)(k-2)(v-k-2)(v-k-1)}{(v-2)(v-3)^2}. \tag{3.56}$$

Therefore, by (3.53), we have

$$\frac{(-k+v-1)}{(v-3)} \leq \frac{4(k-2)(-k+v-2)(-k+v-1)}{(v-3)^2(v-2)},$$

and hence

$$\begin{aligned}(v-2)(v-3) \leq 2(k-2)\cdot 2(v-k-2) &\leq \left(\frac{2(k-2)+2(v-k-2)}{2}\right)^2 \\ &= (v-4)^2,\end{aligned}$$

which is impossible. □

(E) We have $a > 1$ and the following equation holds:

$$v - 2 = \frac{(2ea + a - 1)(2ea + a - 3)}{a^2 - 1}. \tag{3.57}$$

Then we have $v = 4e^2 + 4e - 1 + \frac{4(e-a)(e-a+1)}{a^2-1}$. Hence $\frac{4(e-a)(e-a+1)}{a^2-1}$ is an integer.

Proof. By definition, $a \geq 1$. If $a = 1$, by (3.56), we have $k = 1$, $k = 2$, $v = k+1$, or $v = k+2$, which implies it is a trivial design. So we must have $a > 1$. Then by (3.56), we have

$$v - 2 = \frac{4(k-1)(k-2)(v-k-2)(v-k-1)}{(a^2-1)(v-3)^2}.$$

On the other hand, we have

$$2ea + a - 3 = 2(k-j) + (j-i) - 3 = 2(k-2) - (i+j-1) = 2(k-2)\left(1 - \frac{k-1}{v-3}\right)$$

and

$$2ea + a - 1 = 2(k-j) + (j-i) - 1 = 2(k-1) - (i+j-1) = 2(k-1)\left(1 - \frac{k-2}{v-3}\right).$$

Therefore, (3.57) holds. □

(F) Let $p = v - 2k$. Then $p^2 = (v-2)^2 - 2(2ea + a - 1)(v-3)$.

Proof. By the proof of (E),

$$\begin{aligned} 2ea + a - 1 &= \frac{2(k-1)(v-k-1)}{v-3} = \frac{2(\frac{v-p}{2} - 1)(\frac{v+p}{2} - 1)}{v-3} \\ &= \frac{(v-2)^2 - p^2}{2(v-3)}. \end{aligned} \tag{3.58}$$

□

(G) Let $m = \frac{4(e-a)(e-a+1)}{a^2-1}$. By (E), m is an integer, and the following holds:

$$(2ea + a - 1) | (m-3)(a^2 - 1), \tag{3.59}$$

$$(2ea + a - 1) | (m-3)(2e - a + 1). \tag{3.60}$$

Proof. By (3.54) and (3.52), we have $(v-2)|k(i+j-1)$. On the other hand, by the definitions of e and a, we have

$$i + j - 1 = 2k - 1 - 2(k-j) - (j-i) = v - p - 1 - 2ea - a. \tag{3.61}$$

Since $k = \frac{v-p}{2}$, we have $(v-2)|\frac{(v-p)}{2}(v - p - 1 - 2ea - a)$. Since $(v - p - 1 - 2ea - a) = (v-2) - (p - 1 + 2ea + a)$, we obtain

$$(v-2)|\frac{1}{2}(v-p)(p - 1 + 2ea + a). \tag{3.62}$$

Next, we want to show

$$(v-2)^2|(2ea+a-1)(2ea+a-3)^2. \tag{3.63}$$

Concerning the right-hand side of (3.62), we have

$$\begin{aligned}(v-p)(p-1+2ea+a) &= (v-2-(p-2))(p-1+2ea+a)\\ &= (v-2)(p-1+2ea+a)-(p-2)(p-1+2ea+a).\end{aligned}$$

Moreover, by (F), we have $(p-2)(p-1+2ea+a) = (2ea+a-3)p-(v-2)(-v+2a+4ea)$. Hence (3.62) is rewritten as follows:

$$\begin{aligned}2(v-2)|((v-2)(p-1+2ea+a)\\ +(v-2)(-v+2a+4ea)-(a+2ea-3)p).\end{aligned} \tag{3.64}$$

If $v-2$ is even, by definition, p is even, and by (3.58), $2ea+a-1$ is also even. Therefore, $2(v-2)|(2ea+a-3)p$ and hence $4(v-2)^2|(2ea+a-3)^2p^2$. Again, by (F), we have

$$4(v-2)^2|(2ea+a-3)^2((v-2)^2-2(2ea+a-1)(v-3)).$$

Moreover, since $2ea+a-3$ is even and $v-2$ and $v-3$ are coprime, we obtain the desired condition (3.63). On the other hand, if $v-2$ is odd, we do not have to care about the prime factor 2, and by (3.64), we have $(v-2)|(2ea+a-3)p$. Hence we get $(v-2)^2|(2ea+a-3)^2p^2$. Since $v-2$ and $v-3$ are coprime, by (F), we obtain (3.63). By this condition (3.63) and (E), we have $(2ea+a-1)|(a^2-1)^2$. By expressing $2ea+a-1 = a(2e-a+1)+(a^2-1)$, for the greatest common divisor of $2ea+a-1$, $2e-a+1$ and a^2-1, the following equation holds:

$$\begin{aligned}(2ea+a-1,a^2-1) &= (a(2e-a+1),a^2-1)\\ &= (2e-a+1,a^2-1).\end{aligned} \tag{3.65}$$

Therefore, we have $(2ea+a-1)|(2e-a+1)^2$. Furthermore, since $(2e-a+1)^2 = (4e^2+4e+(a^2-1)-2(2ea+a-1))$, we have

$$(2ea+a-1)|(4e^2+4e+(a^2-1)). \tag{3.66}$$

On the other hand, by definition, we have $(m-3)(a^2-1) = 4e^2+4e+(a^2-1)-4(2ea+a-1)$ and we get (3.59). Since a common divisor of $(2ea+a-1)$ and a^2-1 must divide $(2e-a+1)$, we obtain (3.60). □

(H) We have $2e \le a^2+a-2$.

Proof. By (3.66), we have $(2ea+a-1)|(4e^2+4e+a^2-1)a$. Moreover, since $(4e^2+4e+a^2-1)a = 2e(2ea+a-1)+(2ea+a-1)+2e+a^3-2a+1$, we have $(2ea+a-1)|(2e+a^3-2a+1)$.

Therefore, we have $2ea+a-1 \le 2e+a^3-2a+1$, and $2e(a-1) \le a^3-3a+2 = (a^2+a-2)(a-1)$. Since $a > 1$, we obtain $2e \le a^2 + a - 2$. □

(I) If $m = 0$, the parameters of the design are 4-$(23, 7, 1)$. The cases $m = 1$ and $m = 2$ do not occur. If $m \ge 3$, then $m = 3$, or $2e = a^2 + a - 2$ and $m = a(a - 2)$.

Proof. By (D), we have $m \ge 0$.

Assume $m = 0$. Then $e = a$. By (3.59), we have $(2e - 1)|3(e - 1)$. Since $(2e - 1)$ and $(e - 1)$ are coprime, we have $(2e - 1)|3$. Then $e = a > 1$, and so $e = 2$. Therefore, the parameters of the design are 4-$(23, 7, 1)$.

Assume $m = 1$. We have $\frac{4(e-a)(e-a+1)}{a^2-1} = 1$. Since $e \ge a$, we have $e = \frac{3a-1}{2}$. Then $2ea + a - 1 = 3a^2 - 1$, $(m - 3)(a^2 - 1) = -2a^2 + 2 = -3a^2 + 1 + a^2 + 1$. By (3.59), we have $(3a^2 - 1)|(a^2 + 1)$. Since $a > 1$, this is a contradiction.

Assume $m = 2$. Then by (3.59), we have $(2ea + a - 1)|(a^2 - 1)$. However, we have $(2ea + a - 1) \ge 2a^2 + a - 1 > a^2 - 1$, which is a contradiction.

Assume $m > 3$. We will show that $2e = a^2 + a - 2$ and $m = a(a - 2)$. Let $s = \frac{(m-3)(2e-a+1)}{2ea+a-1}$, $r = \frac{(m-3)(a^2-1)}{2ea+a-1}$. By (3.59), (3.60), both s and r are positive integers. The condition $2e = a^2 + a - 2$ is equivalent to $s = r$. We show $s = r$. By (H), we have $s \le r$. By definition, $sa + r = m - 3$, $\frac{s}{r} = \frac{2e-a+1}{a^2-1}$, and $m = \frac{4(e-a)(e-a+1)}{a^2-1}$. Combining these three equations, we delete m and e. Then we have

$$s^2a^2 - rs(r + 2)a - (s^2 + r^3 + 2r^2) = 0.$$

Therefore

$$a = \frac{2r + r^2 + \sqrt{r^2(r + 4)^2 - 4(r^2 - s^2)}}{2s}.$$

Hence $r^2(r + 4)^2 - 4(r^2 - s^2)$ must be a square. Let $d \ge 0$ and $d^2 = r^2(r + 4)^2 - 4(r^2 - s^2)$. Since $r \ge s \ge 1$, we have $d \le r(r + 4)$. By definition, we have $d^2 = r^2(r + 4)^2 - 4(r^2 - s^2) = (r(r + 4) - 2)^2 + 4s^2 + 16r - 4 > (r(r + 4) - 2)^2$. Therefore, $d = r(r + 4) - 1$ or $d = r(r + 4)$. If $d = r(r+4)-1$, then $s^2 = \frac{1}{4}(2(r^2-4r)+1)$, which contradicts the fact that s is an integer. Therefore, we have $d = r(r + 4)$. Then $s = r$. Thus $2e = a^2 + a - 2$ and $m = a(a - 2)$. □

(J) The case $m = 3$ does not occur.

Proof. Assume $m = 3$. Then by (E) and (G), we have $v = 4e^2 + 4e + 2 = (2e + 1)^2 + 1$. We also have $4(e - a)(e - a + 1) = 3(a^2 - 1)$. Thus we have $(4e + 2 - a)^2 = 3(2e + 1)^2 - 2$. Therefore, we obtain $a = 2(2e + 1) - \sqrt{12e^2 + 12e + 1}$. Then

$$2e - a + 1 = \sqrt{12e^2 + 12e + 1} - (2e + 1),$$
$$2(v - 2) = \left(\sqrt{12e^2 + 12e + 1} - (2e + 1)\right)\left(\sqrt{12e^2 + 12e + 1} + (2e + 1)\right),$$
$$2(2ea + a - 1) = \left(\sqrt{12e^2 + 12e + 1} - (2e + 1)\right)^2.$$

Therefore, by (F), we have

$$\begin{aligned}4p^2 &= 4(v-2)^2 - 8(2ea+a-1)(v-3)\\ &= \left(\sqrt{12e^2+12e+1}-(2e+1)\right)^2\\ &\quad\times\left(\left(\sqrt{12e^2+12e+1}+(2e+1)\right)^2 - 4(4e^2+4e-1)\right).\end{aligned}\tag{3.67}$$

Then $\sqrt{12e^2+12e+1}$ is an integer, and $4p^2$ is a square. Hence by (3.67),

$$\left(\sqrt{12e^2+12e+1}+(2e+1)\right)^2 - 4(4e^2+4e-1)$$

is a square and even. Therefore if we let

$$f = \frac{1}{2}\sqrt{\left(\sqrt{12e^2+12e+1}+(2e+1)\right)^2 - 4(4e^2+4e-1)},$$

then f is an integer. Then the following holds:

$$4f^2 = 2(2e+1)\sqrt{12e^2+12e+1}+6.\tag{3.68}$$

Let $X = 2e+1$, $Y = f$. Then

$$2Y^2 = X\sqrt{3X^2-2}+3.$$

According to Bremner, there does not exist an integral solution of this equation with $X, Y > 0$ except for $(X, Y) = (3, 3)$. This contradicts $X = 2e+1 \geq 2a+1 > 3$. □

(K) Let $2e = a^2+a-2$. Then $a = 2$, $e = 2$, and the parameters of the design are 4-$(23, 7, 1)$.

Proof. We have $2ea+a-1 = (a-1)(a+1)^2$ and $v-2 = (a^3+a^2-a-3)(a+1)$. Therefore, if $a = 2$, then $v = 23$, and by (3.56), we have $k = 7$. Hence $\lambda = 1$ and we obtain a 4-$(23, 7, 1)$-design. From now on, we show the case $a \geq 3$ does not occur. By (F), we have $p^2 = (v-2)^2 - 2(2ea+a-1)(v-3) = (a+1)^3(a^5-a^4-2a^3-2a^2+5a+1)$. Therefore, $(a+1)(a^5-a^4-2a^3-2a^2+5a+1)$ is a square. By putting $x = a+1$, this equation becomes the following:

$$F(x) = x(x^5-6x^4+12x^3-12x^2+12x-6).\tag{3.69}$$

By the assumption, $x = a+1 \geq 4$. Let $x = y\alpha^2$, where y is 1, or the product of distinct primes. Then we have $F(x) = y\alpha^2(y^5\alpha^{10} - 6y^4\alpha^8 + 12y^3\alpha^6 - 12y^2\alpha^4 + 12y\alpha^2 - 6)$. Since $F(x)$ is a square, y is a divisor of 6. That means $x = \alpha^2$, $2\alpha^2$, $3\alpha^2$, or $6\alpha^2$.

(i) The case $x = \alpha^2 \geq 4$.

We have $F(x) = \alpha^2 F_1(\alpha)$, where

$$F_1(\alpha) = \alpha^{10} - 6\alpha^8 + 12\alpha^6 - 12\alpha^4 + 12\alpha^2 - 6.$$

Thus $4F_1(\alpha)$ is a square. Then

$$\begin{aligned}4F_1(\alpha) &= (2\alpha^5 - 6\alpha^3 + 3\alpha)^2 - (12\alpha^4 - 39\alpha^2 + 24)\\ &= (2\alpha^5 - 6\alpha^3 + 3\alpha - 3)^2 + 3(\alpha + 1)(4\alpha^4 - 8\alpha^3 - 4\alpha^2 + 17\alpha - 11).\end{aligned}$$

Therefore, we have $4F_1(\alpha) = (2\alpha^5 - 6\alpha^3 + 3\alpha - 2)^2$ or $(2\alpha^5 - 6\alpha^3 + 3\alpha - 1)^2$. However, there does not exist an integer $\alpha \geq 2$ which satisfies these equations.
(ii) The case $x = 2\alpha^2$.

Since $x \geq 4$, we have $\alpha \geq 2$. We have $F(x) = 4\alpha^2 F_2(\alpha)$, where

$$F_2(\alpha) = (16\alpha^{10} - 48\alpha^8 + 48\alpha^6 - 24\alpha^4 + 12\alpha^2 - 3).$$

Thus $F_2(\alpha)$ must be a square. Then

$$\begin{aligned}4F_2(\alpha) &= (8\alpha^5 - 12\alpha^3 + 3\alpha + 1)^2\\ &\quad - (\alpha + 1)(16\alpha^4 + 8\alpha^3 - 32\alpha^2 - 7\alpha + 13)\\ &= (8\alpha^5 - 12\alpha^3 + 3\alpha - 2)^2\\ &\quad + 32\alpha^5 - 24\alpha^4 - 48\alpha^3 + 39\alpha^2 + 12\alpha - 16.\end{aligned}$$

Therefore $4F_2(\alpha) = (8\alpha^5 - 12\alpha^3 + 3\alpha - 1)^2$ or $(8\alpha^5 - 12\alpha^3 + 3\alpha)^2$. However, there does not exist an integer $\alpha \geq 2$ satisfying these equations.
(iii) The case $x = 3\alpha^2$.

By $x \geq 4$, we have $\alpha \geq 2$. We have $F(x) = 9\alpha^2 F_3(\alpha)$, where

$$F_3(\alpha) = (81\alpha^{10} - 162\alpha^8 + 108\alpha^6 - 36\alpha^4 + 12\alpha^2 - 2).$$

Therefore, $4F_3(\alpha)$ must be a square. Then

$$\begin{aligned}4F_3(\alpha) &= (18\alpha^5 - 18\alpha^3 + 3\alpha)^2 - (36\alpha^4 - 39\alpha^2 + 8)\\ &= (18\alpha^5 - 18\alpha^3 + 3\alpha - 1)^2\\ &\quad + 3(\alpha - 1)(12\alpha^4 - 12\alpha^2 + \alpha + 3).\end{aligned}$$

Therefore, there does not exist an integer $\alpha \geq 2$ satisfying this condition.
(iv) The case $x = 6\alpha^2$.

We have $F(x) = 36\alpha^2 F_4(\alpha)$, where

$$F_4(\alpha) = 1296\alpha^{10} - 1296\alpha^8 + 432\alpha^6 - 72\alpha^4 + 12\alpha^2 - 1.$$

Thus $F_4(\alpha)$ must be a square. We have

$$\begin{aligned}4F_4(\alpha) &= (72\alpha^5 - 36\alpha^3 + 3\alpha)^2 - (72\alpha^4 - 39\alpha^2 + 4)\\ &= (72\alpha^5 - 36\alpha^3 + 3\alpha - 1)^2\\ &\quad + (144\alpha^5 - 72\alpha^4 - 72\alpha^3 + 39\alpha^2 + 6\alpha - 5).\end{aligned}$$

There does not exist an integer α satisfying this condition. □

3.6 Tight t-designs for odd t in Johnson schemes and Hamming schemes

Delsarte proved the Fisher type inequality on the size of a t-design Y in a general Q-polynomial scheme as follows (Theorem 3.16 in Section 3.2):

$$|Y| \geq m_0 + m_1 + \cdots + m_{[\frac{t}{2}]}.$$

He defined that Y is a tight t-design if equality holds in the above inequality. By this definition, it is easily proved that if there exists a tight t-design, t must be even. In this section, we discuss Fisher type inequalities for odd t in $J(v,d)$ and $H(d,q)$. The following inequalities hold.

Let Y be a $(2e+1)$-design in the Johnson scheme $J(v,d)$, where $e \geq 1$. Then the following holds:

$$|Y| \geq \frac{v}{d}\binom{v-1}{e}. \tag{3.70}$$

Let Y be a $(2e+1)$-design in the Hamming scheme $H(d,q)$, where $e \geq 1$. Then the following holds:

$$|Y| \geq 1 + d(q-1) + \binom{d}{2}(q-1)^2 + \cdots + \binom{d}{e}(q-1)^e + \binom{d-1}{e}(q-1)^{e+1}. \tag{3.71}$$

The proofs of the above inequalities are as follows.

Proof of the case of $J(v,d)$. Let $(V,\mathcal{B})$ be a $(2e+1)$-(v,d,λ) design. Fix a point P of V, and let $\mathcal{B}' = \{B\backslash\{P\} \mid B \in \mathcal{B}, P \in B\}$. Then $(V\backslash\{P\},\mathcal{B}')$ becomes a $2e$-$(v-1,d-1,\lambda_{2e})$ design. Hence, by the Fisher type inequality for a $2e$-design, we have $\lambda_1 \geq \binom{v-1}{e}$. On the other hand, we have $|Y| = \lambda_0 = \frac{v}{d}\lambda_1$. We obtain (3.70). □

Proof of the case of $H(d,q)$. Consider the Hamming scheme on $X = F^d$ ($|F| = q$). Let $Y \subset X$ be an orthogonal array with strength $t = 2e+1$; Y can be expressed as a $|Y| \times d$-matrix. Since $t \geq 2e+1 \geq 3$, by definition, we have $|\{i \mid y_{i,j} = a\}| = \frac{\lambda}{q}$ (always a constant) for $a \in F$ and $1 \leq j \leq d$. Next, define a submatrix Y' of Y as follows. Fix $a \in F$. Then there exist exactly $\frac{\lambda}{q}$ a's in the first column of Y. Select the $\frac{\lambda}{q}$ rows whose first entry is a. Then we obtain a $\frac{\lambda}{q} \times d$-matrix whose first column consists of all a's. By deleting the first column, we obtain the $\frac{\lambda}{q} \times (d-1)$-matrix Y'. Then Y' becomes an orthogonal array of strength $t-1 (= 2e)$ in $H(d-1,q)$. Therefore, by the Fisher type inequality for $H(d-1,q)$, we get

$$|Y'| = \frac{|Y|}{q} \geq 1 + (d-1)(q-1) + \binom{d-1}{2}(q-1)^2 + \cdots + \binom{d-1}{e}(q-1)^e.$$

By an elementary calculation, we obtain

$$|Y| \geq 1 + d(q-1) + \binom{d}{2}(q-1)^2 + \cdots + \binom{d}{e}(q-1)^e + \binom{d-1}{e}(q-1)^{e+1}.$$ □

By the above discussion, it seems natural to define tight $(2e+1)$-designs in $J(v,d)$ and $H(d,q)$ as the cases where equality holds in inequalities (3.70) and (3.71), respectively. Most part of the classification of tight $(2e+1)$-designs seems to be essentially reduced to that of tight $2e$-designs.

Unlike the cases of $J(v,d)$ and $H(d,q)$, there is no combinatorial or geometric interpretation of designs in a general Q-polynomial scheme. It is unclear how to obtain the Fisher type inequality for a t-design with odd t in general. This is an open problem. As will be seen later, Delsarte's theory is extended to a finite set of points on a sphere in a Euclidean space, and to a finite set of points on several spheres in a Euclidean space. In this area, the theory of cubature formulas in analysis has been studied ahead of the development of the combinatorial theory, and since some time ago, they have been studied in parallel. In the theory of analysis, lower bounds for the number of nodes in cubature formulas of odd degree t were obtained by Möller [352]. It is most desirable that we can apply a similar method to Q-polynomial schemes.

4 Codes and designs in association schemes (continued)

4.1 The Assmus–Mattson theorem and its extensions (Relative designs in Delsarte theory)

We introduce a famous theorem by Assmus and Mattson on a construction of designs from codes. Combinatorial 5-designs, which are not directly related to 5-transitive Mathieu groups, were first found by this method [6]. Let $N = \{1, 2, \ldots, n\}$ and $F_2 = \{0, 1\}$. Each element of F_2^n corresponds to a subset of N as follows. For $\boldsymbol{u} = (u_1, u_2, \ldots, u_n) \in F_2^n$, the subset $\{i \mid u_i = 1, 1 \le i \le n\}$ of N is called the *support* of $\boldsymbol{u}$ and is denoted by $\overline{\boldsymbol{u}}$. If $\boldsymbol{u}$ has weight m, then $\overline{\boldsymbol{u}}$ is an m-subset of N. By this correspondence, the set of codewords in C of weight m is identified with a subset of the set $N^{(m)}$ consisting of m-subsets of N. This means, in the language of association schemes, a subset in the Hamming scheme $H(n, 2)$ can be described in terms of the Johnson scheme $J(n, m)$.

Theorem 4.1 (Assmus and Mattson (1969) [6]). *Let C be an $[n, k, \delta]$ code over F_2. Namely, C is a k-dimensional subspace of F_2^n with minimum distance δ. Let $t < \delta$. Moreover suppose the following condition holds.*

There exist at most $\delta - t$ positive integers in $\{1, 2, \ldots, n - t\}$ which arise as the weights of codewords in $C^\perp$.

Then the following (1), (2) hold:

(1) $\{\overline{\boldsymbol{u}} \in N^{(m)} \mid \boldsymbol{u} \in C^\perp,\ w(\boldsymbol{u}) = m\}$ *forms a t-design in the Johnson scheme $J(n, m)$;*

(2) $\{\overline{\boldsymbol{u}} \in N^{(m)} \mid \boldsymbol{u} \in C,\ w(\boldsymbol{u}) = m\}$ *forms a t-design in the Johnson scheme $J(n, m)$.*

In (1), (2), we only consider the case that there exists a codeword such that $w(\boldsymbol{u}) = m$.

The original Assmus–Mattson theorem in [6] deals with $[n, k, \delta]$ codes over a general finite field F_q. Here, for the sake of simplicity, we restrict the proof to codes over the binary filed F_2. There are many English books which contain the proof of the Assmus–Mattson theorem. However, there seem to be no Japanese books. So we introduce the proof based on the original paper, by following the proof from the book by van Lint and Wilson [319].

Proof. Fix a t-subset T of N. For $\boldsymbol{u} \in F_2^n$, define a codeword $\boldsymbol{u}'$ of length $n - t$ by $\boldsymbol{u}' = (u_i)_{i \in N \setminus T}$. The mapping $\boldsymbol{u} \longmapsto \boldsymbol{u}'$ is linear. If we let $C' = \{\boldsymbol{u}' \mid \boldsymbol{u} \in C\}$, then C' is a linear code of length $n - t$. Next, let $C_0 = \{\boldsymbol{u}' \mid \boldsymbol{u} \in C,\ u_i = 0 \text{ for all } i \in T\}$. Then C_0 is a subspace of C'. Let $B = C^\perp$. Similarly, for B, define $B' = \{\boldsymbol{u}' \mid \boldsymbol{u} \in B\}$, $B_0 = \{\boldsymbol{u}' \mid \boldsymbol{u} \in B,\ u_i = 0 \text{ for all } i \in T\}$.

For the proof of Theorem 4.1, we show $B_0 = (C')^\perp$ in Step 1, and then in Step 2, we show that the weight enumerators of B_0 and C' do not depend on the choice of T, but only on n, k, t. Moreover, we prove Theorem 4.1 (1) in Step 3 and Theorem 4.1 (2) in Step 4.

https://doi.org/10.1515/9783110630251-004

Step 1: $B_0 = (C')^\perp$.

Since $t < \delta$, the sizes of C and C' are equal. Thus, we have $\dim(C') = \dim(C) = k$. Therefore $\dim((C')^\perp) = n - t - k$. On the other hand, since $\dim(B) - \dim(B_0) \leq t$, we have $\dim(B_0) \geq \dim(B) - t = n - k - t$. Moreover, since $B_0 \subset (C')^\perp$, we obtain $B_0 = (C')^\perp$.

Step 2: The weight enumerators of B_0 and C' do not depend on the choice of T.

Let $\alpha_j = |\{\boldsymbol{u}' \in C' \mid w(\boldsymbol{u}') = j\}|$ and $\beta_i = |\{\boldsymbol{u} \in B_0 \mid w(\boldsymbol{u}) = i\}|$. Assume $\{w(\boldsymbol{u}) \mid \boldsymbol{u} \in B,\ 0 < w(\boldsymbol{u}) \leq n - t\} = \{\ell_1, \ell_2, \dots, \ell_r\}$, where $0 < \ell_1 < \ell_2 < \cdots < \ell_r \leq n - t$. By the assumption of the theorem, $r \leq \delta - t$. Hence if we let $1 \leq j \leq r - 1$, then $1 \leq j \leq \delta - t - 1$, and so if we let $w(\boldsymbol{u}') = j$ for some $\boldsymbol{u} \in C$, then $0 < w(\boldsymbol{u}) \leq t + w(\boldsymbol{u}') = t + j \leq \delta - 1$. Therefore we have $\alpha_j = 0$. Namely, we have $(\alpha_0, \alpha_1, \dots, \alpha_{r-1}) = (1, 0, \dots, 0)$, which is independent of the choice of T. Since $\{w(\boldsymbol{u}') \mid w(\boldsymbol{u}') > 0, \boldsymbol{u}' \in B_0\} \subset \{\ell_1, \ell_2, \dots, \ell_r\}$, by the MacWilliams identity for $(B_0)^\perp = C'$ (Chapter 1, Section 1.6), we obtain

$$|B_0|\alpha_j = \sum_{\ell=0}^{n-t} \beta_\ell Q_j^{(n-t)}(\ell) = \binom{n-t}{j} + \sum_{i=1}^{r} \beta_{\ell_i} Q_j^{(n-t)}(\ell_i), \tag{4.1}$$

where $Q_j^{(n-t)}(x)$ is the Krawtchouk polynomial for $H(n - t, 2)$, that is,

$$Q_j^{(n-t)}(x) = \sum_{\ell=0}^{j} (-1)^\ell \binom{x}{\ell}\binom{n-t-x}{j-\ell}$$

(Chapter 2, Section 2.10.2). With $j = 0, 1, \dots, r - 1$, (4.1) is a system of r linear equations with r unknowns and its coefficient matrix is a non-singular matrix of size r. The reason is that the Krawtchouk polynomial is an orthogonal polynomial and hence the above coefficient matrix is reduced to the Vandermonde matrix with entries $\{\ell_1, \ell_2, \dots, \ell_r\}$. Therefore, $\beta_{\ell_1}, \dots, \beta_{\ell_r}$ are determined by n, k, t only. This implies that the weight enumerator of B_0 does not depend on the choice of T, but only on n, k, t. By using the MacWilliams identity again, it can be shown that the weight enumerator of C' does not depend on the choice of T, but only on n, k, t.

Step 3: Proof of Theorem 4.1 (1).

Let $m > 0$ and $\mathcal{B}_m = \{\overline{\boldsymbol{u}} \subset N^{(m)} \mid \boldsymbol{u} \in C^\perp,\ w(\boldsymbol{u}) = m\}$. It suffices to show $(N, \mathcal{B}_m)$ is a t-design in $J(n, m)$. The proof is as follows.

The case $m \leq n - t$:

By the definition of B_0, we have $|\{\overline{\boldsymbol{u}} \in \mathcal{B}_m \mid \overline{\boldsymbol{u}} \cap T = \emptyset\}| = |\{\boldsymbol{u}' \in B_0 \mid w(\boldsymbol{u}') = m\}| = \beta_m$. As was shown in Step 2, the number β_m is independent of the choice of T. Therefore, the complement $\mathcal{B}_m^c$ of $\mathcal{B}_m$ is a t-design. Hence by Definition 1.34, $\mathcal{B}_m$ is a t-design.

The case $m > n - t$:

Let $t' = n - m (< t)$. Then by the assumption of the theorem, there exist at most $\delta - t'$ positive integers in $\{1, 2, \dots, n - t, n - t + 1, \dots, n - t'\}$ which arise as the weights of codewords in $C^\perp$. If we apply the similar discussion to a t'-subset T' of N, we can show that $\mathcal{B}_m^c$ is a t'-design. Since $m = n - t'$, $\mathcal{B}_m^c$ is an empty set or is equal to $N^{(t')}$. Therefore, $\mathcal{B}_m$ is a trivial t-design as $\mathcal{B}_m = N^{(m)}$.

Step 4: Proof of Theorem 4.1 (2).

Let $\mathcal{C}_m = \{\overline{\boldsymbol{u}} \subset N^{(m)} \mid \boldsymbol{u} \in C, w(\boldsymbol{u}) = m\}$. We show $(N, \mathcal{C}_m)$ is a t-design in $J(n,m)$ by induction on m. Let

$$C_{m,T} = \{\boldsymbol{u} = (u_1, \ldots, u_n) \in C \mid w(\boldsymbol{u}) = m,\ u_i = 1 \text{ for all } i \in T\}.$$

Since $|\{\overline{\boldsymbol{u}} \in \mathcal{C}_m \mid T \subset \overline{\boldsymbol{u}}\}| = |C_{m,T}|$, it suffices to show that $|C_{m,T}|$ depends only on n, m, t. First, consider the case $m = \delta$. Then we have $|C_{\delta,T}| = |\{\boldsymbol{u}' \in C' \mid w(\boldsymbol{u}') = \delta - t\}|$. For, if we let $\boldsymbol{u} \in C_{\delta,T}$, then $u_i = 1$ for any $i \in T$, and so $|\{i \in N \backslash T \mid u_i = 1\}| = \delta - t$. By Step 2, $|C_{\delta,T}|$ depends only on n, t, δ. Next, assume that $m > \delta$ and for any m' with $\delta \le m' < m$, $\mathcal{C}_{m'}$ is a t-design. We show $\mathcal{C}_m$ is a t-design. Note that the map $\boldsymbol{u} \to \boldsymbol{u}'$ is a bijection since $t < \delta$. We have the following:

$$\begin{aligned}
&\big|\{\boldsymbol{u}' \in C' \mid w(\boldsymbol{u}') = m - t\}\big| \\
&\quad = \big|\{\boldsymbol{u} \in C \mid w(\boldsymbol{u}') = m - t,\ w(\boldsymbol{u}) \le m\}\big| \\
&\quad = \sum_{k \le m} \big|\{\boldsymbol{u} \in C \mid w(\boldsymbol{u}) = k,\ \big|\{i \in T \mid u_i = 1\}\big| = k - m + t\}\big| \\
&\quad = C_{m,T} \\
&\qquad + \sum_{k \le m-1} \big|\{\boldsymbol{u} \in C \mid w(\boldsymbol{u}) = k,\ \big|\{i \in T \mid u_i = 1\}\big| = k - m + t\}\big|.
\end{aligned} \tag{4.2}$$

By Step 2, $|\{\boldsymbol{u}' \in C' \mid w(\boldsymbol{u}') = m - t\}|$ does not depend on the choice of T. By induction hypothesis, $\mathcal{C}_k$ $(\delta \le k < m)$ is a t-design, and hence it is an s-design for $s \le t$. It can be easily shown that each term of the summation in (4.2) is a constant which does not depend on the choice of T (see [319, (19.6) on page 220]). Therefore $C_{m,T}$ does not depend on the choice of T. This completes the proof. □

Theorem 4.1 deals with codes over F_2. The same result holds for codes over F_q. Namely, under the same condition, for any m, the set of the supports of codewords in C of weight m forms a t-design in $J(n,m)$.

In the paper [161] of 1977, Delsarte introduced the concept of relative designs in general Q-polynomial schemes and proved an analogue of the Assmus–Mattson theorem. In this section, we introduce his result. First, we give some terminologies. Let $\mathfrak{X} = (X, \{R_i\}_{0 \le i \le d})$ be a Q-polynomial scheme with respect to the ordering of the primitive idempotents $E_0, E_1, \ldots, E_d$ of the Bose–Mesner algebra $\mathfrak{A}$. Let v_i^* $(0 \le i \le d)$ be the polynomials associated with the Q-polynomial structure. We have $Q_j(i) = v_j^*(\theta_i^*)$, where $\theta_i^* = Q_1(i)$. In Chapter 2, Section 2.4, V is the $|X|$-dimensional complex vector space indexed by X. Here, since $\mathfrak{A}$ is a symmetric association scheme, we may identify V with the $|X|$-dimensional real vector space $\mathbb{R}^{|X|}$. For $x \in X$, the vector $\hat{x} \in \mathbb{R}^{|X|}$ is defined by

$$\hat{x}(y) = \begin{cases} 1, & \text{if } y = x, \\ 0, & \text{if } y \neq x. \end{cases}$$

The vector $\hat{x}$ is called the characteristic vector of x. In Chapter 3, we defined the characteristic vector ψ_Y of a subset $Y \subset X$. With this notation, we have $\hat{x} = \psi_{\{x\}} = \psi_x$. For $\chi \in V$ and $x \in X$, the x-entry of χ is expressed as $\chi(x)$, which means V is the space of real valued functions on X. In what follows, we identify $V = \mathbb{R}^{|X|}$ with the space of real valued functions on X. The characteristic vector ψ_Y of a subset Y can be regarded as a function which takes 1 on Y and 0 on $X \backslash Y$. In this way, the concept of designs is naturally extended to functions on X. Before defining designs, we extend the definition of the inner distribution of a subset of X, appearing in Chapter 3, to a function in V. The following definition is a natural extension of the definition of the inner distribution of a subset of X, which was defined in the Ph.D. thesis [159] of Delsarte, to a function in V. Note that the following $\boldsymbol{a}_\chi$ is a positive scalar multiple of the one in Delsarte's paper [161]. For $\chi \in V$, let $\|\chi\|^2 = \sum_{x \in X} \chi(x)^2$. For a function χ satisfying $\|\chi\|^2 \neq 0$, define a $(d+1)$-dimensional row vector $\boldsymbol{a}_\chi = (a_0, a_1, \ldots, a_d)$ so that each entry a_i $(0 \leq i \leq d)$ becomes as follows:

$$a_i = \frac{1}{\|\chi\|^2} \sum_{(x,y) \in R_i} \chi(x)\chi(y). \tag{4.3}$$

The vector $\boldsymbol{a}_\chi$ is called the *inner distribution* of χ. In particular, if χ is the characteristic vector ψ_Y of a subset Y, we have $\|\psi_Y\|^2 = |Y|$ and the definition of the inner distribution of $\chi = \psi_Y$ and that of Y are completely identical. By definition, the following proposition follows (Proposition 3.6 in Chapter 3).

Proposition 4.2. *Let $\chi \neq 0$. Then the following (1) and (2) hold:*
(1) $a_0 = 1$;
(2) $\sum_{i=0}^d a_i = \frac{(\sum_{x \in X} \chi(x))^2}{\|\chi\|^2}$; *in particular, if $\chi = \psi_Y$, we have $\sum_{i=0}^d a_i = |Y|$.*

Next, let

$$\boldsymbol{a}_\chi^* = \boldsymbol{a}_\chi Q \tag{4.4}$$

and write $\boldsymbol{a}_\chi^* = (a_0^*, a_1^*, \ldots, a_d^*)$. Then the following holds (Proposition 3.7 and Theorem 3.10).

Proposition 4.3. *Let $\chi \in V$ and $\chi \neq \mathbf{0}$. Then for any $i \in D$, the following (1), (2), (3), and (4) hold:*
(1) $a_i = \frac{1}{\|\chi\|^2} {}^t\chi A_i \chi$;
(2) $a_i^* = \frac{|X|}{\|\chi\|^2} {}^t\chi E_i \chi$; *in particular,* $a_0^* = \sum_{i=0}^d a_i = \frac{(\sum_{x \in X} \chi(x))^2}{\|\chi\|^2}$;

(3) $a_i^* \geq 0$;
(4) $a_i^* = 0$ *if and only if* $E_i\chi = \mathbf{0}$.

Proof. (1) We have

$$\begin{aligned} a_i &= \frac{1}{\|\chi\|^2} \sum_{(x,y)\in R_i} \chi(x)\chi(y) \\ &= \frac{1}{\|\chi\|^2} \sum_{(x,y)\in X} \chi(x)A_i(x,y)\chi(y) = \frac{1}{\|\chi\|^2} {}^t\chi A_i \chi. \end{aligned}$$

(2) We have

$$\begin{aligned} a_i^* &= \sum_{\ell=0}^{d} a_\ell Q_i(\ell) = \frac{1}{\|\chi\|^2} \sum_{\ell=0}^{d} {}^t\chi A_\ell \chi Q_i(\ell) \\ &= \frac{1}{\|\chi\|^2} \sum_{\ell=0}^{d} {}^t\chi Q_i(\ell) A_\ell \chi = \frac{|X|}{\|\chi\|^2} {}^t\chi E_i \chi. \end{aligned}$$

(3) We have

$$a_i^* = \frac{|X|}{\|\chi\|^2} {}^t\chi E_i \chi = \frac{|X|}{\|\chi\|^2} {}^t(E_i\chi)(E_i\chi) = \frac{|X|}{\|\chi\|^2} \|E_i\chi\|^2 \geq 0.$$

(4) It is obvious by (3). □

For $\chi \in V$, we defined the $(d+1)$-dimensional vectors $\boldsymbol{a}_\chi$ and $\boldsymbol{a}_\chi^*$. Moreover, we give the following.

Definition 4.4. For $\chi \in V$, we define the degree s_χ, the minimum distance δ_χ, the dual degree s_χ^*, and the dual minimum distance δ_χ^* of $\chi \in V$ as follows:

$$\begin{aligned} s_\chi &= \left|\{i \mid a_i \neq 0,\ i \neq 0\}\right|, & \delta_\chi &= \min\{i \mid a_i \neq 0, i \neq 0\}, \\ s_\chi^* &= \left|\{i \mid a_i^* \neq 0,\ i \neq 0\}\right|, & \delta_\chi^* &= \min\{i \mid a_i^* \neq 0, i \neq 0\}. \end{aligned}$$

Definition 4.5. Let $\mathfrak{X} = (X, \{R_i\}_{0 \leq i \leq d})$ be a Q-polynomial scheme. For $\chi \in V$ ($\chi \neq 0$), χ is a t-design in $\mathfrak{X}$ if

$$E_j\chi = \mathbf{0}$$

for $j = 1, 2, \ldots, t$.

We obtain this definition by extending the equivalence condition (3) of designs in Theorem 3.10 to the functions in V (or a pair of a subset and a weight function). Here is an equivalence condition of a t-design $\chi \in V$ in a Q-polynomial scheme $\mathfrak{X}$.

Lemma 4.6. *Let $\mathfrak{X}$ be a Q-polynomial scheme. For $t \in D^{\times}$, $\chi \in V$ $(\chi \neq 0)$ is a t-design if and only if the following equation holds for any $i, j \in D$ satisfying $i + j \leq t$:*

$$E_i\Big(|X|\Delta_\chi E_j - \sum_{x\in X}\chi(x)\delta_{i,j}I\Big) = 0, \tag{4.5}$$

where Δ_χ is the diagonal matrix whose diagonal entries are defined by $\Delta_\chi(x,x) = \chi(x)$ $(x \in X)$.

Proof. In general, we denote the sum of squares of entries of the matrix M by $\|M\|^2$. We prove

$$\Big\|E_i\Big(|X|\Delta_\chi E_j - \sum_{x\in X}\chi(x)\delta_{i,j}I\Big)\Big\|^2 = |X|\sum_{\ell=1}^{n} q_{i,j}^{\ell}\,{}^t\chi E_\ell \chi. \tag{4.6}$$

We compute $\|E_i(|X|\Delta_\chi E_j - \sum_{x\in X}\chi(x)\delta_{i,j}I)\|^2$. If $i \neq j$, we have

$$\begin{aligned}
&\||X|E_i\Delta_\chi E_j\|^2 \\
&= |X|^2 \sum_{x,y\in X}((E_i\Delta_\chi E_j)(x,y))^2 \\
&= |X|^2 \sum_{x,y\in X}\Big(\sum_{z\in X}E_i(x,z)\chi(z)E_j(z,y)\Big)\Big(\sum_{z'\in X}E_i(x,z')\chi(z')E_j(z',y)\Big) \\
&= |X|^2 \sum_{z,z'\in X}\chi(z)\chi(z')\sum_{x\in X}E_i(x,z)E_i(x,z')\sum_{y\in X}E_j(z,y)E_j(z',y) \\
&= |X|^2 \sum_{z,z'\in X}\chi(z)\chi(z')E_i(z,z')E_j(z,z') \\
&= |X|^2 \sum_{z,z'\in X}\chi(z)\chi(z')(E_i\circ E_j)(z,z') \\
&= |X| \sum_{z,z'\in X}\chi(z)\chi(z')\sum_{\ell=0}^{d}q_{i,j}^{\ell}E_\ell(z,z') = |X|\sum_{\ell=1}^{d}q_{i,j}^{\ell}\,{}^t\chi E_\ell\chi
\end{aligned} \tag{4.7}$$

(note that $q_{i,j}^0 = 0$ if $i \neq j$). Next we consider the case $i = j$. Let $\sum_{x\in X}\chi(x) = \alpha$. The (x,y)-entry of the matrix in the left-hand side of (4.5) yields

$$|X|\sum_{z\in X}\chi(z)E_i(x,z)E_i(y,z) - \alpha E_i(x,y). \tag{4.8}$$

Therefore, we have

$$\begin{aligned}
&\||X|E_i\Delta_\chi E_i - \alpha E_i\|^2 \\
&= |X|^2\sum_{x,y\in X}\Big(\sum_{z\in X}\chi(z)E_i(x,z)E_i(y,z)\Big)^2
\end{aligned}$$

$$- 2\alpha|X| \sum_{x,y\in X}\sum_{z\in X} \chi(z)E_i(x,z)E_i(y,z)E_i(x,y) + \alpha^2 \sum_{x,y\in X} E_i(x,y)^2$$

$$= |X| \sum_{\ell=0}^{d} q_{i,i}^{\ell}\, {}^t\chi E_\ell \chi - 2\alpha \sum_{z\in X}\chi(z) \sum_{x\in X} |X|E_i(x,z)^2 + \alpha^2 m_i. \tag{4.9}$$

Since we have $|X|q_{i,i}^{0}\,{}^t\chi E_0\chi = \alpha^2 m_i$ and

$$\begin{aligned}\sum_{z\in X}\chi(z)\sum_{x\in X}|X|E_i(x,z)^2 &= \sum_{z\in X}\chi(z)\sum_{x\in X}\sum_{\ell=0}^{d} q_{i,i}^{\ell}E_\ell(x,z)\\ &= \sum_{z\in X}\chi(z)q_{i,i}^{0} = \alpha m_i,\end{aligned} \tag{4.10}$$

(4.9) is equal to $|X|\sum_{\ell=1}^{d} q_{i,i}^{\ell}\,{}^t\chi E_\ell\chi$. Next suppose χ is a t-design and $i+j \le t$. Then by the property of Q-polynomial schemes, $q_{i,j}^{\ell} = 0$ for all $\ell > t$. Since χ is a t-design, $E_\ell\chi = 0$ for all $1 \le \ell \le t$. Thus by (4.6), we obtain (4.5). Conversely, suppose (4.5) holds for all $i,j \in D$ with $i+j \le t$. By the property of Q-polynomial schemes, $q_{i,j}^{\ell}$ is non-negative real for all $i,j,\ell \in D$ and we have ${}^t\chi E_\ell\chi \ge 0$. Therefore $q_{i,j}^{\ell}\,{}^t\chi E_\ell\chi = 0$. For ℓ with $1 \le \ell \le t$, by choosing $i,j\ (= \ell - i)$ satisfying $i+j = \ell$, by the property of Q-polynomial schemes, we have $q_{i,j}^{\ell} = q_{i,\ell-i}^{\ell} \ne 0$. Hence ${}^t\chi E_\ell\chi = 0$, and so $E_\ell\chi = 0$. That is, χ is a t-design. □

Next, we define relative designs, which are defined for a function in V with respect to $u_0 \in X$.

Definition 4.7. Let $\mathfrak{X} = (X, \{R_i\}_{0\le i\le d})$ be a Q-polynomial scheme. Fix $u_0 \in X$. For a function $\chi \in V$ ($\chi \ne 0$), χ is called a *relative t-design* with respect to u_0 if the vectors $E_j\chi$ and $E_j\hat{u}_0$ are linearly dependent for all j with $1 \le j \le t$.

Remark 4.8. The vectors $E_0\chi$ and $E_0\hat{u}_0$ are always linearly dependent.

The following proposition follows.

Proposition 4.9. *Let $\mathfrak{X}$ be a Q-polynomial scheme. The following holds:*

(1) *A t-design $\chi \in V$ in $\mathfrak{X}$ becomes a relative t-design with respect to every fixed vertex $u_0 \in X$.*
(2) *For $u_0 \in X$, let $X_i = \{x \in X \mid (x,u_0) \in R_i\}$ $(0 \le i \le d)$. Then for every integer i $(0 \le i \le d)$, the characteristic vector (function) ψ_{X_i} of X_i is a relative t-design with respect to u_0.*

Proof. (1) This is clear by the definition.
(2) Let $x \in X_\ell$. Then for every $0 \le j \le d$, we have

$$(E_j\psi_{X_i})(x) = \sum_{z\in X_i} E_j(x,z) = \sum_{v=0}^{d}\sum_{\substack{z\in X_i\\(x,z)\in R_v}} E_j(x,z) = \frac{1}{|X|}\sum_{v=0}^{d} p_{iv}^{\ell}Q_j(v).$$

Hence by Theorem 2.22 (8), we have

$$(E_j\psi_{X_i})(x) = \frac{1}{|X|}P_i(j)Q_j(\ell) = P_i(j)E_j(x, u_0).$$

It turns out that $E_j\psi_{X_i} = P_i(j)E_j\hat{u_0}$ holds for every $j = 1, 2, \ldots, d$. □

The next lemma is very important when we consider an analogue of the Assmus–Mattson theorem. We recall the binary Hamming scheme $H(n, 2)$. Let F_2 be the binary field. Then $H(n, 2)$ has the vertex set $X = F_2^n$, and the relation R_i $(0 \le i \le n)$ is defined by the Hamming distance. Let $\boldsymbol{u}_0 = (0, 0, \ldots, 0) \in X = F_2^n$. Let $X_k = \Gamma_k(\boldsymbol{u}_0) = \{\boldsymbol{x} \in X \mid (\boldsymbol{x}, \boldsymbol{u}_0) \in R_k\}$ be the set of elements in F_2^n with Hamming weight k. Then X_k has the structure of the Johnson scheme $J(n, k)$.

Lemma 4.10. *We retain the notation above. For $1 \le t \le k$, the characteristic function $\psi_Y \in V$ of $Y \subset X_k$ is a relative t-design in $H(n, 2)$ with respect to $\boldsymbol{u}_0$ if and only if Y is a t-design in the Johnson scheme $J(n, k)$.*

Before the proof of Lemma 4.10, we give the following proposition.

Proposition 4.11. *Let Q be the second eigenmatrix of the binary Hamming scheme $H(n, 2)$. Then the following holds:*

$$\sum_{\ell=0}^{m}(-1)^{\ell}\binom{m}{\ell}Q_m(k - m + 2\ell) = 2^{2m}.$$

Proof. By Theorem 2.86 in Chapter 2, Section 2.10, the eigenmatrix Q of $H(n, 2)$ is given in terms of the Krawtchouk polynomials as follows by putting $q = 2$ in (2.27):

$$Q_m(u) = K_m(u) = \sum_{j=0}^{m}(-1)^j\binom{n-u}{m-j}\binom{u}{j} = \sum_{j=0}^{m}(-2)^j\binom{n-j}{m-j}\binom{u}{j}.$$

The generator function is $(1 + x)^{n-u}(1 - x)^u$, and $Q_m(k - m + 2\ell)$ is the coefficient of x^m of the polynomial

$$(1 + x)^{n-(k-m+2\ell)}(1 - x)^{k-m+2\ell}.$$

Hence $\sum_{\ell=0}^{m}(-1)^{\ell}\binom{m}{\ell}Q_m(k - m + 2\ell)$ is the coefficient of x^m of the polynomial

$$\sum_{\ell=0}^{m}(-1)^{\ell}\binom{m}{\ell}(1 + x)^{n-(k-m+2\ell)}(1 - x)^{k-m+2\ell}. \tag{4.11}$$

The polynomial (4.11) can be rewritten as follows:

$$\begin{aligned}
&(1 + x)^{n-k+m}(1 - x)^{k-m}\sum_{\ell=0}^{m}(-1)^{\ell}\binom{m}{\ell}\left(\frac{1-x}{1+x}\right)^{2\ell}\\
&= (1 + x)^{n-k+m}(1 - x)^{k-m}\left(1 - \left(\frac{1-x}{1+x}\right)^2\right)^m\\
&= (1 + x)^{n-k-m}(1 - x)^{k-m}(4x)^m.
\end{aligned} \tag{4.12}$$

Therefore the coefficient of x^m of this polynomial is $4^m = 2^{2m}$. □

Proof of Lemma 4.10. We construct the structure of the Johnson scheme on X_k, by identifying $\boldsymbol{x} \in X_k$ with the support $\overline{\boldsymbol{x}}$ of $\boldsymbol{x}$. Namely, for $\boldsymbol{x} \in X_k$, a k-subset $\overline{\boldsymbol{x}}$ of $\{1, 2, \ldots, n\}$ is defined by $\overline{\boldsymbol{x}} = \{i \mid 1 \le i \le n, x_i = 1\}$. This correspondence becomes the one-to-one correspondence between X_k and the set Ω of k-subsets of the set $\{1, \ldots, n\}$. Then Ω has the structure of $J(n,k) = (\Omega, \{R_i^J\}_{0 \le i \le k})$. We have $(\overline{\boldsymbol{x}}, \overline{\boldsymbol{y}}) \in R_i^J$ for $\overline{\boldsymbol{x}}, \overline{\boldsymbol{y}} \in \Omega$ if and only if $|\overline{\boldsymbol{x}} \cap \overline{\boldsymbol{y}}| = k - i$ if and only if $(\boldsymbol{x}, \boldsymbol{y}) \in R_{2i}$ for vertices $\boldsymbol{x}, \boldsymbol{y} \in X_k$ in $H(n,2)$. Assume $Y \subset X_k$ is a relative t-design in $H(n,2)$ with respect to $\boldsymbol{u}_0$, and show that $\overline{Y} = \{\overline{\boldsymbol{y}} \mid \boldsymbol{y} \in Y\}$ is a t-(n,k,λ) design. Let $1 \le \mu \le t$. For a μ-subset $\overline{\boldsymbol{x}} = \{i_1, i_2, \ldots, i_\mu\}$ of $\{1, 2, \ldots, n\}$, define $\lambda_\mu(\boldsymbol{x}) = |\{\overline{\boldsymbol{y}} \in \overline{Y} \mid \overline{\boldsymbol{x}} \subset \overline{\boldsymbol{y}}\}|$. By induction, we show that $\lambda_\mu(\boldsymbol{x})$ does not depend on the choice of $\overline{\boldsymbol{x}}$, but only on μ. In terms of $H(n,2)$, $\lambda_\mu(\boldsymbol{x})$ is characterized as follows. Let $\boldsymbol{x} = (x_1, x_2, \ldots, x_n) \in X_\mu$ be the element of $X_\mu \subset X$ which corresponds to $\overline{\boldsymbol{x}}$. Then we have the following:

$$\lambda_\mu(\boldsymbol{x}) = \left|\{\boldsymbol{y} \in Y \mid y_{i_1} = x_{i_1}, y_{i_2} = x_{i_2}, \ldots, y_{i_\mu} = x_{i_\mu}\}\right|.$$

Since $\boldsymbol{y} \in Y \subset X_k$, we have

$$\{\boldsymbol{y} \in Y \mid y_{i_1} = x_{i_1}, y_{i_2} = x_{i_2}, \ldots, y_{i_\mu} = x_{i_\mu}\} = Y \cap \Gamma_{k-\mu}(\boldsymbol{x}).$$

So the following holds:

$$\lambda_\mu(\boldsymbol{x}) = \left|Y \cap \Gamma_{k-\mu}(\boldsymbol{x})\right|. \tag{4.13}$$

Furthermore, for an integer ℓ with $1 \le \ell \le \mu$ and $\overline{\boldsymbol{z}} \subset \overline{\boldsymbol{x}}$ with $|\overline{\boldsymbol{z}}| = \ell$, we define a number $a_{\mu,\ell}(\boldsymbol{x}, \boldsymbol{z})$ as follows:

$$a_{\mu,\ell}(\boldsymbol{x}, \boldsymbol{z}) = \left|\{\boldsymbol{y} \in Y \mid \overline{\boldsymbol{y}} \cap \overline{\boldsymbol{x}} = \overline{\boldsymbol{z}}\}\right|. \tag{4.14}$$

We prove $\lambda_\mu(\boldsymbol{x})$ and $a_{\mu,\ell}(\boldsymbol{x}, \boldsymbol{z})$ do not depend on the choices of $\boldsymbol{x}$ and $\boldsymbol{z}$, but depend only on μ and ℓ, by induction. Since ψ_Y is a relative t-design, we have

$$E_j \psi_Y = \alpha_j \psi_{\{\boldsymbol{u}_0\}} \tag{4.15}$$

for $j = 1, 2, \ldots, t$.

The case $\mu = 1$.

Let $\boldsymbol{x} = (x_1, \ldots, x_n) \in X_1$, and let $x_i = 1$, i. e., $\overline{\boldsymbol{x}} = \{i\}$. Since the choices of ℓ and $\boldsymbol{z} \in X_\ell$, which define $a_{\mu,\ell}(\boldsymbol{x}, \boldsymbol{z})$, are $\ell = 1$, $\boldsymbol{z} = \boldsymbol{x}$ only, we have $a_{1,1}(\boldsymbol{x}, \boldsymbol{x}) = \lambda_1(\boldsymbol{x})$. For $j = 1$, compare the $\boldsymbol{x}$-entry of both sides of (4.15). Then since $\boldsymbol{x} \in \Gamma_1(\boldsymbol{u}_0)$, the right-hand side is $\alpha_1 E_1(\boldsymbol{x}, \boldsymbol{u}_0) = \frac{\alpha_1}{|X|} Q_1(1)$, which is independent of i, i. e., the choice of $\overline{\boldsymbol{x}} \in X_1$. Multiplying the left-hand side of (4.13) by $|X|$ yields

$$\begin{aligned}
(|X| E_1 \psi_Y)(\boldsymbol{x}) &= \sum_{\boldsymbol{y} \in Y} |X| E_1(\boldsymbol{x}, \boldsymbol{y}) \\
&= \sum_{\boldsymbol{y} \in Y \cap \Gamma_{k-1}(\boldsymbol{x})} |X| E_1(\boldsymbol{x}, \boldsymbol{y}) + \sum_{\boldsymbol{y} \in Y \cap \Gamma_{k+1}(\boldsymbol{x})} |X| E_1(\boldsymbol{x}, \boldsymbol{y}) \\
&= \left|Y \cap \Gamma_{k-1}(\boldsymbol{x})\right| Q_1(k-1) + \left(|Y| - \left|Y \cap \Gamma_{k-1}(\boldsymbol{x})\right|\right) Q_1(k+1) \\
&= \lambda_1(\boldsymbol{x})\big(Q_1(k-1) - Q_1(k+1)\big) + |Y| Q_1(k+1).
\end{aligned} \tag{4.16}$$

Since $Q_1(k-1) \neq Q_1(k+1)$, $\lambda_1(\boldsymbol{x})$ is independent of i, or the choice of $\boldsymbol{x} \in X_1$. Next, suppose for every integer μ with $1 \le \mu \le m-1$, $\lambda_\mu = \lambda_\mu(\boldsymbol{x})$ does not depend on $\boldsymbol{x}$, but only on μ, and consider the case $\boldsymbol{x} \in X_m$. Let $j = m$, and compute the $\boldsymbol{x}$-entry of both sides of (4.15). The right-hand side is $\frac{\alpha_m}{|X|}Q_m(m)$, which is independent of the choice of $\boldsymbol{x} \in X_m$. The left-hand side is

$$
\begin{aligned}
&\sum_{\boldsymbol{y}\in Y} E_m(\boldsymbol{x},\boldsymbol{y})\\
&\quad= \sum_{\substack{\overline{\boldsymbol{y}}\supset\overline{\boldsymbol{x}}\\ \boldsymbol{y}\in Y}} E_m(\boldsymbol{x},\boldsymbol{y}) + \sum_{\ell=1}^{m}\sum_{\substack{|\overline{\boldsymbol{y}}\cap\overline{\boldsymbol{x}}|=m-\ell\\ \boldsymbol{y}\in Y}} E_m(\boldsymbol{x},\boldsymbol{y})\\
&\quad= \frac{\lambda_m(\boldsymbol{x})}{|X|}Q_m(k-m)\\
&\qquad+\frac{1}{|X|}\sum_{\ell=1}^{m}\sum_{\substack{\overline{\boldsymbol{z}}\subset\overline{\boldsymbol{x}}\\ |\overline{\boldsymbol{z}}|=m-\ell}} \left|\{\boldsymbol{y}\in Y \mid \overline{\boldsymbol{y}}\cap\overline{\boldsymbol{x}}=\overline{\boldsymbol{z}}\}\right| Q_m(k-(m-2\ell)).
\end{aligned} \tag{4.17}
$$

Next, let $\overline{\boldsymbol{z}_{m-\ell}} \subset \overline{\boldsymbol{x}}$, where $|\overline{\boldsymbol{z}_{m-\ell}}| = m-\ell$ and $1 \le \ell \le m-1$. By induction on ℓ, we show the following:

$$
a_{m,m-\ell}(\boldsymbol{x},\boldsymbol{z}_{m-\ell}) = \sum_{i=0}^{\ell-1}(-1)^i\binom{\ell}{i}\lambda_{m-\ell+i} + (-1)^\ell\lambda_m(\boldsymbol{x}). \tag{4.18}
$$

If $\ell = 1$, we have $\{\boldsymbol{y}\in Y \mid \overline{\boldsymbol{y}} \supset \overline{\boldsymbol{z}_{m-1}}\} = \{\boldsymbol{y}\in Y \mid \overline{\boldsymbol{y}}\cap\overline{\boldsymbol{x}} = \overline{\boldsymbol{z}_{m-1}}\} \cup \{\boldsymbol{y}\in Y \mid \overline{\boldsymbol{y}}\supset\overline{\boldsymbol{x}}\}$. So we have

$$
a_{m,m-1}(\boldsymbol{x},\boldsymbol{z}_{m-1}) = \lambda_{m-1} - \lambda_m(\boldsymbol{x}),
$$

and we get (4.18). Next, suppose that (4.18) holds for any integer ν with $1 \le \nu \le \ell-1$ and for $\overline{\boldsymbol{z}_{m-\nu}} \subset \overline{\boldsymbol{x}}$ with $|\boldsymbol{z}_{m-\nu}| = m-\nu$. Let $\overline{\boldsymbol{z}_{m-\ell}} \subset \overline{\boldsymbol{x}}$ and $|\boldsymbol{z}_{m-\ell}| = m-\ell$. Here, we abbreviate as $a_{m,m-\nu}(\boldsymbol{x},\boldsymbol{z}_{m-\nu}) = a_{m,m-\nu}$ for $1 \le \nu \le \ell-1$. By the induction hypothesis, we have

$$
\begin{aligned}
\lambda_{m-\ell} &= \left|\{\boldsymbol{y}\in Y \mid \overline{\boldsymbol{y}} \supset \overline{\boldsymbol{z}_{m-\ell}}\}\right|\\
&= \sum_{j=0}^{\ell-1}\sum_{\substack{\overline{\boldsymbol{u}}\subset\overline{\boldsymbol{x}}\setminus\overline{\boldsymbol{z}_{m-\ell}}\\ |\overline{\boldsymbol{u}}|=j}} \left|\{\boldsymbol{y}\in Y \mid \overline{\boldsymbol{y}}\cap\overline{\boldsymbol{x}} = \overline{\boldsymbol{z}_{m-\ell}}\cup\overline{\boldsymbol{u}}\}\right|\\
&\quad+\left|\{\boldsymbol{y}\in Y \mid \overline{\boldsymbol{y}}\supset\overline{\boldsymbol{x}}\}\right|\\
&= a_{m,m-\ell}(\boldsymbol{x},\boldsymbol{z}_{m-\ell}) + \sum_{j=1}^{\ell-1}\binom{\ell}{j}a_{m,m-\ell+j} + \lambda_m(\boldsymbol{x}).
\end{aligned} \tag{4.19}
$$

Therefore, we have

$$
\begin{aligned}
a_{m,m-\ell}(\boldsymbol{x},\boldsymbol{z}_{m-\ell}) &= \lambda_{m-\ell} - \sum_{j=1}^{\ell-1}\binom{\ell}{j}a_{m,m-\ell+j} - \lambda_m(\boldsymbol{x})\\
&= \lambda_{m-\ell} - \sum_{j=1}^{\ell-1}\binom{\ell}{j}\left(\sum_{i=0}^{\ell-j-1}(-1)^i\binom{\ell-j}{i}\lambda_{m-(\ell-j)+i} + (-1)^{\ell-j}\lambda_m(\boldsymbol{x})\right)
\end{aligned}
$$

$$
\begin{aligned}
&- \lambda_m(\boldsymbol{x}) \\
&= \lambda_{m-\ell} - \sum_{j=1}^{\ell-1} \binom{\ell}{j} \sum_{i=0}^{\ell-j-1} (-1)^i \binom{\ell-j}{i} \lambda_{m-\ell+j+i} + (-1)^\ell \lambda_m(\boldsymbol{x}) \\
&= \lambda_{m-\ell} + \sum_{s=1}^{\ell-1} (-1)^s \binom{\ell}{s} \lambda_{m-\ell+s} + (-1)^\ell \lambda_m(\boldsymbol{x}).
\end{aligned}
\tag{4.20}
$$

This proves (4.18).

By (4.18) and (4.17), the left-hand side of (4.15) becomes as follows:

$$
\begin{aligned}
&\frac{\lambda_m(\boldsymbol{x})}{|X|} Q_m(k-m) \\
&+ \frac{1}{|X|} \sum_{\ell=1}^{m} \binom{m}{\ell} \left(\lambda_{m-\ell} + \sum_{s=1}^{\ell-1} (-1)^s \binom{\ell}{s} \lambda_{m-\ell+s} + (-1)^\ell \lambda_m(\boldsymbol{x}) \right) \\
&\times Q_m(k-m+2\ell).
\end{aligned}
\tag{4.21}
$$

The coefficient of $|X|\lambda_m(\boldsymbol{x})$ in the above polynomial is

$$
\sum_{\ell=0}^{m} (-1)^\ell \binom{m}{\ell} Q_m(k-m+2\ell),
$$

which equals 2^{2m} by Proposition 4.11. Therefore, by (4.15), $\lambda_m(\boldsymbol{x})$ is independent of the choice of $\boldsymbol{x}$. This completes the proof of Lemma 4.10. □

We use the terminology of Terwilliger algebras introduced in Chapter 2, Section 2.6. Let $T = T(u_0)$ be the Terwilliger algebra. For $\chi \in V$, we consider

$$
E_i^* \chi, \quad 0 \le i \le d, \tag{4.22}
$$

where $E_0^*, E_1^*, \ldots, E_d^*$ form a basis of the dual Bose–Mesner algebra. Let $D = \{0, 1, \ldots, d\}$. We define the support $\mathrm{sup}(\chi)$ and $\mathrm{end}(\chi)$ of χ with respect to u_0 as follows:

$$
\mathrm{sup}(\chi) = \{i \in D \mid E_i^* \chi \neq 0,\ i \neq 0\}, \tag{4.23}
$$

$$
\mathrm{end}(\chi) = \min\{i \in D \mid E_i^* \chi \neq 0,\ i \neq 0\}. \tag{4.24}
$$

The support $\mathrm{sup}(\chi)$ and $\mathrm{end}(\chi)$ defined above depend on the choice of $u_0 \in X$. By Proposition 4.3, s_χ^* and δ_χ^*, which was introduced in Definition 4.4, are expressed as follows:

$$
s_\chi^* = |\{i \in D \mid E_i \chi \neq 0,\ i \neq 0\}|, \quad \delta_\chi^* = \min\{i \in D \mid E_i \chi \neq 0,\ i \neq 0\}.
$$

The following theorem is an analogue of the Assmus–Mattson theorem.

Theorem 4.12 (Delsarte ([161], Theorem 8.4)). *Let $\mathfrak{X}$ be a Q-polynomial scheme. Let $\chi \in V$ and fix $u_0 \in X$. We retain the above definitions and notation. Assume $1 \leq |sup(\chi)| \leq \delta_\chi^* - 1$. Then for every $i \in D$ with $E_i^*\chi \neq 0$, $E_i^*\chi$ is a relative $(\delta_\chi^* - |sup(\chi)|)$-design with respect to u_0.*

Proof. Let $i \in D$. If $i = 0$, we have $E_0^*\chi = \chi(u_0)\hat{u_0}$. Hence for every $j \in D$, we have $E_jE_0^*\chi = \chi(u_0)E_j\hat{u_0}$, and $E_0^*\chi$ is a relative d-design with respect to u_0. From now on, fix $i \in \sup(\chi)$. We define a polynomial $f(z)$ as follows:

$$f(z) = \prod_{\substack{j\in\sup(\chi)\\ j\neq i}} \frac{z - \theta_j^*}{\theta_i^* - \theta_j^*}. \tag{4.25}$$

The polynomial $f(z)$ has degree $|\sup(\chi)| - 1$, and for $j \in \sup(\chi)$, we have $f(\theta_j^*) = \delta_{i,j}$. Let Δ_χ be a diagonal matrix indexed by X whose x-th diagonal entry is $\chi(x)$ for $x \in X$. That is, $\Delta_\chi(x,y) = \delta_{x,y}\chi(x)$. Then the following holds. Since $0 \notin \sup(\chi)$, we have

$$\begin{aligned}\left(\Delta_\chi\left(\sum_{j=0}^{d} f(\theta_j^*)A_j\right)\hat{u_0}\right)(x) &= (\Delta_\chi(A_i\hat{u_0} + f(\theta_0^*)A_0\hat{u_0}))(x)\\ &= \chi(x)A_i(x,u_0) + f(\theta_0^*)\chi(x)A_0(x,u_0)\\ &= (E_i^*\chi)(x) + f(\theta_0^*)(E_0^*\chi)(x).\end{aligned} \tag{4.26}$$

Therefore, we have

$$\Delta_\chi\left(\sum_{j=0}^{d} f(\theta_j^*)A_j\right)\hat{u_0} = E_i^*\chi + f(\theta_0^*)\chi(u_0)\hat{u_0}. \tag{4.27}$$

Next, let

$$f(z) = \sum_{\ell=0}^{m-1} f_\ell v_\ell^*(z) \tag{4.28}$$

be the expansion of $f(z)$ in terms of $v_0^*(z), v_1^*(z), \ldots, v_d^*(z)$, where $m = |\sup(\chi)|$. Then

$$\begin{aligned}\sum_{j=0}^{d} f(\theta_j^*)A_j &= \sum_{j=0}^{d}\sum_{\ell=0}^{m-1} f_\ell v_\ell^*(\theta_j^*)A_j\\ &= \sum_{\ell=0}^{m-1} f_\ell \sum_{j=0}^{d} Q_\ell(j)A_j = |X|\sum_{\ell=0}^{m-1} f_\ell E_\ell.\end{aligned} \tag{4.29}$$

Then (4.27) is rewritten as follows:

$$|X|\sum_{\ell=0}^{m-1} f_\ell\Delta_\chi E_\ell\hat{u_0} = E_i^*\chi + f(\theta_0^*)\chi(u_0)\hat{u_0}. \tag{4.30}$$

Note that χ is a $(\delta_\chi^* - 1)$-design by the assumption of the theorem. We multiply both sides of (4.30) by E_j for j with $1 \le j \le \delta_\chi^* - |\mathrm{sup}(\chi)|$. Since $j + \ell \le \delta_\chi^* - 1$ holds for ℓ with $0 \le \ell \le m - 1 = |\mathrm{sup}(\chi)| - 1$, we can apply Lemma 4.6. Since χ is a $(\delta_\chi^* - 1)$-design, by Lemma 4.6, we have

$$|X|E_j\Delta_\chi E_\ell = \sum_{x\in X} \chi(x)\delta_{j,\ell}E_j,$$

and

$$\sum_{x\in X} \chi(x)f_jE_j\hat{u_0} = E_j(E_i^*\chi) + f(\theta_0^*)\chi(u_0)E_j\hat{u_0}.$$

Therefore, we obtain

$$E_j(E_i^*\chi) = \left(f_j \sum_{x\in X} \chi(x) - f(\theta_0^*)\chi(u_0)\right)E_j\hat{u_0},$$

which means that $E_i^*\chi$ is a relative $(\delta_\chi^* - |\mathrm{sup}(\chi)|)$-design with respect to u_0. □

This theorem looks weaker than the original Assmus–Mattson theorem (Theorem 4.1). For example, if we apply this theorem to the Golay [24, 12, 8]-code, it only claims that the set of supports of codewords of a fixed weight forms a 4-design. (The original Assmus–Mattson theorem claims that it becomes a 5-design.) However, as was written in Brouwer, Cohen, and Neumaier [113, pages 61, 62], if a code contains all codewords of a certain weight, say, w, we can ignore w when we evaluate t. So for the case of the Golay [24, 12, 8]-code, we can get a 5-design since it contains the unique codeword of weight 24. In this sense, the above theorem is not so weak.

4.2 t-Designs in regular semilattices

First, we give definitions and basic facts on *semilattices*.

Definition 4.13 (Poset). A *partial order* on a set L is a binary relation $\le$ on L satisfying the following (1)–(3):
(1) reflexivity: $a \le a$;
(2) transitivity: if $a \le b$ and $b \le c$, then $a \le c$;
(3) antisymmetry: if $a \le b$ and $b \le a$, then $a = b$;

where $a, b, c \in L$. If a binary relation $\le$ is a partial order on L, a pair $(L, \le)$ is called a *partially ordered set* or a *poset*.

Definition 4.14 (Meet semilattice). Let L be a poset. For a pair a, b in L, an element of L, denoted by $a \wedge b$, is called the *meet* of a and b if it satisfies the following (1), (2):

(1) $a \wedge b \le a$, $a \wedge b \le b$;
(2) if $c \le a$ and $c \le b$ for $c \in L$, then $c \le a \wedge b$.

Note that the meet $a \wedge b$ is uniquely determined if it exists. The poset L is called a *meet semilattice* if the meet exists for every pair of elements of L.

Definition 4.15 (Join semilattice). Let L be a poset. For a pair a, b in L, an element of L, denoted by $a \vee b$, is called the *join* of a and b if it satisfies the following (1), (2):
(1) $a \vee b \ge a$, $a \vee b \ge b$;
(2) if $c \ge a$ and $c \ge b$ for $c \in L$, then $c \ge a \vee b$.

Note that the join $a \vee b$ is uniquely determined if it exists. The poset L is called a *join semilattice* if the join exists for every pair of elements of L. The poset L is called a *lattice* if it is a meet lattice and a join lattice.

In what follows, a semilattice means a meet semilattice. Assume that for a semilattice L, there exists a unique element u such that $u \le x$ for all x. We denote this element by 0.

Definition 4.16 (Graded semilattice). Let L be a semilattice consisting of finite elements. For $x, y \in L$, we write $x < y$ if $x \le y$ and $x \ne y$. For $x \le y \in L$, a sequence $x_0, \ldots, x_r$ of elements in L is called a *chain* from x to y of length r if it satisfies $x = x_0 < x_1 < \cdots < x_r = y$. A chain $x = x_0 < x_1 < \cdots < x_r = y$ from x to y is said to be *maximal* if there exists no chain from x to y which contains $x_0, \ldots, x_r$. A semilattice L is said to be *graded* if it satisfies the following:
(1) for every $x \le y$, all maximal chains from x to y have the same length;
(2) there exists a map $h : L \longrightarrow \mathbb{Z}_{\ge 0}$;
(3) $h(0) = 0$;
(4) for every $x \le y$, $h(y) - h(x)$ is the length of a maximal chain from x to y;
(5) if we let $n = \max\{h(x) \mid x \in L\}$ and for every s with $0 \le s \le n$ we define

$$L_s = \{x \in L \mid h(x) = s\}, \tag{4.31}$$

then $L_0, L_1, \ldots, L_n$ are not empty.

The map h is called the *height function* of L. Each L_j $(0 \le j \le n)$ is called a *fiber*. In particular, L_n is called the *top fiber*.

For subsets A, B of a semilattice L, define $A \wedge B = \{a \wedge b \mid a \in A,\ b \in B\}$.

Definition 4.17 (Short graded semilattice). Let L be a graded semilattice and L_n the top fiber of L. The semilattice L is called a *short graded semilattice* if $L_n \wedge L_n = L$.

Proposition 4.18. *Let L be a short graded semilattice and L_n the top fiber of L. Then for any integers j and r with $0 \le j \le r \le n$, there exist $a \in L_r$ and $y \in L_n$ such that $h(a \wedge y) = j$.*

Proof. Let $z \in L_j$. Let $x, y \in L_n$ be elements which satisfy $z = x \wedge y$. The length of a maximal chain from z to x is $n - j$. Let $z = a_j < \cdots < a_r < \cdots < a_n = x$ be a maximal chain. Since $y \wedge a_r \le a_r \le x$, $y \wedge a_r \le y$, we have $y \wedge a_r \le x \wedge y = z$. On the other hand, since $z = x \wedge y \le y$, $z = a_j \le a_r$, we obtain $z \le y \wedge a_r$. Therefore, $y \wedge a_r = z \in L_j$. □

Definition 4.19 (Regular semilattice). Let L be a short graded semilattice and L_n the top fiber of L. The semilattice L is called a *regular semilattice* if the following (1), (2), (3) hold:
(1) For $y \in L_n$ and $z \in L_r$ with $z \le y$, the number of $u \in L_s$ such that $z \le u \le y$ is a constant which depends only on r, s. The constant is denoted by $\mu(r, s)$.
(2) For $u \in L_s$, the number of $z \in L_r$ such that $z \le u$ is a constant which depends only on r, s. The constant is denoted by $\nu(r, s)$.
(3) For $a \in L_r$ and $y \in L_n$ with $a \wedge y \in L_j$, the number of pairs $(b, z) \in L_s \times L_n$ such that $b \le z$, $b \le y$, $a \le z$ is a constant which depends only on j, r, s. The constant is denoted by $\pi(j, r, s)$.

Remark 4.20. The constants μ, ν, and π defined in Definition 4.19 are called the parameters of a regular semilattice. If $0 \le r \le s \le n$, then $\mu(r, s)$ and $\nu(r, s)$ are positive integers. We also have $\mu(r, r) = \nu(r, r) = 1$. Note that (3) implies (2), which will be stated in the following proposition.

Proposition 4.21. *In Definition* 4.19, *(3) implies (2). Namely, we have* $\pi(j, n, s) = \nu(s, j)$.

Proof. We assume $r = n$ in (3) of Definition 4.19. By the assumption, we have $a \in L_n$, $a \wedge y \in L_j$. Since $a \le z \in L_n$, we have $z = a$. Hence we have $\pi(j, n, s) = |\{b \mid b \le a,\ b \le y\}| = |\{b \mid b \le a \wedge y\}|$. Recall that $a \wedge y \in L_j$. Conversely, since L is short, for every $c \in L_j$, there exist $a, y \in L_n$ such that $c = a \wedge y$. Thus $\pi(j, n, s) = |\{b \in L_s \mid b \le c\}|$ is a constant which does not depend on the choice of $c \in L_j$, but only on s and j. Hence we have $\pi(j, n, s) = \nu(s, j)$. □

In what follows, let L be a regular semilattice. We denote the top fiber L_n of L by Ω. In the previous section, we identify the vertex set X of an association scheme with the standard basis of the vector space $\mathbb{R}^{|X|}$, and identify the space of functions on X with $\mathbb{R}^{|X|}$. Here, we make similar identifications. Namely, Ω is identified with the standard basis of $V = \mathbb{R}^{|\Omega|}$, and we identify V with the space of functions on Ω. Let $\chi \in V$. For an integer j with $1 \le j \le n$, define a real valued function $\lambda_{j,\chi}$ on L_j as follows. For $z \in L_j$,

$$\lambda_{j,\chi}(z) = \sum_{\substack{x \in \Omega \\ x \ge z}} \chi(x). \tag{4.32}$$

Definition 4.22 (Geometric design). Let $\chi \in V = \mathbb{R}^{|\Omega|}$, $\chi \ne 0$. Let t be a natural number.
(1) If the function $\lambda_{t,\chi}(z)$ takes a constant value λ on L_t, we call χ a *geometric t-design*. The constant λ is called the *index* of a geometric t-design.
(2) If there exists a fixed element x_0 in Ω such that for each j with $0 \le j \le t$, the function $\lambda_{t,\chi}(z)$ takes a constant value $\lambda_{x_0,j}$ on $\{z \in L_t \mid h(z \wedge x_0) = j\}$, we call χ a

geometric relative t-design with respect to x_0. The constant $\lambda_{x_0,j}$ is called the *index* of a geometric relative t-design.

The definition of geometric designs is very natural in the sense that they are closely related to association schemes, as we will see in the following theorems.

To begin with, we explain how to construct an association scheme on the top fiber $\Omega = L_n$ of a regular semilattice L. We define the relation $R_i \subset \Omega \times \Omega$ on Ω as follows:

$$R_i = \{(x,y) \in \Omega \times \Omega \mid h(x \wedge y) = n - i\}, \quad 0 \le i \le n. \tag{4.33}$$

By Proposition 4.18, we have $R_i \neq \emptyset$ $(0 \le i \le n)$. Let $(x,y) \in R_0$. By definition, $h(x \wedge y) = n$. Since $x \wedge y \le x$, the length of a maximal chain from $x \wedge y$ to x is $h(x) - h(x \wedge y) = 0$. That is, $x \wedge y = x$. Similarly, we have $x \wedge y = y$, and hence $x = y$. It follows that $R_0 = \{(x,x) \mid x \in \Omega\}$. If $i \neq j$, then $R_i \cap R_j = \emptyset$. We also have $0 \le h(x \wedge y) \le n$ for $x, y \in \Omega$. It turns out that $\Omega \times \Omega = R_0 \cup R_1 \cup \cdots \cup R_n$ is a partition of $\Omega \times \Omega$. Clearly, we have ${}^tR_i = R_i$ $(0 \le i \le n)$. In the following, we show that $(\Omega, \{R_i\}_{0 \le i \le n})$ becomes a symmetric association scheme. The method of the proof is similar to the way we found the first and second eigenmatrices of the Johnson scheme in Chapter 2, Section 2.10.3. We use a similar matrix as the matrix appearing in Lemma 2.93 (Kantor). In this way, the top fiber of a regular semilattice carries the structure of a symmetric association scheme. As will be discussed later in this chapter, it is well known that Hamming schemes and Johnson schemes appear as the top fibers of certain regular semilattices. In general, it is an open problem whether there exists a regular semilattice whose top fiber is not related to the structure of a Q-polynomial scheme.

Now, we prove two lemmas.

Lemma 4.23 (Delsarte ([160], Lemma 1)). *We retain the above definitions and notation. Let $a \in L_r$. The cardinality of the set $\{z \in \Omega \mid z \ge a\}$ does not depend on the choice of a and equals $\pi(r,r,0)$. We denote $\theta(r) = \pi(r,r,0)$.*

Proof. Take $y \in \Omega$ satisfying $y \ge a$. Then we have $a \wedge y = a \in L_r$. We apply condition (3) of a regular semilattice with $j = r$. If we let $s = 0$, the element b in condition (3) must be the least element 0 of L. Therefore, we have $\pi(r,r,0) = |\{z \in \Omega \mid a \le z\}|$. This proves the lemma. □

Lemma 4.24 (Delsarte ([160], Lemma 2)). *Fix $y \in \Omega$ and $u \in L_s$ with $u \le y$. Then the number of $z \in L_r$ such that $u \wedge z \in L_j$ and $z \le y$ does not depend on the choice of y, u, but only on j, r, s. The number is denoted by $\psi(j,r,s)$.*

Proof. Let $h = \min\{r,s\}$. Fix k with $0 \le k \le h$. We count the cardinality of the set $\{(x,z) \in L_k \times L_r \mid x \le u,\ x \le z \le y\}$ in two ways. First, fix $x \in L_k$. Then we get

$$\begin{aligned}&\{(x,z) \in L_k \times L_r \mid x \le u,\ x \le z \le y\}\\ &= \bigcup_{\substack{x \in L_k\\ x \le u}} \{z \in L_r \mid x \le z \le y\} = \mu(k,r)\nu(k,s).\end{aligned} \tag{4.34}$$

Next, choose $z \in L_r$ first. Let $u \wedge z \in L_j$. Then j takes a value in $\{0, 1, \ldots, h\}$. Since $x \leq u$ and $x \leq z$, it suffices to count the number of x such that $x \leq u \wedge z$. Namely,

$$\begin{aligned}
&\{(x, z) \in L_k \times L_r \mid x \leq u,\ x \leq z \leq y\} \\
&= \sum_{j=0}^{h} \bigcup_{\substack{z \in L_r \\ u \wedge z \in L_j \\ z \leq y}} \{x \in X_k \mid x \leq u \wedge z\} \\
&= \sum_{j=0}^{h} \big|\{z \in L_r \mid u \wedge z \in L_j,\ z \leq y\}\big| \nu(k, j) \\
&= \sum_{j=0}^{h} \psi(j, r, s)\nu(k, j). \qquad (4.35)
\end{aligned}$$

Thus, we obtain a system of $h+1$ linear equations in $h+1$ unknowns $\psi(j, r, s)$ $(0 \leq j \leq h)$ as follows:

$$\sum_{j=0}^{h} \psi(j, r, s)\nu(k, j) = \mu(k, r)\nu(k, s) \qquad (0 \leq k \leq h). \tag{4.36}$$

Since $\nu(j, j) = 1$ $(0 \leq j \leq h)$ (Remark 4.20) and $\nu(k, j) = 0$ for all $0 \leq j < k \leq h$, the coefficient matrix of the above system of linear equations is non-singular. Therefore, the number $\psi(j, r, s)$ depends only on j, r, s. □

The following proposition immediately follows from (4.36).

Proposition 4.25. *Let $(\nu'(r, s))_{\substack{0 \leq r \leq n \\ 0 \leq s \leq n}}$ be the inverse of the matrix $(\nu(r, s))_{\substack{0 \leq r \leq n \\ 0 \leq s \leq n}}$. Then $\psi(j, r, s)$ is given by the following formula:*

$$\psi(j, r, s) = \sum_{i=0}^{n} \nu(i, s)\nu'(j, i)\mu(i, r). \tag{4.37}$$

We denote the adjacency matrix of the relation R_i on Ω by A_i $(0 \leq i \leq n)$. It is easy to show that $A_0 = I$ (= the identity matrix), and $A_0 + A_1 + \cdots + A_n = J$ (= the all 1's matrix). Similarly to the case of the Johnson scheme, we define the matrices $C_0, C_1, \ldots, C_n$ by

$$C_i = \sum_{k=0}^{n} \nu(i, k)A_{n-k} = \sum_{k=i}^{n} \nu(i, k)A_{n-k}. \tag{4.38}$$

Since $\nu(0, 0) = \nu(0, 1) = \cdots = \nu(0, n) = 1$, we have

$$C_0 = A_0 + \cdots + A_n. \tag{4.39}$$

Lemma 4.26. *The vector space spanned by $C_0, C_1, \ldots, C_n$ coincides with the vector space spanned by $A_0, A_1, \ldots, A_n$.*

Proof. Since the transition matrix $(\nu(i,k))_{\substack{0\le k\le n\\0\le i\le n}}$ is a triangular matrix whose diagonal entries are all 1's, it is clear. □

Theorem 4.27. *The pair $(\Omega, \{R_i\}_{0\le i\le n})$ constructed from the top fiber Ω of a regular semilattice L is a symmetric association scheme.*

By definition, $A_0, A_1, \ldots, A_n$ are symmetric matrices. In order to prove Theorem 4.27, it suffices to show that the vector space spanned by $A_0, A_1, \ldots, A_n$ is closed under the ordinary matrix multiplication. We use the basis $C_0, C_1, \ldots, C_n$. To compute C_rC_s, we use a matrix M_i, which is similar to the matrix used by Kantor. Let M_i be the matrix indexed by $L_i \times \Omega$ whose (x,y)-entry is defined by

$$M_i(x,y) = \begin{cases} 1, & \text{if } x \le y, \\ 0, & \text{otherwise.} \end{cases} \tag{4.40}$$

Lemma 4.28. *We have $C_i = {}^tM_iM_i$ for $i = 0, 1, \ldots, n$.*

Proof. Let $(x,y) \in R_j$, i. e., $x \wedge y \in L_{n-j}$. Then we have

$$\begin{aligned}({}^tM_iM_i)(x,y) &= \sum_{z\in L_i} M_i(z,x)M_i(z,y) = \big|\{z \in L_i \mid z \le x,\ z \le y\}\big| \\ &= \big|\{z \in L_i \mid z \le x \wedge y\}\big| = \nu(i, n-j).\end{aligned}$$

Therefore we have

$${}^tM_iM_i = \sum_{j=0}^{n} \nu(i, n-j)A_j = \sum_{k=0}^{n} \nu(i,k)A_{n-k} = C_i. \tag{4.41}$$

□

Proof of Theorem 4.27. It suffices to show the following:

$$C_rC_s = \sum_{k=0}^{n}\left(\sum_{j=0}^{n} \psi(j, r, n-k)\pi(j, r, s)\right)A_k. \tag{4.42}$$

Let $(x,y) \in R_k$, i. e., $x \wedge y \in L_{n-k}$. Then we have

$$\begin{aligned}(C_rC_s)(x,y) &= ({}^tM_rM_r{}^tM_sM_s)(x,y) = \sum_{z\in\Omega} ({}^tM_rM_r)(x,z)({}^tM_sM_s)(z,y) \\ &= \sum_{z\in\Omega}\Big(\sum_{a\in L_r} M_r(a,x)M_r(a,z)\Big)\Big(\sum_{b\in L_s} M_s(b,z)M_s(b,y)\Big) \\ &= \sum_{z\in\Omega} \big|\{a \in L_r \mid a \le x \wedge z\}\big|\big|\{b \in L_s \mid b \le z \wedge y\}\big| \\ &= \big|\{(a,b,z) \in L_r \times L_s \times \Omega \mid a \le x \wedge z,\ b \le y \wedge z\}\big|.\end{aligned} \tag{4.43}$$

First, count the number of possible a's. For each j, let $a \wedge y \in L_j$. Then the number of such a is $|\{a \in L_r \mid a \wedge y \in L_j,\ a \leq x\}|$. Assume $a \in L_r$, $a \wedge y \in L_j$, and $a \leq x$. Let $c = x \wedge y$. Since $\wedge$ is associative and $a \leq x$, we have $a \wedge (x \wedge y) = (a \wedge x) \wedge y = a \wedge y$. So we have $a \wedge c = a \wedge y \in L_j$. Therefore, in Lemma 4.24, by substituting $n-k$ for s and counting the number of c, x, a instead of u, y, z, respectively, we obtain $|\{a \in L_r \mid a \wedge c \in L_j,\ a \leq x\}| = \psi(j, r, n-k)$. Next, fix $a \in L_r$ such that $y \in \Omega$, $a \wedge y \in L_j$, and $a \leq x$. Then the number of pairs (b, z) such that $a \leq z$, $b \leq y \wedge z$ is equal to $\pi(j, r, s)$. Thus we obtain (4.42). □

Let $\mathfrak{A}$ be the Bose–Mesner algebra of the association scheme $(\Omega, \{R_i\}_{0 \leq i \leq n})$ obtained from a regular semilattice L. Let $E_0, E_1, \ldots, E_n$ be the basis of primitive idempotents of $\mathfrak{A}$, where $E_0 = \frac{1}{|\Omega|}J$. The Bose–Mesner algebra $\mathfrak{A}$ is commutative since $(\Omega, \{R_i\}_{0 \leq i \leq n})$ is symmetric. Let $(\nu'(r,s))_{\substack{0 \leq r \leq n \\ 0 \leq s \leq n}}$ be the inverse of the matrix $(\nu(r,s))_{\substack{0 \leq r \leq n \\ 0 \leq s \leq n}}$. Since $(\nu(r,s))_{\substack{0 \leq r \leq n \\ 0 \leq s \leq n}}$ is a non-singular integral upper triangular matrix whose diagonal entries are all 1's, so is $(\nu'(r,s))_{\substack{0 \leq r \leq n \\ 0 \leq s \leq n}}$. Namely, $\nu'(r,s)$ are all integers. By (4.42) and (4.38), we have

$$\begin{aligned} C_r C_s &= \sum_{k=0}^{n} \sum_{j=0}^{n} \sum_{i=0}^{n} \nu(i, n-k) \nu'(j,i) \mu(i,r) \pi(j,r,s) A_k \\ &= \sum_{j=0}^{n} \sum_{i=0}^{n} \nu'(j,i) \mu(i,r) \pi(j,r,s) \sum_{k=0}^{n} \nu(i, n-k) A_k \\ &= \sum_{j=0}^{n} \sum_{i=0}^{n} \nu'(j,i) \mu(i,r) \pi(j,r,s) C_i \\ &= \sum_{i=0}^{r} \mu(i,r) \left(\sum_{j=0}^{n} \nu'(j,i) \pi(j,r,s) \right) C_i. \end{aligned} \tag{4.44}$$

Therefore, $C_r C_s$ is a linear combination of $C_0, C_1, \ldots, C_r$. Similarly, $C_s C_r$ is also a linear combination of $C_0, C_1, \ldots, C_s$. Moreover, since $\mathfrak{A}$ is commutative, $C_r C_s = C_s C_r$ is a linear combination of $C_0, C_1, \ldots, C_{\min\{r,s\}}$. Hence, if we let $\mathfrak{A}_r$ be the subspace of $\mathfrak{A}$ spanned by $C_0, C_1, \ldots, C_r$, then $\mathfrak{A}_r$ is an ideal of $\mathfrak{A}$. By (4.39), it turns out that $\mathfrak{A}_0 = \langle C_0 \rangle = \langle E_0 \rangle$. So by a suitable rearrangement of the ordering of the primitive idempotents of $\mathfrak{A}$, we have $\mathfrak{A}_r = \langle C_0, C_1, \ldots, C_r \rangle = \langle E_0, E_1, \ldots, E_r \rangle$. We denote

$$C_r = \sum_{i=0}^{r} \rho(i,r) E_i. \tag{4.45}$$

If $r \leq s$, we have

$$C_r C_s = \sum_{i=0}^{r} \rho(i,r) E_i \sum_{j=0}^{s} \rho(j,s) E_j \equiv \rho(r,r) \rho(r,s) E_r \pmod{\mathfrak{A}_{r-1}}. \tag{4.46}$$

On the other hand, by (4.44), we have

$$C_r C_s \equiv \sum_{j=0}^{r} \nu'(j,r) \pi(j,r,s) \rho(r,r) E_r \pmod{\mathfrak{A}_{r-1}}. \tag{4.47}$$

Thus we obtain

$$\rho(r,s) = \sum_{j=0}^{r} \nu'(j,r)\pi(j,r,s). \tag{4.48}$$

Then we get the following theorem.

Theorem 4.29. *Each entry of the first eigenmatrix P of $(\Omega, \{R_i\}_{0\le i\le n})$ is a rational integer.*

Proof. By (4.38), (4.45), and (4.48), we have

$$\begin{aligned} A_i &= \sum_{s=0}^{n} \nu'(n-i,s)C_s = \sum_{s=0}^{n} \nu'(n-i,s) \sum_{j=0}^{n} \rho(j,s)E_j \\ &= \sum_{j=0}^{n}\sum_{s=0}^{n} \nu'(n-i,s) \sum_{k=0}^{n} \nu'(k,j)\pi(k,j,s)E_j. \end{aligned} \tag{4.49}$$

Therefore, we have

$$P_i(j) = \sum_{s=0}^{n} \nu'(n-i,s) \sum_{k=0}^{n} \nu'(k,j)\pi(k,j,s). \tag{4.50}$$

These are all rational integers. □

Delsarte [160] did not mention whether the association scheme obtained from a regular semilattice becomes a Q-polynomial scheme. We tried to show that it becomes Q-polynomial with respect to the ordering defined by (4.45), but it does not seem easy. Probably, there may exist a regular semilattice such that the association scheme obtained from it is not Q-polynomial. In the following, for the case $n = 3$, we find a necessary condition for the association scheme obtained from a regular semilattice to be Q-polynomial.

Proposition 4.30. *Let L be a regular semilattice whose top fiber is L_3. Let $(\Omega, \{R\}_{0\le i\le 3})$ be the association scheme obtained from L. Then $(\Omega, \{R\}_{0\le i\le 3})$ is Q-polynomial with respect to the ordering of the primitive idempotents defined by (4.45) if and only if the following (1), (2) hold:*
(1) $\nu(1,3)^2 - \nu(1,3) - \nu(2,3)(\nu(1,2)^2 - \nu(1,2)) = 0$;
(2) $\nu(1,2), \nu(1,3) \ge 2$ *and* $\nu(1,3) \ne \nu(1,2)$.

Proof. By the condition of a regular semilattice, we have $\nu(0,0) = \nu(0,1) = \nu(0,2) = \nu(0,3) = 1$. By (4.39) and (4.45), we have $A_0 + A_1 + A_2 + A_3 = C_0 = \rho(0,0)E_0$, and so $\rho(0,0) = |\Omega|$. By (4.39) and (4.45), we get

$$C_3 = A_0 = \rho(0,3)E_0 + \rho(1,3)E_2 + \rho(2,3)E_2 + \rho(3,3)E_3. \tag{4.51}$$

So we have $\rho(0,3) = \rho(1,3) = \rho(2,3) = \rho(3,3) = 1$. By plugging them in (4.45) and using (4.39), we can express $E_1 \circ E_1$ as a linear combination of E_0, E_1, E_2, E_3, where the coefficient of E_3 is

$$\frac{1}{\rho(1,1)^2}(\nu(1,3)^2 - \nu(1,3) - (\nu(1,2)^2 - \nu(1,2))\nu(2,3)).$$

Hence Proposition 4.30 (1) follows. By (1), if $\nu(1,2) = 1$, then $\nu(1,3) = 1$. Then the assumption that L is short does not hold. Therefore, $\nu(1,2), \nu(1,3) \geq 2$ and $\nu(2,3) = \frac{\nu(1,3)(\nu(1,3)-1)}{\nu(1,2)(\nu(1,2)-1)}$. Next, if we express $E_1 \circ E_1 \circ E_1$ as a linear combination of E_0, E_1, E_2, E_3, then the coefficient of E_3 is

$$\frac{1}{\rho(1,1)^3}(\nu(1,3) - 1)(\nu(1,3) - \nu(1,2))\nu(1,3).$$

The above coefficient is non-zero if and only if E_3 is a polynomial of degree 3 in E_1 with respect to the Hadamard product. □

Research problems

(1) The Johnson scheme $J(v, 3)$ satisfies the conditions of Proposition 4.30. Find other examples of regular semilattices with $n = 3$ which satisfy the conditions of Proposition 4.30.
(2) Find a regular semilattice with $n = 3$ which does not satisfy the conditions of Proposition 4.30, if it exists.

In the rest of this section, we assume the association scheme obtained from a regular semilattice is Q-polynomial with respect to the ordering of the primitive idempotents defined by (4.45), and we show that the concept of t-designs and relative t-designs in a regular semilattice is identical to that of t-designs and relative t-designs in the corresponding Q-polynomial scheme. Firstly, we introduce a basic property of geometric t-designs.

Lemma 4.31 (Delsarte ([160], Lemma 12)). *Let Ω be the top fiber of a regular semilattice L and let t be a natural number with $t \leq n$. If $C_t\chi = \mathbf{0}$ holds for $\chi \in \mathbb{R}^{|\Omega|}$, then $C_i\chi = \mathbf{0}$ holds for all i with $1 \leq i \leq t$. Here, C_i is the matrix defined by (4.38).*

Proof. By the assumption of the lemma and by Lemma 4.28, we have ${}^t\chi\,{}^tM_tM_t\chi = 0$, which implies $M_t\chi = \mathbf{0}$. Fix $a \in L_i$, and compute $\sum_{\substack{x\in\Omega \\ a\leq z\leq x, z\in L_t}} \chi(x)$ in two ways. Then we have

$$\begin{aligned}\sum_{\substack{x\in\Omega \\ a\leq z\leq x, z\in L_t}} \chi(x) &= \sum_{a\leq x, x\in\Omega} \sum_{\substack{a\leq z\leq x \\ z\in L_t}} \chi(x) = \mu(i,t) \sum_{a\leq x, x\in\Omega} \chi(x) \\ &= \mu(i,t)(M_i\chi)(a).\end{aligned} \tag{4.52}$$

On the other hand, we have

$$\begin{aligned}\sum_{\substack{x\in\Omega\\ a\le z\le x, z\in L_t}} \chi(x) &= \sum_{a\le z, z\in L_t} \sum_{z\le x} \chi(x)\\ &= \sum_{a\le z, z\in L_t} (M_t\chi)(z) = 0.\end{aligned} \tag{4.53}$$

Since $\mu(i,t) > 0$, by (4.52) and (4.53), we have $M_i\chi = \mathbf{0}$. Therefore, $C_i\chi = \mathbf{0}$ holds for all i $(1 \le i \le t)$. □

We give one more lemma before the proof of Delsarte's theorem on geometric designs.

Lemma 4.32. *We retain the above notation. The following (1) and (2) hold:*
(1) *$\rho(i,r) > 0$ holds for every integer i, r with $0 \le i \le r \le n$;*
(2) *$C_r\mathfrak{A}_i = \mathfrak{A}_i$ holds for every integer i, r with $0 \le i \le r \le n$.*

Proof. (1) By (4.45), $\rho(i,r)$ is an eigenvalue of $C_r = {}^tM_rM_r$. Since C_r is a positive semidefinite symmetric matrix, we have $\rho(i,r) \ge 0$. If there exist i, r such that $\rho(i,r) = 0$ $(0 \le i \le r \le n)$, then $C_rE_i = \mathbf{0}$. Therefore, by Lemma 4.31, we have $C_iE_i = \mathbf{0}$, and $\rho(i,i) = 0$. Then $C_i \in \langle C_0, C_1, \ldots, C_{i-1}\rangle = \langle E_0, E_1, \ldots, E_{i-1}\rangle$, which is a contradiction. Thus (1) holds.
(2) We have $C_r\mathfrak{A}_i \subset \mathfrak{A}_i$. Moreover, since $C_rE_i = \rho(i,r)E_i$ and $\rho(i,r) > 0$, we have $E_i \in C_r\mathfrak{A}_i$. Thus we have $C_r\mathfrak{A}_i = \mathfrak{A}_i$ $(i \le r)$. □

Theorem 4.33 (Delsarte ([160], Theorem 13)). *Let $L_n = \Omega$ be the top fiber of a regular semilattice L. If $\chi \in V = \mathbb{R}^{|\Omega|}$ is a geometric t-design of index λ, then χ is a geometric i-design of index $\frac{\lambda\theta(i)}{\theta(t)}$ for every integer i with $1 \le i \le t$.*

Proof. We use the above notation. By Definition 4.22 (1) and Definition 4.19 (2), for all x, we have

$$\begin{aligned}({}^tM_tM_t\chi)(x) &= \sum_{z\in L_t} M_t(z,x)(M_t\chi)(z)\\ &= \lambda\sum_{z\in L_t} M_t(z,x) = \lambda v(t,n).\end{aligned} \tag{4.54}$$

Hence, by Lemma 4.28, we obtain

$$C_t\chi = \lambda v(t,n)\psi_\Omega, \tag{4.55}$$

where ψ_Ω is the characteristic function of Ω. Moreover, by Lemma 4.23, we have $|\{x \in \Omega \mid z \le x\}| = \theta(t)$ for arbitrary $z \in L_t$. So we obtain

$$\begin{aligned}({}^tM_tM_t\psi_\Omega)(x) &= \sum_{z\in L_t} M_t(z,x) \sum_{y\in\Omega} M_t(z,y)\psi_\Omega(y)\\ &= \theta(t)v(t,n).\end{aligned} \tag{4.56}$$

Thus by Lemma 4.28, we get

$$C_t\psi_\Omega = \theta(t)\nu(t,n)\psi_\Omega. \tag{4.57}$$

By (4.55) and (4.57), for every integer i with $1 \le i \le t$, we have

$$C_i\left(\chi - \frac{\lambda}{\theta(t)}\psi_\Omega\right) = \mathbf{0}. \tag{4.58}$$

Therefore, we have

$$M_i\left(\chi - \frac{\lambda}{\theta(t)}\psi_\Omega\right) = \mathbf{0}, \tag{4.59}$$

which means

$$(M_i\chi)(z) = \frac{\lambda}{\theta(t)}(M_i\psi_\Omega)(z) = \frac{\lambda\theta(i)}{\theta(t)}. \tag{4.60}$$

Hence χ is a geometric i-design of index $\frac{\lambda\theta(i)}{\theta(t)}$. □

Theorem 4.34 (Delsarte ([161], Theorem 9.8)). *Let $\mathfrak{X} = (\Omega, \{R_i\}_{0\le i\le n})$ be the association scheme obtained from the top fiber $\Omega = L_n$ of a regular semilattice L. Assume $\mathfrak{X}$ is Q-polynomial with respect to the ordering of the primitive idempotents defined by (4.45). Let $\chi \in V = \mathbb{R}^{|\Omega|}$ and let t be an integer with $1 \le t \le n$. Then the following (1) and (2) hold:*

(1) *χ is a t-design in $\mathfrak{X}$ if and only if χ is a geometric t-design in Ω;*
(2) *fix $x_0 \in \Omega$; then χ is a relative t-design in $\mathfrak{X}$ with respect to x_0 if and only if χ is a geometric relative t-design in Ω with respect to x_0.*

Proof. (1) First assume that χ is a geometric t-design in Ω of index λ. By (4.58) and (4.45) in the proof of Theorem 4.33, for every integer j with $1 \le j \le t$, the following holds:

$$\sum_{\ell=0}^{j} \rho(j,\ell)E_\ell\left(\chi - \frac{\lambda}{\theta(t)}\psi_\Omega\right) = \mathbf{0}. \tag{4.61}$$

Multiplying both sides of (4.61) by E_j $(1 \le j \le t)$ yields $\rho(j,j)E_j\chi = \mathbf{0}$. Since $\rho(j,j) \ne 0$, we obtain $E_j\chi = \mathbf{0}$ $(1 \le j \le t)$. Namely, by Definition 4.5, χ is a t-design in $\mathfrak{X}$. Conversely, assume χ is a t-design in $\mathfrak{X}$, and follow the above discussion in reverse order. Then we can prove that χ is a geometric t-design in Ω.

(2) First, we introduce a matrix as follows. Let $M_{k,j}$ be the matrix indexed by $L_k \times \Omega$ whose (z,x)-entry is defined by

$$M_{k,j}(z,x) = \begin{cases} 1, & \text{if } h(z \wedge x) = j, \\ 0, & \text{otherwise.} \end{cases} \tag{4.62}$$

Then for $(z, x) \in L_k \times \Omega$ with $h(z \wedge x) = j$, the following holds:

$$(M_k{}^tM_iM_i)(z,x) = \sum_{y\in\Omega}\sum_{u\in L_i} M_k(z,y)M_i(u,y)M_i(u,x)$$
$$= \pi(j,k,i). \tag{4.63}$$

Therefore we have

$$M_kC_i = \sum_{j=0}^{n} \pi(j,k,i)M_{k,j}. \tag{4.64}$$

Next, we consider the dimensions. We have the following:

$$k+1 \geq \dim(\langle M_{k,i} \mid 0 \leq i \leq k\rangle) \geq \dim(\langle M_kC_i \mid 0 \leq i \leq k\rangle)$$
$$\geq \dim(\langle {}^tM_kM_kC_i \mid 0 \leq i \leq k\rangle) = \dim(\langle C_kC_i \mid 0 \leq i \leq k\rangle). \tag{4.65}$$

By Lemma 4.32, we have $\dim(C_k\mathfrak{A}_k) = \dim(\mathfrak{A}_k) = k+1$, which implies equality holds in (4.65). Therefore, we have

$$\langle M_{k,i} \mid 0 \leq i \leq k\rangle = \langle M_kC_i \mid 0 \leq i \leq k\rangle, \tag{4.66}$$
$$\langle {}^tM_kM_{k,i} \mid 0 \leq i \leq k\rangle = \langle {}^tM_kM_kC_i \mid 0 \leq i \leq k\rangle = \mathfrak{A}_k. \tag{4.67}$$

Thus, the matrix $(\pi(j,k,i))_{\substack{0\leq j\leq k\\ 0\leq i\leq k}}$ of size $k+1$ is non-singular. Now we assume χ is a geometric relative t-design in Ω with respect to x_0 of index $\lambda_{x_0,j}$. Then, for $z \in L_t$ with $h(z \wedge x_0) = j$, we have $(M_t\chi)(z) = \sum_{x\in\Omega} M_t(z,x)\chi(x) = \lambda_{x_0,j}$. So we obtain

$$M_t\chi = \sum_{j=0}^{t} \lambda_{x_0,j}M_{t,j}\psi_{x_0}. \tag{4.68}$$

Hence, by (4.67) and (4.68), we get

$$C_t\chi = {}^tM_tM_t\chi = \sum_{j=0}^{t} \lambda_{x_0,j}{}^tM_tM_{t,j}\psi_{x_0} \in \mathfrak{A}_t\psi_{x_0}. \tag{4.69}$$

Therefore, for every j with $0 \leq j \leq t$, we have

$$\rho(j,t)E_j\chi = C_tE_j\chi \in E_j\mathfrak{A}_t\psi_{x_0} = \langle E_j\psi_{x_0}\rangle. \tag{4.70}$$

Thus $E_j\chi$ and $E_j\psi_{x_0}$ are linearly dependent. That is, χ is a relative t-design in the Q-polynomial scheme $\mathfrak{X}$ with respect to x_0.

Conversely, suppose χ is a relative t-design in a Q-polynomial scheme $\mathfrak{X}$ with respect to x_0. Following the discussion above in reverse order, by $\langle E_j\chi\rangle \in \langle E_j\psi_{x_0}\rangle$ $(0 \leq$

$j \le t$) and (4.67), we have ${}^tM_tM_t\chi = C_t\chi \in \mathfrak{A}_t\psi_{x_0} = \langle {}^tM_tM_{t,i} \mid 0 \le i \le t\rangle\psi_{x_0}$. Therefore, there exists a constant $\lambda_{x_0,j}$ $(0 \le j \le t)$ such that

$$ {}^tM_tM_t\chi = \sum_{j=0}^{t} \lambda_{x_0,j}\, {}^tM_tM_{t,j}\psi_{x_0}. \tag{4.71}$$

This implies

$$(M_t\,{}^tM_t)(M_t\chi) = (M_t\,{}^tM_t)\left(\sum_{j=0}^{t} \lambda_{x_0,j}M_{t,j}\psi_{x_0}\right). \tag{4.72}$$

Hence we obtain

$$M_t\chi = \sum_{j=0}^{t} \lambda_{x_0,j}M_{t,j}\psi_{x_0}. \tag{4.73}$$

Therefore χ is a geometric relative t-design with respect to x_0 of index $\lambda_{x_0,j}$ $(0 \le j \le t)$. □

Corollary 4.35. *If χ is a geometric relative t-design with respect to $x_0 \in \Omega$, it is a geometric i-design with respect to $x_0 \in \Omega$ for every i with $1 \le i \le t$.*

Remark 4.36. For the Hamming scheme $H(n,q)$ and the Johnson scheme $J(v,d)$, there exist regular semilattices such that $H(n,q)$ and $J(v,d)$ become the association schemes obtained from their top fibers. In this sense, Theorem 4.34 is a generalization of Theorem 3.20. We give two examples of regular semilattices.

Example 4.37. Let F be a q-element set, where $q \ge 2$. Let "$*$" be a new symbol for an element which does not belong to F. Let L be the set of words of length n over $F \cup \{*\}$. We consider a partial order in L as follows. For $x = (x_1, x_2, \ldots, x_n)$, $y = (y_1, y_2, \ldots, y_n) \in L$, we define $x \le y$ if the following holds:

$$\text{For every } i\ (1 \le i \le n), \text{ we have } x_i = * \text{ or } x_i = y_i. \tag{4.74}$$

It is known that L becomes a regular semilattice with respect to the above partial order. The regular semilattice L is called the Hamming lattice, whose height function is defined by $h(x) = |\{i \mid x_i \ne *\}|$. The Hamming scheme $H(n,q)$ arises from the top fiber L_n of L.

Example 4.38. Let Ω be a v-element set. Let $d \le \frac{v}{2}$ and $L = \{x \subset \Omega \mid |x| \le d\}$. For $x, y \in L$, we define $x \le y$ if $x \subseteq y$. With this partial order, L becomes a regular semilattice called the truncated Boolean lattice, whose height function is given by $h(x) = |x|$. The Johnson scheme $J(v,d)$ arises from the top fiber L_d of L.

Remark 4.39. The geometric interpretation of t-designs in regular semilattices was started by Delsarte [160]. It seems that he came up with this concept when he considered the geometric interpretation of t-designs in various Q-polynomial schemes.

Munemasa [355] and Stanton [438] also gave geometric characterizations of t-designs in some Q-polynomial schemes (see also [356]). They consider posets whose structures are weaker than regular semilattices. Can we give a geometric interpretation of t-designs in Q-polynomial schemes by considering associated semilattices or posets with weaker structures? Which kinds of Q-polynomial schemes carry the structure of regular semilattices? These problems are very interesting for future study. It is one of the important open problems whether there exists a code which produces a t-design with $t \geq 6$ by the Assmus–Mattson theorem.

Many extensions and generalizations of the Assmus–Mattson theorem are known. There are many references (for example, see Tanaka [451]). (The authors recommend Bachoc [8], Tanabe [449], and Janusz [267].)

At present, the work of Tanaka [451] seems to be a culmination of this area. We close the part on the Assmus–Mattson theorem by taking a quick look at his study. (For details, see Tanaka [451].) The following is a reformulation of the Assmus–Mattson theorem for the case of $H(n,2)$ or $H(n,q)$. (This is essentially the same as Theorem 4.1, which was introduced in the beginning of this section.)

Theorem 4.1′. *Let Y be a linear code in the Hamming scheme $H(n,q)$ with vertex set F_q^n. Suppose the following (a) or (b) holds:*
(a) *There exist at most $\delta - t$ positive integers in $\{1, 2, \ldots, n-t\}$ which arise as the weights of codewords in $Y^\perp$.*
(b) *There exist at most $\delta^* - t$ positive integers in $\{1, 2, \ldots, n-t\}$ which arise as the weights of codewords in Y.*

Then the set of the supports of codewords of weight m in Y forms a t-design in the Johnson scheme $J(n,m)$.

Moreover, Tanaka [451] gave an extension of the Assmus–Mattson theorem to general P- and Q-polynomial schemes. Let $\mathfrak{X} = (X, \{R_{0 \leq i \leq n}\})$ be a P- and Q-polynomial scheme of class n. (Here we use n for the class of an association scheme since we consider codes.) Let $\mathfrak{A}$ be the Bose–Mesner algebra of $\mathfrak{X}$. In Chapter 2, Section 2.6, we defined the dual Bose–Mesner algebra $\mathfrak{A}^* = \mathfrak{A}^*(u_0)$ and the Terwilliger algebra $T = T(u_0)$ with respect to a fixed vertex u_0 of X (Definition 2.33 and Definition 2.35). We say $\chi \in V$ is a non-trivial code if $\chi \notin E_0 V$ and $\chi \notin E_0^* V$. In what follows, fix $u_0 \in X$. We denote the characteristic vector of $x \in X$ by $\hat{x}$. As will be explained in Chapter 6, the Terwilliger algebra acts naturally on $V = \mathbb{C}^{|X|}$. In this sense, V is called the standard module of the Terwilliger algebra. Let $A = A_1$, $A^* = A_1^*$. Then we have $\mathfrak{A} = \langle A \rangle$, $\mathfrak{A}^* = \langle A^* \rangle$. Furthermore, the following hold:

$$AV_i^* \subseteq V_{i-1}^* + V_i^* + V_{i+1}^*,$$
$$AV_0^* \subseteq V_0^* + V_1^*, \quad AV_n^* \subseteq V_{n-1}^* + V_n^*,$$

$$A^*V_i \subseteq V_{i-1} + V_i + V_{i+1},$$
$$A^*V_0 \subseteq V_0 + V_1, \quad A^*V_n \subseteq V_{n-1} + V_n.$$

For details, see Chapter 6, Section 6.2. (The above properties lead to the concept of tridiagonal pairs, which will be given in Chapter 6, Section 6.2.) We define the subspaces $U_{i,j}$ $(0 \le i, j \le n)$ of V as follows:

$$U_{i,j} = (V_0^* + V_1^* + \cdots + V_i^*) \cap (V_j + V_{j+1} + \cdots + V_n).$$

Theorem 4.40. *Let $\mathfrak{X} = (X, \{R_{0\le i\le n}\})$ be a P- and Q-polynomial scheme and let $\chi \in V$ be a non-trivial code. For each integer r with $1 \le r \le t$, assume the following (a′) or (b′) holds:*
(a′) $|\{j \mid r \le j \le n-r,\ E_j\chi \ne 0\}| \le \delta - r$;
(b′) $|\{j \mid r \le j \le n-r,\ E_j^*\chi \ne 0\}| \le \delta^* - r$.

Then the following (1) and (2) hold:
(1) *For any element $F \in T$ of the Terwilliger algebra and for any integer i with $1 \le i \le t$, the vector $F\chi$ is orthogonal to $U_{i,n-i} \cap (\mathfrak{A}\hat{u}_0)^\perp$.*
(2) *For any element $F \in T$ of the Terwilliger algebra and for any integer j with $1 \le j \le t$, the vector $F\chi$ is orthogonal to $U_{n-j,j} \cap (\mathfrak{A}\hat{u}_0)^\perp$.*

Tanaka [451] claims that for the case of $H(n,q)$, if (a) in Theorem 4.1′ holds, then (a′) in Theorem 4.40 holds for every r with $1 \le r \le t$, and the same is true for (b) and (b’). He claims, however, the conclusions (1) and (2) in Theorem 4.40 are much stronger than those of Theorem 4.1′. Namely, for any element F of the Terwilliger algebra, $F\chi$ (in particular, any non-zero $E_m^*\chi$ $(1 \le m \le n)$) is orthogonal to $U_{i,n-i} \cap \mathfrak{A}\psi_{u_0}$, $U_{i,n-i} = (V_0^* + \cdots + V_i^*) \cap (V_{n-i} + \cdots + V_n)$ (Example 5.4 in Tanaka [451]). Therefore, he claims that Theorem 4.40 generalizes the original Assmus–Mattson theorem for $H(n,q)$. He also claims that the proof is valid for any codes which are not necessarily linear (by exchanging the role of weights of $Y^\perp$ by the dual distribution of Y).

For relatively larger t, say, at most 10, many examples of t-(v,k,λ) designs are constructed ([306], [330]). If repeated blocks are allowed, the existence is shown under very weak conditions (Wilson [517], Graver and Jurkat [204]). In the 1960s, there was a vague impression that there exists no non-trivial t-(v,k,λ)-design for $t \ge 6$. However, the situation has changed, and now it is popularly believed that there are many and various types of t-designs for large t.

Concerning other Q-polynomial schemes (especially for P- and Q-polynomial schemes) aside from Johnson schemes, for some cases the existence of t-designs is well known, and for other cases, the existence is not known at all. The situation differs depending on each association scheme. Generally speaking, for the cases of self-dual association schemes, e. g., $H(n,q)$ (q is a prime power), and P- and Q-polynomial schemes coming from classical forms, the construction of examples is easy (the dual of a linear code of large minimum distance will be an example) and many examples are known. On the other hand, for cases such as $J_q(v,k)$, the construction is still not easy. For instance, examples of t-designs in $J_q(v,k)$, except for $t = 2$, are known for $t = 3$ only [110].

For arbitrarily large t, the existence of non-trivial t-(v,k,λ) designs for some v,k,λ are shown by Teirlinck [457] in 1987. In his result, $k = t+1$ and v,λ are explicitly determined[1]. The construction is based on induction. For the case of the Hamming scheme $H(n,q)$, if q is a prime power, by considering the n-dimensional vector space over the finite field F_q, it can be proved that there exists a subspace such that the minimum distance of the dual space is arbitrarily large. By using this fact, it is clear that a t-design exists for any t. However, if q is not a prime power, we cannot use the property of vector spaces since there is no finite field of order q. For this reason, there was no rigorous proof while the existence was believed. For other association schemes, e. g., the association schemes constructed from bilinear forms or classical forms, the existence of t-designs for arbitrary t is trivially true. For the q-Johnson scheme $J_q(v,k)$, the existence of t-designs for large t was left as an open problem. In this situation, Kuperberg, Lovett, and Peled [293] (see also [292]) gave a general method to show the existence of t-(v,k,λ) designs and t-designs in $H(n,q)$ (i. e., orthogonal arrays of strength t). This is the existence theorem based on probability theory, which does not give a construction. It is, however, a very strong theorem. Also, Fazeli, Lovett, and Vardy [184] applied the method of [293] and showed the existence of t-designs for arbitrary t in the q-Johnson scheme $J_q(v,k)$. It seems that this implies the existence of t-designs for arbitrary t in many other Q-polynomial schemes. (We expect further progress.)

We have already discussed the geometric interpretation of t-designs associated with regular semilattices. For other classical known P- and Q-polynomial schemes, we can obtain their geometric interpretation by choosing appropriate posets. See Munemasa [355], Stanton [438], etc. However, it is a subtle problem how well we can achieve the geometric interpretation for general Q-polynomial schemes or P- and Q-polynomial schemes. Song considered the case of the P- and Q-polynomial scheme which has the same parameter as $H(n,4)$, that is, the case of the Doob scheme (unpublished, 1980s), but he could not find good posets. Precisely, he showed the non-existence of such posets if the class is small. For various Q-polynomial schemes or P- and Q-polynomial schemes, the problem whether posets are associated with them needs further study.

The main families of P- and Q-polynomial schemes are related to permutation representations of Chevalley groups, which will be discussed in later chapters. Explicit calculations of the spherical functions for these permutation representations (or the character tables of the corresponding association schemes) were given by Stanton [436, 437, 438], Dunkl [174], etc. In these papers, it was first shown that they

1 In passing, let us mention that Ito [251] may be interesting since it gives a t-transitive set (in the symmetric group S_v) from a combinatorial t-design.

are P- and Q-polynomial schemes. For special cases, there are lots of various studies. Through these studies, together with the work by researchers in representation theory of Chevalley groups, the understanding of orthogonal polynomials and spherical functions has been deepened. Following this development, algebraic combinatorics began. It started with the study of P- and Q-polynomial schemes, and developed to the study of (the character tables of) general commutative association schemes. For an exposition of the historical background, see, for example, Bannai [27], Song and Tanaka [433], or Martin and Tanaka [336]. For the development of the study on character tables of commutative association schemes, there is a series of papers by Bannai, Song and others, for example, Bannai, Hao, and Song [57], Bannai, Song, Hao, and Wei [73], etc. A part of the above study was related to the study of parameters and character tables of various association schemes which was developed in China. Here we take a look at the history of algebraic combinatorics in China.

We briefly explain the trends of the study related to algebraic combinatorics. It is said that the first study of association schemes in China was conducted by L. C. Chang and Pao-Lu Hsu in the 1950s. (For details, see the preface to Wang, Huo, and Ma [504] written by Zhe-Xian Wan [= C.-h. Wan].) The study related to group theory, especially classical groups, was started by L.-k. Hua and Zhe-Xian Wan. Hua and Wan's book, *Classical groups* [241], and Zhe-Xian Wan's books, *Geometry of Matrices* [501] and *Geometry of Classical Groups over Finite Fields* [502], were published. In the 1960s, Zhe-Xian Wan and his group studied association schemes related to classical groups, which were advanced studies at the time. Although the research of this trend was succeeded in spite of the interruption due to the Cultural Revolution, communications with researchers in the West were almost completely cut off. Full-scale resumption of exchanges with the West began in the early 1980s. At that time, there were some leading research groups, such as the groups of Zhe-Xian Wan (Chinese Academy of Sciences), Yangxian Wang and Hongzen Wei (Hebei Normal University), and Hao Shen (Shanghai Jiao Tong University). (Bannai had a personal mathematical connection with these groups and first visited China in 1992.) These groups resumed their study actively, but they might have felt that they lagged behind. The feature of these studies is to find the parameters of association schemes related to classical groups or classical finite geometry, which was discussed in Wang, Huo, and Ma [504] (2010) (Chinese edition, 2006). There are some results on association schemes related to finite geometry by younger researchers, such as Kaishun Wang, Jun Guo, Fenggao Li, Suogang Gao, Rongquan Feng, Changli Ma, Jianmin Ma, etc. (Cf. [186, 423, 505, 506, 507, 510, 511, 512] for example.) (The study by a Japanese researcher, Hirotake Kurihara [295, 296], is also related to their studies.) These studies are heavily based on calculations, and it is a future project to develop the theoretical aspects of these studies. These results of calculations are variable as primary sources, and the authors hope that we will be able to make use of them in order to construct a general theory of designs in association schemes related to classical geometry. The above is a too simple overview. We should mention that there are many other research streams of association schemes, such as the study related to design theory, group theory, and representation theory, and the research of the MIT school. This is beyond the scope of this book.

For many of the known P- and Q-polynomial schemes, the non-existence of perfect e-codes and tight t-designs is shown in a unified manner. In particular, if $q \neq \pm 1$, there are no perfect e-codes and tight t-designs except for special cases of $e = 1$ or $t = 2$ (Chihara [135].) For the proof, we use the Lloyd type theorem. Depending on the cases, we can directly prove it by using a combinatorial method. At present, we need properties of each P- and Q-polynomial scheme for the proof ([135]). Namely, even if $q \neq \pm 1$, the non-existence of perfect codes and tight designs in general P- and Q-polynomial schemes is not proved yet by using Askey–Wilson polynomials only. The case $q = \pm 1$ is more subtle. For example, as was seen in Chapter 3, classification of the perfect codes (and tight designs) in Johnson schemes is not completed.

5 Algebraic combinatorics on spheres and general remarks on algebraic combinatorics

In Chapters 2–4, we treated association schemes and their subsets (codes and designs). In this chapter, we first discuss finite subsets (spherical codes and spherical designs) of spheres in Euclidean spaces. We explain that we can study finite subsets of spheres in a very analogous way as we studied the subsets of association schemes (Delsarte's theory) in Chapters 3 and 4. As for the study of finite subsets of spheres, the first two authors already published the monograph *Algebraic Combinatorics on Spheres* (1999) [32], written in Japanese. In the original Japanese version of the present book, we referred to the corresponding part of [32] in addition to the original source. In this English version, we keep this style. The main purpose of the latter part of this section is to state our own overview on how the study of finite subsets of association schemes and/or spheres are generalized and also in which directions we should proceed. We will mention many research problems that are open at the present stage, hoping that they are attacked by the reader.

5.1 Finite subsets on spheres

In this section, we first explain that similar methods to the study of finite subsets of association schemes are used for the study of finite subsets of spheres. The origin of this study can be found in Delsarte, Goethals, and Seidel [163]. This will be described as Delsarte's theory on spheres. See also the survey article [38].

The main purpose of the study that we call algebraic combinatorics on a sphere is to study "good" finite subsets of the sphere. What "good" means can be regarded as a part of the problem. Actually, what "good" is not unique and there are various viewpoints. Among them, roughly speaking, we will first treat the following two viewpoints: the viewpoint from coding theory and the viewpoint from design theory.

5.1.1 Study of finite sets on the sphere from the viewpoint of coding theory

We divide this subsection into parts (a) to (e) as follows.

(a) Let N be a natural number. Among all the N-element subsets of the sphere, find subsets with the property that the minimum value of the non-zero distances (i. e., the minimum distance) is the largest. Then classify such sets. (Such sets are called "optimal" codes.)

This problem is also called the "*Tammes problem*" [447]. This problem, in botany, originated from the problem to study the locations of the pollen grain on a pistil of flower. The classification of optimal codes for the 2-dimensional sphere S^2 (in the 3-dimensional Euclidean space) had been known for $N \leq 12$ and $N = 24$ until relatively recently (for the details, see Ericson and Zinoviev [180]). This problem for $N = 13$ was

https://doi.org/10.1515/9783110630251-005

solved by Musin and Tarasov (2012) [363] and for $N = 14$ also by Musin and Tarasov (2015) [364].

The following problem (b) is in a similar direction to problem (a).

(b) Suppose that a positive real number is given. Among all the subsets of the sphere, what is the largest size of them having the property that the minimum distance is greater than or equal to that given positive real number? Then also classify those subsets with this largest size.

In the unit sphere S^{n-1} in the real Euclidean space $\mathbb{R}^n$, the problem of finding a subset whose Euclidean distance between the distinct points in it are at least 1 (or equivalently the geodesic distance on the sphere is at least $\pi/3$, or the central angle is at least 60 degrees, or the minimum inner product is at most 1/2) is well known as the problem of finding the kissing number $k(n)$.

Kissing number problem

A sphere is given. How many spheres of the same radius can you put around the given sphere so that they touch (kiss) the given sphere and also do not overlap each other?

It is well known and very obvious that $k(2) = 6$, because we can put exactly 6 coins around a given coin. It is not so easy to find $k(3)$. There was a famous dispute between *Newton and Gregory* in 1694. It is said that Newton claimed the answer to be 12, but Gregory claimed it to be 13, although the details of the discussion are not known. We can put 12 spheres around the given sphere so that the spheres touch at the given sphere in the places corresponding to the 12 vertices of the regular icosahedron inscribed in the given sphere. The central angle of the nearest two points among these 12 points is more than 63 degrees apart. So, these 12 outside spheres are separated. Therefore, if we move these spheres to some directions and keep them touching the central sphere, there might exist a space to put another sphere there. The problem whether this is possible or not is called *the 13 spheres problem*, and various people have tried to settle this problem. Now, the answer is known to be $k(3) = 12$. It is said that the first rigorous proof was obtained by Schütte and van der Waerden in 1953 [416]. There were several proofs before, but they are believed to be not quite rigorous. After 1953, at least about 10 proofs have been published. But none of them is very easy, and good proofs remain wanted ([103, 5, 125, 157, 307, 396, 404, 363, 240, 327]). The paper by Maehara [328] is relatively easy, and may be accessible for undergraduate students. Next, we will discuss kissing number problems for higher-dimensional spaces. (Since this topic was described carefully in [32] until the relatively recent development, we hope that the readers refer to the book. See also [270]) Here we also give a quick review. In 1979, $k(8) = 240$ and $k(24) = 196560$ were proved by Odlyzko and Sloane [389] (see also Conway and Sloane [147, Chapter 13]) in the USA and by Levenshtein [312] in the Soviet Union independently. At that time there was not much free communication between the USA and the Soviet Union. It is said that Odlyzko and Sloane utilized a computer, but Levenshtein did all his calculations by hand.

Their proof is obtained applying the linear programming method. For the details, see Odlyzko and Sloane [389] or Conway and Sloane [147, Chapter 13] or [32, Chapter 1]. (This method is called Delsarte's method.) Since then, there had not been many developments on the kissing number $k(n)$ for $n > 3$ and $k \neq 8, 24$ until Musin announced $k(4) = 24$ by circulating the preprint in 2003. This was a big surprise to this research area. His proof was officially published [360]. Soon afterward, he applied his method in the 4-dimensional space to the 3-dimensional case [359], and gave a new proof of $k(3) = 12$ in a very convincing way. Musin's method to prove $k(3) = 12$ and $k(4) = 24$ was to improve Delsarte's method combining with geometric considerations. As already mentioned in the present book, this method by Delsarte is based on the linear programming method. In the meanwhile, with the development of computers, the *semidefinite programming method* has very much developed in applied mathematics generalizing the method of linear programming. Schrijver [414] made it clear that the method of semidefinite programming can be applied to combinatorics to get stronger results. Then, Bachoc and Vallentin [13] applied the method of Schrijver to the kissing number problem, and then gave another proof of $k(3) = 12$ and $k(4) = 24$. They made crucial use of a computer when they applied semidefinite programming, and they used just standard, well-established software packages. So, the correctness of their result is very convincing. The correctness of $k(3) = 12$ is also evidenced by the work of Musin and Tarasov [363], as they gave the complete classification of optimal codes of 13 points. (Here, Musin and Tarasov also used the computer substantially.)

We would like to emphasize that (i) the solution of Kepler's conjecture by Hales [207], (ii) the progress on the sphere packing problems in 8- and 24-dimensional Euclidean spaces by Cohn and Elkies [139] and by Cohn and Kumar [142], and (iii) the determination of the lattice packings of the spheres in these two dimensions by Cohn and Kumar [142] are very important breakthroughs together with the work of Musin [360]. More recently, further breakthroughs of the sphere packing problems in 8- and 24-dimensional Euclidean spaces were obtained by Viazovska [496] and Cohn, Kumar, Miller, Radchenko, and Viazovska [143].

We mention another viewpoint related to the coding theoretical viewpoint, namely, s-distance sets.

(c) A subset $X \subset S^{n-1}$ is called an s-distance set if there are exactly s non-zero distances between 2 elements of X. (Some people say that it is an s-distance set if there are at most s non-zero distances.) Note that the same concept was already discussed for subsets of association schemes. The typical problem here is to find the maximum size of s-distance sets on S^{n-1}.

The paper by Delsarte, Goethals, and Seidel [163] is one of the starting points for the study of s-distance subsets on spheres. Namely, their main result is that if X is an s-distance set on S^{n-1}, then the following holds:

$$|X| \leq \binom{n-1+s}{s} + \binom{n-1+s-1}{s-1}.$$

Although the details will be explained later, the number on the right-hand side is exactly the number appearing in the tight spherical $2s$-designs.

There have been some new developments in the study of s-distance subsets on spheres and of the Euclidean spaces. The most notable ones are due to Musin [361], Nozaki and Shinohara [388], and Musin and Nozaki [362] on the refinement of the evaluation of the bound (under some additional conditions) on 2-distance sets, and generalizations on s-distance sets. In particular, Nozaki's generalization of Larman–Roger–Seidel's theorem to s-distance sets [386] and Kurihara and Nozaki's study on the spherical embedding and their characterization of Q-polynomial association schemes [297] seem to be very important and useful results (see also [298]). See also Barg and Yu [77], Yu [524], and Glazyrin and Yu [195] for more recent results on spherical 2-distance sets. In particular, Glazyrin and Yu [195] solved infinitely many cases of maximum spherical 2-distance sets. See also Barg and Musin [76].

Here is another problem related to the coding theoretical viewpoint.

(d) *Coulomb–Thomson problem*

Let N be a natural number. Among all the N-element subsets $X = \{x_1, x_2, \dots, x_N\}$ of S^{n-1}, find subsets that minimize the following value:

$$\sum_{1\le i<j\le N} \frac{1}{\|x_i - x_j\|}.$$

Also, determine this minimum value as well as the structure of such sets.

It is possible to consider a similar problem for other energy functions, i. e., to find the minimum energy subset $X = \{x_1, x_2, \dots, x_N\}$, the minimum energy value, as well as the structure of such X. In this direction, the concept of *universally optimal codes* is very interesting. We will discuss it later in this book.

(e) *The covering problem*

Let N be a natural number. Among all the N-element subsets $X = \{x_1, x_2, \dots, x_N\}$ of S^{n-1}, find subsets that minimize the following value:

$$\max_{x\in S^{n-1}} \Big\{ \min_{1\le i\le N} (d(x, x_i)) \Big\}.$$

Also, determine this minimum value, as well as the structures of such sets. (Here $d(x, y)$ denotes the distance between x and y.)

For other closely related problems, we refer the reader to Conway and Sloane [147, Chapter 2, Section 1.3 (page 36)].

5.1.2 Design theoretical study of finite subsets on the sphere

We consider the following problem. What are the good finite subsets that approximate the sphere very well as a whole? One very interesting condition is the following concept of *spherical designs*. We consider non-empty finite subsets on the unit sphere $S^{n-1} = \{(x_1, x_2, \dots, x_n) \mid x_1^2 + x_2^2 + \dots + x_n^2 = 1\}$.

Definition 5.1 (Spherical t-designs, Delsarte–Goethals–Seidel [163]). For a finite subset X of the sphere S^{n-1}, if the equality

$$\frac{1}{|S^{n-1}|}\int_{S^{n-1}} f(x)\,d\sigma(x) = \frac{1}{|X|}\sum_{x\in X} f(x)$$

is attained for all polynomials $f(x) = f(x_1.x_2, \dots, x_n)$ of degree at most t, then X is called a *spherical t-design* on S^{n-1}, where $|S^{n-1}|$ denotes the surface area of the sphere S^{n-1}.

As the definition shows, "X is a spherical t-design" means that the value of the surface integral over the sphere (divided by the surface area) of any polynomial of degree at most t can be calculated as the average value of the polynomial at the points of the finite set X. In this sense, we can say that X is a good finite subset that approximates the sphere well. It is obvious from the definition that if X is a spherical t-design, then it is an i-design for any integer i with $1 \le i \le t$. There are many equivalent conditions that X is a spherical t-design on S^{n-1}. In the following we will mention equivalent conditions (i)–(iv). For more details and the proof, the reader is referred to [163, 32].

(i) The equation

$$\sum_{x\in X} \varphi(x) = 0$$

holds for any homogeneous harmonic polynomial $\varphi(x)$ of degree ℓ with $1 \le \ell \le t$.

(ii) The equation

$$\sum_{x\in X} f(x^g) = \sum_{x\in X} f(x)$$

holds for any polynomial $f(x)$ of degree at most t and for any orthogonal transformation $g \in O(d)$.

An alternative, but essentially equal expression of (ii) is stated as (ii′).

(ii′) Any moment of degree at most t of X is invariant under any orthogonal transformation. (Namely, this is the case where we use the monomials $x_1^{i_1}x_2^{i_2}\cdots x_n^{i_n}$ with $0 \le i_1 + i_2 + \cdots + i_n \le t$.)

Furthermore, obviously from (ii) we have the following equivalent condition.

(iii) The equation

$$\frac{1}{|X|}\sum_{x\in X} f(x^g) = \frac{1}{|S^{n-1}|}\int_{S^{n-1}} f(x)\,d\sigma(x)$$

holds for any polynomial $f(x)$ of degree at most t and for any orthogonal transformation $g \in O(d)$.

Let $\mathcal{P}(\mathbb{R}^n)$ be the space of all the polynomials in n variables $x_1, x_2, \dots, x_n$. Let $\mathrm{Hom}_\ell(\mathbb{R}^n)$ be the space of homogeneous polynomials of degree ℓ. Let $\mathcal{H}(\mathbb{R}^n)$ be the subspace of $\mathcal{P}(\mathbb{R}^n)$ consisting of all the harmonic polynomials. If we denote the Laplacian by $\Delta = (\frac{\partial}{\partial x_1})^2 + (\frac{\partial}{\partial x_2})^2 + \cdots + (\frac{\partial}{\partial x_n})^2$, then we have

$$\mathcal{H}(\mathbb{R}^n) = \{f(x) \in \mathcal{P}(\mathbb{R}^n) \mid \Delta f(x) = 0\}.$$

Let $\mathrm{Harm}_\ell(\mathbb{R}^n) = \mathcal{H}(\mathbb{R}^n) \cap \mathrm{Hom}_\ell(\mathbb{R}^n)$ be the subspace consisting of all the homogeneous harmonic polynomials of degree ℓ. Furthermore, when we restrict the domain of definition to the sphere, we denote $\mathcal{P}(S^{n-1})$, $\mathrm{Hom}_\ell(S^{n-1})$, $\mathcal{H}(S^{n-1})$, and $\mathrm{Harm}_\ell(S^{n-1})$, respectively. Now, we define the inner product of two polynomials $f(x)$ and $g(x)$ in $\mathcal{P}(S^{n-1})$ by

$$\langle f, g\rangle = \frac{1}{|S^{n-1}|} \int_{S^{n-1}} f(x)g(x)\, d\sigma(x).$$

Let $h_\ell = \dim(\mathrm{Harm}_\ell(\mathbb{R}^n))$ and let $\varphi_{\ell,1}(x), \varphi_{\ell,2}(x), \ldots, \varphi_{\ell,h_\ell}(x)$ be the orthonormal basis with respect to the inner product on $\mathrm{Harm}_\ell(\mathbb{R}^n)$ given above.

Let us recall Gegenbauer polynomials. We denote the Gegenbauer polynomial of degree ℓ by $G_\ell^{(n)}(x)$. It is known that Gegenbauer polynomials satisfy the following recursive relations:

$$\frac{k_\ell}{k_{\ell+1}} = \frac{\ell+1}{n+2\ell}, \tag{5.1}$$

$$\frac{k_\ell}{k_{\ell+1}} G_{\ell+1}^{(n)}(x) = xG_\ell^{(n)}(x) - \left(1 - \frac{k_{\ell-2}}{k_{\ell-1}}\right) G_{\ell-1}^{(n)}(x), \tag{5.2}$$

where $k_{-1} = 0$, $k_0 = 1$, $G_{-1}^{(n)} \equiv 0$, $G_0^{(n)} \equiv 1$.

Remark 5.2. Gegenbauer polynomials $G_\ell^{(n)}(x)$ ($\ell = 0, 1, \ldots$) are orthogonal polynomials on the interval $[-1, 1]$ with respect to the integral with the weight function $\omega(x) = (1-x^2)^{\frac{n-3}{2}}$. Namely, $\int_{-1}^{1} G_\ell^{(n)}(x)G_m^{(n)}(x)w(x)\, dx = \delta_{\ell,m} \frac{|S^{n-1}|h_\ell}{|S^{n-2}|}$ holds ([163], [32], [445], etc.).

The following facts are well known.

(α) We have $\dim(\mathrm{Harm}_\ell(\mathbb{R}^n)) = \dim(\mathrm{Harm}_\ell(S^{n-1})) = G_\ell^{(n)}(1) = \binom{n-1+\ell}{\ell} - \binom{n-1+\ell-2}{\ell-2}$.

(β) We have $\mathcal{P}(S^{n-1}) = \mathrm{Harm}_0(S^{n-1}) \perp \mathrm{Harm}_1(S^{n-1}) \perp \cdots \perp \mathrm{Harm}_\ell(S^{n-1}) \perp \cdots\cdots$.
(This is the orthogonal decomposition with respect to the above inner product.)

(γ) For any x and y in S^{n-1}, any non-negative integer ℓ, and any orthonormal basis $\varphi_{\ell,1}(x), \varphi_{\ell,2}(x), \ldots, \varphi_{\ell,h_\ell}(x)$ of $\mathrm{Harm}_\ell(\mathbb{R}^n)$, the following "addition formula" holds:

$$\sum_{i=1}^{h_\ell} \varphi_{\ell,i}(x)\varphi_{\ell,i}(y) = G_\ell^{(n)}(x \cdot y). \tag{5.3}$$

Here $x \cdot y$ denotes the standard inner product in $\mathbb{R}^n$.

(δ) For any non-negative integers i and j, we have

$$G_i^{(n)}(x)G_j^{(n)}(x) = \sum_{k=0}^{i+j} q_k(i,j)G_k^{(n)}(x), \quad q_k(i,j) \geq 0. \tag{5.4}$$

Moreover, $q_k(i,j) > 0$ holds if and only if $|i-j| \le k \le i+j$ and $k \equiv i+j \pmod 2$ hold. Here, note that $q_0(i,i) = G_i^{(n)}(1) = h_i$.

To discuss further equivalent conditions for X to be a spherical t-design, we introduce the following notation. We consider the following matrix H_ℓ of size $|X| \times h_\ell$. Here the rows are indexed by the elements in X and the columns are indexed by the elements of a fixed orthonormal basis $\varphi_{\ell,1}(x), \varphi_{\ell,2}(x), \ldots, \varphi_{\ell,h_\ell}(x)$ of $\mathrm{Harm}_\ell(\mathbb{R}^n)$. Namely, the (x,i)-entry is defined by $H_\ell(x,i) = \varphi_{\ell,i}(x)$ for $x \in X$ and $1 \le i \le h_\ell$. Note that H_0 is the column vector of size $|X|$ whose entries are all 1. These matrices are called the characteristic matrices. Then each of the following conditions (iv)–(vii) is equivalent to the condition for X to be a spherical t-design.

(iv) For $\ell = 1, 2, \ldots, t$, ${}^tH_\ell H_0 = 0$.

(v) For any non-negative integers k and ℓ, ${}^tH_k H_\ell = |X|\Delta_{k,\ell}$ holds, with $0 \le k+\ell \le t$. Here we define $\Delta_{\ell,\ell}$ to be the identity matrix and $\Delta_{k,\ell}$[1] to be the zero matrix if $k \ne \ell$.

(vi) We have ${}^tH_e H_e = |X|I$ and ${}^tH_e H_r = 0$, where $e = [\frac{t}{2}]$ and $r = e - (-1)^t$.

(vii) We have $\sum_{(x,y)\in X\times X} G_\ell^{(n)}(x \cdot y) = 0$ for $\ell = 1, 2, \ldots, t$.

The reader is referred to [163, 32] for the proof of these claims.

The next two theorems are also important for the purpose of characterizing spherical t-designs.

Theorem 5.3 (Sidelnikov's inequality, cf. Venkov [503]). *Let X be any finite subset of the sphere S^{n-1}. Then the following inequality holds:*

$$\frac{1}{|X|^2} \sum_{(x,y)\in X\times X} (x \cdot y)^\ell \ge \begin{cases} 0, & \text{for odd integer } \ell, \\ \frac{(\ell-1)!!(n-2)!!}{(n+\ell-2)!!}, & \text{for even integer } \ell. \end{cases} \tag{5.5}$$

Moreover, X is a spherical t-design if and only if the equalities in (5.5) hold for all integers ℓ with $1 \le \ell \le t$.

Theorem 5.4 (Venkov's fundamental equation [493, 494]). *Let X be an antipodal finite set. Then X is a spherical t-design if and only if the equality*

$$\frac{1}{|X|} \sum_{x\in X} (a \cdot x)^{2\ell} = \frac{1 \cdot 3 \cdot \cdots \cdot (2\ell-1)}{n(n+2)\cdots(n+2\ell-2)} (a \cdot a)^\ell \tag{5.6}$$

holds for any integer with $1 \le \ell \le [\frac{t}{2}]$ and for any vector a in $\mathbb{R}^n$. Moreover, it is also equivalent to the condition that (5.6) holds just for $\ell = [\frac{t}{2}]$.

Now, for a finite subset X of S^{n-1} we define

$$A(X) = \{x \cdot y \mid x, y \in X,\ x \ne y\}, \quad A'(X) = \{1\} \cup A(X).$$

1 Exactly speaking, $\Delta_{k,\ell}$ is a matrix of size $h_k \times h_\ell$.

In addition, for each $\alpha \in A'(X)$, let us define the matrix D_α whose rows and columns are indexed by the elements of X. Namely,

$$D_\alpha(x,y) = \begin{cases} 1, & \text{for } x \cdot y = \alpha, \\ 0, & \text{for } x \cdot y \neq \alpha, \end{cases} \tag{5.7}$$

and for each α we define

$$d_\alpha = \sum_{(x,y)\in X\times X} D_\alpha(x,y). \tag{5.8}$$

(Note that $D_1 = I$ and $d_1 = |X|$.)

Lemma 5.5 (Theorem 5.10 in [163], Lemma 4.2.2 in [32]). *Suppose the Gegenbauer expansion of a real coefficient polynomial $F(x)$ is given by*

$$F(x) = \sum_{k=0}^{(\infty)} f_k G_k^{(n)}(x) \quad \textit{(finite sum)}, \tag{5.9}$$

and the following conditions are satisfied:

$$F(\alpha) \geq 0 \quad \textit{for any } \alpha \in [-1,1],\ F(1) > 0,\ f_0 > 0,\ \textit{and } f_k \leq 0 \textit{ for any } k \geq t+1.$$

Then, for a spherical t-design X on S^{n-1}, the following condition holds:

$$|X| \geq \frac{F(1)}{f_0}. \tag{5.10}$$

Moreover, equality holds in (5.10) if and only if $F(\alpha) = 0$ for any $\alpha \in A(X)$ and $f_k {}^tH_kH_0 = 0$ for any integer k with $k > t$.

Proof. By using the addition formula (5.3) and (5.8) we have

$$\begin{aligned} \|{}^tH_kH_0\|^2 &= \sum_{i=1}^{h_k}(({}^tH_kH_0)(i))^2 = \sum_{i=1}^{h_k}\Big(\sum_{x\in X} H_k(x,i)\Big)^2 \\ &= \sum_{i=1}^{h_k}\sum_{x\in X}\varphi_{k,i}(x)\sum_{y\in X}\varphi_{k,i}(y) = \sum_{(x,y)\in X\times X}\sum_{i=1}^{h_k}\varphi_{k,i}(x)\varphi_{k,i}(y) \\ &= \sum_{(x,y)\in X\times X} G_k^{(n)}(x\cdot y) = \sum_{\alpha\in A'(X)} d_\alpha G_k^{(n)}(\alpha). \end{aligned} \tag{5.11}$$

Therefore, we obtain

$$\begin{aligned} \sum_{k=0}^{(\infty)} f_k\|{}^tH_kH_0\|^2 &= \sum_{k=0}^{(\infty)} f_k \sum_{\alpha\in A'(X)} d_\alpha G_k^{(n)}(\alpha) = \sum_{\alpha\in A'(X)} d_\alpha F(\alpha) \\ &= |X|F(1) + \sum_{\alpha\in A(X)} d_\alpha F(\alpha). \end{aligned} \tag{5.12}$$

Moreover, since X is a t-design, by condition (v) in Section 5.1.2, we have

$$f_0\|{}^tH_0H_0\|^2 - |X|F(1) = \sum_{\alpha\in A(X)} d_\alpha F(\alpha) - \sum_{k=t+1}^{(\infty)} f_k\|{}^tH_kH_0\|^2. \tag{5.13}$$

Thus, we get $f_0|X|^2 - |X|F(1) \geq 0$ by the assumption. Also, if equality holds in (5.10), then the right-hand side of (5.13) must be equal to 0. Since $d_\alpha > 0$ for any $\alpha \in A(X)$ and $f_k \leq 0$ for any $k \geq t+1$, we have $F(\alpha) = 0$ for any $\alpha \in A(X)$ and $f_k\|{}^tH_kH_0\|^2 = 0$ for any $k \geq t+1$. □

The existence of spherical t-designs on S^{n-1} for any n and t was proved by Seymour and Zaslavsky [422]. Namely, they proved that for any n and t, and for a sufficiently large positive integer N, there exists a spherical t-design on S^{n-1} whose cardinality is N. However, their proof is an existence proof, and the explicit constructions were not given. Therefore, our interest goes to the problem how small-sized spherical t-designs on S^{n-1} can exist. The next theorem gives a very natural lower bound for the size of a spherical t-design on S^{n-1}.

Theorem 5.6 (Theorem 5.11 in [163], Theorem 4.2.3 in [32]). *Let X be a spherical $2e$-design on S^{n-1}. Then the following inequality holds:*

$$|X| \geq \binom{n-1+e}{e} + \binom{n-1+e-1}{e-1} = R_e(1), \tag{5.14}$$

where $R_e(x) = G_0^{(n)}(x) + G_1^{(n)}(x) + \cdots + G_e^{(n)}(x)$. Moreover, equality holds in (5.14) if and only if the set $A(X)$ is equal to the set of zeros of $R_e(x)$.

Remark 5.7. The polynomials $R_i(x)$ $(i = 0, 1, 2, \ldots)$ appearing in Theorem 5.6 form a system of orthogonal polynomials called the Jacobi polynomials.

Proof of Theorem 5.6. Apply Lemma 5.5 for the function $F(x) = R_e(x)^2$. Then the previous statement in (δ) about the non-negativity of the coefficients of Gegenbauer expansions of the product of two Gegenbauer polynomials is applied. Then $F(x)$ satisfies all the assumptions of Lemma 5.5 for $t = 2e$. Therefore, we have the following:

$$|X| \geq \frac{F(1)}{f_0}.$$

By (5.4) we have

$$F(1) = R_e(1)^2, \quad f_0 = \sum_{i=0}^{e} q_0(i,i) = \sum_{i=0}^{e} G_i^{(n)}(1) = R_e(1).$$

Then we can show, by an easy calculation, that $R_e(1) = \sum_{i=0}^{e} h_i = \binom{n-1+e}{e} + \binom{n-1+e-1}{e-1}$. □

In the case that t is an odd integer, we have the following bound.

Theorem 5.8 (Theorem 5.12 in [163], Theorem 4.2.5 in [32]). *Let X be a spherical $(2e+1)$-design on S^{n-1}. Then we have the following inequality:*

$$|X| \geq 2\binom{n-1+e}{e} = 2C_e(1), \tag{5.15}$$

where $C_e(x) = G_e^{(n)}(x) + G_{e-2}^{(n)}(x) + \cdots + G_{e-2[\frac{e}{2}]}^{(n)}(x)$. Moreover, equality holds in (5.15) if and only if $A(X)$ consists of -1 and the set of all zeros of $C_e(x)$.

For the proof, we refer the reader to [163, 32].

Remark 5.9. The polynomials C_e in Theorem 5.8 are also Jacobi polynomials. If equality holds in Theorem 5.8 (equation (5.15)), then X is shown to be an antipodal set.

Definition 5.10 (Tight spherical designs). If equality holds either in (5.14) of Theorem 5.6 or in (5.15) of Theorem 5.8, then X is called a *tight spherical t-design*. (Note that a tight spherical t-design for odd t is antipodal.)

A tight spherical t-design, if it exists, makes it possible to calculate the integral of any polynomial f of degree at most t as the average value of f on finitely many points with the smallest possible number, so it is a very good finite subset that approximates the sphere most efficiently. In addition, tight spherical t-designs have very good combinatorial structures, as is explained below. In Section 5.1.1 (c), we mentioned s-distance sets in view of coding theory. Delsarte's theory, which deals with codes and designs in Q-polynomial association schemes in a unified way, has similarly developed in the theory of codes and designs on the sphere, which is observed in Theorem 5.6 and Theorem 5.8. The polynomials $R_e(x)$ and $C_e(x)$ appearing in Theorem 5.6 and Theorem 5.8 are orthogonal polynomials. Moreover, the upper bounds on the sizes of e-distance sets (on the sphere) and antipodal e-distance sets (on the sphere) are given by $R_e(1)$ and $2C_e(1)$, respectively. We have already seen in Theorem 3.16 and Theorem 3.19 that in the case of Q-polynomial association schemes, the orthogonal polynomials $v_i^*(x)$ $(1 \leq i \leq d)$ that define the second eigenmatrix Q play a very similar role as Gegenbauer polynomials play in the case of spheres.

Now, we want to study the structures of tight spherical designs and spherical designs close to tight designs. For a finite subset X on S^{n-1}, we define parameters $(n, |X|, s, t)$. Here, n is the dimension of the space $S^{n-1} \subset \mathbb{R}^n$, $s = |A(X)|$, and t is the largest integer for X to be a t-design. We call s the *strength* of X, as we used that word in the same way for subsets of Q-polynomial association schemes. (So, it means X is not a $(t+1)$-design.) For a subset $A = [-1, \alpha]$ of $[-1, 1)$, we say that X is an A-code if $A(X) \subset A$. We may also call X an α-code. If a polynomial $F(x)$ satisfies $F(x) = 0$ for any x in $A \subset [-1, 1)$, then $F(x)$ is called an *annihilator* of A. Moreover, if $F(1) = 1$, then $F(x)$ is called a *normalized annihilator*.

Lemma 5.11. *Let $A \subset [-1, 1)$. Let $F(x)$ be a polynomial and let $F(x) = \sum_{k=0}^{(\infty)} f_k G_k^{(n)}(x)$ (finite sum) be the Gegenbauer expansion. Suppose that $F(\alpha) \leq 0$ for any $\alpha \in A$, $f_0 > 0$,* *and $f_k \geq 0$ for any natural number k. If there exists an A-code X on S^{n-1}, the following holds:*

$$|X| \leq \frac{F(1)}{f_0}. \tag{5.16}$$

Moreover, equality holds in (5.16) if and only if $F(\alpha) = 0$ for any $\alpha \in A(X)$ and $f_k \|{}^t H_k H_0\| = 0$ for any integer $k \geq 1$. Note that $A(X) = \{x \cdot y \mid x, y \in X, x \neq y\}$.

Proof. By using formula (5.12) in the proof of Lemma 5.5, we get

$$\sum_{k=0}^{(\infty)} f_k \|{}^t H_k H_0\|^2 = |X|F(1) + \sum_{\alpha \in A(X)} d_\alpha F(\alpha), \tag{5.17}$$

and from the assumption, we have

$$|X|F(1) - f_0 \|{}^t H_0 H_0\|^2 = \sum_{k=1}^{(\infty)} f_k \|{}^t H_k H_0\|^2 - \sum_{\alpha \in A(X)} d_\alpha F(\alpha) \geq 0. \tag{5.18}$$

So, we get (5.16). Moreover, equality holds in (5.16) if and only if the right-hand side of (5.18) becomes 0. Since each term in the right-hand side is a non-negative real number, we get the conclusion of Lemma 5.11. □

Let A be a finite subset of $[-1, 1)$ with $|A| = s$. Let us define

$$F_A(x) = \prod_{\alpha \in A} \frac{x - \alpha}{1 - \alpha}. \tag{5.19}$$

Then $F_A(x)$ is the normalized annihilator polynomial of A of degree s.

Theorem 5.12 (Theorem 6.4 in [163], Theorem 4.3.4 in [32]). *We use the notation above. Let X be an A-code on S^{n-1}, and let*

$$F_A(x) = \sum_{k=0}^{s} f_k G_k^{(n)}(x)$$

be the Gegenbauer expansion of $F_A(x)$. Then the following (1) and (2) hold:

(1) *If $f_i \geq 0$ holds for any integer i with $0 \leq i \leq s$, then $f_i \leq \frac{1}{|X|}$ for any i with $0 \leq i \leq s$.*
(2) *If there exists an integer j with $0 \leq j \leq s$ satisfying $f_j = \frac{1}{|X|}$, then for any A-code Y, we have $|Y| \leq |X|$.*

Proof. (1) We fix j with $0 \leq j \leq s$. We set $Z(x) = \frac{G_j^{(n)}(x)}{G_j^{(n)}(1)} F_A(x)$. Then $Z(x)$ is a normalized annihilator of A. Let $Z(x) = \sum_{k=0}^{s+j} g_k G_k^{(n)}(x)$ be the Gegenbauer expansion. Then we can see that $g_0 = f_j$ and that $g_k \geq 0$ for all integer $k \geq 1$, by (iii) (in Section 5.1.2) and also by using the property (5.4) of Gegenbauer polynomials given in Remark 5.2 (δ) and $f_i \geq 0$ for $0 \leq i \leq s$. Therefore, by applying formula (5.12) in the proof of Lemma 5.5 to the function $Z(x)$, and by the fact that X is an A-code, we have

$$\sum_{k=0}^{s+j} g_k \|{}^t H_k H_0\|^2 = |X| Z(1) = |X|. \tag{5.20}$$

Therefore, we have

$$|X| - g_0 \|{}^t H_0 H_0\|^2 = \sum_{k=1}^{s+j} g_k \|{}^t H_k H_0\|^2 \geq 0. \tag{5.21}$$

So, we get $f_j = g_0 \leq \frac{1}{|X|}$.

(2) Assume $f_j = \frac{1}{|X|}$ holds for an integer j, satisfying $0 \leq j \leq s$. Apply Lemma 5.11 for the polynomial $Z(x)$ given in the proof of (1) with this j. Then for any $\alpha \in A$ we obtain $Z(\alpha) = 0$, $g_0 = f_j = \frac{1}{|X|} > 0$, $g_k \geq 0$ for $k \geq 1$. Hence we obtain $|Y| \leq \frac{Z(1)}{g_0} = |X|$. □

Theorem 5.13 (Theorem 6.5 in [163], Theorem 4.3.5 in [32]). *We keep the notation used in Theorem 5.12. For an A-code X, we have the following:*

(1) *If X is a t-design and $t \geq s$, then we have $f_0 = f_1 = \cdots = f_{t-s} = \frac{1}{|X|}$.*

(2) *Suppose that for an integer r with $0 \leq r \leq s$ we have $f_0 = f_1 = \cdots = f_r = \frac{1}{|X|}$, and for all integers i with $r < i \leq s$ we have $f_i > 0$. Then X is an $(r + s)$-design.*

Proof. (1) As in the proof of Theorem 5.12, we fix an integer j with $0 \leq j \leq t - s$. Then we consider the normalized annihilator $Z(x) = \frac{G_j^{(n)}(x)}{G_j^{(n)}(1)} F_A(x)$ of degree $s + j$. Since X is a t-design, we have $\|{}^t H_k H_0\|^2 = 0$ for any k with $1 \leq k \leq s + j$. Therefore, by (5.20), we get $s_j = g_0 = \frac{1}{|X|}$.

(2) We use the function $Z(x) = x^r F_A(x)$. First, let us consider the Gegenbauer expansion of $xF_A(x)$. Then,

$$\begin{aligned}
xF_A(x) &= \sum_{k=0}^{s} f_k x G_k^{(n)}(x) = \sum_{k=0}^{s} f_k \left(\frac{n+k-3}{n+2k-4} G_{k-1}^{(n)}(x) + \frac{k+1}{n+2k} G_{k+1}^{(n)}(x) \right) \\
&= \frac{1}{|X|} G_0^{(n)}(x) + \sum_{k=1}^{s-1} \left(f_{k+1} \frac{n+k-2}{n+2k-2} + f_{k-1} \frac{k}{n+2k-2} \right) G_k^{(n)}(x) \\
&\quad + \frac{s}{n+2s-2} f_{s-1} G_s^{(n)}(x) + \frac{s+1}{n+2s} f_s G_{s+1}^{(n)}(x) \\
&= \frac{1}{|X|} \sum_{k=0}^{r-1} G_k^{(n)}(x) + \sum_{k=r}^{s+1} f_k' G_k^{(n)}(x).
\end{aligned} \tag{5.22}$$

Here, $f_k' > 0$ for any k with $r \leq k \leq s+1$. Repeating this process r times, the Gegenbauer expansion of $Z(x)$ becomes $Z(x) = x^r F_A(x) = \frac{1}{|X|} G_0^{(n)}(x) + \sum_{k=1}^{s+r} g_k G_k^{(n)}(x)$. Moreover, we know that $g_k > 0$, for any k with $1 \leq k \leq s + r$. Then applying formula (5.12) for the function $Z(x)$, since X is an A-code and $Z(x)$ is a normalized annihilator of A, we get $\sum_{k=1}^{s+r} g_k \|{}^t H_k H_0\|^2 = |X| - g_0 \|{}^t H_0 H_0\|^2 = 0$. Therefore, ${}^t H_k H_0 = 0$ holds. So, X is proved to be an $(s + r)$-design. □

Theorem 5.14 (Theorem 6.6 in [163], Theorem 4.3.6 in [32]). *For a finite subset X of S^{n-1} with degree s and strength t, the following hold:*

(1) $t \le 2s$;

(2) $|X| \le R_s(1) = \binom{n+s-1}{s} + \binom{n+s-2}{s-1}$;

(3) *if* $|X| = R_s(1)$ *holds, then* $t = 2s$ *holds and* X *is a tight* $2s$*-design;*

(4) *if* $t = 2s$*, then* $|X| = R_s(1)$ *holds, and so* X *is a tight* t*-design.*

Proof. We first prove (2). Let $A = A(X)$ and let $F_A(x)$ be the normalized annihilator of A of degree s, and let $F_A(x) = \sum_{k=0}^{s} f_k G_k^{(n)}(x)$ be the Gegenbauer expansion of $F(x)$. Then for $x, y \in X$, we have

$$\sum_{k=0}^{s} f_k(H_k{}^t H_k)(x,y) = \sum_{k=0}^{s} f_k \sum_{i=1}^{h_k} \varphi_{k,i}(x)\varphi_{k,i}(y) = \sum_{k=0}^{s} f_k G_k^{(n)}(x \cdot y). \tag{5.23}$$

So, we get

$$\sum_{k=0}^{s} f_k(H_k{}^t H_k) = \sum_{k=0}^{s} f_k \sum_{\alpha \in A'(X)} G_k^{(n)}(\alpha) D_\alpha = \sum_{\alpha \in A'(X)} F_A(\alpha) D_\alpha = I. \tag{5.24}$$

Let T be the diagonal matrix of size $R_e(1)$ whose diagonal entries are

$$f_0, \underbrace{f_1, \ldots, f_1}_{G_1^{(n)}(1)}, \ldots, \underbrace{f_i, \ldots, f_i}_{G_i^{(n)}(1)}, \ldots, \underbrace{f_s, \ldots, f_s}_{G_s^{(n)}(1)}.$$

Now, let $H = [H_0, H_1, \ldots, H_s]$ be the $|X| \times R_s(1)$-matrix. Then we get

$$H\, T\, {}^tH = \sum_{k=0}^{s} f_k(H_k{}^t H_k) = I \text{ (the identity matrix of size } |X|).$$

Therefore, we have $\operatorname{rank}(H) \ge |X|$, and so we have $|X| \le R_s(1)$.

Now we prove (1). Let $t \ge s$. Then by Theorem 5.13 (1), we get $f_0 = f_1 = \cdots = f_{t-s} = \frac{1}{|X|}$. Since $F_A(x)$ is a polynomial of degree s, we have $f_i = 0$ for $i > s$. Therefore, we have $t - s \le s$. Namely, we get $t \le 2s$.

(3) Suppose $|X| = R_e(1)$. Then the matrix H defined above is a square matrix, and also non-singular. Therefore, the matrix T is also non-singular, and so $T^{-1} = {}^tHH$. Since ${}^tHH$ is a positive definite matrix, each entry of T must be a positive real number. Namely, $f_0, f_1, \ldots, f_s$ are all positive real numbers. Applying Theorem 5.12 (1), we get $0 < f_i \le \frac{1}{|X|} = \frac{1}{R_s(1)}$ for any i with $0 \le i \le s$. Since $1 = F_A(1) = \sum_{k=0}^{s} f_k G_k^{(n)}(1)$ and $\sum_{k=0}^{s} G_k^{(n)}(1) = R_s(1)$, we have $f_i = \frac{1}{R_e(1)}$ for each i with $0 \le i \le s$. Namely, X is a $2s$-design by Theorem 5.13 (2). And so, it is a tight $2s$-design.

(4) Suppose $t = 2s$. Then $|X| \ge R_s(1)$ by Theorem 5.6. Therefore, by (2), we get $|X| = R_e(1)$. □

The following theorem is proved by essentially the same argument as in the proof of Theorem 5.14. However, the details are a bit more complicated. (The reader is referred to [163, 32] for the details.)

Theorem 5.15 (Theorem 6.8 in [163], Theorem 4.3.7 in [32]). *Let X be an s-distance set and also a t-design on S^{n-1}. Let $A = A(X) = \{x \cdot y \mid x, y \in X, x \neq y\}$, and $A' = A'(X) = \{1\} \cup A$. Suppose that $A' = -A'(= \{-\alpha \mid \alpha \in A'\})$ holds. Then the following hold:*
(1) $t \leq 2s - 1$;
(2) $|X| \leq 2C_{s-1}(1)$;
(3) *if* $|X| = 2C_{s-1}(1)$ *holds, then* $t = 2s - 1$, *and X is a tight* $(2s - 1)$*-design;*
(4) *if* $t = 2s - 1$ *holds, then* $|X| = 2C_{s-1}(1)$, *and X is a tight* $(2s - 1)$*-design.*

The next theorem is not explicitly mentioned as a theorem in [163], but it is clear from Theorem 5.6, Theorem 5.8, Theorem 5.14, and Theorem 5.15.

Theorem 5.16 (Theorem 4.3.8 in [32]).
(1) *Suppose* $|X| = R_s(1)$. *Then X is a $2s$-design if and only if X is an s-distance set.*
(2) *Suppose* $|X| = 2C_{s-1}(1)$. *Then X is a* $(2s - 1)$*-design if and only if X is an antipodal s-distance set.*

As we have seen above, the arguments in algebraic combinatorics on spheres are developed parallel to the case of association schemes.

Association schemes obtained from tight spherical t-designs

The contents here are due to Delsarte, Goethals, and Seidel [163]. We faithfully follow [163] in the discussion below (see also [32, Section 7.2]).

Let A be a subset of $[-1, 1)$ and let X be an A-code on S^{n-1}. For $\alpha, \beta \in A'(= A \cup \{1\})$ and for $x, y \in X$, we define

$$v_\alpha(x) = \big|\{z \in X \mid x \cdot z = \alpha\}\big|,$$
$$p_{\alpha,\beta}(x, y) = \big|\{z \in X \mid x \cdot z = \alpha, z \cdot y = \beta\}\big|.$$

By definition, we have $p_{\alpha,\alpha}(x, x) = v_\alpha(x)$ and $v_1(x) = 1$. If for each $\alpha \in A'$, $v_\alpha(x)$ is a constant independent of the choice of $x \in X$, then the A-code X is said to be *distance-invariant*. Moreover, for any $\alpha, \beta, \gamma \in A'$, if $p_{\alpha,\beta}(x, y)$ is a constant independent of the choice of $x, y \in X$ such that $x \cdot y = \gamma$, an association scheme is attached to the A-code X. To be more precise, let $A' = \{\alpha_0(= 1), \alpha_1, \ldots, \alpha_s\}$, and let $R_i = \{(x, y) \in X \times X \mid x \cdot y = \alpha_i\}$. Then

$$X \times X = R_0 \cup R_1 \cup \cdots \cup R_s$$

holds, and $(X, \{R_i\}_{0 \leq i \leq s})$ becomes an association scheme. The reference [163] gives a condition when an A-code X on S^{n-1} has this good property that an association scheme is attached to it (see also [32]). Now, let us define some more notation. Let

$$x^i = \sum_{\ell=0}^{i} f_{i,\ell} G_\ell^{(n)}(x) \tag{5.25}$$

be the Gegenbauer expansion of the monomial x^i. Also, let

$$F_{i,j}(x) = \sum_{\ell=0}^{\min\{i,j\}} f_{i,\ell} f_{j,\ell} G_\ell^{(n)}(x). \tag{5.26}$$

Then the next lemma follows.

Lemma 5.17. *Let X be a finite subset of S^{n-1} which is an s-distance set and a t-design. Let $x, y \in X$ satisfy $x \cdot y = \gamma$, and let $\alpha, \beta \in A'(X)$. For non-negative integers i, j satisfying $i + j \leq t$, we have the following linear equations:*

$$\sum_{\alpha \in A(X)} \alpha^{i+j} p_{\alpha,\alpha}(x,x) = |X| F_{i,j}(1) - 1 \quad (\text{for } \gamma = 1), \tag{5.27}$$

$$\sum_{\alpha,\beta \in A(X)} \alpha^i \beta^j p_{\alpha,\beta}(x,y) = |X| F_{i,j}(\gamma) - \gamma^i - \gamma^j \quad (\text{for } \gamma \neq 1). \tag{5.28}$$

(Note that if $\gamma = 1$, then $x = y$, and we have

$$p_{\alpha,\beta}(x,x) = \delta_{\alpha,\beta} p_{\alpha,\alpha}(x,x).$$

If $x \neq y$, then $p_{1,\beta}(x,y) = \delta_{\beta,\gamma}$ and $p_{\alpha,1}(x,y) = \delta_{\alpha,\gamma}$ hold.)

Proof. We use the equivalent condition (v), given in Section 5.1.2, for X to be a spherical t-design. Namely, we use the condition that ${}^tH_i H_j = |X| \Delta_{i,j}$ holds. Then, we have the following matrix equation (whose rows and columns are indexed by the elements of X):

$$\left(\sum_{k=0}^{i} f_{i,k} H_k {}^tH_k \right) \left(\sum_{\ell=0}^{j} f_{j,\ell} H_\ell {}^tH_\ell \right) = |X| \sum_{k=0}^{\min\{i,j\}} f_{i,k} f_{j,k} H_k {}^tH_k. \tag{5.29}$$

The (x, y)-entry of the left-hand side of (5.29), where $x \cdot y = \gamma$, becomes as follows, by a similar calculation to that we did in (5.24):

$$\begin{aligned}
&\left(\left(\sum_{k=0}^{i} f_{i,k} H_k {}^tH_k \right) \left(\sum_{\ell=0}^{j} f_{j,\ell} H_\ell {}^tH_\ell \right) \right)(x,y) \\
&= \left(\left(\sum_{k=0}^{i} f_{i,k} \sum_{\alpha \in A'(X)} G_k^{(n)}(\alpha) D_\alpha \right) \left(\sum_{\ell=0}^{j} f_{j,\ell} \sum_{\beta \in A'(X)} G_\ell^{(n)}(\beta) D_\beta \right) \right)(x,y) \\
&= \left(\sum_{\alpha \in A'(X)} \alpha^i D_\alpha \sum_{\beta \in A'(X)} \beta^j D_\beta \right)(x,y) \\
&= \sum_{\alpha,\beta \in A'(X)} \alpha^i \beta^j p_{\alpha,\beta}(x,y).
\end{aligned} \tag{5.30}$$

Meanwhile, the (x, y)-entry of the right-hand side of (5.29) is given by

$$\left(|X| \sum_{k=0}^{\min\{i,j\}} f_{i,k} f_{j,k} H_k {}^t H_k\right)(x, y) = |X| \sum_{k=0}^{\min\{i,j\}} f_{i,k} f_{j,k} G_k^{(n)}(x \cdot y)$$
$$= |X| F_{i,j}(x \cdot y) = |X| F_{i,j}(\gamma). \tag{5.31}$$

Since $p_{1,1}(x, y) = \delta_{1,\gamma}$ and $p_{\alpha,1}(x, y) = \delta_{\alpha,\gamma}$, $p_{1,\beta}(x, y) = \delta_{\beta,\gamma}$ for $\alpha, \beta \neq 1$, we complete the proof of (5.27) and (5.29). □

This lemma leads to the following theorem.

Theorem 5.18. *Let X be a finite subset of S^{n-1} which is an s-distance set and a t-design. Then we have the following assertions:*
(1) *If $t \geq s-1$, then X is distance-invariant.*
(2) *If $t \geq 2s-2$, then a Q-polynomial scheme of class s is attached to X.*

Proof. Let $A'(X) = \{\alpha_0(=1), \alpha_1, \ldots, \alpha_s\}$.
(1) In formula (5.27) in Lemma 5.17, remarking that $v_{\alpha_i}(x) = p_{\alpha_i,\alpha_i}(x, x)$ for $0 \leq i \leq s-1$ and $j = 0$, we have

$$\sum_{\ell=1}^{s} \alpha_\ell{}^i v_{\alpha_\ell}(x) = |X| F_{i,0}(1) - 1 \tag{5.32}$$

for $i = 0, 1, \ldots, s-1$. Namely, we have s linear equations with s variables $v_{\alpha_1}(x), v_{\alpha_2}(x), \ldots, v_{\alpha_s(x)}$. The coefficient matrix of this system of linear equations is the Vandermonde matrix $W = (\alpha_\ell^i)_{\substack{0 \leq i \leq s-1 \\ 1 \leq \ell \leq s}}$ and non-singular. Therefore, $v_{\alpha_1}(x), v_{\alpha_2}(x), \ldots, v_{\alpha_s(x)}$ are constant numbers independent of the choice of $x \in X$.
(2) Similarly to (1), regarding that $i + j \leq 2s - 2 \leq t$ for $0 \leq i, j \leq s-1$, we have s^2 linear equations in s^2 variables $p_{\alpha_i,\alpha_j}(x, y)$ by formula (5.29) in Lemma 5.17. The coefficient matrix is exactly $W \otimes W$, and hence non-singular. Therefore, if $x \cdot y = \gamma$ then $p_{\alpha_i,\alpha_j}(x, y)$ is a constant number independent of the choice of x and y, and only depends on $\alpha_1, \alpha_2, \ldots, \alpha_s$ and γ. This means that X has the structure of an association scheme. Now we want to show that this association scheme is a Q-polynomial scheme. We set $A_i = D_{\alpha_i}$ for $0 \leq i \leq s$ and let $\mathfrak{A}$ be the Bose–Mesner algebra of the association scheme. Let us set $E_j = \frac{1}{|X|} H_j {}^t H_j$ $(0 \leq i \leq s-1)$, $E_s = I - \sum_{j=0}^{s-1} E_j$. Then, using formula (5.24) used in the proof of Theorem 5.14, we get $E_j = \frac{1}{|X|} \sum_{i=0}^{s} G_j^{(n)}(\alpha_i) A_i$. Therefore, $E_0, E_1, \ldots, E_s$ are contained in $\mathfrak{A}$.

By condition (v) in Section 5.1.2, we have $E_i E_j = \delta_{i,j} E_i$ for $0 \leq i, j \leq s-1$. We also have $E_i E_s = \delta_{i,s} E_s$. Namely, $E_0, E_1, \ldots, E_s$ form the basis of the primitive idempotents of $\mathfrak{A}$. Let $Q = (Q_j(i))_{0 \leq i \leq s, 0 \leq j \leq s}$ be the second eigenmatrix. For j with $0 \leq j \leq s-1$, $Q_j(i)$ is expressed by using the Gegenbauer polynomial as $Q_j(i) = G_j^{(n)}(\alpha_i)$. Meanwhile, since

$$E_s = I - \sum_{j=0}^{s-1} E_j$$
$$= \frac{1}{|X|}\left(|X| - \sum_{j=0}^{s-1} Q_j(0) \right) A_0 - \frac{1}{|X|} \sum_{i=1}^{s} \left(\sum_{j=0}^{s-1} Q_j(i) \right) A_i, \tag{5.33}$$

we have

$$Q_s(0) = |X| - \sum_{j=0}^{s-1} Q_j(0) = |X| - \sum_{j=0}^{s-1} G_j^{(n)}(1), \tag{5.34}$$

$$Q_s(i) = -\sum_{j=0}^{s-1} Q_j(i) = -\sum_{j=0}^{s-1} G_j^{(n)}(\alpha_i). \tag{5.35}$$

Thus $\theta^* = Q_1(i) = G_1^{(n)}(\alpha_i) = n\alpha_i$, and $Q_j(i) = G_j^{(n)}(\alpha_i) = G_j^{(n)}(\frac{\theta_i^*}{n})$ for $0 \le i \le s$, $0 \le j \le s-1$. We need some more works to consider $Q_s(i)$. Namely, we consider the annihilator polynomial $F(x) = \prod_{i=1}^{s} \frac{x-\alpha_i}{1-\alpha_i}$ of $A(X)$. Since $F(\alpha_i) = 0$ for $1 \le i \le s$, we have

$$Q_s(i) = F(\alpha_i) - \sum_{j=0}^{s-1} G_j^{(n)}(\alpha_i) = F\left(\frac{\theta_i^*}{n}\right) - \sum_{j=0}^{s-1} G_j^{(n)}\left(\frac{\theta_i^*}{n}\right)$$

for $i = 1, 2, \ldots, s$. Moreover, we get

$$Q_s(0) = |X| - \sum_{j=0}^{s-1} G_j^{(n)}(1) = |X|F(1) - \sum_{j=0}^{s-1} G_j^{(n)}(1)$$
$$= |X|F\left(\frac{\theta_0^*}{n}\right) - \sum_{j=0}^{s-1} G_j^{(n)}\left(\frac{\theta_0^*}{n}\right). \tag{5.36}$$

Therefore condition (3) of Proposition 2.81 (3) in Chapter 2 is verified. This implies that this association scheme is Q-polynomial. □

We get the following theorem immediately from Theorem 5.6, Theorem 5.8, and Theorem 5.18.

Theorem 5.19. *For any tight spherical t-design X on S^{n-1}, a Q-polynomial association scheme of class $[\frac{t+1}{2}]$ is associated with it.*

As for Theorem 5.18 (2), if we additionally assume that X is antipodal, then $t \ge 2s-2$ is replaced by $t \ge 2s-3$ ([36]). See also Suda [441].

Tight t-designs are, if they exist, the most extremal t-designs among all the t-designs. So, we are very much interested in finding and classifying tight t-designs and designs close to them. We now explain the current situation of the classification of tight spherical t-designs.

Theorem 5.20 (Some theorems and explanations).

(1) *Tight t-designs on S^1 are completely classified. They are the $(t+1)$ vertices of a regular $(t+1)$-gon inscribed in S^1. (So, in what follows, we assume that $n \geq 3$ unless otherwise stated.)*

(2) *For $n \geq 3$, tight t-designs on S^{n-1} are completely classified for $t = 1, 2, 3$. Namely:*
 (2a) *If $t = 1$, then X is a pair of two antipodal points $\{x, -x\}$ on S^{n-1}.*
 (2b) *If $t = 2$, then X is the $(n+1)$ vertices of a regular simplex inscribed in S^{n-1}.*
 (2c) *If $t = 3$, then it is the $2n$ vertices of a generalized regular octahedron (also called a cross polytope) inscribed in S^{n-1}.*

(3) *For $n \geq 3$, tight spherical t-designs on S^{n-1} exist only for $t \leq 5$, $t = 7$, or $t = 11$ ([163], [55, 56]).*

(4) *For $n \geq 3$, tight spherical 11-designs on S^{n-1} exist only when $n = 24$. Moreover, it must be isomorphic (i. e., similar) to the set of 196,560 minimum vectors (normalized as unit vectors) of the Leech lattice ([72]).*

(5) *For $t = 4, 5, 7$, the classification of tight spherical t-designs is still open.*
 (5a) *For $n \geq 3$, if there exists a tight spherical 4-design on S^{n-1}, then it is known that n must be of the form*

$$n = (2m+1)^2 - 3,$$

where m is an integer. The first two numbers for n satisfying this condition are $n = 6$ and $n = 22$. For these two numbers, there are tight spherical 4-designs of sizes 27 and 275, and each of them is unique for each of the dimensions given here. On the other hand, no other tight spherical 4-designs are known for any other values of n. Bannai, Munemasa, and Venkov [71] proved the non-existence of tight spherical 4-designs for some infinitely many values of n satisfying the condition $n = (2m+1)^2 - 3$, where m is an integer, which is given above. In the case of $n = 46$, there is a non-existence theorem on tight spherical 4-designs by Makhnev only using purely graph theoretical arguments ([331]).

 (5b) *Let $n \geq 3$. It is shown that a tight spherical 5-design on S^{n-1} exists if and only if a tight spherical 4-design on S^{n-2} (the sphere of 1 dimension less) exists. Namely, if there is one, then another is constructed and vice versa. Therefore, if a tight spherical 5-design exists on S^{n-1}, then n must be either $n = 3$ or $(2m+1)^2 - 2$, where m is an integer by (5a). Note that for $n = 3$, there exists a tight spherical 5-design, namely, the 12 vertices of a regular icosahedron inscribed in S^2. (The corresponding tight 4-design is the set of vertices of a regular pentagon in $S^{n-2} = S^1$.) So, this case (5b) is essentially reduced to case (5a).*
 Moreover, tight spherical 5-designs do exist for $n = 3, 7, 23$ and the sizes are 12, 56, and 552, respectively. Each of them is known to be unique up to the orthogonal transformation.

 (5c) *For $n \geq 3$, it is known that if there exists a tight spherical 7-design on S^{n-1}, then n must be of the form*

$$n = 3m^2 - 4,$$

where m is an integer. The first two such numbers n are 8 and 23, and there are tight 7-designs of sizes 240 and 4600, respectively. Each of them is known to be unique for $n = 8$ and $n = 23$. No other tight spherical 7-designs are known, and Bannai, Munemasa, and Venkov [71] proved the non-existence for certain infinitely many values of n satisfying the above condition $n = 3m^2 - 4$.

The reader is referred to the reference [163, 55, 56, 72, 71] for the details of the proof as well as further explanations. The basic idea of the proof is that $A(X)$ must be the zeros of certain orthogonal polynomials (Lloyd type polynomials, i. e., something like Wilson polynomials in tight combinatorial t-designs) ([55, 56]) and that these inner product values must be rational numbers in general. The fact that Q-polynomial association schemes are associated with tight designs ([163]) and some number theoretical arguments are used to get (3). For (5), we can get the restriction on n relatively easily by a similar consideration, but to eliminate the remaining n is not so easy. See [71] for the details. For the classification of tight t-designs for $t = 4, 5, 7$, Nebe and Venkov [374] proved further non-existence results beyond [71][2]. However, it seems that the complete classification of tight spherical t-designs for $t = 4, 5, 7$ is not easy and not expected to be solved very soon.

At the beginning of this chapter, we mentioned optimal codes. The stronger concept "*universally optimal*" seems to be a very interesting concept. So, here we will discuss this topic.

Definition 5.21 (Absolutely monotonic function). A function $\alpha : [-1, 1) \to \mathbb{R}$ is called an *absolute monotonic function*, if α is of class C^∞ and if the k-th derivative $\alpha^{(k)}$ of α satisfies $\alpha^{(k)}(x) \geq 0$ for any integer $k \geq 0$ and for any $x \in (-1, 1)$.

Definition 5.22 (Universally optimal code). A finite subset X on S^{n-1} is called a *universally optimal code*, if for any subset Y on S^{n-1} with $|Y| = |X|$ and for any absolutely monotonic function α, the following condition holds:

$$\sum_{x,y \in X, x \neq y} \alpha(x \cdot y) \leq \sum_{x,y \in Y, x \neq y} \alpha(x \cdot y).$$

Remark 5.23. By considering the function $\alpha(x) = (x + 1)^m$ on $x \in [-1, 1)$ $(m = 1, 2, \ldots)$ and letting m go to infinity, we can see that a universally optimal code is an optimal code. Universally optimal codes are interpreted as codes which satisfy the energy minimizing conditions not only for very special classes of potential functions but for general wide classes of potential functions. These subsets can be regarded as very good finite point-sets on the sphere, even from the viewpoint of physics or chemistry.

2 For an explanation of the work of Boris Venkov who passed away on Nov 10, 2011, see Nebe [369].

The following concept is essentially equivalent to that of universally optimal codes. This definition might be intuitively understandable from the viewpoint of physics.

Definition 5.24 (Completely monotonic function). A function $f : (0,4] \to \mathbb{R}$ is called a *completely monotonic function* if f is of class C^∞ and if the k-th derivative $f^{(k)}$ of f satisfies $(-1)^k f^{(k)}(x) \geq 0$ for any integer $k \geq 0$.

The following definition is in fact known to be equivalent to Definition 5.22.

Definition 5.25 (Universally optimal code). A finite subset X on S^{n-1} is called a *universally optimal code* if for any subset Y on S^{n-1} with $|Y| = |X|$ and for any completely monotonic function f, the following condition holds:

$$\sum_{x,y\in X, x\neq y} f(\|x-y\|^2) \leq \sum_{x,y\in Y, x\neq y} f(\|x-y\|^2).$$

Before Cohn and Kumar defined and studied universally optimal codes, there had been some preceding studies to find energy minimizing finite subsets on S^{n-1} for some special classes of potential functions by Yudin, Kolushov, Andreev, and others ([525, 283, 284, 2, 3], etc.).

Cohn and Kumar [140] tried to classify universally optimal designs. For the sphere S^2 in $\mathbb{R}^3$, they proved relatively simply that there are only three kinds of universally optimal codes, i. e., the regular tetrahedron, the regular octahedron, and the regular icosahedron, by using the results of Leech [308]. However, for other dimensions $n \geq 4$, the problem is completely open. They conjecture that for each $n \geq 4$, there are only finitely many universally optimal codes, although this is also still open. The list of known universally optimal codes is given in Cohn and Kumar [140]. Except one example which is the set of 120 vertices of a regular polytope (600-cell) on $S^3 \subset \mathbb{R}^4$ with $t = 11$ and $s = 8$, all other known universally optimal codes satisfy the condition that $t \geq 2s-1$. At the present stage, it is not clear[3] whether there are any other universally optimal codes with $t < 2s-1$. Ballinger et al. give some candidates of universally optimal codes ([18]). Most interesting and promising candidates are (i) $n = 10$ and $|X| = 40$ and (ii) $n = 16$ and $|X| = 64$. These two examples have the structures of association schemes ([43]). Further candidates that also have the structures of association schemes can be seen in Abdukhalikov, Bannai, and Suda [1]. As is seen from these examples, universally optimal codes are usually associated to association schemes. However, it is still an interesting open question whether there is a universally optimal code which is not associated with an association scheme. Anyway, we would like to say that the classification of universally optimal codes, or more specifically the classification of those X with $t \geq 2s-1$, or more generally the classification of those X with $t \geq 2s-2$ (then association schemes are always attached) may be very interesting open problems that we should try to attack.

---

3 Personally, the author (Eiichi Bannai) thinks that there should exist some such examples.

In particular when we try to study a finite subset X on S^{n-1} with strength t and degree s satisfying $t \geq 2s - 1$, the series of work by Levenshtein [315, 313, 314] are extremely important. So, let us comment on the crucial points of these works. The idea of his proof is to "try to find good test functions." For the classical Jacobi polynomials $P_i^{a+\frac{n-3}{2},b+\frac{n-3}{2}}(x)$, where $a, b \in \{0, 1\}$, let us define

$$T_k^{1,\varepsilon}(x,y) = \sum_{i=0}^{k} r_i^{1,\varepsilon} P_i^{\frac{n-1}{2},\varepsilon+\frac{n-3}{2}}(x) P_i^{\frac{n-1}{2},\varepsilon+\frac{n-3}{2}}(y), \tag{5.37}$$

$$r_i^{1,\varepsilon} = \left(\frac{n+2i-1+\varepsilon}{n+\varepsilon-1}\right)^{2-\varepsilon} \binom{n+i-2-\varepsilon}{i}, \quad \varepsilon \in \{0,1\}. \tag{5.38}$$

Now, let $t_k^{1,1}$ and $t_k^{1,0}$ be the largest zeros of polynomials $P_k^{\frac{n-1}{2},\frac{n-1}{2}}(x)$ and $P_k^{\frac{n-1}{2},\frac{n-3}{2}}(x)$, respectively. Here, we remark that if we set $P_k^{(n)}(x) = P_k^{\frac{n-1}{2},\frac{n-1}{2}}(x)$, then we get

$$P_k^{(n)}(x) = \frac{G_k^{(n)}(x)}{G_k^{(n)}(1)}.$$

(Namely, $P_k^{(n)}(x)$ are the Gegenbauer polynomials normalized as $P_k^{(n)}(1) = 1$.) Based on these polynomials, Levenshtein defined the polynomials $f_m^{(y)}(x)$ by

$$f_m^{(y)}(x) = (x+1)^{\varepsilon}(x-1)\left(T_{k-1}^{1,\varepsilon}(x,y)\right)^2, \tag{5.39}$$

with $m = 2k + 1 - \varepsilon$ and $t_{k-1+\varepsilon}^{1,1-\varepsilon} \leq y \leq t_k^{1,\varepsilon}$.

Theorem 5.26 (Levenshtein [313, 314, 315]). *Let X be a finite set on S^{n-1} with $|X| = N$. Let $A(X) \subset [-1, y]$ (so X is a y-code). Then the following inequality holds:*

$$N \leq \begin{cases} L_{2k-1}(n,y), & \text{for } t_{k-1}^{1,1} \leq y < t_k^{1,0}, \\ L_{2k}(n,y), & \text{for } t_k^{1,0} \leq y < t_k^{1,1}. \end{cases}$$

Here, we define

$$L_{2k-1}(n,y) = \binom{k+n-3}{k-1}\left(\frac{2k+n-3}{n-1} - \frac{P_{k-1}^{(n)}(y) - P_k^{(n)}(y)}{(1-y)P_k^{(n)}(y)}\right), \tag{5.40}$$

$$L_{2k}(n,y) = \binom{k+n-2}{k}\left(\frac{2k+n-1}{n-1} - \frac{(1+y)(P_k^{(n)}(y) - P_{k+1}^{(n)}(y))}{(1-y)(P_k^{(n)}(y) + P_{k+1}^{(n)}(y))}\right). \tag{5.41}$$

Moreover, if $N = L_m(n,y)$, then X is an m-design on S^{n-1} and $A(X)$ coincides with the set of the zeros of the polynomial $f_m^{(y)}(x)$.

The importance of the results of Levenshtein lies in the fact that the converse of the claim holds. Namely, if X is a spherical t-design with $t \geq 2s - 1$, then the following holds:

$$|X| = L_t(n, \alpha).$$

Here, α is the largest element in $A(X)$. There are many works already done to find the lower bounds of the sizes of t-designs, using either linear programming or semidefinite programming. In the work of Levenshtein, he tried to find the lower bound under the condition $t \geq 2s - 1$ by using polynomials as test functions. Besides, there are works of Boyvalenkov [105] and his school to improve the Fisher type lower bound for t-designs ([106, 107, 108, 104, 378]). Meanwhile, Yudin [526] gives a method to improve the Fisher type lower bound. He uses not only polynomials but also other functions as test functions. The paper of Cohn and Kumar [140] uses this idea of Yudin. There is more recent work by Bondarenko and Viazovska [98] in this direction. We will mention further extensions by Bondarenko, Radchenko, and Viazovska later in this chapter in between Theorem 5.32 and Definition 5.33. In that work it is shown that there are many successes to improve the lower bound when the dimension n is fixed and t goes to infinity. On the other hand, when t is fixed and n goes to infinity, much progress has not been made yet, except for the non-existence of tight t-designs in some extreme cases. (This is partly seen from an unpublished computer experiment by Sikirić, Schürmann, and Vallentin [168] that searches for the lower bounds by using either linear programming or semidefinite programming. This may imply that even semidefinite programming has some limitations, which seems interesting.)

5.1.3 Connections of spherical designs with group theory, number theory, modular forms

(a) t-Designs obtained as orbits of finite groups

Let us consider what kinds of spherical t-designs there are. The most natural way of the construction is to consider orbits of a finite group G in the orthogonal group $\mathrm{O}(n)$. Namely, for $x \in S^{n-1}$ we consider the orbit x^G of x by G as follows:

$$X = x^G = \{x^g \mid g \in G\} \subset S^{n-1}.$$

There are many possibilities for G. We expect that if we take larger finite subgroups G in $O(n)$, then we may get better designs. This topic was already treated in [32, Chapter 6] in a detailed way, so here we just mention the points that we think important, leaving the details to the book [32]. The most important finite groups are: real reflection groups (including Weyl groups and more generally Coxeter groups), the Conway group Co.0 in the 24-dimensional space and their various subgroups, and Clifford groups. The research in this direction was started by Sobolev in the 1960s (cf. [412]). See also Sidelnikov [428, 429]. There are some works on properties of finite groups such that orbits become t-designs (for details, see [32]). Here we just mention some important facts that we believe interesting.

For each $n \geq 3$, among spherical t-designs that are obtained as an orbit of a finite group G of $\mathrm{O}(n)$, those with large t are not yet found. We remark the following.

Theorem 5.27 (Bannai [25] (1984)). *For $x_1, x_2 \in S^{n-1}$ suppose that an orbit x_1^G of G is a t_1-design (but not a (t_1+1)-design) and also the orbit x_2^G of G is a t_2-design (but not a (t_2+1)-design). Then we get $t_2 \le 2t_1 + 1$ (hence symmetrically, we also get $t_1 \le 2t_2 + 1$).*

This result implies that the property that a finite subgroup G of $O(n)$ is *t-homogeneous*, namely, every orbit of G is a t-design, has an important meaning.

Group theoretically, the condition that a finite subgroup G of $O(n)$ is t-homogeneous is equivalent to the condition that the representation of the i-th spherical representation ρ_i (i. e., the representation of $O(n)$ acting on the space $\mathrm{Harm}_i(\mathbb{R}^n)$ of homogeneous harmonic polynomials of degree i), if restricted to G, does not contain the identity representation as an irreducible component. This concept was introduced as an analogue of t-homogeneous permutation groups. Let G be a subgroup of the symmetric group S_v on the set $V = \{1, 2, \dots, v\}$. We say G is a t-homogeneous permutation group if G acts transitively on the set of t-element subsets of V. This is equivalent to the condition that for any k-element subset of V, its orbit by G makes a t-(v, k, λ) combinatorial design. This condition is slightly weaker than the condition G being t-transitive, and representation theoretically this is equivalent to the following condition. Let $\chi_{(n-i,i)}$ be the representation of S_v corresponding to the Young diagram of type $(n-i, i)$. Then the restriction of $\chi_{(n-i,i)}$ to the subgroup G does not contain the identity representation as an irreducible component for all i with $1 \le i \le t$. So, this concept of t-homogeneous (linear) group for a subgroup of $O(n)$ was obtained as an analogue of t-homogeneous permutation groups.

The same statement, replacing x_1, x_2 by k-element subsets of V, and also the same proof as for Theorem 5.27, holds for classical combinatorial t-designs. This is also mentioned in Cameron and Praeger [123].

For small n, finite subgroups of $O(n)$ are classified. The results for $n = 3$ are well known. Conway and Smith [148] give the classification for $n = 4$. Miezaki [346] managed to prove that finite subgroups of $O(4)$ are at most 11-homogeneous, by using the result of Conway and Smith [148]. So, from Theorem 5.27, it turns out that for any spherical t-design on S^3 that is obtained as an orbit of a finite group, we have $t \le 23$. It is known that there exist some 19-designs as an orbit of the real reflection group $W(H_4)$ (Goethals and Seidel [201] or [32]). Meanwhile, it is known that for $n = 8$, $W(E_8)$ is 7-homogeneous and there are some 11-designs as its orbits. For $n = 24$, $\mathrm{Co.0}$ is 11-homogeneous, and there are some 15-designs as its orbits.

Theorem 5.28 (Bannai (1984) [26]). *There exists a function $f(n)$ such that for $n \ge 3$ if there exists a spherical t-design that is an orbit of a finite group G in $O(n)$, then $t \le f(n)$. (The explicit form of $f(n)$ is not yet known.)*

As far as looking at known examples, we know only up to 11 homogeneous groups G in $O(n)$ for $n \ge 3$. Also only up to 19-designs, which are orbits of a finite group, are known. It is conjectured that the function that appears in Theorem 5.28 may be replaced by an absolute constant, say, by 19, but it is still an open problem. We wonder whether we can use the classification of finite simple groups to prove this claim. (We encourage the reader to try to prove this.)

As mentioned on the previous page it is shown that if there is a t-(v, k, λ) combinatorial design on whose blocks a subgroup G of S_v acts transitively, then $t \leq 7$, by using the classification of ℓ-homogeneous finite permutation groups (with $\ell \geq 2$), which is proved in turn by using the classification of finite simple groups. Cameron and Praeger [123] conjecture that $t \leq 5$.

(b) Spherical t-designs that are obtained as shells of a lattice

In addition to the spherical t-designs which are obtained as orbits of a finite subgroup G in $O(n)$, there is another natural method for the construction of the spherical t-designs obtained as shells of a lattice.

Let us fix some notation. A subset L of $\mathbb{R}^n$ is called a *lattice* if there exists a basis $\{v_1, v_2, \ldots, v_n\}$ of the real vector space $\mathbb{R}^n$ and L is the set of linear combinations of $\{v_1, v_2, \ldots, v_n\}$ with integral coefficients. We call $\{v_1, v_2, \ldots, v_n\}$ a generator of L. A lattice L is a free Abelian group of rank n. For the important examples of lattices, the reader is referred to basic references on lattices, Conway and Sloane [147] or Ebeling [175]. (In what follows, $\{e_1, e_2, \ldots, e_n\}$ is the standard orthonormal basis of L.)

- Let $\mathbb{Z}^n$ be the set of all integer points: $\{(x_1, x_2, \ldots, x_n) \mid x_i \in \mathbb{Z}, 1 \leq i \leq n\}$.
- The A_n-lattice is $\{(a_1, a_2, \ldots, a_{n+1}) \mid a_i \in \mathbb{Z}, a_1 + \cdots + a_n + a_{n+1} = 2\}$.
  (Note that for the hyperplane H, we have $H = \{(a_1, \ldots, a_n, a_{n+1}) \in \mathbb{R}^{n+1} \mid a_1 + \cdots + a_n + a_{n+1} = 2\} \cong \mathbb{R}^n$.) In particular, the A_2-lattice is called the hexagonal lattice, as shown below.

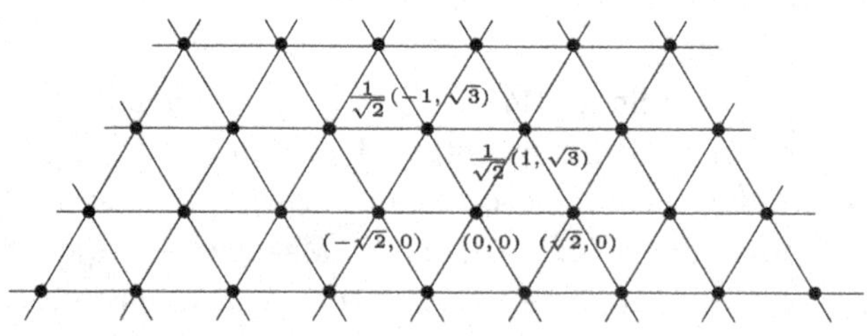

- The D_n-lattice is the free Abelian group generated by the set $\{\pm e_i \pm e_j \mid 1 \leq i, j \leq n, i \neq j$, with all the choices of $\pm\} = \{(a_1, \ldots, a_n) \in \mathbb{R}^n \mid a_1 + \cdots + a_n \equiv 0 \pmod 2, a_i \in \mathbb{Z}, 1 \leq i \leq n\}$.
- The E_n-lattices ($n = 6, 7, 8$).
  The E_8-lattice is defined as the set of integral linear combinations of the vectors $\{\pm e_i \pm e_j \mid 1 \leq i, j \leq 8, i \neq j$ with all possible choices of signs$\} \cup \{\frac{1}{2}(\pm e_1 \pm e_2 \pm \cdots \pm e_8) \mid$ with an even number of minus signs$\}$. This is equivalent to the D_8^+ lattice

defined as the union of the D_8-lattice and the set $\{D_8 + \frac{1}{2}(e_1 + e_2 + \cdots + e_8)\}$. So, it is $\{(a_1, a_2, \ldots, a_8) \mid a_1 + a_2 + \cdots + a_8 \equiv 0 \pmod 2, a_i \in \mathbb{Z} \text{ or } a_i \in \mathbb{Z} + \frac{1}{2}, 1 \le i \le 8\}$.
The E_7-lattice is isomorphic to the subset (of the E_8-lattice) consisting of the vectors perpendicular to any fixed vector of length $\sqrt{2}$ of the E_8-lattice. (So, it is a lattice in $\mathbb{R}^7$.)
The E_6-lattice is isomorphic to the lattice (in $\mathbb{R}^6$) which is the subset of the E_8-lattice consisting of the vectors perpendicular to a pair of two vectors of length $\sqrt{2}$, 120 degrees apart in the E_8-lattice.
- The Leech lattice Λ_{24} is defined as $\Lambda_{24} = \{\frac{1}{\sqrt{8}}(\mathbf{0} + 2\mathbf{c} + 4\mathbf{x})\} \cup \{\frac{1}{\sqrt{8}}(\mathbf{1} + 2\mathbf{c} + 4\mathbf{y})\}$. Here, $\mathbf{0} = (0, 0, \ldots, 0)$, $\mathbf{1} = (1, 1, \ldots, 1)$, and $\mathbf{c}$ is in the Golay code (regarding the components of $\mathbf{c}$ as real numbers 0 and 1 rather than the elements of the finite field F_2). Also, $\mathbf{x} = (x_1, x_2, \ldots, x_{24})$ and $\mathbf{y} = (y_1, y_2, \ldots, y_{24})$ are vectors in $\mathbb{R}^{24}$ satisfying $x_i, y_i \in \mathbb{Z}$, $1 \le i \le 24$, $x_1 + x_2 + \cdots + x_{24} \equiv 0 \pmod 2$, $y_1 + y_2 + \cdots + y_{24} \equiv 1 \pmod 2$.

We say that a lattice L is an *integral lattice* if the ordinary Euclidean inner product $x \cdot y$ is an integer for any $x, y \in L$. If $x \cdot y$ is always an even integer, then L is called an *even lattice*. It is immediate from the definition that an even lattice is an integral lattice. For any lattice L, we define the dual lattice L^* by

$$L^* = \{x \in \mathbb{R}^n \mid x \cdot y \in \mathbb{Z}, \forall y \in L\}.$$

So, the lattice L is an integral lattice if and only if $L \subset L^*$. We say that an integral lattice L is *unimodular* if $L = L^*$ holds. For an integral lattice L, $|L^*/L|$ is the determinant of the Gram matrix of a basis $\{v_1, v_2, \ldots, v_n\}$ of L. This quantity is also equal to the square of the determinant of the $n \times n$-matrix whose column vectors are a basis $\{v_1, v_2, \ldots, v_n\}$ of L.
- The lattice $\mathbb{Z}^n$ is a unimodular integral lattice, but is not an even lattice.
- The A_n-lattice is an even lattice, but not a unimodular lattice; $|L^*/L| = n + 1$ and L^*/L is isomorphic to $\mathbb{Z}_{n+1}$, the cyclic group of order $n + 1$.
- The D_n-lattice is an integral lattice, but not a unimodular lattice; $|L^*/L| = 4$. Moreover, $L^*/L \cong \mathbb{Z}_4$ if n is odd, and $L^*/L \cong \mathbb{Z}_2 \times \mathbb{Z}_2$ if n is even.
- The E_6-lattice is an even lattice, but not a unimodular lattice; $L^*/L \cong \mathbb{Z}_3$.
- The E_7-lattice is an even lattice, but not a unimodular lattice; $L^*/L \cong \mathbb{Z}_2$.
- The E_8-lattice is an *even unimodular* lattice.
- The Leech lattice Λ_{24} is an even unimodular lattice.

As we mentioned in Chapter 1, Section 1.6, it is known that for a Type II code, the length n must be a multiple of 8. Similarly, it is known that for an even unimodular lattice L in $\mathbb{R}^n$, n must be a multiple of 8. Now, for an integral lattice L and a positive integer m, let us define

$$L_m = \{x \in L \mid x \cdot x = m\}.$$

Namely, L_m is the set of points of L with the constant length $\sqrt{m}$ from the origin. We call L_m a *shell* of L. Then $\frac{1}{\sqrt{m}}L_m$ is a finite subset of the unit sphere S^{n-1}. We can expect that there are some good spherical t-designs among them.[4]

One of the very important results is the following theorem by Venkov (Theorem 5.30, discussed later). Before stating Venkov's theorem, let us discuss the concept of *extremal lattices*. Let L be an even unimodular lattice in $\mathbb{R}^n$. (Such a code is also called a Type II code.) Then as we mentioned already n must be a multiple of 8. It is known that for an even unimodular lattice L in $\mathbb{R}^n$, we have

$$\min\{x \cdot x \mid x \in L, x \neq 0\} \leq 2\left[\frac{n}{24}\right] + 2.$$

If equality holds in the above inequality, L is called an *extremal even unimodular lattice*. (This situation is very similar to the case of extremal Type II doubly even codes.)

The problem of finding (or constructing) and classifying extremal even unimodular lattices is a very interesting problem, similarly to the case of extremal type II codes. Many works have been done on this problem.

The current situation of the study of extremal even unimodular lattices is as follows. (In the original Japanese version of this book, we described the situation of October 2015. It seems that the situation has not changed in June 2020, when we wrote this English version.)

- For $n = 8$, the E_8-lattice is the unique such lattice.
- For $n = 16$, there are exactly two such lattices: $E_8 \oplus E_8$ and D_{16}^+.
- For $n = 24$, the Leech lattice Λ_{24} is the unique such lattice.
- For $n = 32$, there are many examples known. King (2003) [280] proved the following result. The mass of the even unimodular lattices with no roots, i. e., vectors of length $\sqrt{2}$, is $\frac{1310037331282023326658917}{238863431761920000} = 5.48 \times 10^6$. This implies that there are at least 10^7 such lattices. Here the mass means the quantity $\sum_{L \in Y} \frac{1}{|\mathrm{Aut}(L)|}$, where $\mathrm{Aut}(L)$ is the automorphism group of L and Y consists of all the representatives of the isomorphism classes of extremal even unimodular lattices with no roots.
- For $n = 40$, there are some (many) examples known. The first example was given by McKay [340]. King [279] proved that there are at least 12,579 extremal Type II codes of length 40. Now it is known that the exact value is 16,470, as proved by Betsumiya, Harada, and Munemasa [89]. On the other hand, Kitazume, Kondo, and Miyamoto [281, Theorem 3] proved that the extremal unimodular lattices which are obtained from extremal type II codes by so-called Construction C are not isomorphic to each other. So, there are at least as many non-isomorphic extremal unimodular lattices as them.

4 As far as all known examples of lattices are concerned, for any $n \geq 2$, there is no known shell L_m (of any lattice) which is a 12-design. (Moreover, for $n = 2$, no 6-design is known; for $n = 3$, no 4-design is known.) It is an interesting open question whether there is an absolute constant t_0 such that if L_m is a t-design, then $t \leq t_0$ for all n, m, and L. (So, we may be able to take $t_0 = 11$.)

- For $n = 48$, there have been known three examples: P_{48n}, P_{48p}, and P_{48q} (Conway and Sloane [147], Nebe (1998) [367], and others). Nebe found a new example [370, 371].
- For $n = 56$ and 64, it is known that there is at least one for each. Nebe, Ozeki, and Quebbemann constructed some examples, but isomorphism classes among them are not completely determined yet.
- For $n = 72$, in 2010 Nebe announced that she constructed one example. Her paper was published in 2012 [368].
- For $n = 80$, two examples were known by Bachoc and Nebe [12]. Two more examples have been discovered by Stehlé and Watkins [439] and Watkins [509].
- For $n \geq 88$, no example has been discovered yet.

The existence of an extremal even unimodular lattice of dimension 72 was the long-standing famous open problem, and Nebe affirmatively solved it in 2010. It is very interesting whether a similar thing will happen for other values of n, in particular for values that are multiples of 24. It is known that if there exists an extremal even unimodular lattice of dimension n, then n is bounded above by about $40,000$ (Mallow, Odlyzko, and Sloane (1975) [332]) (see also Gaborit (2004) [188, Table 3]). Miezaki informed us that the above number $40,000$ is not correct and must be replaced by $163,264$ (Nebe [372] and Jenkins and Rouse [268]).

Remark 5.29. Griess constructed the Leech lattice Λ_{24} from the E_8-lattice in the following way. Let M and N be the lattices in $\mathbb{R}^8$ isomorphic to the E_8-lattice and $M \cap N \cong \sqrt{2}E_8$. There are in fact many such choices. Let L be defined by

$$L = \{(w + x, w + y, w + z) \mid w \in M,\ x, y, z \in N,\ x + y + z \in M \cap N\}.$$

Then L is a lattice in $\mathbb{R}^{24}$ and the square of the minimum distance equals 4. So, it is an extremal even unimodular lattice in $\mathbb{R}^{24}$. Griess [205] tried to construct even unimodular lattices in $\mathbb{R}^{72}$ by a similar method to the above case, starting from the Leech lattice Λ_{24}. Namely, he tried to construct L using M, N isomorphic to Λ_{24} with $M \cap N = \sqrt{2}\Lambda_{24}$. Since there are too many choices of such M and N, Griess could not find an extremal even unimodular lattice in $\mathbb{R}^{72}$. However, Nebe could, in fact, find such good M and N with the property that the square of the minimum distance of L is 8, namely, L is an extremal even unimodular lattice in $\mathbb{R}^{72}$.

Now, let us return to the work of Venkov.

Theorem 5.30 (Venkov, 1984). *Let L be an extremal even unimodular lattice in $\mathbb{R}^n$. Then $n \equiv 0 \pmod 8$. Then for each even integer $2m$, we consider the shell L_{2m}. Then the following assertions hold:*
(1) *If $n \equiv 0 \pmod{24}$, then L_{2m} becomes an 11-design.*
(2) *If $n \equiv 8 \pmod{24}$, then L_{2m} becomes a 7-design.*
(3) *If $n \equiv 16 \pmod{24}$, then L_{2m} becomes a 3-design.*

(Note that since L is an even lattice, the square of the distance of any point from the origin is an even integer.)

The proof of this theorem is seen in Venkov [493]. See also Venkov [494]. The proof written in Japanese is also available in [32, Chapter 10, Section 2]. It is crucially important that the proof uses the theory of modular forms.

The relation between modular forms (theta series of lattices) and spherical designs through lattices is very deep, and the study was started by Boris Venkov. As mentioned in Section 1.6 of Chapter 1 in this book, this relation is an analogue of the relation between polynomials (weight enumerators of codes) and combinatorial designs, through codes. As for the general references on this topic, we refer the reader to Broué and Enguehard [112], Conway and Sloane [147], Runge [409], Ozeki [391], Nebe, Rains, and Sloane [373], Ebeling [175], and Huffman and Pless [397]. Broué and Enguehard (1972) [112] showed that modular forms are obtained from the weight enumerators of codes (Theorem 5.31 below). The concept of modular forms is very important and appears in number theory and related areas. We refer the reader to books on number theory for the details. Roughly speaking, a modular form is a good holomorphic function on the complex upper half-plane satisfying the good property (automorphic property) that by the action of the modular group $\mathrm{SL}(2,\mathbb{Z})$ it is just changed to a certain factor times the original function. The weight of a modular form is related to this factor. Here, we only consider modular forms of even weight and for the full modular group $\mathrm{SL}(2,\mathbb{Z})$.

Theorem 5.31 (Broué–Enguehard (1972) [112]). *Let C be a doubly even self-dual code of length n over the binary field F_2, and let $W_C(x,y) = \sum_{c\in C} x^{n-wt(c)}y^{wt(c)}$ be the weight enumerator of C. Then if we put $\theta_3(2\tau,0)$ to x and put $\theta_2(2\tau,0)$ to y, then $W_C(\theta_3(2\tau,0),\theta_2(2\tau,0))$ is a modular form of weight $n/2$. (Exactly speaking, these are modular forms with respect to the full modular group $\mathrm{SL}(2,\mathbb{Z})$.) Here, $\theta_2(2\tau,0)$ and $\theta_3(2\tau,0)$ are defined as follows:*

$$\theta_2(\tau,z) = \sum_{n\in\mathbb{Z}} n e^{\pi i(n+1/2)^2\tau+(2n+1)\pi iz},$$

$$\theta_3(\tau,z) = \sum_{n\in\mathbb{Z}} e^{\pi in^2\tau+2n\pi iz}.$$

Here, τ is an element in the upper half-plane, and z is an element of the complex number field $\mathbb{C}$. In the proof of this theorem, the variable τ is not needed, but it is sometimes better to use theta series in this form, so we keep this expression.

We call the map that assigns $\theta_3(2\tau,0)$ to x and $\theta_2(2\tau,0)$ to y the Broué–Enguehard map. This Broué-Enguehard map gives the isomorphism between the algebras generated by the weight enumerators of binary self-dual doubly even codes over F_2. This algebra gives an isomorphism from the invariant ring $\mathbb{C}[x,y]^G$, for the 2-dimensional complex reflection group No. 9 (of the Shephard–Todd classification) of order 192, to the subspace of modular forms, exactly speaking the subalgebra $\mathbb{C}[E_4, \Delta_{12}]$ generated by the Eisenstein series E_4 of weight 4 and Δ_{12}, a cusp form of weight 12. By this Broué–Enguehard map, the weight enumerator $W_{e_8}(x, y)$ for the Hamming code e_8 corresponds to E_4, but note that $W_{g_{24}}$, the weight enumerator of the Golay code g_{24}, does not corresponds to Δ_{12} directly, nor to the theta series of the Leech lattice directly.

The essence of the proof of the theorem of Broué and Enguehard is as follows. For a code C of length n over F_2, the *Construction A* to construct a lattice L_C in $\mathbb{R}^n$ is known. Namely, let φ be the natural homomorphism from $\mathbb{Z}^n$ to $(\mathbb{Z}/2\mathbb{Z})^n = F_2^n$. For a code C which is a subset of F_2^n, let $\varphi^{-1}(C)$ be the full inverse image of C and let $L_C = \frac{1}{\sqrt{2}}\varphi^{-1}(C)$. Then L_C is the desired lattice in $\mathbb{R}^n$. For example, for the Hamming code e_8, L_{e_8} becomes the E_8-lattice in $\mathbb{R}^8$ generated by the roots of type E_8. Generally, for a self-dual doubly even code C, L_C is an even unimodular lattice. Then from the theta series of the lattice L_C, we get a modular form of weight $n/2$. It is well known that the space of modular forms is given as $\mathbb{C}[E_4, E_6]$, where E_6 is the modular form called Eisenstein series of weight 6. Also, it is known that the image of the Broué–Enguehard map is the subspace $\mathbb{C}[E_4, \Delta_{12}]$ and not the whole space of modular forms. However, by considering in the following way, we can get all the space of modular forms. Let

$$\sigma_1 = \frac{1}{\sqrt{2}}\begin{bmatrix} 1 & 1 \\ 1 & -1 \end{bmatrix}, \; x \longmapsto \frac{1}{\sqrt{2}}(x+y), \; y \longmapsto \frac{1}{\sqrt{2}}(x-y),$$

$$\sigma_2 = \begin{bmatrix} 1 & 0 \\ 0 & \sqrt{-1} \end{bmatrix}, \; x \longmapsto x, \; y \longmapsto \sqrt{-1}y.$$

(Note that there is some difference in the notation given in Chapter 1, Section 1.6.) The group $G = \langle \sigma_1, \sigma_2 \rangle$ is the complex reflection group No. 9 of order 192. Then the map defined by $\sigma_1 \to -1$ and $\sigma_2 \to 1$ gives a homomorphism from G to the cyclic group of order 2. The kernel H of this homomorphism is a subgroup of index 2 of G and is also a complex reflection group, No. 8 and of order 96. Then

$$H = \left\langle \frac{1+i}{2}\begin{bmatrix} 1 & 1 \\ 1 & -1 \end{bmatrix}, \sigma_2 = \begin{bmatrix} 1 & 0 \\ 0 & \sqrt{-1} \end{bmatrix} \right\rangle.$$

The invariant ring $\mathbb{C}[x, y]^H$ for H is isomorphic to a polynomial ring and the Broué–Enguehard map gives the isomorphism between $\mathbb{C}[x, y]^H$ and the space of full modular forms $\mathbb{C}[E_4, E_6]$. Here, $\mathbb{C}[x, y]^H$ is generated by f_1, the weight enumerator of the Hamming $[8, 4, 4]$ code, and the homogeneous polynomial $f_2 = x^{12} - 33x^8y^4 - 33x^4y^8 + y^{12}$. Now, note that the Broué–Enguehard map maps f_2 to E_6. (As the negative sign appears, we can see that f_2 cannot be the weight enumerator of a code.)

The theorem of Broué–Enguehard means that the whole space of modular forms is understood to be isomorphic to the invariant ring of a certain finite group. There are many generalizations of this situation that the space of certain modular forms can essentially be understood as the space of the invariant ring of a certain finite group. (This isomorphism may not be completely isomorphic but at least we can grasp the essential part as the space of invariants by a finite group.) For example, Siegel modular forms of genus g (the case of $g = 1$ corresponds to the usual modular forms for $\mathrm{SL}(2, \mathbb{Z})$) can be understood close to the space of certain multiweight enumerators, and in turn the polynomial invariants by the Clifford group $G = 4 * 2^{1+2g} Sp(2g, 2)$ that is a subgroup of $\mathrm{GL}(2^g, \mathbb{C})$ (Duke, Runge); and certain Hilbert modular forms are related to the Lee weight enumerators of certain codes over F_p (Hirzebruch, van der Geer); and Jacobi modular forms are related to the invariants of the groups H or G of order 96 and 192 that appeared already, acting as the simultaneous diagonal action (Bannai–Ozeki, Runge). (In the Jacobi case, the variable z in the theta series matters.) In these cases, although all the modular forms are not obtained by this method, it seems that the essential parts are actually obtained. Also, many more generalizations are expected. In coding theory, codes not only over the finite field but also certain finite rings, as well as certain p-adic rings have been studied. And there are close connections between these general codes and various modular forms. Anyway, it seems to be very interesting to see that finite objects (like codes, weight enumerators, or invariant rings of finite groups) describe the essence of the infinite objects (like modular forms).

Considering the shells of a lattice as spherical designs is similar to considering the subsets of a fixed weight of a code as combinatorial designs. In this sense, we can regard Venkov's theorem as a spherical analogue of the Assmus–Mattson theorem.

All known examples of spherical t-designs obtained from shells of a lattice satisfy $t \leq 11$. This is very similar to the situation that all known combinatorial t-designs obtained as shells (we tentatively call a subset of a fixed size a shell) satisfy $t \leq 5$. Whether these restrictions, i. e., $t \leq 11$ for such spherical t-designs, or $t \leq 5$ for such combinatorial t-designs, are absolute restrictions or not seems to be a very important question at this stage.[5]

The method of using the orbits of a finite subgroup of $O(n)$ for $n \geq 3$ and the method of using the shells of a lattice in $\mathbb{R}^n$ are good methods to find spherical t-designs. But it seems difficult to get t-designs for large t. Then, a question arises. Do there exist spherical t-designs for large t? The following theorem of Seymour and Zaslavsky [422] answered this question decisively.

Theorem 5.32 (Seymour–Zaslavsky (1984) [422]). *For any positive integers n and t, spherical t-designs on S^{n-1} exist.*

One proof of this important theorem is given in [32], so we will not repeat that proof. The existence of spherical t-designs and interval t-designs (that will be explained later) was first proved by P. D. Seymour and T. Zaslavsky (1984) [422]. A proof

5 As a generalization of relations between codes and combinatorial designs, or lattices and spherical designs, there is a relation between vertex operator algebras and quantum designs ([236, 338, 347]).

of the interval t-designs which is relatively easy to understand was obtained by Arias de Reyna (1988) [4]. Meanwhile, G. Wagner (1991) [499], P. Rabau and B. Bajnok (1991) [398], and Bajnok (1992) [15] proved that if there exists an interval t-design X on the interval $[-1, 1]$, then t-designs of size $(t + 1)|X|$ exist on S^2. Moreover, they proved that if there exists an interval t-design on the interval $[-1, 1]$ with Gegenbauer weight $w(x) = (1-x^2)^{\frac{n-3}{2}}$, then spherical t-designs on S^{n-1} are constructed. As is seen from the proof of Proposition 5.35 of the simple case, the separation of variables of integrals is utilized here. So, we would like to start with the discussion of the existence and the construction of interval t-designs on $[-1, 1]$. The proofs in the original papers [422, 4] were just the existence proofs. So, the information on the size $|X|$ was not obtained. On the other hand, in the proof of [499] and [398], the existence of interval t-design was obtained in a way that the size $|X|$ is bounded from above. The proofs for the existence of spherical t-designs on S^2 are done through constructions by using interval designs whose sizes are of order $O(t^3)$. This result was best possible at the stage of the early 1990s. Also, as for the spherical t-designs on S^{n-1}, the existence of designs whose sizes are of order $O(t^{\frac{n^2-n}{2}})$, which was proved by Korevaar and Meyers [287], was the best result at that time.

For interval t-designs X on $[-1, 1]$, Bernstein (1937) [86] showed that if $t \to \infty$, then $|X|$ must be at least of order $O(t^2)$. Moreover, Bernstein (1937) [87] obtained the result that if we allow multiple points, i. e., if x_i and x_j are not necessarily different, then there are interval designs whose sizes are of order $O(t^2)$. (However, this result was not known widely.) Kuijlaars [290] excavated this paper of Bernstein and succeeded in further proving that slightly moving (deforming) Bernstein's multiple point interval t-design, one can separate multiple points and can get a real interval t-design of order $O(t^2)$ whose points are all distinct.

On the other hand, it seems that many researchers have conjectured that there should be spherical t-designs on S^{n-1} of order $O(t^{n-1})$. There was some progress toward this conjecture by Bondarenko and Viazovska [98, 99] and then the conjecture was finally fully settled by Bondarenko, Radchenko, and Viazovska [100, 101]. The Fisher type lower bound of sizes of spherical t-designs on S^{n-1} implies that if n is fixed and t goes to ∞, then it is of order $O(t^{n-1})$. Therefore no essential improvement is possible, except for the exact determination of the coefficients. On the other hand, it seems to be still unknown what will happen if t is fixed and n goes to ∞.

So far, we have discussed the existence problem of spherical t-designs. Besides the results already mentioned, there are some related works, for example, by Hardin and Sloane [212], Mhaskar, Narcowich, and Ward [345], Chen, Frommer, and Lang [133], Chen and Wormersley [134], and Gröf and Pott [203].

In most of the papers mentioned above, including Seymour and Zaslavsky, their existence proofs were either analytic or topological, and use the continuity of real numbers crucially. Also, in most cases, the proofs were existence proofs, and so they were not useful for the explicit construction of the t-designs. The following paper by

Kuperberg [291] exceptionally gives an explicit construction of interval t-designs. Let us discuss it below. (However, it is very delicate to what extent the constructions are explicit.) Using this result of Kuperberg, Okuda [390] gives an explicit construction of spherical t-designs on S^3 (see also Cohn, Conway, Elkies, and Kumar [138] for another approach of this result). Reznick [403] may be interesting in connection with spherical designs.

Definition 5.33 (Interval t-design). Let t be a positive integer, and let $w(x)$ be a weight function on the interval $[-1,1]$. A subset $\{x_1,x_2,\dots,x_N\}$ of $[-1,1]$ is called an *interval t-design* on $[-1,1]$ with respect to the weight function $w(x)$, if the following equality holds for any polynomial $f(x)$ of degree at most t:

$$\frac{1}{\int_{-1}^{1} w(x)dx}\int_{-1}^{1} f(x)w(x)\,dx = \sum_{i=1}^{N} w(x_i)f(x_i).$$

Kuperberg discovered a method to construct interval t-designs on the interval $[-1,1]$ for $w(x)\equiv 1$.

Theorem 5.34 (Kuperberg [291]). *Let $Q_s(x)$ be the polynomial of degree s defined by*

$$Q_s(x) = x^s - \frac{x^{s-1}}{3} + \frac{x^{s-2}}{3\cdot 15} - \cdots + \frac{(-1)^s}{1\cdot 3\cdot 15\cdot\,\cdots\,\cdot(4^s-1)}.$$

Then the following assertions hold:

1. *The polynomial $Q_s(x)$ has s positive real zeros $\alpha_1,\alpha_2,\dots,\alpha_s$ which are distinct from each other.*
2. *If we define $Z=\{\pm\sqrt{\alpha_1}\pm\sqrt{\alpha_2}\cdots\pm\sqrt{\alpha_s}\mid$ take all the choices of signs$\}$, then $|Z|=2^s$ and $Z\subset(-1,1)$. (Note that in the original paper [291] and also in [32], there are typos that the square roots in $\sqrt{\alpha_i}$'s are missing.)*
3. *The set Z gives a Chebyshev type quadrature formula of degree $2s+1$. Namely, the 2^s points give an interval $(2s+1)$-design on $[-1,1]$ with respect to the weight function $w(x)\equiv 1$.*

As mentioned already, if there is an interval t-design on $[-1,1]$ with respect to the weight function $w(x)\equiv 1$, then a spherical t-design on S^2 is constructed, and if there is an interval t-design on $[-1,1]$ with respect to the Gegenbauer weight function $w(x)=(1-x^2)^{\frac{n-3}{2}}$, then a spherical t-design on S^{n-1} is constructed, as mentioned in Wagner [499], Rabau and Bajnok [398], and Bajnok [14]. Now we will see how spherical $(2s+1)$-designs on S^2 are constructed from the interval $(2s+1)$-design, or Chebyshev type quadrature, $Z=\{z_1,z_2,\dots,z_{2^s}\}$ on $[-1,1]$ which is constructed by Kuperberg (by Theorem 5.34). Now, for an integer $m\ge 2s+2$, let Y be the set of the m vertices of a regular m-gon inscribed in S^1. For example, Y is given as $Y=\{y_k=(\cos\frac{2\pi k}{m},\sin\frac{2\pi k}{m})\mid k=0,1,\dots,m-1\}$. Since $m\ge 2s+2$, Y is a $(2s+1)$-design. Now we define the subset X

on S^2 as follows:

$$X = \{(r_i y_k, z_i) \mid 1 \le i \le 2^s, 0 \le k \le m-1\}.$$

Here, $r_i = \sqrt{1 - z_i^2}$ $(1 \le i \le 2^s)$. So, X is a subset of size $2^s m$ on S^2.

Proposition 5.35. *The set X defined above becomes a spherical* $(2s+1)$*-design on* S^2.

Proof. We consider monomials $x_1^{\lambda_1} x_2^{\lambda_2} x_3^{\lambda_3}$ $(\lambda_1 + \lambda_2 + \lambda_3 \le 2s+1)$ of degree at most $2s+1$ on $S^2 \subset \mathbb{R}^3$. Since Y is an $(m-1)$-design on S^2 and since $\lambda_1 + \lambda_2 \le 2s+1 \le m-1$, we get

$$\frac{1}{2\pi} \int_{S^1} x_1^{\lambda_1} x_2^{\lambda_2} \, d\sigma x = \frac{1}{m} \sum_{k=0}^{m-1} \left(\cos \frac{2\pi k}{m}\right)^{\lambda_1} \left(\sin \frac{2\pi k}{m}\right)^{\lambda_2}. \tag{5.42}$$

On the other hand, we have

$$\begin{aligned}
&\frac{1}{|X|} \sum_{x \in X} x_1^{\lambda_1} x_2^{\lambda_2} x_3^{\lambda_3} \\
&= \frac{1}{2^s m} \sum_{i=1}^{2^s} \sum_{k=0}^{m-1} \left(r_i \cos \frac{2\pi k}{m}\right)^{\lambda_1} \left(r_i \sin \frac{2\pi k}{m}\right)^{\lambda_2} z_i^{\lambda_3} \\
&= \frac{1}{2^s m} \sum_{i=1}^{2^s} \left(\sqrt{1 - z_i^2}\right)^{\lambda_1 + \lambda_2} z_i^{\lambda_3} \sum_{k=0}^{m-1} \left(\cos \frac{2\pi k}{m}\right)^{\lambda_1} \left(\sin \frac{2\pi k}{m}\right)^{\lambda_2}.
\end{aligned} \tag{5.43}$$

First, let us consider the case when $\lambda_1 + \lambda_2$ is odd. Then either λ_1 or λ_2 is odd. Therefore, we get

$$\frac{1}{4\pi} \int_{S^2} x_1^{\lambda_1} x_2^{\lambda_2} x_3^{\lambda_3} \, d\sigma x = 0, \quad \frac{1}{2\pi} \int_{S^1} x_1^{\lambda_1} x_2^{\lambda_2} \, d\sigma x = 0.$$

Therefore, by (5.42) and (5.43), we have

$$\frac{1}{4\pi} \int_{S^2} x_1^{\lambda_1} x_2^{\lambda_2} x_3^{\lambda_3} \, d\sigma x = \frac{1}{|X|} \sum_{x \in X} x_1^{\lambda_1} x_2^{\lambda_2} x_3^{\lambda_3} \; (= 0).$$

Next, let us consider the case where λ_1 or λ_2 is even. Then, since Z is a quadrature formula of degree $2s+1$ (i. e., interval t-design) on the interval $[-1, 1]$, we have

$$\frac{1}{2} \int_{-1}^{1} x^{\lambda} \, dx = \frac{1}{2^s} \sum_{i=1}^{2^s} z_i^{\lambda}$$

for any integer λ with $0 \le \lambda \le 2s+1$. In particular, since λ_1 or λ_2 is even, we get

$$\begin{aligned}
\frac{1}{2^s} \sum_{i=1}^{2^s} (1 - z_i^2)^{\frac{\lambda_1 + \lambda_2}{2}} z_i^{\lambda_3} &= \frac{1}{2} \int_{-1}^{1} (1 - x^2)^{\frac{\lambda_1 + \lambda_2}{2}} x^{\lambda_3} \, dx \\
&= \frac{1}{2} \int_{0}^{\pi} \sin^{1+\lambda_1+\lambda_2} \psi \cos^{\lambda_3} \psi \, d\psi.
\end{aligned} \tag{5.44}$$

Also, using the polar coordinates $x_1 = \cos\theta\sin\psi$, $x_2 = \sin\theta\sin\psi$, $z = \cos\psi$, the integral over S^2 is represented as follows:

$$\begin{aligned}
&\frac{1}{4\pi}\int\limits_{S^2} x_1^{\lambda_1}x_2^{\lambda_2}x_3^{\lambda_3}\,d\sigma x \\
&= \frac{1}{4\pi}\int\limits_0^{2\pi}\int\limits_0^{\pi}(\cos\theta\sin\psi)^{\lambda_1}(\sin\theta\sin\psi)^{\lambda_2}(\cos\psi)^{\lambda_3}\sin\psi\,d\psi\,d\theta \\
&= \left(\frac{1}{2\pi}\int\limits_0^{2\pi}\cos^{\lambda_1}\theta\sin^{\lambda_2}\theta\,d\theta\right)\left(\frac{1}{2}\int\limits_0^{\pi}\sin^{1+\lambda_1+\lambda_2}\psi\cos^{\lambda_3}\psi\,d\psi\right).
\end{aligned}$$

Therefore, by (5.42), (5.43), and (5.44), we obtain

$$\frac{1}{4\pi}\int\limits_{S^2} x_1^{\lambda_1}x_2^{\lambda_2}x_3^{\lambda_3}\,d\sigma x = \frac{1}{|X|}\sum_{x\in X} x_1^{\lambda_1}x_2^{\lambda_2}x_3^{\lambda_3}. \tag{5.45}$$

□

The construction of spherical t-designs on S^{n-1} using interval t-designs on $[-1,1]$ with respect to the Gegenbauer weight is proved in a similar way. The following two questions seem to be interesting as research problems.

Problem 1: Find a method to concretely construct interval t-designs on $[-1,1]$ with respect to the Gegenbauer weight. Is it possible to extend Kuperberg's method to this case?

Problem 2: Is it possible to find concrete constructions of interval t-designs without using the continuity of real numbers? In particular, are there interval t-designs whose points are all rational numbers[6]?

Now, let us return to the spherical t-designs obtained from shells of a lattice.

1. When we consider spherical t-designs obtained from orbits of a finite group $G \subset O(n)$, we mentioned in Theorem 5.27 that if one is a t_1-design and the other is a t_2-design, then one is not extremely large compared with the other, i. e., we get $t_2 \le 2t_1 + 1$. We wonder whether a similar situation can happen for two spherical t-designs obtained as shells of the same lattice. It is shown easily that this does not hold unconditionally. Namely, we can find an example of a lattice with one shell having a t_1-design (although $t_1 \le 11$ anyway) and another shell is only a 1-design. But under some restricted conditions such as the condition that both designs generate the whole lattice, can we say that t_1 is not too large compared with t_2?

6 This was answered affirmatively by Cui, Xia, and Xiang [152] in 2019.

2. When we consider the shells of the E_8-lattice L, each shell of L is a spherical 7-design. This was seen from the theorem of Venkov (Theorem 5.30), or from the fact that the shell is a union of some orbits of the reflection group $W(E_8)$. This, of course, means that each shell is a 7-design, but this does not exclude the possibility that it becomes an 8-design or t-design for bigger t. This is a very simple and natural question. Actually, this has an unexpected important meaning, as we will see next.

Theorem 5.36 (Venkov, de la Harpe, Pache [392, 165, 166]). *Let L_{2m} be a shell of the E_8-lattice. Then L_{2m} becomes an 8-design if and only if $\tau(m) = 0$, where Ramanujan's tau function τ is defined by*

$$\eta(q)^{24} = q\prod_{i\geq 1}(1-q^i)^{24} = \sum_{k\geq 1}\tau(k)q^k. \tag{5.46}$$

Here η denotes the Dedekind eta function.

The reason why this theorem is interesting is closely related to the famous conjecture in number theory called Lehmer's conjecture (Lehmer [309], Serre [420, 421]).

Lehmer's conjecture. For each integer $k \geq 1$, $\tau(k) \neq 0$.

So, Lehmer's conjecture was reformulated in terms of spherical t-designs. However, there are still great difficulties to prove Lehmer's conjecture, and this still remains an open problem at the present stage. Many analogues or extensions of Lehmer's conjecture are known. (Pache [392] gives a detailed explanation about this.) The most simple analogue of Lehmer's conjecture is reformulated as follows.

Problem I: It is easy to see that any shell L_m of the lattice $L = \mathbb{Z}^2$ in $\mathbb{R}^2$ is a spherical 3-design. (Because the cyclic group of order 4 of 90-degree rotation makes L_m invariant.) Is there any 4-design among these shells L_m?

Problem II: It is easy to see that any shell L_{2m} of the A_2-lattice (hexagonal lattice) L in $\mathbb{R}^2$ is a spherical 5-design. (Because the cyclic group of order 6 of 60-degree rotation makes L_{2m} invariant.) Is there any 6-design among these shells L_{2m}?

We call these problems *toy models* of Lehmer's conjecture. These two problems were solved negatively by Bannai and Miezaki (2010) [67]. Here the theory of modular forms also played an important role.

Theorem 5.37 (Bannai–Miezaki (2010) [67]). *The following hold:*
(1) *In the integer lattice $L = \mathbb{Z}^2$ in $\mathbb{R}^2$, any shell L_m cannot be a 4-design.*
(2) *In the A_2-lattice in $\mathbb{R}^2$, any shell L_m cannot be a 6-design.*

A generalization for other 2-dimensional lattices can be found in [69]. A proof of the above theorem that does not use modular forms can be found in [70]. Let us recall again that for any 2-dimension lattice, no shell that is a 6-design is known. Meanwhile, for any 3-dimensional lattice, no shell that is a 4-design is known. In addition, in any lattice in any dimension, no shell that is a 12-design is known. These are very important and interesting open problems. We propose another open problem that was proposed by Yudin.

Problem. Let us consider the lattice $\mathbb{Z}^2$. We have explained that the shell of the lattice, namely, the points on the circle whose center is at the origin, cannot be a 4-design. Is there any 4-design on a circle whose center is not necessarily at the origin? We can consider similar problems for other lattices.

The following result is not directly related to spherical t-designs. But we would like to mention it in order to advertise this result.

Theorem 5.38 (Bannai–Miezaki (2012) [68]). *Let L be any integral lattice in $\mathbb{R}^2$. For any positive integer N, there is a circle that contains exactly N points of L.*

For the proof of this theorem, we need the theory of 2-dimensional integral lattices and some number theoretical method related to class field theory. As for the background of this problem, see Maehara [329]. In addition, if we use this result for 2-dimensional lattices, it is possible to show that for any positive integer N and for any integral lattice in $\mathbb{R}^n$ ($n \geq 3$), there is a sphere that contains exactly N points in L, although this extension for higher dimensions is rather formal (and not substantial).

We would like to remark about the shells of extremal even unimodular lattices. As we have seen in Theorem 5.30 (Venkov's theorem), if $n \equiv 0 \pmod{24}$, then any shell is an 11-design. Also, for $n \equiv 8 \pmod{24}$ and $n \equiv 16 \pmod{24}$, then any shell is a 7-design and a 3-design, respectively. In the case of $n = 24$, the smallest shell of the Leech lattice has 196, 560 points and is a tight 11-design, and so it cannot be a 12-design. In the case of $n = 8$, the smallest shell of the E_8-lattice has 240 points and is a tight 7-design and so it cannot be an 8-design. In the case of $n \equiv 0 \pmod{24}$, can the smallest shell of an extremal even unimodular lattice be a 12-design? Or more strongly, is there any case where all the shells are 12-designs? Also, in the cases $n \equiv 8 \pmod{24}$ and $n \equiv 16 \pmod{24}$, is there any possibility that all the shells are 8-designs and 4-designs, respectively? These problems are simple, but the answer is not yet obtained completely. We can see a partial answer in [66]. (This might be also interesting from the viewpoint of modular forms.) Note that similar questions can be made on the shells of extremal Type II codes. It is interesting that in case of extremal codes, all the shells have the same strength (Janusz [267], [66]). However, whether there is actually an extremal code whose shells have a larger strength t than the value of t determined by Venkov's theorem is still an open problem. (No such examples are known.) For some partial results, please see [66]. Horiguchi, Miezaki, and Nakasora [239] and Miezaki and Nakasora [349] give some partial results improving previously known results.

(c) Spherical designs appearing in physics and chemistry

Now, we list what we think interesting in an informal way. Frankly speaking, the contents here are the authors' personal impression and also meant to be a guidance for the readers what references to look at. Of course there are many other interesting and deep relations between spherical designs and physics or chemistry that we are not aware of. So, we hope and expect that the contents could be supplemented by other experts. We will list eight references or topics (i)–(viii), and make some comments on them. The first two expository articles were already mentioned in [32]. The rest arises after the publication of [32].

(i) Saff and Kuijlaars, "Distributing many points on a sphere" [410]. (Many references mentioned there are also very useful.)
(ii) Smale, "Mathematical problems for the next century" [430].
(iii) M. Atiyah and P. Sutcliffe, "Polyhedra in Physics, Chemistry and Geometry" [7]. Milano J. math. 71 (2003), 33–58.
(iv) H. Cohn, "Order and disorder in energy minimization" [137]. Proceedings of the International Congress of Mathematics, Volume IV, Hindustan Book Agency, New Delhi (2010), 2416–2443 (see also [141]).

The main topic of papers (i) and (ii) is to find finite subsets of the sphere S^{n-1} for a given potential function $f : [0,2] \to \mathbb{R}$, which have the property that among all the subsets of the same size, the energy for f is minimum. For example, the most important problem is the next problem, called the Coulomb–Thomson problem. Let the potential function be $f(r) = \frac{1}{r}$ and let the energy (for the potential function f) of a subset Y be defined by

$$\sum_{x,y\in Y, x\neq y} \frac{1}{\|x-y\|}.$$

Here, we fix a positive integer N, and to try to find and/or determine a subset X of size N with the smallest energy among all subsets Y of the same size N. If we take a different potential function f, of course the minimum energy subsets are generally different. The following question:

"Is there a finite subset that is the minimum energy subset for all good potential functions?"

is the key to the concept of universally optimal codes proposed by Cohn and Kumar. Of course, there are many choices of potential functions, and these potential functions usually come from the requirement from physics and/or chemistry. We explain some results that we think interesting in the following (α) to (ζ). (For the notation, the reader is referred to the original source.)

(α) There is no universally optimal code of size 5 on S^2. However, for any given completely monotonic potential function f, it seems and is conjectured that the energy minimizing sets are either (1) 5 points consisting of three points of the equilateral triangle on the equator and north and south poles, or (2) 4 points on the square slightly south of the equator and the north pole (although the exact shape depends on the potential function f). (Many experiments confirm this claim, but it is not rigorously proved yet. See [144].)

(β) The 8 points of a cube and the 12 points of a regular dodecahedron are not universally optimal, and not even optimal either. However, it is shown that there are some potential functions that these two sets are energy minimizing sets for those potential functions. Examples of these potential functions are also explicitly given ([142]).

(γ) As will be discussed in Section 5.2.1 in this chapter, the concept of universally optimal codes is extended to compact symmetric spaces of rank 1 (projective spaces over $\mathbb{R}, \mathbb{C}, \mathbb{H}, \mathbb{O}$). For example, $\mathbb{PR}^2$ is the 2-dimensional projective space, and this is the space where the lines through the origin in $\mathbb{R}^3$ are regarded as points of $\mathbb{PR}^2$. Cohn and Woo [144] classified universally optimal codes on $\mathbb{PR}^2$ completely. Namely, they are described as follows:
  (1) less than 3 lines that are orthogonal to each other;
  (2) the 4 lines that connect antipodal points of the cube;
  (3) the 6 lines that connect antipodal points of a regular icosahedron;
  (4) the $3 + 4 = 7$ points that connect antipodal points of the union of a cube and its dual (the regular octahedron).

  It is interesting to note that in example (4), the automorphism group of the universally optimal code cannot be transitive on the set of points. This might suggest that a similar situation might happen for universally optimal codes on the sphere.

(δ) Although the details will not be explained here, the concept of universally optimal codes is defined for lattices (or more generally for periodic infinite sets). Some study on this topic has already started. The A_2-lattice, the D_4-lattice, the E_8-lattice, the Leech lattice Λ_{24}, and possibly D_{16}^+ are expected to be universally optimal codes, although these are not yet verified. Coulangeon and Schürman [150] have shown that if a lattice has the property that all the shells are spherical 4-designs, then the lattice is "locally" universally optimal.

(ϵ) Good finite point sets that appear in physics and chemistry are not limited on the sphere. One method is to consider the energy function such as the one appearing in [7],

$$\sum_{x,y\in X, x\neq y} \frac{1}{||x-y||} + \sum_{x\in X} ||x||^2,$$

or generally, to consider the energy function

$$\sum_{x,y\in X, x\neq y} f(||x-y||) + \sum_{x\in X} ||x||^2,$$

which involves the distances from the origin. More generally, the energy functions like

$$\sum_{x,y\in X, x\neq y} \left(\frac{1}{\|x-y\|^{12}} - \frac{1}{||x-y||^6} \right)$$

are considered. Here, we are interested in grasping the cluster of particles in the Euclidean space $\mathbb{R}^n$ as an energy minimizing set for certain energy functions.

(ζ) The following is a very casual idea of the first author (Eiichi Bannai). We want to propose a definition of universally optimal code in $\mathbb{R}^n$. We would like to hear the feedback from the readers whether this definition is appropriate or not. (Please contact us if you have any comments.)

Definition. For a subset X in $\mathbb{R}^n$, the center of X is defined by

$$x_0 = \frac{1}{|X|} \sum_{x \in X} x.$$

The deviation $m(X)$ of X is defined by

$$m(X) = \frac{1}{|X|} \sum_{x \in X} \|x - x_0\|^2.$$

Then X is called optimal, if the condition

$$\min_{\substack{x,y \in X \\ x \neq y}} d(x, y) \geq \min_{\substack{x,y \in Y \\ x \neq y}} d(x, y)$$

holds for any finite subsets Y in $\mathbb{R}^n$ with $|Y| = |X|$ and $m(Y) = m(X)$. Moreover, X is called universally optimal if

$$\sum_{x,y \in X, x \neq y} f(\|x - y\|^2) \leq \sum_{x,y \in Y, x \neq y} f(\|x - y\|^2)$$

holds for any finite subsets Y in $\mathbb{R}^n$ with $|Y| = |X|$ and $m(Y) = m(X)$ and for any complete monotonic function $f(x)$ (Definition 5.24).

We hope that this concept could be related to the concept of Euclidean t-designs that we will discuss in the next subsection. From a very personal attachment of the first author, the concept of rigid spherical t-designs (namely, those t-designs that cannot be deformed keeping the property of being t-designs) should reflect some physical and/or chemical properties well, and their classification problem seems interesting.

(v) There are many work on fullerenes.

It is beyond our ability to discuss this topic. This started with the discovery of C_{60} and the study of such objects on spheres or on some layers of spheres should be closely related with the research described in this book. We expect that the study of finite sets on the layers of several spheres should closely be related with the study of Euclidean designs, which we will discuss in the next subsection.

There are researchers with a strong interest in the areas of algebraic combinatorics and related topics; there are works by Deza, Shpectrov, Klin, Kerber, and others. (We hope that the interested reader can find the details of these relevant works for themselves.)

(vi) The "Optimal Configurations on the Sphere and Other Manifolds (Optimal 2010)" conference was held in Nashville, Tennessee in May 2010.

At this conference, optimal objects (in various senses) in mathematics and other related areas including physics and chemistry were studied. In the sense of studying many good finite point sets on the earth (the sphere), there were some talks from the geocentrism viewpoint. From the viewpoint of analysis, the cubature formulas were very much studied (as will be mentioned in the next subsection). There were many interesting studies. One question that was presented was: What is the best density of filling the Euclidean space with the ellipsoid of the same size (instead of the sphere of the same size). Interestingly, in the 2-dimensional plane, it is the same as in the sphere case. But in dimensions $n \geq 3$, the density can get smaller (Bezdek and Kuperberg [90] and Schürmann [415]). (In Japan there are some works by the group of Odagaki on this topic.) Anyway, it seems that there are many interesting new directions. This conference was subsequently followed up by the following conferences: Vienna in October 2014, ICERC, Brown University in May 2018, and Vienna again in January 2021.

As for the relation of spherical t-designs and physics, one of the authors is specifically interested in the paper of Crann, Pereira, and Kribs [151].

(vii) Spherical t-designs seem to be closely related to anticoherent spin states in physics.

Conjecture (Conjecture 1 (page 8) in [151]). *A spin state is anticoherent if and only if its Majorana representation is a spherical t-design.*

Unfortunately, counter-examples were obtained for both directions by Bannai and Tagami. It would be very interesting if we could say anything positive by modifying this conjecture.

(viii) The book [442] of Sunada, *Why do diamonds look so beautiful?*

We are very much interested in and excited about this book. Unfortunately, it seems that this book is published only in Japanese. (Please also see the article of Sunada in the Notice of AMS [443].) Sunada [442] is based on the series of papers of Kotani and Sunada ([288], etc.). The part that is closely related to spherical designs is as follows. For any finite graph, which is not necessarily a simple graph, construct a finite set X of vectors in $\mathbb{R}^n$ satisfying condition (2) of Theorem 3.3 in [442, page 107]:

$$\sum_{x \in X} (x \cdot u)^2 = \text{constant}. \tag{5.47}$$

Here, u is an arbitrary unit vector in $\mathbb{R}^n$. Then, using this set X, we can construct a crystal lattice. Note that the condition (5.47) is one of the important parts of X to be a spherical 2-design. Generally, these points of X are not necessarily on one sphere. However, we observed that if we take a strongly regular graph (or some good graph), then X is on a single sphere (see some Master's theses at Kyushu University and Junichi Shigezumi's report [425]). Although it was not easy to prove this, Hirotake Kurihara [294], in his Master's thesis, proved this result in a very general form. Namely, he succeeded in proving this for any graph attached to an association scheme. (We believe this result is interesting, although it is not published in a journal.) It would be an important problem to consider which of those X obtained in this way is interesting as spherical designs. (However, it is seen that spherical t-designs for large t cannot appear in this way.) We also would like to mention that the condition (5.47) is the condition that X is a *tight frame*. It seems that there are many researchers interested in tight frames (Miezaki and Tagami [348], Kotelina and Pevny [289], and Bachoc and Ehler [11]; see also Waldron [500]).

In concluding this subsection, we would like to mention one more thing. Larman, Rogers, and Seidel [305] studied 2-distance sets in the real Euclidean space $\mathbb{R}^{n-1}$. They proved that if the size of a 2-distance set is not small, then the ratio of the squared lengths of the 2 distances are the ratio $\frac{k-1}{k}$ of some consecutive positive integers $k-1$ and k. This result of Larman, Rogers, and Seidel is often used efficiently in the study of spherical or Euclidean t-designs. It was hoped to generalize their result to 3-distance sets, and this was solved by Nozaki [386] in a very beautiful way. Bannai and Bannai [34] found that this integer k can be read from the character table if the 2-distance set comes from a strongly regular graph. Kurihara and Nozaki [297] generalized this for Q-polynomial association schemes. The study in this direction is showing a very interesting and deep progress.

5.2 Study of finite subsets on other spaces

5.2.1 Finite subsets on projective spaces (compact symmetric spaces of rank 1)

The study of finite subsets on the sphere is generalized to the study of finite subsets on the real projective space. The real projective space is obtained by identifying two antipodal points of the sphere. Therefore, we can regard the study of finite subsets on the real projective space as the study of antipodal subsets on the sphere. In this sense, some of the results mentioned in the previous sections can be regarded as the results on the real projective space. The projective spaces over the real and complex spaces are identified as the space of 1-dimensional subspaces of the vector space over each of these fields. In this sense, the study of equiangular lines has been made. See Delsarte, Goethals, and Seidel (1975) [162] and Koornwinder (1976) [286]. Besides real and complex fields, there are projective spaces over the quaternion (skew) field and the Cayley octonion. (Only the projective plane exists over the Cayley octonion.) These spaces are, together with the sphere, called compact symmetric spaces of rank 1. It is shown by Cartan [124] (work in the 1920s) that these are the only connected compact symmetric spaces of rank 1. Also, H.-C. Wang (1952) [503] showed that the compact symmetric spaces of rank 1 are characterized as compact 2-point homogeneous symmetric spaces. Note that this last property essentially corresponds to the distance-transitive property, and therefore in the case of graphs, to the property of distance-regular graphs or P-polynomial association schemes.

Some attempts to look at these spaces in a more general framework have been made. Neumaier (1981) [375] introduced the concept of "Delsarte spaces" to consider t-designs on both Q-polynomial association schemes and compact symmetric spaces of rank 1. Hoggar (1982) [233] studied finite subsets on projective spaces, by developing general theory and discussing many examples of t-designs, and then made it clear that a very similar theory to those on the sphere is developed on these projective spaces. The study of tight t-designs is also made on these projective spaces (Bannai and Hoggar (1989) [58], Hoggar (1989) [234], Hoggar (1990) [235]). The details will not be explained here, but the reader is referred to original papers of Hoggar. (A short explanation in Japanese is given in [32, Chapter 13].)

Levenshtein called these spaces "polynomial spaces" and it looks that his formalization is essentially similar to Neumaier [375] and Godsil (1988) [198]. Levenshtein studied finite subsets on these spaces in a very thorough and yet very useful way. The reader is referred to the original papers of Levenshtein [313, 315]. As subsequent developments, there are some works by the group of Boyvalenkov [107, 108, 109], Lyubich and Vaserstein [321], Lyubich [322], and others. We think it is safe to say that the study of finite subsets on compact symmetric spaces of rank 1 is as well developed as that on spheres. It is interesting to remark that the existence of tight 2-designs on the complex projective space is equivalent to the existence of Symmetric Informationally Complete Projective Operators (SIC-POVMs), and the existence of Mutually Unbiased Bases (MUBs) is closely related to the existence of a very special kind of 2-designs. The existence of these objects is an important topic in quantum physics, in particular, quantum information theory. For more details, the reader is referred to Scott [418] and Roy and Scott [407].

5.2.2 Finite subsets on compact symmetric spaces of general ranks

The compact symmetric spaces of a general rank were classified by E. Cartin around 1920 (Helgason (1962) [213], Wolf [519], mathematical dictionaries, etc.).

Now, we will see that finite subsets (codes and designs) on these spaces are very much studied. Typical compact symmetric spaces of rank greater than 1 are real Grassmann spaces. Finite subsets there have been studied extensively by Shor, Sloane, Calderbank, Hardin, Rains, and others. This work came from the study of error correcting codes in quantum computers, indeed a very interesting topic. The reader is referred to [427, 118]. For a simple explanation in Japanese, please see [32, Chapter 15, Section 15.6].

The study of finite subsets of real Grassmann spaces from the viewpoint of algebraic combinatorics was started by Bachoc, Coulangeon, and Nebe (2002) [10], and then further developed in Bachoc, Bannai, and Coulangeon (2004) [9]. There, the concepts of t-designs and tight t-designs were defined, and a similar theory to the case of those on the spheres has been developed, although the classification of tight t-designs is not yet within reach. In the case of compact rank 1 symmetric spaces, 1-variable Jacobi polynomials appear as spherical functions. In the case of real Gassmann space, multivariable orthogonal polynomials appear. In fact, in the general compact symmetric rank r, certain generalized Jacobi polynomials of r variables appear (the reader is referred to Vretare (1984) [498] or further more general work by Koornwinder (cf. [285])).

Roughly speaking, there are two kinds of compact symmetric spaces. One class consists of the homogeneous spaces G/H of simple Lie groups G by subgroups H, and another class consists of the compact simple Lie groups G themselves. (The second class is also interpreted as $(G \times G)/G$ by the diagonal subgroup isomorphic to G.) The real Grassmann space is one in the first class, and the work on the real Grassmann case is generalized for other symmetric spaces in the first class. (For the complex Grassmann case, see Y. Miura [350] [the Master's thesis of Kyushu University (2004)] and Roy (2010) [406]. There was an unpublished generalization by Takanori Yasuda on this topic.)

In the case of a simple Lie group G itself, the concept of t-designs is defined for finite subsets X of G. In particular, for the unitary group $U(d)$, this is given in Roy and Scott [407]. The concept of tight t-designs is also defined there, but the classification problem is still difficult to handle, and the problem still seems to be open. It seems that the situation is the same for other simple Lie groups. (We wonder whether any work has been done already on this topic.) Meyer [344] considered certain designs in $U(d)$. It seems that these designs are not the same as the t-designs in Roy and Scott [407], but related to the Hermite constants of lattices, and so it may have a number theoretical interest. (It seems that the work of Gularnich and Tiep [206] and Tiep [481] might be related with the work of Meyer [344].) In these definitions of finite subsets of $U(d)$, designs are defined basically for certain subsets of irreducible representations of $U(d)$. We also remark that there are some (unpublished) works of Takayuki Okuda that consider designs for subsets of irreducible representations of any compact group. This situation is a kind of analogy of considering T-designs in an association scheme $\mathfrak{X} = (X, \{R_i\}_{0 \le i \le d})$ for a subset $T \subset \{1, 2, \ldots, d\}$. Since there are many possible subsets T, it would be important and interesting to consider what subsets T should be taken.

Another generalization of finite subsets on the sphere would be the study of finite sets on complex spheres. This is related to the study of finite sets on the complex projective space or on the unitary group $U(d)$, but there are some differences. For this study, please see Roy and Suda [408]. This study is also related to MUBs, quantum information theory, and quantum physics. The following are some papers related to design theory in quantum information theory, e.g., [401, 419, 520].

For the discussion of unitary t-designs in $U(d)$ as well as spherical t-designs, we would like to add the following recent papers.

- E. Bannai, G. Navaro, N. Rizo, and P. H. Tiep, Unitary t-groups, J. Math. Soc. Japan 72 (2020), 909–921;

- E. Bannai, M. Nakahara, D. Zhao, and Y. Zhu, On the explicit construction of certain unitary t-designs, J. Phys. A: Math. Theor., 52 (49): 495301, Nov. 2019;
- E. Bannai, Y. Nakata, T. Okuda, and D. Zhao, On explicit construction of exact unitary designs, arXiv:2009.11170.

(The last paper gives explicit constructions of unitary t-designs as well as spherical t-designs by using an induction.)

5.2.3 Finite subsets in Euclidean spaces

So far we have discussed subsets of association schemes, finite point sets on unit spheres in Euclidean spaces, finite subsets in projective spaces, and more generally finite subsets in compact symmetric spaces. In this section, we would like to discuss finite sets in Euclidean spaces. The fact that a Euclidean space is not a compact space causes a quite different situation compared with the cases listed above. In particular, when we consider designs in Euclidean spaces, we have some difficulties because in design theory we usually consider how to approximate the given set with a subset which consists of finite elements. On the other hand, there are cases, such as s-distance sets, where the property of non-compactness does not become any obstruction to the study. Here, at first, let us consider design theory in Euclidean spaces, which seems to have big differences from design theory on finite sets.

Among the researchers of combinatorial theory, Neumaier and Seidel introduced the idea of t-designs to Euclidean spaces ([377], 1988). They also introduced the Fisher type inequality for $2e$-designs and the idea of tight $2e$-designs in Euclidean spaces. On the other hand, when we reconsider t-designs in Euclidean spaces, we find that there is already a similar concept, which is called a "cubature formula" in approximation theory and numerical analysis. It is almost exactly the same concept as the idea of Euclidean designs. The theory of cubature formulas developed without any contact with combinatorial theory. The study of rotatable designs in statistics is also a very similar research subject to Euclidean designs and the history is far longer than that of combinatorial study. However, from a personal viewpoint, at this stage, the authors feel that the researches of them in analysis or combinatorics have some differences and in some parts are developed much more deeply compared with that in statistics.

Here we give the definition of t-designs in $\mathbb{R}^n$ due to Neumaier and Seidel [377]. The concept that "a finite set $X \subset \mathbb{R}^n$ is a Euclidean design" is a two-step generalization of a spherical design in the following sense.

(1) Relax the condition $X \subset S^{n-1}$.
(2) Consider a weight function on X, that is, X is a weighted set.

First we introduce some symbols. Let X be a subset of $\mathbb{R}^n$ which consists of a finite number of points; X intersects with some spheres centered at the origin $0 \in \mathbb{R}^n$. Let p be the number of such spheres and let $r_1, r_2, \ldots, r_p$ be the radii of these spheres. Then

$\{r_1, r_2, \ldots, r_p\} = \{\|x\| \mid x \in X\}$, where $x = (x_1, x_2, \ldots, x_n)$ and $\|x\| = \sqrt{x_1^2 + x_2^2 + \cdots + x_n^2}$. If X contains the origin 0 of $\mathbb{R}^n$, then $0 \in \{\|x\| \mid x \in X\}$ holds. So we consider 0 as a special case of a sphere, i. e., this corresponds to the relaxation (1). As for (2), a t-design X in the Euclidean space $\mathbb{R}^n$ is a finite set weighted by a positive real valued weight function w defined on X. Let $S^{n-1}(r_p)$ be the sphere of radius r_p centered at the origin 0. Let $S = S^{n-1}(r_1) \cup \cdots \cup S^{n-1}(r_p)$ and call S the support of X, i. e., X is supported by p concentric spheres.

Definition 5.39 (Neumaier–Seidel (1988)). The notation is as given above. Let t be a natural number. A weighted subset (X, w) in $\mathbb{R}^n$ is called a *Euclidean t-design* if the condition

$$\sum_{i=1}^{p} \frac{w(X_i)}{|S^{n-1}(r_i)|} \int_{S^{n-1}(r_i)} f(x)\, d\sigma_i(x) = \sum_{x \in X} w(x) f(x)$$

is satisfied for any polynomial $f(x) = f(x_1, \ldots, x_n) \in \mathcal{P}(\mathbb{R}^n)$ of degree at most t.

The following conditions are known to be equivalent to the condition given in Definition 5.39.

(1) Let t be a natural number. For any integers k and j satisfying $1 \le k$, $0 \le j$, $k + 2j \le t$ and for any harmonic polynomial $\varphi(x) \in \mathrm{Harm}_k(\mathbb{R}^n)$ the following condition holds:

$$\sum_{x \in X} w(x) \|x\|^{2j} \varphi(x) = 0.$$

(2) Any moment of X with degree at most t is invariant under the action of orthogonal transformations.

Condition (2) given above is equivalent to the following condition:

(2′) $\sum_{x \in X} w(x) f(x) = \sum_{x \in X} w(x) f(x^\sigma)$ holds for any polynomial $f(x) = f(x_1, \ldots, x_n)$ of degree at most t and any $\sigma \in O(n)$.

Remark 5.40.

(1) When we consider t-designs on S^{n-1}, we can use exactly the same condition (2′) above. Hence the definition of t-design using (2′) is a good definition which puts together designs on the sphere and designs in the Euclidean space as a whole.

(2) A Euclidean t-design on exactly one sphere can be consider as a weighted spherical t-design. Moreover if the weight is constant, then we can regard it as an ordinary spherical t-design. In that sense the concept of Euclidean designs is regarded as two-step generalization of the concept of spherical designs. Euclidean designs without weight functions are also interesting research objects; however, if we put weights on them we can find many more examples. So we think it is interesting to consider weighted designs. In analysis we usually consider cubature formulas

with weight functions. In particular, designs with constant weights are called cubature formulas of Chebyshev type.

When we study Euclidean designs, the fundamental strategy is the same as the one we took when we sought for spherical designs. That is, we seek for designs X with smaller cardinalities $|X|$. Neumaier and Seidel [377] and Delsarte and Seidel [164] obtained the following Fisher type inequality for $t = 2e$.

Theorem 5.41. *Let (X, w) be a Euclidean 2e-design in $\mathbb{R}^n$ and let $S = S^{n-1}(r_1) \cup \cdots \cup S^{n-1}(r_p)$ be the support of X. Then the following inequality holds:*

$$|X| \geq \dim(\mathcal{P}_e(S)), \tag{5.48}$$

where $\mathcal{P}_e(\mathbb{R}^n) = \bigoplus_{j=0}^{e} \operatorname{Hom}_j(\mathbb{R}^n)$ and $\mathcal{P}_e(S) = \{f|_S \mid f \in \mathcal{P}_e(\mathbb{R}^n)\}$.

Remark 5.42.

(1) For a positive odd integer $t = 2e + 1$, Delsarte and Seidel obtained the following lower bound for $|X|$ under the condition that X is antipodal and the weight function is symmetric with respect to the origin 0, i. e., $w(-x) = w(x)$ ([164]):

$$|X| \geq 2\dim(\mathcal{P}_e^*(S)).$$

Here $\mathcal{P}_e^*(\mathbb{R}^n) = \bigoplus_{j=0}^{[\frac{e}{2}]} \operatorname{Hom}_{e-2j}(\mathbb{R}^n)$ and $\mathcal{P}_e^*(S) = \{f|_S \mid f \in \mathcal{P}_e^*(\mathbb{R}^n)\}$.

(2) It is well known that $\dim(\mathcal{P}_e(\mathbb{R}^n)) = \binom{n+e}{e}$. On the other hand $\dim(\mathcal{P}_e(S))$ depends on the number p of the spheres in the support S of X. Let $\varepsilon_S = 0$ if $0 \notin S$ and $\varepsilon_S = 1$ if $0 \in S$. Then, it is also known that the following holds ([164, 74], etc.).
For $p \leq [\frac{e+\varepsilon_S}{2}]$,

$$\dim(\mathcal{P}_e(S)) = \varepsilon_S + \sum_{i=0}^{2(p-\varepsilon_S)-1} \binom{n+e-i-1}{e-i} < \dim(\mathcal{P}_e(\mathbb{R}^n)),$$

and for $p \geq [\frac{e+\varepsilon_S}{2}] + 1$,

$$\dim(\mathcal{P}_e(S)) = \sum_{i=0}^{e} \binom{n+e-i-1}{e-i} = \dim(\mathcal{P}_e(\mathbb{R}^n)).$$

We will give a neat definition of the tight design of a Euclidean space later, however Delsarte, Seidel, and Neumaier called a Euclidean design (X, w) tight when equality holds in the inequality given in Theorem 5.41 or Remark 5.42 (1). One of the authors, Eiichi Bannai, has a memory related to this subject. He had known Seidel well at a personal level, and there were several occasions to discuss mathematics with him. Every time Seidel suggested him to study the existence (non-existence) problem of tight Euclidean t-designs. He even said to him that "You are the most right person to work on this problem." At that time, unfortunately we had not reached any good idea to attack that problem. It was only after his death (in 2001) that we could get some results on that problem. We believe that Seidel would have been very happy if we could let him know these results, and we deeply regret that we cannot report them to him.

The authors had not been taking the study of Euclidean designs into consideration before they really started their work around 2002–2003. The first result was the paper [35] published in 2006. This paper may not be easy to read; however, as a first step we took up the study of tight Euclidean designs for the case $t = 4$ and $p = 2$. Delsarte, Seidel, and Neumaier were pessimistic about the existence of tight Euclidean designs, and they conjectured that there is no tight t-design for $t \geq 4$ except those trivially obtained. However, against expectations, we could find tight 4-designs and tight 5-designs ([35, 74]). We had been thinking that the lower bound of the cardinalities of Euclidean t-designs X with odd integer t should be the same as given in Remark 5.42 (1) without assuming that X are symmetric with respect to the origin 0. However we could not give a proof of it. Around that time we had some interaction with researchers in analysis (M. Hirao and M. Sawa). They introduced us to the work of Möller (1979) [352] on cubature formulas and we could find out that his theorem had already solved our problem. Hence for odd integer $t = 2e + 1$, we are also able to give the definition of tight $(2e + 1)$-designs. In the following, we give theorems on the natural lower bound of the cardinalities of Euclidean $(2e + 1)$-designs and the definition of tight Euclidean t-designs. Please refer to [352, 45] for the detailed story.

Theorem 5.43 (Möller [352]). *Let (X, w) be a Euclidean $(2e + 1)$-design in $\mathbb{R}^n$. Let $S = S^{n-1}(r_1) \cup \cdots \cup S^{n-1}(r_p)$ be the support of X. Then the following inequality holds:*

$$|X| \geq \begin{cases} 2\dim(\mathcal{P}_e^*(S)) - 1, & \text{if } e \text{ is even and } 0 \in X, \\ 2\dim(\mathcal{P}_e^*(S)), & \text{otherwise,} \end{cases}$$

where $\mathcal{P}_e^*(\mathbb{R}^n) = \bigoplus_{j=0}^{[\frac{e}{2}]} \mathrm{Hom}_{e-2j}(\mathbb{R}^n)$ and $\mathcal{P}_e^*(S) = \{f|_S \mid f \in \mathcal{P}_e^*(\mathbb{R}^n)\}$.

Here we gave the theorem in the form that we can apply for Euclidean t-designs. However the original theorem by Möller is a more general theorem on cubature formulas.

Definition 5.44 (Tight Euclidean design [377, 164, 45]).

(1) Let (X, w) be a Euclidean t-design in $\mathbb{R}^n$. If equality holds in the inequality of Theorem 5.41 or Theorem 5.43, then (X, w) is called a tight Euclidean design on p concentric spheres.

(2) Moreover, according to t being even or odd, if $\dim(\mathcal{P}_{[\frac{t}{2}]}(S)) = \dim(\mathcal{P}_{[\frac{t}{2}]}(\mathbb{R}^n))$ or $\dim(\mathcal{P}^*_{[\frac{t}{2}]}(S)) = \dim(\mathcal{P}^*_{[\frac{t}{2}]}(\mathbb{R}^n))$ holds, then (X, w) is a tight Euclidean t-design of $\mathbb{R}^n$.

Remark 5.45. Here we would like to explain the subtle differences between arguing cubature formulas in analysis and arguing Euclidean designs as combinatorial objects. As an example let us take cubature formulas with respect to the Gaussian measure. We define the integration of a polynomial $f(x)$ on $\mathbb{R}^n$ by $\int_{\mathbb{R}^n} f(x)e^{-\|x\|^2}\,dx$. A set $\{u_1, u_2, \ldots, u_N\}$ of points in $\mathbb{R}^n$ is called a cubature formula in $\mathbb{R}^n$ of degree t with respect to the Gaussian measure if there exist positive real numbers $w_1, w_2, \ldots, w_N$ satisfying the following condition:

$$\frac{1}{\int_{\mathbb{R}^n} e^{-\|x\|^2}\,dx} \int_{\mathbb{R}^n} f(x)e^{-\|x\|^2}\,dx = w_1 f(u_1) + \cdots + w_N f(u_N) \tag{5.49}$$

for any polynomial $f(x)$ of degree at most t. We can show that this cubature formula of degree t with respect to Gaussian measure is actually a Euclidean t-design of $\mathbb{R}^n$. We can show that any cubature formula of degree t with respect to a radially symmetric measure is a Euclidean t-design of $\mathbb{R}^n$, not only with respect to the Gaussian measure. Now get back to the theorem of Möller. Then we have the following lower bound for N.

1. Case $t = 2e$:

$$N \geq \dim(\mathcal{P}_e(\mathbb{R}^n)) = \binom{n+e}{e}. \tag{5.50}$$

2. Case $t = 2e + 1$:

$$N \geq \begin{cases} 2\dim(\mathcal{P}_e^*(\mathbb{R}^n)) - 1, & \text{if } e \text{ is even and } u_i = 0 \text{ for some } i, \\ 2\dim(\mathcal{P}_e^*(\mathbb{R}^n)), & \text{otherwise.} \end{cases} \tag{5.51}$$

Let us explain the idea of tight designs in Euclidean spaces, for example, using a tight 9-design (X, w) in $\mathbb{R}^2$. Let us consider the possibility that the support S of X satisfies $\dim(\mathcal{P}_4^*(S)) = \dim(\mathcal{P}_4^*(\mathbb{R}^2)) = 9$. For example, in the picture on the left of the figure given below, we have $S = \{0\} \cup S^1(r_1) \cup S^1(r_2)$. In the picture on the right, we have $S = S^1(r_1) \cup S^1(r_2) \cup S^1(r_3)$. Also, there are several kinds. In these two cases 17 points and 18 points are the natural lower bounds for the 9-designs in $\mathbb{R}^2$, respectively. So we say each of them is a tight 9-design of $\mathbb{R}^2$. According to the idea of Möller, the 17 points on the left is the cubature formula with the smallest number of points ([352]).

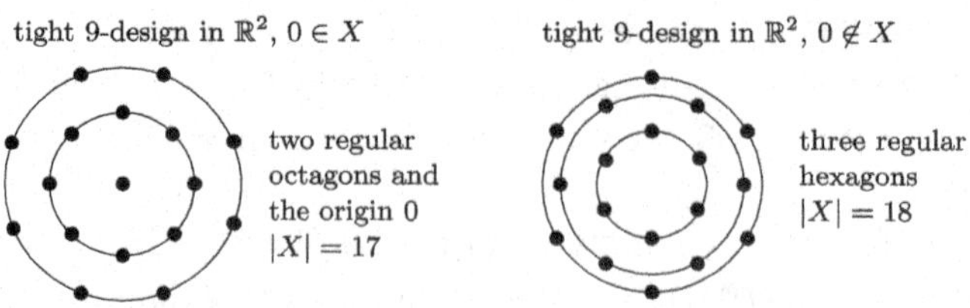

The circles are centered at the origin.
Consecutive two regular polygons are in the position of rotation by half the central angle.

Using the results obtained by Möller we can prove the following theorem.

Theorem 5.46 ([352, 45]). *Let (X, w) be a tight $(2e+1)$-design on a union of p concentric spheres. Then the following hold:*
(1) *If e is odd, then X is antipodal and the weight function w is symmetric with respect to the origin 0.*
(2) *If e is even and $0 \in X$, then X is antipodal and the weight function w is symmetric with respect to 0.*
(3) *Assume e is even and $0 \notin X$. Moreover, assume for each line l passing through 0, $X \cap l$ contains at most $\frac{e}{2}+1$ points which are not antipodal to each other (i. e., $Y \subset X \cap l$ and $x \neq -y$ for any distinct two points $x, y \in Y$, then $|Y| \leq \frac{e}{2}+1$). Then X is antipodal and the weight function w is symmetric with respect to 0.*

Remark 5.47.
(1) Theorem 5.46 shows that for the case where e is even, if the number p of concentric spheres is at most $\frac{e}{2}+1$, then tight $(2e+1)$-designs are antipodal and the weight functions are symmetric with respect to the origin. Hence all the examples of Euclidean 5-designs obtained in [74] are tight 5-designs on two concentric spheres. Please refer to [74] for further details.
(2) For odd integer t, up to the present, no example of a non-antipodal tight t-design of $\mathbb{R}^n$ nor of a non-antipodal tight t-design on p concentric spheres has been reported to exist so far. In general, for a union S of p concentric spheres, it is known that if $p \leq [\frac{e}{2}]$, then $\dim(\mathcal{P}_e^*(S)) < \dim(\mathcal{P}_e^*(\mathbb{R}^n))$ holds, and if $p \geq [\frac{e}{2}]+1$, then $\dim(\mathcal{P}_e^*(S)) = \dim(\mathcal{P}_e^*(\mathbb{R}^n))$ holds. It is also an interesting problem whether we can prove the same thing without the assumption of Theorem 5.46 (3).

The following facts on t-designs in $\mathbb{R}^n$ are known, although they are not necessarily listed in chronological order.

It is easy to see that (X, w), with $0 \notin X$, is a t-design if and only if $(X \cup \{0\}, w)$ is a t-design, where $w(0)$ could be any positive real number. Therefore when we consider tight t-designs X on p concentric spheres, assuming $0 \notin X$, we only have to check carefully whether $(X \cup \{0\}, w)$ is a tight t-design on $p+1$ concentric spheres or not.

We introduce some more notations. Let $S = S^{n-1}(r_1) \cup \cdots \cup S^{n-1}(r_p)$ be the support of the finite set X of a t-design (X, w). For each pair $X_i = X \cap S^{n-1}(r_i) \neq \{0\}$ and $X_j = X \cap S^{n-1}(r_j) \neq \{0\}$ we define $A(X_i, X_j) = \{\frac{x \cdot y}{r_i r_j} \mid x \in X_i, y \in X_j\}$. In particular for $i = j$ we write $A(X_i) = A(X_i, X_i)$.

Theorem 5.48 ([35, 74]). *Let (X, w) be a tight t-design on p-concentric spheres in $\mathbb{R}^n$. Assume $0 \notin X$. Then the following (1) and (2) hold:*
(1) *Let $t = 2e$. Then the following (i)–(iv) hold:*
  (i) *The weight function w is constant on X_i, i. e., there exists a positive real number w_i satisfying $w(x) = w_i$ for any $x \in X_i$ and i with $1 \leq i \leq p$.*
  (ii) *Each X_i is an at most e-distance set, i. e., $|A(X_i)| \leq e$ holds for $i = 1, 2, \ldots, p$.*
  (iii) *We have $|A(X_i, X_j)| \leq e$ for $1 \leq i \neq j \leq p$.*
  (iv) *If $e \geq p$, then X_i is a spherical $2(e-p+1)$-design on S^{n-1}.*

(2) *Let $t = 2e+1$. If X is antipodal and w is symmetric with respect to 0, then the following (i)–(iv) hold:*
   (i) *The weight function w is constant on each X_i, i. e., there exists a positive constant w_i and $w(x) = w_i$ holds for any $x \in X_i$ and $i = 1, 2, \ldots, p$.*
   (ii) *We have $|A(X_i)| \leq e + 1$ for $i = 1, 2, \ldots, p$.*
   (iii) *We have $|A(X_i, X_j)| \leq e$ for $1 \leq i \neq j \leq p$.*
   (iv) *For $e \geq p$, X_i is a spherical $(2(e - p + 1) + 1)$-design on S^{n-1}.*

Remark 5.49. By Theorem 5.46, if $p \leq \frac{e}{2} + 1$, then the condition in Theorem 5.48 (2) that X is antipodal and w is symmetric with respect to 0 is automatically satisfied.

As we have seen in Theorem 5.18 and Theorem 5.19, tight spherical t-designs or almost tight spherical t-designs have structures of Q-polynomial association schemes. Then, for the Euclidean designs, what kind of property do they have? In particular, consider a tight t-design (X, w) on a union of two concentric spheres. Let $X = X_1 \cup X_2$. Then Theorem 5.48 implies that X_1 and X_2 are spherical $(t - 2)$-designs and at most $[\frac{t+1}{2}]$-distance sets. Hence Theorem 5.18 and Theorem 5.19 imply that they have the structure of Q-polynomial schemes. Also Theorem 5.48 implies $|A(X_1, X_2)| \leq [\frac{t}{2}]$. On the other hand there is a concept of a coherent configuration which is one of the generalizations of association schemes which are very important research objects in algebraic combinatorics. The concept of coherent configurations was defined by D. G. Higman around 1970 ([216, 217, 218, 219, 220]). There are several ways to define coherent configurations. Since we want to describe the connection with Euclidean designs we give the following definition.

Definition 5.50 (Coherent configuration). Let X be a finite set; X is divided into p subsets, $X = X_1 \cup \cdots \cup X_p$ with $X_\lambda \cap X_\mu = \emptyset$ $(1 \leq \lambda \neq \mu \leq p)$. For each $1 \leq \lambda, \mu \leq p$, we consider the following partitions:

$$X_\lambda \times X_\mu = \begin{cases} R_{\lambda,\lambda}^{(0)} \cup R_{\lambda,\lambda}^{(1)} \cup \cdots \cup R_{\lambda,\lambda}^{(s_{\lambda,\lambda})}, & \text{for } \lambda = \mu, \\ R_{\lambda,\mu}^{(1)} \cup R_{\lambda,\mu}^{(2)} \cup \cdots \cup R_{\lambda,\mu}^{(s_{\lambda,\mu})}, & \text{for } \lambda \neq \mu. \end{cases}$$

If X and the partition $\{R_{\lambda,\mu}^{(l)}\}_{1-\delta_{\lambda,\mu} \leq l \leq s_{\lambda,\mu}}$ of $X_\lambda \times X_\mu$ satisfy the following conditions (1)–(3), then $(X, \{R_{\lambda,\mu}^{(l)}\}_{1-\delta_{\lambda,\mu} \leq l \leq s_{\lambda,\mu}})$ is called a *coherent configuration*:

(1) $R_{\lambda,\lambda}^{(0)} = \{(x, x) \mid x \in X_\lambda\}$ for $1 \leq \lambda \leq p$;
(2) $s_{\lambda,\mu} = s_{\mu,\lambda}$ and there exists an integer ℓ' satisfying ${}^t R_{\lambda,\mu}^{(\ell)} = R_{\mu,\lambda}^{(\ell')}$ and $1 - \delta_{\lambda,\mu} \leq \ell' \leq s_{\lambda,\mu}$, where ${}^t R_{\lambda,\mu}^{(\ell)} = \{(x, y) \in X_\mu \times X_\lambda \mid (y, x) \in R_{\lambda,\mu}^{(\ell)}\}$;
(3) for integers λ, μ, ν, k satisfying $1 \leq \lambda, \mu, \nu \leq p$, $1 - \delta_{\lambda,\mu} \leq k \leq s_{\lambda,\mu}$, the cardinality of the set

$$\{z \in X_\nu \mid (x, z) \in R_{\lambda,\nu}^{(i)}, (z, y) \in R_{\nu,\mu}^{(j)}\}$$

is independent of the choice of $(x, y) \in R_{\lambda,\mu}^{(k)}$ and depends only on the choices of the integers $\lambda, \mu, \nu, i, j, k$.

Remark 5.51.

(1) From the definition it is clear that $(X_\lambda, \{R^{(l)}_{\lambda,\lambda}\}_{0\le\ell\le s_{\lambda,\lambda}})$ is an association scheme of class $s_{\lambda,\lambda}$ for $\lambda = 1, 2, \ldots, p$. In particular if $p = 1$ this is an association scheme itself.

(2) The nature of coherent configurations is purely combinatorial objects which are obtained from the orbits of a permutation group (not necessarily transitive) of X acting on $X \times X$. When G acts transitively it gives an association scheme. (Example 2.3 in Section 2.1 in Chapter 2). In this sense coherent configurations can be considered as a generalization of association schemes. Higman called association schemes homogeneous coherent configurations. Even now sometimes we come across this expression.

(3) For $1 \le \lambda, \mu \le p$, $1 - \delta_{\lambda,\mu} \le i \le s_{\lambda,\mu}$ define a matrix $A^{(i)}_{\lambda,\mu}$ indexed by $X \times X$ by

$$A^{(i)}_{\lambda,\mu}(x,y) = \begin{cases} 1, & \text{for } (x,y) \in R^{(i)}_{\lambda,\mu}, \\ 0, & \text{for } (x,y) \notin R^{(i)}_{\lambda,\mu}. \end{cases}$$

The vector space $\mathfrak{A}$ spanned by $\{A^{(i)}_{\lambda,\mu} \mid 1 \le \lambda, \mu \le p,\ 1 - \delta_{\lambda,\mu} \le i \le s_{\lambda,\mu}\}$ is closed under the ordinary matrix multiplication, so it is an algebra not necessarily commutative. This algebra is called the *coherent algebra*.

Let (X, w) be a t-design in $\mathbb{R}^n$ supported by the union $S = S^{n-1}(r_1) \cup \cdots \cup S^{n-1}(r_p)$ of p concentric spheres with positive radii $r_1, \ldots, r_p$. Also we assume that the weight w is constant on each X_ν for $\nu = 1, \ldots, p$. Let $w = w_\nu$ on X_ν for $\nu = 1, \ldots, p$. Then each X_ν $(1 \le \nu \le p)$ is a spherical $(t - 2(p - 1))$-design. We define $A(X_\lambda, X_\mu) = \{\alpha^{(i)}_{\lambda,\mu} \mid 1 - \delta_{\lambda,\mu} \le i \le s_{\lambda,\mu}\}$. In this situation, if the number of points in

$$\left\{ z \in X_\nu \mid \frac{x \cdot z}{r_\lambda r_\nu} = \alpha^{(i)}_{\lambda,\nu},\ \frac{z \cdot y}{r_\nu r_\mu} = \alpha^{(j)}_{\nu,\mu} \right\}$$

depends only on the choice of $\lambda, \mu, \nu, i, j, k$, then the finite set X is associated by the structure of a coherent configuration. Unfortunately if $p \ge 3$, tight t-designs on p concentric spheres hardly exist. Actually the existence of a tight 5-design on 3 concentric spheres in $\mathbb{R}^4$ which does not have a structure of a coherent configuration is known [230]. However, it is proved that tight t-designs on 2 concentric spheres always have structures of coherent configurations [39]. This fact makes the possibility of classifying tight t-designs on 2 concentric spheres large.

Next, before we finish this subsection, let us discuss the current status of the study on the classification problem of tight t-designs in $\mathbb{R}^n$.

For $t = 1, 2, 3$, the classification problem on tight Euclidean t-designs is settled in general ([74, 50]).

(1) A tight 1-design in $\mathbb{R}^n$ is $X = \{x, -x\}$, where x is an arbitrary vector in $\mathbb{R}^n$.

(2) Any tight 2-design in $\mathbb{R}^n$ is similar to (X, w), with $|X| = n + 1$, $\{x \cdot y \mid x, y \in X,\ x \ne y\} = \{-1\}$ (i. e., X is a 1-inner product set) and $w(x) = \frac{1}{1+\|x\|^2}$. For any integer $p \le n+1$

there exists X satisfying this property with $|\{\|x\| \mid x \in X\}| = p$ ([50]). The Fisher type inequality for the 1-inner product sets is obtained by Deza and Frankl [169]. Also Nozaki [385] gave a more algebraic combinatorial proof for this problem.

(3) Any tight 3-design in $\mathbb{R}^n$ is similar to (X, w) satisfying $|X| = 2n$ and

$$X = \{\pm r_i e_i \mid 1 \le i \le n\},$$

with $w(\pm r_i e_i) = \frac{1}{nr_i^2}$ $(1 \le i \le n)$, where $\{e_1, \ldots, e_n\}$ is the standard basis of $\mathbb{R}^n$ and $r_1, \ldots, r_n$ are arbitrary positive real numbers ([74, 16]).

For $t \ge 4$, in particular if p is not small, the complete classification seems extremely difficult. As mentioned in [50], it is expected that there are many such kinds of tight t-designs. This is a quite new development considering that non-existence was conjectured in [377, 164], etc. For $n = 2$, Cools and Schmid [149], Bajnok [16, 17], and others listed all the known examples for the case $p \le [\frac{e}{2}]+1$ and classification was essentially possible. In the following we give the result without proof. Please refer to [45] for a precise proof.

(4) Any tight t design on p concentric circles in $\mathbb{R}^2$ for $p \le [\frac{t}{4}] + 1$ is similar to one of the following:
   (i) Case $0 \in X$: $X = \{0\} \cup X_1 \cup \cdots \cup X_{p-1}$, where each X_i is the regular $(t - 2p + 5)$-gon on the circle of radius $r_i > 0$. Let $r_1 > r_2 > \cdots > r_{p-1} > r_p = 0$. Then consecutive X_i and X_{i+1} are in the position of rotating half of the central angle. Let $w_i = w(x)$, $x \in X_i$. Then $\frac{w_i}{w_1}$ is a constant which depends only on $r_1, \ldots, r_{p-1}$. If $p = [\frac{t}{4}] + 1$, then X is a tight t-design of $\mathbb{R}^2$.
   (ii) Case $0 \notin X$: $X = X_1 \cup \cdots \cup X_p$, where each X_i is the regular $(t-2p+3)$-gon on the circle of radius $r_i > 0$. Let $r_1 > r_2 > \cdots > r_p > 0$. Then consecutive X_i and X_{i+1} are in the position of rotating half of the central angle. Let $w_i = w(x)$, $x \in X_i$. Then $\frac{w_i}{w_1}$ is a constant which depends only on $r_1, \ldots, r_p$. If $p = [\frac{t}{4}] + 1$, then X is a tight t-design of $\mathbb{R}^2$.

(5) If a tight t-design X of $\mathbb{R}^2$ contains 0, then $t = 4k$ or $t = 4k + 1$ and $p = k + 1$ holds. (This is also true for tight t-designs of $\mathbb{R}^n$.) It is expected that there are many tight t-designs of $\mathbb{R}^n$ on p concentric spheres which do not contain 0 for $p \ge [\frac{t}{4}] + 2$. However except the case for $n = 2$, $t = 4$, it is not studied so far (in the case $n = 2$, $t = 4$, it was proved [45] that infinitely many of them exist).

(6) For the case $n \ge 3$, the existence of tight t-designs of $\mathbb{R}^n$ with $p \ge 4$ is not known yet. In the case $p = 3$, the following two examples are known so far.
A tight 7-design of $\mathbb{R}^3$: $X = X_1 \cup X_2 \cup X_3$, where $|X_1| = 6$, $|X_2| = 8$, $|X_3| = 12$, X_1 is the regular octahedron, X_2 is a regular hexahedron, and X_3 consists of the mid points of the 12 edges of the regular hexahedron (for the details, see [17]).
A tight 5-design of $\mathbb{R}^4$: $X = X_1 \cup X_2 \cup X_3$, where $|X_1| = 2$, $|X_2| = 8$, $|X_3| = 12$ (for the details, see [230]).

In the following we mainly discuss tight t-designs (X, w) supported by 2 concentric spheres for the case $t \geq 4$. If $0 \in X$, then $X \backslash \{0\}$ is a spherical tight t-design. Therefore we consider the case $0 \notin X$. For $t = 4, 5, 6, 7$, $\dim \mathcal{P}_{[\frac{t}{2}]}(S) = \dim \mathcal{P}_{[\frac{t}{2}]}(\mathbb{R}^n)$ and $\dim \mathcal{P}^*_{[\frac{t}{2}]}(S) = \dim \mathcal{P}^*_{[\frac{t}{2}]}(\mathbb{R}^n)$ hold and it is a tight t-design of $\mathbb{R}^n$. It is known that a tight t-design on 2 concentric spheres has the structure of a coherent configuration ([39]) and this fact plays a big role on the classification problems and discoveries of concrete examples we mention in the following.

Let (X, w) be a t-design on two concentric spheres and $X = X_1 \cup X_2$, $2 \leq |X_1| \leq |X_2|$.

(1) For the case $t = 4$, there is a remarkable fact that they are related to tight designs on Johnson schemes or Hamming schemes. The classification of tight 4-designs is not yet completed, and so far the following are the best results known at present (as of 2020) ([75, 39]) ($|X_1| \geq n + 1$ holds for $t = 4$).
   (a) If (X, w) is tight and $|X_1| = n + 1$, then we must have $n = 2, 4, 5, 6, 22$. For $n = 4, 5, 6$, X_2 has the structure of the Johnson scheme $J(n + 1, 2)$. In particular for $n = 22$, X_2 has the structure of the tight 4-$(23, 7, 1)$ design of the Johnson scheme $J(23, 7)$ ([75]).
   (b) If (X, w) is tight and $|X_1| = n+2$, then $n = 4$ holds. Moreover X_2 has the structure of the Hamming scheme $H(2, 3)$ ([75]).
   (c) For the case where (X, w) is tight and $|X_1| \geq n + 3$, the classification problem is not yet settled; however, for $n = 22$ and $|X_1| = 33$ the existence is shown. In this case X_2 is a tight 4-design on the Hamming scheme $H(11, 3)$ ([75]).
   (d) There is a series of parameters of tight 4-designs for $n = (6k^2 - 3)^2 - 3$ $(k \geq 1)$ in which $|X_i|$ and the values in $A(X_i, X_j)$ $(i, j = 1, 2)$ depend only on $k(n)$. The case $k = 1$ is exactly the case (a) with $n = 6$ and actually it exists. However, in general, the existence or non-existence is an unsolved problem ([39]).
   (e) If the weight function w of (X, w) is constant on each X_i $(i = 1, 2)$, then X has the structure of a coherent configuration and (X, w) is a tight 4-design or $n = 2$, or $n = (2k - 1)^2 - 4$ $(k \geq 2)$. If $n = 2$, then it is a tight 5-design which consists of two squares. If $n = (2k - 1)^2 - 4$, $k \geq 2$, then $|X_i|$, $A(X_i, X_j)$ $(i, j = 1, 2)$ depend only on k. These parameters satisfy all the integral conditions of coherent configurations; however, the existence is not known yet ([39]).

   The structure of the coherent configurations associated with examples of the tight 4-designs in (a), (b), and (c) and candidates of 4-designs given in (d) are unique for each dimension and the ratio of the radii is also a constant (without any freedom). On the other hand the coherent configuration associated with the family of candidates of 4-designs in (e) which are not tight is unique for each dimension but the ratio of the radii is arbitrary and the weight function becomes constant on X by a suitable choice of them.

(2) The classification of tight 5-designs on 2 concentric spheres is done. They exist only for $n = 2, 3, 5, 6$. The ratio of the radii is flexible; however, the structure of the coherent configuration is unique ([74]).

(3) Only one example has been found for the case of a tight 6-design on the union of 2 concentric spheres. It is in the case $n = 22$ and $|X_1| = 275$, $|X_2| = 2025$, where X_1 is a tight 4-design on S^{21}. Both X_1 and X_2 are orbits of the McLaughlin group, which is known as a simple group, acting on $\mathbb{R}^{22}$ as an orthogonal transformation ([48]). In this case the ratio of the radii of two concentric spheres is a constant. It is known that a tight 4-design on the unit sphere S^{21} is unique. If we choose this design for X_1, then X_2 has to be on the sphere of radius $\sqrt{11}$. It is still unknown whether there exists any other tight 6-design on 2 concentric spheres.

(4) Tight 7-designs on two concentric spheres exist only for $n = 2, 4, 7$. In each of the cases the ratio of the radii is flexible; however, the structure of the corresponding coherent configuration is unique. In the case $n = 4$, we have $|X_1| = |X_2| = 24$ and both of them have the structure of the kissing configuration. In the case $n = 7$, we have $|X_1| = 56$ and $|X_2| = 126$, and X_1 is a spherical tight 5-design. In this case we can make the weight function w constant on X by a suitable choice of the ratio of the radii of the spheres ([37]).

(5) If $n \geq 3$, there is no tight 9-design on a union of 2 concentric spheres ([40]). On the other hand it is proved that if a tight 9-design (X, w) has the property $0 \in X$, then we must have $p = 3$. Hence $(X \backslash \{0\}, w)$ must be a tight 9-design on a union of 2 concentric spheres. This fact implies that the radially symmetric minimal cubature formula mentioned by Möller does not exist for $n \geq 3$.

(6) We expect that there is no tight t-design with $t \geq 11$ on the union of 2 concentric spheres except for $n = 2$. The authors (Ei. Bannai and Et. Bannai) have been working on this using the properties of the corresponding coherent configuration. However, we could not obtain the final solution so far (cf. [42]).

Verlinden and Cools [495] proved that a minimal cubature formula of degree $t = 4k+1$ in $\mathbb{R}^2$, in the sense of Möller, does not exist except for some small values of k. Hirao and Sawa [229] and Bannai, Bannai, Hirao, and Sawa [46] extended their result to the case without the condition $t = 4k + 1$, by using the properties of tight t-designs of $\mathbb{R}^2$.

When we study tight t-designs of $\mathbb{R}^n$, tight t-designs on p concentric spheres, or sometimes designs having a property that is close to tightness, it is very interesting to consider under what kind of situation structures of coherent configurations appear.

5.2.4 Connections with analysis (in particular, numerical analysis, approximation theory, orthogonal polynomials, and cubature formulas)

Cubature formulas

Let D be a domain (or subset) in $\mathbb{R}^n$, X a subset of D, and w a positive real valued function on X. Now, let μ be a measure on D.

We say (X, w) is a *cubature formula* of degree t if

$$\frac{1}{\mu(D)}\int_D f(x)\mu(x)\,dx = \sum_{x\in X} w(x)f(x)$$

holds for any polynomial $f(x) = f(x_1, x_2, \dots, x_n)$ of degree at most t.
(We sometimes consider cubature formulas for other classes of functions, instead of polynomials of degree at most t. In some cases, the function w is allowed to take negative values. However, here we consider the case of $f(x)$ being polynomials of degree at most t and w being positive real valued functions unless otherwise stated. If $n = 1$, the cubature formula is also called the "quadrature formula.")

Anyway, the cubature formula means that it gives the approximate value (more exactly speaking, it gives exact values for polynomials $f(x)$ of degree at most t), by taking the average value at the finitely many points in X. The cubature formulas have been studied extensively in analysis, in particular in numerical analysis, approximation theory, and the theory of orthogonal polynomials. Spherical designs and Euclidean designs which we just considered are special but very important cases of the study of cubature formulas and t-designs. The case we consider here is the case where D and μ are radially symmetric. We say the subset D in $\mathbb{R}^n$ is *radially symmetric* if $x \in D$ implies any $y \in \mathbb{R}^n$ with $|x| = |y|$ is in D. (Namely, D is a union of spheres whose centers are at the origin. The number of such concentric spheres can be finite or infinite.) We also say μ is *radially symmetric* if the measure μ is constant on each concentric sphere.

If $D = S^{n-1}$ and μ is a constant, then the cubature formula corresponds to weighted spherical designs. If D is the union of finitely many concentric spheres and μ is constant on each concentric sphere, then this case corresponds to Euclidean designs we just considered in the previous section. Other than these cases, there are many interesting cases given as follows:

$$\begin{aligned}
&D = \mathbb{R}^n \quad \text{and} \quad \mu(x) = e^{-||x||},\\
&D = \mathbb{R}^n \quad \text{and} \quad \mu(x) = e^{-||x||^2},\\
&D = \mathbb{R}^n \quad \text{and} \quad \mu(x) = e^{-||x||^k},\\
&D = \mathbb{B}^n \quad \text{(the unit ball) and} \quad \mu(x) = \text{(a constant)},\\
&D = \mathbb{B}^n \quad \text{(the unit ball) and} \quad \mu(x) = (1 - ||x||^2)^\alpha,
\end{aligned}$$

and so on. Now we want to give our explanation taking the example of $D = \mathbb{R}^n$ and $\mu(x) = e^{-||x||^2}$. Let

$$\frac{1}{\int_{\mathbb{R}^n} e^{-||x||^2}\,dx}\int_{\mathbb{R}^n} f(x)e^{-||x||^2}\,dx = \sum_{x\in X} w(x)f(x)$$

be satisfied for any polynomial $f(x) = f(x_1, x_2, \dots, x_n)$ of degree at most t. This (X, w) is called a Gaussian t-design. As is shown in [33], we have the following Fisher type inequality for Gaussian $2e$-designs:

$$|X| \geq \binom{n+e}{e}.$$

If the equality

$$|X| = \binom{n+e}{e}$$

holds, then such (X, w) is called a *tight Gaussian* $2e$*-design*. We are interested in the problem whether such tight Gaussian $2e$-designs exist. Actually, this problem is not yet settled. This problem is still open, even if we assume that $w(x)$ is a constant function. On the other hand, we know the following facts:

(1) a Gaussian $2e$-design is a Euclidean $2e$-design;
(2) X is on at least $[e/2] + 1$ concentric spheres;
(3) if X is a tight Gaussian $2e$-design, then $w(x)$ is a constant function for each concentric sphere;
(4) if X is a tight Gaussian $2e$-design, then each X_i (the set of points of X on the i-th concentric sphere) is an e-distance set.

By utilizing this information, Bannai and Bannai [33] give the classification of tight Gaussian 4-designs.

It would be very desirable if we could classify tight Gaussian $2e$-designs for $e \geq 3$, but we have no good idea how to attack this problem in general. The case that might be attackable is the case where the number of concentric spheres is $[e/2] + 1$ or very close to that.

In the study of cubature formulas, some combinatorial concepts are very useful. For example, in the study of tight Euclidean 4-designs, tight 4-designs in Johnson schemes and/or Hamming schemes are very relevant ([75]). It is interesting to note that Victoir [497] has succeeded in drastically decreasing the number of points in the cubature formula, by replacing the part of the points that is in a certain orbit of a group, by t-designs in an association scheme (Victoir [497], Sawa, Hirao, and Kageyama [412]). See also Nozaki and Sawa [387] and Sawa and Xu [413], where cubature formulas whose numbers of points are relatively small are explicitly constructed and studied. Many further developments are expected. Besides using t-designs in Johnson or Hamming schemes, regular t-wise balanced designs are also used (Section 5.2.5 of the present book).

5.2.5 Analogy between Euclidean or hyperbolic t-designs and relative t-designs in association schemes

As we have seen in the previous section, the Euclidean space $\mathbb{R}^n$ is a non-compact symmetric space of rank 1. Non-compact symmetric spaces are also classified ([213],

[519]) and the real hyperbolic space $\mathbb{H}^n$ is one of them:

$$\mathbb{H}^n = \{(x_0, x_1, \ldots, x_n) \in \mathbb{R}^{n+1} \mid x_0^2 - x_1^2 - \cdots - x_n^2 = 1,\ x_0 > 0\}.$$

For 2 elements $x = (x_0, x_1, \ldots, x_n)$ and $y = (y_0, y_1, \ldots, y_n)$ the distance between them is defined by

$$d(x, y) = \operatorname{arccosh}(x_0 y_0 - x_1 y_1 - \cdots - x_n y_n).$$

Here, $\cosh(z) = \frac{e^z + e^{-z}}{2}$. This space is also known as the famous non-Euclidean space called Bolyai–Lobachevsky space. This space is also expressed as the homogeneous space

$$\mathbb{H}^n = \mathrm{SO}^1(n+1)/\,\mathrm{SO}(n) (= O^1(n+1)/O(n)).$$

(For the notation, see [213, 519], etc.) This space is a Riemannian symmetric space and 2-point homogeneous, like spheres and Euclidean spaces. Roughly speaking, this space is described as in the figure below.

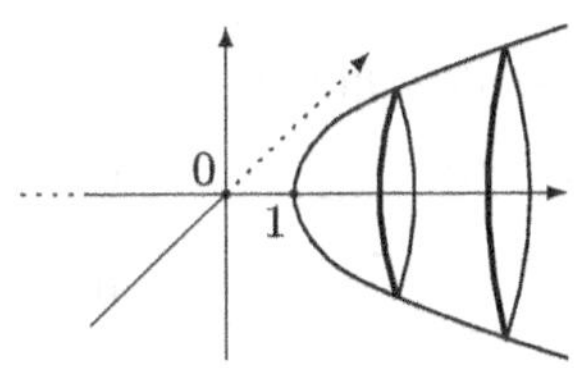

As we have already explained in [32], one of the authors was eager to find a reasonable definition of t-designs in $\mathbb{H}^n$. In Bannai, Blockhuis, Delsarte, and Seidel [54], we have succeeded in finding an addition theorem in harmonic analysis, and proved that $|X| \leq \binom{n+s}{s}$ for an s-distance set in $\mathbb{H}^n$. Then we hoped to define t-designs in $\mathbb{H}^n$, but we were not successful at all. Relatively recently, we proposed in [41] the definition of t-designs in $\mathbb{H}^n$. By the definition of $\mathbb{H}^n$, $\mathrm{Isom}(\mathbb{H}^n) \cong O^1(n+1)$, and for $z_0 = (1, 0, 0, \ldots, 0) \in \mathbb{H}^n$ the stabilizer of z_0 is isomorphic to $O(n)$. On the other hand, $\mathrm{Isom}(\mathbb{R}^n) \cong E(n)(= \mathbb{R}^n \cdot O(n))$, and the stabilizer of $z_0 = (0, 0, \ldots, 0) \in \mathbb{R}^n$ is isomorphic to the group $O(n)$. Therefore, the following definition seems to be natural as the definition of t-designs on $\mathbb{H}^n$.

Definition 5.52 (t-Designs on real hyperbolic spaces). Let X be a finite subset of the real hyperbolic space $\mathbb{H}^n = \{(x_0, x_1, \ldots, x_n) \in \mathbb{R}^{n+1} \mid x_0^2 - (x_1^2 + \cdots + x_n^2) = 1\}$ and let $w : X \to \mathbb{R}_{>0}$ be a positive real valued weight function on X. Fix a special point $z_0 = (1, 0, \ldots, 0) \in \mathbb{H}^n$. We call (X, w_0) a t-design on $\mathbb{H}^n$ with respect to z_0 if any moment of degree at most t in $x_1, x_2, \ldots, x_n$ is invariant by any element $\sigma \in O(n)$ where $O(n)$ is isomorphic to the stabilizer of z_0 by $\mathrm{Isom}(\mathbb{H}^n)$. In other words,

$$\sum_{x \in X} f(x_1, \ldots, x_n) = \sum_{x \in X^\sigma} f(x_1, \ldots, x_n)$$

holds for any polynomial $f(x_1, x_2, \ldots, x_n)$ of degree at most t.

Let us add several comments (a) to (d).

(a) The similarity with t-designs in Euclidean spaces would be very clear. (Compare with condition (2′) given right before Remark 5.40.)

(b) For the sphere $S^n = \{(x_0, x_1, \dots, x_n) \in \mathbb{R}^{n+1} \mid x_0^2 + x_1^2 + \dots + x_n^2 = 1\}$ by taking a special point $z_0 \in S^{n-1}$ we can define a t-design on S^n with respect to the special point $z_0 \in S^n$ as follows.
Let X be a finite subset of S^n, let $w : X \to \mathbb{R}_{>0}$ be a positive real valued weight function, and let $z_0 = (1, 0, \dots, 0)$ be the special point. Then (X, w) is called a t-design on S^n with respect to the special point z_0 if any moment of degree at most t in the variables $x_1, x_2, \dots, x_n$ is invariant by any element of $O(n)$ (the stabilizer of z_0 in $O(n+1) \cong \mathrm{Isom}(S^n)$).

(c) This definition of a t-design on S^n with respect to the special point z_0 is very weak when we compare it with the definition of a usual t-design on S^n, because the latter has the stronger property that any moment of degree at most t in the variables $x_0, x_1, \dots, x_n$ is invariant by the element in $O(n+1)$.

(d) Now we want to mention the relations among t-designs on the three spaces S^n, $\mathbb{R}^n$ and $\mathbb{H}^n$. Now let X be one of the following three spaces, and let $w : X \to \mathbb{R}_{>0}$.

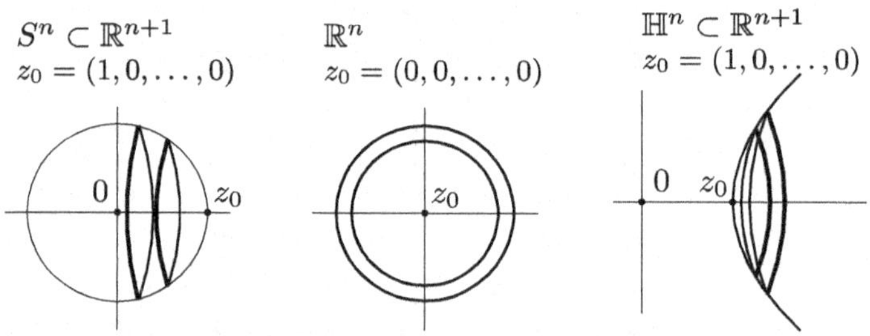

If there exists a t-design in this sense, then there also "essentially" exist t-designs in two other spaces. This is explained in the following:

$$x = (x_0, x_1, \dots, x_n) \to (x_1, \dots, x_n) \leftarrow x = (x_0, x_1, \dots, x_n),$$

$$\left(\sqrt{1 - x_1^2 - \dots - x_n^2}, x_1, \dots, x_n\right) \leftarrow (x_1, \dots, x_n) \to x = \left(\sqrt{1 + x_1^2 + \dots + x_n^2}, x_1, \dots, x_n\right).$$

The reason why we used the term "essentially" is that in $S^n (\subset \mathbb{R}^{n+1})$ the condition

$$x_1{}^2 + x_2{}^2 + \dots + x_n{}^2 \le 1$$

may not hold. In the above map, it may not hold indeed, but in $\mathbb{R}^n$ we can expand or shrink X, keeping it as (isomorphic) Euclidean t-designs. (So, by shrinking appropriately if necessary, we can get the isomorphism as mentioned above.) This fact implies that the Fisher type inequality for t-designs, tight t-designs, and their classifications should go exactly parallel to each of these cases. This situation that whole things go parallel to three cases might be regarded as uninteresting by some people, but at the personal level, I was very much convinced by the correctness of the definition of Euclidean t-designs for the first time. The designs we considered here on $\mathbb{R}^n$ and $\mathbb{H}^n$ are considered as kinds of self-centralized or "ptolemaic" designs.

The idea of this design is based on a very similar viewpoint of t-designs and relative t-designs of Q-polynomial schemes $\mathfrak{X} = (X, \{R_i\}_{0 \le i \le d})$ defined in Definition 4.5 and Definition 4.7 in Chapter 4. More precisely, we defined t-designs as elements of the vector space $V = \mathbb{R}^{|X|}$, i. e., the space of the functions on X. We choose an arbitrary point $u_0 \in X$ and we decompose X into $(d+1)$ shells (layers), $X_i = \{x \in X \mid (x, u_0) \in R_i\}$ $(0 \le i \le d)$. Here note that considering the element in V is equivalent to considering the weight function on X. For a non-negative weight function ϕ on X, we consider $Y = \{x \in X \mid \phi(x) \neq 0\}$, the support of ϕ. Then considering (Y, ϕ) is equivalent to considering a positive real valued weight function on a finite subset in the case of Euclidean designs.

When we consider a Euclidean t-design and if the support contains only one sphere, then the theory corresponds to that of spherical t-designs. Similarly, when we consider a relative t-design on an association scheme and if the support of Y is just on one layer (namely, if $Y \subset X_i$ for one i), then a similar situation holds. For example, in the association scheme $H(n, 2)$, for a subset $Y \subset X_k$ and the characteristic vector ϕ_Y of Y, (Y, ϕ_Y) is a relative t-design with respect to $u_0 = (0, 0, \ldots, 0)$ if and only if Y (as a subset of X_k) is a t-design, i. e., a t-(n, k, λ) design of the Johnson scheme $J(n, k) \cong X_k$. In this way, the special point u_0 of the Q-polynomial scheme $\mathfrak{X} = (X, \{R_i\}_{0 \le i \le d})$ corresponds to the origin of $\mathbb{R}^n$ and it is natural to regard each layer X_i, determined by u_0, as each of the concentric spheres in the Euclidean case. Now, we want to give a formulation of a relative t-design on Q-polynomial scheme $\mathfrak{X} = (X, \{R_i\}_{0 \le i \le d})$ with respect to the point $u_0 \in X$ in terms of the function space. We use the notation that we used already. Let $L(X)$ be the vector space of real valued functions on X. We use this notation instead of $V = \mathbb{R}^{|X|}$. For $u \in X$, let ϕ_u be the characteristic function of $\{u\}$, so $\phi_u = \phi_{\{u\}}$. (Previously we used the term characteristic vector, but here we use the term characteristic function.) Then, $\{\phi_u \mid u \in X\}$ is a basis of $L(X)$. For each j with $0 \le j \le d$, let $L_j(X)$ be the subspace of $L(X)$ generated by $\{E_j \phi_u \mid u \in X\}$. If we regard $L_j(X)$ as the subspace of V, then $L_j(X)$ is isomorphic to the subspace spanned by the column vectors of E_j. Then, we can define the natural inner product on $L(X)$:

$$f \cdot g = \sum_{x \in X} f(x) g(x) \quad \text{for } f, g \in L(X).$$

If we recall the definition of the primitive idempotents $E_0, E_1, \ldots, E_d$, we get the following proposition.

Proposition 5.53.
(1) *We have* $\dim(L_j(X)) = m_j$ *for* $j = 0, 1, \ldots, d$.
(2) *With respect to the inner product defined above,* $L(X)$ *has the following orthogonal decomposition:*

$$L(X) = L_0(X) \perp L_1(X) \perp \cdots \perp L_d(X).$$

Now, we give the definition of a relative t-design on a Q-polynomial scheme, imitating the definition of Euclidean t-designs. We use the following notation. Let X_i denote the i-th layer of X with respect to the special point u_0 as defined before. We consider a subset Y of X. Let $p = |\{\mu \mid Y \cap X_\mu \neq \emptyset\}|$ and set $\{\mu_1, \mu_2, \ldots, \mu_p\} = \{\mu \mid Y \cap X_\mu \neq \emptyset\}$. We call $S = X_{\mu_1} \cup X_{\mu_2} \cup \cdots \cup X_{\mu_p}$ the *support* of Y. Set $Y_{\mu_i} = Y \cap X_{\mu_i}$, for $1 \leq i \leq p$. We consider a positive real valued function w defined on Y, and define $w(Y_{\mu_i}) = \sum_{y \in Y_{\mu_i}} w(y)$ for $1 \leq i \leq p$.

Definition 5.54. Under the above notation, the pair (Y, w) of a subset $Y \subset X$ and a positive real valued weight function w on Y is called a relative t-design with respect to the special point u_0 if the condition

$$\sum_{i=1}^{p} \frac{w(Y_{\mu_i})}{|X_{\mu_i}|} \sum_{x \in X_{\mu_i}} f(x) = \sum_{y \in Y} w(y) f(y) \tag{5.52}$$

holds for any $f \in L_0(X) + L_1(X) + \cdots + L_t(X)$.

In Definition 4.7 in Chapter 4, we defined a relative t-design for a non-zero function $\psi \in L(X)$; here we define a relative t-design for a positive real valued weight function on a subset Y of X. These definitions are equivalent, as we will see below.

Theorem 5.55. *Let* $\psi \in L(X)$ *be a non-negative function. Define* $\bar{\psi} \in L(X)$ *by* $\bar{\psi}(x) = \frac{1}{X_i} \sum_{y \in X_i} \psi(y)$ *for any* i $(0 \leq i \leq d)$ *and* $x \in X_i$. *(Namely,* $\bar{\psi}$ *becomes a constant function on each* X_i.*) Define* $Y = \{y \in X \mid \psi(y) \neq 0\}$ *and* $w = \psi|_Y$. *Then the following conditions (1), (2), (3) are equivalent:*
(1) ψ *is a relative t-design with respect to the point* u_0 *in the sense of Definition 4.7;*
(2) *for each* j *with* $1 \leq j \leq t$, $E_j\psi$ *and* $E_j\bar{\psi}$ *are linearly dependent;*
(3) (Y, w) *is a relative t-design with respect to the point* u_0 *in the sense of Definition 5.54.*

Proof. Let ψ_{u_0} and ψ_{X_i} be the characteristic functions of u_0 and each layer X_i, respectively. For $x \in X_\ell$, $(E_j\psi_{u_0})(x) = E_j(x, u_0) = \frac{1}{|X|} Q_j(\ell)$. Since the second eigenmatrix Q is a non-singular matrix, $E_j(\psi_{u_0}) \neq 0$. Let the support of Y be $S = X_{\mu_1} \cup \cdots \cup X_{\mu_p}$.

(1) ⇔ (2). As we have seen in Proposition 4.9, ψ_X is a relative t-design, and as we have seen in that proof, $E_j\psi_{X_i} = P_i(j)\psi_{u_0}$ holds for $j = 1, 2, \ldots, d$. Meanwhile, from the definition, we have

$$E_j\bar{\psi} = E_j \sum_{i=0}^{d} \frac{1}{|X_i|} \sum_{y \in X_i} \psi(y)\psi_{X_i} = \left(\sum_{i=0}^{d} \frac{1}{|X_i|} \sum_{y \in X_i} \psi(y) P_i(j) \right) E_j\psi_{u_0}.$$

Therefore, $E_j\bar{\psi}$ and $E_j\psi_{u_0}$ are linearly dependent.

In order to show (1) ⇔ (3), we need some preparation. The condition (5.52) is the condition on any integer j with $0 \le j \le t$ and $f \in L_j(S)$. Since $L_j(X)$ is generated by $\{E_j\psi_u, u \in X\}$, it suffices to consider the case of $f = E_j\psi_u|_S$. Let $(u_0, u) \in R_\ell$. By Theorem 2.22 (8), we have

$$\sum_{x\in X_v} E_j(x,u) = \frac{1}{|X|}\sum_{k=0}^{d}\sum_{\substack{x\in X_v\\(x,u)\in R_k}} Q_j(k) = \frac{1}{|X|}\sum_{k=0}^{d} p_{v,k}^{\ell} Q_j(k)$$
$$= \frac{1}{|X|}P_v(j)Q_j(\ell). \tag{5.53}$$

Therefore, the left-hand side of (5.52) becomes

$$\sum_{i=1}^{p}\frac{w(Y_{v_i})}{|X_{v_i}|}\sum_{x\in X_{v_i}}(E_j\psi_u)(x)$$
$$= \frac{1}{|X|}\sum_{i=1}^{p}\frac{w(Y_{v_i})}{|X_{v_i}|}P_{v_i}(j)Q_j(\ell) = \sum_{i=1}^{p}\frac{w(Y_{v_i})}{|X_{v_i}|}P_{v_i}(j)E_j(u,u_0)$$
$$= \sum_{i=1}^{p}\frac{w(Y_{v_i})}{|X_{v_i}|}P_{v_i}(j)(E_j\psi_{u_0})(u). \tag{5.54}$$

From $|X_{v_i}| = k_{v_i}$ and Theorem 2.22 (3), we have

$$\sum_{i=1}^{p}\frac{w(Y_{v_i})}{|X_{v_i}|}P_{v_i}(j) = \frac{1}{m_j}\sum_{i=1}^{p} w(Y_{v_i})Q_j(v_i). \tag{5.55}$$

For v satisfying $v \in \{v_1, v_2, \dots, v_p\}$ we have $\psi(y) \ne 0$ for $y \in X_v$, and for $x \notin Y$ we have $\psi(x) = 0$. Therefore, we have

$$\frac{1}{m_j}\sum_{i=1}^{p} w(Y_{v_i})Q_j(v_i) = \frac{1}{m_j}\sum_{i=1}^{p}\sum_{y\in Y_{v_i}}\psi(y)Q_j(v_i)$$
$$= \frac{1}{m_j}\sum_{v=0}^{d}\sum_{x\in X_v}\psi(x)Q_j(v) = \frac{|X|}{m_j}\sum_{v=0}^{d}\sum_{x\in X_v}E_j(u_0,x)\psi(x)$$
$$= \frac{|X|}{m_j}\sum_{x\in X}E_j(u_0,x)\psi(x) = \frac{|X|}{m_j}(E_j\psi)(u_0). \tag{5.56}$$

Thus, by (5.54), (5.55), and (5.56), we see that the left-hand side of (5.52) is given by

$$\frac{|X|}{m_j}(E_j\psi)(u_0)(E_j\psi_{u_0})(u). \tag{5.57}$$

(1) ⇒ (3): By the assumption, for each integer j with $0 \le j \le t$, there exists a real number α_j such that $E_j\psi = \alpha_j E_j\psi_{u_0}$. So, by (5.57), the left-hand side of (5.52) becomes

$$\frac{|X|}{m_j}(\alpha_j E_j\psi_{u_0})(u_0)(E_j\psi_{u_0})(u) = \frac{|X|}{m_j}\alpha_j E_j(u_0,u_0)E_j(u_0,u) = \frac{\alpha_j}{|X|}Q_j(\ell).$$

On the other hand, the right-hand side of (5.52) becomes

$$\sum_{y\in Y} w(y)(E_j\psi_u)(y) = \sum_{x\in X} \psi(x)E_j(x,u) = (E_j\psi)(u)$$
$$= (\alpha_j E_j\psi_{u_0})(u) = \alpha_j E_j(u,u_0) = \frac{\alpha_j}{|X|}Q_j(\ell). \tag{5.58}$$

(3) ⇒ (1): Using (5.57), we can modify equality (5.52) as

$$\frac{|X|}{m_j}(E_j\psi)(u_0)(E_j\psi_{u_0})(u) = (E_j\psi)(u). \tag{5.59}$$

Therefore, $E_j\psi = (\frac{|X|}{m_j}(E_j\psi)(u_0))E_j\psi_{u_0}$ holds. Namely, ψ is a relative t-design with respect to the point u_0. □

From Theorem 5.55 and Proposition 4.9, we have the following.

Proposition 5.56. *Fix $u_0 \in X$. For any integer i with $0 \le i \le d$ let $X_i = \{u \in X \mid (u,u_0) \in R_i\}$. Also, let ψ_{X_i} be the characteristic function of X_i. Then (X_i, ψ_{X_i}) is a relative d-design.*

We call (X_i, ψ_{X_i}) the trivial relative design. When we define spherical or Euclidean t-designs, we described them as properties with respect to the polynomials of degree at most t. The relative t-design described in this chapter is described by using the indices of the primitive idempotents. The next proposition shows that a similar property holds.

Proposition 5.57. *Let $0 \le j \le \frac{d}{2}$. Then*

$$f^2 \in \sum_{i=0}^{2j} L_i(X)$$

holds for any $f \in \sum_{i=0}^{j} L_i(X)$.

Proof. It suffices to show that for non-negative integers ν, μ with $\mu+\nu \le 2j$ and arbitrary $u, v \in X$ the relation

$$(E_\nu\psi_u)(E_\mu\psi_v) \in \sum_{i=0}^{2j} L_i(X) \tag{5.60}$$

holds. As explained in Proposition 5.53 (2), each $L_i(X)$ $(0 \le i \le d)$ is a factor of the orthogonal decomposition of $L(X)$. Therefore, the proof is completed if we show that $(E_\nu\psi_u)(E_\mu\psi_v) \perp E_k\psi_z$ for any k $(0 \le k \le d)$ with $k > \mu+\nu$ and $z \in X$. Now, the following holds:

$$|X|\sum_{x\in X} E_\nu(x,u)E_\mu(x,v)E_k(x,z)$$
$$= |X|\sum_{i=0}^{d}\sum_{\substack{x\in X\\(x,v)\in R_i}} E_\nu(x,u)E_\mu(x,v)E_k(x,z)$$

$$= \sum_{i=0}^{d} \sum_{\substack{x \in X \\ (x,v) \in R_i}} E_v(x,u) Q_\mu(i) E_k(x,z)$$

$$= \sum_{i=0}^{d} \sum_{x \in X} E_v(x,u) Q_\mu(i) E_i^*(x,x) E_k(x,z)$$

$$= \sum_{x \in X} E_v(x,u) A_\mu^*(x,x) E_k(x,z) = (E_v A_\mu^* E_k)(u,z). \tag{5.61}$$

By Corollary 2.37 (2) in Chapter 2, Section 2.6, the condition $E_v A_\mu^* E_k = 0$ is equivalent to the condition $q_{v,\mu}^k = 0$. Also, by the property of Q-polynomial association schemes (Proposition 2.83 (1) in Chapter 2, Section 2.9), if $0 \le v, \mu, k \le d$ and $v + \mu < k$, then $q_{v,\mu}^k = 0$. Therefore we have $(E_v \psi_u)(E_\mu \psi_v) \perp E_k \psi_z$. □

By using Proposition 5.57, we can obtain the lower bound of the size $|Y|$ of a relative $2e$-design (Y, w) with expect to the point u_0 as follows.

Theorem 5.58. *For the size $|Y|$ of a relative $2e$-design (Y, w) on a Q-polynomial association scheme $\mathfrak{X} = (X, \{R_i\}_{0 \le i \le d})$ with respect to the point $u_0 \in X$, we have*

$$|Y| \ge \dim(L_0(S) + L_1(S) + \cdots + L_e(S)), \tag{5.62}$$

where $S = X_{v_1} \cup X_{v_2} \cup \cdots \cup X_{v_p}$ is the support of Y and $L_i(S) = \{f|_S \mid f \in L_i(X)\}$ $(0 \le i \le d)$.

Proof. Let $\rho : L_0(S) + L_1(S) + \cdots + L_e(S) \longrightarrow L(Y)$ be defined by $\rho(f) = f|_Y$ for $f \in L_0(S) + L_1(S) + \cdots + L_e(S)$. If we prove that ρ is injective, then we can see that $|Y| = \dim(L(Y)) \ge \dim(L_0(S) + L_1(S) + \cdots + L_e(S))$ holds. For that purpose, we have only to show that if $f \in L_0(S) + L_1(S) + \cdots + L_e(S)$ satisfies $f(y) = 0$ for any $y \in Y$, then $f(x) = 0$ for any $x \in S$. Since $f^2 \in L_0(X) \perp L_1(X) \perp \cdots \perp L_{2e}(X)$ holds by Proposition 5.57, from the defining relation (5.52) we get

$$\sum_{i=1}^{p} \frac{w(Y_{v_i})}{|X_{v_i}|} \sum_{x \in X_{v_i}} (f(x))^2 = \sum_{y \in Y} w(y)(f(y))^2 = 0. \tag{5.63}$$

Therefore we get $f(x) = 0$ for any $x \in X_{v_i} (1 \le i \le p)$ and we get $f \equiv 0$ as a function in $L_0(S) + L_1(S) + \cdots + L_e(S)$. Therefore, ρ is injective. □

Definition 5.59. If equality holds in the inequality in Theorem 5.58, then (Y, w) is called a *tight relative $2e$-design* with respect to the point u_0.

While continuing the study of Euclidean designs, we noticed that a similar concept can be defined on Q-polynomial association schemes, and then realized that it is exactly the concept of relative t-designs that was already defined and studied by Delsarte (1977) [161]. The following theorem suggests a similarity.

Theorem 5.60 ([44]). *Let $\mathfrak{X} = (X, \{R_i\}_{0 \le i \le d})$ be a Q-polynomial scheme, and let (Y, w) be a tight relative $2e$-design with respect to the point u_0. Suppose the stabilizer subgroup*

G_{u_0} of u_0 of the automorphism group G of $\mathfrak{X}$ acts transitively on each X_{v_i}. Then the weight function w becomes a constant function on each $Y_{v_i} = X_{v_i} \cap Y$ $(1 \le i \le p)$.

We will not discuss the proof here. (Please see [44].) There are many Q-polynomial association schemes that satisfy this assumption, including the Hamming scheme $H(d,q)$, the Johnson scheme $J(v,d)$, and many others. For a tight Euclidean t-design on two concentric spheres, it is known that the structure of a coherent configuration is generally attached ([75, 39]). We think it is natural to conjecture that the same conclusion (a coherent configuration is attached for a tight relative t-design) if the assumptions in Theorem 5.60 is satisfied and if $p = 2$. Actually, it is shown ([44]) that this is true for tight relative 2-designs on $H(d,2)$ if $p = 2$. The concept of relative t-designs on Q-polynomial schemes was first defined and studied by Delsarte in the late 1970s ([161]). There are many problems that are to be studied more, including finding more explicit examples.

Remark 5.61.
(1) In the Hamming scheme, for the fixed $u_0 = (0,0,\dots,0)$, we consider the layer X_k and a subset Y in X_k. Let ψ_Y be the characteristic function of Y. Then (Y,ψ_Y) is a tight relative $2e$-design with respect to the point u_0 if and only if Y is a classical tight $2e$-(n,k,λ) design.
(2) For the designs in Euclidean spaces, Möller obtained a natural lower bound for the number of points of a (centrally symmetric) t-design for odd $t = 2e+1$. It is expected to find a natural lower bound for the size of a relative $(2e+1)$-design on a Q-polynomial scheme with respect to the fixed point u_0.

Remark 5.62. Delsarte and Seidel [164] defined the new concept of designs on the Hamming scheme $H(n,2)$ which removes the condition that the block sizes are constant, as an analogue of Euclidean designs. Let $F_2 = \{0,1\}$ and $H(n,2)$ be the Hamming scheme on the set $X = F_2^n$. For $x = (x_1, x_2, \dots, x_n) \in X$, we define $\bar{x} = \{i \mid x_i = 1, 1 \le i \le n\}$.

Definition (Delsarte–Seidel, Definition 6.1 in [164]). Let (Y,w) be a weighted subset of X, where Y is a subset of X and w is a positive real valued weight function on Y. We say that (Y,w) *admits the index* j if the quantity

$$\sum_{\substack{y \in Y \\ \bar{z} \subset \bar{y}}} w(y)$$

does not depend on the choice of $z \in X_j$ and is determined only by j.

A pair (Y,w) that admits the index j is called a *j-wise balanced design*. If (Y,w) admits all integers j with $0 \le j \le t$, then it is called a *regular t-wise balanced design*.

Moreover, Delsarte and Seidel used the following space as the space corresponding to the space of polynomials of degree at most t in the real Euclidean space. First we define, for each $z \in X$, the function $f_z : X \longrightarrow \mathbb{R}$ by

$$f_z(x) = \begin{cases} 1, & \text{if } \overline{z} \subset \overline{x} \text{ holds,} \\ 0, & \text{otherwise.} \end{cases}$$

(For z with $|\overline{z}| = j$, f_z corresponds to a column vector of the matrix M_j defined in Chapter 2.) For each j, we define $\mathrm{Hom}_j(X) = \langle f_z \mid z \in X, |\overline{z}| = j \rangle$ as the subspace of the space of the real valued functions on X. Note that by Lemma 2.93, $\dim(\mathrm{Hom}_j(X)) = \binom{n}{j}$. Dersarte and Seidel proved the following two theorems.

Theorem (Delsarte–Seidel, Theorem 6.2 in [164]). *In $H(n,2)$, (Y,w) admits the index j if and only if*

$$\sum_{i=1}^{p} \frac{w(Y_{v_i})}{|X_{v_i}|} \sum_{x \in X_{v_i}} f(x) = \sum_{y \in Y} w(y) f(y) \tag{5.64}$$

holds for any $f \in \mathrm{Hom}_j(X)$.

Theorem (Delsarte–Seidel, Theorem 6.3 in [164]). *If (Y,w) admits all the indices j with $0 \le j \le 2e$, then*

$$|Y| \ge \dim(\mathrm{Hom}_0(S) + \mathrm{Hom}_1(S) + \cdots + \mathrm{Hom}_e(S)) \tag{5.65}$$

holds, where $S = X_{v_1} \cup X_{v_2} \cup \cdots \cup X_{v_p}$.

Therefore, it becomes a very interesting problem to find the dimension of the right-hand side of (5.65) in general. In Definition 5.54 in this chapter, we gave a definition of relative design (Y,w) (in any Q-polynomial scheme) that is equivalent to Definition 4.7 of relative t-design in the sense of Delsarte. Also, in Theorem 5.58, we already obtained the lower bound (5.62):

$$|Y| \ge \dim(L_0(S) + L_1(S) + \cdots + L_e(S)),$$

which very much looks like the lower bound (5.65) mentioned above.

Although these two formulas look very much alike, the spaces of functions appearing in (5.65) and (5.62) are mathematically different objects in general. In Chapter 4, we have explained that the top fiber of a regular semilattice Ω has the structure of an association scheme. For the semilattice constructed in Example 4.37, for $q = 2$, the top fiber becomes the association scheme $H(n,2)$. Let w be a non-negative real valued function on $\Omega = F_2^n$ and let $Y = \{y \mid w(y) \neq 0\}$. If (Y,w) is a relative t-design with respect to the point $x_0 = (0,0,\ldots,0)$, then $w(y)$ becomes a geometric t-design on Ω, and so if we consider the function defined by (4.45), the pair (Y,w) admits j for any integer with $0 \le j \le t$. Namely, (Y,w) becomes a j-wise balanced design for any integer j with $0 \le j \le t$. By these considerations, in $H(n,2)$ the value in (5.65) is smaller than the value in (5.62) in general. Many studies are ongoing, including the study of exact values of (5.62) as well as the case where equality holds, i. e., the study of tight relative $2e$-designs.

Remark 5.63. The research of relative $2e$-designs has restarted with the joint work by Li, Bannai, and Bannai [316]. Some progress has been made since the authors wrote the manuscript of the original version of this book around 2011. In particular, Ziqing Xiang [521] (who was an undergraduate student at Shanghai Jiao Tong University at that time) succeeded in proving that in $H(n, 2)$ the right-hand side of (5.65) (under some reasonable conditions like $e \leq v_i \leq n - e$) is equal to $\binom{n}{e} + \binom{n}{e-1} + \cdots + \binom{n}{e-p+1}$. Moreover, Bannai, Bannai, Suda, and Tanaka [49] proved that as a space over $H(n, 2)$, we get

$$\begin{aligned} &\mathrm{Hom}_0(X) + \mathrm{Hom}_1(X) + \cdots + \mathrm{Hom}_e(X) \\ &\quad = L_0(X) + L_1(X) + \cdots + L_e(X). \end{aligned} \tag{5.66}$$

So, the result of Xiang ([521]) implies that the value of the right-hand side of (5.65) has the same value as that of (5.62), because of (5.57). However, this property (5.57) does not hold in general for P- and Q-polynomial schemes even if $P = Q$ (due to a remark by Suda and Tanaka). Therefore it is very desirable to calculate the right-hand side of (5.62) for general Q-polynomial schemes. Also, it would be very interesting to calculate the right-hand side of (5.65) for general P-polynomial schemes. The reader is referred to [49] for these discussions, as well as Bannai, Bannai, Tanaka, and Zhu [51].

The study of tight relative 2-designs on two shells of $H(n, 2)$ and $J(v, k)$ ([44, 528]) as well as tight relative 4-designs on two shells on $H(n, 2)$ ([53]) is ongoing. The research on tight relative $2e$-designs in general has started. For updates, see Bannai, Bannai, Tanaka, and Zhu (2020) [52].

6 P- and Q-polynomial schemes

6.1 P-polynomial/Q-polynomial schemes revisited

6.1.1 Distance-regular graphs revisited

We begin with the definition of a distance-transitive graph. Let $\Gamma = (X, R)$ be a finite simple graph (Chapter 1, Section 1.1), where X is the vertex set and R is the edge set. For $x, y \in X$, let $\partial(x, y)$ be the length of a shortest path connecting x and y. If there is no such path, then we set $\partial(x, y) = \infty$. We let $d = \mathrm{Max}\{\partial(x, y) \mid x, y \in X\}$ denote the maximum of the distances between vertices in X, and call d the *diameter* of Γ. In what follows, we assume that the diameter satisfies $d < \infty$, that is, Γ is a connected graph. Let S^X be the symmetric group on X. An element $\sigma \in S^X$ is said to be an automorphism of Γ if $(x^\sigma, y^\sigma) \in R$ for any edge $(x, y) \in R$. The set of all automorphisms of Γ will be denoted by $\mathrm{Aut}(\Gamma)$. The set $\mathrm{Aut}(\Gamma)$ forms a subgroup of S^X, and is called the *automorphism group* of the graph Γ. It acts on $X \times X$ by $(x, y)^\sigma = (x^\sigma, y^\sigma)$ ($\sigma \in \mathrm{Aut}(\Gamma)$). For $0 \le i \le d$, we set

$$R_i = \{(x, y) \in X \times X \mid \partial(x, y) = i\}. \tag{6.1}$$

Each R_i is invariant under the action of $\mathrm{Aut}(\Gamma)$. Namely, $\mathrm{Aut}(\Gamma)$ acts on each R_i. We say that Γ is a *distance-transitive graph* if the action of $\mathrm{Aut}(\Gamma)$ on R_i is *transitive*. That is, we call Γ a distance-transitive graph if there exists $\sigma \in \mathrm{Aut}(\Gamma)$ such that $x' = x^\sigma$, $y' = y^\sigma$ for any $x, y, x', y' \in X$ satisfying $\partial(x, y) = \partial(x', y')$.

We now forget about the group action, and return again to a finite connected simple graph $\Gamma = (X, R)$. Recall that d denotes the diameter of Γ. For $x, y \in X$ and $i, j \in \{0, 1, \ldots, d\}$, let

$$p_{i,j}(x, y) = \big|\{z \in X \mid (x, z) \in R_i, (z, y) \in R_j\}\big|. \tag{6.2}$$

If Γ is a distance-transitive graph, then for any $i, j, k \in \{0, 1, \ldots, d\}$ there exists a constant $p_{i,j}^k$ such that

$$p_{i,j}^k = p_{i,j}(x, y) \quad ((x, y) \in R_k). \tag{6.3}$$

In other words, $p_{i,j}(x, y)$ is determined only by i, j, and k, and is independent of the choice of $(x, y) \in R_k$. A finite connected simple graph for which (6.3) holds is called a *distance-regular graph*. A distance-transitive graph is a distance-regular graph. Usually, a distance-regular graph is defined by using the following condition, which is weaker than (6.3):

$$p_{1,j}^k = p_{1,j}(x, y) \quad ((x, y) \in R_k, |k - j| \le 1). \tag{6.4}$$

https://doi.org/10.1515/9783110630251-006

Namely, for $x_0 \in X$, let

$$\Gamma_i(x_0) = \{x \in X \mid \partial(x_0, x) = i\}.$$

If the numbers

$$\begin{aligned} b_i &= |\Gamma_{i+1}(x_0) \cap \Gamma_1(x)| \quad (0 \le i \le d-1),\\ a_i &= |\Gamma_i(x_0) \cap \Gamma_1(x)| \quad (0 \le i \le d),\\ c_i &= |\Gamma_{i-1}(x_0) \cap \Gamma_1(x)| \quad (1 \le i \le d) \end{aligned} \tag{6.5}$$

are constant and independent of the choice of $x_0 \in X$, $x \in \Gamma_i(x_0)$ (i. e., if $b_i = p^i_{1,i+1} = p_{1,i+1}(x, x_0)$, $a_i = p^i_{1,i} = p_{1,i}(x, x_0)$, $c_i = p^i_{1,i-1} = p_{1,i-1}(x, x_0)$), we define Γ to be a distance-regular graph.

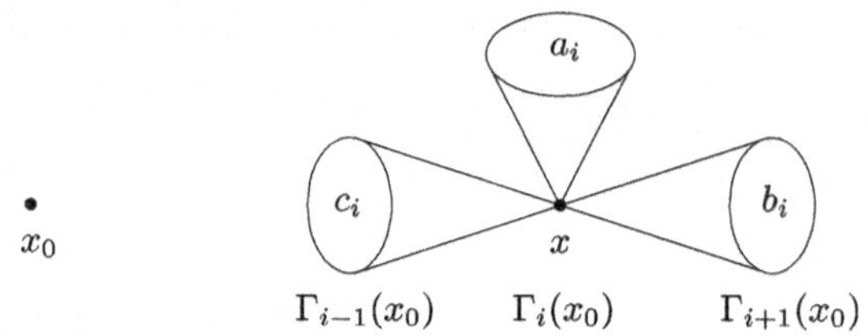

A distance-regular graph is regular with degree $k = |\Gamma_1(x_0)| = b_0$ and the following holds:

$$k = a_0 + b_0 = c_i + a_i + b_i = c_d + a_d \quad (1 \le i \le d-1). \tag{6.6}$$

Clearly, we have $a_0 = 0$, $c_1 = 1$. In the following, we will check that (6.4) implies (6.3).

Assume (6.4). Let A_i be the adjacency matrix of R_i:

$$A_i(x, y) = \begin{cases} 1, & \text{if } \partial(x, y) = i,\\ 0, & \text{otherwise.} \end{cases} \tag{6.7}$$

By the triangle inequality of the distance function $\partial(x, y)$, the condition (6.4) can be written as follows. We have

$$\text{(i)} \qquad A_1A_j = b_{j-1}A_{j-1} + a_jA_j + c_{j+1}A_{j+1} \quad (0 \le j \le d),$$

and $b_{i-1}c_i \neq 0$ $(1 \le i \le d)$, where b_i, a_i, c_i are defined in (6.5). Here, we let b_{-1} be an indeterminate and $c_{d+1} = 1$, $A_{-1} = 0$, $A_{d+1} = 0$. We define a polynomial $v_i(x)$ of degree i by the following three-term recurrence:

$$xv_j(x) = b_{j-1}v_{j-1}(x) + a_jv_j(x) + c_{j+1}v_{j+1}(x), \tag{6.8}$$

where $v_{-1}(x) = 0$, $v_0(x) = 1$. Then we have

$$\text{(ii)} \qquad A_i = v_i(A_1) \quad (0 \le i \le d).$$

By (ii), we have

$$J = v_0(A_1) + v_1(A_1) + v_2(A_1) + \cdots + v_d(A_1), \tag{6.9}$$

where J is the all 1's matrix. The matrices $v_0(A_1) = I, v_1(A_1), v_2(A_1), \ldots, v_d(A_1)$ are linearly independent. Hence, $A_1^0 = I, A_1, A_1^2, \ldots, A_1^d$ are linearly independent. Since A_1 is the adjacency matrix of a graph Γ of degree k, every row sum of A_1 is k, and hence we have $(A_1 - kI)J = 0$. Therefore, the polynomial

$$(x-k)(v_0(x) + v_1(x) + v_2(x) + \cdots + v_d(x)) \tag{6.10}$$

is a scalar multiple of the minimal polynomial of A_1. In particular, the minimal polynomial of A_1 has degree $d+1$. Note that we only need the condition that the polynomial $v_i(x)$ has degree i in the above discussion. This is important because it matters when we define P-polynomial schemes. By using the fact that $v_i(x)$ is given by (6.8), we can show that $v_{d+1}(A_1) = 0$ and that the minimal polynomial of A_1 has degree $d+1$. In fact, $v_{d+1}(x)$ equals the polynomial (6.10). Let $\mathfrak{A}$ be the subalgebra of $M_X(\mathbb{C})$ generated by A_1. Since the degree of the minimal polynomial of A_1 is $d+1$, the dimension of $\mathfrak{A}$ is $d+1$ and $\{A_i = v_i(A_1) \mid 0 \le i \le d\}$ forms a basis of $\mathfrak{A}$. Therefore, there exists a non-negative integer $p_{i,j}^k$ such that

$$\text{(iii)} \qquad A_iA_j = \sum_{k=0}^{d} p_{i,j}^k A_k.$$

As we defined in (6.7), A_i is the adjacency matrix of R_i, so (iii) is equivalent to (6.3). Hence, (6.4) implies (6.3). Combinatorial conditions (6.4), (6.3) correspond to algebraic conditions (i), (iii), respectively, and algebraically we have (i) ⇒ (ii) ⇒ (iii) ⇒ (i).

The combinatorial condition (6.3) or the equivalent algebraic condition (iii) implies $(X, \{R_i\}_{0\le i\le d})$ is a symmetric association scheme, which will be denoted by $\mathfrak{X}$. Namely, the distance-regular graph defined by the combinatorial condition (6.4) or the equivalent algebraic condition (i) induces a symmetric association scheme defined by (iii). By the triangle inequality of the distance function $\partial(x,y)$ appearing in (6.7), the intersection numbers in (iii) satisfy

$$\text{(iii)}' \qquad p_{i,j}^k \begin{cases} = 0, & \text{if } i < |k-j| \text{ or } i > k+j, \\ \neq 0, & \text{if } i = |k-j| \text{ or } i = k+j, \end{cases}$$

for $i, j, k \in \{0, 1, \ldots, d\}$. The intersection numbers in (i) satisfy

$$\text{(i)}' \qquad p_{1,j}^k \begin{cases} = 0, & \text{if } 1 < |k-j|, \\ \neq 0, & \text{if } 1 = |k-j|, \end{cases}$$

for $j, k \in \{0, 1, \ldots, d\}$. Assume a symmetric association scheme $\mathfrak{X} = (X, \{R_i\}_{0\le i\le d})$ satisfies (i)′. Similarly to the discussion which shows (i) implies (ii), the following holds:

$$\text{(ii)}' \qquad A_i \text{ is a polynomial in } A_1 \text{ of degree } i \quad (0 \le i \le d).$$

By (ii)′, it is clear that for $i, j, k \in \{0, 1, \dots, d\}$,

$$\text{(iii)}'' \qquad p_{i,j}^k \begin{cases} = 0, & \text{if } i + j < k, \\ \neq 0, & \text{if } i + j = k. \end{cases}$$

On the other hand, since $k_\gamma p_{\alpha,\beta}^\gamma = k_\beta p_{\gamma,\alpha}^\beta = k_\alpha p_{\beta,\gamma}^\alpha$ (Chapter 2, Proposition 2.17), (iii)″ implies (iii)′. Therefore we have (i)′ ⇒(ii)′ ⇒(iii)′ ⇒(i)′. A symmetric association scheme $\mathfrak{X} = (X, \{R_i\}_{0\le i\le d})$ is called a P-polynomial scheme if it satisfies one of (i)′, (ii)′, (iii)′ with respect to the ordering $R_0, R_1, \dots, R_d$. Then the polynomial of degree i in (ii)′ coincides with the polynomial $v_i(x)$ given by (6.8), and the minimal polynomial of A_1 is a scalar multiple of $v_{d+1}(x)$. Moreover, $v_{d+1}(x)$ coincides with the polynomial $(x - k)(v_0(x) + v_1(x) + \cdots + v_d(x))$ in (6.10).

In this way, P-polynomial schemes arise from distance-regular graphs. Conversely, if we assume $\mathfrak{X} = (X, \{R_i\}_{0\le i\le d})$ is P-polynomial with respect to the ordering $R_0, R_1, \dots, R_d$, it is clear by (i)′ that in the graph $\Gamma = (X, R_1)$, R_i is the distance i relation given by (6.1). So Γ is a distance-regular graph. In this sense, distance-regular graphs are often identified with P-polynomial schemes. There is another definition for P-polynomial schemes in the framework of Terwilliger algebras. Although the Terwilliger algebra first appeared in the papers [468, 469, 470] published in 1992–1993, Terwilliger had been considering the idea of this algebra since late 1980s. He obtained the idea inspired by Bannai and Ito [60], and his paper [468] was a kind of answer to the book, that is, his paper shows how he interpreted it. Terwilliger himself calls the Terwilliger algebra the *subconstituent algebra*. What Terwilliger is interested in is distance-regular graphs or P- and Q-polynomial schemes, and the subconstituent algebra is invented in order to study them. It seems that this is a key to the success of the theory. Since the idea of the subconstituent algebra is natural, there must have been several people who had similar ideas. However, these ideas lacked the objects like distance-regular graphs or P- and Q-polynomial schemes, and this is why these ideas could not come to light as fruitful theories.

Let $\mathfrak{X} = (X, \{R_i\}_{0\le i\le d})$ be a symmetric association scheme. We review the Terwilliger algebra of $\mathfrak{X}$. Let A_i be the adjacency matrix of the relation R_i and $V = \bigoplus_{x\in X} \mathbb{C}x$ the standard module of the Bose–Mesner algebra $\mathfrak{A}$. Fix a base vertex $x_0 \in X$, and set

$$V_i^* = V_i^*(x_0) = \bigoplus_{x\in\Gamma_i(x_0)} \mathbb{C}x, \tag{6.11}$$

where $\Gamma_i(x_0) = \{x \in X \mid (x_0, x) \in R_i\}$. Let V_i^* be the projection determined by the direct sum decomposition $V = \sum_{j=0}^d V_j^*$ as follows:

$$E_i^* = E_i^*(x_0) : V \longrightarrow V_i^*. \tag{6.12}$$

The subalgebra

$$\mathfrak{A}_i^* = \mathfrak{A}_i^*(x_0) = \langle E_0^*, E_1^*, \dots, E_d^* \rangle \tag{6.13}$$

of End(V) is called the *dual Bose–Mesner algebra*. Moreover, the subalgebra of End(V) generated by $\{\mathfrak{A}^*, \mathfrak{A}\}$, i. e.,

$$T = T(x_0) = \langle \mathfrak{A}^*, \mathfrak{A} \rangle, \tag{6.14}$$

is called the Terwilliger algebra. By Corollary 2.37 in Chapter 2, we have

$$p^{\gamma}_{\alpha,\beta} = 0 \Longleftrightarrow E^*_{\alpha} A_{\beta} E^*_{\gamma} = 0. \tag{6.15}$$

Hence, by $p^k_{1,j} = p^k_{j,1}$, condition (i)′ is equivalent to

$$\text{(iv)} \qquad E^*_j A E^*_k \begin{cases} = 0, & \text{if } 1 < |k-j|, \\ \neq 0, & \text{if } 1 = |k-j|, \end{cases}$$

where $A = A_1$. Namely, (iv) is the fourth definition for $\mathfrak{X}$ to be a P-polynomial scheme. Since $V^*_i = E^*_i V$, for a P-polynomial scheme, we have

$$AV^*_i \subseteq V^*_{i-1} + V^*_i + V^*_{i+1} \quad (0 \le i \le d), \tag{6.16}$$

where $V^*_{-1} = V^*_{d+1} = 0$. Comparing the above with the definition of distance-regular graphs (6.5) or (i), although the contents are the same, we can see the framework changes from the representation of $\mathfrak{A}$ to that of T.

In what follows, let $\mathfrak{X} = (X, \{R_i\}_{0 \le i \le d})$ be a P-polynomial scheme. In Section 6.1.4, we formally reframe several facts on the Bose–Mesner algebra $\mathfrak{A} = \langle A_0, A_1 \dots, A_d \rangle$ of $\mathfrak{X}$ in the framework of orthogonal polynomials. For this purpose, we review basic facts from this point of view. The adjacency matrix A_i of R_i is given by the equation $A_i = v_i(A_1)$ in (ii), where $v_i(x)$ is the polynomial in (6.8). The standard module $V = \bigoplus_{x \in X} \mathbb{C}x$ of $\mathfrak{A}$ is decomposed into the direct sum of the eigenspaces V_i of A_1:

$$V = \bigoplus_{i=0}^{d} V_j, \quad A_1|_{V_i} = \theta_i. \tag{6.17}$$

Since A_1 is a real symmetric matrix, the eigenvalues θ_i $(0 \le i \le d)$ are real. Let $E_i : V \longrightarrow V_i$ be the projection onto V_i defined by the above decomposition. Then we have

$$E_i = \prod_{j \neq i} \frac{A_1 - \theta_j}{\theta_i - \theta_j}. \tag{6.18}$$

The Bose–Mesner algebra $\mathfrak{A}$ has two bases, $\{A_0, A_1, \dots, A_d\}$ and $\{E_0, E_1, \dots, E_d\}$. Let $P = (P_j(i))$ be the transition matrix (first eigenmatrix) between them:

$$A_j = \sum_{i=0}^{d} P_j(i) E_i. \tag{6.19}$$

Namely, $P_j(i)$ is the eigenvalue of A_j on V_i. The eigenvalue of A_1 on V_i is $P_1(i) = \theta_i$ and $A_j = v_j(A_1)$. So we have

$$P_j(i) = v_j(\theta_i). \tag{6.20}$$

Here we set

$$k_i = |\Gamma_i(x_0)|, \tag{6.21}$$

where $\Gamma_i(x_0) = \{x \in X \mid (x_0, x) \in R_i\}$. The number k_i is the row sum of A_i. By Proposition 2.21 in Chapter 2, we have $k_i = P_i(0)$. Hence, by (6.20), we have

$$k_i = v_i(\theta_0). \tag{6.22}$$

In particular, if we let k be the degree of the distance-regular graph $\Gamma = (X, R_1)$, we get

$$k = k_1 = \theta_0. \tag{6.23}$$

Next, let m_i be the multiplicity of the eigenvalue θ_i of A_1:

$$m_i = \dim V_i = \operatorname{tr} E_i. \tag{6.24}$$

By rewriting the first and second orthogonality relations in Theorem 2.22 in Chapter 2 using (6.20), we have

$$\sum_{\nu=0}^{d} v_\nu(\theta_i)v_\nu(\theta_j)\frac{1}{k_\nu} = \delta_{i,j}\frac{n}{m_i}, \tag{6.25}$$

$$\sum_{\nu=0}^{d} v_i(\theta_\nu)v_j(\theta_\nu)m_\nu = \delta_{i,j}nk_i, \tag{6.26}$$

where $n = |X| = \sum_{i=0}^{d} k_i = \sum_{i=0}^{d} m_i$. Note that $P_j(i) = v_j(\theta_i)$ is real since it is an eigenvalue of the real symmetric matrix A_j. By setting $i = j$ in (6.25), we obtain the Biggs formula on the multiplicity m_i as follows.

Theorem 6.1 (Multiplicity formula).

$$m_i = \frac{n}{\sum_{\nu=0}^{d} v_\nu(\theta_i)^2/k_\nu}. \tag{6.27}$$

Let

$$\rho : \mathfrak{A} \longrightarrow M_{d+1}(\mathbb{C}) \cong \operatorname{End}(\mathfrak{A}) \tag{6.28}$$

be the matrix expression of the regular representation $\mathfrak{A} \longrightarrow \operatorname{End}(\mathfrak{A})$ of the Bose–Mesner algebra $\mathfrak{A}$ with respect to the basis $A_0, A_1, \ldots, A_d$ of $\mathfrak{A}$. Set $\rho(A_i) = {}^tB_i$ and we call B_i the intersection matrix of A_i (Chapter 2, Section 2.5). Since $A_iA_j = \sum_{k=0}^{d} p_{i,j}^k A_k$, the (k,j)-entry of ${}^tB_i$ is $p_{i,j}^k$. If we set $\mathfrak{B} = \langle B_0, B_1, \ldots, B_d\rangle$, since $\mathfrak{A}$ is commutative, by the correspondence $A_i \mapsto B_i$, we have

$$\mathfrak{A} \cong \mathfrak{B} \quad \text{(as algebras)}.$$

Let $B = {}^tB_1$. By (i), B is the tridiagonal matrix

$$B = \begin{bmatrix} a_0 & b_0 & & & & \\ c_1 & a_1 & b_1 & & 0 & \\ & \ddots & \ddots & \ddots & & \\ & & c_{d-1} & a_{d-1} & b_{d-1} \\ & 0 & & c_d & a_d \end{bmatrix}, \tag{6.29}$$

where $b_{i-1}c_i \neq 0$ $(1 \le i \le d)$. A tridiagonal matrix is said to be *irreducible* if the superdiagonal entries and the subdiagonal entries are all non-zero. So the above matrix is an irreducible tridiagonal matrix. The minimal polynomials of A_1 and $B = {}^tB_1$ are identical, which are equal to the polynomial of degree $d+1$ given by (6.10), and hence B has $d+1$ distinct eigenvalues $\theta_0, \theta_1, \ldots, \theta_d$. In particular, B is diagonalizable. Furthermore, by (6.6), (6.23), every row sum of B is θ_0. Such B, that is, a diagonalizable irreducible tridiagonal matrix whose row sums are constant, will be discussed in Section 6.1.4. Since $k_i = |\Gamma_i(x_0)|$ in (6.21) satisfies $k_{i-1}b_{i-1} = k_ic_i$, we have

$$k_i = \frac{b_{i-1}}{c_i}k_{i-1} = \frac{b_0b_1\cdots b_{i-1}}{c_1c_2\cdots c_i} \quad (0 \le i \le d). \tag{6.30}$$

Therefore, $\{k_i\}_{i=0}^d$ are determined by B only. Moreover, by (6.8), the polynomials $\{v_i(x)\}_{i=0}^d$ depend only on B, and $\{\theta_i\}_{i=0}^d$ are distinct eigenvalues of B. By the above discussion, the multiplicities $\{m_i\}_{i=0}^d$ determined by (6.27) in Theorem 6.1 are constants depending only on B.

For an irreducible tridiagonal matrix B which has the form of (6.29) to coincide with the transpose ${}^tB_1$ of the intersection matrix of a distance-regular graph, it is necessary that it is diagonalizable, it has a constant row sum, and m_i $(0 \le i \le d)$, which are determined by (6.27), and k_i $(0 \le i \le d)$, which are determined by (6.30), are positive integers. The above necessary condition is called the *feasibility condition* and B is said to be *feasible* if it satisfies the feasibility condition. There are various other conditions for the irreducible tridiagonal matrix B to coincide with the transpose of the intersection matrix of a distance-regular graph ([94, 113, 459, 461]). These conditions are also called feasibility conditions, and it depends on the text what the feasibility condition refers to. The conditions appearing in the following proposition are typical examples.

Proposition 6.2. *The intersection matrix of a distance-regular graph* $\Gamma = (X, R)$ *satisfies the following:*

(1) $k = b_0 \geq b_1 \geq \cdots \geq b_{d-1}$;
(2) $1 = c_1 \leq c_2 \leq \cdots \leq c_d$;
(3) *if* $i + j \leq d$, *then* $c_i \leq b_j$;
(4) *if* $a_1 \neq 0$, *then* $a_i \neq 0$ $(2 \leq i \leq d-1)$.

Remark 6.3. Assume the Terwilliger algebra T is 1-thin. Namely, every irreducible T-module of endpoint 1 is thin (Section 6.2, Remark 6.30). Then $a_1 = 0$ implies $a_i = 0$ $(2 \leq i \leq d-1)$ [170, 172].

As a preparation for the proof of the above proposition, we define the *distance distribution diagram* of a distance-regular graph $\Gamma = (X, R)$. Fix 2 vertices $x, y \in X$ at distance 1. For x, y and $i, j \in \{0, 1, \ldots, d\}$, set

$$D_i^j = \Gamma_i(x) \cap \Gamma_j(y) = \{z \in X \mid \partial(x,z) = i, \partial(y,z) = j\}. \tag{6.31}$$

If D_i^j is not empty, clearly we have $|j - i| \leq 1$. (There are cases such that D_i^j is empty even if the inequality holds.) There are edges joining vertices in D_i^j and those in D_k^ℓ only if D_i^j and D_k^ℓ are non-empty and $|j - \ell| \leq 1$, $|i - k| \leq 1$ hold. (There are cases such that there exists no edge even if the inequalities hold.) The undirected graph with vertex set $\{D_i^j \mid |j - i| \leq 1\}$ and edge set $\{D_i^j D_k^\ell \mid |j - \ell| \leq 1, |i - k| \leq 1\}$ is called the distance distribution diagram.

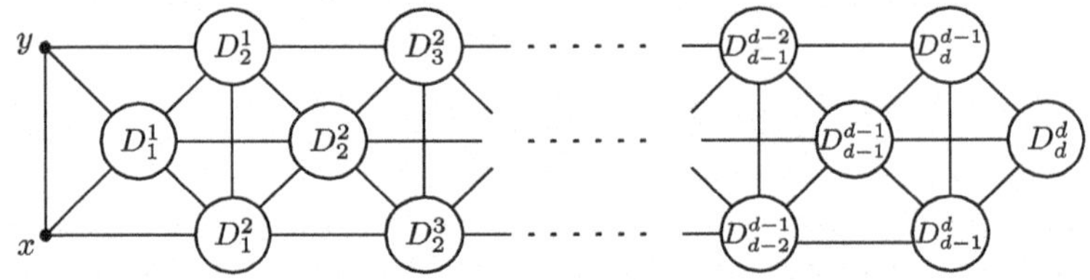

Lemma 6.4. *The sets D_i^{i-1} $(1 \leq i \leq d)$ are non-empty. There are b_i edges joining any vertex $z \in D_i^{i-1}$ and vertices in D_{i+1}^i, and there are c_i edges joining any vertex $z \in D_{i+1}^i$ and vertices in D_i^{i-1}. Similarly, the sets D_{i-1}^i $(1 \leq i \leq d)$ are non-empty. There are b_i edges joining any vertex $z \in D_{i-1}^i$ and vertices in D_i^{i+1}, and there are c_i edges joining any vertex $z \in D_i^{i+1}$ and vertices in D_{i-1}^i.*

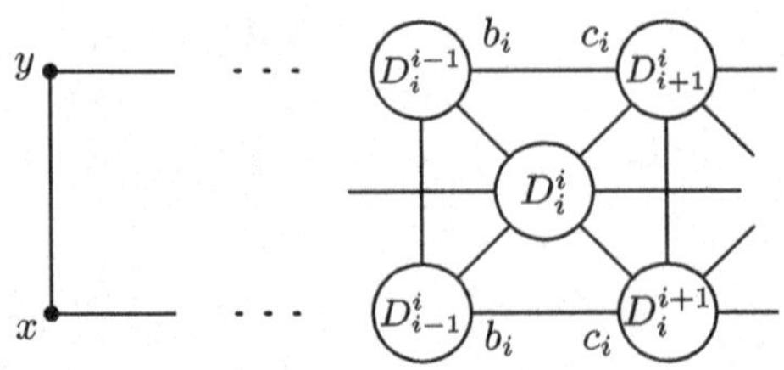

Proof. We prove by induction on i. When $i = 1$, $D_1^0 = \{y\}$ is non-empty. If we let x be a base vertex, we have $D_1^0 \subset \Gamma_1(x)$. There are b_1 edges joining $z = y \in D_1^0$ and vertices in $\Gamma_2(x)$, and each of these edges has another endpoint in D_2^1 since $\Gamma_2(x) = D_2^1 \cup D_2^2 \cup D_2^3$. Therefore there are b_1 edges joining $z \in D_1^0$ and vertices in D_2^1. Next assume that D_i^{i-1} is non-empty and that there are b_i edges joining any vertex in D_i^{i-1} and vertices in D_{i+1}^i. Since $b_i \neq 0$, D_{i+1}^i is non-empty. If we let x be a base vertex and $z \in D_{i+1}^i \subset \Gamma_{i+1}(x)$, then there are b_{i+1} edges joining z and vertices in $\Gamma_{i+2}(x)$, and each of these edges has another endpoint in D_{i+2}^{i+1} since $\Gamma_{i+2}(x) = D_{i+2}^{i+1} \cup D_{i+2}^{i+2} \cup D_{i+2}^{i+3}$. Therefore there are b_{i+1} edges joining $z \in D_{i+1}^i$ and vertices in D_{i+2}^{i+1}. The remaining part of the lemma is similarly proved. □

Proof of Proposition 6.2. (1) Fix $z \in D_i^{i-1}$. By Lemma 6.4, there are b_i edges joining z and vertices in D_{i+1}^i. On the other hand, we have $z \in D_i^{i-1} \subset \Gamma_{i-1}(y)$, and the b_i edges joining z and vertices in D_{i+1}^i are contained in the set of the b_{i-1} edges joining z and vertices in $\Gamma_i(y)$. Therefore, we get $b_{i-1} \geq b_i$. Similarly, we obtain (2), i. e., $c_i \leq c_{i+1}$.
(3) By Lemma 6.4, we can choose a path $y_0 = y, y_1, \dots, y_{d-1}$ in Γ such that $y_{i-1} \in D_i^{i-1}$ $(1 \leq i \leq d)$. Note that $y_{i-1} \in \Gamma_{i-1}(y_0)$. Choose any y_d in $\Gamma_d(y_0) = D_{d-1}^d \cup D_d^d$ which is joined with y_{d-1} by an edge. Then we have $\partial(y_i, y_j) = |i - j|$ $(i, j = 0, 1, \dots, d)$ since $\partial(y_0, y_d) = d$. By Lemma 6.4, there are c_i edges joining $y_i \in D_{i+1}^i$ and vertices in D_i^{i-1}. Since $\partial(y_d, y_i) = d-i$, we have $y_i \in \Gamma_{d-i}(y_d)$. By the distance distribution diagram, we can easily verify that the endpoints of these edges other than y_i are contained in $\Gamma_{d-i+1}(y_d)$. Therefore we obtain $b_{d-i} \geq c_i$. If $i + j \leq d$, by (1), we obtain $b_j \geq b_{d-i} \geq c_i$.
(4) We use the path $y_0 = y, y_1, \dots, y_{d-1}$ in the above proof of (3). Since $a_1 \neq 0$, there exists a triangle which contains the edge $y_{i-1}y_i$. Let $y_{i-1}y_iw$ be one of such triangles. The vertex w is adjacent to y_{i-1}, y_i, and $y_{i-1} \in D_i^{i-1}$, $y_i \in D_{i+1}^i$. By the distance distribution diagram, we have $w \in D_i^{i-1} \cup D_i^i \cup D_{i+1}^i$. If $w \in D_i^{i-1} \cup D_i^i$, then the edge $y_{i-1}w$ is contained in $\Gamma_i(x)$. Therefore $a_i \neq 0$. If $w \in D_{i+1}^i$, then the edge y_iw is contained in $\Gamma_i(y)$. Hence $a_i \neq 0$. □

The condition that the intersection numbers $p_{i,j}^\ell$ are non-negative integers and the condition that the Krein numbers $q_{i,j}^\ell$ are real and non-negative can also be called feasibility conditions. By Theorem 2.23, Theorem 2.22 (3) in Chapter 2, and (6.20), we have

$$p_{i,j}^\ell = \frac{1}{k_\ell |X|} \sum_{\nu=0}^{d} m_\nu v_i(\theta_\nu) v_j(\theta_\nu) v_\ell(\theta_\nu), \tag{6.32}$$

$$q_{i,j}^\ell = \frac{m_i m_j}{|X|} \sum_{\nu=0}^{d} \frac{1}{k_\nu^2} v_\nu(\theta_i) v_\nu(\theta_j) v_\nu(\theta_\ell). \tag{6.33}$$

So the right-hand sides of (6.32) and (6.33) are constants determined only by the intersection matrix of the distance-regular graph. Hence, the condition that the right-hand side of (6.32) is a non-negative integer and the condition that the right-hand side of (6.33) is real and non-negative are necessary conditions for the irreducible tridiagonal matrix B of (6.29) to be the transpose of the intersection matrix of a distance-regular graph. In this way, there are various conditions which can be called feasibility conditions. Usually, the feasibility condition refers to the condition that m_i are integers as in (6.27). The proof of the non-existence of Moore graphs in Chapter 1, Section 1.2 uses the feasibility condition (6.27). The technique of using the integrality of m_i was established by the study of generalized polygons by Feit and Higman [185] and the study of permutation groups by D. G. Higman [215]. It spread widely after the study of Moore graphs by Bannai and Ito [59] and Damerell [156] and the publication of the textbook by Biggs [94]. This technique has now become a standard tool. Beyond these studies, there is a far-reaching goal to classify distance-regular graphs with large diameters. Roughly speaking, the classification problem of distance-regular graphs consists of the following two parts:

(A) To narrow down the possibilities of parameters of distance-regular graphs.
(B) Characterization of distance-regular graphs by parameters.

For the latest information, we refer the reader to the survey paper by Van Dam, Koolen, and Tanaka [155].

Recently, the Bannai–Ito conjecture [60, page 237] was solved by Bang, Dubickas, Koolen, and Moulton [19].

The Bannai–Ito Conjecture. *Let* Γ *be a distance-regular graph of degree* k *and diameter* d $(k \geq 3)$. *Then there exists a function* $f(k)$ *in* k *which is independent of* Γ *such that* $d \leq f(k)$. *In particular, if we fix the degree* $k \geq 3$, *then there are finitely many distance-regular graphs.*

For details, we refer the reader to [61, Section 2] and [19]. The reason why it is called the Bannai–Ito conjecture is that the serious challenge to this conjecture began with a series of papers, "On distance-regular graphs with fixed valency I, II, III, IV" by Bannai and Ito [63, 64, 62, 65]. The conjecture itself existed in Biggs [97] in the 1970s (see also Terwilliger [458]). There is a same conjecture for distance-transitive graphs, which was solved by Cameron [121] and Cameron, Praeger, Saxl, and Seitz [122] by using the classification of finite simple groups in the early 1980s. (After that, the proof without using the classification of finite simple groups was obtained by Weiss [513].) In [61, Section 2], there is a detailed explanation how the study of the conjecture changes by the transition of objects from distance-transitive graphs to distance-regular graphs.

There are two streams of the study of the Bannai–Ito conjecture. One is based on the technique of using multiplicities of eigenvalues, which originates from Ito [250], and through Bannai and Ito [63, 64, 62, 65], ends at Bang, Dubickas, Koolen, and Moulton [19]. The other is based on the technique of using distance distribution diagrams, which originates from Ivanov [262], and through Biggs, Boshier, and Shawe-Taylor [96], arrives at Hiraki [221, 222, 223, 224, 225, 226, 227, 228]. The technique of using multiplicities of eigenvalues culminated in [19]. On the other hand, for the technique of using distance distribution diagrams, we expect various developments beyond the study of Hiraki.

6.1.2 Q-polynomial schemes revisited

As a preparation for Section 6.4, where P- and Q-polynomial schemes will be discussed, we review the definition of Q-polynomial schemes briefly following the manner of the previous subsection, where P-polynomial schemes are discussed. Especially, the definition in terms of Terwilliger algebras will play an important role later. Let $\mathfrak{X} = (X, \{R_i\}_{0\le i\le d})$ be a symmetric association scheme and let $E_0, E_1, \ldots, E_d$ be the primitive idempotents of the Bose–Mesner algebra $\mathfrak{A}$ of $\mathfrak{X}$. By (4″) in Chapter 2, Section 2.3, we have

$$E_i \circ E_j = \frac{1}{|X|} \sum_{k=0}^{d} q_{i,j}^k E_k$$

with respect to the Hadamard product of $\mathfrak{A}$. Note that Krein numbers $q_{i,j}^k$ are nonnegative real (Theorem 2.26, Corollary 2.37). Then the following conditions (i)′, (ii)′, and (iii)′ on $q_{i,j}^k$ are equivalent, and a symmetric association scheme satisfying these conditions is called a Q-polynomial scheme. For the proof of the equivalence, the proof in the previous subsection is valid if we replace the ordinary matrix product by the Hadamard product and the intersection number $p_{i,j}^k$ by the Krein number $q_{i,j}^k$. We have:
(i)′

$$q_{1,j}^k \begin{cases} = 0, & \text{if } 1 < |k-j|, \\ \neq 0, & \text{if } 1 = |k-j|; \end{cases}$$

(ii)′ E_i is a polynomial of degree i in E_1 with respect to the Hadamard product ($0 \le i \le d$);

(iii)′

$$q_{i,j}^k \begin{cases} = 0, & \text{if } i < |k-j| \text{ or } i > k+j, \\ \neq 0, & \text{if } i = |k-j| \text{ or } i = k+j. \end{cases}$$

If we set $b_i^* = q_{1,i+1}^i$, $a_i^* = q_{1,i}^i$, $c_i^* = q_{1,i-1}^i$, then (i)′ is rewritten as
(i)

$$E_1 \circ E_j = \frac{1}{|X|}(b_{j-1}^* E_{j-1} + a_j^* E_j + c_{j+1}^* E_{j+1}) \quad (0 \le j \le d), \tag{6.34}$$

and $b_{i-1}^* c_i^* \neq 0$ ($1 \le i \le d$), where b_{-1}^* is indeterminate and $c_{d+1}^* = 1$, $E_{-1} = 0$, $E_{d+1} = 0$. Moreover, if we set $m = b_0^* = q_{1,1}^0$, we have

$$m = a_0^* + b_0^* = c_i^* + a_i^* + b_i^* = c_d^* + a_d^* \quad (1 \le i \le d-1) \tag{6.35}$$

(Proposition 2.24).

We will discuss the duality of P-polynomial schemes and Q-polynomial schemes again in Section 6.1.4 from the aspect of orthogonal polynomials.

Here we discuss the duality of P-polynomial schemes and Q-polynomial schemes in the framework of Terwilliger algebras. Let $\mathfrak{A}^*$ be the dual Bose–Mesner algebra of the association scheme $\mathfrak{X}$ and let $E_0^*, E_1^*, \ldots, E_d^*$ be the primitive idempotents of $\mathfrak{A}^*$. As in Chapter 2, Section 2.6, we set

$$A_i^* = \sum_{\alpha=0}^{d} Q_i(\alpha)E_\alpha^*,$$

where $Q_i(\alpha)$ is the (α, i)-entry of the second eigenmatrix of $\mathfrak{X}$. Let $V = \bigoplus_{x\in X} \mathbb{C}x$ be the standard module of $\mathfrak{X}$, E_i $(0 \le i \le d)$ the primitive idempotents of $\mathfrak{A}$, and V_i $(0 \le i \le d)$ the image of V by E_i: $V = \bigoplus_{i=0}^{d} V_i$, where $V_i = E_iV$. Let $A^* = A_1^*$. Then the symmetric association scheme $\mathfrak{X}$ is a Q-polynomial scheme if and only if

$$A^*V_i \subset V_{i-1} + V_i + V_{i+1} \quad (0 \le i \le d),$$

where $V_{-1} = V_{d+1} = 0$. To be precise, condition (i)$'$ of Q-polynomial schemes is equivalent to

(iv)

$$E_jA^*E_k \begin{cases} = 0, & \text{if } 1 < |k-j|, \\ \neq 0, & \text{if } 1 < |k-j|. \end{cases}$$

(This follows directly from Corollary 2.37.) In the previous subsection, we adopted the following as the fourth definition of P-polynomial schemes:

$$AV_i^* \subset V_{i-1}^* + V_i^* + V_{i+1}^* \quad (0 \le i \le d),$$

where $V_{-1}^* = V_{d+1}^* = 0$. More precisely,

$$E_j^*AE_k^* \begin{cases} = 0, & \text{if } 1 < |k-j|. \\ \neq 0, & \text{if } 1 = |k-j|. \end{cases}$$

The dual of this definition is the above condition (iv) of a Q-polynomial scheme.

For P-polynomial schemes, as was seen in Section 6.1.1, there is a graph theoretic interpretation (distance-regularity). There is no known graph theoretic interpretation for Q-polynomial schemes. Since the origin of Q-polynomial schemes is design theory (Delsarte's theory) which is the dual of coding theory, it seems to be hard to interpret Q-polynomial schemes in the framework of graph theory. Q-polynomial schemes have a connection with spherical geometry through the spherical representation. It is known that Q-polynomial schemes arise from various combinatorial objects through the spherical representation. For instance, Q-polynomial schemes arise from spherical tight designs or similar objects (e. g., Theorem 5.19). By the discussion in Chapter 2,

Section 2.11, it is important that the balanced condition of the spherical representation gives a condition for a P-polynomial scheme to be a Q-polynomial scheme. This discussion is a prototype of the theory to derive the TD relation in the classification of L-pairs in the next section.

As was seen in Chapter 2, Section 2.5, let

$$\rho^* : \mathfrak{A}^\circ \longrightarrow M_{d+1}(\mathbb{C}) \cong \mathrm{End}(\mathfrak{A}^\circ)$$

be the matrix expression of the regular representation $\mathfrak{A}^\circ \longrightarrow \mathrm{End}(\mathfrak{A}^\circ)$ of the algebra $\mathfrak{A}^\circ$ with the Hadamard product with respect to the basis $|X|E_0, |X|E_1, \ldots, |X|E_d$, and let $\rho^*(|X|E_i) = {}^tB_i^*$. The (k, j)-entry of ${}^tB_i^*$ is $q_{i,j}^k$. We call B_i^* the dual intersection matrix. Let $\mathfrak{B}^* = \langle B_0^*, B_1^*, \ldots, B_d^* \rangle$. Since $\mathfrak{A}^\circ$ is commutative, by the correspondence $|X|E_i \longrightarrow B_i^*$, we have

$$\mathfrak{A}^\circ \cong \mathfrak{B}^* \quad \text{(as algebras)}.$$

Let $B^* = {}^tB_1^*$. By (6.34), B^* is the irreducible tridiagonal matrix

$$B^* = \begin{bmatrix} a_0^* & b_0^* & & & & \\ c_1^* & a_1^* & b_1^* & & 0 & \\ & \ddots & \ddots & \ddots & & \\ & 0 & & c_{d-1}^* & a_{d-1}^* & b_{d-1}^* \\ & & & & c_d^* & a_d^* \end{bmatrix}, \tag{6.36}$$

and $b_{i-1}^* c_i^* \neq 0$ $(1 \le i \le d)$. By (6.35), every row sum of B^* is $m = b_0^*$. Since the Krein numbers $q_{i,j}^k$ are real and non-negative, the matrix B^* is a real matrix and $b_{i-1}^* c_i^* > 0$ $(1 \le i \le d)$ holds. In particular, B^* is diagonalizable and has $d + 1$ distinct real eigenvalues (Corollary 6.11 and Proposition 6.12).

What is the necessary condition for the irreducible tridiagonal matrix B^* in (6.36) to coincide with the transpose ${}^tB_1^*$ of the dual intersection matrix of a Q-polynomial scheme? The problem of the so-called *feasibility* of B^* is not deeply studied compared to the case of P-polynomial schemes, for the study of Q-polynomial schemes itself just started recently.

In what follows, assume that the irreducible tridiagonal matrix B^* in (6.36) coincides with the transpose ${}^tB_1^*$ of the dual intersection matrix. Since $a_0^* = q_{1,0}^0$, $c_1^* = q_{1,0}^1$, we have $a_0^* = 0$, $c_1^* = 1$. Moreover, as was seen before, the row sums of B^* are equal to $m = b_0^*$. Furthermore, if we set

$$m_i = \frac{b_0^* b_1^* \cdots b_{i-1}^*}{c_1^* c_2^* \cdots c_i^*} \quad (1 \le i \le d), \tag{6.37}$$

we can verify m_i $(1 \le i \le d)$ are positive integers as follows.

Let $v_i(x)$ $(0 \le i \le d+1)$ be the polynomial of degree i defined by the following three-term recurrence:

$$xv_j^*(x) = b_{j-1}^* v_{j-1}^*(x) + a_j^* v_j^*(x) + c_{j+1}^* v_{j+1}^*(x), \tag{6.38}$$

where $v_{-1}^*(x) = 0$, $v_0^*(x) = 1$, b_{-1}^* is indeterminate, and $c_{d+1}^* = 1$. Since we assume B^* coincides with the transpose ${}^tB_1^*$ of the dual intersection matrix of a Q-polynomial scheme, by (6.34), we have

$$|X|E_j = v_j^*(|X|E_1) \quad (0 \le j \le d)$$

with respect to the Hadamard product. Taking the Hadamard product of the above equation and A_i, for the (i,j)-entry $Q_j(i)$ of the second eigenmatrix, we obtain

$$Q_j(i) = v_j^*(\theta_i^*), \quad \theta_i^* = Q_1(i) \tag{6.39}$$

(Proposition 2.31). Here $\theta_i^* = Q_1(i)$ $(0 \le i \le d)$ are the eigenvalues of $B^* = {}^tB_1^*$ (Proposition 2.31), and θ_0^* is the Perron–Frobenius eigenvalue of B^* (since every row sum of B^* is b_0^* and $b_0^* = q_{1,1}^0 = Q_1(0) = \theta_0^*$). Note that the set $\{\theta_i^*\}_{i=0}^d$ and θ_0^* are determined by B^* only. If we set $i = 0$ in (6.38),

$$m_j = Q_j(0) = \operatorname{tr} E_j \tag{6.40}$$

is a positive integer (Proposition 2.20, Proposition 2.21). If we set $x = \theta_0^*$ in (6.39), by (6.39), (6.40), we obtain

$$\theta_0^* m_j = b_{j-1}^* m_{j-1} + a_j^* m_j + c_{j+1}^* m_{j+1} \tag{6.41}$$

$(0 \le j \le d-1)$, where $m_{-1} = 0$, $m_0 = 1$. As a solution of the recurrence (6.41), we obtain (6.37). (We use the fact that every row sum of B^* is $b_0^* = \theta_0^*$.) The right-hand side of (6.37) is a positive integer.

Next, we set $n = \sum_{i=0}^d m_i$, where $m_0 = 1$ and m_i is determined by (6.37). We also set

$$k_i = \frac{n}{\sum_{v=0}^d v_v^*(\theta_i^*)^2/m_v} \tag{6.42}$$

for the eigenvalues θ_i^* $(0 \le i \le d)$ of B^* and the polynomials $v_v^*(x)$ $(0 \le v \le d)$ determined by (6.38). Then we can verify that k_i $(0 \le i \le d)$ are positive integers as follows.

Since we assume $B^* = {}^tB_1^*$, (6.39) holds. The second orthogonality relation (Theorem 2.22 (5))

$$\sum_{v=0}^d P_i(v)P_j(v)m_v = \delta_{i,j}|X|k_i$$

is rewritten as follows by using $Q_j(i)/m_j = P_i(j)/k_i$:

$$\sum_{\nu=0}^{d} Q_\nu(i)Q_\nu(j)\frac{1}{m_\nu} = \delta_{i,j}|X|\frac{1}{k_j}.$$

Here m_i is given by (6.40) (hence by (6.37)), and k_i is given by

$$k_i = P_i(0) = (\text{the row sum of } A_i) \tag{6.43}$$

(Propositions 2.20 and 2.21). Moreover, by (6.39) and $|X| = \sum_{i=0}^{d} m_i = n$, we obtain

$$\sum_{\nu=0}^{d} v_\nu^*(\theta_i^*)v_\nu^*(\theta_j^*)\frac{1}{m_\nu} = \delta_{i,j}\frac{n}{k_j}. \tag{6.44}$$

If we set $i = j$, we obtain (6.42), and by (6.43), k_i are positive integers.

By Theorem 2.23, we have

$$p_{i,j}^{\ell} = \frac{k_i k_j}{|X|}\sum_{\nu=0}^{d}\frac{1}{m_\nu{}^2}v_\nu^*(\theta_i^*)v_\nu^*(\theta_j^*)v_\nu^*(\theta_\ell^*), \tag{6.45}$$

$$q_{i,j}^{\ell} = \frac{1}{m_\ell|X|}\sum_{\nu=0}^{d} k_\nu v_i^*(\theta_\nu^*)v_j^*(\theta_\nu^*)v_\ell^*(\theta_\nu^*). \tag{6.46}$$

So the right-hand sides of (6.45) and (6.46) depend only on the dual intersection matrix of the Q-polynomial scheme. Hence the conditions that the right-hand side of (6.45) is a non-negative integer and that the right-hand side of (6.46) is real and non-negative are necessary conditions for the irreducible tridiagonal matrix in (6.36) to coincide with the transpose ${}^tB_1^*$ of the intersection matrix of a Q-polynomial scheme.

However, it is not known well how strong they are as necessary conditions.

For the intersection matrices of P-polynomial schemes, Proposition 6.2 (1)–(4) hold, but for Q-polynomial schemes, we cannot expect the same properties except for Proposition 6.2 (4). As a Q-polynomial version of Proposition 6.2 (4), we have the following.

Proposition 6.5 ([170]). *Assume the irreducible tridiagonal matrix B^* in (6.36) coincides with the transpose ${}^tB_1^*$ of the dual intersection matrix for a Q-polynomial scheme. Then $a_1^* \neq 0$ implies $a_i^* \neq 0$ $(2 \le i \le d-1)$.*

Remark 6.6. Assume that the Terwilliger algebra T is dual 1-thin. Namely, we assume every irreducible T-module of dual endpoint 1 is thin (Remark 6.30 in Section 6.2). Then $a_1^* = 0$ implies $a_i^* = 0$ $(2 \le i \le d-1)$ [170, 172].

By weakening Proposition 6.2 (1)–(3) and considering the Q-polynomial version of them, we obtain the following conjectures.

(1) The conjecture by Bannai and Ito [60, III-1]: There exists i ($1 \leq i \leq d$) such that

$$m_0 = 1 < m_1 \leq m_2 \leq \cdots \leq m_i \geq m_{i+1} \geq \cdots \geq m_d.$$

(2) The conjecture by D. Stanton: For $i < \frac{d}{2}$, the following hold:

$$m_i \leq m_{i+1} \quad \text{and} \quad m_i \leq m_{d-i}.$$

Obviously, conjecture (2) is stronger than (1). For a conditional approach to conjecture (2), we refer the reader to [411, 393].

As a dual version of the Bannai–Ito conjecture for distance-regular graphs (Section 6.1.1), the following theorem holds for Q-polynomial schemes.

Theorem 6.7 (Martin–Williford [337]). *Let $\mathfrak{X} = (X, \{R_i\}_{0\leq i\leq d})$ be a Q-polynomial scheme with respect to the ordering $E_0, E_1, \ldots, E_d$ of the primitive idempotents. Let $m = \operatorname{rank}(E_1)$ and assume $m \geq 3$. Then there exists a function $f(m)$ in m which is independent of $\mathfrak{X}$ such that $d \leq f(m)$. In particular, if we fix $m \geq 3$, there are finitely many Q-polynomial schemes with $\operatorname{rank}(E_1) = m$.*

In [337], they first show that the splitting field of a Q-polynomial scheme $\mathfrak{X}$ is an at most degree 2 extension of the rationals, and this is the key to the proof. Here the splitting field of $\mathfrak{X}$ means the smallest extension of the rationals containing all the entries $P_j(i)$ of the first eigenmatrix P.

6.1.3 P-polynomial schemes and Q-polynomial schemes

In this subsection, we present several additional basic facts on P-polynomial schemes and Q-polynomial schemes without proofs. For details, see, for example, [60].

Imprimitive P-polynomial schemes

Let $\mathfrak{X} = (X, \{R_i\}_{0\leq i\leq d})$ be a P-polynomial scheme. Let A_i be the adjacency matrix of R_i. Then we have

$$A_1 A_i = b_{i-1}A_{i-1} + a_i A_i + c_{i+1}A_{i+1} \quad (0 \leq i \leq d),$$

where $A_{-1} = 0$, $A_{d+1} = 0$. The first intersection matrix $B_1 = (p^k_{1,j})$ of $\mathfrak{X}$ has the following form:

$$
{}^tB_1 = \begin{bmatrix} a_0 & b_0 & & & \\ c_1 & a_1 & b_1 & & 0 \\ & \ddots & \ddots & \ddots & \\ 0 & & c_{d-1} & a_{d-1} & b_{d-1} \\ & & & c_d & a_d \end{bmatrix}. \tag{6.47}
$$

Assume $b_0 > 2$. (If $b_0 = 2$, the graph (X, R_1) is an n-gon.) In what follows, assume $\mathfrak{X}$ is imprimitive. Namely, there exists Ω such that $\{0\} \subsetneq \Omega \subsetneq \{0,1,\dots,d\}$ and

$$A_iA_j = \sum_{k\in\Omega} p_{i,j}^k A_k \quad (\forall i,j \in \Omega)$$

(Chapter 2, Section 2.7.3). Then one of the following holds:

$$\textbf{Case 1:} \quad \Omega = \{0,2,4,\dots,[d/2]\}, \tag{6.48}$$

$$\textbf{Case 2:} \quad \Omega = \{0,d\}. \tag{6.49}$$

Moreover, a P-polynomial scheme $\mathfrak{X}$ is imprimitive with Case 1 if and only if the following holds in (6.47):

$$a_i = 0, \quad 0 \le i \le d. \tag{6.50}$$

Similarly, a P-polynomial scheme $\mathfrak{X}$ is imprimitive with Case 2 if and only if the following holds in (6.47):

$$b_i = c_{d-i}, \quad 0 \le i \le d-1 \quad (i \ne [d/2]). \tag{6.51}$$

For an imprimitive P-polynomial scheme $\mathfrak{X}$, if Case 1 holds, $\mathfrak{X}$ is called *bipartite*, and if Case 2 holds, $\mathfrak{X}$ is called *antipodal*.

Let $\mathfrak{X} = (X, \{R_i\}_{0\le i\le d})$ be a bipartite P-polynomial scheme. Define a relation $\sim$ on X as follows by using Ω in (6.48):

$$x \sim y \Longleftrightarrow (x,y) \in \bigcup_{i\in\Omega} R_i, \quad \Omega = \{0,2,4,\dots,[d/2]\}.$$

Then the relation $\sim$ becomes an equivalence relation and X is partitioned into exactly two equivalence classes X_1, X_2:

$$X = X_1 \cup X_2, \quad X_1 \cap X_2 = \emptyset.$$

We call X_i $(i = 1,2)$ a *bipartite half*. Let $Y = X_1$ or $Y = X_2$. Then the subscheme $\mathfrak{Y} = (Y, \{R_{2i}\}_{0\le i\le [\frac{d}{2}]})$ on the bipartite half Y arises (Chapter 2, Section 2.7.4), and $\mathfrak{Y}$ is a P-polynomial scheme (the graph (Y, R_2) is a distance-regular graph). Let B_1' be the intersection matrix $\mathfrak{Y}$ with respect to R_2. Then by using (6.47), we have

$${}^tB_1' = \begin{bmatrix} a_0' & b_0' & & & \\ c_1' & a_1' & b_1' & & 0 \\ & \ddots & \ddots & \ddots & \\ 0 & & c_{d'-1}' & a_{d'-1}' & b_{d'-1}' \\ & & & c_{d'}' & a_{d'}' \end{bmatrix}, \quad d' = \left[\frac{d}{2}\right],$$

where

$$b'_i = \frac{b_{2i}b_{2i+1}}{c_2}, \quad c'_{i+1} = \frac{c_{2i+1}c_{2i+2}}{c_2}, \quad 0 \le i \le d' - 1. \tag{6.52}$$

We denote the subscheme $\mathfrak{Y}$ on the bipartite half Y by $\frac{1}{2}\mathfrak{X}$.

Remark 6.8. If we set $Y = X_2$ instead of $Y = X_1$, we obtain the same intersection matrix B'_1 of the subscheme $\mathfrak{Y}$ on Y, but the corresponding distance-regular graphs are not necessarily isomorphic.

Next, let $\mathfrak{X} = (X, \{R_i\}_{0\le i\le d})$ be an antipodal P-polynomial scheme. Define a relation $\sim$ on X as follows by using $\Omega = \{0, d\}$ in (6.49):

$$x \sim y \iff (x, y) \in R_0 \cup R_d.$$

Then the relation $\sim$ becomes an equivalence relation, and X is partitioned into r equivalence classes, where $r = |X|/(1 + k_d)$:

$$X = X_1 \cup X_2 \cup \cdots \cup X_r, \quad X_i \cap X_j = \emptyset \ (i \ne j).$$

Here, $|X_i| = 1 + k_d$ $(1 \le i \le r)$, $k_d = b_0 b_1 \cdots b_{d-1}/c_1 c_2 \cdots c_d$. By (6.51), we have

$$k_d = \frac{b_{d'}}{c_{d-d'}} \quad \left(d' = \left[\frac{d}{2}\right]\right). \tag{6.53}$$

The set $\Sigma = \{X_1, X_2, \ldots, X_r\}$ of equivalence classes is called a system of imprimitivity. By Proposition 2.66, there exists a partition

$$\{0, 1, \ldots, d\} = \Omega_0 \cup \Omega_1 \cup \cdots \cup \Omega_t, \quad \Omega_i \cap \Omega_j = \emptyset \ (i \ne j)$$

of the index set $\{0, 1, \ldots, d\}$, where $0 \in \Omega_0$, and if we let

$$R_{\Omega_i} = \bigcup_{j\in\Omega_i} R_j,$$

then R_{Ω_i} $(0 \le i \le t)$ becomes a union of some of $X_\alpha \times X_\beta = \{(x, y) \mid x \in X_\alpha, y \in X_\beta\}$. Therefore, we can regard R_{Ω_i} as a relation on the system Σ of imprimitivity. By this identification, $\overline{\mathfrak{X}} = (\Sigma, \{R_{\Omega_i}\}_{0\le i\le t})$ becomes an association scheme. Namely, $\overline{\mathfrak{X}}$ is a quotient scheme of $\mathfrak{X}$ (Definition 2.68 in Chapter 2, Section 2.7.4). If $\mathfrak{X}$ is an antipodal P-polynomial scheme, then the quotient scheme $\overline{\mathfrak{X}}$ also becomes a P-polynomial scheme. (In this case, the class t of $\overline{\mathfrak{X}}$ satisfies $t = [\frac{d}{2}]$.) The quotient scheme $\overline{\mathfrak{X}}$ is called the *antipodal quotient scheme* of $\mathfrak{X}$. Let $\tilde{B}_1$ be the intersection matrix of the antipodal quotient scheme $\overline{\mathfrak{X}} = (\Sigma, \{R_{\Omega_i}\}_{0\le i\le [\frac{d}{2}]})$ with respect to R_{Ω_1}, where $1 \in \Omega_1$. Then by using (6.47), we have

$${}^t\tilde{B}_1 = \begin{bmatrix} \tilde{a}_0 & \tilde{b}_0 & & & \\ \tilde{c}_1 & \tilde{a}_1 & \tilde{b}_1 & & 0 \\ & \ddots & \ddots & \ddots & \\ 0 & & \tilde{c}_{d'-1} & \tilde{a}_{d'-1} & \tilde{b}_{d'-1} \\ & & & \tilde{c}_{d'} & \tilde{a}_{d'} \end{bmatrix}, \quad d' = \left[\frac{d}{2}\right],$$

where

$$\tilde{b}_i = b_i \ (0 \le i \le d' - 1), \quad \tilde{c}_i = c_i \ (1 \le i \le d' - 1),$$
$$\tilde{c}_{d'} = \begin{cases} (1 + k_d)c_{d'}, & \text{if } d \text{ is even}, \\ c_{d'}, & \text{if } d \text{ is odd}. \end{cases} \tag{6.54}$$

The above results are due to Biggs, Gardiner, and Smith ([431]). (The proofs are relatively easy if we consider the definition of distance-regular graphs. We recommend the reader to prove them as exercises.) It seems that Biggs regarded the primitivity and imprimitivity as important problems as soon as he obtained the idea of distance-regular graphs, for permutation groups acting on distance-transitive graphs are prototypes of distance-regular graphs, and the primitivity and imprimitivity of permutation groups naturally arise as important problems.

Imprimitive Q-polynomial schemes

Let $\mathfrak{X} = (X, \{R_i\}_{0 \le i \le d})$ be a Q-polynomial scheme. Let $\mathfrak{A}$ be the Bose–Mesner algebra of $\mathfrak{X}$, and let $E_0, E_1, \dots, E_d$ be the primitive idempotents of $\mathfrak{A}$. Then the following holds with respect to the Hadamard product:

$$E_1 \circ E_i = \frac{1}{|X|}(b^*_{i-1}E_{i-1} + a^*_i E_i + c^*_{i+1}E_{i+1}), \quad 0 \le i \le d,$$

where $E_{-1} = 0, E_{d+1} = 0$, and the first dual intersection matrix $B^*_1 = (q^k_{1,j})$ of $\mathfrak{X}$ is given by

$${}^tB^*_1 = \begin{bmatrix} a^*_0 & b^*_0 & & & & \\ c^*_1 & a^*_1 & b^*_1 & & 0 & \\ & \ddots & \ddots & \ddots & & \\ 0 & & c^*_{d-1} & a^*_{d-1} & b^*_{d-1} \\ & & & c^*_d & a^*_d \end{bmatrix}. \tag{6.55}$$

Remark 6.9. In what follows, we assume $b^*_0 > 2$. Since $b^*_0 = m_1 = \operatorname{rank}(E_1)$, by considering the spherical representation ρ_{E_1} in Chapter 2, Section 2.11 (Definition 2.96), the case $b^*_0 = 1$ does not occur, and if $b^*_0 = 2$, the set $\{\rho_{E_1}(x) \mid x \in X\}$ of points is located on the circle as the regular $(d+1)$-gon (Lemma 2.97).

Assume $\mathfrak{X}$ is imprimitive. Namely, there exists Λ such that $\{0\} \subsetneq \Lambda \subsetneq \{0, 1, \dots, d\}$ and

$$E_i \circ E_j = \frac{1}{|X|} \sum_{k \in \Lambda} q^k_{i,j} E_k \quad (\forall i, j \in \Lambda)$$

(Chapter 2, Section 2.7.3). Then one of the following holds:

$$\textbf{Case 1:} \quad \Lambda = \{0, 2, 4, \dots, [d/2]\}, \tag{6.56}$$

$$\textbf{Case 2:} \quad \Lambda = \{0, d\}. \tag{6.57}$$

Moreover, a Q-polynomial scheme $\mathfrak{X}$ is imprimitive with Case 1 if and only if the following holds in (6.55):

$$a_i^* = 0, \quad 0 \le i \le d. \tag{6.58}$$

Also, a Q-polynomial scheme $\mathfrak{X}$ is imprimitive with Case 2 if and only if the following holds in (6.55):

$$b_i^* = c_{d-i}^*, \quad 0 \le i \le d-1 \quad (i \ne [d/2]). \tag{6.59}$$

For an imprimitive Q-polynomial scheme $\mathfrak{X}$, if Case 1 holds, $\mathfrak{X}$ is called *dual bipartite*, and if Case 2 holds, $\mathfrak{X}$ is called *dual antipodal*.

Let $\mathfrak{X} = (X, \{R_i\}_{0 \le i \le d})$ be a dual bipartite Q-polynomial scheme. By using Λ in (6.56), define the subalgebra $\mathfrak{A}_\Lambda$ of the Bose–Mesner algebra $\mathfrak{A}$ of $\mathfrak{X}$ as follows:

$$\mathfrak{A}_\Lambda = \langle E_{2i} \mid 0 \le i \le [d/2] \rangle.$$

Since $\mathfrak{A}_\Lambda$ is closed under the Hadamard product, a partition of the index set, which corresponds to the primitive idempotents $A_{\Omega_i} = \sum_{j \in \Omega_i} A_j$ $(0 \le i \le [d/2])$ of $\mathfrak{A}_\Lambda$ with respect to the Hadamard product, arises as follows:

$$\{0, 1, \ldots, d\} = \Omega_0 \cup \Omega_1 \cup \cdots \cup \Omega_{[d/2]}, \quad \Omega_i \cap \Omega_j = \emptyset \ (i \ne j).$$

Let $0 \in \Omega_0$. Then $R_{\Omega_0} = \bigcup_{j \in \Omega_0} R_j$ is an equivalence relation on X. (This follows from the fact that the identity element of $\mathfrak{A}_\Lambda$ is a scalar multiple of A_{Ω_0}.) Let $\Sigma = \{X_1, X_2, \ldots, X_r\}$ be the set of equivalence classes. Then Σ is a system of imprimitivity of $\mathfrak{X}$. If we set $R_{\Omega_i} = \bigcup_{j \in \Omega_i} R_j$, then $\mathfrak{A}_\Lambda$ is isomorphic to the Bose–Mesner algebra of the association scheme $\overline{\mathfrak{X}} = (\Sigma, \{R_{\Omega_i}\}_{0 \le i \le [d/2]})$ on Σ (Proposition 2.66 in Chapter 2). The association scheme $\overline{\mathfrak{X}}$ is a quotient scheme of $\mathfrak{X}$ (Definition 2.68 in Chapter 2). Then $\overline{\mathfrak{X}}$ is a Q-polynomial scheme. Let $\hat{B}_1^*$ be the dual intersection matrix of $\overline{\mathfrak{X}}$ for E_2. Then by using (6.55), we have

$$^t\hat{B}_1^* = \begin{bmatrix} \hat{a}_0^* & \hat{b}_0^* & & & & \\ \hat{c}_1^* & \hat{a}_1^* & \hat{b}_1^* & & 0 & \\ & \ddots & \ddots & \ddots & & \\ & & \hat{c}_{d'-1}^* & \hat{a}_{d'-1}^* & \hat{b}_{d'-1}^* & \\ & 0 & & \hat{c}_{d'}^* & \hat{a}_{d'}^* & \end{bmatrix}, \quad d' = [d/2], \tag{6.60}$$

where

$$\hat{b}_i^* = \frac{b_{2i}^* b_{2i+1}^*}{c_2^*}, \quad \hat{c}_{i+1}^* = \frac{c_{2i+1}^* c_{2i+2}^*}{c_2^*}, \quad 0 \le i \le d'-1. \tag{6.61}$$

Next, let $\mathfrak{X} = (X, \{R_i\}_{0\le i\le d})$ be a dual antipodal Q-polynomial scheme. By using Λ in (6.57), define the subalgebra $\mathfrak{A}_\Lambda$ of the Bose–Mesner algebra $\mathfrak{A}$ of $\mathfrak{X}$ as follows:

$$\mathfrak{A}_\Lambda = \langle E_0, E_d\rangle.$$

Since $\mathfrak{A}_\Lambda$ is closed under the Hadamard product, by the similar discussion above, a partition of the index set arises as follows:

$$\{0,1,\dots,d\} = \Omega_0 \cup \Omega_1, \quad \Omega_0 \cap \Omega_1 = \emptyset.$$

Let $0 \in \Omega_0$. Then $R_{\Omega_0} = \bigcup_{j\in\Omega_0} R_j$ is an equivalence relation on X. (In particular, $A_{\Omega_0} = \sum_{j\in\Omega_0} A_j$ is a scalar multiple of the identity element $E_\Lambda = E_0 + E_d$ of $\mathfrak{A}_\Lambda$.) If we let $Y \subsetneq X$ be an equivalence class, then a subscheme $\mathfrak{Y} = (Y, \{R_i\}_{i\in\Omega_0})$ on Y arises. The Bose–Mesner algebra of $\mathfrak{Y}$ is $\mathfrak{A}_{\Omega_0} = \langle A_i \mid i \in \Omega_0\rangle$, which is a subalgebra of the Bose–Mesner algebra $\mathfrak{A}$ of $\mathfrak{X}$. Corresponding to the primitive idempotent $E_{\Lambda_i} = \sum_{j\in\Lambda_i} E_j$ of $\mathfrak{A}_{\Omega_0}$, a partition of the index set arises as follows:

$$\{0,1,\dots,d\} = \Lambda_0 \cup \Lambda_1 \cup \cdots \cup \Lambda_s, \quad \Lambda_i \cap \Lambda_j = \emptyset \ (i \ne j)$$

$(1 + s = |\Omega_0|)$. Let $0 \in \Lambda_0$. Then we have $\Lambda_0 = \Lambda = \{0, d\}$ (since the identity element $A_{\Omega_0} = \sum_{j\in\Omega_0} A_j$ of $\mathfrak{A}_{\Omega_0}$ with respect to the Hadamard product is a scalar multiple of $E_{\Lambda_0} = \sum_{j\in\Lambda_0} E_j$ ($E_0 = \frac{1}{|X|}J$)). Since $\mathfrak{X}$ is a Q-polynomial scheme, we may set

$$\Lambda_i = \{i, d-i\}, \quad i = 0,1,2,\dots,$$

and we obtain $s = [\frac{d}{2}]$. The subscheme $\mathfrak{Y} = (Y, \{R_i\}_{i\in\Omega_0})$ on Y becomes a Q-polynomial scheme. Let $\tilde{B}_1^*$ be the dual intersection matrix of $\mathfrak{Y}$ for E_{Λ_1}. Then by using (6.55), we have

$${}^t\tilde{B}_1^* = \begin{bmatrix} \tilde{a}_0^* & \tilde{b}_0^* & & & \\ \tilde{c}_1^* & \tilde{a}_1^* & \tilde{b}_1^* & & 0 \\ & \ddots & \ddots & \ddots & \\ & & \tilde{c}_{d'-1}^* & \tilde{a}_{d'-1}^* & \tilde{b}_{d'-1}^* \\ 0 & & & \tilde{c}_{d'}^* & \tilde{a}_{d'}^* \end{bmatrix}, \quad d' = [d/2],$$

where

$$\tilde{b}_i^* = b_i^* \ (0 \le i \le d'-1), \quad \tilde{c}_i^* = c_i^* \ (1 \le i \le d'-1),$$
$$\tilde{c}_{d'}^* = \begin{cases} (1+m_d)c_{d'}^*, & \text{if } d \text{ is even}, \\ c_{d'}^*, & \text{if } d \text{ is odd}. \end{cases} \tag{6.62}$$

Here, $m_d = \operatorname{tr} E_d$.

The above results are due to Suzuki [444]. The proofs are not easy unlike the case of P-polynomial schemes. It is necessary to refine the technique on the Krein numbers $q_{i,j}^k$ used in the Ph. D. thesis of Garth Dickie. Moreover, in [444], there were exceptional cases of $d = 4, 6$, which were eliminated later by Cerzo and Suzuki [132] and Tanaka and Tanaka [455], respectively. Readers are also referred to [335].

Double P-polynomial structures

Let $\mathfrak{X} = (X, \{R_i\}_{0\le i\le d})$ be a symmetric association scheme. When we define a P-polynomial scheme, the ordering of the relations $\{R_i\}_{0\le i\le d}$ matters. Precisely, if

$$A_1A_i = b_{i-1}A_{i-1} + a_iA_i + c_{i+1}A_{i+1}, \quad 0 \le i \le d,$$

holds with respect to the ordering $R_0, R_1, \ldots, R_d$, $\mathfrak{X}$ is defined to be a P-polynomial scheme, where A_i is the adjacency matrix of R_i and $A_{-1} = 0$, $A_{d+1} = 0$. Usually, unless otherwise stated, we understand that the ordering of $\{R_i\}_{0\le i\le d}$ is fixed for a P-polynomial scheme $\mathfrak{X}$, and it is P-polynomial with respect to this ordering. However, there is a case that a symmetric association scheme $\mathfrak{X} = (X, \{R_i\}_{0\le i\le d})$ is P-polynomial with respect to the ordering $R_0, R_1, \ldots, R_d$ and is also P-polynomial with respect to another ordering $R_0, R_{i_1}, \ldots, R_{i_d}$. In this case, $\mathfrak{X}$ is said to have a *double P-polynomial structure*. Since the ordering $R_0, R_1, \ldots, R_d$ is fixed, we call another ordering $R_0, R_{i_1}, \ldots, R_{i_d}$ the second P-polynomial ordering. There are only four patterns of second P-polynomial orderings (note that $\mathfrak{X}$ cannot have a triple P-polynomial structure, namely, $\mathfrak{X}$ cannot be P-polynomial with respect to three orderings of $\{R_i\}_{0\le i\le d}$):

(I) $R_0, R_2, R_4, R_6, \ldots, R_5, R_3, R_1$;

(II) $R_0, R_d, R_1, R_{d-1}, R_2, R_{d-2}, R_3, R_{d-3} \cdots$;

(III) $R_0, R_d, R_2, R_{d-2}, R_4, R_{d-4}, \ldots, R_{d-5}, R_5, R_{d-3}, R_3, R_{d-1}, R_1$;

(IV) $R_0, R_{d-1}, R_2, R_{d-3}, R_4, R_{d-5}, \ldots, R_5, R_{d-4}, R_3, R_{d-2}, R_1, R_d$.

If we regard ordering (I) as a standard ordering, then the ordering $R_0, R_1, \ldots, R_d$ coincides with pattern (II), and if we regard ordering (II) as a standard ordering, then the ordering $R_0, R_1, \ldots, R_d$ coincides with pattern (I). In this sense, pattern (I) and pattern (II) are dual to each other. Similarly, (III) and (IV) are self-dual. Namely, the dual of (III) is (III) itself, and the dual of (IV) is (IV) itself.

We can determine whether a P-polynomial scheme has a double P-polynomial structure or not by the intersection numbers. Let $\mathfrak{X} = (X, \{R_i\}_{0\le i\le d})$ be a symmetric association scheme which is P-polynomial with respect to the ordering $R_0, R_1, \ldots, R_d$, and also let A_i be the adjacency matrix of R_i and $p_{i,j}^k$ the intersection numbers:

$$A_iA_j = \sum_{k=0}^{d} p_{i,j}^k A_k.$$

The first intersection matrix $B_1 = (p_{1,j}^k)$ is given by (6.47). Then $\mathfrak{X}$ has a double P-polynomial structure of pattern (I) if and only if

$$a_i = 0 \quad (0 \le i \le d-1) \quad \text{and} \quad a_d \ne 0, \tag{6.63}$$

where $a_i = p_{1,i}^i$. A P-polynomial scheme satisfying condition (6.63) is called *almost bipartite*.

The P-polynomial scheme $\mathfrak{X}$ has a double P-polynomial structure of pattern (II) if and only if

$$p_{j,d}^{d} = 0 \quad (2 \leq j \leq d). \tag{6.64}$$

The necessary and sufficient conditions for $\mathfrak{X}$ to have a double P-polynomial structure of pattern (III) are as follows. In the case $d = 2\ell + 1 \geq 5$,

$$a_i = 0 \quad (i \neq \ell + 1), \quad p_{j,d}^{d} = 0\ (j \neq 0, 2), \quad \text{and} \quad p_{2,d}^{d} \neq 0, \tag{6.65}$$

where $a_{\ell+1}$ is arbitrary. In the case $d = 2\ell \geq 6$,

$$a_i = 0\ (i \neq \ell, \ell + 1), \quad a_\ell \neq 0, \quad a_{\ell+1} \neq 0, \quad p_{j,d}^{d} = 0\ (j \neq 0, 2), \quad \text{and} \quad p_{2,d}^{d} \neq 0. \tag{6.66}$$

In the case $d = 3$, $p_{1,3}^{3} = 0$ and the graph (X, R_3) is connected. In the case $d = 4$, $p_{1,4}^{4} = 0$, $p_{1,4}^{3}(p_{1,1}^{1} + p_{1,2}^{2} - p_{1,3}^{3}) = p_{1,2}^{2}$, and the graph (X, R_4) is connected.

The necessary and sufficient conditions for $\mathfrak{X}$ to have a double P-polynomial structure of pattern (IV) are as follows. In the case $d = 2\ell$,

$$b_i = c_{d-i}\ (i \neq \ell) \quad \text{and} \quad a_i = 0\ (i \neq \ell), \tag{6.67}$$

where a_ℓ is arbitrary. In the case $d = 2\ell + 1$,

$$b_i = c_{d-i}\ (i \neq \ell), \quad a_i = 0\ (i \neq \ell,\ \ell + 1), \quad \text{and} \quad a_\ell = a_{\ell+1} \neq 0. \tag{6.68}$$

The above results are due to Bannai and Bannai [31]. Note that there are several errors in [60, Theorem 4.2 in Section 4 of Chapter III] about double P-polynomial structures. In the above list, errors are corrected.

Note that if a P-polynomial scheme $\mathfrak{X}$ has a double P-polynomial structure of pattern (I), the *bipartite doubling* of $\mathfrak{X}$ also becomes a P-polynomial scheme. For details, see [60, III. 6, Remark (5)].

Double Q-polynomial structures

Let $\mathfrak{X} = (X, \{R_i\}_{0\leq i\leq d})$ be a symmetric association scheme. Let $\mathfrak{A}$ be the Bose–Mesner algebra of $\mathfrak{X}$ and $\{E_i\}_{0\leq i\leq d}$ the primitive idempotents of $\mathfrak{A}$. When we define a Q-polynomial scheme, the ordering of $\{E_i\}_{0\leq i\leq d}$ matters. Precisely, $\mathfrak{X}$ is a Q-polynomial scheme if

$$E_1 \circ E_i = \frac{1}{|X|}(b_{i-1}^{*}E_{i-1} + a_i^{*}E_i + c_{i+1}^{*}E_{i+1}), \quad 0 \leq i \leq d,$$

holds with respect to the ordering $E_0, E_1, \dots, E_d$, where $E_{-1} = E_{d+1} = 0$. However, there is a case where $\mathfrak{X}$ becomes a Q-polynomial scheme with respect to another ordering $E_0, E_{i_1}, \dots, E_{i_d}$. In this case, $\mathfrak{X}$ is said to have a *double Q-polynomial structure*, and another ordering $E_0, E_{i_1}, \dots, E_{i_d}$ is called the second Q-polynomial ordering. There are only four patterns of second Q-polynomial orderings (note that $\mathfrak{X}$ cannot have a triple Q-polynomial structure, namely, $\mathfrak{X}$ cannot be Q-polynomial with respect to three orderings of $\{E_i\}_{0\leq i\leq d}$):

(I) $E_0, E_2, E_4, , E_6, \ldots, E_5, E_3, E_1$;
(II) $E_0, E_d, E_1, E_{d-1}, E_2, E_{d-2}, E_3, E_{d-3}, \ldots$;
(III) $E_0, E_d, E_2, E_{d-2}, E_4, E_{d-4}, \ldots, E_{d-5}, E_5, E_{d-3}, E_3, E_{d-1}, E_1$;
(IV) $E_0, E_{d-1}, E_2, E_{d-3}, E_4, E_{d-5}, \ldots, E_5, E_{d-4}, E_3, E_{d-2}, E_1, E_d$.

Similarly to the case of P-polynomial schemes, (I) and (II) are dual to each other, and (III) and (IV) are self-dual.

We can determine whether a Q-polynomial scheme has a double Q-polynomial structure by the Krein numbers. Let $\mathfrak{X} = (X, \{R_i\}_{0\le i\le d})$ be a Q-polynomial scheme with respect to the ordering $E_0, E_1, \ldots, E_d$ of the primitive idempotents $\{E_i\}_{0\le i\le d}$, and let $q_{i,j}^k$ be the Krein numbers:

$$E_i \circ E_j = \frac{1}{|X|} \sum_{k=0}^{d} q_{i,j}^k E_k.$$

The first dual intersection matrix $B_1^* = (q_{1,j}^k)$ is given by (6.55). Then $\mathfrak{X}$ has a double Q-polynomial structure of pattern (I) if and only if

$$a_i^* = 0\ (0 \le i \le d-1) \quad \text{and} \quad a_d^* \ne 0, \tag{6.69}$$

where $a_i^* = q_{1,i}^i$. A Q-polynomial scheme satisfying condition (6.69) is called *almost dual bipartite*. The necessary and sufficient condition for $\mathfrak{X}$ to have a double Q-polynomial structure of pattern (II) is

$$q_{j,d}^d = 0\ (2 \le j \le d) \quad \text{and} \quad q_{1,d}^d \ne 0. \tag{6.70}$$

The necessary and sufficient conditions for $\mathfrak{X}$ to have a double Q-polynomial structure of pattern (III) are the following. In the case $d \ge 5$,

$$q_{j,d}^d = 0\ (j \ne 0, 2) \quad \text{and} \quad q_{2,d}^d \ne 0. \tag{6.71}$$

(In this case, if $d = 2\ell + 1$, we have $a_j^* = 0$ $(j \ne \ell + 1)$, and if $d = 2\ell$, we have $a_j^* = 0$ $(j \ne \ell, \ell + 1)$, $a_\ell^* \ne 0$, $a_{\ell+1}^* \ne 0$.) In the case $d = 3$, $q_{1,3}^3 = 0$, $q_{2,3}^3 \ne 0$, and in the case $d = 4$, $q_{1,4}^4 = q_{3,4}^4 = 0$ and $q_{2,4}^4 \ne 0$, $q_{2,3}^4 \ne 0$. The necessary and sufficient conditions for $\mathfrak{X}$ to have a double Q-polynomial structure of Pattern (IV) are the following. In the case $d \ge 4$,

$$q_{j,d}^{d-1} = 0 \quad (j \ne 0, 1). \tag{6.72}$$

(In this case, if $d = 2\ell$, we have $a_j^* = 0$ $(j \ne \ell)$, and if $d = 2\ell+1$, we have $a_j^* = 0$ $(j \ne \ell, \ell+1)$ and $a_\ell^* \ne 0$, $a_{\ell+1}^* \ne 0$.) In the case $d = 3$, $q_{1,2}^2 \ne 0$, $q_{3,2}^2 = 0$.

The above results are due to Suzuki [444]. In [444], there was an exceptional case of $d = 5$, which was eliminated by Ma and Wang [323] later. Note that if a Q-polynomial scheme $\mathfrak{X}$ has a double Q-polynomial structure of pattern (I), the *bipartite doubling* of $\mathfrak{X}$ is also a Q-polynomial scheme. For details, see [60, III. 6, Remark (5)].

6.1.4 Orthogonal polynomials

In this section, we consider an irreducible tridiagonal matrix in general:

$$B = \begin{bmatrix} a_0 & b_0 & & & & 0 \\ c_1 & a_1 & b_1 & & & \\ & \ddots & \ddots & \ddots & & \\ & & & c_{d-1} & a_{d-1} & b_{d-1} \\ 0 & & & & c_d & a_d \end{bmatrix}, \quad b_{i-1}c_i \neq 0 \ (1 \le i \le d). \tag{6.73}$$

We set the complex field $\mathbb{C}$ as the base field. The intersection matrix B_1 of a P-polynomial scheme and the dual intersection matrix B_1^* become irreducible tridiagonal matrices. If we consider P-polynomial schemes and Q-polynomial schemes, since B_1 and B_1^* are real matrices and their eigenvalues are also real, it is enough to consider the real field as a base field. If we consider L-pairs later, we need to consider complex matrices.

For B, we define the polynomial $v_i(x)$ $(0 \le i \le d+1)$ of degree i by the following three-term recurrence:

$$xv_i(x) = b_{i-1}v_{i-1}(x) + a_iv_i(x) + c_{i+1}v_{i+1}(x) \quad (0 \le i \le d), \tag{6.74}$$

where $v_{-1}(x) = 0$, $v_0(x) = 1$, b_{-1} is an indeterminate, and $c_{d+1} = 1$. We also define a sequence $k_0, k_1, \dots, k_{d+1}$ by $k_0 = 1$, and

$$k_i = \frac{b_{i-1}}{c_i}k_{i-1} = \frac{b_0b_1\cdots b_{i-1}}{c_1c_2\cdots c_i} \quad (1 \le i \le d+1), \tag{6.75}$$

where $b_d = c_{d+1} = 1$. Then we have

$$x\frac{v_i(x)}{k_i} = c_i\frac{v_{i-1}(x)}{k_{i-1}} + a_i\frac{v_i(x)}{k_i} + b_i\frac{v_{i+1}(x)}{k_{i+1}} \quad (0 \le i \le d), \tag{6.76}$$

where c_0, k_{-1} are indeterminate. We call $\{v_i(x)\}_{i=0}^{d+1}$ and $\{k_i\}_{i=0}^{d+1}$ a *system of polynomials* and a *sequence of degrees* determined by B, respectively. Conversely, if polynomials $v_i(x)$ $(0 \le i \le d+1)$ of degree i satisfy the three-term recurrence (6.74) for $i = 0, 1, \dots, d$, the corresponding tridiagonal matrix B is uniquely determined.

Proposition 6.10. *The minimal polynomial of B is $c_1c_2\cdots c_dv_{d+1}(x)$.*

Proof. Let $\mathbf{e}_0 = {}^t(1, 0, \dots, 0)$, $\mathbf{e}_1 = {}^t(0, 1, \dots, 0)$, $\cdots$, $\mathbf{e}_{d+1} = {}^t(0, 0, \dots, 0, 1)$ form the standard basis of $\mathbb{C}^{d+1}$. We have

$$B\mathbf{e}_i = b_{i-1}\mathbf{e}_{i-1} + a_i\mathbf{e}_i + c_{i+1}\mathbf{e}_{i+1}, \tag{6.77}$$

where $\mathbf{e}_{-1} = \mathbf{0}$, $\mathbf{e}_{d+1} = \mathbf{0}$.

Consider the polynomial ring $\mathbb{C}[x]$, and let $\mathcal{I}$ be the *ideal* generated by $v_{d+1}(x)$. By comparing (6.74) and (6.77), we can see that the action of B on $\mathbb{C}^{d+1}$ is isomorphic to the natural action of x on the quotient ring $\mathbb{C}[x]/\mathcal{I}$ by the correspondence $\boldsymbol{e}_i \mapsto v_i(x)$ $(0 \leq i \leq d)$. Since the leading coefficient of $v_{d+1}(x)$ is $\frac{1}{c_1 c_2 \cdots c_d}$, $c_1 c_2 \cdots c_d v_{d+1}(x)$ is the minimal polynomial of the action of x on $\mathbb{C}[x]/\mathcal{I}$. □

Allowing repetition, let $\theta_0, \theta_1, \ldots, \theta_d$ be the zeros of $v_{d+1}(x)$:

$$c_1 c_2 \cdots c_d v_{d+1}(x) = (x - \theta_0)(x - \theta_1) \cdots (x - \theta_d). \tag{6.78}$$

Corollary 6.11. *The irreducible tridiagonal matrix B is diagonalizable if and only if the zeros $\theta_0, \theta_1, \ldots, \theta_d$ of $v_{d+1}(x)$ are mutually distinct. In this case, the eigenvalues of B are $\theta_0, \theta_1, \ldots, \theta_d$. Namely, B has $d + 1$ distinct eigenvalues.*

In what follows, we assume that B is
(I) diagonalizable.

In this case, $\{v_i(x)\}_{i=0}^d$ are called a *system of orthogonal polynomials*. The set of zeros of $v_{d+1}(x)$ is called the *support* of the system of orthogonal polynomials. Note that $\theta_i \neq \theta_j$ $(i \neq j)$.

Proposition 6.12. *Suppose a tridiagonal matrix B is*
(I)′ *a real matrix and $b_{i-1}c_i > 0$ $(1 \leq i \leq d)$.*

Then B is diagonalizable and the eigenvalues of B are all real.

Proof. If we let $\lambda_0 = 1$, $\lambda_i = \sqrt{\frac{c_i}{b_{i-1}}}\lambda_{i-1}$ $(1 \leq i \leq d)$, then λ_i is real and we have $\frac{\lambda_i}{\lambda_{i-1}} b_{i-1} = \frac{\lambda_{i-1}}{\lambda_i} c_i$ $(1 \leq i \leq d)$. Let $\Lambda = \operatorname{diag}(\lambda_0, \lambda_1, \ldots, \lambda_d)$ be the diagonal matrix whose diagonal entries are $\lambda_0, \lambda_1, \ldots, \lambda_d$. Then $\Lambda^{-1}B\Lambda$ is a real symmetric matrix. Therefore $\Lambda^{-1}B\Lambda$ is diagonalizable and the eigenvalues are all real. □

In the following, we fix an ordering $\theta_0, \theta_1, \ldots, \theta_d$ of zeros of v_{d+1}, and set diagonal matrices D and K as follows:

$$D = \operatorname{diag}(\theta_0, \theta_1, \ldots, \theta_d), \quad K = \operatorname{diag}(k_0, k_1, \ldots, k_d). \tag{6.79}$$

Let

$$P = (v_j(\theta_i)) \tag{6.80}$$

be the matrix whose (i, j)-entry is $v_j(\theta_i)$. We also let

$$P' = \left(\frac{v_i(\theta_j)}{k_i}\right) = K^{-1}{}^tP. \tag{6.81}$$

Note that the (i, j)-entry of P' is $\frac{v_i(\theta_j)}{k_i}$.

Proposition 6.13. *The matrices P and P' are non-singular matrices and the following hold:*
(1) $PBP^{-1} = D$;
(2) $(P')^{-1}BP' = D$.

Proof. By (6.74) and (6.78), the i-th row $(v_0(\theta_i), v_1(\theta_i), \ldots, v_d(\theta_i))$ of P is a left eigenvector of B with eigenvalue θ_i. Therefore $PB = DP$. Moreover, since $\theta_i \neq \theta_j$ $(i \neq j)$, the 0-th, 1-th, …, d-th rows belong to distinct eigenspaces, and so they are linearly independent. Hence, P is non-singular and (1) holds. By (6.76), (6.78), the j-th column ${}^t(\frac{v_0(\theta_j)}{k_0}, \frac{v_1(\theta_j)}{k_0}, \ldots, \frac{v_d(\theta_j)}{k_0})$ of P' is a right eigenvector of B with eigenvalue θ_j. Therefore $BP' = P'B$. Moreover, similarly to (1), since $\theta_i \neq \theta_j$ $(i \neq j)$, the 0-th, 1-th, …, d-th columns belong to distinct eigenspaces of P', and so they are linearly independent. Hence P' is non-singular and (2) holds. □

Theorem 6.14. *There exists a non-singular diagonal matrix $M = \mathrm{diag}(\mu_0, \mu_1, \ldots, \mu_d)$ $(\mu_i \neq 0,\ 0 \le i \le d)$ such that the following (1), (2) hold:*
(1) $M^{-1} = PK^{-1}{}^tP$;
(2) $K = {}^tPMP$.

Proof. By Proposition 6.13, we have $P^{-1}DP = B = P'D(P')^{-1}$. Therefore PP' and D are commutative. Since D is diagonal and their diagonal entries $\theta_0, \theta_1, \ldots, \theta_d$ are distinct, a matrix which commutes with D must be a diagonal matrix. Hence, there exists a non-singular diagonal matrix M such that $M^{-1} = PP'$. By (6.81), we have $P' = K^{-1}{}^tP$ and we obtain (1); (2) follows directly from (1). □

Computing the (i,j)-entries of both sides of (1) and (2) in Theorem 6.14 yields the following orthogonality relations.
The first orthogonality relation:

$$\sum_{\nu=0}^{d} v_\nu(\theta_i) v_\nu(\theta_j) \frac{1}{k_\nu} = \delta_{i,j} \frac{1}{\mu_i}. \tag{6.82}$$

The second orthogonality relation:

$$\sum_{\nu=0}^{d} v_i(\theta_\nu) v_j(\theta_\nu) \mu_\nu = \delta_{i,j} k_i. \tag{6.83}$$

Theorem 6.15. *The following holds:*

$$\mu_i = \frac{k_d}{v'_{d+1}(\theta_i) v_d(\theta_i)} \qquad (0 \le i \le d), \tag{6.84}$$

where $v'_{d+1}(x)$ is the derived function of $v_{d+1}(x)$.

The number μ_i is called the *Christoffel number*. As a preparation for the proof of Theorem 6.15, we prove the *Christoffel–Darboux formula* in the following proposition.

Proposition 6.16 (Christoffel–Darboux formula). *For $i \in \{0,1,\dots,d\}$, the following equation holds:*

$$\sum_{\nu=0}^{i} v_\nu(x)v_\nu(y)\frac{1}{k_\nu} = \frac{c_{i+1}}{k_i}\frac{v_{i+1}(x)v_i(y) - v_i(x)v_{i+1}(y)}{x-y}. \tag{6.85}$$

Proof. By (6.74), we have

$$\begin{aligned} &v_{\nu+1}(x)v_\nu(y) - v_\nu(x)v_{\nu+1}(y) \\ &\quad = \frac{1}{c_{\nu+1}}((x-a_\nu)v_\nu(x) - b_{\nu-1}v_{\nu-1}(x))v_\nu(y) \\ &\qquad - \frac{1}{c_{\nu+1}}((y-a_\nu)v_\nu(y) - b_{\nu-1}v_{\nu-1}(y))v_\nu(x) \\ &\quad = \frac{1}{c_{\nu+1}}(x-y)v_\nu(x)v_\nu(y) + \frac{b_{\nu-1}}{c_{\nu+1}}(v_\nu(x)v_{\nu-1}(y) - v_{\nu-1}(x)v_\nu(y)). \end{aligned}$$

Since $b_{\nu-1} = \frac{c_\nu}{k_{\nu-1}}k_\nu$ by (6.75), we have

$$\begin{aligned} \frac{1}{k_\nu}(x-y)v_\nu(x)v_\nu(y) &= \frac{c_{\nu+1}}{k_\nu}(v_{\nu+1}(x)v_\nu(y) - v_\nu(x)v_{\nu+1}(y)) \\ &\quad - \frac{c_\nu}{k_{\nu-1}}(v_\nu(x)v_{\nu-1}(y) - v_{\nu-1}(x)v_\nu(y)). \end{aligned}$$

Taking the summation through $\nu = 0,1,\dots,i$ yields the desired equation. □

Corollary 6.17. *For $i \in \{0,1,\dots,d\}$, the following holds:*

$$\sum_{\nu=0}^{i} v_\nu(x)^2\frac{1}{k_\nu} = \frac{c_{i+1}}{k_i}(v'_{i+1}(x)v_i(x) - v'_i(x)v_{i+1}(x)). \tag{6.86}$$

Proof. Note that

$$\begin{aligned} &v_{i+1}(x)v_i(y) - v_i(x)v_{i+1}(y) \\ &\quad = (v_{i+1}(x) - v_{i+1}(y))v_i(y) - (v_i(x) - v_i(y))v_{i+1}(y). \end{aligned}$$

Taking the limit of (6.85) as $x - y \longrightarrow 0$ yields the desired equation. □

Proof of Theorem 6.15. Set $i = d$, $x = \theta_j$ in (6.86). Since $v_{d+1}(\theta_j) = 0$, we get

$$\sum_{\nu=0}^{d} v_\nu(\theta_j)^2\frac{1}{k_\nu} = \frac{1}{k_d}v'_{d+1}(\theta_j)v_d(\theta_j).$$

(Note that $c_{d+1} = 1$.) On the other hand, let $i = j$ in (6.82). We obtain

$$\sum_{\nu=0}^{d} v_\nu(\theta_j)^2 \frac{1}{k_\nu} = \frac{1}{\mu_j}.$$

Hence we get $\frac{1}{\mu_j} = \frac{1}{k_d} v'_{d+1}(\theta_j) v_d(\theta_j)$. □

Proposition 6.18. *We have $\mu_0 + \mu_1 + \cdots + \mu_d = 1$.*

Proof. By Theorem 6.14, we have $P^{-1} = K^{-1}\,{}^tPM$. Hence the 0-th row of P^{-1} coincides with the 0-th row $(\mu_0, \mu_1, \ldots, \mu_d)$ of $K^{-1}\,{}^tPM$. On the other hand, since the 0-th row of P is ${}^t(1, 1, \ldots, 1)$, by comparing with the $(0,0)$-entry, we obtain $1 = (P^{-1}P)(0,0) = \mu_0 + \mu_1 + \cdots + \mu_d$. □

The above is a general theory on diagonalizable irreducible tridiagonal matrices. Among the diagonalizable irreducible tridiagonal matrices of size $d + 1$, the subset of non-negative real matrices forms an important class. For any matrix which belongs to this class, we can choose an eigenvector such that all components are real and positive as a basis of the (1-dimensional) eigenspace corresponding to the Perron–Frobenius eigenvalue θ_0. Since an eigenvector corresponding to θ_0 is ${}^t(\frac{v_0(\theta_0)}{k_0}, \frac{v_1(\theta_0)}{k_1}, \ldots, \frac{v_d(\theta_0)}{k_d})$, this property can be rephrased as $v_i(\theta_0) > 0$ $(0 \le i \le d)$. (Conversely, the Perron–Frobenius eigenvalue θ_0 can be characterized by the positivity of a corresponding eigenvector.) For a general diagonalizable irreducible tridiagonal matrix of size $d + 1$, we consider the weaker condition:

$(\mathrm{II})_1$ There exists an eigenvalue θ_0 such that $v_i(\theta_0) \neq 0$ $\quad(0 \le i \le d)$.

For two diagonalizable irreducible tridiagonal matrices B, B' of size $d + 1$, we define B and B' to be equivalent if there exists a non-singular diagonal matrix Λ of size $d + 1$ such that $B' = \Lambda^{-1} B \Lambda$. Let θ_0 be an eigenvalue of B. Condition $(\mathrm{II})_1$ holds, i. e., any component of any eigenvector for the eigenvalue θ_0 is non-zero, if and only if there exists a matrix B' whose row sum is θ_0, where B' is equivalent to B (i. e., ${}^t(1, 1, \ldots, 1)$ becomes an eigenvector of B' with eigenvalue θ_0). Moreover, such matrix B' is unique in the equivalence class which contains B. Therefore by replacing matrices in the same equivalence class, the above condition $(\mathrm{II})_1$ can be replaced by the following:

$(\mathrm{II})_2$ There exists an eigenvalue θ_0 such that $v_i(\theta_0) = k_i$ $(0 \le i \le d)$.

Namely, the condition $(\mathrm{II})_1$ can be replaced by the following:

$(\mathrm{II})_3$ Every row sum is θ_0.

In this case, θ_0 is an eigenvalue of B and ${}^t(1, 1, \ldots, 1)$ is an eigenvector for the eigenvalue θ_0. From the above discussion, in the following, we consider irreducible tridiagonal matrices of the form (6.73) satisfying the following:

(I) they are diagonalizable;

(II) every row sum is θ_0.

(We denote condition $(\mathrm{II})_3$ by condition (II) as a separate condition from (I).)

Proposition 6.19. *The following condition is the necessary and sufficient condition for condition (II):*

$$v_{d+1}(x) = (x - \theta_0)(v_0(x) + v_1(x) + \cdots + v_d(x)). \tag{6.87}$$

Proof. Taking the summation of the three-term recurrence (6.74) through $i = 0, 1, \ldots, d$ yields (6.87) by condition (II). Conversely, by (6.87), we obtain condition (II). □

Proposition 6.20.
(1) *We have $k_i = v_i(\theta_0)$ $(0 \le i \le d)$.*
(2) *Set $n = k_0 + k_1 + \cdots + k_d$. Then $n \neq 0$.*

Proof. (1) By (6.76) and (6.78), the vector ${}^t(\frac{v_0(\theta_0)}{k_0}, \frac{v_1(\theta_0)}{k_1}, \ldots, \frac{v_d(\theta_0)}{k_d})$ is a right eigenvector for the eigenvalue θ_0. On the other hand, by (I) and Corollary 6.11, the eigenspace of B for the eigenvalue θ_0 is 1-dimensional. By (II), the vector ${}^t(1, 1, \ldots, 1)$ is an eigenvector of B for the eigenvalue θ_0. So $v_0(\theta_0) = 1$, $k_0 = 1$, and hence $1 = \frac{v_0(\theta_0)}{k_0} = \frac{v_1(\theta_0)}{k_1} = \cdots = \frac{v_d(\theta_0)}{k_d}$.
(2) By condition (I) and Corollary 6.11, $v_{d+1}(x) = 0$ has no multiple root. Hence by Proposition 6.19, θ_0 is not a zero of $v_0(x) + v_1(x) + \cdots + v_d(x)$. By (1), we have $v_i(\theta_0) = k_i$, and hence we obtain $k_0 + k_1 + \cdots + k_d \neq 0$. □

We normalize the Christoffel number μ_i in (6.84) and set

$$m_i = n\mu_i \quad (0 \le i \le d). \tag{6.88}$$

We call m_i the *multiplicity*. In terms of the degree k_i and the multiplicity m_i, we rewrite the orthogonality relations as follows.
The first orthogonality relation:

$$\sum_{\nu=0}^{d} v_\nu(\theta_i) v_\nu(\theta_j) \frac{1}{k_\nu} = \delta_{i,j} \frac{n}{m_i}. \tag{6.89}$$

The second orthogonality relation:

$$\sum_{\nu=0}^{d} v_i(\theta_\nu) v_j(\theta_\nu) m_\nu = \delta_{i,j} n k_i. \tag{6.90}$$

Moreover, setting $j = i$ in (6.89) and rewriting (6.84) in terms of m_i yields the following multiplicity formula.

Theorem 6.21 (Multiplicity formula). *We have*

$$m_i = \frac{n}{\sum_{\nu=0}^{d} v_\nu(\theta_i)^2 / k_\nu} = \frac{n k_d}{v'_{d+1}(\theta_i) v_d(\theta_i)}. \tag{6.91}$$

Proposition 6.22. *The following hold:*

(1) $m_0 = 1$;
(2) $m_0 + m_1 + \cdots + m_d = n$.

Proof. (1) Set $i = 0$ in (6.91) and note that $v_\nu(\theta_0) = k_\nu$ and $n = \sum_{\nu=0}^{d} k_\nu$. Then we get $m_0 = 1$.
(2) By Proposition 6.18, we have $\mu_0 + \mu_1 + \cdots + \mu_d = 1$. By noting that $m_i = n\mu_i$, the result follows. □

Other than the irreducible tridiagonal matrix B in (6.73), assume another irreducible tridiagonal matrix B^* is given as follows:

$$B^* = \begin{bmatrix} a_0^* & b_0^* & & & \\ c_1^* & a_1^* & b_1^* & & 0 \\ & \ddots & \ddots & \ddots & \\ & & c_{d-1}^* & a_{d-1}^* & b_{d-1}^* \\ 0 & & & c_d^* & a_d^* \end{bmatrix}, \quad b_{i-1}^* c_i^* \neq 0 \ (1 \le i \le d). \tag{6.92}$$

Suppose B and B^* are (I) diagonalizable and (II) their row sums are θ_0 and θ_0^*, respectively. Let $\{v_i(x)\}_{i=0}^{d+1}$, $\{k_i\}_{i=0}^{d+1}$, and $\{v_i^*(x)\}_{i=0}^{d+1}$, $\{k_i^*\}_{i=0}^{d+1}$ be the systems of polynomials and the sequences of degrees of B and B^*, respectively. Note that B and B^* have $d+1$ distinct eigenvalues, respectively, and they are the roots of $v_{d+1}(x)$ and $v_{d+1}^*(x)$, respectively. Note also that the row sums θ_0, θ_0^* are the eigenvalues of B, B^*. If the ordering $\theta_0, \theta_1, \ldots, \theta_d$ of eigenvalues of B and the ordering $\theta_0^*, \theta_1^*, \ldots, \theta_d^*$ of eigenvalues of B^* are given, define square matrices P, P', Q, Q' of size $d+1$ indexed by i, j with $0 \le i, j \le d$ as follows:

$$\begin{aligned} P &= (v_j(\theta_i)), \quad P' = \left(\frac{v_i(\theta_j)}{k_i}\right), \\ Q &= (v_j^*(\theta_i^*)), \quad Q' = \left(\frac{v_i^*(\theta_j^*)}{k_i^*}\right). \end{aligned} \tag{6.93}$$

By Proposition 6.13, P, P', Q, Q' are non-singular and the following hold:

$$\begin{aligned} P^{-1}DP &= B = P'D(P')^{-1}, \\ Q^{-1}D^*Q &= B^* = Q'D^*(Q')^{-1}, \end{aligned} \tag{6.94}$$

where D, D^* are diagonal matrices defined by

$$\begin{aligned} D &= \operatorname{diag}(\theta_0, \theta_1, \ldots, \theta_d), \\ D^* &= \operatorname{diag}(\theta_0^*, \theta_1^*, \ldots, \theta_d^*). \end{aligned} \tag{6.95}$$

Definition 6.23. If there exist an ordering $\theta_0, \theta_1, \dots, \theta_d$ of the eigenvalues of B (the zeros of $v_{d+1}(x)$) and an ordering $\theta_0^*, \theta_1^*, \dots, \theta_d^*$ of the eigenvalues of B^* (the zeros of $v_{d+1}^*(x)$) such that ${}^tP' = Q'$, i. e.,

$$\text{(III)} \quad \frac{v_j(\theta_i)}{k_j} = \frac{v_i^*(\theta_j^*)}{k_i^*} \tag{6.96}$$

holds for $i, j \in \{0, 1, \dots, d\}$, we call $\{v_i(x)\}_{i=0}^d, \{v_i^*(x)\}_{i=0}^d$ a *dual system of orthogonal polynomials.*

When $\{v_i(x)\}_{i=0}^d, \{v_i^*(x)\}_{i=0}^d$ form a dual system of orthogonal polynomials, we assume an ordering $\theta_0, \theta_1, \dots, \theta_d$ of the eigenvalues of B (the zeros of $v_{d+1}(x)$) and an ordering $\theta_0^*, \theta_1^*, \dots, \theta_d^*$ of the eigenvalues of B^* (the zeros of $v_{d+1}^*(x)$) such that (III) holds are fixed. Note that by (6.87), the zeros of $v_{d+1}(x)$ are determined by the row sum θ_0 and the system of orthogonal polynomials $\{v_i(x)\}_{i=0}^d$, and the zeros of $v_{d+1}^*(x)$ are determined by the row sum θ_0^* and the system of orthogonal polynomials $\{v_i^*(x)\}_{i=0}^d$.

Let μ_i and m_i $(0 \le i \le d)$ be the Christoffel numbers and the multiplicities of a system of orthogonal polynomials $\{v_i(x)\}_{i=0}^d$. Also, let μ_i^* and m_i^* $(0 \le i \le d)$ be the Christoffel numbers and the multiplicities of a system of orthogonal polynomials $\{v_i^*(x)\}_{i=0}^d$. By (6.88) and Proposition 6.22, we have

$$\begin{aligned} m_i &= n\mu_i, \quad n = \sum_{i=0}^d k_i = \sum_{i=0}^d m_i \neq 0, \\ m_i^* &= n^*\mu_i^*, \quad n^* = \sum_{i=0}^d k_i^* = \sum_{i=0}^d m_i^* \neq 0. \end{aligned} \tag{6.97}$$

Let $K = \operatorname{diag}(k_0, k_1, \dots, k_d)$, $M = \operatorname{diag}(\mu_0, \mu_1, \dots, \mu_d)$, $K^* = \operatorname{diag}(k_0^*, k_1^*, \dots, k_d^*)$, and $M^* = \operatorname{diag}(\mu_0^*, \mu_1^*, \dots, \mu_d^*)$. By Theorem 6.14, we have

$$\begin{aligned} M^{-1} &= PK^{-1}{}^tP, \quad K = {}^tPMP, \\ (M^*)^{-1} &= Q(K^*)^{-1}{}^tQ, \quad K^* = {}^tQM^*Q. \end{aligned} \tag{6.98}$$

Note also that

$$P' = K^{-1}{}^tP, \quad Q' = (K^*)^{-1}{}^tQ. \tag{6.99}$$

Proposition 6.24. *Let $\{v_i(x)\}_{i=0}^d, \{v_i^*(x)\}_{i=0}^d$ form a dual system of orthogonal polynomials. Then the following (1)–(3) hold:*

(1) $n = n^*$;

(2) $m_i = k_i^*$, $m_i^* = k_i$ $(0 \le i \le d)$;

(3) $PQ = QP = nI$.

Proof. By using (6.99), we rewrite the dual condition (III) ${}^tP' = Q'$ as $PK^{-1} = (K^*)^{-1t}Q$, $KP^{-1} = {}^t Q^{-1} K^*$. By (6.98), we have $KP^{-1} = {}^tPM$, ${}^tQ^{-1}K^* = M^*Q$. Hence we get

$$ {}^tPM = M^*Q. \tag{6.100}$$

By Proposition 6.20, we have $k_i = v_i(\theta_0)$, $k_i^* = v_i^*(\theta_0^*)$. So we obtain

$$P = \begin{bmatrix} 1 & k_1 & \cdots & k_d \\ 1 & & & \\ \vdots & & * & \\ 1 & & & \end{bmatrix}, \quad Q = \begin{bmatrix} 1 & k_1^* & \cdots & k_d^* \\ 1 & & & \\ \vdots & & * & \\ 1 & & & \end{bmatrix}. \tag{6.101}$$

Therefore, the 0-th row of ${}^tPM$ is $(\mu_0, \mu_1, \ldots, \mu_d)$, and the 0-th column is ${}^t(\mu_0, \mu_0 k_1, \ldots, \mu_0 k_d)$. Moreover, the 0-th row of M^*Q is $(\mu_0^*, \mu_0^* k_1^*, \ldots, \mu_0^* k_d^*)$, and the 0-th column is ${}^t(\mu_0^*, \mu_1^*, \ldots, \mu_d^*)$. Therefore, by (6.100), we get $\mu_i = \mu_0^* k_i^*$, $\mu_0 k_i = \mu_i^*$ $(0 \le i \le d)$. If we let $i = 0$, we have $\mu_0 = \mu_0^*$, and by Proposition 6.22, we have $m_0 = 1 = m_0^*$. Thus $n\mu_0 = m_0 = 1 = m_0^* = n^*\mu_0^*$, which implies $n = n^*$. We also get $m_i = n\mu_i = n\mu_0^* k_i^* = n^*\mu_0^* k_i^* = m_0^* k_i^* = k_i^*$, and $m_i^* = n^*\mu_i^* = n^*\mu_0 k_i = n\mu_0 k_i = m_0 k_i = k_i$.

Finally, we show $PQ = QP = nI$. By (1), (2), we have

$$nM = K^*, \quad nM^* = K.$$

As we stated in the beginning, the dual condition (III) ${}^tP' = Q'$ is rewritten as $PK^{-1} = (K^*)^{-1t}Q$. By using $nM = K^*$, we rewrite $PK^{-1} = n^{-1}M^{-1t}Q$ again. Substituting this for $M^{-1} = PK^{-1t}P$ in (6.98) yields $M^{-1} = n^{-1}M^{-1t}Q^tP$. Therefore $PQ = nI$. ☐

Theorem 6.25. *Let B, B^* be irreducible tridiagonal matrices given by (6.73), (6.92), respectively. Assume B, B^* are (I) diagonalizable and (II) every row sum of B is θ_0 and every row sum of B^* is θ_0^*. Let $\{v_i(x)\}_{i=0}^d$, $\{v_i^*(x)\}_{i=0}^d$ be the systems of orthogonal polynomials determined by B, B^*, respectively. Then with respect to an ordering $\theta_0, \theta_1, \ldots, \theta_d$ of the eigenvalues of B and an ordering $\theta_0^*, \theta_1^*, \ldots, \theta_d^*$ of the eigenvalues of B^*, the following are equivalent:*

(i) *There exists a non-singular matrix S such that*

$$SBS^{-1} = \operatorname{diag}(\theta_0, \theta_1, \ldots, \theta_d),$$
$$S^{-1}B^*S = \operatorname{diag}(\theta_0^*, \theta_1^*, \ldots, \theta_d^*).$$

(ii) *For any $i, j \in \{0, 1, \ldots, d\}$,*

$$\frac{v_j(\theta_i)}{k_j} = \frac{v_i^*(\theta_j^*)}{k_i^*},$$

where $\{k_i\}_{i=0}^d$, $\{k_i^*\}_{i=0}^d$ are the sequences of degrees determined by B, B^*, respectively. Namely, $\{v_i(x)\}_{i=0}^d$, $\{v_i^*(x)\}_{i=0}^d$ form a dual system of orthogonal polynomials.

If (i) holds, by setting $P = (v_j(\theta_i))$, $Q = (v_j^*(\theta_i^*))$, there exists a non-zero constant $\lambda \in \mathbb{C}$ such that $S = \lambda P$, $S^{-1} = \frac{1}{\lambda n}Q$, where $n = \sum_{i=0}^{d} k_i = \sum_{i=0}^{d} k_i^*$.

Proof. Following (6.93), set $P' = (\frac{v_i(\theta_j)}{k_i})$, $Q' = (\frac{v_i^*(\theta_j^*)}{k_i^*})$.

(i)⇒(ii): Let $D = \operatorname{diag}(\theta_0, \theta_1, \ldots, \theta_d)$, $D^* = \operatorname{diag}(\theta_0^*, \theta_1^*, \ldots, \theta_d^*)$. By (6.94), we have $S^{-1}DS = B = P^{-1}DP$, and so SP^{-1} and D are commutative. By Corollary 6.11, the eigenvalues $\theta_0, \theta_1, \ldots, \theta_d$ of B are distinct, which means that if a matrix commutes with D, it must be a diagonal matrix. Therefore there exists non-zero constants λ_i $(0 \le i \le d)$ such that

$$S = \operatorname{diag}(\lambda_0, \lambda_1, \ldots, \lambda_d)P. \tag{6.102}$$

Similarly, by (6.94), we have $SD^*S^{-1} = B^* = Q'D^*(Q')^{-1}$. Hence $(Q')^{-1}S$ and D are commutative, and so there exists non-zero constants λ_i^* $(0 \le i \le d)$ such that

$$S = Q' \operatorname{diag}(\lambda_0^*, \lambda_1^*, \ldots, \lambda_d^*). \tag{6.103}$$

By (6.101), the 0-th row and the 0-th column of the right-hand side of (6.102), i. e., $\operatorname{diag}(\lambda_0, \lambda_1, \ldots, \lambda_d)P$, are $\lambda_0(1, k_1, \ldots, k_d)$ and ${}^t(\lambda_0, \lambda_1, \ldots, \lambda_d)$. The 0-th row and the 0-th column of the right-hand side $Q' \operatorname{diag}(\lambda_0^*, \lambda_1^*, \ldots, \lambda_d^*)$ of (6.103) are $(\lambda_0^*, \lambda_1^*, \ldots, \lambda_d^*)$ and $\lambda_0^{*t}(1, 1, \ldots, 1)$. Therefore, we obtain $\lambda_0 k_i = \lambda_i^*$, $\lambda_i = \lambda_0^*$ $(0 \le i \le d)$. If we let $\lambda = \lambda_0 = \lambda_0^*$, since $\lambda_i = \lambda$, $\lambda_i^* = \lambda k_i$ $(0 \le i \le d)$, by (6.102), (6.103), we obtain $\lambda P = S = \lambda Q'K$, where $K = \operatorname{diag}(k_0, k_1, \ldots, k_d)$. In particular, we get $P = Q'K$, $PK^{-1} = Q'$. On the other hand, since ${}^tP' = PK^{-1}$ by (6.99), we have ${}^tP' = Q'$, i. e., (ii) holds. Moreover, since $P^{-1} = \frac{1}{n}Q$ by Proposition 6.24, by $\lambda P = S$, we get $S^{-1} = \frac{1}{\lambda n}Q$.

(ii)⇒(i): By the assumption, $Q' = {}^tP'$ holds. By (6.99), we have $P' = K^{-1t}P$, and hence $Q' = {}^tP' = PK^{-1}$. By (6.94), we have $D^* = (Q')^{-1}B^*Q'$. Substituting $Q' = PK^{-1}$ yields $D^* = KP^{-1}B^*PK^{-1}$. Therefore, $D^* = P^{-1}B^*P$. On the other hand, since $D = PBP^{-1}$ by (6.94), by setting $S = P$, we have (i). □

Remark 6.26. The matrices $B, B^*, D = \operatorname{diag}(\theta_0, \theta_1, \ldots, \theta_d)$, and $D^* = \operatorname{diag}(\theta_0^*, \theta_1^*, \ldots, \theta_d^*)$ in Theorem 6.25 which satisfy (i) are called an L-pair (precisely, an L-pair associated with an L-system) (Section 6.1.3). By Theorem 6.25, if we look at dual systems of orthogonal polynomials in another way, we get L-pairs. In Section 6.3 in this chapter, we classify L-pairs and show that dual systems of orthogonal polynomials are identical to dual systems of Askey–Wilson polynomials. In [60], the direct proof of the above fact was given.

6.2 Tridiagonal pairs (TD-pairs)

In the previous section, we stated some properties on irreducible representations of the Terwilliger algebra of a P- and Q-polynomial scheme. In this section, we axiomatize the essential part of them and introduce the concept of tridiagonal pairs (TD-pairs). Some TD-pairs do not arise from irreducible representations of Terwilliger algebras. However, this wider framework of TD-pairs is suitable for the study of irreducible representations of Terwilliger algebras.

The classification of TD-pairs implies the determination of all the irreducible representations of Terwilliger algebras. The classification of TD-pairs is almost completed [260, 253]. In this book, we deal with so-called *L-pairs*, a special class of TD-pairs. Using the classification of L-pairs we can determine the *principal representations* of Terwilliger algebras. This implies, if we ignore combinatorial structures, we can determine the Bose–Mesner algebras of P- and Q-polynomial schemes at the algebraic level.

In Section 6.2.1, we discuss weight space decompositions of TD-pairs, and in Section 6.2.2, we discuss TD-relations following [256]. We also introduce Askey–Wilson parameters (AW-parameters). The classification of L-pairs will be discussed in Section 6.3.

We prepare the notation. Let V be a finite-dimensional vector space over the complex field $\mathbb{C}$, where $\dim(V) \geq 1$, and let $\mathrm{End}(V)$ be the algebra over $\mathbb{C}$ consisting of the linear transformations on V. In this section, A, A^* denote diagonalizable linear transformations on V. We denote the eigenvalues of A, A^* by θ_i $(0 \leq i \leq d)$, θ_i^* $(0 \leq i \leq d^*)$, respectively, and denote the eigenspace of A for θ_i by V_i and the eigenspace of A^* for θ_i^* by V_i^*. Note that $\theta_i \neq \theta_j$, $\theta_i^* \neq \theta_j^*$ whenever $i \neq j$. Let E_i be the projection onto V_i and E_i^* the projection onto V_i^*:

$$E_i : V = \oplus_{j=0}^{d} V_j \longrightarrow V_i, \quad A|_{V_i} = \theta_i,$$
$$E_i^* : V = \oplus_{j=0}^{d^*} V_j^* \longrightarrow V_i^*, \quad A^*|_{V_i^*} = \theta_i^*.$$

Note that the following hold:

$$E_i = \prod_{j \neq i} \frac{A - \theta_j}{\theta_i - \theta_j}, \quad E_i^* = \prod_{j \neq i} \frac{A^* - \theta_j^*}{\theta_i^* - \theta_j^*}. \tag{6.104}$$

In particular, E_i is contained in the subalgebra $\langle A \rangle$ of $\mathrm{End}(V)$ generated by A, and E_i^* is contained in the subalgebra $\langle A^* \rangle$ of $\mathrm{End}(V)$ generated by A^*. The subalgebra of $\mathrm{End}(V)$ generated by A, A^* is denoted by $\langle A, A^* \rangle$. We keep the above notation throughout this section unless otherwise stated.

Definition 6.27 (Tridiagonal pairs (TD-pairs)). The pair A, A^* is called a *TD-pair* if the following (1), (2), (3) hold:

(1) there exists an ordering $V_0^*, V_1^*, \ldots, V_{d^*}^*$ of the eigenspaces V_i^* of A^* satisfying the following:

$$AV_i^* \subseteq V_{i-1}^* + V_i^* + V_{i+1}^* \quad (1 \leq i \leq d^* - 1), \tag{6.105}$$
$$AV_0^* \subseteq V_0^* + V_1^*, \quad AV_{d^*}^* \subseteq V_{d^*-1}^* + V_{d^*}^*; \tag{6.106}$$

(2) there exists an ordering $V_0, V_1, \dots, V_d$ of the eigenspaces V_i of A satisfying the following:

$$A^* V_i \subseteq V_{i-1} + V_i + V_{i+1} \quad (1 \le i \le d-1), \tag{6.107}$$

$$A^* V_0 \subseteq V_0 + V_1, \quad A^* V_d \subseteq V_{d-1} + V_d; \tag{6.108}$$

(3) V is irreducible as an $\langle A, A^* \rangle$-module; namely, a subspace of V which is A-invariant and A^*-invariant is $\{0\}$ or V.

Two TD-pairs $A, A^* \in \mathrm{End}(V)$ and $B, B^* \in \mathrm{End}(V')$ are said to be isomorphic if there exists an isomorphism $\varphi : V \longrightarrow V'$ as vector spaces such that $\varphi A = B\varphi$, $\varphi A^* = B^* \varphi$.

$$\begin{array}{ccc} V & \xrightarrow{\varphi} & V' \\ {\scriptstyle A, A^*}\downarrow & \circlearrowright & \downarrow{\scriptstyle B, B^*} \\ V & \xrightarrow{\varphi} & V' \end{array}$$

Remark 6.28. Let A, A^* be a TD-pair. Then:

(1) There are exactly two orderings of the eigenspaces V_i^* of A^* satisfying Definition 6.27 (1) if $d^* \ge 1$, and if one of them is $V_0^*, V_1^*, \dots, V_{d^*}^*$, then another ordering is the reversed ordering $V_{d^*}^*, \dots, V_1^*, V_0^*$. The same property holds for orderings of the eigenspaces V_i of A satisfying Definition 6.27 (2).

(2) As will be proved in Corollary 6.33 in the next subsection, we have $d = d^*$. We call d the *diameter* of a TD-pair A, A^*. A TD-pair with $d = 0$ is called a trivial TD-pair.

If a pair $A, A^* \in \mathrm{End}(V)$ is a TD-pair, we assume an ordering $V_0, V_1, \dots, V_d$ of eigenspaces of A and an ordering $V_0^*, V_1^*, \dots, V_d^*$ of eigenspaces of A^* satisfying Definition 6.27 are given unless otherwise stated. Such a quadruple $(A, A^*; \{V_i\}_{i=0}^d, \{V_i^*\}_{i=0}^{d^*})$ is called a *TD-system*. Instead of eigenspaces V_i, V_i^*, we sometime use eigenvalues θ_i, θ_i^* or projections E_i, E_i^* to denote a TD-system by $(A, A^*; \{\theta_i\}_{i=0}^d, \{\theta_i^*\}_{i=0}^{d^*})$ or $(A, A^*; \{E_i\}_{i=0}^d, \{E_i^*\}_{i=0}^{d^*})$. Sometimes we abbreviate A, A^* to denote a TD-system by $(\{V_i\}_{i=0}^d, \{V_i^*\}_{i=0}^{d^*})$, $(\{\theta_i\}_{i=0}^d, \{\theta_i^*\}_{i=0}^{d^*})$, or $(\{E_i\}_{i=0}^d, \{E_i^*\}_{i=0}^{d^*})$.

By Remark 6.28 (1), a TD-pair $A, A^* \in \mathrm{End}(V)$ is associated with the following four TD-systems:

$$(A, A^*; \{V_i\}_{i=0}^d, \{V_i^*\}_{i=0}^{d^*}), \quad (A, A^*; \{V_{d-i}\}_{i=0}^d, \{V_i^*\}_{i=0}^{d^*}),$$
$$(A, A^*; \{V_i\}_{i=0}^d, \{V_{d^*-i}^*\}_{i=0}^{d^*}), \quad (A, A^*; \{V_{d-i}\}_{i=0}^d, \{V_{d^*-i}^*\}_{i=0}^{d^*}).$$

Remark 6.29. If $A, A^* \in \mathrm{End}(V)$ form a TD-pair, for any complex numbers $\lambda, \mu, \lambda^*, \mu^*$ ($\lambda \ne 0, \lambda^* \ne 0 \in \mathbb{C}$), $\lambda A + \mu, \lambda^* A^* + \mu^* \in \mathrm{End}(V)$ also form a TD-pair. Let $(A, A^*; \{V_i\}_{i=0}^d, \{V_i^*\}_{i=0}^{d^*})$ be a TD-system associated with a TD-pair A, A^*. Then $(\lambda A + \mu, \lambda^* A^* + \mu^*; \{V_i\}_{i=0}^d, \{V_i^*\}_{i=0}^{d^*})$ is a TD-system associated with a TD-pair $\lambda A + \mu, \lambda^* A^* + \mu^*$.

Remark 6.30. Let $\mathfrak{X} = (X, \{R_i\}_{0\le i\le D})$ be a P- and Q-polynomial scheme of diameter D. Let $A_0, A_1, \dots, A_D$ be the adjacency matrices and let $E_0, E_1, \dots, E_D$ be the primitive idempotents of the Bose–Mesner algebra $\mathfrak{A}$ of $\mathfrak{X}$, each of which follows the P-polynomial ordering and the Q-polynomial ordering, respectively. Let $A_0^*, A_1^*, \dots, A_D^*$ be the dual adjacency matrices and $E_0^*, E_1^*, \dots, E_D^*$ the dual primitive idempotents of the dual Bose–Mesner algebra $\mathfrak{A}^*$ of $\mathfrak{X}$, where $E_i^* = E_i^*(x_0)$ for a fixed base vertex $x_0 \in X$, each of which follows the Q-polynomial ordering and the P-polynomial ordering, respectively. The Terwilliger algebra is generated by $A = A_1$ and $A^* = A_1^*$:

$$T = T(x_0) = \langle A, A^* \rangle \subset \mathrm{End}(V), \quad V = \bigoplus_{x\in X} \mathbb{C}x,$$

where V is the standard T-module. By Proposition 2.39 in Chapter 2, Section 2.6, the standard module V is decomposed into the direct sum of irreducible T-modules. Let W be an irreducible T-module of V. By condition (iv) for P-polynomial schemes and condition (iv) for Q-polynomial schemes in Sections 6.1.1 and 6.1.2, and by the irreducibility of W, there exist integers r, t, d, d^* such that

$$\{i \mid E_i^* W \ne 0,\ 0 \le i \le D\} = \{i \mid r \le i \le r + d^*\},$$
$$\{i \mid E_i W \ne 0,\ 0 \le i \le D\} = \{i \mid t \le i \le t + d\}.$$

The restrictions $A|_W, A^*|_W \in \mathrm{End}(W)$ of A, A^* on W form a TD-pair. In particular, $d = d^*$. Moreover, if we let $W_i = E_i W$, $W_i^* = E_i^* W$, by the general theory of TD-pairs which will be shown later (Corollary 6.37), we have $\dim W_{r+i} = \dim W_{t+i}^*$ $(0 \le i \le d)$. Furthermore, by Theorem 6.39, which will be shown later, the sequence of dimensions $\{\dim W_{r+i}^*\}_{i=0}^d$ are unimodal and symmetric. We call $d = d^*$ the *diameter* of an irreducible T-module W, and r and t the *endpoint* and the *dual endpoint* of W, respectively. If $\dim W_{r+i}^* = 1$ $(0 \le i \le d)$ holds, an irreducible T-module W is called *thin*. If every irreducible T-module of the standard T-module V is thin, T is called *thin*. In this case, we say a P- and Q-polynomial scheme $\mathfrak{X}$ is thin with respect to a base vertex x_0. If it is thin with respect to any base vertex x_0, we say $\mathfrak{X}$ is thin. If an irreducible T-module W is thin, a TD-pair $A|_W, A^*|_W$ becomes an L-pair. (The definition of an L-pair will be introduced in the next section.) Note that the primary T-module is thin (Definition 2.41 in Chapter 2, Section 2.6). For the primary T-module, we have $r = t = 0$, $d = D$. Conversely, if one of $r = 0$, $t = 0$, $d = D$ holds, such irreducible T-modules coincide with the primary T-module.

6.2.1 Weight space decompositions

Let $A, A^* \in \mathrm{End}(V)$ be a TD-pair and $(A, A^*; \{V_i\}_{i=0}^d, \{V_i^*\}_{i=0}^{d^*})$ an associated TD-system. For a pair i, j of integers with $0 \le i \le d^*$ and $0 \le j \le d$, we set

$$U_{i,j} = (V_0^* + \cdots + V_i^*) \cap (V_j + \cdots + V_d). \tag{6.109}$$

For a pair i, j of integers with $i \notin \{0, 1, \dots, d^*\}$ or $j \notin \{0, 1, \dots, d\}$, we set $U_{i,j} = 0$.

Lemma 6.31. *The following hold:*
(1) $(A - \theta_j)U_{i,j} \subseteq U_{i+1,j+1}$, *where* θ_j *is an eigenvalue of* A *on* V_j;
(2) $(A^* - \theta_i^*)U_{i,j} \subseteq U_{i-1,j-1}$, *where* θ_i^* *is an eigenvalue of* A^* *on* V_i^*.

Proof. Statement (1) follows directly from $(A - \theta_j)(V_0^* + \cdots + V_i^*) \subseteq V_0^* + \cdots + V_{i+1}^*$, $(A - \theta_j)(V_j + \cdots + V_d) \subseteq V_{j+1} + \cdots + V_d$. Statement (2) is similarly proved. □

Proposition 6.32. *We have* $U_{i,j} = 0$ $(0 \le i < j \le d)$.

Proof. Let $W = U_{0,j-i} + U_{1,j-i+1} + \cdots + U_{i,j} + \cdots + U_{d^*,j-i+d^*}$. By Lemma 6.31, W is $\langle A, A^* \rangle$-invariant. On the other hand, since $W \subseteq V_{j-i} + \cdots + V_d \neq V$ and V is an irreducible $\langle A, A^* \rangle$-module, we have $W = 0$. In particular, $U_{i,j} = 0$. □

Corollary 6.33. *We have* $d = d^*$.

Proof. Assume $d^* < d$. We have $U_{d^*,d} = (V_0^* + V_1^* + \cdots + V_{d^*}^*) \cap V_d = V \cap V_d = V_d$. Therefore, $U_{d^*,d} \neq 0$, which contradicts Proposition 6.32. Hence $d \le d^*$. Next, exchanging the roles of A and A^*, applying the above discussion to a TD-system $(A^*, A; \{V_i^*\}_{i=0}^{d^*}, \{V_i\}_{i=0}^{d})$ yields $d^* \le d$. □

We call d the *diameter* of a TD-pair A, A^*.

Definition 6.34. Let $U_i = U_{i,i} = (V_0^* + \cdots + V_i^*) \cap (V_i + \cdots + V_d)$. We call U_i the *weight space* of a TD-system $(A, A^*; \{V_i\}_{i=0}^{d}, \{V_i^*\}_{i=0}^{d^*})$. If $i \notin \{0, 1, \ldots, d\}$, let $U_i = 0$. Note that $U_0 = V_0^*$, $U_d = V_d$.

By applying Lemma 6.31 to the weight space U_i, we obtain the following lemma.

Lemma 6.35. *The following hold:*
(1) $(A - \theta_i)U_i \subseteq U_{i+1}$ $(0 \le i \le d)$;
(2) $(A^* - \theta_i^*)U_i \subseteq U_{i-1}$ $(0 \le i \le d)$.

Theorem 6.36. *The following hold:*
(1) $V = U_0 \oplus U_1 \oplus \cdots \oplus U_d$ *(a direct sum)*;
(2) $U_0 + U_1 + \cdots + U_i = V_0^* + V_1^* + \cdots + V_i^*$ $(0 \le i \le d)$, $U_i + U_{i+1} + \cdots + U_d = V_i + V_{i+1} + \cdots + V_d$ $(0 \le i \le d)$.

The direct sum decomposition $V = U_0 \oplus U_1 \oplus \cdots \oplus U_d$ *is called the weight space decomposition (or split decomposition).*

Proof. (1) If we let $V = U_0 + U_1 + \cdots + U_d$, by Lemma 6.35, W is $\langle A, A^* \rangle$-invariant. On the other hand, since $W \supseteq U_0 = V_0^* \neq 0$, we have $W \neq 0$. Since V is an irreducible $\langle A, A^* \rangle$-module, we have $W = V$. Moreover, since $U_0 + U_1 + \cdots + U_i \subseteq V_0^* + V_1^* + \cdots + V_i^*$ and $U_{i+1} \subseteq V_{i+1} + \cdots + V_d$, we have

$$(U_0 + U_1 + \cdots + U_i) \cap U_{i+1} \subseteq U_{i,i+1}.$$

By Proposition 6.32, we have $U_{i,i+1} = 0$. So $V = U_0 + U_1 + \cdots + U_d$ is a direct sum.

(2) We prove the former claim only. The latter is similarly proved. First, clearly we have $U_0 + U_1 + \cdots + U_i \subseteq V_0^* + V_1^* + \cdots + V_i^*$. On the other hand,

$$\begin{aligned} V_0^* + V_1^* + \cdots + V_i^* &= (A^* - \theta_{i+1}^*) \cdots (A^* - \theta_{d^*}^*)V \\ &= \sum_{j=0}^{d} (A^* - \theta_{i+1}^*) \cdots (A^* - \theta_{d^*}^*)U_j. \end{aligned}$$

By Lemma 6.35 (2), if $i+1 \le j$, we have $(A^* - \theta_{i+1}^*) \cdots (A^* - \theta_j^*)U_j \subseteq U_i$, and for arbitrary j, we have $A^*U_j \subseteq U_{j-1} + U_j$. Hence if $j \le i$, for any natural number k, we have $A^{*k}U_j \subseteq U_0 + U_1 + \cdots + U_i$. Therefore for any j we have $(A^* - \theta_{i+1}^*) \cdots (A^* - \theta_{d^*}^*)U_j \subseteq U_0 + \cdots + U_i$. Thus we obtain

$$V_0^* + V_1^* + \cdots + V_i^* \subseteq U_0 + U_1 + \cdots + U_i.$$

□

Let the map F_i be the projection from V onto the weight space U_i:

$$F_i : V = \bigoplus_{j=0}^{d} U_j \longrightarrow U_i. \tag{6.110}$$

Corollary 6.37.
(1) *The maps* $E_i|_{U_i} : U_i \longrightarrow V_i$ *and* $F_i|_{V_i} : V_i \longrightarrow U_i$ *are isomorphisms as vector spaces, and one is the inverse of the other.*
(2) *The maps* $E_i^*|_{U_i} : U_i \longrightarrow V_i^*$ *and* $F_i|_{V_i^*} : V_i^* \longrightarrow U_i$ *are isomorphisms as vector spaces, and one is the inverse of the other.*
(3) *We have* $\dim(V_i) = \dim(U_i) = \dim(V_i^*)$ $(0 \le i \le d)$ *and* $\dim(V_i) = \dim(V_{d-i})$ $(0 \le i \le d)$. *Hence* $\dim(V_i^*) = \dim(V_{d-i}^*)$, $\dim(U_i) = \dim(U_{d-i})$ $(0 \le i \le d)$.

Proof. (1) Let $W_j = V_j + \cdots + V_d$ $(0 \le j \le d)$. By Theorem 6.36, we have the direct sum decomposition $W_j = U_j + \cdots + U_d$. Therefore we obtain an isomorphism $V_i \cong W_i/W_{i+1} \cong U_i$ as vector spaces. By this isomorphism, if $v \in V_i$ maps to $u \in U_i$, we have $v + W_{i+1} = u + W_{i+1}$, and $E_i u = v$, $F_i v = u$.
(2) By Theorem 6.36, we have the direct sum $V_0^* + \cdots + V_j^* = U_0 + \cdots + U_j$. So we obtain the desired result similarly to (1).
(3) By (1), (2), we get $\dim(V_i) = \dim(U_i) = \dim(V_i^*)$. In particular, if we consider as the TD-system $(A, A^*; \{V_i\}_{i=0}^d, \{V_i^*\}_{i=0}^d)$, we get $\dim(V_i) = \dim(V_i^*)$. For the TD-system $(A, A^*; \{V_{d-i}\}_{i=0}^d, \{V_i^*\}_{i=0}^d)$, the same thing must hold. Namely, we have $\dim(V_{d-i}) = \dim(V_i^*)$, which proves (3). □

Definition 6.38. Let $R, F \in \mathrm{End}(V)$ be the maps defined as follows:

$$R = A - \sum_{i=0}^{d} \theta_i F_i, \quad L = A^* - \sum_{i=0}^{d} \theta_i^* F_i. \tag{6.111}$$

We call R the *raising map* and L the *lowering map*. By Lemma 6.35, we have

$$RU_i \subseteq U_{i+1}, \quad LU_i \subseteq U_{i-1} \quad (0 \le i \le d). \tag{6.112}$$

Theorem 6.39. *Let d be the diameter of a TD-pair. For a pair i, j of integers with $0 \le i \le j \le d$, R and L satisfy the following:*

(1) *The map $R^{j-i}|_{U_i} : U_i \longrightarrow U_j$ is injective if $i + j \le d$, surjective if $i + j \ge d$, and bijective if $i + j = d$.*
(2) *The map $L^{j-i}|_{U_j} : U_j \longrightarrow U_i$ is injective if $i + j \ge d$, surjective if $i + j \le d$, and bijective if $i + j = d$.*
(3) *We have*

$$\dim(U_0) \le \dim(U_1) \le \cdots \le \dim(U_{[\frac{d}{2}]}),$$
$$\dim(U_d) \le \dim(U_{d-1}) \le \cdots \le \dim(U_{d-[\frac{d}{2}]}).$$

In particular, $U_i \cong U_{d-i}$, $i = 0, 1, \ldots, [\frac{d}{2}]$.

Proof. We prove (1) only since (2) is similarly proved and (3) is clear from (1) and (2). First, we show that R^{j-i} is injective under the assumption that $i + j \le d$. Assume $u \in U_i$, $R^{j-i}u = 0$. Since $R^{j-i}|_{U_i} = (A - \theta_{j-1})(A - \theta_{j-2}) \cdots (A - \theta_i)|_{U_i}$, we have $u \in \operatorname{Ker}(A - \theta_{j-1})(A - \theta_{j-2}) \cdots (A - \theta_i) = V_{j-1} + V_{j-2} + \cdots + V_i \subseteq V_{j-1} + V_{j-2} + \cdots V_0$. On the other hand, by Theorem 6.36, we have $u \in U_i \subseteq V_0^* + V_1^* + \cdots + V_i^*$, which implies $u \in (V_0^* + V_1^* + \cdots + V_i^*) \cap (V_{j-1} + V_{j-2} + \cdots + V_0)$. By applying Proposition 6.32 to the TD-system $(A, A^*; \{V_{d-i}\}_{i=0}^d, \{V_i^*\}_{i=0}^d)$, since $i < d-j+1$, the weight space for indices $i, d-j+1$ is $(V_0^* + V_1^* + \cdots + V_i^*) \cap (V_{d-(d-j+1)} + V_{d-(d-j+2)} + \cdots + V_{d-d}) = 0$. Hence $u = 0$. This means $R^{j-i}|_{U_i}$ is injective.

Next, we consider the case $i + j = d$. By the above proof, $R^{j-i} : U_i \longrightarrow U_j$ is injective. Moreover, by Corollary 6.37 (3), we have $\dim(U_i) = \dim(U_{d-i}) = \dim(U_j)$, and so $R^{j-i}|_{U_i}$ is surjective.

Finally, we show R^{j-i} is surjective under the condition $i+j \ge d$. Note that $d-j \le i \le j$ and apply the above discussion to $d - j$ and j. Since $d - j + j = d$, $R^{2j-d} : U_{d-j} \longrightarrow U_j$ is a bijection. Therefore, $U_j = R^{2j-d}(U_{d-j}) = R^{j-i}(R^{i+j-d}U_{d-j})$. Since $R^{i+j-d}U_{d-j} \subseteq U_i$, we have $U_j \subseteq R^{j-i}(U_i)$ and $R^{j-i} : U_i \longrightarrow U_j$ is surjective. □

Lemma 6.40. *Let d denote the diameter of a TD-pair A, A^* and let $i, j, k, \ell, m \in \{0, 1, 2, \ldots, d\}$.*

(1) *If $m < |i - j|$, then $E_i^* A^m E_j^* = 0$.*
(2) *Let $\ell + m = |i - j|$. Then*

$$E_i^* A^\ell E_k^* A^m E_j^* = \begin{cases} E_i^* A^{\ell+m} E_j^*, & \text{if } i + \ell = k = j - m \text{ or } i - \ell = k = j + m, \\ 0, & \text{otherwise.} \end{cases}$$

(3) *We have*

$$E_i^* A^\ell A^* A^m E_j^* = \begin{cases} \theta_{i+\ell}^* E_i^* A^{\ell+m} E_j^*, & \text{if } \ell + m = j - i, \\ \theta_{i-\ell}^* E_i^* A^{\ell+m} E_j^*, & \text{if } \ell + m = i - j, \\ 0, & \text{if } \ell + m < |i - j|. \end{cases}$$

The above (1), (2), and (3) hold if we replace A by A^*, E_v^* by E_v, and θ_v^* by θ_v ($v \in \{0,1,\dots,d\}$) at the same time.

Proof. (1) This is clear by the definition of a TD-pair.
(2) By the assumption $\ell+m = |i-j|$ and (1), if $E_i^* A^\ell E_k^* A^m E_j^* \neq 0$, we have $i+\ell = k = j-m$ or $i-\ell = k = j+m$. If $i+\ell = k = j-m$ or $i-\ell = k = j+m$, for any $v \neq k$, we have $E_i^* A^\ell E_v^* A^m E_j^* = 0$. Namely, $\sum_{v=0}^d E_v^* = I$ is the identity map, and hence we get $E_i^* A^{\ell+m} E_j^* = E_i^* A^\ell (\sum_{v=0}^d E_v^*) A^m E_j^* = E_i^* A^\ell E_k^* A^m E_j^*$.
(3) For each case of (1), (2), substituting $A^* = \sum_{v=0}^d \theta_v^* E_v^*$ yields the desired result. The proof is similar for the cases in (1), (2), and (3), replacing A by A^*, E_v^* by E_v, and θ_v^* by θ_v ($v \in \{0,1,\dots,d\}$) at the same time. □

Proposition 6.41. *Let $i,j,k \in \{0,1,\dots,d\}$. If $k = |i-j|$, we have $E_i^* A^k E_j^* \neq 0$ and $E_i A^{*k} E_j \neq 0$.*

Proof. We prove $E_i^* A^k E_j^* \neq 0$ only. The proof of $E_i A^{*k} E_j \neq 0$ is similar.
If $i \geq j$, we have $k = i-j$ and by Lemma 6.40, we have

$$\begin{aligned} E_d^* A^{d-i} (E_i^* A^k E_j^*) A^j E_0^* &= (E_d^* A^{d-i} E_i^* A^{i-j} E_j^*) A^j E_0^* \\ &= E_d^* A^{d-j} E_j^* A^j E_0^* = E_d^* A^d E_0^*. \end{aligned} \tag{6.113}$$

If $i \leq j$, we have $k = j-i$, and similarly by Lemma 6.40, we have

$$\begin{aligned} E_0^* A^i (E_i^* A^k E_j^*) A^{d-j} E_d^* &= E_0^* A^i (E_i^* A^{j-i} E_j^* A^{d-j} E_d^*) \\ &= E_0^* A^d E_d^*. \end{aligned} \tag{6.114}$$

Therefore it suffices to show $E_d^* A^d E_0^* \neq 0$ and $E_0^* A^d E_d^* \neq 0$. If we show $E_d^* A^d E_0^* \neq 0$, by the similar discussion on a TD-system $(A, A^*; \{V_i\}_{i=0}^d, \{V_{d-i}^*\}_{i=0}^d)$, we can show $E_0^* A^d E_d^* \neq 0$. Hence it suffices to show $E_d^* A^d E_0^* \neq 0$. By Lemma 6.40 (1), for any m with $0 \leq m \leq d-1$, we have $E_d^* A^m E_0^* = 0$. Thus we have $E_d^* A^d E_0^* = E_d^* (A-\theta_{d-1})(A-\theta_{d-2}) \cdots (A-\theta_0) E_0^*$. Moreover, since $E_0^* V = V_0^* = U_0$, by the fact that $R^d|_{U_0} : U_0 \longrightarrow U_d$ is bijective in Theorem 6.39 (1), we obtain

$$\begin{aligned} &E_d^* (A-\theta_{d-1})(A-\theta_{d-2}) \cdots (A-\theta_0) E_0^* V \\ &= E_d^* (A-\theta_{d-1})(A-\theta_{d-2}) \cdots (A-\theta_0) U_0 = E_d^* U_d. \end{aligned} \tag{6.115}$$

Then by Corollary 6.37 (2), we have $E_d^* U_d = V_d^*$. Thus we obtain $E_d^* A^d E_0^* \neq 0$. □

Remark 6.42 (Character formula conjecture, Leonard pairs). Let $A, A^* \in \mathrm{End}(V)$ be a TD-pair and $V = U_0 \oplus U_1 \oplus \cdots \oplus U_d$ its weight space decomposition. The following inequality is proved in [257, 383, 384]:

$$\dim(U_i) \leq \binom{d}{i} \quad (0 \leq i \leq d). \tag{6.116}$$

In particular, $\dim(U_0) = \dim(U_d) = 1$. Moreover, define the character of a TD-pair A, A^* by

$$ch(t) = \sum_{i=0}^{d} (\dim U_i)t^i. \tag{6.117}$$

It is conjectured that there exist natural numbers $\ell_1, \ell_2, \ldots, \ell_n$ such that the following character formula holds [256]:

$$ch(t) = \prod_{i=1}^{n} \frac{1 - t^{\ell_i+1}}{1 - t}. \tag{6.118}$$

TD-pairs are classified into Types I, II, and III in the following subsection. The character formula for Type I is proved in [260]. It is announced that the character formulas for Type II and III are proved in [253].

A TD-pair A, A^* is called a Leonard pair (L-pair) if the weight space U_i is 1-dimensional for any i with $0 \le i \le d$. This condition $\dim(U_0) = \dim(U_1) = \cdots = \dim(U_d) = 1$ holds if and only if the character satisfies

$$ch(t) = \frac{1 - t^{d+1}}{1 - t}. \tag{6.119}$$

Note that TD-pairs arising from Terwilliger algebras of P- and Q-polynomial schemes become L-pairs (Remark 6.29).

Let A, A^* be an arbitrary TD-pair and $\ell_1, \ell_2, \ldots, \ell_n$ the natural numbers determined by the character formula of A, A^*. It is conjectured that for each i $(1 \le i \le n)$, there exists an L-pair A_i, A_i^* of diameter ℓ_i such that A, A^* are a kind of tensor product of A_i, A_i^*. The conjecture is true for TD-pairs of Type I [260], and it is announced that the cases for Type II and III are proved in [253].

6.2.2 TD-relations

Let $A, A^* \in \mathrm{End}(V)$ be a TD-pair and $(A, A^*; \{V_i\}_{i=0}^d, \{V_i^*\}_{i=0}^d)$ an associated TD-system.

Lemma 6.43. *For a subalgebra $\langle A \rangle$ of $\mathrm{End}(V)$ generated by A, define a subspace $\mathcal{L}$ of $\langle A, A^* \rangle$ as follows:*

$$\mathcal{L} = Span\{XA^*Y - YA^*X \mid X, Y \in \langle A \rangle\}.$$

Then each of the following sets (1), (2) forms a basis of $\mathcal{L}$:
(1) $\{E_iA^*E_{i+1} - E_{i+1}A^*E_i \mid 0 \le i \le d-1\}$;
(2) $\{A^iA^* - A^*A^i \mid 1 \le i \le d\}$.

The above claim holds if we replace A by A^* and E_i by E_i^* $(0 \le i \le d-1)$ at the same time.

Proof. Since the proof is similar to that of Lemma 2.104 in Chapter 2, we check the points only. The algebra $\langle A\rangle$ is spanned by $\{E_i \mid 0 \le i \le d\}$. By Lemma 6.40 (1), for i, j with $|i-j| > 1$, we have $E_iA^*E_j = 0$, and so it is clear that the set (1) spans $\mathcal{L}$. Therefore we have $\dim(\mathcal{L}) \le d$. If we show the set (2) is linearly independent, we get $\dim(\mathcal{L}) \ge d$ and hence $\dim(\mathcal{L}) = d$. Namely, each of the sets (1), (2) forms a basis of $\mathcal{L}$. We show the set (2) is linearly independent. We assume the set (2) is linearly dependent to get the contradiction. There exists an integer r $(1 \le r \le d)$ such that $\sum_{i=1}^{r} \lambda_i(A^iA^* - A^*A^i) = 0$. By Lemma 6.40, we obtain

$$\begin{aligned}E_r^*\left(\sum_{i=1}^{r} \lambda_i(A^iA^* - A^*A^i)\right)E_0^* &= \sum_{i=1}^{r} \lambda_i(E_r^*A^iA^*E_0^* - E_r^*A^*A^iE_0^*)\\ &= \lambda_r(\theta_0^* - \theta_r^*)E_r^*A^rE_0^*.\end{aligned}$$

Hence $E_r^*A^rE_0^* = 0$, which contradicts Proposition 6.41. □

Theorem 6.44. *There exist complex numbers $\beta, \gamma, \delta, \gamma^*, \delta^* \in \mathbb{C}$ such that the following relations (TD) hold:*

$$(TD)\begin{cases} A^3A^* - (\beta+1)(A^2A^*A - AA^*A^2) - A^*A^3\\ \quad = \gamma(A^2A^* - A^*A^2) + \delta(AA^* - A^*A),\\ A^{*3}A - (\beta+1)(A^{*2}AA^* - A^*AA^{*2}) - AA^{*3}\\ \quad = \gamma^*(A^{*2}A - AA^{*2}) + \delta^*(A^*A - AA^*).\end{cases}$$

The eigenvalues $\{\theta_i\}_{i=0}^d$ of A and the eigenvalues $\{\theta_i^*\}_{i=0}^d$ of A^* satisfy the following recurrence relations:

$$\begin{cases} \delta = \theta_{i+1}^2 - \beta\theta_{i+1}\theta_i + \theta_i^2 - \gamma(\theta_{i+1} + \theta_i) \quad (0 \le i \le d-1),\\ \delta^* = \theta^{*2}_{i+1} - \beta\theta_{i+1}^*\theta_i^* + \theta^{*2}_i - \gamma^*(\theta_{i+1}^* + \theta_i^*) \quad (0 \le i \le d-1),\end{cases} \tag{6.120}$$

$$\begin{cases} \gamma = \theta_{i+1} - \beta\theta_i + \theta_{i-1} \quad (1 \le i \le d-1),\\ \gamma^* = \theta_{i+1}^* - \beta\theta_i^* + \theta_{i-1}^* \quad (1 \le i \le d-1),\end{cases} \tag{6.121}$$

$$\beta = \frac{\theta_{i+1} - \theta_i + \theta_{i-1} - \theta_{i-2}}{\theta_i - \theta_{i-1}} = \frac{\theta_{i+1}^* - \theta_i^* + \theta_{i-1}^* - \theta_{i-2}^*}{\theta_i^* - \theta_{i-1}^*} \quad (2 \le i \le d-1). \tag{6.122}$$

If $d \ge 3$, $\beta, \gamma, \gamma^*, \delta, \delta^*$ are uniquely determined. If $d = 2$, we can choose arbitrary β, and then $\gamma, \gamma^*, \delta, \delta^*$ are uniquely determined. If $d = 1$, we can choose arbitrary β, γ, γ^*, and then δ, δ^* are uniquely determined. If $d = 0$, we can choose arbitrary $\beta, \gamma, \gamma^*, \delta, \delta^*$.

The relations (TD) in Theorem 6.44 are called the *TD-relations*. Moreover, $\{\theta_i\}_{i=0}^d$, $\{\theta_i^*\}_{i=0}^d$ satisfying the recurrence relations (6.120) are called a (β, γ, δ)-sequence and a $(\beta^*, \gamma^*, \delta^*)$-sequence, respectively.

Proof. We show the first equation of the relations (TD). The second one is similarly proved by replacing β by β^* at first, and showing $\beta = \beta^*$ later.

Since $A^2A^*A - AA^*A^2 \in \mathcal{L}$, it is expressed as a linear combination of the basis (2) in Lemma 6.43. Namely, we have the unique expression as follows:

$$A^2A^*A - AA^*A^2 = \sum_{i=1}^{r} \lambda_i(A^iA^* - A^*A^i) \quad (\lambda_r \neq 0). \tag{6.123}$$

Note that $A^2A^*A - AA^*A^2 \neq 0$. For, by Lemma 6.40 (3), Proposition 6.41, we have

$$E_3^*(A^2A^*A - AA^*A^2)E_0^* = (\theta_1^* - \theta_2^*)E_3^*A^3E_0^* \neq 0,$$

where $d \geq 3$. We deal with the case $d \leq 2$ separately. Similarly to the proof of Lemma 6.43, we have

$$E_r^*(A^2A^*A - AA^*A^2)E_0^* = \lambda_r(\theta_0^* - \theta_r^*)E_r^*A^rE_0^* \neq 0.$$

By Lemma 6.40 (3), $E_r^*(A^2A^*A - AA^*A^2)E_0^* \neq 0$ only if $r \leq 3$.

If we multiply both sides of (6.123) by E_3^* from the left and by E_0^* from the right, by Lemma 6.40 (3), the right-hand side must be zero if $r \leq 2$. However, as we noted before, the left-hand side becomes $E_3^*(A^2A^*A - AA^*A^2)E_0^* = (\theta_1^* - \theta_2^*)E_3^*A^3E_0^* \neq 0$, which is a contradiction. Therefore if $d \geq 3$, we have $r = 3$. Then the right-hand side of (6.123) is $E_3^*(\sum_{i=1}^{3} \lambda_i(A^iA^* - A^*A^i))E_0^* = \lambda_3(\theta_0^* - \theta_3^*)E_3^*A^3E_0^*$, and so we get $\lambda_3 = \frac{\theta_1^*-\theta_2^*}{\theta_0^*-\theta_3^*}$. Setting

$$\beta + 1 = \frac{1}{\lambda_3}, \quad \gamma = -\frac{\lambda_2}{\lambda_3}, \quad \delta = -\frac{\lambda_1}{\lambda_3} \tag{6.124}$$

yields the first equation of the relations (TD). Since $\lambda_3, \lambda_2, \lambda_1$ are uniquely determined, so are β, γ, δ.

The case $d = 2$: Given arbitrary β in the first equation in (TD). Since $A^2A^* - A^*A^2$ and $AA^* - A^*A$ form a basis of $\mathcal{L}$, γ and δ are uniquely determined.

The cases $d = 1$ and $d = 0$ are similar.

Next we show $\{\theta_i\}_{i=0}^{d}$ is a (β, γ, δ)-sequence. Let $0 \leq i \leq d-1$. Multiplying the first equation in (TD) by E_{i+1} from the left and by E_i from the right yields

$$\begin{aligned}(\theta_{i+1}^3 - (\beta+1)(\theta_{i+1}^2\theta_i - \theta_{i+1}\theta_i^2) - \theta_i^3)E_{i+1}A^*E_i \\ = (\gamma(\theta_{i+1}^2 - \theta_i^2) + \delta(\theta_{i+1} - \theta_i))E_{i+1}A^*E_i.\end{aligned} \tag{6.125}$$

Since $E_{i+1}A^*E_i \neq 0$ and $\theta_{i+1} - \theta_i \neq 0$, we have

$$\theta_{i+1}^2 - \beta\theta_{i+1}\theta_i + \theta_i^2 = \gamma(\theta_{i+1} + \theta_i) + \delta \quad (0 \leq i \leq d-1).$$

Taking the difference between the i-th and $(i-1)$-th of these d equations and deleting δ yields

$$\theta_{i+1} + \theta_{i-1} - \beta\theta_i = \gamma \quad (1 \leq i \leq d-1).$$

Again, taking the difference between the i-th and the $(i-1)$-th of the $d-1$ equations and deleting γ yields

$$\theta_{i+1} - \theta_i + \theta_{i-1} - \theta_{i-2} - \beta(\theta_i - \theta_{i-1}) = 0 \quad (2 \le i \le d-1).$$

We obtain the first equation of β in (6.122). Similarly, we can show that $\{\theta_i^*\}_{i=0}^d$ is a $(\beta^*, \gamma^*, \delta^*)$-sequence by the second equation of (TD).

Finally, we show $\beta = \beta^*$. If $d \le 2$, we can choose arbitrary β, β^* so we may set $\beta = \beta^*$. Next, let $d \ge 3$. By the second equation in (TD), we can show $\{\theta_i^*\}_{i=0}^d$ is a $(\beta^*, \gamma^*, \delta^*)$-sequence, which means $\beta^* = \frac{\theta_{i+1}^* - \theta_i^* + \theta_{i-1}^* - \theta_{i-2}^*}{\theta_i^* - \theta_{i-1}^*}$ holds for $i = 2, 3, \ldots, d-1$. In particular, if we let $i = 2$, we get $\beta^* + 1 = \frac{\theta_3^* - \theta_0^*}{\theta_2^* - \theta_1^*}$. On the other hand, since $\beta + 1 = \frac{\theta_0^* - \theta_3^*}{\theta_1^* - \theta_2^*}$ in (6.124), we have $\beta^* = \beta$. □

Next, in order to express the eigenvalues $\{\theta_i\}_{i=0}^d$, $\{\theta_i^*\}_{i=0}^d$ of a TD-pair A, A^* by the AW-parameters, we consider a general (β, γ, δ)-sequence.

A sequence $\{x_i\}_{i=0}^d$ $(d \ge 3)$ is called a β-sequence if for a constant $\beta \in \mathbb{C}$, it satisfies

$$\beta(x_i - x_{i-1}) = x_{i+1} - x_i + x_{i-1} - x_{i-2} \quad (2 \le i \le d-1).$$

If $\{x_i\}_{i=0}^d$ $(d \ge 3)$ is a β-sequence,

$$\gamma = x_{i+1} - \beta x_i + x_{i-1} \quad (1 \le i \le d-1)$$

is a constant which is independent of i. Such a sequence $\{x_i\}_{i=0}^d$ $(d \ge 2)$ is called a (β, γ)-sequence. Conversely, if $d \ge 3$, a (β, γ)-sequence is a β-sequence. Moreover, if $\{x_i\}_{i=0}^d$ is a (β, γ)-sequence,

$$\delta = x_{i+1}^2 - \beta x_{i+1} x_i + x_i^2 - \gamma(x_{i+1} + x_i) \quad (0 \le i \le d-1)$$

is a constant which is independent of i. Such a sequence $\{x_i\}_{i=0}^d$ $(d \ge 1)$ is called a (β, γ, δ)-sequence. Conversely, if $d \ge 2$, a (β, γ, δ)-sequence satisfying $x_{i+1} \ne x_{i-1}$ $(1 \le i \le d-1)$ is a (β, γ)-sequence.

Lemma 6.45. *Let $\{x_i\}_{i=0}^d$ be a (β, γ)-sequence. Then the following hold:*
(1) $x_i^2 - \gamma x_i - \delta = x_{i+1} x_{i-1}$ $(1 \le i \le d-1)$;
(2) $(\beta - 2)x_i^2 + 2\gamma x_i + \delta = (x_{i+1} - x_i)(x_i - x_{i-1})$ $(1 \le i \le d-1)$.

Proof. Substituting (1) and (2) by $\delta = x_{i+1}^2 - \beta x_{i+1} x_i + x_i^2 - \gamma(x_{i+1} + x_i)$ and $\beta x_i = x_{i+1} + x_{i-1} - \gamma$ yields the desired equations. □

By

$$\beta(x_i - x_{i-1}) = x_{i+1} - x_i + x_{i-1} - x_{i-2},$$

we expand a β-sequence $\{x_i\}_{i=0}^{d}$ to $\{x_i\}_{i\in\mathbb{Z}}$. Fix a complex number $\beta \in \mathbb{C}$. The set of β-sequences forms a 3-dimensional vector space with respect to standard sums and scalar multiples of sequences. Let

$$\beta = q + q^{-1}. \tag{6.126}$$

A β-sequence is called Type I if $q \neq \pm 1$, Type II if $q = 1$, and Type III if $q = -1$.

The vector space formed by the set of β-sequences of Type I has the following basis:

$$\{1\}_{i\in\mathbb{Z}}, \quad \{q^i\}_{i\in\mathbb{Z}}, \quad \{q^{-i}\}_{i\in\mathbb{Z}}.$$

For the case of Type II, it has the basis

$$\{1\}_{i\in\mathbb{Z}}, \quad \{i\}_{i\in\mathbb{Z}}, \quad \{i^2\}_{i\in\mathbb{Z}},$$

and for the case of Type III, it has the basis

$$\{1\}_{i\in\mathbb{Z}}, \quad \{(-1)^i\}_{i\in\mathbb{Z}}, \quad \{(-1)^i i\}_{i\in\mathbb{Z}}.$$

Therefore β-sequences $\{x_i\}_{i\in\mathbb{Z}}$ are expressed as follows:

Type I: $x_i = a + bq^i + cq^{-i}$;

Type II: $x_i = a + bi + ci^2$;

Type III: $x_i = a + (-1)^i(b + ci)$.

The constants γ, δ are expressed as follows:

Type I	Type II	Type III
$\gamma = (2-\beta)a$	$\gamma = 2c$	$\gamma = 4a$
$\delta = (2-\beta)((2+\beta)bc - a^2)$	$\delta = b^2 - c^2 - 4ac$	$\delta = c^2 - 4a^2$

For the case of Type I, by exchanging q and q^{-1} if necessary, we have the following expression:

$$x_i = x_0 + h\frac{1}{q^i}(1-q^i)(1-sq^{i+1}),$$

Taking limits $q \longrightarrow 1$, $(1-q)^2 h \longrightarrow h'$, $\frac{1-s}{1-q} \longrightarrow s'$ and replacing h', s' by h, s yields the following expression of Type II:

$$x_i = x_0 + hi(i+1+s).$$

Moreover, taking limits $h \longrightarrow 0$, $hs \longrightarrow h'$ and replacing h' by h yields the following expression of Type II:

$$x_i = x_0 + hi.$$

These two expressions of Type II exhaust all cases of Type II. This means that all Type II cases appear in the limiting cases of Type I. For the case of Type I, taking limits $q \longrightarrow -1$, $(1+q)h \longrightarrow h'$, $\frac{1-s}{1+q} \longrightarrow s'$ and replacing h', s' by h, s yields the following expression of Type III:

$$x_i = \begin{cases} x_0 + 2hi, & \text{if } i \text{ is even,} \\ x_0 - 2h(i+1+s), & \text{if } i \text{ is odd.} \end{cases}$$

This expression exhausts all cases of Type III. Namely, all cases of Type III appear in the limiting case of Type I. The above expressions are called the expressions by AW-parameters.

Remark 6.46.

(1) For the expression of the case of Type I by AW-parameters, taking limits $h \longrightarrow 0$, $hsq \longrightarrow h'$ and replacing h' by h yields the following expression of the case of Type I:

$$x_i = x_0 - h(1 - q^i).$$

This coincides with the expression obtained by replacing q by q^{-1} and setting $s = 0$ in the expression of the case of Type I by AW-parameters. However, if it is not allowed to replace q by q^{-1}, we have to use the following two expressions by AW-parameters:

$$x_i = x_0 + h\frac{1}{q^i}(1 - q^i)(1 - sq^{i+1}),$$
$$x_i = x_0 - h(1 - q^i).$$

These two expressions exhaust all cases of Type I.

(2) As was stated before, we extend a β-sequence $\{x_i\}_{i=0}^d$ $(d \geq 3)$ to β-sequence $\{x_i\}_{i\in\mathbb{Z}}$. If $d = 2$, we extend a sequence x_0, x_1, x_2 to a β-sequence $\{x_i\}_{i\in\mathbb{Z}}$, where β is arbitrary. If $d = 1$, we extend a sequence x_0, x_1 to a (β, γ)-sequence $\{x_i\}_{i\in\mathbb{Z}}$, where β, γ are arbitrary. If $d = 0$, we extend to a β-sequence $x_i = x_0$ $(i \in \mathbb{Z})$. A (β, γ)-sequence extended to $i \in \mathbb{Z}$ is a β-sequence, and conversely, a β-sequence is a (β, γ)-sequence. In this sense, for any $d \geq 0$, we mention a β-sequence $\{x_i\}_{i=0}^d$ or a (β, γ)-sequence $\{x_i\}_{i=0}^d$, and express $\{x_i\}_{i=0}^d$ by AW-parameters.

(3) When we express a β-sequence $\{x_i\}_{i=0}^d$ by AW-parameters, the condition for $x_i = x_j$ in each case is given as follows.

Type I and $h \neq 0$:

$$x_i = x_j \Longleftrightarrow q^{i-j} = 1 \text{ or } sq^{1+i+j} = 1.$$

The limiting case of Type I: $h \longrightarrow 0$, $hsq \longrightarrow h'$, and $h' \neq 0$:

$$x_i = x_j \Longleftrightarrow q^{i-j} = 1.$$

Type II and $h \neq 0$:

$$x_i = x_j \Longleftrightarrow i = j \quad \text{or} \quad s + 1 + i + j = 0.$$

The limiting case of Type II: $h \longrightarrow 0$, $hs \longrightarrow h'$, and $h' \neq 0$:

$$x_i = x_j \Longleftrightarrow i = j.$$

Type III and $h \neq 0$:

$$x_i = x_j \Longleftrightarrow i = j, \quad \text{or} \quad s + 1 + i + j = 0 \text{ and } 1 + i + j \text{ is even.}$$

Let $(A, A^*, \{V_i\}_{i=0}^d, \{V_i^*\}_{i=0}^d)$ be a TD-system. By Theorem 6.44, since the eigenvalues $\{\theta_i\}_{i=0}^d$ and $\{\theta_i^*\}_{i=0}^d$ of A, A^* are β-sequences, they can be expressed by AW-parameters. Let

$$\beta = q + q^{-1}.$$

Note that for the case of Type I, we need a restriction if we exchange q and q^{-1} since we express $\{\theta_i\}_{i=0}^d$ and $\{\theta_i^*\}_{i=0}^d$ by AW-parameters at the same time. The following list exhausts all of the possible cases of $\{\theta_i\}_{i=0}^d$ and $\{\theta_i^*\}_{i=0}^d$. The cases (I), (II), and (III) are most general expressions of Type I, Type II, and Type III, the case (IA) is the limiting case of (I), and the cases (IIA), (IIB), and (IIC) are the limiting case of (II). The symbols $\Lambda_{\mathrm{I}}, \Lambda_{\mathrm{IA}}, \Lambda_{\mathrm{II}}, \Lambda_{\mathrm{IIA}}, \Lambda_{\mathrm{IIB}}, \Lambda_{\mathrm{IIC}}$, and Λ_{III} denote the sets of AW-parameters. At each end, the condition for $\theta_i \neq \theta_j$ $(i \neq j)$, $\theta_i^* \neq \theta_j^*$ $(i \neq j)$ is given.

(I) We have

$$\begin{cases} \theta_i = \theta_0 + h\frac{1}{q^i}(1 - q^i)(1 - sq^{i+1}) & (0 \leq i \leq d), \\ \theta_i^* = \theta_0^* + h^*\frac{1}{q^i}(1 - q^i)(1 - s^*q^{i+1}) & (0 \leq i \leq d), \end{cases}$$

$$\Lambda_{\mathrm{I}} = (s, s^*; h, h^*, \theta_0, \theta_0^* \mid q, d),$$

$$h, h^* \neq 0; \quad q^i \neq 1\ (1 \leq i \leq d); \quad s, s^* \notin \{q^{-2}, q^{-3}, \ldots, q^{-2d}\}.$$

(IA) In (I), set $s^* = 0$, and take limits $h \longrightarrow 0$, $hsq \longrightarrow h'$. Then we have

$$\begin{cases} \theta_i = \theta_0 - h(1 - q^i) & (0 \leq i \leq d), \\ \theta_i^* = \theta_0^* - h^*(1 - q^{-i}) & (0 \leq i \leq d), \end{cases}$$

$$\Lambda_{\mathrm{IA}} = (-, -; h, h^*, \theta_0, \theta_0^* \mid q, d),$$

$$h, h^* \neq 0; \quad q^i \neq 1\ (1 \leq i \leq d).$$

(II) In (I), take limits $q \longrightarrow 1$, $(1-q)^2 h \longrightarrow h'$, $(1-q)^2 h^* \longrightarrow (h^*)'$, $\frac{1-s}{1-q} \longrightarrow s'$, $\frac{1-s^*}{1-q} \longrightarrow (s^*)'$. Then we have

$$\begin{cases} \theta_i = \theta_0 + hi(i + 1 + s) & (0 \leq i \leq d), \\ \theta_i^* = \theta_0^* + h^*i(i + 1 + s^*) & (0 \leq i \leq d), \end{cases}$$

$$\Lambda_{\mathrm{II}} = (s, s^*; h, h^*, \theta_0, \theta_0^* \mid d),$$
$$h, h^* \neq 0; \quad s, s^* \notin \{-2, -3, \ldots, -2d\}.$$

(IIA) In (II), take limits $h^* \longrightarrow 0$, $h^*s^* \longrightarrow (h^*)'$. Then we have

$$\begin{cases} \theta_i = \theta_0 + hi(i+1+s) & (0 \leq i \leq d), \\ \theta_i^* = \theta_0^* + h^*i & (0 \leq i \leq d), \end{cases}$$
$$\Lambda_{\mathrm{IIA}} = (s, -; h, h^*, \theta_0, \theta_0^* \mid d),$$
$$h, h^* \neq 0; \quad s \notin \{-2, -3, \ldots, -2d\}.$$

(IIB) In (II), take limits $h \longrightarrow 0$, $hs \longrightarrow h'$. Then we have

$$\begin{cases} \theta_i = \theta_0 + hi & (0 \leq i \leq d), \\ \theta_i^* = \theta_0^* + h^*i(i+1+s^*) & (0 \leq i \leq d), \end{cases}$$
$$\Lambda_{\mathrm{IIB}} = (-, s^*; h, h^*, \theta_0, \theta_0^* \mid d),$$
$$h, h^* \neq 0; \quad s^* \notin \{-2, -3, \ldots, -2d\}.$$

(IIC) In (II), take limits $h \longrightarrow 0$, $h^* \longrightarrow 0$, $hs \longrightarrow h'$, $h^*s^* \longrightarrow (h^*)'$. Then we have

$$\begin{cases} \theta_i = \theta_0 + hi & (0 \leq i \leq d), \\ \theta_i^* = \theta_0^* + h^*i & (0 \leq i \leq d), \end{cases}$$
$$\Lambda_{\mathrm{IIC}} = (-, -; h, h^*, \theta_0, \theta_0^* \mid d),$$
$$h, h^* \neq 0.$$

(III) In (I), take limits $q \longrightarrow -1$, $(1+q)h \longrightarrow h'$, $(1+q)h^* \longrightarrow (h^*)'$, $\frac{1-s}{1+q} \longrightarrow s'$, $\frac{1-s^*}{1+q} \longrightarrow (s^*)'$. Then we have

$$\theta_i = \begin{cases} \theta_0 + 2hi, & \text{if } i \text{ is even} \quad (0 \leq i \leq d), \\ \theta_0 - 2h(i+1+s), & \text{if } i \text{ is odd} \quad (0 \leq i \leq d), \end{cases}$$
$$\theta_i^* = \begin{cases} \theta_0^* + 2h^*i, & \text{if } i \text{ is even} \quad (0 \leq i \leq d), \\ \theta_0^* - 2h^*(i+1+s^*), & \text{if } i \text{ is odd} \quad (0 \leq i \leq d), \end{cases}$$
$$\Lambda_{\mathrm{III}} = (s, s^*;\ h, h^*, \theta_0, \theta_0^* \mid d),$$
$$h, h^* \neq 0; \quad s, s^* \notin \{-2, -4, \ldots, -2(d-1), -2d\}.$$

By Theorem 6.44, a TD-pair $A, A^* \in \mathrm{End}(V)$ satisfies the TD-relations. The following theorem shows the converse is true [474].

Theorem 6.47. *Let $A, A^* \in \mathrm{End}(V)$ be diagonalizable linear transformations. Assume V is irreducible as an $\langle A, A^* \rangle$-module. If there exist complex numbers $\beta, \gamma, \delta \in \mathbb{C}$ such* *that the first equation of (TD)*

$$\begin{aligned} & A^3A^* - (\beta+1)(A^2A^*A - AA^*A^2) - A^*A^3 \\ & = \gamma(A^2A^* - A^*A^2) + \delta(AA^* - A^*A), \end{aligned}$$

holds, then there exists an ordering $V_0, V_1, \ldots, V_d$ of the eigenspaces V_i of A such that

$$A^*V_i \subseteq V_{i-1} + V_i + V_{i+1} \quad (0 \le i \le d).$$

Here, we let $V_{-1} = V_d$, $V_{d+1} = V_0$ if $\beta = q + q^{-1}$, $q \neq \pm 1$, $q^{d+1} = 1$, and let $V_{-1} = 0$, $V_{d+1} = 0$ otherwise.

Moreover, if we exchange the roles of A and A^*, the same holds.

Proof. If $d = 0, 1$, it is clear. Suppose $d \ge 2$. For $\theta \in \mathbb{C}$, let $V(\theta) = \{x \in V \mid Av = \theta v\}$. If $V(\theta) \neq \{0\}$, θ is an eigenvalue of A. Let $v \in V(\theta)$ and let $A^3A^* - (\beta+1)(A^2A^*A - AA^*A^2) - A^*A^3 - \gamma(A^2A^* - A^*A^2) - \delta(AA^* - A^*A)$ act on v. Then we get

$$(A-\theta)(A^2 - \beta\theta A + \theta^2 - \gamma(A+\theta) - \delta)A^*v = 0.$$

Let θ', θ'' be the roots of the quadratic equation $x^2 - \beta\theta x + \theta^2 - \gamma(x+\theta) - \delta = 0$. Then

$$(A-\theta)(A-\theta')(A-\theta'')A^*v = 0$$

holds for any $v \in V(\theta)$. Therefore, we get

$$A^*V(\theta) \subseteq V(\theta) + V(\theta') + V(\theta''). \tag{6.127}$$

Next, define a relation of an undirected graph on the set $\{\theta_i\}_{i=0}^d$ of distinct eigenvalues of A as follows:

$$\theta_i \sim \theta_j \iff \theta_i^2 - \beta\theta_i\theta_j + \theta_j^2 - \gamma(\theta_i + \theta_j) - \delta = 0, \quad i \neq j. \tag{6.128}$$

By (6.127), we obtain

$$A^*V(\theta_i) \subseteq V(\theta_i) + \sum_{\theta_j \sim \theta_i} V(\theta_j).$$

Since V is an irreducible $\langle A, A^*\rangle$-module by the assumption, the graph must be connected. By (6.128), the number of edges coming from each vertex is at most 2. Hence, the graph is a path or a cycle. Namely, there exists an ordering $\theta_0, \theta_1, \ldots, \theta_d$ of eigenvalues such that

$$\begin{aligned} & \theta_i \sim \theta_{i+1} \quad (0 \le i \le d-1), \\ & \theta_0 \sim \theta_d \quad \text{if the graph is a cycle.} \end{aligned}$$

Namely, the eigenvalues $\{\theta_i\}_{i=0}^d$ of A form a (β,γ,δ)-sequence. Since $d \geq 2$, we extend the (β,γ)-sequence $\{\theta_i\}_{i=0}^d$ to the (β,γ)-sequence $\{\theta_i\}_{i\in\mathbb{Z}}$. If we extend the graph to $\{\theta_i\}_{i\in\mathbb{Z}}$ by (6.128), since a (β,γ)-sequence is a (β,γ,δ)-sequence, we have

$$\theta_i \sim \theta_{i+1} \quad (i \in \mathbb{Z}).$$

Therefore, if $\theta_0 \sim \theta_d$, we have $\theta_{-1} = \theta_d$, $\theta_{d+1} = \theta_0$. Hence by Remark 6.46 (3), $\theta_{-1} = \theta_d$, $\theta_{d+1} = \theta_0$ and $\theta_0, \theta_1, \ldots, \theta_d$ are distinct only if $q \neq \pm 1$, $q^{d+1} = 1$. □

Remark 6.48. Let $\mathfrak{X} = (X, \{R_i\}_{0\leq i\leq d})$ be a P- and Q-polynomial scheme and $\Gamma = (X, R_1)$ the associated distance-regular graph. Let $T = T(x_0)$ be the Terwilliger algebra of $\mathfrak{X}$, and let A, A^* be standard generators of T (A is the adjacency matrix of Γ). Let $V = \bigoplus_{x\in X} \mathbb{C}x$ be the standard module, and let W $(\subset V)$ be the principal T-module (Definition 2.41). In Remark 6.42, we stated that $A|_W, A^*|_W \in \mathrm{End}(V)$ form an L-pair, which consists of a special class of TD-pairs. Let $\{\theta_i \mid 0 \leq i \leq d\}$ be the set of distinct eigenvalues of $A|_W$. Then $\{\theta_i \mid 0 \leq i \leq d\}$ is the set of eigenvalues of A (Lemma 2.40).

(1) If $d \geq 5$ and Γ is not the n-gon, then $\theta_0, \theta_1, \ldots, \theta_d$ are all rational numbers [60, 170]. In particular, in the expression of θ_i by AW-parameters, $\beta = q + q^{-1}$ is a rational number.
(2) If the pair $A|_W, A^*|_W \in \mathrm{End}(W)$ is of Type III, one of the following holds [463] (Section 6.4):
   (i) Γ is the Hamming graph $H(d, 2)$;
   (ii) Γ is the odd graph O_{d+1};
   (iii) Γ is the antipodal quotient of $H(2d + 1, 2)$.
   The classification of Γ is completed for the cases of Type II, Type IIA, Type IIB, and Type IIC [462], [460], [465], [466], [176].
(3) For the core part of known P- and Q-polynomial schemes, in the expression of θ_i^* by AW-parameters, we have $s^* = 0$ (Section 6.4). AW-parameters are called classical if $s^* = 0$. It is conjectured that except for polygons, P- and Q-polynomial schemes are expressed by classical AW-parameters or are "relatives" of such schemes (Section 6.4).

6.3 Leonard pairs (L-pairs)

We already mentioned the definition of an L-pair in Remark 6.42. Here we give the definition again.

Definition 6.49. Let $A, A^* \in \mathrm{End}(V)$ be a TD-pair, and let $(A, A^*; \{V_i\}_{i=0}^d, \{V_i^*\}_{i=0}^d)$ be an associated TD-system. A pair A, A^* is called a *Leonard pair* or *L-pair* and $(A, A^*; \{V_i\}_{i=0}^d, \{V_i^*\}_{i=0}^d)$ is called a *Leonard system* or *L-system* if $\dim(V_i) = \dim(V_i^*) = 1$ for all i $(0 \leq i \leq d)$. If we emphasize the space which A, A^* act on, we say an L-pair on V, or an L-system on V. Note that the definition of an L-pair does not depend on the choice of the associated TD-systems. If $d \geq 1$, for a given L-pair, there are four associated L-systems.

In this section, we classify L-pairs (precisely, L-systems). In Section 6.3.1, we introduce the standard basis and the dual system of orthogonal polynomials. There is a one-to-one correspondence between L-systems and dual systems of orthogonal polynomials. In Sections 6.3.2–6.3.5, we classify L-systems. In Section 6.3.6, it will be shown that dual systems of orthogonal polynomials are identical to the dual systems of AW-polynomials. Historically, the concept of dual systems of orthogonal polynomials arose from the algebraic properties of character tables of P- and Q-polynomial schemes (the base field is the real field $\mathbb{R}$), and the classification was made [310, 60]. The concept of L-systems was introduced later to understand dual systems of orthogonal polynomials in the framework of representation theory of Terwilliger algebras [475]. Readers are referred to [479] as an expository article.

In this section, L-pairs will be classified by the following procedure. First, in Section 6.3.2, we introduce pre-L-pairs. It is very hard to construct L-pairs since the constraints are strong. The advantage of pre-L-pairs is that they are freely constructed since there are no constraints. In Section 6.3.3, we derive a necessary condition for a pre-L-pair to become an L-pair. The key is the lemma by Terwilliger [475, Corollary 11.4]. In Section 6.3.4, we show an L-pair satisfies AW-relations. AW-relations are stronger than TD-relations. In Section 6.3.5, we show the necessary condition which is obtained in Section 6.3.3 is also a sufficient condition to be an L-pair in terms of AW-relations. In this way, the classification of L-pairs (precisely, L-pairs associated with L-systems) is completed.

Finally, in Section 6.3.6, by using the classification of L-pairs, we describe dual systems of orthogonal polynomials in terms of a q-analogue of hypergeometric series ${}_4\phi_3$ (and its limit). By this, it turns out that dual systems of orthogonal polynomials coincide with dual systems of AW-polynomials.

We begin with preliminary facts on L-pairs. From the following theorem, it turns out that an L-pair A, A^* on V generates $\mathrm{End}(V)$. (By Burnside's theorem, any TD-pair $A, A^* \in \mathrm{End}(V)$ generates $\mathrm{End}(V)$.) Note that the following theorem on L-pairs is much stronger than Burnside's theorem.

Theorem 6.50. *Let $A, A^* \in \mathrm{End}(V)$ form an L-pair, and let $(A, A^*; \{V_i\}_{i=0}^d, \{V_i^*\}_{i=0}^d)$ be an associated L-system. Let $E_i : V = \bigoplus_{j=0}^d V_j \longrightarrow V_i$, $E_i^* : V = \bigoplus_{j=0}^d V_j^* \longrightarrow V_i^*$ be the projections onto the eigenspaces. Then each of the following (1)–(4) forms a basis of $\mathrm{End}(V)$ as a vector space:*

(1) $A^i E_0^* A^j$ $(0 \le i \le d, 0 \le j \le d)$;

(2) $(A^*)^i E_0 (A^*)^j$ $(0 \le i \le d, 0 \le j \le d)$;

(3) $E_i E_0^* E_j$ $(0 \le i \le d, 0 \le j \le d)$;

(4) $E_i^* E_0 E_j^*$ $(0 \le i \le d, 0 \le j \le d)$.

In particular, A, A^* generate $\mathrm{End}(V)$ as an algebra: $\mathrm{End}(V) = \langle A, A^* \rangle$.

Proof. We prove (1) only. We can prove (2) similarly to (1). Immediately, (3) follows from (1), and (4) follows from (2).

Since $\dim(\mathrm{End}(V)) = (d+1)^2$, it suffices to show $A^i E_0^* A^j$ $(0 \le i, j \le d)$ are linearly independent. Note that by the definition of L-pairs, we have

$$E_k^* A^i E_0^* A^j E_\ell^* \begin{cases} = 0, & \text{if } k > i \text{ or } \ell > j, \\ \neq 0, & \text{if } k = i \text{ and } \ell = j. \end{cases}$$

(We can show the above property for general TD-pairs similarly to the proof of Lemma 6.40 and Proposition 6.41.) Assume $\sum_{i,j=0}^{d} c_{i,j} A^i E_0^* A^j = 0$. Fix k, ℓ and multiply both sides of the equation by E_k^* from the left and by E_ℓ^* from the right. By the property noted above, we get

$$c_{k,\ell} E_k^* A^k E_0^* A^\ell E_\ell^* + \sum_{\substack{k \le i \le d, \ell \le j \le d \\ (i,j) \neq (k,\ell)}} c_{i,j} E_k^* A^i E_0^* A^j E_\ell^* = 0. \tag{6.129}$$

Hence, if $c_{i,j} = 0$ for all (i,j) satisfying $(i,j) \neq (k,\ell)$, $k \le i \le d, \ell \le j \le d$, we have $c_{k,\ell} = 0$. If we set $(k,\ell) = (d,d)$, since there is no (i,j) satisfying the condition, we have $c_{d,d} = 0$. Therefore, we obtain $c_{d,d-1} = 0, c_{d,d-2} = 0, \ldots, c_{d,0} = 0$ sequentially. Similarly, we obtain $c_{d-1,d} = 0, c_{d-2,d} = 0, \ldots, c_{0,d} = 0$. Likewise, by the double induction, we obtain $c_{k,\ell} = 0$ for all k, ℓ. □

Let $A, A^* \in \mathrm{End}(V)$ be an L-pair, and let $(A, A^*; \{V_i\}_{i=0}^d, \{V_i^*\}_{i=0}^d)$ be an associated L-system. Choose vectors v_i^*, v_i $(0 \le i \le d)$ such that $v_i^* \in V_i$, $v_i \in V_i^*$, $v_i^* \neq 0$, $v_i \neq 0$. The vectors $v_0, v_1, \ldots, v_d$ form a basis of V and so do $v_0^*, v_1^*, \ldots, v_d^*$. With respect to the basis $v_0, v_1, \ldots, v_d$, the representation matrix of A is an irreducible tridiagonal matrix

$$B = \begin{bmatrix} a_0 & b_0 & & & & \\ c_1 & a_1 & b_1 & & 0 & \\ & \ddots & \ddots & \ddots & & \\ & & c_{d-1} & a_{d-1} & b_{d-1} \\ 0 & & & c_d & a_d \end{bmatrix}, \quad b_{i-1}c_i \neq 0 \ (1 \le i \le d), \tag{6.130}$$

and the representation matrix of A^* is a diagonal matrix

$$D^* = \begin{bmatrix} \theta_0^* & & & 0 \\ & \theta_1^* & & \\ & & \ddots & \\ 0 & & & \theta_d^* \end{bmatrix}. \tag{6.131}$$

Namely, we have

$$Av_i = b_{i-1}v_{i-1} + a_i v_i + c_{i+1} v_{i+1} \quad (0 \le i \le d), \tag{6.132}$$

$$A^* v_i = \theta_i^* v_i \quad (0 \le i \le d), \tag{6.133}$$

where $v_{-1} = 0$, $v_{d+1} = 0$, $c_{d+1} = 1$, and b_{-1} is an indeterminate. Similarly, with respect to the basis $v_0^*, v_1^*, \ldots, v_d^*$, the representation matrix of A^* is an irreducible tridiagonal matrix,

$$B^* = \begin{bmatrix} a_0^* & b_0^* & & & \\ c_1^* & a_1^* & b_1^* & & 0 \\ & \ddots & \ddots & \ddots & \\ & & c_{d-1}^* & a_{d-1}^* & b_{d-1}^* \\ 0 & & & c_d^* & a_d^* \end{bmatrix}, \quad b_{i-1}^* c_i^* \neq 0 \ (1 \le i \le d), \tag{6.134}$$

and the representation matrix of A is a diagonal matrix

$$D = \begin{bmatrix} \theta_0 & & & 0 \\ & \theta_1 & & \\ & & \ddots & \\ 0 & & & \theta_d \end{bmatrix}. \tag{6.135}$$

Namely, we have

$$A^* v_i^* = b_{i-1}^* v_{i-1}^* + a_i^* v_i^* + c_{i+1}^* v_{i+1}^* \quad (0 \le i \le d), \tag{6.136}$$

$$A v_i^* = \theta_i v_i^* \quad (0 \le i \le d), \tag{6.137}$$

where $v_{-1}^* = 0$, $v_{d+1}^* = 0$, $c_{d+1}^* = 1$, and b_{-1}^* is an indeterminate. If we let P be the transition matrix, i. e.,

$$(v_0, v_1, \dots, v_d) = (v_0^*, v_1^*, \dots, v_d^*)P, \tag{6.138}$$

then we have

$$PBP^{-1} = D, \quad PD^*P^{-1} = B^*. \tag{6.139}$$

The pair B, D^* is an L-pair on $\mathbb{C}^{d+1}$, and the pair B^*, D is an L-pair which is isomorphic to B, D^*.

Conversely, let B, B^* be irreducible tridiagonal matrices given by (6.130), (6.134), and let D, D^* be diagonal matrices given by (6.135), (6.131). Suppose there exists a nonsingular matrix P such that (6.139) holds. Due to the following proposition, the pair B, D^* becomes an L-pair on $\mathbb{C}^{d+1}$, and the pair B^*, D becomes an L-pair which is isomorphic to B, D^*. Therefore, every abstract L-pair is isomorphic to an L-pair B, D^*. In fact, the original definition of L-pairs by Terwilliger consists of such irreducible tridiagonal matrices B, B^* and diagonal matrices D, D^* [475].

Proposition 6.51. *Let B, B^*, D, D^*, and P be the matrices satisfying the above conditions. Then the following (1)–(3) hold:*

(1) *the eigenvalues $\{\theta_0, \theta_1, \dots, \theta_d\}$ of B are distinct;*
(2) *the eigenvalues $\{\theta_0^*, \theta_1^*, \dots, \theta_d^*\}$ of B^* are distinct;*
(3) *$\mathbb{C}^{d+1}$ is irreducible as a $\langle B, D^* \rangle$-module. Namely, if a subspace W of $\mathbb{C}^{d+1}$ is B-invariant and D^*-invariant, then $W = \{0\}$ or $W = \mathbb{C}^{d+1}$.*

Proof. Statements (1) and (2) follow directly from Corollary 6.11. However, to make the point clear, we discuss it again. Choose column vectors $\boldsymbol{e}_0 = {}^t(1,0,\dots,0)$, $\boldsymbol{e}_1 = {}^t(0,1,\dots,0)$, $\cdots$, $\boldsymbol{e}_{d+1} = {}^t(0,0,\dots,0,1)$ as a basis of $\mathbb{C}^{d+1}$. Note that

$$B\boldsymbol{e}_i = b_{i-1}\boldsymbol{e}_{i-1} + a_i\boldsymbol{e}_i + c_{i+1}\boldsymbol{e}_{i+1} \quad (0 \le i \le d), \tag{6.140}$$

where $\boldsymbol{e}_{-1} = \boldsymbol{e}_{d+1} = 0$.

Since the proof of (2) is similar to that of (1), we prove (1) only. By (6.140), for $0 \le i \le d$, the vector $B^i\boldsymbol{e}_0 - c_1c_2\cdots c_i\boldsymbol{e}_i$ is a linear combination of $\boldsymbol{e}_0, \boldsymbol{e}_1, \dots, \boldsymbol{e}_{i-1}$. Therefore, $\boldsymbol{e}_0, B\boldsymbol{e}_0, B^2\boldsymbol{e}_0, \dots, B^d\boldsymbol{e}_0$ are linearly independent. In the full matrix algebra $M_{d+1}(\mathbb{C})$, $I = B^0, B, B^2, \dots, B^d$ are linearly independent, where I is the identity matrix. Hence, the minimal polynomial of B has degree $d+1$. On the other hand, since B is diagonalizable, the eigenvalues $\theta_0, \theta_1, \dots, \theta_d$ of B are distinct.

(3) For each i $(0 \le i \le d)$, define the diagonal matrix E_i^* in $M_{d+1}(\mathbb{C})$ as follows:

$$E_i^*(\ell, j) = \begin{cases} 1, & \text{if } (\ell, j) = (i, i), \\ 0, & \text{otherwise.} \end{cases}$$

By (2), the diagonal entries $\theta_0^*, \theta_1^*, \dots, \theta_d^*$ of D^* are distinct. Therefore, $E_i^* \in \langle D^* \rangle$ $(0 \le i \le d)$. Namely, if $W \ne \{0\}$ is D^*-invariant, W is E_i^*-invariant for $i = 0, 1, \dots, d$. Hence, for non-zero $\boldsymbol{u} = \sum_{\ell=0}^{d} \alpha_\ell \boldsymbol{e}_\ell \in W$ with $\alpha_i \ne 0$, we have $E_i^*\boldsymbol{u} = \alpha_i\boldsymbol{e}_i \in W$. So there exists $\boldsymbol{e}_i$ such that $\boldsymbol{e}_i \in W$. On the other hand, since (6.140) holds for B, we can sequentially show that $\boldsymbol{e}_0, \boldsymbol{e}_1, \dots, \boldsymbol{e}_d$ are contained in W, and obtain $W = \mathbb{C}^{d+1}$. □

6.3.1 Standard bases, dual systems of orthogonal polynomials

Let $A, A^* \in \mathrm{End}(V)$ be an L-pair, and let $(A, A^*; \{V_i\}_{i=0}^d, \{V_i^*\}_{i=0}^d)$ be an associated L-system. As we have seen before, if we choose $v_i \in V_i^*, v_i \ne 0$, then $v_0, v_1, \dots, v_d$ form a basis of V, and with respect to this basis the representation matrix of A is an irreducible tridiagonal matrix B given by (6.130) and that of A^* is a diagonal matrix D^* given by (6.131). Moreover, if we choose $v_i^* \in V_i, v_i^* \ne 0$, then $v_0^*, v_1^*, \dots, v_d^*$ form a basis of V, and with respect to this basis, the representation matrix of A^* is an irreducible tridiagonal matrix B^* given by (6.134) and that of A is a diagonal matrix D given by (6.135).

Definition 6.52.

(1) If B has the constant row sum θ_0, i. e.,

$$a_0 + b_0 = c_i + a_i + b_i = c_d + a_d = \theta_0 \quad (1 \le i \le d-1), \tag{6.141}$$

we call $v_0, v_1, \dots, v_d$ a *standard basis*. Here, θ_0 is the eigenvalue of A on V_0.

(2) If B^* has the constant row sum θ_0^*, i. e.,

$$a_0^* + b_0^* = c_i^* + a_i^* + b_i^* = c_d^* + a_d^* = \theta_0^* \quad (1 \le i \le d-1), \tag{6.142}$$

we call $v_0^*, v_1^*, \dots, v_d^*$ a *dual standard basis*. Here, θ_0^* is the eigenvalue of A^* on V_0^*.

Proposition 6.53. *Let $E_i : V = \bigoplus_{j=0}^{d} V_j \longrightarrow V_i$, $E_i^* : V = \bigoplus_{j=0}^{d} V_j^* \longrightarrow V_i^*$ be the projections onto the eigenspaces V_i, V_i^*, respectively.*

(1) *If we let $v_i = E_i^* v \in V_i^*$ $(0 \le i \le d)$ for $v \in V_0$, $v \neq 0$, then $v_0, v_1, \ldots, v_d$ form a standard basis. Moreover, for any standard basis $v_0', v_1', \ldots, v_d'$, there exists a non-zero constant $\lambda \in \mathbb{C}$ such that $v_i' = \lambda v_i$ $(0 \le i \le d)$. In this sense, a standard basis uniquely exists.*

(2) *If we let $v_i^* = E_i v^* \in V_i$ $(0 \le i \le d)$ for $v^* \in V_0^*$, $v^* \neq 0$, then $v_0^*, v_1^*, \ldots, v_d^*$ form a dual standard basis. Moreover, for any dual standard basis $v_0^{*\prime}, v_1^{*\prime}, \ldots, v_d^{*\prime}$, there exists a non-zero constant $\lambda^* \in \mathbb{C}$ such that $v_i^{*\prime} = \lambda^* v_i^*$ $(0 \le i \le d)$. In this sense, a dual standard basis uniquely exists.*

Proof. Since the proof of (2) is similar to that of (1), we prove (1) only. First, we show $v_i \neq 0$. Let $I \in \operatorname{End}(V)$ be the identity map. Since $\sum_{j=0}^{d} E_j^* = I$, we have $E_i^* E_0 = \sum_{j=0}^{d} E_i^* E_0 E_j^*$. By Theorem 6.50, $E_i^* E_0 E_j^*$ $(0 \le j \le d)$ are linearly independent, and so we have $E_i^* E_0 \neq 0$. Hence $0 \neq E_i^* E_0 V = E_i^* V_0 \subseteq V_i^*$ and $\dim(V_i^*) = 1$. So we have $V_i^* = E_i^* V_0$. Substituting $V_0 = \mathbb{C} v$, we have $V_i^* = \mathbb{C} E_i^* v$, which implies $v_i = E_i^* v \neq 0$. Next we show the representation matrix B of A with respect to the basis $v_0, v_1, \ldots, v_d$ has the constant row sum θ_0. Since $\sum_{j=0}^{d} E_j^* = I$, we have $v = (\sum_{j=0}^{d} E_j^*) v = \sum_{j=0}^{d} E_j^* v$. Hence, we have

$$v = \sum_{j=0}^{d} v_j. \tag{6.143}$$

Since $v \in V_0$, we have $Av = \theta_0 v = \sum_{i=0}^{d} \theta_0 v_i$. On the other hand, if we let $B = (b_{i,j})$, we obtain

$$Av = \sum_{j=0}^{d} Av_j = \sum_{j=0}^{d} \sum_{i=0}^{d} b_{i,j} v_i = \sum_{i=0}^{d} \left(\sum_{j=0}^{d} b_{i,j} \right) v_i \quad (0 \le i \le d).$$

Therefore, $\sum_{j=0}^{d} b_{i,j} = \theta_0$ for $i = 0, 1, \ldots, d$. Namely, B has the constant row sum θ_0. Finally, we show that for any standard basis $v_0', v_1', \ldots, v_d'$, there exists a constant $\lambda \neq 0$ such that $v_i' = \lambda v_i$, $(0 \le i \le d)$. Clearly, for each i, there exists $\lambda_i \in \mathbb{C}$, $\lambda_i \neq 0$ such that $v_i' = \lambda_i v_i$. We show $\lambda_0 = \lambda_i$ $(0 \le i \le d)$. Let B' be the representation matrix of A with respect to the basis $v_0', v_1', \ldots, v_d'$. Then we have

$$B' = \begin{bmatrix} a_0 & \frac{\lambda_1}{\lambda_0} b_0 & & & \\ \frac{\lambda_0}{\lambda_1} c_1 & a_1 & \frac{\lambda_2}{\lambda_1} b_1 & & 0 \\ & \ddots & \ddots & \ddots & \\ & & \frac{\lambda_{d-2}}{\lambda_{d-1}} c_{d-1} & a_{d-1} & \frac{\lambda_d}{\lambda_{d-1}} b_{d-1} \\ & 0 & & \frac{\lambda_{d-1}}{\lambda_d} c_d & a_d \end{bmatrix}.$$

Since $v_0', v_1', \ldots, v_d'$ form a standard basis, B' has constant row sum θ_0. Therefore, by (6.141), we have $1 = \frac{\lambda_1}{\lambda_0} = \frac{\lambda_2}{\lambda_1} = \cdots = \frac{\lambda_d}{\lambda_{d-1}}$. □

In what follows, for an L-pair $A, A^* \in \mathrm{End}(V)$, we fix an associated L-system $(A, A^*; \{V_i\}_{i=0}^d, \{V_i^*\}_{i=0}^d)$, and let $v_0, v_1, \ldots, v_d$ be a standard basis and $v_0^*, v_1^*, \ldots, v_d^*$ a dual standard basis. With respect to the standard basis $v_0, v_1, \ldots, v_d$, let B, D^* be the representation matrices of A, A^*, respectively. Then B is an irreducible tridiagonal matrix given by (6.130) and satisfies the condition (6.141) on the constant row sum. On the other hand, D^* is a diagonal matrix given by (6.131). Moreover, with respect to the dual standard basis $v_0^*, v_1^*, \ldots, v_d^*$, let D, B^* be the representation matrices of A, A^*, respectively. Then B^* is an irreducible tridiagonal matrix given by (6.134) and satisfies the condition (6.142) on the constant row sum. On the other hand, D is a diagonal matrix given by (6.135). By the uniqueness of the standard basis and the dual standard basis, B, D^*, D, B^* are uniquely determined. Following (6.138), let P be the transition matrix between the standard basis and the dual standard basis. By (6.139), we have

$$PBP^{-1} = D, \quad P^{-1}B^*P = D^*.$$

Let $\{v_i(x)\}_{i=0}^d$ and $\{v_i^*(x)\}_{i=0}^d$ be the dual system of orthogonal polynomials determined by B, B^*, and let $\{k_i\}_{i=0}^d$ and $\{k_i^*\}_{i=0}^d$ be the sequences of degrees. By Theorem 6.25, for any $i, j \in \{0, 1, \ldots, d\}$, we have

$$\frac{v_j(\theta_i)}{k_j} = \frac{v_i^*(\theta_j^*)}{k_i^*}.$$

Namely, $\{v_i(x)\}_{i=0}^d$ and $\{v_i^*(x)\}_{i=0}^d$ form a dual system of orthogonal polynomials. Conversely, by Theorem 6.25, a dual system of orthogonal polynomials induces an L-pair associated with an L-system. By Theorem 6.25, without loss of generality, we can set the transition matrix P as

$$P = (v_j(\theta_i)).$$

Moreover, by Proposition 6.24, if we let

$$Q = (v_j^*(\theta_i^*)),$$

then we have

$$PQ = QP = nI, \quad n = \sum_{i=0}^d k_i = \sum_{i=0}^d k_i^*.$$

6.3.2 Pre-L-pairs

In this section, we define pre-L-pairs and study their parameters.

Let V be a finite-dimensional vector space over $\mathbb{C}$. Assume a direct sum decomposition $V = \bigoplus_{i=0}^d U_i$ and linear transformations $R : V \longrightarrow V$ and $L : V \longrightarrow V$ such that

$$RU_i \subseteq U_{i+1}, \quad LU_i \subseteq U_{i-1} \ (0 \le i \le d) \tag{6.144}$$

are given, where $U_{-1} = U_{d+1} = 0$. Moreover, distinct numbers $\theta_0, \theta_1, \ldots, \theta_d \in \mathbb{C}$ and distinct numbers $\theta_0^*, \theta_1^*, \ldots, \theta_d^* \in \mathbb{C}$ are given. Then the pair of linear transformations

$$A = R + \sum_{i=0}^{d} \theta_i F_i, \quad A^* = L + \sum_{i=0}^{d} \theta_i^* F_i \tag{6.145}$$

is called a *pre-TD-pair*, where F_i denotes the projection

$$F_i : V = \bigoplus_{j=0}^{d} U_j \longrightarrow U_i. \tag{6.146}$$

We call U_i the weight space of a pre-TD-pair $A, A^* \in \mathrm{End}(V)$ and $V = \bigoplus_{j=0}^{d} U_j$ the weight space decomposition. Note that a TD-pair is a pre-TD-pair by (6.111), (6.112) in Definition 6.38 in Section 6.2.

If a pair $A, A^* \in \mathrm{End}(V)$ is a pre-TD-pair, by (6.145), both A and A^* are diagonalizable and their eigenvalues are $\{\theta_i\}_{i=0}^{d}$ and $\{\theta_i^*\}_{i=0}^{d}$, respectively. Let $V = \bigoplus_{i=0}^{d} V_i$ $(A|_{V_i} = \theta_i)$, $V^* = \bigoplus_{i=0}^{d} V_i^*$ $(A^*|_{V_i^*} = \theta_i^*)$ be the eigenspace decompositions of A, A^*, respectively, and let $V = \bigoplus_{i=0}^{d} U_i$ be the weight space decomposition. Then we have

$$\begin{aligned} &U_0 + U_1 + \cdots + U_i = V_0^* + V_1^* + \cdots + V_i^*, \\ &U_i + U_{i+1} + \cdots + U_d = V_i + V_{i+1} + \cdots + V_d, \\ &U_i = (V_0^* + V_1^* + \cdots + V_i^*) \cap (V_i + \cdots V_{i+1} + \cdots + V_d). \end{aligned} \tag{6.147}$$

Moreover, if we let

$$\begin{aligned} E_i &: V = \bigoplus_{j=0}^{d} V_j \longrightarrow V_i, \\ E_i^* &: V = \bigoplus_{j=0}^{d} V_j^* \longrightarrow V_i^* \end{aligned} \tag{6.148}$$

be the projections onto the eigenspaces V_i, V_i^* of A, A^*, respectively, then we have

$$V_i \cong U_i \cong V_i^* \quad \text{(isomorphic as vector spaces)} \tag{6.149}$$

and we obtain pairs of isomorphisms such that the one is the inverse of the other as follows:

$$F_i|_{V_i} : V_i \longrightarrow U_i, \quad E_i|_{U_i} : U_i \longrightarrow V_i, \tag{6.150}$$

and

$$F_i|_{V_i^*} : V_i^* \longrightarrow U_i, \quad E_i^*|_{U_i} : U_i \longrightarrow V_i^*, \tag{6.151}$$

where F_i is the projection given by (6.146). Note that by (6.148), the weight space decomposition of a pre-TD-pair $A, A^* \in \mathrm{End}(V)$ is determined by the eigenspaces $\{V_i\}_{i=0}^d$ of A and the eigenspaces $\{V_i^*\}_{i=0}^d$ of A^*. We call $(A, A^*; \{V_i\}_{i=0}^d, \{V_i^*\}_{i=0}^d)$ a *pre-TD-system*. We sometimes abbreviate A, A^* and call $(\{V_i\}_{i=0}^d, \{V_i^*\}_{i=0}^d)$ a pre-TD-system. The orderings of the eigenspaces of A, A^* are fixed as $V_0, V_1, \dots, V_d$ and $V_0^*, V_1^*, \dots, V_d^*$.

Let $V = \bigoplus_{i=0}^d U_i$ be the weight space decomposition of a pre-TD-pair $A, A^* \in \mathrm{End}(V)$, and let $(\{V_i\}_{i=0}^d, \{V_i^*\}_{i=0}^d)$ be a pre TD-system. By (6.149), we have $\dim(V_i) = \dim(U_i) = \dim(V_i^*)$ in general. If $\dim(V_i) = \dim(U_i) = \dim(V_i^*) = 1$ $(0 \le i \le d)$, we call A, A^* a *pre-L-pair* and $(\{V_i\}_{i=0}^d, \{V_i^*\}_{i=0}^d)$ a *pre-L-system*. In this case, if we choose a non-zero u_i from each U_i, the vectors $u_0, u_1, \dots, u_d$ form a basis of V, and the matrix representation of A, A^* with respect to $u_0, u_1, \dots, u_d$ are

$$A = \begin{bmatrix} \theta_0 & & & 0 \\ x_1 & \theta_1 & & \\ & \ddots & \ddots & \\ 0 & & x_d & \theta_d \end{bmatrix}, \tag{6.152}$$

$$A^* = \begin{bmatrix} \theta_0^* & y_0 & & 0 \\ & \theta_1^* & \ddots & \\ & & \ddots & y_{d-1} \\ 0 & & & \theta_d^* \end{bmatrix}. \tag{6.153}$$

If we choose a non-zero vector v_i from each V_i^*, then $v_0, v_1, \dots, v_d$ form a basis of V, and the matrix representation of A with respect to $v_0, v_1, \dots, v_d$ is

$$A = \begin{bmatrix} a_0 & & & * \\ c_1 & a_1 & & \\ & \ddots & \ddots & \\ 0 & & c_d & a_d \end{bmatrix}, \tag{6.154}$$

and the matrix representation of A^* is a diagonal matrix $A^* = \mathrm{diag}(\theta_0^*, \theta_1^*, \dots, \theta_d^*)$. Here, letting E_i^* be the projection given by (6.148), we have

$$a_i = \mathrm{tr}(E_i^* A E_i^*) \quad (0 \le i \le d). \tag{6.155}$$

Furthermore, if we choose a non-zero vector v_i^* from each V_i, then $v_0^*, v_1^*, \dots, v_d^*$ form a basis of V, the matrix representation of A^* with respect to $v_0^*, v_1^*, \dots, v_d^*$ is

$$A^* = \begin{bmatrix} a_0^* & b_0^* & & 0 \\ & a_1^* & \ddots & \\ & * & \ddots & b_{d-1}^* \\ & & & a_d^* \end{bmatrix}, \tag{6.156}$$

and the matrix representation of A is a diagonal matrix $A = \mathrm{diag}(\theta_0, \theta_1, \ldots, \theta_d)$. Here, letting E_i be the projection given by (6.148), we have

$$a_i^* = \mathrm{tr}(E_i A^* E_i) \quad (0 \le i \le d). \tag{6.157}$$

We also have

$$\begin{aligned}\sum_{i=0}^{d} a_i &= \sum_{i=0}^{d} \theta_i,\\ \sum_{i=0}^{d} a_i^* &= \sum_{i=0}^{d} \theta_i^*.\end{aligned} \tag{6.158}$$

Set $\lambda_i = x_{i+1}y_i$ $(0 \le i \le d-1)$ for x_{i+1}, y_i in (6.152), (6.153). Then by (6.145), we obtain

$$\lambda_i = \mathrm{tr}(LR|_{U_i}) = \mathrm{tr}(RL|_{U_{i+1}}) \quad (0 \le i \le d-1) \tag{6.159}$$

and it turns out that λ_i does not depend on the choice of a basis $u_0, u_1, \ldots, u_d$ $(u_i \in U_i)$. By (6.159), we extend the definition of λ_i to $i = -1, d$ so that $\lambda_{-1} = \mathrm{tr}(RL|_{U_0}) = 0, \lambda_d = \mathrm{tr}(LR|_{U_d}) = 0$. We call a triple $\{\theta_i\}_{i=0}^d, \{\theta_i^*\}_{i=0}^d, \{\lambda_i\}_{i=0}^{d-1}$ the data of a pre-L-pair A, A^*. We define the isomorphism of pre-L-pairs similarly to that of TD-pairs by the commutative diagram in Definition 6.27 in Section 6.2. Note that if $\lambda_i \ne 0$ $(0 \le i \le d-1)$, the data $\{\theta_i\}_{i=0}^d, \{\theta_i^*\}_{i=0}^d, \{\lambda_i\}_{i=0}^{d-1}$ determine an isomorphism class of a pre-L-pair. Note also that $\{\theta_i\}_{i=0}^d, \{\theta_i^*\}_{i=0}^d, \{\lambda_i\}_{i=0}^{d-1}$ form the data of a pre-L-pair only if $\theta_i \ne \theta_j$ $(i \ne j)$, $\theta_i^* \ne \theta_j^*$ $(i \ne j)$, and if this condition holds, we can construct a pre-L-pair with data $\{\theta_i\}_{i=0}^d, \{\theta_i^*\}_{i=0}^d, \{\lambda_i\}_{i=0}^{d-1}$.

In what follows, let $A, A^* \in \mathrm{End}(V)$ be a pre-L-pair, $V = \bigoplus_{i=0}^d U_i$ the weight space decomposition, $(\{V_i\}_{i=0}^d, \{V_i^*\}_{i=0}^d)$ a pre-L-system, and $\{\theta_i\}_{i=0}^d, \{\theta_i^*\}_{i=0}^d, \{\lambda_i\}_{i=0}^{d-1}$ its data. The linear transformations R, L determined by (6.144), (6.145), and (6.146) are called the *raising map* and the *lowering map*, respectively.

Lemma 6.54. *The following hold:*

(1) *For $u_i \in U_i$, let $u_{i+j} = R^j u_i \in U_{i+j}$ $(j = 1, 2, \ldots, d-i)$. Then*

$$\sum_{j=0}^{d-i} \frac{1}{(\theta_i - \theta_{i+1})(\theta_i - \theta_{i+2}) \cdots (\theta_i - \theta_{i+j})} u_{i+j} \tag{6.160}$$

is contained in V_i.

(2) *For $u_i \in U_i$, let $u_{i-j} = L^j u_i \in U_{i-j}$ $(j = 1, 2, \ldots, i)$. Then*

$$\sum_{j=0}^{i} \frac{1}{(\theta_i^* - \theta_{i-1}^*)(\theta_i^* - \theta_{i-2}^*) \cdots (\theta_i^* - \theta_{i-j}^*)} u_{i-j} \tag{6.161}$$

is contained in V_i^*.

Proof. We prove (1) only. The proof of (2) is similar. Let $v = \sum_{j=0}^{d-i} c_j u_{i+j}$. By (6.145), we have $Au_{i+j} = u_{i+j+1} + \theta_{i+j}u_{i+j}$. So we obtain $Av = \sum_{j=0}^{d-i}(c_{j-1} + \theta_{i+j}c_j)u_{i+j}$, where $c_{-1} = 0$. Hence $Av = \theta_i v$ only if $c_{j-1}+\theta_{i+j}c_j = \theta_i c_j$ $(0 \le j \le d-i)$, that is, $c_j = \frac{1}{\theta_i-\theta_{i+j}}c_{j-1}$ $(1 \le j \le d-i)$. If we let $c_0 = 1$, we obtain (6.160). □

Let E_i, E_i^* be the projections of the eigenspaces of A, A^* onto V_i, V_i^* as in (6.148), and define $\{a_i\}_{i=0}^d, \{a_i^*\}_{i=0}^d$ by $a_i = \mathrm{tr}(E_i^* A E_i^*)$ in (6.155) and by $a_i^* = \mathrm{tr}(E_i A^* E_i)$ in (6.157).

Theorem 6.55. *The following hold:*

(1)

$$\frac{\lambda_i}{\theta_{i+1}^* - \theta_i^*} - \frac{\lambda_{i-1}}{\theta_i^* - \theta_{i-1}^*} = \theta_i - a_i \quad (0 \le i \le d),$$

where for $i = 0, d$, *we set* $\frac{\lambda_{-1}}{\theta_0^*-\theta_{-1}^*} = 0$, $\frac{\lambda_d}{\theta_{d+1}^*-\theta_d^*} = 0$;

(2)

$$\frac{\lambda_i}{\theta_{i+1} - \theta_i} - \frac{\lambda_{i-1}}{\theta_i - \theta_{i-1}} = \theta_i^* - a_i^* \quad (0 \le i \le d),$$

where for $i = 0, d$, *we set* $\frac{\lambda_{-1}}{\theta_0-\theta_{-1}} = 0$, $\frac{\lambda_d}{\theta_{d+1}-\theta_d} = 0$.

In particular, we have

$$\frac{\lambda_i}{\theta_{i+1}^* - \theta_i^*} = \sum_{j=0}^{i} \theta_j - \sum_{j=0}^{i} a_j \quad (0 \le i \le d),$$

$$\frac{\lambda_i}{\theta_{i+1} - \theta_i} = \sum_{j=0}^{i} \theta_j^* - \sum_{j=0}^{i} a_j^* \quad (0 \le i \le d).$$

Proof. We prove (1) only. Statement (2) is similarly proved. The last statement in the theorem follows immediately from (1), (2).

Choose a non-zero vector u_i from U_i. Following Lemma 6.54, we set $u_{i+j} = R^j u_i \in U_{i+j}$ $(1 \le j \le d-i)$, $u_{i-j} = L^j u_i \in U_{i-j}$ $(1 \le j \le i)$. Since

$$v = u_i + \frac{1}{\theta_i^* - \theta_{i-1}^*}u_{i-1} + \cdots \tag{6.162}$$

is contained in V_i^*, by the definition of a_i, we have

$$E_i^* A E_i^* v = a_i v. \tag{6.163}$$

Since $E_i^* v = v$, the left-hand side of (6.163) equals $E_i^* Av$. Therefore, by (6.162), we have

$$Av = Au_i + \frac{1}{\theta_i^* - \theta_{i-1}^*}Au_{i-1} + \cdots$$

and by (6.145), we have

$$Au_i = Ru_i + \theta_i u_i = u_{i+1} + \theta_i u_i,$$
$$Au_{i-1} = Ru_{i-1} + \theta_{i-1}u_{i-1} = RLu_i + \theta_{i-1}u_{i-1} = \lambda_{i-1}u_i + \theta_{i-1}u_{i-1}.$$

It follows that

$$Av = u_{i+1} + \left(\theta_i + \frac{\lambda_{i-1}}{\theta_i^* - \theta_{i-1}^*}\right)u_i + \cdots. \tag{6.164}$$

On the other hand, by (6.148), we have $U_{i-1} + \cdots + U_0 = V_{i-1}^* + \cdots + V_0^*$. Hence, by multiplying both sides of (6.164) by E_i^*, the terms after u_{i-1} vanish as follows:

$$E_i^* Av = E_i^* u_{i+1} + \left(\theta_i + \frac{\lambda_{i-1}}{\theta_i^* - \theta_{i-1}^*}\right)E_i^* u_i. \tag{6.165}$$

Since E_i^* is the projection onto the eigenspace V_i^* of the diagonalizable matrix A^*, we have

$$E_i^* = \prod_{j\neq i} \frac{A^* - \theta_j^*}{\theta_i^* - \theta_j^*}. \tag{6.166}$$

By (6.145), we have

$$\begin{aligned}
\frac{A^* - \theta_{i+1}^*}{\theta_i^* - \theta_{i+1}^*}u_{i+1} &= \frac{1}{\theta_i^* - \theta_{i+1}^*}Lu_{i+1} = \frac{1}{\theta_i^* - \theta_{i+1}^*}LRu_i = \frac{\lambda_i}{\theta_i^* - \theta_{i+1}^*}u_i, \\
\frac{A^* - \theta_j^*}{\theta_i^* - \theta_j^*}u_i &= \frac{A^* - \theta_i^* + \theta_i^* - \theta_j^*}{\theta_i^* - \theta_j^*}u_i = \frac{1}{\theta_i^* - \theta_j^*}Lu_i + u_i \\
&= u_i + \frac{1}{\theta_i^* - \theta_j^*}u_{i-1}, \\
\frac{A^* - \theta_j^*}{\theta_i^* - \theta_j^*}u_k &= \frac{\theta_k^* - \theta_j^*}{\theta_i^* - \theta_j^*}u_k + \frac{1}{\theta_i^* - \theta_j^*}u_{k-1} \quad (k = i-1, \ldots, 0).
\end{aligned}$$

So by (6.165), (6.166), we obtain

$$E_i^* Av = \left(\frac{\lambda_i}{\theta_i^* - \theta_{i+1}^*} + \theta_i + \frac{\lambda_{i-1}}{\theta_i^* - \theta_{i-1}^*}\right)u_i + (\text{the terms after } u_{i-1}). \tag{6.167}$$

Since $E_i^* Av = a_i v$ by (6.163), by comparing the coefficient of u_i, we get

$$a_i = \frac{\lambda_i}{\theta_i^* - \theta_{i+1}^*} + \theta_i + \frac{\lambda_{i-1}}{\theta_i^* - \theta_{i-1}^*}.$$

If $i = 0$, we should set $u_{-1} = 0$ in (6.162) and $\frac{\lambda_{-1}}{\theta_0^* - \theta_{-1}^*} = 0$ in (6.164), and if $i = d$, we should set $u_{d+1} = 0$ in (6.165) and $\frac{\lambda_d}{\theta_d^* - \theta_{d+1}^*} = 0$ in (6.167). □

Let $A, A^* \in \mathrm{End}(V)$ be an L-pair. There are four L-systems associated with A, A^*. For each L-system, A, A^* form a pre-L-pair. Using new symbols $\hat{\lambda}_i, \check{\lambda}_i, \tilde{\lambda}_i$, we give a list of the data of each pre-L-pair as below. We also give for each pre-L-pair $\{a_i\}_{i=0}^d$ determined by $a_i = \mathrm{tr}(E_i^* A E_i^*)$ in (6.155) and $\{a_i^*\}_{i=0}^d$ determined by $a_i^* = \mathrm{tr}(E_i A^* E_i)$ in (6.157).

L-system	Data of the pre-L-pair A, A^*	
$(\{V_i\}_{i=0}^d, \{V_i^*\}_{i=0}^d)$	$\{\theta_i\}_{i=0}^d, \{\theta_i^*\}_{i=0}^d, \{\lambda_i\}_{i=0}^{d-1}$	$\{a_i\}_{i=0}^d, \{a_i^*\}_{i=0}^d$
$(\{V_{d-i}\}_{i=0}^d, \{V_i^*\}_{i=0}^d)$	$\{\theta_{d-i}\}_{i=0}^d, \{\theta_i^*\}_{i=0}^d, \{\hat{\lambda}_i\}_{i=0}^{d-1}$	$\{a_i\}_{i=0}^d, \{a_{d-i}^*\}_{i=0}^d$
$(\{V_i\}_{i=0}^d, \{V_{d-i}^*\}_{i=0}^d)$	$\{\theta_i\}_{i=0}^d, \{\theta_{d-i}^*\}_{i=0}^d, \{\check{\lambda}_i\}_{i=0}^{d-1}$	$\{a_{d-i}\}_{i=0}^d, \{a_i^*\}_{i=0}^d$
$(\{V_{d-i}\}_{i=0}^d, \{V_{d-i}^*\}_{i=0}^d)$	$\{\theta_{d-i}\}_{i=0}^d, \{\theta_{d-i}^*\}_{i=0}^d, \{\tilde{\lambda}_i\}_{i=0}^{d-1}$	$\{a_{d-i}\}_{i=0}^d, \{a_{d-i}^*\}_{i=0}^d$

If we let A, A^* be an L-pair as in the above table, then A^*, A also form an L-pair and $(\{V_i^*\}_{i=0}^d, \{V_i\}_{i=0}^d)$ is an associated L-system. If we regard A^*, A as a pre-L-pair with respect to this L-system, we let $\{\theta_i^*\}_{i=0}^d$, $\{\theta_i\}_{i=0}^d$ $\{\lambda_i^*\}_{i=0}^{d-1}$ be its data:

L-system	Data of the pre-L-pair A^*, A	
$(\{V_i^*\}_{i=0}^d, \{V_i\}_{i=0}^d)$	$\{\theta_i^*\}_{i=0}^d, \{\theta_i\}_{i=0}^d, \{\lambda_i^*\}_{i=0}^{d-1}$	$\{a_i^*\}_{i=0}^d, \{a_i\}_{i=0}^d$

Lemma 6.56. *We have*

$$\frac{\lambda_i - \hat{\lambda}_i}{\theta_{i+1}^* - \theta_i^*} = \sum_{j=0}^{i} (\theta_j - \theta_{d-j}) \quad (0 \le i \le d-1).$$

Proof. By Theorem 6.55, we have

$$\frac{\lambda_i}{\theta_{i+1}^* - \theta_i^*} = \sum_{j=0}^{i} \theta_j - \sum_{j=0}^{i} a_j,$$

$$\frac{\hat{\lambda}_i}{\theta_{i+1}^* - \theta_i^*} = \sum_{j=0}^{i} \theta_{d-j} - \sum_{j=0}^{i} a_j.$$

Taking the difference between the above equations yields the desired result. □

Lemma 6.57. *The following hold:*
(1) $\tilde{\lambda}_i = \lambda_{d-i-1}$ $(0 \le i \le d-1)$;
(2) $\check{\lambda}_i = \hat{\lambda}_{d-i-1}$ $(0 \le i \le d-1)$;
(3) $\lambda_i^* = \lambda_i$ $(0 \le i \le d-1)$.

Proof. (1) By Theorem 6.55, we have

$$\frac{\tilde{\lambda}_i}{\theta_{d-i-1}^* - \theta_{d-i}^*} = \sum_{j=0}^{i} \theta_{d-j} - \sum_{j=0}^{i} a_{d-j} = \sum_{j=d-i}^{d} \theta_j - \sum_{j=d-i}^{d} a_j.$$

Since $\sum_{j=0}^{d} a_j = \sum_{j=0}^{d} \theta_j$ by (6.158), we have

$$\frac{\tilde{\lambda}_i}{\theta^*_{d-i-1} - \theta^*_{d-i}} = -\sum_{j=0}^{d-i-1} \theta_j + \sum_{j=0}^{d-i-1} a_j.$$

Again by Theorem 6.55, we have

$$\frac{\lambda_{d-i-1}}{\theta^*_{d-i} - \theta^*_{d-i-1}} = \sum_{j=0}^{d-i-1} \theta_j - \sum_{j=0}^{d-i-1} a_j.$$

Hence we have $\tilde{\lambda}_i = \lambda_{d-i-1}$.
(2) Since $\check{\lambda}_i = \hat{\tilde{\lambda}}_i$, by (1), we obtain $\check{\lambda}_i = \hat{\lambda}_{d-i-1}$.
(3) By Theorem 6.55, we have

$$\frac{\lambda^*_i}{\theta_{i+1} - \theta_i} = \sum_{j=0}^{i} \theta^*_j - \sum_{j=0}^{i} a^*_j = \frac{\lambda_i}{\theta_{i+1} - \theta_i}.$$ □

6.3.3 Terwilliger's lemma

Let $A, A^* \in \mathrm{End}(V)$ be a pre-L-pair, $V = \bigoplus_{i=0}^{d} U_i$ the weight space decomposition, $(\{V_i\}_{i=0}^{d}, \{V_i^*\}_{i=0}^{d})$ an associated pre-L-pair, and $\{\theta_i\}_{i=0}^{d}$, $\{\theta^*_i\}_{i=0}^{d}$, $\{\lambda_i\}_{i=0}^{d-1}$ its data. Let E_i, E^*_i be the projections onto V_i, V^*_i, respectively, given by (6.148). Following Lemma 6.56, we define $\{\hat{\lambda}_i\}_{i=0}^{d-1}$ by

$$\hat{\lambda}_i = \lambda_i - (\theta^*_{i+1} - \theta^*_i) \sum_{j=0}^{i} (\theta_j - \theta_{d-j}). \tag{6.168}$$

In this subsection, we consider several necessary conditions for $(A, A^*; \{V_i\}_{i=0}^{d}, \{V^*_i\}_{i=0}^{d})$ to be an L-system.

Proposition 6.58. *The sequence $(A, A^*; \{V_i\}_{i=0}^{d}, \{V^*_i\}_{i=0}^{d})$ is an L-system only if the following (1)–(4) hold:*
(1) $\lambda_i \neq 0\ (0 \leq i \leq d-1)$;
(2) $\hat{\lambda}_i \neq 0\ (0 \leq i \leq d-1)$;
(3) $A(V^*_2 + V^*_3 + \cdots + V^*_d) \subseteq V^*_1 + V^*_2 + \cdots + V^*_d$;
(4) $A^*(V_0 + V_1 + \cdots + V_{d-2}) \subseteq V_0 + V_1 + \cdots + V_{d-1}$.

Proof. (1) If we let $\lambda_i = 0$, since $\lambda_i = \mathrm{tr}(LR|_{U_i}) = \mathrm{tr}(RL|_{U_{i+1}})$, we have $RU_i = 0$ or $LU_{i+1} = 0$. If $RU_i = 0$, then $U_0 + U_1 + \cdots + U_i$ is $\langle A, A^*\rangle$-invariant. If $LU_{i+1} = 0$, then $U_{i+1} + \cdots + U_d$ is $\langle A, A^*\rangle$-invariant. In any case, it contradicts the fact that V is irreducible as an $\langle A, A^*\rangle$-module.

(2) If $(A, A^*; \{V_i\}_{i=0}^d, \{V_i^*\}_{i=0}^d)$ is an L-pair, so is $(A, A^*; \{V_{d-i}\}_{i=0}^d, \{V_i^*\}_{i=0}^d)$. Therefore, by (1), the data $\{\theta_i\}_{i=0}^d, \{\theta_i^*\}_{i=0}^d, \{\hat{\lambda}_i\}_{i=0}^{d-1}$ must satisfy $\hat{\lambda}_i \neq 0$.
Statements (3) and (4) are clear by the definition of an L-system. □

The next lemma appears as Corollary 11.4 in the paper [475] by Terwilliger. This is a key lemma for the classification of L-pairs, which we call Terwilliger's lemma in this book.

Lemma 6.59 (Terwilliger's lemma).
(1) *The condition that* $E_0^*(A - a_0)(A^* - \theta_1^*) = 0$ *is a necessary condition for*

$$A(V_2^* + V_3^* + \cdots + V_d^*) \subseteq V_1^* + V_2^* + \cdots + V_d^*,$$

where $a_0 = \operatorname{tr}(E_0^* A E_0^*)$.
(2) *The condition that* $E_d(A^* - a_d^*)(A - \theta_{d-1}) = 0$ *is a necessary condition for*

$$A^*(V_0 + V_1 + \cdots + V_{d-2}) \subseteq V_0 + V_1 + \cdots + V_{d-1},$$

where $a_d^* = \operatorname{tr}(E_d A^* E_d)$.

Proof. We prove (1) only since (2) is similarly proved.
Note that

$$A(V_2^* + V_3^* + \cdots + V_d^*) \subseteq V_1^* + V_2^* + \cdots + V_d^*$$

is equivalent to

$$E_0^* A E_i^* = 0 \quad (2 \leq i \leq d). \tag{6.169}$$

Moreover,

$$E_0^*(A - a_0)(A^* - \theta_1^*) = 0$$

and

$$E_0^*(A - a_0)(A^* - \theta_1^*)E_i^* = 0 \quad (0 \leq i \leq d) \tag{6.170}$$

are equivalent. Therefore, since $(A^* - \theta_1^*)E_i^* = (\theta_i^* - \theta_1^*)E_i^*$, we have

$$\begin{aligned} E_0^*(A - a_0)(A^* - \theta_1^*)E_i^* &= (\theta_i^* - \theta_1^*)E_0^*(A - a_0)E_i^* \\ &= (\theta_i^* - \theta_1^*)(E_0^* A E_i^* - a_0 \delta_{0,i} E_0^*). \end{aligned}$$

Hence (6.170) always holds for $i = 0, 1$. Therefore, since $\theta_i^* - \theta_1^* \neq 0$ $(2 \leq i \leq d)$, (6.170) is equivalent to

$$E_0^* A E_i^* = 0 \quad (2 \leq i \leq d).$$

Namely, (6.170) is equivalent to (6.169). □

Theorem 6.60. *Let $A, A^* \in \mathrm{End}(V)$ be a pre-L-pair, $(\{V_i\}_{i=0}^d, \{V_i^*\}_{i=0}^d)$ an associated pre-L-system, and $\{\theta_i\}_{i=0}^d, \{\theta_i^*\}_{i=0}^d, \{\lambda_i\}_{i=0}^{d-1}$ its data. Assume $d \geq 2$, $\lambda_i \neq 0$ $(0 \leq i \leq d-1)$. Define $\{\lambda_i'\}_{i=0}^{d-1}$ by*

$$\lambda_i' = \lambda_i - (\theta_i - \theta_d)(\theta_{i+1}^* - \theta_0^*) \tag{6.171}$$

and for $i = -1, d$, we set $\lambda_{-1}' = 0$, $\lambda_d' = 0$. Then the following (1), (2) hold:

(1) *The condition*

$$\lambda_{i-1}' = \frac{\theta_i^* - \theta_1^*}{\theta_{i+1}^* - \theta_0^*}\lambda_i' + \lambda_{d-1}' \quad (0 \leq i \leq d-1)$$

is a necessary condition for $A(V_2^* + V_3^* + \cdots + V_d^*) \subseteq V_1^* + V_2^* + \cdots + V_d^*$.

(2) *The condition*

$$\lambda_i' = \frac{\theta_i - \theta_{d-1}}{\theta_{i-1} - \theta_d}\lambda_{i-1}' + \lambda_0' \quad (1 \leq i \leq d)$$

is a necessary condition for $A^*(V_0 + V_1 + \cdots + V_{d-2}) \subseteq V_0 + V_1 + \cdots + V_{d-1}$.

Proof. Let $V = \bigoplus_{i=0}^d U_i$ be the weight space decomposition. Choose a non-zero vector u_0 from U_0 and set $u_i = R^i u_0$ $(0 \leq i \leq d)$. We also let $u_{-1} = u_{d+1} = 0$. We have $Lu_{i+1} = \lambda_i u_i$ $(0 \leq i \leq d-1)$. Since $\lambda_i \neq 0$ $(0 \leq i \leq d-1)$, we have $u_i \neq 0$ $(0 \leq i \leq d)$. Therefore $u_0, u_1, \ldots, u_d$ form a basis of V.

We prove (1) only. The proof of (2) is similar. By Terwilliger's lemma, it suffices to show that condition (1) is a necessary condition for

$$E_0^*(A - a_0)(A^* - \theta_1^*)u_i = 0 \quad (0 \leq i \leq d). \tag{6.172}$$

Noting that $Ru_j = u_{j+1}$, $Lu_j = LRu_{j-1} = \lambda_{j-1}u_{j-1}$, by (6.145), we have

$$\begin{aligned} Au_j &= u_{j+1} + \theta_j u_j, \\ A^* u_j &= \theta_j^* u_j + \lambda_{j-1}u_{j-1}. \end{aligned} \tag{6.173}$$

Therefore we obtain

$$\begin{aligned} (A - a_0)(A^* - \theta_1^*)u_i &= (A - a_0)((\theta_i^* - \theta_1^*)u_i + \lambda_{i-1}u_{i-1}) \\ &= (\theta_i^* - \theta_1^*)u_{i+1} + ((\theta_i - a_0)(\theta_i^* - \theta_1^*) + \lambda_{i-1})u_i \\ &\quad + \lambda_{i-1}(\theta_{i-1} - a_0)u_{i-1}. \end{aligned} \tag{6.174}$$

By (6.166), we have

$$E_0^* = \prod_{j=1}^d \frac{A^* - \theta_j^*}{\theta_0^* - \theta_j^*}.$$

So, by using

$$(A^* - \theta_1^*) \cdots (A^* - \theta_k^*) u_k = \lambda_0 \cdots \lambda_{k-1} u_0,$$
$$\frac{A^* - \theta_j^*}{\theta_0^* - \theta_j^*} u_0 = u_0,$$

we get

$$E_0^* u_k = \frac{\lambda_0 \cdots \lambda_{k-1}}{(\theta_0^* - \theta_1^*) \cdots (\theta_0^* - \theta_k^*)} u_0. \tag{6.175}$$

By (6.174), (6.175), we obtain

$$\begin{aligned} &E_0^*(A - a_0)(A^* - \theta_1^*) u_i \\ &= \frac{\lambda_0 \cdots \lambda_{i-1}}{(\theta_0^* - \theta_1^*) \cdots (\theta_0^* - \theta_i^*)} \\ &\times \left\{ \frac{\theta_i^* - \theta_1^*}{\theta_0^* - \theta_{i+1}^*} \lambda_i + ((\theta_i - a_0)(\theta_i^* - \theta_1^*) + \lambda_{i-1}) + (\theta_{i-1} - a_0)(\theta_0^* - \theta_i^*) \right\} u_0. \end{aligned}$$

Therefore, (6.172) holds only if

$$\frac{\theta_i^* - \theta_1^*}{\theta_0^* - \theta_{i+1}^*} \lambda_i + (\theta_i - a_0)(\theta_i^* - \theta_1^*) + \lambda_{i-1} + (\theta_{i-1} - a_0)(\theta_0^* - \theta_i^*) = 0 \tag{6.176}$$

holds for any i with $0 \le i \le d$. For $i = 0$, since the term containing u_{-1} does not appear in (6.174), the third term and the fourth term in (6.176) do not appear. For $i = d$, since the term containing u_{d+1} does not appear in (6.174), the first term in (6.176) does not appear. Namely, in (6.176), we should regard $\theta_{d+1}^*, \theta_{-1}$ as indeterminates, and set $\lambda_d = \lambda_{-1} = 0$. Under this rule, we may extend (6.171) to $-1 \le i \le d$ and set

$$\lambda_i = \lambda_i' + (\theta_i - \theta_d)(\theta_{i+1}^* - \theta_0^*). \tag{6.177}$$

If we substitute (6.177) for (6.176), we obtain

$$\frac{\theta_i^* - \theta_1^*}{\theta_0^* - \theta_{i+1}^*} \lambda_i' + \lambda_{i-1}' + (\theta_d - a_0)(\theta_0^* - \theta_1^*) = 0. \tag{6.178}$$

Namely, (6.172) holds if and only if (6.178) holds for any i with $0 \le i \le d$. For $i = 0$, (6.178) becomes

$$\lambda_0' = (\theta_d - a_0)(\theta_1^* - \theta_0^*). \tag{6.179}$$

By Theorem 6.55 (1), we have

$$\lambda_0 = (\theta_0 - a_0)(\theta_1^* - \theta_0^*). \tag{6.180}$$

Hence we have

$$\lambda'_0 = \lambda_0 - (\theta_0 - \theta_d)(\theta^*_1 - \theta^*_0) = (\theta_d - a_0)(\theta^*_1 - \theta^*_0)$$

and (6.179) always holds. Therefore, (6.178) is equivalent to

$$\lambda'_{i-1} = \frac{\theta^*_i - \theta^*_1}{\theta^*_{i+1} - \theta^*_0}\lambda'_i + \lambda'_0. \tag{6.181}$$

Equation (6.181) trivially holds for $i = 0$, and for $i = d$, we have

$$\lambda'_{d-1} = \lambda'_0. \tag{6.182}$$

In (6.181), replacing λ'_0 by λ'_{d-1} yields

$$\lambda'_{i-1} = \frac{\theta^*_i - \theta^*_1}{\theta^*_{i+1} - \theta^*_0}\lambda'_i + \lambda'_{d-1}. \tag{6.183}$$

Equation (6.183) trivially holds for $i = d$, and we have (6.182) for $i = 0$. Therefore, (6.181) holds for any i with $1 \le i \le d$ if and only if (6.183) holds for any i with $0 \le i \le d-1$. □

For a pre-L-system $(A, A^*; \{V_i\}_{i=0}^d, \{V^*_i\}_{i=0}^d)$ to be an L-system, the data $\{\theta_i\}_{i=0}^d$, $\{\theta^*_i\}_{i=0}^d$, $\{\lambda_i\}_{i=0}^{d-1}$ must be quite restricted. First of all, since an L-pair is a TD-pair, by Theorem 6.44 in Section 6.2, $\{\theta_i\}_{i=0}^d$, $\{\theta^*_i\}_{i=0}^d$ are β-sequences and are expressed by AW-parameters, as was seen in Section 6.2.2. Moreover, by Proposition 6.58 and Theorem 6.60, we have

$$\begin{cases} \lambda'_{i-1} = \frac{\theta^*_i - \theta^*_1}{\theta^*_{i+1} - \theta^*_0}\lambda'_i + \lambda'_{d-1} & (0 \le i \le d-1), \\ \lambda'_i = \frac{\theta_i - \theta_{d-1}}{\theta_{i-1} - \theta_d}\lambda'_{i-1} + \lambda'_0 & (1 \le i \le d). \end{cases} \tag{6.184}$$

By solving (6.184), $\{\lambda'_i\}_{i=0}^{d-1}$ and $\{\lambda_i\}_{i=0}^{d-1}$ are expressed by AW-parameters and λ'_0, as will be shown below.

Case (I): We have

$$\begin{cases} \theta_i = \theta_0 + hq^{-i}(1 - q^i)(1 - sq^{i+1}) & (0 \le i \le d), \\ \theta^*_i = \theta^*_0 + h^*q^{-i}(1 - q^i)(1 - s^*q^{i+1}) & (0 \le i \le d), \end{cases} \tag{6.185}$$
$$h, h^* \ne 0; \quad q^i \ne 1\ (1 \le i \le d); \quad s, s^* \notin \{q^{-2}, q^{-3}, \ldots, q^{-2d}\}.$$

In this case, by inserting

$$\begin{cases} \theta_i - \theta_j = hq^{-i}(1 - q^{i-j})(1 - sq^{i+j+1}), \\ \theta^*_i - \theta^*_j = h^*q^{-i}(1 - q^{i-j})(1 - s^*q^{i+j+1}), \end{cases} \tag{6.186}$$

(6.184) becomes

$$\begin{cases} \lambda'_{i-1} = \frac{q(1-q^{i-1})}{1-q^{i+1}}\lambda'_i + \lambda'_{d-1} & (0 \le i \le d-1), \\ \lambda'_i = \frac{q^{-1}(1-q^{i-d+1})}{1-q^{i-d-1}}\lambda'_{i-1} + \lambda'_0 & (1 \le i \le d). \end{cases} \tag{6.187}$$

By solving (6.187), we obtain the solution as follows:

$$\lambda'_i = \frac{(1-q^{d-i})(1-q^{i+1})}{(1-q)(1-q^d)}\lambda'_0 \quad (-1 \le i \le d). \tag{6.188}$$

By substituting (6.188) and (6.186) into (6.177), we get

$$\begin{aligned} \lambda_i &= hh^* q^{-2i-1}(1-q^{i-d})(1-q^{i+1}) \\ &\quad \times \left\{ -\frac{q^{d+i+1}\lambda'_0}{hh^*(1-q)(1-q^d)} + (1-sq^{i+d+1})(1-s^*q^{i+2}) \right\}. \end{aligned} \tag{6.189}$$

If we set $i = 0$ in (6.189), we have

$$\lambda'_0 = \lambda_0 - hh^* q^{-1}(1-q^{-d})(1-q)(1-sq^{d+1})(1-s^*q^2). \tag{6.190}$$

Substituting (6.190) into (6.189) yields

$$\begin{aligned} \lambda_i &= hh^* q^{-2i-1}(1-q^{i-d})(1-q^{i+1}) \\ &\quad \times \left\{ -\frac{q^{d+i+1}\lambda_0}{hh^*(1-q)(1-q^d)} + (1-q^i)(1-ss^*q^{i+d+3}) \right\}. \end{aligned} \tag{6.191}$$

Here, there exist constants r_1, r_2 which are independent of i such that

$$(1-r_1q^{i+1})(1-r_2q^{i+1}) = (1-q^i)(1-ss^*q^{i+d+3}) - \frac{q^{d+i+1}\lambda_0}{hh^*(1-q)(1-q^d)}. \tag{6.192}$$

Namely, r_1, r_2 are the solution of

$$r_1r_2 = ss^*q^{d+1}, \tag{6.193}$$

$$r_1 + r_2 = q^{-1} + ss^*q^{d+2} + \frac{q^d\lambda_0}{hh^*(1-q)(1-q^d)}. \tag{6.194}$$

Here, we regard r_1, r_2 as parameters satisfying (6.193) and consider that λ_0 is given by (6.194). Then by (6.191), (6.192), λ_i is expressed by AW-parameters and r_1, r_2 as follows:

$$\lambda_i = hh^* q^{-2i-1}(1-q^{i-d})(1-q^{i+1})(1-r_1q^{i+1})(1-r_2q^{i+1}) \quad (0 \le i \le d-1). \tag{6.195}$$

Next, we find $\hat{\lambda}_i$ $(0 \le i \le d-1)$ given by (6.168). Substituting (6.186) and (6.195) into (6.168) yields

$$\begin{aligned}\hat{\lambda}_i &= hh^* q^{-2i-1}(1-q^{i+1})(1-q^{i-d}) \\ &\quad \times \{(1-r_1q^{i+1})(1-r_2q^{i+1}) - (1-s^*q^{2i+2})(1-sq^{d+1})\}.\end{aligned}$$

By (6.193), we have $r_1r_2 = ss^*q^{d+1}$. Hence we have

$$\begin{aligned}&(1-r_1q^{i+1})(1-r_2q^{i+1}) - (1-s^*q^{2i+2})(1-sq^{d+1}) \\ &\quad = s^*q^{2i+2} - (r_1+r_2)q^{i+1} + sq^{d+1} \\ &\quad = \begin{cases} \frac{1}{s^*}(s^*q^{i+1}-r_1)(s^*q^{i+1}-r_2), & \text{if } s^* \neq 0, \\ -(r_1+r_2)q^{i+1} + sq^{d+1}, & \text{if } s^* = 0. \end{cases}\end{aligned}$$

Therefore, if $s^* \neq 0$, we have

$$\hat{\lambda}_i = \frac{hh^*}{s^*}q^{-2i-1}(1-q^{i+1})(1-q^{i-d})(r_1-s^*q^{i+1})(r_2-s^*q^{i+1}) \quad (0 \le i \le d-1). \tag{6.196}$$

If $s^* = 0$, by (6.193), we have $r_1r_2 = ss^*q^{d+1} = 0$. So without loss of generality we may assume $r_2 = 0$. Then we obtain

$$\hat{\lambda}_i = hh^*q^{-2i-1}(1-q^{i+1})(1-q^{i-d})(-r_1q^{i+1} + sq^{d+1}),$$

that is,

$$\hat{\lambda}_i = hh^*q^{d-2i}(1-q^{i+1})(1-q^{i-d})(s-r_1q^{i-d}), \quad (0 \le i \le d-1). \tag{6.197}$$

Case (IA): In (I), we set $s^* = 0$ and take the limits $h \to 0$, $hsq \to h'$ $(h' \neq 0)$. Then the limit of (6.191) is

$$\lambda_i = -h^*q^{-2i-1}(1-q^{i-d})(1-q^{i+1})\frac{q^{d+i+1}\lambda_0}{h^*(1-q)(1-q^d)}.$$

Let

$$r_1' = \frac{q^d\lambda_0}{h'h^*(1-q)(1-q^d)}.$$

Then we get

$$\lambda_i = -h'h^*r_1'q^{-i}(1-q^{i-d})(1-q^{i+1}). \tag{6.198}$$

By (6.193), we have $r_1r_2 = ss^*q^{d+1} = 0$. So without loss of generality, we may assume $r_2 = 0$ in (6.195). In (6.195), if we set

$$hr_1 \to h'r_1', \quad r_2 = 0, \tag{6.199}$$

then we get (6.198). Therefore, by (6.198), (6.197), we obtain

$$\hat{\lambda}_i = h'h^* q^{d-2i-1}(1 - q^{i+1})(1 - q^{i-d})(1 - r_1' q^{i-d+1}). \tag{6.200}$$

Case (II): In (I), take the limits $q \to 1$, $(1-q)^2 h \to h'$, $(1-q)^2 h^* \to (h^*)'$, $\frac{1-s}{1-q} \to s'$, $\frac{1-s^*}{1-q} \to (s^*)'$ ($h' \neq 0$, $(h^*)' \neq 0$). The limit of (6.191) is

$$\lambda_i = h'(h^*)'(i-d)(i+1)\left\{-\frac{\lambda_0}{h'(h^*)'d} + i(i+d+3+s'+(s^*)')\right\}. \tag{6.201}$$

We regard the limits of parameters r_1, r_2 as

$$\frac{1-r_1}{1-q} \to r_1', \quad \frac{1-r_2}{1-q} \to r_2'. \tag{6.202}$$

By taking the limits as

$$\frac{1-r_1 r_2}{1-q} \to r_1' + r_2', \quad \frac{1-ss^*q^{d+1}}{1-q} \to s' + (s^*)' + d + 1,$$

(6.193) becomes

$$r_1' + r_2' = s' + (s^*)' + d + 1. \tag{6.203}$$

By (6.193), (6.194) becomes

$$q(q^{-1} - r_1)(q^{-1} - r_2) + \frac{q^d \lambda_0}{hh^*(1-q)(1-q^d)} = 0.$$

Dividing by $(1-q)^2$ and taking the limit $q \to 1$, we obtain

$$(r_1' + 1)(r_2' + 1) + \frac{\lambda_0}{h'(h^*)'d} = 0.$$

By (6.203), we have

$$r_1' r_2' = -s' - (s^*)' - d - 2 - \frac{\lambda_0}{h'(h^*)'d}. \tag{6.204}$$

Namely, we introduce new parameters r_1', r_2' satisfying (6.203) and consider that λ_0 is given by (6.204). Then by (6.203), (6.204), we have

$$(i+1+r_1')(i+1+r_2') = -\frac{\lambda_0}{h'(h^*)'d} + i(i+d+3+s'+(s^*)').$$

So (6.201) becomes

$$\lambda_i = h'(h^*)'(i-d)(i+1)(i+1+r_1')(i+1+r_2'). \tag{6.205}$$

Note that (6.205) is the limit of (6.195) under (6.202). Therefore, by (6.168), (6.196), (6.202), we get

$$\hat{\lambda}_i = h'(h^*)'(i+1)(i-d)(i+1+(s^*)'-r_1')(i+1+(s^*)'-r_2'). \tag{6.206}$$

Case (IIA): In (II), we take the limits $(h^*)' \to 0$, $(h^*)'(s^*)' \to (h^*)''$ ($(h^*)'' \neq 0$). Then (6.201) becomes

$$\lambda_i = h'(i-d)(i+1)\left\{-\frac{\lambda_0}{h'd} + i(h^*)''\right\}. \tag{6.207}$$

Rewrite (6.203) as

$$(s^*)' + 1 - r_2' = r_1' - s' - d, \tag{6.208}$$

and consider both sides of (6.208) to be finite under the limits $(h^*)' \to 0$, $(h^*)'(s^*)' \to (h^*)''$. Namely, $(h^*)'((s^*)' - r_2') \to 0$, and hence

$$(h^*)'r_2' \to (h^*)''. \tag{6.209}$$

Multiplying both sides of (6.204) by $(h^*)'$ and taking the limit $(h^*)' \to 0$ yields

$$r_1' = -1 - \frac{\lambda_0}{h'(h^*)''d}. \tag{6.210}$$

By (6.207), (6.210), we get

$$\lambda_i = h'(h^*)''(i-d)(i+1)(i+1+r_1'). \tag{6.211}$$

In (6.205), if we take the limit $(h^*)'r_2' \to (h^*)''$ as in (6.209), we obtain (6.211). Therefore, by (6.208), (6.168), (6.206), we get

$$\hat{\lambda}_i = h'(h^*)''(i+1)(i-d)(i+r_1'-s'-d). \tag{6.212}$$

Case (IIB): In (II), we take the limits $h' \to 0$, $h's' \to h''$ ($h'' \neq 0$).
Similarly to Case (IIA), we let $h'(s'-r_2') \to 0$, that is,

$$h'r_2' \to h''. \tag{6.213}$$

By (6.205), (6.206), as the limiting case of Case (II), we get

$$\begin{aligned} \lambda_i &= h''(h^*)'(i-d)(i+1)(i+1+r_1'), \\ \hat{\lambda}_i &= -h''(h^*)'(i+1)(i-d)(i+1+(s^*)'-r_1'). \end{aligned} \tag{6.214}$$

Case (IIC): In (II), we take the limits $h' \to 0$, $(h^*)' \to 0$, $h's' \to h''$, and $(h^*)'(s^*)' \to (h^*)''$ ($h'' \neq 0$, $(h^*)'' \neq 0$).

The limit of (6.201) is

$$\lambda_i = r(i-d)(i+1), \quad r = -\frac{\lambda_0}{d}. \tag{6.215}$$

Following the definition of (6.168), we find $\hat{\lambda}_i$. By $\theta_i = \theta_0 + h''i$, $\theta_i^* = \theta_0^* + (h^*)''i$, we have

$$\begin{aligned}
\hat{\lambda}_i &= \lambda_i - (\theta_{i+1}^* - \theta_i^*)\sum_{j=0}^{i}(\theta_j - \theta_{d-j}) \\
&= \lambda_i - (h^*)''\sum_{j=0}^{i} h''(2j-d) \\
&= \lambda_i - h''(h^*)''(i+1)(i-d).
\end{aligned}$$

By (6.215), we get

$$\hat{\lambda}_i = (r - h''(h^*)'')(i+1)(i-d). \tag{6.216}$$

We interpret (6.215), (6.216) as limits of (6.205), (6.206). Namely, as the limits of r_1', r_2', we have

$$h'r_1' \to r_1'', \quad (h^*)'r_2' \to r_2''.$$

Multiplying both sides of (6.204) by $h'(h^*)'$ and taking the limit yields

$$r_1''r_2'' = -\frac{\lambda_0}{d} = r.$$

So as the limits of (6.205), we get (6.215). By (6.203), we rewrite (6.206) as follows:

$$\begin{aligned}
\hat{\lambda}_i &= h'(h^*)'(i+1)(i-d) \\
&\quad \times \{(i+1+(s^*)')^2 - (r_1'+r_2')(i+1+(s^*)') + r_1'r_2'\} \\
&= h'(h^*)'(i+1)(i-d)\{(i+1+(s^*)')(i+1+(s^*)' - r_1' - r_2') + r_1'r_2'\} \\
&= h'(h^*)'(i+1)(i-d)\{(i+1+(s^*)')(i-s'-d) + r_1'r_2'\}.
\end{aligned}$$

Moreover, by taking the limit, we get

$$\hat{\lambda}_i = (i+1)(i-d)(-(h^*)''h'' + r_1''r_2''),$$

that is, we get (6.216).

Case (III): in (I), we take the limits $q \to -1$, $(1+q)h \to h'$, $(1+q)h^* \to (h^*)'$, $\frac{1-s}{1+q} \to s'$, and $\frac{1-s^*}{1+q} \to (s^*)'$ $(h' \neq 0, (h^*)' \neq 0)$.

Next we write (6.191) as follows:

$$\lambda_i = \frac{(1-q^{d-i})(1-q^{i+1})}{(1-q)(1-q^d)}\lambda_0 + hh^* q^{-2i-1}(1-q^{i-d})(1-q^{i+1})(1-q^i)(1-ss^* q^{i+d+3}). \tag{6.217}$$

Here we introduce the following symbol:

$$\varepsilon(j) = \begin{cases} 1, & \text{if } j \text{ is even,} \\ 0, & \text{if } j \text{ is odd.} \end{cases} \tag{6.218}$$

The limit of the first term of (6.217) is

$$\frac{(1-q^{d-i})(1-q^{i+1})}{(1-q)(1-q^d)}\lambda_0 \to \begin{cases} \frac{\lambda_0}{d}(d-i)^{\varepsilon(i)}(i+1)^{\varepsilon(i+1)}, & \text{if } d \text{ is even,} \\ \varepsilon(i)\lambda_0, & \text{if } d \text{ is odd.} \end{cases}$$

The limit of the second term of (6.217) is

$$\begin{aligned} &hh^* q^{-2i-1}(1-q^{i+1})(1-q^i)(1-q^{i-d})(1-ss^* q^{i+d+3}) \\ &\to -4h'(h^*)'(i+1)^{\varepsilon(i+1)} i^{\varepsilon(i)} (i-d)^{\varepsilon(i-d)} (i+d+3+s'+(s^*)')^{\varepsilon(i-d+1)}. \end{aligned}$$

Therefore, the limits of (6.217) is

$$\lambda_i = \begin{cases} (d-i)(\frac{\lambda_0}{d} + 4h'(h^*)'i), & \text{if } d \text{ and } i \text{ are even,} \\ (i+1)(\frac{\lambda_0}{d} - 4h'(h^*)'(i+d+3+s'+(s^*)')), & \text{if } d \text{ is even and } i \text{ is odd,} \\ \lambda_0 - 4h'(h^*)'i(i+d+3+s'+(s^*)'), & \text{if } d \text{ is odd and } i \text{ is even,} \\ -4h'(h^*)'(i+1)(i-d), & \text{if } d \text{ and } i \text{ are odd.} \end{cases} \tag{6.219}$$

We regard the limits of parameters r_1, r_2 as

$$\frac{1+r_1}{1+q} \to r_1', \quad \frac{1+r_2 q^{d+1}}{1+q} \to r_2' + d + 1. \tag{6.220}$$

We write (6.193) as $r_1 r_2 q^{-d-1} = ss^*$. Since we have

$$\begin{aligned} \frac{1-r_1 r_2 q^{-d-1}}{1+q} &\to r_1' + r_2' - d - 1, \\ \frac{1-ss^*}{1+q} &\to s' + (s^*)', \end{aligned}$$

(6.193) becomes

$$r_1' + r_2' = s' + (s^*)' + d + 1. \tag{6.221}$$

By (6.193), we rewrite (6.194) as

$$(r_1 - q^{-1})(r_2 - q^{-1})(1 - q^d) + \frac{q^{d-1}\lambda_0}{hh^*(1-q)} = 0.$$

Since we have

$$\frac{r_1 - q^{-1}}{1+q} \to r_1' + 1,$$

$$\frac{(r_2 - q^{-1})(1-q^d)}{1+q} \to \begin{cases} 2d, & \text{if } d \text{ is even,} \\ 2(r_2'+1), & \text{if } d \text{ is odd,} \end{cases}$$

(6.194) becomes

$$\frac{\lambda_0}{4h'(h^*)'} = \begin{cases} (r_1'+1)d, & \text{if } d \text{ is even,} \\ -(r_1'+1)(r_2'+1), & \text{if } d \text{ is odd.} \end{cases} \tag{6.222}$$

Therefore, by (6.221), (6.222), we rewrite (6.219) as

$$\lambda_i = \begin{cases} -4h'(h^*)'(i-d)(i+1+r_1'), & \text{if } d \text{ and } i \text{ are even,} \\ -4h'(h^*)'(i+1)(i+1+r_2'), & \text{if } d \text{ is even and } i \text{ is odd,} \\ -4h'(h^*)'(i+1+r_1')(i+1+r_2'), & \text{if } d \text{ is odd and } i \text{ is even,} \\ -4h'(h^*)'(i+1)(i-d), & \text{if } d \text{ and } i \text{ are odd.} \end{cases} \tag{6.223}$$

By using the symbol given by (6.218), we rewrite (6.223) as

$$\begin{aligned} \lambda_i = {} & -4h'(h^*)'(i+1)^{\varepsilon(i+1)}(i-d)^{\varepsilon(i-d)} \\ & \times (i+1+r_1')^{\varepsilon(i)}(i+1+r_2')^{\varepsilon(i-d+1)}. \end{aligned} \tag{6.224}$$

Note that (6.224) is the limit of (6.195) under (6.220). Hence, by (6.220), (6.168), (6.196), we get

$$\begin{aligned} \hat{\lambda}_i = {} & (-1)^d 4h'(h^*)'(i+1)^{\varepsilon(i+1)}(i-d)^{\varepsilon(i-d)} \\ & \times (i+1+(s^*)'-r_1')^{\varepsilon(i)}(i+1+(s^*)'-r_2')^{\varepsilon(i-d+1)}. \end{aligned} \tag{6.225}$$

We summarize the above discussion as the following theorem.

Theorem 6.61. *A pre-L-system $(A, A^*; \{V_i\}_{i=0}^d, \{V_i^*\}_{i=0}^d)$ becomes an L-system only if the data $\{\theta_i\}_{i=0}^d$, $\{\theta_i^*\}_{i=0}^d$, $\{\lambda_i\}_{i=0}^{d-1}$ have the expression by AW-parameters as the following table.*

In the table, we give the data $\{\theta_i\}_{i=0}^d$, $\{\theta_i^*\}_{i=0}^d$, $\{\lambda_i\}_{i=0}^{d-1}$, and $\{\hat{\lambda}_i\}_{i=0}^{d-1}$ in (6.168). The symbols Λ_I, Λ_{IA}, Λ_{II}, Λ_{IIA}, Λ_{IIB}, Λ_{IIC}, and Λ_{III} refer to the sets of AW-parameters. At the bottom of each part, we give the condition for $\theta_i \neq \theta_j$, $\theta_i^* \neq \theta_j^*$ $(i \neq j)$, $\lambda_i \neq 0$, $\hat{\lambda}_i \neq 0$ $(0 \leq i \leq d-1)$.

(I)

$$\begin{array}{|ll|}\hline \theta_i = \theta_0 - h(1-q^{-i})(1-sq^{i+1}) & (0 \le i \le d) \\ \theta^*_i = \theta^*_0 - h^*(1-q^{-i})(1-s^*q^{i+1}) & (0 \le i \le d) \\ \lambda_i = hh^*q^{-2i-1}(1-q^{i+1})(1-q^{i-d}) & \\ \quad \times (1-r_1q^{i+1})(1-r_2q^{i+1}) & (0 \le i \le d-1) \\ \textit{If } s^* \ne 0, & \\ \hat{\lambda}_i = \frac{1}{s^*}hh^*q^{-2i-1}(1-q^{i+1})(1-q^{i-d}) & \\ \quad \times (r_1 - s^*q^{i+1})(r_2 - s^*q^{i+1}) & (0 \le i \le d-1) \\ \textit{If } s^* = 0, & \\ \hat{\lambda}_i = hh^*q^{d-2i}(1-q^{i+1})(1-q^{i-d})(s-r_1q^{i-d}) & (0 \le i \le d-1) \\ \hline \end{array}$$

$\Lambda_I = (r_1, r_2, s, s^*; h, h^*, \theta_0, \theta^*_0 \mid q, d)$, $r_1r_2 = ss^*q^{d+1}$:
 $h, h^* \ne 0$; $q^i \ne 1$ $(1 \le i \le d)$; $s, s^* \notin \{q^{-2}, q^{-3}, \ldots, q^{-2d}\}$;
 If $s^* \ne 0$, $r_1, r_2 \notin \{q^{-1}, q^{-2}, \ldots, q^{-d}\} \cup \{s^*q, s^*q^2, \ldots, s^*q^d\}$.
 If $s^* = 0$, $r_2 = 0$, $r_1 \notin \{q^{-1}, q^{-2}, \ldots, q^{-d}\} \cup \{sq, sq^2, \ldots, sq^d\}$.

(IA) *In (I), set* $s^* = 0$, $r_2 = 0$, *and take the limits* $h \to 0$, $hsq \to h'$, $hr_1 \to h'r'_1$.

$$\begin{array}{|ll|}\hline \theta_i = \theta_0 - h(1-q^{i}) & (0 \le i \le d) \\ \theta^*_i = \theta^*_0 - h^*(1-q^{-i}) & (0 \le i \le d) \\ \lambda_i = -hh^*r_1q^{-i}(1-q^{i+1})(1-q^{i-d}) & (0 \le i \le d-1) \\ \hat{\lambda}_i = hh^*q^{d-2i-1}(1-q^{i+1})(1-q^{i-d})(1-r_1q^{i-d+1}) & (0 \le i \le d-1) \\ \hline \end{array}$$

$\Lambda_{IA} = (r_1, -, -, -; h, h^*, \theta_0, \theta^*_0 \mid q, d)$:
 $h, h^* \ne 0$; $q^i \ne 1$ $(1 \le i \le d)$; $r_1 \ne 0, q^i$ $(0 \le i \le d-1)$.

(II) *In (I), take the limits* $q \to 1$, $(1-q)^2h \to h'$, $(1-q)^2h^* \to (h^*)'$, $\frac{1-s}{1-q} \to s'$, $\frac{1-s^*}{1-q} \to (s^*)'$, $\frac{1-r_1}{1-q} \to r'_1$, $\frac{1-r_2}{1-q} \to r'_2$.

$$\begin{array}{|ll|}\hline \theta_i = \theta_0 + hi(i+1+s) & (0 \le i \le d) \\ \theta^*_i = \theta^*_0 + h^*i(i+1+s^*) & (0 \le i \le d) \\ \lambda_i = hh^*(i+1)(i-d)(i+1+r_1)(i+1+r_2) & (0 \le i \le d-1) \\ \hat{\lambda}_i = hh^*(i+1)(i-d) & \\ \quad \times (i+1+s^*-r_1)(i+1+s^*-r_2) & (0 \le i \le d-1) \\ \hline \end{array}$$

$\Lambda_{II} = (r_1, r_2, s, s^*; h, h^*, \theta_0, \theta^*_0 \mid d)$, $r_1 + r_2 = s + s^* + d + 1$:
 $h, h^* \ne 0$; $s, s^* \notin \{-2, -3, \ldots, -2d\}$;
 $r_1, r_2 \notin \{-1, -2, \ldots, -d\} \cup \{s^*+1, s^*+2, \ldots, s^*+d\}$.

(IIA) *In (II), take the limits* $h^* \to 0$, $h^*s^* \to (h^*)'$, $h^*r_2 \to (h^*)'$.

$$\begin{array}{|ll|}\hline \theta_i = \theta_0 + hi(i+1+s) & (0 \le i \le d) \\ \theta^*_i = \theta^*_0 + h^*i & (0 \le i \le d) \\ \lambda_i = hh^*(i+1)(i-d)(i+1+r_1) & (0 \le i \le d-1) \\ \hat{\lambda}_i = hh^*(i+1)(i-d)(i+r_1-s-d) & (0 \le i \le d-1) \\ \hline \end{array}$$

$\Lambda_{IIA} = (r_1, -, s, -; h, h^*, \theta_0, \theta^*_0 \mid d)$:
 $h, h^* \ne 0$; $s \notin \{-2, -3, \ldots, -2d\}$;
 $r_1 \notin \{-1, -2, \ldots, -d\} \cup \{s+1, s+2, \ldots, s+d\}$.

(**IIB**) *In (II), take the limits $h \to 0$, $hs \to h'$, $hr_2 \to h'$.*

$\theta_i = \theta_0 + hi$	$(0 \le i \le d)$
$\theta_i^* = \theta_0^* + h^* i(i+1+s^*)$	$(0 \le i \le d)$
$\lambda_i = hh^*(i+1)(i-d)(i+1+r_1)$	$(0 \le i \le d-1)$
$\hat{\lambda}_i = -hh^*(i+1)(i-d)(i+1+s^*-r_1)$	$(0 \le i \le d-1)$

$\Lambda_{IIB} = (r_1, -, -, s^*; h, h^*, \theta_0, \theta_0^* \mid d)$:

$h, h^* \neq 0$; $s^* \notin \{-2, -3, \dots, -2d\}$;

$r_1 \notin \{-1, -2, \dots, -d\} \cup \{s^*+1, s^*+2, \dots, s^*+d\}$.

(**IIC**) *In (II), take the limits $h \to 0$, $h^* \to 0$, $hs \to h'$, $h^*s^* \to (h^*)'$, $hr_1 \to r_1'$, $h^*r_2 \to r_2'$. Set $r = r_1'r_2'$.*

$\theta_i = \theta_0 + hi$	$(0 \le i \le d)$
$\theta_i^* = \theta_0^* + h^* i$	$(0 \le i \le d)$
$\lambda_i = r(i+1)(i-d)$	$(0 \le i \le d-1)$
$\hat{\lambda}_i = (r-hh^*)(i+1)(i-d)$	$(0 \le i \le d-1)$

$\Lambda_{IIC} = (r_1, r_2, -, -; h, h^*, \theta_0, \theta_0^* \mid d)$:

$r = r_1 r_2$; $h, h^* \neq 0$; $r \neq 0, hh^*$.

(**III**) *In (I), take the limits $q \to -1$, $(1+q)h \to h'$, $(1+q)h^* \to (h^*)'$, $\frac{1-s}{1+q} \to s'$, $\frac{1-s^*}{1+q} \to (s^*)'$, $\frac{1+r_1}{1+q} \to r_1'$, $\frac{1+r_2 q^{d+1}}{1+q} \to r_2' + d + 1$. Set $\varepsilon(i) = 1$ if i is even, and $\varepsilon(i) = 0$ if i is odd.*

$\theta_i = \theta_0 + (-1)^i 2hi^{\varepsilon(i)}(i+1+s)^{\varepsilon(i+1)}$	$(0 \le i \le d)$
$\theta_i^* = \theta_0^* + (-1)^i 2h^* i^{\varepsilon(i)}(i+1+s^*)^{\varepsilon(i+1)}$	$(0 \le i \le d)$
$\lambda_i = -4hh^*(i+1)^{\varepsilon(i+1)}(i-d)^{\varepsilon(i-d)} \times (i+1+r_1)^{\varepsilon(i)}(i+1+r_2)^{\varepsilon(i-d+1)}$	$(0 \le i \le d-1)$
$\hat{\lambda}_i = (-1)^d 4hh^*(i+1)^{\varepsilon(i+1)}(i-d)^{\varepsilon(i-d)} \times (i+1+s^*-r_1)^{\varepsilon(i)}(i+1+s^*-r_2)^{\varepsilon(i-d+1)}$	$(0 \le i \le d-1)$

$\Lambda_{III} = (r_1, r_2, s, s^*; h, h^*, \theta_0, \theta_0^* \mid d)$:

$r_1 + r_2 = s + s^* + d + 1$; $h, h^* \neq 0$; $s, s^* \notin \{-2, -4, \dots, -2(d-1), -2d\}$;

$r_1 \neq -1-2j, s^*+1+2j$ $(0 \le 2j \le d-1)$,

$r_2 \neq -d+2j, s^*+d-2j$ $(0 \le 2j \le d-1)$.

6.3.4 AW-relations

The relations of A, A^* in $\mathrm{End}(V)$

$$
\text{(AW)} \quad \begin{cases} A^2A^* - \beta AA^*A + A^*A^2 \\ \quad = \gamma(AA^* + A^*A) + \delta A^* + \gamma^* A^2 + \omega A + \eta^*, \\ (A^*)^2 A - \beta A^*AA^* + A(A^*)^2 \\ \quad = \gamma^*(A^*A + AA^*) + \delta^* A + \gamma(A^*)^2 + \omega A^* + \eta \end{cases}
$$

are called *Askey–Wilson relations (AW-relations)*. Here, $\beta, \gamma, \gamma^*, \delta, \delta^*, \omega, \eta, \eta^*$ are complex constants determined for each AW-relation. In this section, we first show that

L-pairs satisfy AW-relations. Then, we consider the condition for pre-TD-pairs and pre-L-pairs to satisfy AW-relations. As an application, we show that if a TD-pair satisfies AW-relations, it becomes an L-pair.

Before the discussion of the main theory, we look at another expression of AW-relations (AW). In the first equation of (AW), replace ω by ω^* and denote it as $(\mathrm{AW})_1$, and denote the second equation of (AW) as $(\mathrm{AW})_2$. Namely, if we set

$$C = A^2A^* - \beta AA^*A + A^*A^2 - \gamma(AA^* + A^*A) - \delta A^*, \tag{6.226}$$

$$C^* = (A^*)^2A - \beta A^*AA^* + A(A^*)^2 - \gamma^*(A^*A + AA^*) - \delta^* A, \tag{6.227}$$

then we have

$$(\mathrm{AW})_1 : \quad C = \gamma^* A^2 + \omega^* A + \eta^*, \tag{6.228}$$

$$(\mathrm{AW})_2 : \quad C^* = \gamma(A^*)^2 + \omega A^* + \eta. \tag{6.229}$$

If we let $\omega = \omega^*$, (AW) holds if and only if $(\mathrm{AW})_1$ and $(\mathrm{AW})_2$ hold. Define the bracket product $[\ ,\ ]$ by $[X, Y] = XY - YX$. We rewrite TD-relations as follows:

$$(\mathrm{TD}) \quad \begin{cases} [A, C] = 0, \\ [A^*, C^*] = 0. \end{cases} \tag{6.230}$$

Hence, if both $(\mathrm{AW})_1$ and $(\mathrm{AW})_2$ hold, then (TD) holds. In particular, if (AW)-relations hold, then (TD)-relations hold.

Lemma 6.62. *Let $[A, A^*] \neq 0$. If $(AW)_1$ and $(AW)_2$ hold, we have $\omega = \omega^*$, that is, (AW) holds.*

Proof. By the definition (6.226), (6.227) of C, C^*, we have

$$[C, A^*] + [C^*, A] = \gamma[(A^*)^2, A] + \gamma^*[A^2, A^*]. \tag{6.231}$$

On the other hand, if $(\mathrm{AW})_1$, $(\mathrm{AW})_2$ hold, by (6.228), (6.229), we get

$$[C, A^*] + [C^*, A] = \gamma^*[A^2, A^*] + \gamma[(A^*)^2, A] + (\omega^* - \omega)[A, A^*]. \tag{6.232}$$

By (6.231), (6.232), and $[A, A^*] \neq 0$, we obtain $\omega^* - \omega = 0$. □

The next lemma is checked by the direct calculation.

Lemma 6.63. *For $A, A^* \in \mathrm{End}(V)$ and $\alpha, \alpha^* \in \mathbb{C}$, we let*

$$\tilde{A} = A + \alpha, \quad \tilde{A}^* = A^* + \alpha^*.$$

If A, A^* satisfy $(AW)_1$ for constants $\beta, \gamma, \delta, \gamma^*, \omega^*, \eta^*$, then $\tilde{A}, \tilde{A}^*$ satisfy $(AW)_1$ for constants β and

$$\tilde{\gamma} = \gamma - \alpha(\beta - 2),$$
$$\tilde{\delta} = \delta - \alpha(\gamma + \tilde{\gamma}),$$

$$\begin{aligned}
\tilde{\gamma}^* &= \gamma^* - \alpha^*(\beta - 2),\\
\tilde{\omega}^* &= \omega^* - 2(\alpha^*\gamma + \alpha\tilde{\gamma}^*) = \omega^* - 2(\alpha^*\gamma + \alpha\gamma^* - (\beta - 2)\alpha\alpha^*),\\
\tilde{\eta}^* &= \eta^* - \alpha^*\delta - \alpha(\alpha\tilde{\gamma}^* + \tilde{\omega}^*) = \eta^* - \alpha^*\tilde{\delta} - \alpha(-\alpha\gamma^* + \omega^*).
\end{aligned}$$

If A, A^* satisfy $(AW)_2$ for constants $\beta, \gamma^*, \delta^*, \gamma, \omega, \eta$, then $\tilde{A}, \tilde{A}^*$ satisfy $(AW)_2$ for constants β and

$$\begin{aligned}
\tilde{\gamma}^* &= \gamma^* - \alpha^*(\beta - 2),\\
\tilde{\delta}^* &= \delta^* - \alpha^*(\gamma^* + \tilde{\gamma}^*),\\
\tilde{\gamma} &= \gamma - \alpha(\beta - 2),\\
\tilde{\omega} &= \omega - 2(\alpha\gamma^* + \alpha^*\tilde{\gamma}) = \omega - 2(\alpha\gamma^* + \alpha^*\gamma - (\beta - 2)\alpha^*\alpha),\\
\tilde{\eta} &= \eta - \alpha\delta^* - \alpha^*(\alpha^*\tilde{\gamma} + \tilde{\omega}) = \eta - \alpha\tilde{\delta}^* - \alpha^*(-\alpha^*\gamma + \omega).
\end{aligned}$$

Let $A, A^* \in \mathrm{End}(V)$ be an L-pair, and let $(\{V_i\}_{i=0}^d, \{V_i^*\}_{i=0}^d)$ be an associated L-system. Let $\{\theta_i\}_{i=0}^d$ be the eigenvalues of A, and let $\{\theta_i^*\}_{i=0}^d$ be the eigenvalues of A^*. Let E_i, E_i^* be the projections $E_i : V = \bigoplus_{j=0}^d V_j \longrightarrow V_i$, $E_i^* : V = \bigoplus_{j=0}^d V_j^* \longrightarrow V_i^*$ and let $a_i = \mathrm{tr}(E_i^* A E_i^*)$, $a_i^* = \mathrm{tr}(E_i A^* E_i)$ $(0 \le i \le d)$. Since an L-pair is a TD-pair, by Theorem 6.44 in Section 6.2, there exist constants $\beta, \gamma, \gamma^*, \delta, \delta^*$ such that $\{\theta_i\}_{i\in\mathbb{Z}}$ is a (β, γ, δ)-sequence and $\{\theta_i^*\}_{i\in\mathbb{Z}}$ is a $(\beta, \gamma^*, \delta^*)$-sequence in the sense of Remark 6.46 (2). Then the following theorem holds [480].

Theorem 6.64. *An L-pair A, A^* satisfies AW-relations. Namely, there exist constants $\omega, \eta, \eta^* \in \mathbb{C}$ such that C, C^* in (6.226), (6.227) satisfy*

$$(AW) \quad \begin{cases} C = \gamma^* A^2 + \omega A + \eta^*,\\ C^* = \gamma (A^*)^2 + \omega A^* + \eta. \end{cases}$$

Moreover, the following hold:

$$\begin{aligned}
&\gamma^*\theta_i^2 + \omega\theta_i + \eta^* = a_i^*(\theta_i - \theta_{i-1})(\theta_i - \theta_{i+1}) \quad (0 \le i \le d),\\
&\gamma(\theta_i^*)^2 + \omega\theta_i^* + \eta = a_i(\theta_i^* - \theta_{i-1}^*)(\theta_i^* - \theta_{i+1}^*) \quad (0 \le i \le d),\\
&\omega = a_i^*(\theta_i - \theta_{i+1}) + a_{i-1}^*(\theta_{i-1} - \theta_{i-2}) - \gamma^*(\theta_i + \theta_{i-1})\\
&\quad = a_i(\theta_i^* - \theta_{i+1}^*) + a_{i-1}(\theta_{i-1}^* - \theta_{i-2}^*) - \gamma(\theta_i^* + \theta_{i-1}^*) \quad (1 \le i \le d).
\end{aligned}$$

Proof. First we show there exist $\omega^*, \eta^* \in \mathbb{C}$ satisfying $(\mathrm{AW})_1$ in (6.228). Since an L-pair $A, A^* \in \mathrm{End}(V)$ is a TD-pair, by Theorem 6.44 in Section 6.2, it satisfies TD-relations (6.230). Therefore, C and A are commutative. Since A is diagonalizable and each eigenspace is 1-dimensional, an element in $\mathrm{End}(V)$ which commutes with A must be a polynomial in A. Therefore, there exist $\alpha_0 \neq 0, \alpha_1, \dots, \alpha_d \in \mathbb{C}$ such that

$$C = \alpha_0 A^n + \alpha_1 A^{n-1} + \cdots + \alpha_{n-1}A + \alpha_n. \tag{6.233}$$

Here, without loss of generality, we may assume the diameter d of the L-pair A, A^* satisfies $d \geq 2$. For, the minimal polynomial of A has degree $d+1$ and if $d \leq 1$, the right-hand side of (6.233) is a polynomial in A of degree at most 1 and the right-hand side of (6.228) is also a polynomial in A of degree at most 1.

Following the definition (6.226) of C, we find $E_i^* C E_0^*$ $(i \geq 2)$. By Lemma 6.40 in Section 6.2, we have

$$E_i^* C E_0^* = \begin{cases} (\theta_0^* - \beta\theta_1^* + \theta_2^*) E_2^* A^2 E_0^*, & \text{if } i = 2, \\ 0, & \text{if } i \geq 3. \end{cases} \tag{6.234}$$

On the other hand, following (6.233), we find $E_n^* C E_0^*$. By Lemma 6.40 and Proposition 6.41 in Section 6.2, we have

$$E_n^* C E_0^* = \alpha_0 E_n^* A^n E_0^* \neq 0. \tag{6.235}$$

Comparing (6.234) and (6.235) yields $n \leq 2$. Hence, there exist $\alpha^*, \omega^*, \eta^* \in \mathbb{C}$ such that

$$C = \alpha^* A^2 + \omega^* A + \eta^*. \tag{6.236}$$

By (6.236), we have

$$E_2^* C E_0^* = \alpha^* E_2^* A^2 E_0^*. \tag{6.237}$$

By Proposition 6.41, we have $E_2^* A^2 E_0^* \neq 0$. So, by (6.234), (6.237), we have $\alpha^* = \theta_0^* - \beta\theta_1^* + \theta_2^*$. Since $\{\theta_i^*\}_{i \in \mathbb{Z}}$ is a (β, γ^*)-sequence, we have $\theta_0^* - \beta\theta_1^* + \theta_2^* = \gamma^*$. Thus, we get $\alpha^* = \gamma^*$ and $(\mathrm{AW})_1$ in (6.228) holds.

Similarly, there exist $\omega, \eta \in \mathbb{C}$ such that $(\mathrm{AW})_2$ in (6.228) holds. We show $\omega = \omega^*$. Without loss of generality, we may assume the diameter d of the L-pair A, A^* satisfies $d \geq 1$. If $d = 0$, A, A^* are constants, and by a suitable choice of η^*, η in (6.228), (6.229), we may set $\omega = \omega^*$. Let $d \geq 1$. We have $E_1^*[A, A^*]E_0^* = (\theta_0^* - \theta_1^*)E_1^* A E_0^*$. Since $E_1^* A E_0^* \neq 0$ by Proposition 6.41, we have $E_1^*[A, A^*]E_0^* \neq 0$. Hence, $[A, A^*] \neq 0$. By Lemma 6.62, we get $\omega = \omega^*$. Thus (AW) holds.

Following the definition (6.226) of C, we find $E_i C E_i$. We have $E_i C E_i = (\theta_i^2 - \beta\theta_i^2 + \theta_i^2 - 2\gamma\theta_i - \delta) E_i A^* E_i$. By $\mathrm{tr}(E_i A^* E_i) = a_i^*$ and $\dim(V_i) = 1$, we have $E_i A^* E_i = a_i^* E_i$. Moreover, by Lemma 6.45 in Section 6.2, for a (β, γ, δ)-sequence $\{\theta_i\}_{i \in \mathbb{Z}}$, we have $(\beta - 2)\theta_i^2 + 2\gamma\theta_i + \delta = (\theta_{i+1} - \theta_i)(\theta_i - \theta_{i-1})$, and we obtain

$$E_i C E_i = -(\theta_{i+1} - \theta_i)(\theta_i - \theta_{i-1}) a_i^* E_i \quad (0 \leq i \leq d). \tag{6.238}$$

On the other hand, following $C = \gamma^* A^2 + \omega A + \eta^*$, we find $E_i C E_i$. Then we get

$$E_i C E_i = (\gamma^* \theta_i^2 + \omega\theta_i + \eta^*) E_i \quad (0 \leq i \leq d). \tag{6.239}$$

By (6.238), (6.239), we obtain

$$\gamma^* \theta_i^2 + \omega\theta_i + \eta^* = a_i^* (\theta_i - \theta_{i-1})(\theta_i - \theta_{i+1}) \quad (0 \leq i \leq d). \tag{6.240}$$

Similarly, by $E_i^* C^* E_i^*$, we have

$$\gamma(\theta_i^*)^2 + \omega\theta_i^* + \eta = a_i(\theta_i^* - \theta_{i-1}^*)(\theta_i^* - \theta_{i+1}^*) \quad (0 \le i \le d). \tag{6.241}$$

In (6.240), subtracting the $(i-1)$-th equation from the i-th equation and deleting η^* yields the desired equation on ω. Moreover, in (6.241), subtracting the $(i-1)$-th equation from the i-th equation and deleting η yields the other equation on ω. □

Next, let $A, A^* \in \mathrm{End}(V)$ be a pre-TD-pair. Following the definition (6.145) in Section 6.3.2, we express A, A^* as follows. A direct sum decomposition $V = \bigoplus_{i=0}^d U_i$ and linear transformations $R, L \in \mathrm{End}(V)$ such that

$$RU_i \subseteq U_{i+1}, \quad LU_i \subseteq U_{i-1} \quad (0 \le i \le d), \tag{6.242}$$

where $U_{-1} = U_{d+1} = \{0\}$, are given. Moreover, distinct numbers $\theta_0, \theta_1, \ldots, \theta_d \in \mathbb{C}$ and $\theta_0^*, \theta_1^*, \ldots, \theta_d^* \in \mathbb{C}$ are given. For the projection $F_i : V = \bigoplus_{j=0}^d U_j \longrightarrow U_i$, let

$$F = \sum_{i=0}^d \theta_i F_i, \quad F^* = \sum_{i=0}^d \theta_i^* F_i. \tag{6.243}$$

Then A, A^* are given by

$$A = R + F, \quad A^* = L + F^*. \tag{6.244}$$

It is checked by the direct calculation that the relations $(\mathrm{AW})_1$, $(\mathrm{AW})_2$ are rewritten as the following lemma.

Lemma 6.65. *The following hold:*

(1) *We have*

$$A^2A^* - \beta AA^*A + A^*A^2 = X_2 + X_1 + X_0 + X_{-1},$$

where

$$\begin{aligned}
X_2 &= R^2F^* - \beta RF^*R + F^*R^2, \\
X_1 &= R^2L + RFF^* + FRF^* - \beta(RLR + RF^*F + FF^*R) \\
&\quad + LR^2 + F^*FR + F^*RF, \\
X_0 &= RFL + FRL + F^2F^* - \beta(RLF + FLR + FF^*F) \\
&\quad + LFR + LRF + F^*F^2, \\
X_{-1} &= F^2L - \beta FLF + LF^2.
\end{aligned}$$

We also have

$$\gamma(AA^* + A^*A) + \delta A^* + \gamma^* A^2 + \omega^* A + \eta^* = Y_2 + Y_1 + Y_0 + Y_{-1},$$

where

$$\begin{aligned}
Y_2 &= \gamma^* R^2,\\
Y_1 &= \gamma(RF^* + F^*R) + \gamma^*(RF + FR) + \omega^* R,\\
Y_0 &= \gamma(RL + LR + FF^* + F^*F) + \delta F^* + \gamma^* F^2 + \omega^* F + \eta^*,\\
Y_{-1} &= \gamma(FL + LF) + \delta L.
\end{aligned}$$

In particular,

$$\begin{aligned}
&A^2A^* - \beta AA^*A + A^*A^2\\
&\quad = \gamma(AA^* + A^*A) + \delta A^* + \gamma^* A^2 + \omega^* A + \eta^*
\end{aligned}$$

holds if and only if the following (i)–(iv) *hold:*
(i) $X_{-1}|_{U_i} = Y_{-1}|_{U_i}$ $(1 \le i \le d)$;
(ii) $X_2|_{U_i} = Y_2|_{U_i}$ $(0 \le i \le d-2)$;
(iii) $X_0|_{U_i} = Y_0|_{U_i}$ $(0 \le i \le d)$;
(iv) $X_1|_{U_i} = Y_1|_{U_i}$ $(0 \le i \le d-1)$.

(2) *We have*

$$\begin{aligned}
(A^*)^2A - \beta A^*AA^* + A(A^*)^2 &= X'_{-2} + X'_{-1} + X'_0 + X'_1,\\
\gamma^*(A^*A + AA^*) + \delta^* A + \gamma(A^*)^2 + \omega A^* + \eta &= Y'_{-2} + Y'_{-1} + Y'_0 + Y'_1,
\end{aligned}$$

where $X'_{-2}, X'_{-1}, X'_0, X'_1; Y'_{-2}, Y'_{-1}, Y'_0, Y'_1$ *are obtained by exchanging as*

$$X_2, X_1, X_0, X_{-1}; Y_2, Y_1, Y_0, Y_{-1},$$

and applying the following operations:

$$\begin{aligned}
(R, F) &\leftrightarrow (L, F^*),\\
(\gamma, \delta, \omega^*, \eta^*) &\leftrightarrow (\gamma^*, \delta^*, \omega, \eta).
\end{aligned}$$

In particular,

$$\begin{aligned}
&(A^*)^2A - \beta A^*AA^* + A(A^*)^2\\
&\quad = \gamma^*(A^*A + AA^*) + \delta^* A + \gamma(A^*)^2 + \omega A^* + \eta
\end{aligned}$$

holds if and only if the following (i′)–(iv′) *hold:*
(i′) $X'_1|_{U_i} = Y'_1|_{U_i}$ $(0 \le i \le d-1)$;
(ii′) $X'_{-2}|_{U_i} = Y'_{-2}|_{U_i}$ $(2 \le i \le d)$;
(iii′) $X'_0|_{U_i} = Y'_0|_{U_i}$ $(0 \le i \le d)$;
(iv′) $X'_{-1}|_{U_i} = Y'_{-1}|_{U_i}$ $(1 \le i \le d)$.

Proposition 6.66. *Let $d \geq 1$, and assume $R^2U_i \neq 0$ $(0 \leq i \leq d-2)$ and $L^2U_i \neq 0$ $(2 \leq i \leq d)$, and if $d = 1$, assume $RU_0 \neq 0$, $LU_1 \neq 0$. Then the following (1), (2) hold:*

(1) *We have*

$$\begin{aligned} &A^2A^* - \beta AA^*A + A^*A^2 \\ &= \gamma(AA^* + A^*A) + \delta A^* + \gamma^*A^2 + \omega^*A + \eta^* \end{aligned}$$

if and only if the following (i)–(iv) hold:

(i) $\delta = \theta_i^2 - \beta\theta_i\theta_{i-1} + \theta_{i-1}^2 - \gamma(\theta_i + \theta_{i-1})$ $(1 \leq i \leq d)$, $\gamma = \theta_i - \beta\theta_{i-1} + \theta_{i-2}$ $(2 \leq i \leq d)$, *namely, $\{\theta_i\}_{i=0}^d$ is a (β, γ, δ)-sequence;*

(ii) $\gamma^* = \theta_i^* - \beta\theta_{i-1}^* + \theta_{i-2}^*$ $(2 \leq i \leq d)$, *namely, $\{\theta_i^*\}_{i=0}^d$ is a (β, γ^*)-sequence;*

(iii) $((\theta_i - \theta_{i-1})LR - (\theta_{i+1} - \theta_i)RL)|_{U_i} = (\theta_{i+1} - \theta_i)(\theta_i - \theta_{i-1})\theta_i^* + \gamma^*\theta_i^2 + \omega^*\theta_i + \eta^*$ $(0 \leq i \leq d)$;

(iv) $(R^2L - \beta RLR + LR^2)|_{U_i} = ((\theta_{i+2} - \theta_{i+1})\theta_{i+1}^* - (\theta_i - \theta_{i-1})\theta_i^* + \gamma^*(\theta_i + \theta_{i+1}) + \omega^*)R|_{U_i}$ $(0 \leq i \leq d-1)$.

Here, we interpret that θ_i are extended to the (β, γ)-sequence $\{\theta_i\}_{i\in\mathbb{Z}}$, and θ_i^* are extended to the (β, γ^*)-sequence $\{\theta_i^*\}_{i\in\mathbb{Z}}$ in the sense of Remark 6.46 (2) in Section 6.2.

(2) *We have*

$$\begin{aligned} &(A^*)^2A - \beta A^*AA^* + A(A^*)^2 \\ &= \gamma^*(A^*A + AA^*) + \delta^*A + \gamma(A^*)^2 + \omega A^* + \eta \end{aligned}$$

if and only if the following (i′)–(iv′) hold:

(i′) $\delta^* = (\theta_i^*)^2 - \beta\theta_i^*\theta_{i+1}^* + (\theta_{i+1}^*)^2 - \gamma^*(\theta_i^* + \theta_{i+1}^*)$ $(0 \leq i \leq d-1)$, $\gamma^* = \theta_i^* - \beta\theta_{i+1}^* + \theta_{i+2}^*$ $(0 \leq i \leq d-2)$, *namely, $\{\theta_i^*\}_{i=0}^d$ is a $(\beta, \gamma^*, \delta^*)$-sequence;*

(ii′) $\gamma = \theta_i - \beta\theta_{i+1} + \theta_{i+2}$ $(0 \leq i \leq d-2)$, *namely, $\{\theta_i\}_{i=0}^d$ is a (β, γ)-sequence;*

(iii′) $((\theta_i^* - \theta_{i+1}^*)RL - (\theta_{i-1}^* - \theta_i^*)LR)|_{U_i} = (\theta_{i-1}^* - \theta_i^*)(\theta_i^* - \theta_{i+1}^*)\theta_i + \gamma(\theta_i^*)^2 + \omega\theta_i^* + \eta$ $(0 \leq i \leq d)$.

(iv′) $(L^2R - \beta LRL + RL^2)|_{U_i} = ((\theta_{i-2}^* - \theta_{i-1}^*)\theta_{i-1} - (\theta_i^* - \theta_{i+1}^*)\theta_i + \gamma(\theta_i^* + \theta_{i-1}^*) + \omega)L|_{U_i}$ $(1 \leq i \leq d)$.

Here, we interpret that θ_i are extended to the (β, γ)-sequence $\{\theta_i\}_{i\in\mathbb{Z}}$, and θ_i^* are extended to the (β, γ^*)-sequence $\{\theta_i^*\}_{i\in\mathbb{Z}}$ in the sense of Remark 6.46 (2) in Section 6.2.

Proof. We prove (1) only. The proof of (2) is similar. Corresponding to conditions (i), (ii), (iii), (iv) in Lemma 6.65 (1), we obtain (i), (ii), (iii), and (iv) of the proposition as follows.

(i) Since we have

$$X_{-1}|_{U_i} = (\theta_{i-1}^2 - \beta\theta_{i-1}\theta_i + \theta_i^2)L|_{U_i},$$
$$Y_{-1}|_{U_i} = (\gamma(\theta_{i-1} + \theta_i) + \delta)L|_{U_i},$$

by the assumption that $L|_{U_i} \neq 0$ $(1 \le i \le d)$, we get

$$\delta = \theta_i^2 - \beta\theta_i\theta_{i-1} + \theta_{i-1}^2 - \gamma(\theta_i + \theta_{i-1}) \quad (1 \le i \le d).$$

Subtracting the $(i-1)$-th equation from the i-th equation and deleting δ yields

$$\gamma = \theta_i - \beta\theta_{i-1} + \theta_{i-2} \quad (2 \le i \le d).$$

If $d = 1$, we regard $\gamma = \theta_1 - \beta\theta_0 + \theta_{-1}$ as the definition of θ_{-1}.
(ii) Since we have

$$X_2|_{U_i} = (\theta_i^* - \beta\theta_{i+1}^* + \theta_{i+2}^*)R^2|_{U_i}, \quad Y_2|_{U_i} = \gamma^* R^2|_{U_i},$$

by the assumption that $R^2|_{U_i} \neq 0$ $(0 \le i \le d-2)$, we get

$$\gamma^* = \theta_{i+2}^* - \beta\theta_{i+1}^* + \theta_i^* \quad (0 \le i \le d-2).$$

If $d = 1$, we regard $\gamma^* = \theta_2^* - \beta\theta_1^* + \theta_0^*$ as the definition of θ_2^*.
(iii) Since we have

$$X_0|_{U_i} = ((\theta_{i-1} + \theta_i - \beta\theta_i)RL + (\theta_{i+1} + \theta_i - \beta\theta_i)LR + (2-\beta)\theta_i^2\theta_i^*)|_{U_i},$$
$$Y_0|_{U_i} = (\gamma RL + \gamma LR + 2\gamma\theta_i\theta_i^* + \delta\theta_i^* + \gamma^*\theta_i^2 + \omega^*\theta_i + \eta^*)|_{U_i},$$

we obtain

$$\begin{aligned}&((\theta_{i-1} + \theta_i - \beta\theta_i - \gamma)RL + (\theta_{i+1} + \theta_i - \beta\theta_i - \gamma)LR)|_{U_i}\\ &\quad = ((\beta - 2)\theta_i^2 + 2\gamma\theta_i + \delta)\theta_i^* + \gamma^*\theta_i^2 + \omega^*\theta_i + \eta^* \quad (0 \le i \le d).\end{aligned}$$

By (i), $\{\theta_i\}_{i\in\mathbb{Z}}$ is a (β, γ, δ)-sequence. So we have $\gamma + \beta\theta_i = \theta_{i+1} + \theta_{i-1}$. Moreover, by Lemma 6.45 in Section 6.2, we have

$$(\beta - 2)\theta_i^2 + 2\gamma\theta_i + \delta = (\theta_{i+1} - \theta_i)(\theta_i - \theta_{i-1}).$$

Hence, we obtain (iii).
(iv) Since we have

$$\begin{aligned}X_1|_{U_i} &= (R^2L - \beta RLR + LR^2\\ &\quad + ((\theta_i + \theta_{i+1} - \beta\theta_i)\theta_i^* + (\theta_i + \theta_{i+1} - \beta\theta_{i+1})\theta_{i+1}^*)R)|_{U_i},\\ Y_1|_{U_i} &= (\gamma(\theta_i^* + \theta_{i+1}^*) + \gamma^*(\theta_i + \theta_{i+1}) + \omega^*)R|_{U_i},\end{aligned}$$

if we note $\gamma + \beta\theta_j = \theta_{j+1} + \theta_{j-1}$, we obtain (iv). □

Theorem 6.67. *Let $A, A^* \in \mathrm{End}(V)$ be a pre-L-pair and $\{\theta_i\}_{i=0}^d$, $\{\theta_i^*\}_{i=0}^d$, $\{\lambda_i\}_{i=0}^{d-1}$ the data. Let $d \geq 1$ and $\lambda_i \neq 0$ $(0 \leq i \leq d-1)$. Then the following (1), (2), (3) hold:*

(1) *The relation (AW)$_1$*

$$A^2A^* - \beta AA^*A + A^*A^2 = \gamma(AA^* + A^*A) + \delta A^* + \gamma^* A^2 + \omega^* A + \eta^* \tag{6.245}$$

holds if and only if the following (i)–(iv) hold:

(i) $\delta = \theta_i^2 - \beta\theta_i\theta_{i-1} + \theta_{i-1}^2 - \gamma(\theta_i + \theta_{i-1})$ $(1 \leq i \leq d)$, $\gamma = \theta_i - \beta\theta_{i-1} + \theta_{i-2}$ $(2 \leq i \leq d)$, *namely, $\{\theta_i\}_{i=0}^d$ is a (β, γ, δ)-sequence;*

(ii) $\gamma^* = \theta_i^* - \beta\theta_{i-1}^* + \theta_{i-2}^* (2 \leq i \leq d)$, *namely, $\{\theta_i^*\}_{i=0}^d$ is a (β, γ^*)-sequence;*

(iii) $(\theta_i - \theta_{i-1})\lambda_i - (\theta_{i+1} - \theta_i)\lambda_{i-1} = (\theta_{i+1} - \theta_i)(\theta_i - \theta_{i-1})\theta_i^* + \gamma^*\theta_i^2 + \omega^*\theta_i + \eta^*$ $(0 \leq i \leq d)$;

(iv) $\lambda_{i-1} - \beta\lambda_i + \lambda_{i+1} = (\theta_{i+2} - \theta_{i+1})\theta_{i+1}^* - (\theta_i - \theta_{i-1})\theta_i^* + \gamma^*(\theta_i + \theta_{i+1}) + \omega^*$ $(0 \leq i \leq d-1)$.
Here, we interpret that θ_i are extended to the (β, γ)-sequence $\{\theta_i\}_{i\in\mathbb{Z}}$, and θ_i^* are extended to the (β, γ^*)-sequence $\{\theta_i^*\}_{i\in\mathbb{Z}}$ in the sense of Remark 6.46 (2) in Section 6.2. We also let $\lambda_{-1} = 0$, $\lambda_d = 0$. Moreover, (iii) holds for some $\eta^* \in \mathbb{C}$ if and only if (iv) holds under condition (i).

(2) *The relation (AW)$_2$*

$$(A^*)^2A - \beta A^*AA^* + A(A^*)^2 = \gamma^*(A^*A + AA^*) + \delta^* A + \gamma(A^*)^2 + \omega A^* + \eta \tag{6.246}$$

holds if and only if the following (i′)–(iv′) hold:

(i′) $\delta^* = (\theta_i^*)^2 - \beta\theta_i^*\theta_{i+1}^* + (\theta_{i+1}^*)^2 - \gamma^*(\theta_i^* + \theta_{i+1}^*)$ $(0 \leq i \leq d-1)$, $\gamma^* = \theta_i^* - \beta\theta_{i+1}^* + \theta_{i+2}^*$ $(0 \leq i \leq d-2)$,
namely, $\{\theta_i^*\}_{i=0}^d$ is a $(\beta, \gamma^*, \delta^*)$-sequence;

(ii′) $\gamma = \theta_i - \beta\theta_{i+1} + \theta_{i+2}$ $(0 \leq i \leq d-2)$,
namely, $\{\theta_i\}_{i=0}^d$ is a (β, γ)-sequence;

(iii′) $(\theta_i^* - \theta_{i+1}^*)\lambda_{i-1} - (\theta_{i-1}^* - \theta_i^*)\lambda_i = (\theta_{i-1}^* - \theta_i^*)(\theta_i^* - \theta_{i+1}^*)\theta_i + \gamma(\theta_i^*)^2 + \omega\theta_i^* + \eta$ $(0 \leq i \leq d)$;

(iv′) $\lambda_i - \beta\lambda_{i-1} + \lambda_{i-2} = (\theta_{i-2}^* - \theta_{i-1}^*)\theta_{i-1} - (\theta_i^* - \theta_{i+1}^*)\theta_i + \gamma(\theta_i^* + \theta_{i-1}^*) + \omega$ $(1 \leq i \leq d)$.
Here, we interpret that θ_i are extended to the (β, γ)-sequence $\{\theta_i\}_{i\in\mathbb{Z}}$, and θ_i^* are extended to the (β, γ^*)-sequence $\{\theta_i^*\}_{i\in\mathbb{Z}}$ in the sense of Remark 6.46(2) in Section 6.2. We also let $\lambda_{-1} = 0$, $\lambda_d = 0$. Moreover, (iii′) holds for some $\eta \in \mathbb{C}$ if and only if (iv′) holds under condition (i′).

(3) *If the relation (AW)$_1$ in (6.245) holds, then the relation (AW)$_2$ in (6.246) holds where $\omega = \omega^*$. Conversely, if the relation (AW)$_2$ in (6.246) holds, then the relation (AW)$_1$ in (6.245) holds where $\omega^* = \omega$. Moreover, if (AW)$_1$ and (AW)$_2$ hold, then we have $\omega = \omega^*$.*

Proof. Note that since $\lambda_i \neq 0$ $(0 \leq i \leq d-1)$, the assumption of Proposition 6.66, i. e., $R^2U_i \neq 0$ $(0 \leq i \leq d-2)$, $L^2U_i \neq 0$ $(2 \leq i \leq d)$, hold.

(1) Since $RL|_{U_i} = \lambda_{i-1}$, $LR|_{U_i} = \lambda_i$, conditions (i), (ii), (iii), and (iv) in Proposition 6.66 (1) are identical to conditions (i), (ii), (iii), and (iv) of the theorem.

Assume (i). Divide condition (iii) into the i-th equation and the $(i+1)$-th equation as follows:

$$\begin{aligned}&(\theta_{i+1}-\theta_i)\lambda_{i-1}-(\theta_i-\theta_{i-1})\lambda_i\\ &\quad=-(\theta_{i+1}-\theta_i)(\theta_i-\theta_{i-1})\theta_i^*-\gamma^*\theta_i^2-\omega^*\theta_i-\eta^*\quad(0\le i\le d-1),\end{aligned}\tag{6.247}$$

$$\begin{aligned}&(\theta_{i+1}-\theta_i)\lambda_{i+1}-(\theta_{i+2}-\theta_{i+1})\lambda_i\\ &\quad=(\theta_{i+2}-\theta_{i+1})(\theta_{i+1}-\theta_i)\theta_{i+1}^*+\gamma^*\theta_{i+1}^2+\omega^*\theta_{i+1}+\eta^*\quad(0\le i\le d-1).\end{aligned}\tag{6.248}$$

By (i), $\{\theta_i\}_{i\in\mathbb{Z}}$ is a β-sequence, and so we have

$$(\theta_{i+1}-\theta_i)\beta=\theta_{i+2}-\theta_{i+1}+\theta_i-\theta_{i-1},$$

and thus

$$\begin{aligned}&(\theta_{i+1}-\theta_i)(\lambda_{i-1}-\beta\lambda_i+\lambda_{i+1})\\ &\quad=(\theta_{i+1}-\theta_i)\lambda_{i-1}-(\theta_i-\theta_{i-1})\lambda_i+(\theta_{i+1}-\theta_i)\lambda_{i+1}-(\theta_{i+2}-\theta_{i+1})\lambda_i.\end{aligned}$$

Therefore, the i-th equation of (iv) can be written as follows:

$$\begin{aligned}&((\theta_{i+1}-\theta_i)\lambda_{i-1}-(\theta_i-\theta_{i-1})\lambda_i)+((\theta_{i+1}-\theta_i)\lambda_{i+1}-(\theta_{i+2}-\theta_{i+1})\lambda_i)\\ &\quad=(\theta_{i+1}-\theta_i)((\theta_{i+2}-\theta_{i+1})\theta_{i+1}^*-(\theta_i-\theta_{i-1})\theta_i^*+\gamma^*(\theta_i+\theta_{i+1})+\omega^*)\\ &\quad(0\le i\le d-1).\end{aligned}\tag{6.249}$$

Hence, if two of (6.247), (6.248), (6.249) hold, then so does the rest. In particular, if (iii) holds, we have (iv). Moreover, if we assume (iv), for i with $0\le i\le d-1$, the i-th equation in (iii) holds if and only if the $(i+1)$-th equation holds. Therefore, if we determine η^* so that (iii) holds for $i=0$, then (iii) holds for i with $0\le i\le d$.
(2) This is similar to the proof of (1).
(3) Since a (β,γ)-sequence is a (β,γ,δ)-sequence and a (β,γ^*)-sequence is a $(\beta,\gamma^*,\delta^*)$-sequence, (i), (ii) hold if and only if (i′), (ii′) hold. We write (iv), (iv′) as follows:

$$\begin{aligned}&\lambda_{i+1}-\beta\lambda_i+\lambda_{i-1}-\omega^*\\ &\quad=(\theta_{i+2}-\theta_{i+1})\theta_{i+1}^*-(\theta_i-\theta_{i-1})\theta_i^*+\gamma^*(\theta_i+\theta_{i+1})\quad(0\le i\le d-1),\end{aligned}\tag{6.250}$$

$$\begin{aligned}&\lambda_{i+1}-\beta\lambda_i+\lambda_{i-1}-\omega\\ &\quad=(\theta_{i+2}^*-\theta_{i+1}^*)\theta_{i+1}-(\theta_i^*-\theta_{i-1}^*)\theta_i+\gamma(\theta_i^*+\theta_{i+1}^*)\quad(0\le i\le d-1).\end{aligned}\tag{6.251}$$

We find the difference between the right-hand sides of (6.250) and (6.251). Note that $\{\theta_i\}_{i\in\mathbb{Z}}$ is a (β,γ)-sequence and $\{\theta_i^*\}_{i\in\mathbb{Z}}$ is a (β,γ^*)-sequence. Then we get

$$\begin{aligned}&(\theta_{i+2}-\gamma)\theta_{i+1}^*+(\theta_{i-1}-\gamma)\theta_i^*-(\theta_{i+2}^*-\gamma^*)\theta_{i+1}-(\theta_{i-1}^*-\gamma^*)\theta_i\\ &\quad=(\beta\theta_{i+1}-\theta_i)\theta_{i+1}^*+(\beta\theta_i-\theta_{i+1})\theta_i^*-(\beta\theta_{i+1}^*-\theta_i^*)\theta_{i+1}-(\beta\theta_i^*-\theta_{i+1}^*)\theta_i\\ &\quad=0.\end{aligned}$$

Thus, the right-hand sides of (6.250) and (6.251) are equal. So $\omega=\omega^*$ if and only if (iv) and (iv′) are equivalent. □

Theorem 6.68. *Let $A, A^* \in \mathrm{End}(V)$ be a TD-pair. Then A, A^* satisfy AW-relations if and only if A, A^* form an L-pair.*

Proof. If A, A^* form an L-pair, by Theorem 6.64, they satisfy AW-relations. Conversely, suppose A, A^* satisfy AW-relations. Since A, A^* form a TD-pair, if we let $V = \bigoplus_{i=0}^{d} U_i$ be the weight space decomposition and R, L the raising map and the lowering map, condition (iv) in Proposition 6.66 holds. In particular, for $u \in U_i$ $(0 \le i \le d-1)$, we have

$$LR^2u \in \mathrm{Span}\{R^2Lu, RLRu, Ru\}. \tag{6.252}$$

Let $u_0 \in U_0$ be an eigenvector of LR, and let $u_i = R^i u_0$ $(0 \le i \le d)$. Note that $u_i \in U_i$. First, we have $Lu_0 = 0 \in \mathbb{C}u_{-1}$ $(u_{-1} = 0)$, $Lu_1 = LRu_0 \in \mathbb{C}u_0$. Assume $Lu_{i-1} \in \mathbb{C}u_{i-2}$, $Lu_i \in \mathbb{C}u_{i-1}$ $(i \ge 1)$. By (6.252), we have

$$Lu_{i+1} = LR^2u_{i-1} \in \mathrm{Span}\{R^2Lu_{i-1}, RLRu_{i-1}, Ru_{i-1}\}.$$

Since $R^2Lu_{i-1} \in \mathbb{C}R^2u_{i-2} \subseteq \mathbb{C}u_i$, $RLRu_{i-1} = RLu_i \in \mathbb{C}Ru_{i-1} = \mathbb{C}u_i$, $Ru_{i-1} = u_i$, we obtain $Lu_{i+1} \in \mathbb{C}u_i$. By induction, if we let

$$W = \bigoplus_{i=0}^{d} \mathbb{C}u_i,$$

then W is L-invariant. Since W is R-invariant and it is also invariant under the projection $F_i : V = \bigoplus_{j=0}^{d} U_j \longrightarrow U_i$, it is A-invariant and A^*-invariant. Since V is irreducible as an $\langle A, A^*\rangle$-module, we have $W = V$. Therefore, A, A^* is an L-pair. □

6.3.5 Classification

In this section, we prove the following theorem.

Theorem 6.69. *A pre-L-system $(A, A^*; \{V_i\}_{i=0}^{d}, \{V_i^*\}_{i=0}^{d})$ becomes an L-system if and only if the data $\{\theta_i\}_{i=0}^{d}, \{\theta_i^*\}_{i=0}^{d}, \{\lambda_i\}_{i=0}^{d-1}$ are expressed by AW-parameters as in the table of Theorem 6.61.*

By the above theorem, isomorphism classes of L-systems are in one-to-one correspondence with the data in the table of Theorem 6.61, and in this sense, the classification of L-systems is completed. Note that if $\lambda_i \ne 0$ $(0 \le i \le d-1)$, a pre-L-pair with given data $\{\theta_i\}_{i=0}^{d}$, $\{\theta_i^*\}_{i=0}^{d}$, $\{\lambda_i\}_{i=0}^{d-1}$ uniquely exists up to isomorphism.

In order to prove Theorem 6.69, it suffices to show the following proposition.

Proposition 6.70. *Suppose the data $\{\theta_i\}_{i=0}^{d}$, $\{\theta_i^*\}_{i=0}^{d}$, $\{\lambda_i\}_{i=0}^{d-1}$ of a pre-L-system $(A, A^*; \{V_i\}_{i=0}^{d}, \{V_i^*\}_{i=0}^{d})$ are expressed by AW-parameters as in the table of Theorem 6.61. Then the following (1), (2) hold:*

(1) *A, A^* satisfy AW-relations;*
(2) *$V = \bigoplus_{i=0}^{d} V_i = \bigoplus_{i=0}^{d} V_i^*$ is irreducible as an $\langle A, A^* \rangle$-module.*

If a pre-L-system $(A, A^*; \{V_i\}_{i=0}^d, \{V_i^*\}_{i=0}^d)$ becomes an L-system, the data must be expressed by AW-parameters as in the table of Theorem 6.61, which was shown in Theorem 6.61. Conversely, suppose the data of a pre-L-system $(A, A^*; \{V_i\}_{i=0}^d, \{V_i^*\}_{i=0}^d)$ are expressed by AW-parameters as in the table of Theorem 6.61. Since A, A^* satisfy AW-relations by Proposition 6.70 (1), A, A^* satisfy TD-relations by (6.230). Therefore, by Theorem 6.47 in Section 6.2, we have

$$\begin{aligned} AV_i^* &\subseteq V_{i-1}^* + V_i^* + V_{i+1}^* \quad (0 \le i \le d), \\ A^* V_i &\subseteq V_{i-1} + V_i + V_{i+1} \quad (0 \le i \le d), \end{aligned} \tag{6.253}$$

where $V_{-1}^* = 0$ or $V_{-1}^* = V_d^*$, $V_{d+1}^* = 0$ or $V_{d+1}^* = V_0^*$, $V_{-1} = 0$ or $V_{-1} = V_d$, $V_{d+1} = 0$ or $V_{d+1} = V_0$. In (6.253), if we show we may set $V_{-1}^* = V_{d+1}^* = V_{-1} = V_{d+1} = 0$, then A, A^* satisfy conditions (1), (2) in Definition 6.27 in Section 6.2, and moreover, by Proposition 6.70 (2), they satisfy condition (3) of Definition 6.27, and it turns out that A, A^* form a TD-pair. Furthermore, since $\dim(V_i) = 1$ $(0 \le i \le d)$, A, A^* form an L-pair.

First, if $d = 0, 1$, clearly, we may assume $V_{-1}^* = V_{d+1}^* = V_{-1} = V_{d+1} = 0$. Let $d \ge 2$. Since A, A^* form a pre-L-pair, if we let $V = \bigoplus_{i=0}^{d} U_i$ be the weight space decomposition, we have $V_0^* = U_0$, $V_d = U_d$, and

$$\begin{aligned} AV_0^* &= AU_0 \subseteq U_0 + U_1 = V_0^* + V_1^*, \\ A^* V_d &= A^* U_d \subseteq U_{d-1} + U_d = V_{d-1} + V_d. \end{aligned}$$

Therefore, in (6.253), we may set $V_{-1}^* = V_{d+1} = 0$. On the other hand, since the data in the table of Theorem 6.61 satisfy the condition of Theorem 6.60, we have

$$\begin{aligned} A(V_2^* + V_3^* + \cdots + V_d^*) &\subseteq V_1^* + V_2^* + \cdots + V_d^*, \\ A^*(V_0 + V_1 + \cdots + V_{d-2}) &\subseteq V_0 + V_1 + \cdots + V_{d-1}. \end{aligned}$$

In particular, we have

$$\begin{aligned} AV_d^* &\subseteq V_1^* + V_2^* + \cdots + V_d^*, \\ A^* V_0 &\subseteq V_0 + V_1 + \cdots + V_{d-1}. \end{aligned}$$

On the other hand, by (6.253), clearly, we have

$$\begin{aligned} AV_d^* &\subseteq V_0^* + V_{d-1}^* + V_d^*, \\ AV_0 &\subseteq V_0 + V_1 + V_d. \end{aligned}$$

So in (6.253), we may set $V_{d+1}^* = V_{-1} = 0$. In this way, Theorem 6.69 is reduced to Proposition 6.70.

Proof of Proposition 6.70 *(1)*. Let $d \geq 1$. It suffices to check conditions (i), (ii), and (iii) in Theorem 6.67. By the assumption, the data are expressed by AW-parameters in Theorem 6.61. Therefore, there exist constants $\beta, \gamma, \gamma^*, \delta, \delta^*$ such that $\{\theta_i\}_{i=0}^d$ is a (β, γ, δ)-sequence and $\{\theta_i^*\}_{i=0}^d$ is a $(\beta, \gamma^*, \delta^*)$-sequence. Hence, (i), (ii) hold. Thus it suffices to check that there exists a constant ω^* such that (iv) holds:

$$\begin{aligned}&\lambda_{i-1} - \beta\lambda_i + \lambda_{i+1}\\&\quad = (\theta_{i+2} - \theta_{i+1})\theta_{i+1}^* - (\theta_i - \theta_{i-1})\theta_i^* + \gamma^*(\theta_i + \theta_{i+1}) + \omega^* \quad (0 \leq i \leq d-1).\end{aligned}$$

Namely, it suffices to check

$$\lambda_{i-1} - \beta\lambda_i + \lambda_{i+1} - (\theta_{i+2} - \theta_{i+1})\theta_{i+1}^* + (\theta_i - \theta_{i-1})\theta_i^* - \gamma^*(\theta_i + \theta_{i+1}) \tag{6.254}$$

does not depend on i $(0 \leq i \leq d-1)$. Note that θ_i and θ_i^* are extended to the (β, γ)-sequence $\{\theta_i\}_{i\in\mathbb{Z}}$ and the (β, γ^*)-sequence $\{\theta_i^*\}_{i\in\mathbb{Z}}$, respectively. Without loss of generality, we may assume the expression of the data by AW-parameters is of Type I, since the other expressions are obtained as the limiting case of Type I. In the following, we write $\beta = q + q^{-1}$, $q \neq \pm 1$. By (6.171), (6.188), we have

$$\begin{aligned}\lambda_i &= \lambda_i' + (\theta_i - \theta_d)(\theta_{i+1}^* - \theta_0^*), \quad (-1 \leq i \leq d),\\ \lambda_i' &= \frac{\lambda_0'}{(1-q)(1-q^d)}(1 + q^{d+1} - q^{i+1} - q^{d-i}) \quad (-1 \leq i \leq d).\end{aligned}$$

Since $q^{j-1} - \beta q^j + q^{j+1} = 0$ $(j \in \mathbb{Z})$,

$$\lambda_{i-1}' - \beta\lambda_i' + \lambda_{i+1}' = \frac{(2-\beta)(1+q^{d+1})\lambda_0'}{(1-q)(1-q^d)}$$

does not depend on i. Moreover, since we have

$$\begin{aligned}&(\theta_{i-1} - \theta_d)(\theta_i^* - \theta_0^*) - \beta(\theta_i - \theta_d)(\theta_{i+1}^* - \theta_0^*) + (\theta_{i+1} - \theta_d)(\theta_{i+2}^* - \theta_0^*)\\&\quad = \theta_{i-1}\theta_i^* - \beta\theta_i\theta_{i+1}^* + \theta_{i+1}\theta_{i+2}^* - (\theta_{i-1} - \beta\theta_i + \theta_{i+1})\theta_0^*\\&\qquad - \theta_d(\theta_i^* - \beta\theta_{i+1}^* + \theta_{i+2}^* - (2-\beta)\theta_0^*)\\&\quad = \theta_{i-1}\theta_i^* - \beta\theta_i\theta_{i+1}^* + \theta_{i+1}\theta_{i+2}^* - \gamma\theta_0^* - \theta_d(\gamma^* - (2-\beta)\theta_0^*),\end{aligned}$$

we obtain

$$\begin{aligned}\lambda_{i-1} - \beta\lambda_i + \lambda_{i+1} &= \theta_{i-1}\theta_i^* - \beta\theta_i\theta_{i+1}^* + \theta_{i+1}\theta_{i+2}^* + c,\\ c &= \frac{(2-\beta)(1+q^{d+1})\lambda_0'}{(1-q)(1-q^d)} - \gamma\theta_0^* - \theta_d(\gamma^* - (2-\beta)\theta_0^*).\end{aligned}$$

Hence, in order to check (6.254) does not depend on i, it suffices to check

$$\begin{aligned}&\theta_{i-1}\theta_i^* - \beta\theta_i\theta_{i+1}^* + \theta_{i+1}\theta_{i+2}^* - (\theta_{i+2} - \theta_{i+1})\theta_{i+1}^*\\&\quad + (\theta_i - \theta_{i-1})\theta_i^* - \gamma^*(\theta_i + \theta_{i+1})\end{aligned} \tag{6.255}$$

does not depend on i. Substituting (6.255) by $\beta\theta^*_{i+1} = \theta^*_i + \theta^*_{i+2} - \gamma^*$ yields

$$(\theta_{i+1} - \theta_i)\theta^*_{i+2} - (\theta_{i+2} - \theta_{i+1})\theta^*_{i+1} - \gamma^*\theta_{i+1}. \tag{6.256}$$

By using $\theta_{i+2} - \theta_{i+1} + \theta_i - \theta_{i-1} = (\theta_{i+1} - \theta_i)\beta$, subtracting the $(i-1)$-th equation from the i-th equation in (6.256) yields

$$(\theta_{i+1} - \theta_i)(\theta^*_{i+2} - \beta\theta^*_{i+1} + \theta^*_i - \gamma^*) = 0.$$

Therefore, (6.256) and (6.255) do not depend on i. □

From the above discussion, we show that if the data $\{\theta_i\}_{i=0}^d$, $\{\theta^*_i\}_{i=0}^d$, $\{\lambda_i\}_{i=0}^{d-1}$ of a pre-L-system $(A, A^*; \{V_i\}_{i=0}^d, \{V^*_i\}_{i=0}^d)$ are expressed by AW-parameters in the table of Theorem 6.61, A, A^* satisfy AW-relations. Therefore, (6.253) holds where $V^*_{-1} = V^*_{d+1} = V_{-1} = V_{d+1} = 0$. This implies conditions (1), (2) in Definition 6.27 in Section 6.2 hold. In the following, we show condition (3) in Definition 6.27 holds. If conditions (1), (2), (3) in Definition 6.27 hold, A, A^* form an L-pair. In particular, $(A, A^*; \{V_{d-i}\}_{i=0}^d, \{V^*_i\}_{i=0}^d)$ is a pre-L-system and if we let $\{\theta_{d-i}\}_{i=0}^d$, $\{\theta^*_i\}_{i=0}^d$, $\{\hat{\lambda}_i\}_{i=0}^{d-1}$ be the data, then $\hat{\lambda}_i$ is given by (6.168) by Lemma 6.56. For V to be an irreducible $\langle A, A^*\rangle$-module, it is necessary that $\lambda_i \neq 0$, $\hat{\lambda}_i \neq 0$ $(0 \leq i \leq d-1)$. Note that this condition is included in the expression by AW-parameters of Theorem 6.61.

Let $V = \bigoplus_{i=0}^d U_i$ be the weight space decomposition, and let R, L be the raising map and the lowering map. Choose a non-zero vector u_0 in U_0 and let $u_i = R^i u_0$. Then since $\lambda_i \neq 0$ $(0 \leq i \leq d-1)$, we get the following inductively:

$$Lu_i = \lambda_{i-1}u_{i-1} \neq 0 \quad (1 \leq i \leq d). \tag{6.257}$$

In particular, we have $u_i \neq 0$, $U_i = \mathbb{C}u_i$ $(0 \leq i \leq d)$. Let E^*_0 be the projection $V = \bigoplus_{i=0}^d V^*_i \longrightarrow V^*_0$. The following holds.

Proposition 6.71. *If and only if $E^*_0 V_0 \neq 0$, V is irreducible as an $\langle A, A^*\rangle$-module.*

Proof. (Sufficiency) Assume $E^*_0 V_0 \neq 0$. Assume $W \neq 0$ is an $\langle A, A^*\rangle$-submodule of V to show $W = V$. Since A is diagonalizable, W is a direct sum of eigenspaces of A. Choose the minimum i such that $W \cap V_i \neq 0$. Since $\dim(V_i) = 1$, we have $V_i \subseteq W$. We show $i = 0$. Assume $i \geq 1$ to get a contradiction. By Lemma 6.54,

$$v = u_i + \frac{1}{\theta_i - \theta_{i+1}}u_{i+1} + \cdots$$

belongs to V_i. Then by (6.257), we have

$$(A^* - \theta^*_i)v = \lambda_{i-1}u_{i-1} + (\text{the terms of } u_i, \ldots, u_d) \notin U_i + \cdots + U_d = V_i + \cdots + V_d.$$

By the minimality of i, we have

$$(A^* - \theta^*_i)v \in W \subseteq V_i + \cdots + V_d,$$

which is a contradiction. Therefore, we get $i = 0$, and $V_0 \subseteq W$. Since $E_0^* V_0 \neq 0$ and $\dim(V_0^*) = 1$, we have $E_0^* V_0 = V_0^*$. Hence $V_0^* \subseteq W$. Since

$$V_0^* = U_0, \quad U_i = (A - \theta_{i-1}) \cdots (A - \theta_0) U_0,$$

we get $U_i \subseteq W$ $(0 \leq i \leq d)$. Therefore $W = V$.

(Necessity) Assume V is irreducible as an $\langle A, A^* \rangle$-module. Assume $E_0^* V_0 = 0$ to get a contradiction. Define the subspace $V_{i,i+1}$ $(0 \leq i \leq d-1)$ of V by

$$V_{i,i+1} = (V_0 + \cdots + V_i) \cap (V_{i+1}^* + \cdots + V_d^*).$$

Let $V_{-1,0} = 0$, $V_{d,d+1} = 0$. Since it was shown that A, A^* satisfy (1), (2) in Definition 6.27 in Section 6.2, we have

$$(A - \theta_i) V_{i,i+1} \subseteq V_{i-1,i},$$
$$(A^* - \theta_i^*) V_{i,i+1} \subseteq V_{i+1,i+2}.$$

Therefore, if we let

$$W = V_{0,1} + V_{1,2} + \cdots V_{d-1,d},$$

W is $\langle A, A^* \rangle$-invariant. On the other hand, since $V_0 \subseteq V_1^* + \cdots + V_d^*$ by $E_0^* V_0 = 0$, we have

$$V_0 \subseteq V_0 \cap (V_1^* + \cdots + V_d^*) = V_{0,1}.$$

In particular, $W \supseteq V_0 \neq 0$. Moreover, since

$$V_{i,i+1} \subseteq V_{i+1}^* + \cdots + V_d^* \subseteq V_1^* + \cdots + V_d^* \quad (0 \leq i \leq d-1),$$

we have $W \subseteq V_1^* + \cdots + V_d^* \neq V$. This contradicts the irreducibility of V as an $\langle A, A^* \rangle$-module. □

Lemma 6.72. *The following condition is a necessary and sufficient condition for* $E_0^* V_0 \neq 0$*:*

$$\sum_{i=0}^{d} \frac{\lambda_0 \lambda_1 \cdots \lambda_{i-1}}{(\theta_0 - \theta_1) \cdots (\theta_0 - \theta_i)(\theta_0^* - \theta_1^*) \cdots (\theta_0^* - \theta_i^*)} \neq 0.$$

Proof. By Lemma 6.54,

$$v = \sum_{i=0}^{d} \frac{1}{(\theta_0 - \theta_1) \cdots (\theta_0 - \theta_i)} u_i$$

belongs to V_0. Since $\dim(V_0) = 1$, we have $V_0 = \mathbb{C}v$. Since we have

$$E_0^* = \prod_{j=1}^{d} \frac{A^* - \theta_j^*}{\theta_0^* - \theta_j^*},$$

by noting that

$$\begin{aligned}&\frac{A^* - \theta_j^*}{\theta_0^* - \theta_j^*}u_0 = u_0 \quad (1 \le j \le d),\\&(A^* - \theta_j^*)u_j = \lambda_{j-1}u_{j-1} \quad (1 \le j \le d),\end{aligned}$$

we obtain

$$\begin{aligned}E_0^* u_i &= \frac{(A^* - \theta_{i+1}^*)\cdots(A^* - \theta_d^*)}{(\theta_0^* - \theta_{i+1}^*)\cdots(\theta_0^* - \theta_d^*)}\frac{(A^* - \theta_1^*)\cdots(A^* - \theta_i^*)}{(\theta_0^* - \theta_1^*)\cdots(\theta_0^* - \theta_i^*)}u_i\\&= \frac{\lambda_0\cdots\lambda_{i-1}}{(\theta_0^* - \theta_1^*)\cdots(\theta_0^* - \theta_i^*)}u_0.\end{aligned}$$

Therefore we get

$$E_0^* v = \sum_{i=0}^{d} \frac{\lambda_0\cdots\lambda_{i-1}}{(\theta_0 - \theta_1)\cdots(\theta_0 - \theta_i)(\theta_0^* - \theta_1^*)\cdots(\theta_0^* - \theta_i^*)}u_0. \qquad \square$$

As will be stated below, by using the following theorem, we complete the proof of Proposition 6.70 (2).

Theorem 6.73. *We have*

$$\begin{aligned}&\sum_{i=0}^{d} \frac{\lambda_0\cdots\lambda_{i-1}}{(\theta_0 - \theta_1)\cdots(\theta_0 - \theta_i)(\theta_0^* - \theta_1^*)\cdots(\theta_0^* - \theta_i^*)}\\&\quad = \frac{\hat{\lambda}_0\cdots\hat{\lambda}_{d-1}}{(\theta_0 - \theta_1)\cdots(\theta_0 - \theta_d)(\theta_0^* - \theta_1^*)\cdots(\theta_0^* - \theta_d^*)}.\end{aligned}$$

Proof of Proposition 6.70 *(2).* Since $\hat{\lambda}_i \neq 0$ $(0 \le i \le d-1)$, by Theorem 6.73, we have

$$\sum_{i=0}^{d} \frac{\lambda_0\cdots\lambda_{i-1}}{(\theta_0 - \theta_1)\cdots(\theta_0 - \theta_i)(\theta_0^* - \theta_1^*)\cdots(\theta_0^* - \theta_i^*)} \neq 0.$$

By Lemma 6.72, $E_0^* V_0 \neq 0$. Therefore, by Proposition 6.71, V is irreducible as an $\langle A, A^*\rangle$-module. $\square$

For the proof of Theorem 6.73, it suffices to show the case that the data are expressed by AW-parameters of Type I, since the other cases are the limiting cases of

(I). As a preparation, we introduce new symbols and give two lemmas. For $a, a_1, \ldots, a_r \in \mathbb{C}$, let

$$(a;q)_i = \begin{cases} (1-a)\cdots(1-aq^{i-1}) & (i=1,2,\ldots), \\ 1 & (i=0), \end{cases}$$
$$(a_1,\ldots,a_r;q)_i = (a_1;q)_i\cdots(a_r;q)_i.$$

We define a *basic hypergeometric series* ${}_{r+1}\phi_r$ by the following:

$${}_{r+1}\phi_r\left(\begin{matrix} a_1,\ldots,a_{r+1} \\ b_1,\ldots,b_r \end{matrix};q,x\right) = \sum_{i=0}^{\infty} \frac{(a_1,\ldots,a_{r+1};q)_i}{(b_1,\ldots,b_r;q)_i}\frac{x^i}{(q;q)_i}. \tag{6.258}$$

Lemma 6.74. *If the data are expressed by AW-parameters of Type I, we have*

$$\sum_{i=0}^{d} \frac{\lambda_0\cdots\lambda_{i-1}}{(\theta_0-\theta_1)\cdots(\theta_0-\theta_i)(\theta_0^*-\theta_1^*)\cdots(\theta_0^*-\theta_i^*)}$$
$$= {}_3\phi_2\left(\begin{matrix} r_1q, r_2q, q^{-d} \\ sq^2, s^*q^2 \end{matrix};q,q\right).$$

Proof. We have the following:

$$\theta_0-\theta_j = -hq^{-j}(1-q^j)(1-sq^{j+1}),$$
$$(\theta_0-\theta_1)\cdots(\theta_0-\theta_i)(\theta_0^*-\theta_1^*)\cdots(\theta_0^*-\theta_i^*)$$
$$= (hh^*)^i q^{-i(i+1)}(q;q)_i(q;q)_i(sq^2;q)_i(s^*q^2;q)_i,$$
$$\lambda_j = hh^*q^{-2j-1}(1-q^{j+1})(1-q^{j-d})(1-r_1q^{j+1})(1-r_2q^{j+1}),$$
$$\lambda_0\cdots\lambda_{i-1} = (hh^*)^i q^{-i^2}(q;q)_i(q^{-d};q)_i(r_1q;q)_i(r_2q;q)_i.$$

Therefore, we obtain

$$\frac{\lambda_0\cdots\lambda_{i-1}}{(\theta_0-\theta_1)\cdots(\theta_0-\theta_i)(\theta_0^*-\theta_1^*)\cdots(\theta_0^*-\theta_i^*)}$$
$$= \frac{q^i(q^{-d};q)_i(r_1q;q)_i(r_2q;q)_i}{(q;q)_i(sq^2;q)_i(s^*q^2;q)_i}.$$

Since $(q^{-d};q)_i = 0$ $(i = d+1, d+2, \ldots)$, in the summation ${}_3\phi_2$, i goes from 0 to d, and hence the lemma holds. □

Lemma 6.75. *We have*

$${}_3\phi_2\left(\begin{matrix} r_1q, r_2q, q^{-d} \\ sq^2, s^*q^2 \end{matrix};q,q\right)$$
$$= (-1)^d q^{-\frac{1}{2}d(d-1)} \frac{(r_1-s^*q)\cdots(r_1-s^*q^d)(r_2-s^*q)\cdots(r_2-s^*q^d)}{(s^*)^d(sq^2;q)_d(s^*q^2;q)_d}.$$

Proof. We assume $s^* \neq 0$. If $s^* = 0$, by noting that $(r_1 - s^*q^i)(r_2 - s^*q^i)/s^* = sq^{d+1} - (r_1 + r_2)q^i + s^*q^{2i}$, we take the limit $s^* \to 0$.
In the *q-analogue of the Pfaff–Saalschütz formula* [189],

$$ {}_3\phi_2\left(\begin{matrix} a, b, q^{-n} \\ c, abc^{-1}q^{1-n} \end{matrix} ; q, q \right) = \frac{(c/a, c/b; q)_n}{(c, c/ab; q)_n} \quad (n = 0, 1, 2, \ldots), \tag{6.259} $$

if we let $a = r_1 q$, $b = r_2 q$, $c = s^* q^2$, by $r_1 r_2 = ss^* q^{d+1}$, we get

$$ {}_3\phi_2\left(\begin{matrix} r_1 q, r_2 q, q^{-d} \\ sq^2, s^* q^2 \end{matrix} ; q, q \right) = \frac{(s^* q/r_1, s^* q/r_2; q)_d}{(s^* q^2, s^*/r_1 r_2; q)_d}. $$

Since we have

$$ (s^* q/r_i; q)_d = \frac{1}{r_i^d}(r_i - s^* q) \cdots (r_i - s^* q^d) \quad (i = 1, 2), $$

$$ (s^*/r_1 r_2; q)_d = (-1)^d \left(\frac{s^*}{r_1 r_2} \right)^d q^{\frac{1}{2}d(d-1)} (sq^2; q)_d, $$

we get the desired equation. □

Proof of Theorem 6.73. We prove the case that the data are of Type I and $s^* \neq 0$. The other cases are the limiting cases of the above case. We have

$$ \hat{\lambda}_i = \frac{hh^*}{s^*} q^{-2i-1}(1 - q^{i+1})(1 - q^{i-d})(r_1 - s^* q^{i+1})(r_2 - s^* q^{i+1}), $$

$$ \begin{aligned} \hat{\lambda}_0 \cdots \hat{\lambda}_{d-1} &= \left(\frac{hh^*}{s^*} \right)^d q^{-d^2} (q; q)_d (q^{-d}; q)_d \prod_{i=0}^{d-1} (r_1 - s^* q^{i+1})(r_2 - s^* q^{i+1}) \\ &\quad \times (\theta_0 - \theta_1) \cdots (\theta_0 - \theta_d)(\theta_0^* - \theta_1^*) \cdots (\theta_0^* - \theta_d^*) \\ &= (hh^*)^d q^{-d(d+1)} (q; q)_d (q; q)_d (sq^2; q)_d (s^* q^2; q)_d. \end{aligned} $$

Note that

$$ (q^{-d}; q)_d = (-1)^d q^{-\frac{1}{2}d(d+1)} (q; q)_d. $$

Then we get

$$ \begin{aligned} &\frac{\hat{\lambda}_0 \cdots \hat{\lambda}_{d-1}}{(\theta_0 - \theta_1) \cdots (\theta_0 - \theta_d)(\theta_0^* - \theta_1^*) \cdots (\theta_0^* - \theta_d^*)} \\ &= (-1)^d q^{-\frac{1}{2}d(d-1)} \frac{(r_1 - s^* q) \cdots (r_1 - s^* q^d)(r_2 - s^* q) \cdots (r_2 - s^* q^d)}{(s^*)^d (sq^2; q)_d (s^* q^2; q)_d}. \end{aligned} $$

By Lemma 6.74 and Lemma 6.75, the theorem holds. □

Remark 6.76.

1. For $a \in \mathbb{C}$, let

$$(a)_i = \begin{cases} a(a+1)\cdots(a+i-1), & \text{if } i = 1, 2, \ldots, \\ 1, & \text{if } i = 0. \end{cases}$$

Define a hypergeometric series ${}_{r+1}F_r$ by

$${}_{r+1}F_r\left(\begin{matrix} a_1, \ldots, a_{r+1} \\ b_1, \ldots, b_r \end{matrix} ; x \right) = \sum_{i=0}^{\infty} \frac{(a_1)_i \cdots (a_{r+1})_i}{(b_1)_i \cdots (b_r)_i} \frac{x^i}{i!}. \tag{6.260}$$

Note that ${}_{r+1}\phi_r$ is a q-analogue of ${}_{r+1}F_r$. Moreover, (6.259) is a q-analogue of the *Pfaff–Saalschütz formula*:

$${}_3F_2\left(\begin{matrix} a, b, -n \\ c, 1+a+b-c-n \end{matrix} ; 1 \right) = \frac{(c-a)_n (c-b)_n}{(c)_n (c-a-b)_n} \quad (n = 0, 1, \ldots). \tag{6.261}$$

2. In the right-hand side of the equation of Lemma 6.75, the following holds:

$$\begin{aligned} &\frac{(r_1 - sq)\cdots(r_1 - sq^d)(r_2 - sq)\cdots(r_2 - sq^d)}{s^d} \\ &= \frac{(r_1 - s^*q)\cdots(r_1 - s^*q^d)(r_2 - s^*q)\cdots(r_2 - s^*q^d)}{(s^*)^d}. \end{aligned}$$

Proof. Since $r_1 r_2 = ss^* q^{d+1}$, we have

$$\frac{r_1 r_2}{s} - (r_1 q^{d-i+1} + r_2 q^i) + sq^{d+1} = \frac{r_1 r_2}{s^*} - (r_1 q^{d-i+1} + r_2 q^i) + s^* q^{d+1}.$$

Therefore,

$$\frac{(r_1 - sq^i)(r_2 - sq^{d-i+1})}{s} = \frac{(r_1 - s^* q^i)(r_2 - s^* q^{d-i+1})}{s^*}$$

and we get the desired equation. □

3. In the proof of Theorem 6.73, we make use of the q-analogue of the Pfaff–Saalschütz formula. In fact, TD-pairs are related to Drinfeld polynomials, and Theorem 6.73 is the special case of the product formula for them [260]. Hence, conversely we can obtain the q-analogue of the Pfaff–Saalschütz formula from Theorem 6.73.

6.3.6 Dual systems of AW-polynomials

Let $(A, A^*; \{V_i\}_{i=0}^d, \{V_i^*\}_{i=0}^d)$ be an L-system and let $\{\theta_i\}_{i=0}^d$, $\{\theta_i^*\}_{i=0}^d$, $\{\lambda_i\}_{i=0}^{d-1}$ be the data. Let $v_0, v_1, \ldots, v_d$ be the standard basis and $v_0^*, v_1^*, \ldots, v_d^*$ the dual standard basis. Then there exist constants $b_{i-1}, a_i, c_{i+1}; b_{i-1}^*, a_i^*, c_{i+1}^*$ $(0 \le i \le d)$ such that

$$Av_i = b_{i-1}v_{i-1} + a_i v_i + c_{i+1}v_{i+1} \quad (0 \le i \le d), \tag{6.262}$$

$$A^* v_i^* = b_{i-1}^* v_{i-1}^* + a_i^* v_i^* + c_{i+1}^* v_{i+1}^* \quad (0 \le i \le d), \tag{6.263}$$

where $v_{-1} = v_{d+1} = 0$, $v_{-1}^* = v_{d+1}^* = 0$, $c_{d+1} = c_{d+1}^* = 1$, and b_{-1}, b_{-1}^* are indeterminate. Note that since we choose the standard basis and the dual standard basis, the following hold:

$$a_i + b_i + c_i = \theta_0, \tag{6.264}$$

$$a_i^* + b_i^* + c_i^* = \theta_0^*, \tag{6.265}$$

where $c_0 = c_0^* = 0$, $b_d = b_d^* = 0$ (Definition 6.52). Define polynomials $v_i(x)$, $v_i^*(x)$ $(-1 \le i \le d+1)$ of degree i by $v_{-1}(x) = v_{-1}^*(x) = 0$, $v_0(x) = v_0^*(x) = 1$, and

$$xv_i(x) = b_{i-1}v_{i-1}(x) + a_i v_i(x) + c_{i+1}v_{i+1}(x) \quad (0 \le i \le d), \tag{6.266}$$

$$xv_i^*(x) = b_{i-1}^* v_{i-1}^*(x) + a_i^* v_i^*(x) + c_{i+1}^* v_{i+1}^*(x) \quad (0 \le i \le d). \tag{6.267}$$

Then $\{v_i(x)\}_{i=0}^d$, $\{v_i^*(x)\}_{i=0}^d$ form a dual system of orthogonal polynomials (Theorem 6.25 in Section 6.1.4), and the following hold (equation (6.78) in Section 6.1.4):

$$c_1 c_2 \cdots c_d v_{d+1}(x) = (x-\theta_0)(x-\theta_1)\cdots(x-\theta_d), \tag{6.268}$$

$$c_1^* c_2^* \cdots c_d^* v_{d+1}^*(x) = (x-\theta_0^*)(x-\theta_1^*)\cdots(x-\theta_d^*). \tag{6.269}$$

In this section, we first prove the following two theorems.

Theorem 6.77.

(1) *Let $\{\theta_{d-i}\}_{i=0}^d$, $\{\theta_i^*\}_{i=0}^d$, $\{\hat{\lambda}_i\}_{i=0}^{d-1}$ be the data of the L-system $(A, A^*; \{V_{d-i}\}_{i=0}^d, \{V_i^*\}_{i=0}^d)$. Let a_i $(0 \le i \le d)$ be the numbers determined by (6.264). Then the following hold:*

$$b_i = \lambda_i \frac{(\theta_i^* - \theta_{i-1}^*)\cdots(\theta_i^* - \theta_0^*)}{(\theta_{i+1}^* - \theta_i^*)\cdots(\theta_{i+1}^* - \theta_0^*)} \quad (0 \le i \le d-1),$$

$$c_i = \hat{\lambda}_{i-1} \frac{(\theta_i^* - \theta_{i+1}^*)\cdots(\theta_i^* - \theta_d^*)}{(\theta_{i-1}^* - \theta_i^*)\cdots(\theta_{i-1}^* - \theta_d^*)} \quad (1 \le i \le d).$$

(2) *Let $\{\theta_i\}_{i=0}^d$, $\{\theta_{d-i}^*\}_{i=0}^d$, $\{\check{\lambda}_i\}_{i=0}^{d-1}$ be the data of the L-system $(A, A^*; \{V_i\}_{i=0}^d, \{V_{d-i}^*\}_{i=0}^d)$. Let a_i^* $(0 \le i \le d)$ be the numbers determined by (6.265). Then the following hold:*

$$b_i^* = \lambda_i \frac{(\theta_i - \theta_{i-1})\cdots(\theta_i - \theta_0)}{(\theta_{i+1} - \theta_i)\cdots(\theta_{i+1} - \theta_0)} \quad (0 \le i \le d-1),$$

$$c_i^* = \check{\lambda}_i \frac{(\theta_i - \theta_{i+1})\cdots(\theta_i - \theta_d)}{(\theta_{i-1} - \theta_i)\cdots(\theta_{i-1} - \theta_d)} \quad (1 \le i \le d).$$

Remark 6.78.

(1) By Lemma 6.57 (2), we have $\check{\lambda}_i = \hat{\lambda}_{d-i-1}$ $(0 \le i \le d-1)$.

(2) The data of the L-system $(A^*, A; \{V_{d-i}^*\}_{i=0}^d, \{V_i\}_{i=0}^d)$ are $\{\theta_{d-i}^*\}_{i=0}^d$, $\{\theta_i\}_{i=0}^d$, $\{\check{\lambda}_i\}_{i=0}^{d-1}$ (Lemma 6.57 (3)).

Theorem 6.79. *The following hold:*

(1)

$$\frac{v_i(x)}{k_i} = \sum_{j=0}^{i} \frac{(\theta_i^* - \theta_0^*)\cdots(\theta_i^* - \theta_{j-1}^*)}{\lambda_0 \cdots \lambda_{j-1}}(x - \theta_0)\cdots(x - \theta_{j-1}) \quad (0 \le i \le d),$$

where $k_i = \frac{b_0 \cdots b_{i-1}}{c_1 \cdots c_i}$ $(0 \le i \le d)$;

(2)

$$\frac{v_i^*(x)}{k_i^*} = \sum_{j=0}^{i} \frac{(\theta_i - \theta_0)\cdots(\theta_i - \theta_{j-1})}{\lambda_0 \cdots \lambda_{j-1}}(x - \theta_0^*)\cdots(x - \theta_{j-1}^*) \quad (0 \le i \le d),$$

where $k_i^* = \frac{b_0^* \cdots b_{i-1}^*}{c_1^* \cdots c_i^*}$ $(0 \le i \le d)$.

After proving these theorems, by using them, we express constants b_i, c_i, b_i^*, c_i^* and polynomials $\frac{v_i(x)}{k_i}$, $\frac{v_i^*(x)}{k_i^*}$ by AW-parameters appearing in Theorem 6.61 and Theorem 6.69. For the case of Type I, the dual system of orthogonal polynomials $\{v_i(x)\}_{i=0}^d$, $\{v_i^*(x)\}_{i=0}^d$ becomes a dual system of AW-polynomials. The other cases are the limiting cases of Type I.

Proof of Theorem 6.77. (1) As we choose the standard basis $v_0, v_1, \ldots, v_d$, by Proposition 6.53, there exists a non-zero vector v in V_0 such that

$$v_i = E_i^* v \in V_i^*, \tag{6.270}$$

where E_i^* is the projection $V = \bigoplus_{j=0}^d V_j^* \longrightarrow V_i^*$ expressed as follows:

$$E_i^* = \frac{A^* - \theta_0^*}{\theta_i^* - \theta_0^*} \cdots \frac{A^* - \theta_{i-1}^*}{\theta_i^* - \theta_{i-1}^*} \frac{A^* - \theta_{i+1}^*}{\theta_i^* - \theta_{i+1}^*} \cdots \frac{A^* - \theta_d^*}{\theta_i^* - \theta_d^*}. \tag{6.271}$$

Now we consider the L-system $(A, A^*; \{V_{d-i}\}_{i=0}^d, \{V_i^*\}_{i=0}^d)$ and let $V = \bigoplus_{i=0}^d \hat{U}_i$ be the weight space decomposition:

$$\hat{U}_i = (V_0^* + \cdots + V_i^*) \cap (V_{d-i} + \cdots + V_0).$$

Note that $V_0 = \hat{U}_d$. Let $\hat{F}_i$ be the projection $V = \bigoplus_{j=0}^d \hat{U}_j \longrightarrow \hat{U}_i$. Then

$$\hat{R} = A - \sum_{i=0}^{d} \theta_{d-i}\hat{F}_i, \quad \hat{L} = A^* - \sum_{i=0}^{d} \theta_i^* \hat{F}_i$$

are rgw associated raising map and lowering map, respectively, and we have

$$\hat{R}\hat{U}_i \subseteq \hat{U}_{i+1}, \quad \hat{L}\hat{U}_i \subseteq \hat{U}_{i-1} \quad (0 \le i \le d),$$

where $\hat{U}_{-1} = \hat{U}_{d+1} = 0$. If we choose a non-zero vector $\hat{u}_0$ in $\hat{U}_0$ and let $\hat{u}_i = \hat{R}^i\hat{u}_0$, then we have $\hat{u}_i \in \hat{U}_i$, and

$$\hat{L}\hat{u}_i = (A^* - \theta_i^*)\hat{u}_i = \hat{\lambda}_{i-1}\hat{u}_{i-1} \in \hat{U}_{i-1}. \tag{6.272}$$

Since $\hat{\lambda}_{i-1} \neq 0$ $(1 \leq i \leq d)$, we inductively get $\hat{u}_i \neq 0$ $(0 \leq i \leq d)$. In particular, $\hat{u}_d \neq 0$. Since $\hat{U}_d = V_0$ and $\dim(\hat{U}_d) = 1$, without loss of generality, we may let $v = \hat{u}_d$ in (6.270). Thus by (6.270), (6.271), (6.272), we get

$$v_i = E_i^*\hat{u}_d = \frac{A^* - \theta_0^*}{\theta_i^* - \theta_0^*} \cdots \frac{A^* - \theta_{i-1}^*}{\theta_i^* - \theta_{i-1}^*} \frac{\hat{\lambda}_i \cdots \hat{\lambda}_{d-1}}{(\theta_i^* - \theta_{i+1}^*) \cdots (\theta_i^* - \theta_d^*)}\hat{u}_i.$$

By (6.272), note that we have

$$\frac{A^* - \theta_j^*}{\theta_i^* - \theta_j^*}\hat{u}_i = \hat{u}_i + \frac{\hat{\lambda}_{i-1}}{\theta_i^* - \theta_j^*}\hat{u}_{i-1}$$

for $j \neq i$. Then we get

$$v_i - \frac{\hat{\lambda}_i \cdots \hat{\lambda}_{d-1}}{(\theta_i^* - \theta_{i+1}^*) \cdots (\theta_i^* - \theta_d^*)}\hat{u}_i \in \hat{U}_{i-1} + \cdots + \hat{U}_0, \tag{6.273}$$

where we set $\frac{\hat{\lambda}_i \cdots \hat{\lambda}_{d-1}}{(\theta_i^* - \theta_{i+1}^*) \cdots (\theta_i^* - \theta_d^*)} = 1$ if $i = d$. Since we have

$$A\hat{u}_i = \hat{R}\hat{u}_i + \theta_{d-i}\hat{u}_i = \hat{u}_{i+1} + \theta_{d-i}\hat{u}_i,$$

by (6.273), we have

$$Av_i - \frac{\hat{\lambda}_i \cdots \hat{\lambda}_{d-1}}{(\theta_i^* - \theta_{i+1}^*) \cdots (\theta_i^* - \theta_d^*)}\hat{u}_{i+1} \in \hat{U}_i + \cdots + \hat{U}_0. \tag{6.274}$$

Applying (6.273) to v_{i-1}, v_i, v_{i+1} in the right-hand side of (6.262), we get

$$Av_i - c_{i+1}\frac{\hat{\lambda}_{i+1} \cdots \hat{\lambda}_{d-1}}{(\theta_{i+1}^* - \theta_{i+2}^*) \cdots (\theta_{i+1}^* - \theta_d^*)}\hat{u}_{i+1} \in \hat{U}_i + \cdots + \hat{U}_0. \tag{6.275}$$

Comparing the coefficients of $\hat{u}_{i+1}$ in (6.274) and (6.275) yields

$$c_{i+1} = \hat{\lambda}_i \frac{(\theta_{i+1}^* - \theta_{i+2}^*) \cdots (\theta_{i+1}^* - \theta_d^*)}{(\theta_i^* - \theta_{i+1}^*) \cdots (\theta_i^* - \theta_d^*)} \quad (0 \leq i \leq d-1), \tag{6.276}$$

where we set $(\theta_{i+1}^* - \theta_{i+2}^*) \cdots (\theta_{i+1}^* - \theta_d^*) = 1$ if $i = d-1$.

Next, we consider the L-pair $(A, A^*; \{V_i\}_{i=0}^d, \{V_{d-i}^*\}_{i=0}^d)$ and let $\{\theta_i\}_{i=0}^d$, $\{\theta_{d-i}^*\}_{i=0}^d$, $\{\check{\lambda}_i\}_{i=0}^{d-1}$ be the data. For a non-zero vector v in V_0, let $\check{v}_i = E_{d-i}^* v$. Then $\check{v}_0, \check{v}_1, \ldots, \check{v}_d$ form the standard basis of the L-system. If we let

$$A\check{v}_i = \check{b}_{i-1}\check{v}_{i-1} + \check{a}_i\check{v}_i + \check{c}_{i+1}\check{v}_{i+1}, \tag{6.277}$$

by (6.276), we get

$$\check{c}_{i+1} = \tilde{\lambda}_i \frac{(\theta^*_{d-i-1} - \theta^*_{d-i-2}) \cdots (\theta^*_{d-i-1} - \theta^*_0)}{(\theta^*_{d-i} - \theta^*_{d-i-1}) \cdots (\theta^*_{d-i} - \theta^*_0)} \quad (0 \le i \le d-1), \tag{6.278}$$

where $\tilde{\lambda}_i = \hat{\check{\lambda}}_i$. Namely, let $\{\theta_{d-i}\}_{i=0}^d$, $\{\theta^*_{d-i}\}_{i=0}^d$, $\{\tilde{\lambda}_i\}_{i=0}^{d-1}$ be the data of the L-system $(A, A^*; \{V_{d-i}\}_{i=0}^d, \{V^*_{d-i}\}_{i=0}^d)$. By (6.270), we have $\check{v}_i = E^*_{d-i}v = v_{d-i}$. Hence (6.277) becomes

$$Av_{d-i} = \check{b}_{i-1}v_{d-i+1} + \check{a}_i v_{d-i} + \check{c}_{i+1}v_{d-i-1},$$

and by comparing with (6.262), we get $\check{c}_{i+1} = b_{d-i-1}$. Therefore, by (6.278) and by $\tilde{\lambda}_i = \lambda_{d-i-1}$ in Lemma 6.57 (1), we obtain

$$b_{d-i-1} = \lambda_{d-i-1} \frac{(\theta^*_{d-i-1} - \theta^*_{d-i-2}) \cdots (\theta^*_{d-i-1} - \theta^*_0)}{(\theta^*_{d-i} - \theta^*_{d-i-1}) \cdots (\theta^*_{d-i} - \theta^*_0)} \quad (0 \le i \le d-1).$$

(2) We consider the L-system $(A^*, A; \{V^*_i\}_{i=0}^d, \{V_i\}_{i=0}^d)$. By Lemma 6.57 (3), the data of the L-system are $\{\theta^*_i\}_{i=0}^d$, $\{\theta_i\}_{i=0}^d$, $\{\lambda_i\}_{i=0}^{d-1}$. On the other hand, the data of the L-system $(A^*, A; \{V^*_{d-i}\}_{i=0}^d, \{V_i\}_{i=0}^d)$ are, by Lemma 6.57 (3), $\{\theta^*_{d-i}\}_{i=0}^d$, $\{\theta_i\}_{i=0}^d$, $\{\check{\lambda}_i\}_{i=0}^{d-1}$. Replacing θ_i with θ^*_i, and $\hat{\lambda}_{i-1}$ with $\check{\lambda}_{i-1}$ and applying the part of the proof of (1) yields the equations on b^*_i, c^*_i. □

In order to prove Theorem 6.79, we give a proposition. Let $v_0, v_1, \ldots, v_d$ be the standard basis, $v^*_0, v^*_1, \ldots, v^*_d$ the dual standard basis, and $v_i(x)$, $v^*_i(x)$ the polynomials of degree i determined by (6.262), (6.263). Then we have the following.

Proposition 6.80. *We have:*
(1) $v_i(A)v_0 = v_i$ $(0 \le i \le d)$;
(2) $v^*_i(A^*)v^*_0 = v^*_i$ $(0 \le i \le d)$.

Proof. We prove (1) only since the proof of (2) is similar. By (6.266), we have

$$Av_i(A) = b_{i-1}v_{i-1}(A) + a_i v_i(A) + c_{i+1}v_{i+1}(A). \tag{6.279}$$

Since $v_{-1}(A) = 0$, $v_0(A) = I$ (= identity matrix), (1) holds for $i = -1, 0$. Assume (1) holds for any integer i with $0 \le i \le k$. Let both sides of (6.279) act on the vector v_0 of the standard basis and substitute $v_i(A)v_0 = v_i$ $(0 \le i \le k)$. Then we get

$$Av_k = b_{k-1}v_{k-1} + a_k v_k + c_{k+1}v_{k+1}(A)v_0. \tag{6.280}$$

Since $c_{k+1} \ne 0$, by (6.262) and (6.280), we obtain $v_{k+1}(A)v_0 = v_{k+1}$. □

Proof of Theorem 6.79. (1) Let $\frac{v_i(x)}{k_i}$ be the polynomial of degree i defined as follows:

$$\frac{v_i(x)}{k_i} = \sum_{j=0}^{i} t_j(x - \theta_0) \cdots (x - \theta_{j-1}). \tag{6.281}$$

We find the coefficient $t_j \in \mathbb{C}$, where we set $(x - \theta_0) \cdots (x - \theta_{j-1}) = 1$ for $j = 0$. By Proposition 6.20, we have $k_i = v_i(\theta_0)$. Therefore we get

$$t_0 = \frac{v_i(\theta_0)}{k_i} = 1. \tag{6.282}$$

Let $V = \bigoplus_{j=0}^{d} U_j$ be the weight space decomposition where $U_j = (V_0^* + \cdots + V_j^*) \cap (V_j + \cdots + V_d)$, and let R be the raising map. For the standard basis $v_0, v_1, \ldots, v_d$, let $u_j = R^j v_0$. Since $v_0 \in V_0^* = U_0$, we have $u_0 = v_0 \in U_0$, and

$$u_j = R^j v_0 = (A - \theta_{j-1}) \cdots (A - \theta_0) v_0 \in U_j.$$

Therefore we have

$$\sum_{j=0}^{i} t_j u_j = \sum_{j=0}^{i} t_j (A - \theta_{j-1}) \cdots (A - \theta_0) v_0,$$

and by (6.281), the right-hand side is equal to $\frac{v_i(A)}{k_i} v_0$. So we get

$$\sum_{j=0}^{i} t_j u_j = \frac{v_i(A)}{k_i} v_0.$$

By Proposition 6.80, we have $v_i(A) v_0 = v_i$ and so

$$\sum_{j=0}^{i} t_j u_j = \frac{1}{k_i} v_i \in V_i^*.$$

Hence we have

$$(A^* - \theta_i^*) \sum_{j=0}^{i} t_j u_j = 0. \tag{6.283}$$

Let L be the lowering map. Since $L u_j = \lambda_{j-1} u_{j-1}$ and $L u_j = (A^* - \theta_j^*) u_j$, we have

$$(A^* - \theta_i^*) u_j = (L + (\theta_j^* - \theta_i^*)) u_j = \lambda_{j-1} u_{j-1} + (\theta_j^* - \theta_i^*) u_j.$$

Thus by (6.283), we have

$$\sum_{j=0}^{i} t_j (\lambda_{j-1} u_{j-1} + (\theta_j^* - \theta_i^*) u_j) = 0,$$

which implies

$$\sum_{j=0}^{i}(\Lambda_j t_{j+1} + (\theta_j^* - \theta_i^*)t_j)u_j = 0,$$

where $t_{i+1} = 0$. Hence we get

$$\Lambda_j t_{j+1} + (\theta_j^* - \theta_i^*)t_j = 0 \quad (0 \le j \le i-1).$$

By (6.282), we have $t_0 = 1$. So we have

$$t_{j+1} = \frac{\theta_i^* - \theta_j^*}{\Lambda_j} t_j = \frac{(\theta_i^* - \theta_j^*)\cdots(\theta_i^* - \theta_1^*)(\theta_i^* - \theta_0^*)}{\Lambda_j \cdots \Lambda_1 \Lambda_0}.$$

(2) By Lemma 6.57 (3), the data of the L-system$(A^*, A; \{V_i^*\}_{i=0}^d, \{V_i\}_{i=0}^d)$ are $\{\theta_i^*\}_{i=0}^d$, $\{\theta_i\}_{i=0}^d$, $\{\Lambda_i\}_{i=0}^{d-1}$. So if we use (1) of this theorem, we obtain the equation on $\frac{v_i^*(x)}{k_i^*}$ immediately. □

Since a dual system of orthogonal polynomials and an L-system can be identified by Theorem 6.25, Proposition 6.51, and Proposition 6.53, we can classify dual systems of orthogonal polynomials by the classification of L-systems (Theorem 6.69) and by Theorem 6.79.

In the following, we express b_i, c_i, b_i^*, c_i^* in Theorem 6.77 and the polynomials $\frac{v_i(x)}{k_i}, \frac{v_i^*(x)}{k_i^*}$ in Theorem 6.79 by AW-parameters in the table of Theorem 6.61. There are special cases on b_0, b_0^*, c_d, c_d^* because if $i = 0, d$, zero appears in numerators or denominators of b_i, b_i^*, c_i, c_i^* in the general formulas. In Theorem 6.77, if $i = 0$, we set $(\theta_i^* - \theta_{i-1}^*)\cdots(\theta_i^* - \theta_0^*) = 1$, $(\theta_i - \theta_{i-1})\cdots(\theta_i - \theta_0) = 1$, and if $i = d$, we set $(\theta_i^* - \theta_{i+1}^*)\cdots(\theta_i^* - \theta_d^*) = 1$, $(\theta_i - \theta_{i+1})\cdots(\theta_i - \theta_d) = 1$ although in the general formulas it is not necessarily the case if $i = 0, d$.

In the table of the theorem below, ${}_{r+1}\phi_r$ denotes a basic hypergeometric series in (6.258), and ${}_{r+1}F_r$ denotes a hypergeometric series in (6.260). Moreover, θ_i, θ_i^* are given by $\theta_i = \xi(i)$, $\theta_i^* = \xi^*(i)$, where for the case of Type III, if i is even, we set $\theta_i = \xi(i)$, $\theta_i^* = \xi^*(i)$, and if i is odd, we set $\theta_i = \xi(-i-1-s)$, $\theta_i^* = \xi^*(-i-1-s^*)$.

Theorem 6.81. *Let $\{v_i(x)\}_{i=0}^d, \{v_i^*(x)\}_{i=0}^d$ be a dual system of orthogonal polynomials in the sense of Definition 6.23. It is one of the pairs of polynomials in the table below. Conversely, any pair $\{v_i(x)\}_{i=0}^d, \{v_i^*(x)\}_{i=0}^d$ of polynomials in the table below is a dual system of orthogonal polynomials.*

(**I**)

(1) *We have*

$$\frac{v_i(x)}{k_i} = {}_4\phi_3\left(\begin{matrix} q^{-i}, s^*q^{i+1}, q^{-y}, sq^{y+1} \\ q^{-d}, r_1q, r_2q \end{matrix}; q, q\right) \quad (0 \le i \le d),$$

where $x = \xi(y) = \theta_0 + h\frac{1}{q^y}(1 - q^y)(1 - sq^{y+1})$. *Also,*

$$b_i = \frac{h(1-q^{i-d})(1-s^*q^{i+1})(1-r_1q^{i+1})(1-r_2q^{i+1})}{(1-s^*q^{2i+2})(1-s^*q^{2i+1})} \quad (1 \le i \le d-1),$$
$$b_0 = h(1-q^{-d})(1-r_1q)(1-r_2q)/(1-s^*q^2).$$

If $s^* \neq 0$,

$$c_i = \frac{hq}{s^*}\frac{(1-q^i)(q^{-d-1}-s^*q^i)(r_1-s^*q^i)(r_2-s^*q^i)}{(1-s^*q^{2i})(1-s^*q^{2i+1})} \quad (1 \le i \le d-1),$$
$$c_d = h(1-q^d)(r_1-s^*q^d)(r_2-s^*q^d)/s^*q^d(1-s^*q^{2d}).$$

If $s^* = 0$,

$$c_i = hq(1-q^i)(s-r_1q^{i-1-d}) \quad (1 \le i \le d).$$

(2) *We have*

$$\frac{v_i^*(x)}{k_i^*} = {}_4\phi_3\left(\begin{matrix} q^{-i}, sq^{i+1}, q^{-y}, s^*q^{y+1} \\ q^{-d}, r_1q, r_2q \end{matrix}; q, q\right) \quad (0 \le i \le d),$$

where $x = \xi^*(y) = \theta_0^* + h^*\frac{1}{q^y}(1-q^y)(1-s^*q^{y+1})$. *Also,*

$$b_i^* = \frac{h^*(1-q^{i-d})(1-sq^{i+1})(1-r_1q^{i+1})(1-r_2q^{i+1})}{(1-sq^{2i+2})(1-sq^{2i+1})} \quad (1 \le i \le d-1),$$
$$b_0^* = h^*(1-q^{-d})(1-r_1q)(1-r_2q)/(1-sq^2).$$

If $s \neq 0$,

$$c_i^* = \frac{h^*q}{s}\frac{(1-q^i)(q^{-d-1}-sq^i)(r_1-sq^i)(r_2-sq^i)}{(1-sq^{2i})(1-sq^{2i+1})} \quad (1 \le i \le d-1),$$
$$c_d^* = h^*(1-q^d)(r_1-sq^d)(r_2-sq^d)/sq^d(1-sq^{2d}).$$

If $s = 0$,

$$c_i^* = h^*q(1-q^i)(s^*-r_1q^{i-1-d}) \quad (1 \le i \le d).$$

(IA)

(1) *We have*

$$\frac{v_i(x)}{k_i} = {}_2\phi_1\left(\begin{matrix} q^{-i}, q^{-y}, \\ q^{-d} \end{matrix}; q, \frac{hq^y}{r_1}\right) \quad (0 \le i \le d),$$

where $x = \xi(y) = \theta_0 - h(1-q^y)$. *Also,*

$$b_i = -r_1q^{i+1}(1-q^{i-d}) \quad (0 \le i \le d-1),$$
$$c_i = (1-q^i)(h-r_1q^{i-d}) \quad (1 \le i \le d).$$

(2) *We have*

$$\frac{v_i^*(x)}{k_i^*} = {}_2\phi_1\left(\begin{matrix} q^{-i}, q^{-y}, \\ q^{-d} \end{matrix} ; q, \frac{hq^i}{r_1} \right) \quad (0 \le i \le d),$$

where $x = \xi^*(y) = \theta_0^* - h^*(1 - q^{-y})$. *Also,*

$$b_i^* = \frac{h^* r_1}{h} q^{-2i}(1 - q^{i-d}) \quad (0 \le i \le d-1),$$

$$c_i^* = h^* q^{-2i}(1 - q^i)\left(\frac{qr_1}{h} - q^i\right) \quad (1 \le i \le d).$$

(II)

(1) *We have*

$$\frac{v_i(x)}{k_i} = {}_4F_3\left(\begin{matrix} -i, +1 + s^*, -y, y + 1 + s \\ -d, r_1 + 1, r_2 + 1 \end{matrix} ; 1 \right) \quad (0 \le i \le d),$$

where $x = \xi(y) = \theta_0 + hy(y + 1 + s)$. *Also,*

$$b_i = \frac{h(i - d)(i + 1 + s^*)(i + 1 + r_1)(i + 1 + r_2)}{(2i + 2 + s^*)(2i + 1 + s^*)} \quad (1 \le i \le d-1),$$

$$b_0 = -hd(1 + r_1)(1 + r_2)/(2 + s^*),$$

$$c_i = \frac{hi(i + s^* + d + 1)(i + s^* - r_1)(i + s^* - r_2)}{(2i + s^*)(2i + 1 + s^*)} \quad (1 \le i \le d-1),$$

$$c_d = hd(d + s^* - r_1)(d + s^* - r_2)/(2d + s^*).$$

(2) *We have*

$$\frac{v_i^*(x)}{k_i^*} = {}_4F_3\left(\begin{matrix} -i, i + 1 + s, -y, y + 1 + s^* \\ -d, r_1 + 1, r_2 + 1 \end{matrix} ; 1 \right) \quad (0 \le i \le d),$$

where $x = \xi^*(y) = \theta_0^* + h^* y(y + 1 + s^*)$. *Also,*

$$b_i^* = \frac{h^*(i - d)(i + 1 + s)(i + 1 + r_1)(i + 1 + r_2)}{(2i + 2 + s)(2i + 1 + s)} \quad (1 \le i \le d-1),$$

$$b_0^* = -h^* d(1 + r_1)(1 + r_2)/(2 + s),$$

$$c_i^* = \frac{h^* i(i + s + d + 1)(i + s - r_1)(i + s - r_2)}{(2i + s)(2i + 1 + s)} \quad (1 \le i \le d-1),$$

$$c_d^* = h^* d(d + s - r_1)(d + s - r_2)/(2d + s).$$

(IIA)

(1) *We have*

$$\frac{v_i(x)}{k_i} = {}_3F_2\left(\begin{matrix} -i, -y, y + 1 + s \\ -d, r_1 + 1 \end{matrix} ; 1 \right) \quad (0 \le i \le d),$$

where $x = \xi(y) = \theta_0 + hy(y+1+s)$. *Also,*

$$b_i = h(i-d)(i+1+r_1) \quad (0 \le i \le d-1),$$
$$c_i = hi(i+r_1-s-d-1) \quad (1 \le i \le d).$$

(2) *We have*

$$\frac{v_i^*(x)}{k_i^*} = {}_3F_2\left(\begin{matrix} -i, i+1+s, -y \\ -d, r_1+1 \end{matrix}; 1\right) \quad (0 \le i \le d),$$

where $x = \xi^*(y) = \theta_0^* + h^* y$. *Also,*

$$b_i^* = \frac{h^*(i-d)(i+1+s)(i+1+r_1)}{(2i+2+s)(2i+1+s)} \quad (1 \le i \le d-1),$$
$$b_0^* = -h^* d(1+r_1)/(2+s),$$
$$c_i^* = -\frac{h^* i(i+s+d+1)(i+s-r_1)}{(2i+s)(2i+1+s)} \quad (1 \le i \le d-1),$$
$$c_d^* = -h^* d(d+s-r_1)/(2d+s).$$

(IIB)

(1) *We have*

$$\frac{v_i(x)}{k_i} = {}_3F_2\left(\begin{matrix} -i, i+1+s^*, -y \\ -d, r_1+1 \end{matrix}; 1\right) \quad (0 \le i \le d),$$

where $x = \xi(y) = \theta_0 + hy$. *Also,*

$$b_i = \frac{h(i-d)(i+1+s^*)(i+1+r_1)}{(2i+2+s^*)(2i+1+s^*)} \quad (1 \le i \le d-1),$$
$$b_0 = -hd(1+r_1)/(2+s^*),$$
$$c_i = -\frac{hi(i+s^*+d+1)(i+s^*-r_1)}{(2i+s^*)(2i+1+s^*)} \quad (1 \le i \le d-1),$$
$$c_d = -hd(d+s^*-r_1)/(2d+s^*).$$

(2) *We have*

$$\frac{v_i^*(x)}{k_i^*} = {}_3F_2\left(\begin{matrix} -i, -y, y+1+s^* \\ -d, r_1+1 \end{matrix}; 1\right) \quad (0 \le i \le d),$$

where $x = \xi^*(y) = \theta_0^* + h^* y(y+1+s^*)$. *Also,*

$$b_i^* = h^*(i-d)(i+1+r_1) \quad (0 \le i \le d-1),$$
$$c_i^* = h^* i(i+r_1-s^*-d-1) \quad (1 \le i \le d).$$

(**IIC**)

(1) *We have*

$$\frac{v_i(x)}{k_i} = {}_2F_1\left(\begin{matrix} -i, -y \\ -d \end{matrix}; \frac{hh^*}{r}\right) \quad (0 \le i \le d),$$

where $x = \xi(y) = \theta_0 + hy$. *Also,*

$$b_i = \frac{r}{h^*}(i-d) \quad (0 \le i \le d-1),$$
$$c_i = \frac{r - hh^*}{h^*} i \quad (1 \le i \le d).$$

(2) *We have*

$$\frac{v_i^*(x)}{k_i^*} = {}_2F_1\left(\begin{matrix} -i, -y \\ -d \end{matrix}; \frac{hh^*}{r}\right) \quad (0 \le i \le d),$$

where $x = \xi^*(y) = \theta_0^* + h^* y$. *Also,*

$$b_i^* = \frac{r}{h}(i-d) \quad (0 \le i \le d-1),$$
$$c_i^* = \frac{r - hh^*}{h} i \quad (1 \le i \le d).$$

(**III**) *For a real number* t, *we denote the largest integer not exceeding* t *by* $\lfloor t \rfloor$. *Namely,* $\lfloor t \rfloor = j$ *is the integer satisfying* $j \le t < j+1$. *Moreover, for an integer* j, *we set* $\varepsilon(j) = 1$ *if* j *is even, and we set* $\varepsilon(j) = 0$ *if* j *is odd.*

(1) *We have*

$$\begin{aligned}\frac{v_i(x)}{k_i} &= {}_4F_3\left(\begin{matrix} -\lfloor\frac{i}{2}\rfloor, \lfloor\frac{i+2}{2}\rfloor + \frac{s^*}{2}, -\frac{y}{2}, \frac{y+2+s}{2} \\ -\lfloor\frac{d}{2}\rfloor, \frac{1+r_1}{2}, \frac{\varepsilon(d)+1+r_2}{2} \end{matrix}; 1\right) \\ &\quad + (-1)^{i-d} \frac{i^{\varepsilon(i)}(i+1+s^*)^{\varepsilon(i+1)}}{d^{\varepsilon(d)}(1+r_2)^{\varepsilon(d+1)}} \frac{y}{1+r_1} \\ &\quad \times {}_4F_3\left(\begin{matrix} -\lfloor\frac{i-1}{2}\rfloor, \lfloor\frac{i+3}{2}\rfloor + \frac{s^*}{2}, \frac{2-y}{2}, \frac{y+2+s}{2} \\ -\lfloor\frac{d-1}{2}\rfloor, \frac{3+r_1}{2}, \frac{-\varepsilon(d)+3+r_2}{2} \end{matrix}; 1\right) \quad (0 \le i \le d),\end{aligned}$$

where $x = \xi(y) = \theta_0 + 2hy$. *Also,*

$$b_i = \frac{2h(i-d)^{\varepsilon(i-d)}(i+1+s^*)^{\varepsilon(i+1)}(i+1+r_1)^{\varepsilon(i)}(i+1+r_2)^{\varepsilon(i-d+1)}}{2i+2+s^*}$$
$$(0 \le i \le d-1),$$
$$c_i = \frac{-2hi^{\varepsilon(i)}(i+s^*+d+1)^{\varepsilon(i-d+1)}(i+s^*-r_1)^{\varepsilon(i+1)}(i+s^*-r_2)^{\varepsilon(i-d)}}{2i+s^*}$$
$$(1 \le i \le d).$$

(2) *We have*

$$\frac{v_i^*(x)}{k_i^*} = {}_4F_3\left(\begin{matrix} -\lfloor\frac{i}{2}\rfloor, \lfloor\frac{i+2}{2}\rfloor + \frac{s}{2}, -\frac{y}{2}, \frac{y+2+s^*}{2} \\ -\lfloor\frac{d}{2}\rfloor, \frac{1+r_1}{2}, \frac{\varepsilon(d)+1+r_2}{2}\end{matrix}; 1\right)$$
$$+ (-1)^{i-d}\frac{i^{\varepsilon(i)}(i+1+s)^{\varepsilon(i+1)}}{d^{\varepsilon(d)}(1+r_2)^{\varepsilon(d+1)}}\frac{y}{1+r_1}$$
$$\times {}_4F_3\left(\begin{matrix} -\lfloor\frac{i-1}{2}\rfloor, \lfloor\frac{i+3}{2}\rfloor + \frac{s}{2}, \frac{2-y}{2}, \frac{y+2+s^*}{2} \\ -\lfloor\frac{d-1}{2}\rfloor, \frac{3+r_1}{2}, \frac{-\varepsilon(d)+3+r_2}{2}\end{matrix}; 1\right) \quad (0 \le i \le d),$$

where $x = \xi^*(y) = \theta_0^* + 2h^*y$. *Also,*

$$b_i^* = \frac{2h^*(i-d)^{\varepsilon(i-d)}(i+1+s)^{\varepsilon(i+1)}(i+1+r_1)^{\varepsilon(i)}(i+1+r_2)^{\varepsilon(i-d+1)}}{2i+2+s}$$
$$(0 \le i \le d-1),$$
$$c_i^* = \frac{-2h^* i^{\varepsilon(i)}(i+s+d+1)^{\varepsilon(i-d+1)}(i+s-r_1)^{\varepsilon(i+1)}(i+s-r_2)^{\varepsilon(i-d)}}{2i+s}$$
$$(1 \le i \le d).$$

Remark 6.82.

(1) We call a pair of polynomials $\{v_i(x)\}_{i=0}^d$, $\{v_i^*(x)\}_{i=0}^d$ of Type I a dual system of AW-polynomials. We call $v_i(x)$, $v_i^*(x)$ AW-polynomials.

(2) For the limiting case, the notation for AW-parameters is slightly different from the one used in [60]. In our notation, $\theta_0, \theta_0^*, h, h^*$ are called *affine parameters*. If A, A^* form an L-pair, $h^{-1}(A - \theta_0)$, $(h^*)^{-1}(A^* - \theta_0^*)$ also form an L-pair. Hence, note that a general L-pair is obtained by applying an affine transformation to an L-pair such that $\theta_0 = \theta_0^* = 0$, $h = h^* = 1$. For the notation of AW-parameters of Type III, we modify the notation used in [60] so that we can easily compare with the notation of other limiting cases.

6.4 Known P- and Q-polynomial schemes

In this section, we give a list of known P- and Q-polynomial schemes $\mathfrak{X} = (X, \{R_i\}_{0\le i\le d})$ with *sufficiently large diameter d*. We give parameters for principal T-modules which are expressed by AW-parameters. Note that the intersection numbers $p_{i,j}^k$ and the Krein numbers $q_{i,j}^k$ of $\mathfrak{X}$ are determined by the principal T-modules of $\mathfrak{X}$ (Proposition 2.43). Moreover, the principal T-modules of $\mathfrak{X}$ can be described by L-pairs (Proposition 2.43), and L-pairs are given by dual systems of orthogonal polynomials (Remark 6.26). Therefore, by substituting a dual system of orthogonal polynomials in Theorem 6.81 by AW-parameters of Theorem 6.61, we can determine the first eigenmatrix $P = (v_j(\theta_i))$ and the second eigenmatrix $Q = (v_j^*(\theta_i^*))$ of $\mathfrak{X}$ and $b_i = p_{1,i+1}^i$, $a_i = p_{1,i}^i$, $c_i = p_{1,i-1}^i$, $b_i^* = q_{1,i+1}^i$, $a_i^* = q_{1,i}^i$, $c_i^* = q_{1,i-1}^i$.

Given an association scheme, it is relatively easy to check whether it is P-polynomial or not. As was discussed in Section 6.1.1, it suffices to check whether the relation R_1 defines a distance-regular graph. We can check whether a given P-polynomial scheme becomes a Q-polynomial scheme by the following proposition.

Proposition 6.83. *A P-polynomial scheme $\mathfrak{X}$ with intersection numbers $b_i = p^i_{1,i+1}$, $a_i = p^i_{1,i}$, $c_i = p^i_{1,i-1}$ is a Q-polynomial scheme if and only if b_i, c_i are expressed by AW-parameters of Theorem* 6.81.

Proof. If $\mathfrak{X}$ is Q-polynomial, by Theorem 6.81, b_i, c_i are expressed by AW-parameters. Conversely, suppose b_i, c_i of $\mathfrak{X}$ are expressed by AW-parameters of Theorem 6.81. Define b_i^*, c_i^* by using these AW-parameters as in Theorem 6.81. Then irreducible tridiagonal matrices B, B^* with constant row sums are determined, and so is the dual system of orthogonal polynomials $\{v_i(x)\}_{i=0}^d$, $\{v_i^*(x)\}_{i=0}^d$. Here by using the notation in Theorem 6.81, we set $P = (v_j(\theta_i))$, $Q = (v_j^*(\theta_i^*))$. Then we have $PQ = QP = nI$. On the other hand, the intersection matrix of $\mathfrak{X}$ coincides with ${}^tB$, and by $A_j = v_j(A_1)$, the first eigenmatrix is given by $P = (v_j(\theta_i))$. If we let Q' be the second eigenmatrix, since $PQ' = Q'P = nI$, we get $Q' = Q = (v_j^*(\theta_i^*))$. This means that $\mathfrak{X}$ has a Q-polynomial ordering. The expression of eigenvalues θ_i by AW-parameters determines the ordering of primitive idempotents E_i, and $\mathfrak{X}$ is Q-polynomial with respect to this ordering. □

Remark 6.84. When b_i, c_i are expressed by AW-parameters of Theorem 6.81, there is a case that even if the diameter d is large, the expression is not unique, i. e., they are expressed in two ways by AW-parameters (unless $\mathfrak{X}$ is an n-gon, there are at most two ways). In this case, as is seen from the proof of Proposition 6.83, there are two Q-polynomial orderings.

Now we observe how AW-parameters are determined by b_i, c_i of a distance-regular graph by giving an example of Type I. First, we check that b_i is expressed by AW-parameters of Type I as follows (Theorem 6.81):

$$b_i = \frac{h(1-q^{i-d})(1-s^*q^{i+1})(1-r_1q^{i+1})(1-r_2q^{i+1})}{(1-s^*q^{2i+2})(1-s^*q^{2i+1})} \quad (1 \le i \le d-1),$$
$$b_0 = h(1-q^{-d})(1-r_1q)(1-r_2q)/(1-s^*q^2).$$

(Parameters q, r_1, r_2, s^*, h are determined.) As for these parameters, we check c_i coincides with the expression by AW-parameters of Type I. If $s^* = 0$, the expression of c_i determines s, and if $s^* \neq 0$, s is determined by $r_1r_2 = ss^*q^{d+1}$. At this point, we can verify that $\mathfrak{X}$ is a Q-polynomial scheme by Proposition 6.83. Next, using the properties $c_1 = 1$, $b_0 = k_1 = \theta_0$ of a P-polynomial scheme and the properties $c_1^* = 1$, $b_0^* = m_1 = \theta_0^*$ of a Q-polynomial scheme, we determine the affine parameters $h, h^*, \theta_0, \theta_0^*$ as follows. The parameter h is already given. We can obtain h^* by putting $c_1^* = 1$ in the expression of c_i^* by AW-parameters. From the above discussion, we specify b_i^*, c_i^*. The parameters θ_0, θ_0^* are determined by $\theta_0 = b_0$, $\theta_0^* = b_0^*$.

The list of P- and Q-polynomial schemes is divided into two parts: One is the core part and the other is the part which is derived from the core part. Moreover, we classify the latter part into four cases considering how they are derived from the core part:

(a) $\mathfrak{X}$ is imprimitive;
(b) $\mathfrak{X}$ has the second P- or Q-polynomial ordering;
(c) $\mathfrak{X}$ has the extended bipartite double or a fusion scheme;
(d) there is an association scheme which has the same parameters as $\mathfrak{X}$ but is not isomorphic to $\mathfrak{X}$.

Remark 6.85. At the end of each item of the list of the core part in Section 6.4.1, we give comments on the above (a)–(d).

We discuss P- and Q-polynomial schemes of the core part in Section 6.4.1, and we discuss P- and Q-polynomial schemes derived from the core part in Section 6.4.2. As for (d), there is an old important problem on the characterization in terms of parameters, which will be discussed in Section 6.4.2. In Section 6.4.3, we state the current situation on the study of irreducible T-modules.

Now we summarize how to look at the list. For each P- and Q-polynomial scheme $\mathfrak{X} = (X, \{R_l\}_{0\le i\le d})$, let $R_0, R_1, \ldots, R_d$ be the P-polynomial ordering unless otherwise stated. We first list $b_i = p^i_{1,i+1}$, $c_i = p^i_{1,i-1}$ of the distance-regular graph on the relation R_1. The relation R_i equals the distance i of this distance-regular graph. There are cases where $\mathfrak{X}$ has the second P-polynomial ordering ($\mathfrak{X}$ does not have the third P-polynomial ordering unless it is an n-gon). We consider the second P-polynomial ordering in Section 6.4.2.

Next, we list AW-parameters with their types (Theorem 6.61). The expression by AW-parameters determines the Q-polynomial ordering as we have seen in the proof of Proposition 6.83. There are cases that the ordering of AW-parameters is not uniquely determined and $\mathfrak{X}$ has the second Q-polynomial ordering (Remark 6.84). We discuss the second Q-polynomial ordering in Section 6.4.2.

At the end, we list the following parameters:

θ_i: the eigenvalues of the intersection matrix B_1 corresponding to the Q-polynomial ordering;

θ^*_i: the eigenvalues of the dual intersection matrix B^*_1 corresponding to the P-polynomial ordering;

$\lambda_i, \hat{\lambda}_i$: the data of the pre-L-pair (Theorem 6.61);

b^*_i, c^*_i: the entries of the dual intersection matrix B^*_1, i. e., $b^*_i = q^i_{1,i+1}$, $c^*_i = q^i_{1,i-1}$;

k_i: the valency determined by the intersection matrix B_1, i. e., $k_i = b_0 b_1 \cdots b_{i-1} / c_1 c_2 \cdots c_i$;

m_i: the multiplicity of the eigenvalue θ_i, which is equal to the valency k^*_i determined by the dual intersection matrix B^*_1;

$k^*_i = b^*_0 b^*_1 \cdots b^*_{i-1} / c^*_1 c^*_2 \cdots c^*_i$ (Proposition 6.24).

We denote the binomial coefficient and its q-binomial coefficient by $\binom{n}{i}$ and $\binom{n}{i}_q$, respectively:

$$\binom{n}{i} = \frac{n(n-1)\cdots(n-i+1)}{i(i-1)\cdots 1},$$
$$\binom{n}{i}_q = \frac{(q^n-1)\cdots(q^{n-i+1}-1)}{(q^i-1)(q^{i-1}-1)\cdots(q-1)}.$$

Moreover, an isomorphism of an association scheme $\mathfrak{X} = (X, \{R_i\}_{0\le i\le d})$ is a permutation on X preserving each relation R_i, and the set of isomorphisms is denoted by $\mathrm{Aut}(\mathfrak{X})$. If $\mathfrak{X}$ is a P-polynomial scheme, $\mathrm{Aut}(\mathfrak{X})$ is identical to the set of isomorphisms of the distance-regular graph determined by the relation R_1.

If a P- and Q-polynomial scheme is of Type I and satisfies $s = s^* = r_2 = 0$, it is said to have *formally self-dual classical parameters*. Note that in this case, we have $b_i = b_i^*, c_i = c_i^*, h = h^*, \theta_0 = \theta_0^*$. Incidentally, classical parameters are AW-parameters of Type I with $s^* = 0$ (Remark 6.48 (3)). Except for n-gons, known P- and Q-polynomial schemes have classical parameters, and moreover, if they have a non-thin irreducible T-module, i. e., a TD-pair which is not an L-pair, they have formally self-dual classical parameters except for twisted Grassmann schemes. (In fact, they have geometric duality as well as algebraic formal self-duality.) We will make a remark on this point in Section 6.4.1 again.

Finally, we give the notation on (a)–(d). An imprimitive association scheme $\mathfrak{X}$ is called *antipodal* if $R_0 \cup R_d$ becomes an equivalence relation. In this cases, the quotient scheme with respect to $R_0 \cup R_d$ is called the *antipodal quotient* of $\mathfrak{X}$ which is denoted by $\overline{\mathfrak{X}}$. A P-polynomial scheme $\mathfrak{X}$ is said to be *bipartite* if the distance-regular graph determined by the relation R_1 is a bipartite graph. A bipartite P-polynomial scheme is also imprimitive. The association scheme derived on a *bipartite half* is denoted by $\frac{1}{2}\mathfrak{X}$. Note that $\overline{\mathfrak{X}}$ and $\frac{1}{2}\mathfrak{X}$ of P- and Q-polynomial schemes also become P- and Q-polynomial schemes (Section 6.1.3).

6.4.1 The core of P- and Q-polynomial schemes

In this section, we give a list of the core part of known P- and Q-polynomial schemes $\mathfrak{X} = (X, \{R_i\}_{0\le i\le d})$. We divide them into two families. The second family consists of P- and Q-polynomial schemes with formally self-dual (classical) parameters, and the other belong to the first family. For most cases of P- and Q-polynomial schemes $\mathfrak{X}$, the automorphism group $\mathrm{Aut}(\mathfrak{X})$ acts transitively on each relation R_i, and hence the distance-regular graph determined on the relation R_1 is a distance-transitive graph. We list the group which acts distance-transitively on X, and in general $\mathrm{Aut}(\mathfrak{X})$ is a little larger than it. For the details, we refer the reader to [113].

For the cases which belong to the second family, X itself has the structure of an abelian group and acts regularly on X as a subgroup of $\mathrm{Aut}(\mathfrak{X})$. This means the character group $\hat{X}$ of an abelian group X has the structure of an association scheme ($\hat{X}$ is a

Schur ring), and the first eigenmatrix of the association scheme on $\hat{X}$ equals the second eigenmatrix of $\mathfrak{X}$. If we consider examples of the second family, the second eigenmatrix and the first eigenmatrix of $\mathfrak{X}$ are equal, and hence the association scheme on $\hat{X}$ is isomorphic to $\mathfrak{X}$ via the isomorphism of X and $\hat{X}$ as groups. For details, we refer the reader to [60].

The association schemes of the second family has duality as well as formal duality. In Delsarte's theory, duality of codes and designs is algebraic and formal, in general. For P- and Q-polynomial schemes of the second family, geometric duality holds in the sense that codes are regarded as designs and vice versa. This holds for P- and Q-polynomial schemes in the list, and it does not mean that $P = Q$ is derived from a general discussion.

P- and Q-polynomial schemes of the first family have a remarkable property that every irreducible T-module is thin, i. e., it can be described as an L-pair. We will discuss this in Section 6.4.3 again.

Now we give the list.

The first family

(i) The Johnson scheme $J(v,d)$ $(1 \le d \le \frac{v}{2})$

Let V be a v-element set and let $X = \binom{V}{d}$ be the set of d-subsets of V. Assume $1 \le d \le \frac{v}{2}$. Define the relation R_i on X by

$$(x,y) \in R_i \Longleftrightarrow |x \cap y| = d - i$$

(Chapter 2, Section 2.10.3). The symmetric group S_v acts transitively on each R_i. The ordering $R_0, R_1, \ldots, R_d$ gives a P-polynomial ordering of $J(v,d) = (X, \{R_i\}_{0 \le i \le d})$. The parameters $b_i = p^i_{1,i+1}$, $c_i = p^i_{1,i-1}$ of the distance-regular graph determined by R_1 are given as follows:

$$\begin{aligned} b_i &= (d-i)(v-d-i), \quad 0 \le i \le d-1, \\ c_i &= i^2, \quad 1 \le i \le d. \end{aligned}$$

The Johnson scheme $J(v,d)$ is a Q-polynomial scheme with the following AW-parameters:

$$\begin{aligned} \Lambda_{\mathrm{IIA}} &= (r_1, -, s, -; h, h^*, \theta_0, \theta_0^* \mid d), \\ r_1 &= -v + d - 1, \quad s = -v - 2, \\ h &= 1, \quad h^* = -v(v-1)/d(v-d), \\ \theta_0 &= d(v-d), \quad \theta_0^* = v - 1. \end{aligned}$$

The other parameters are as follows:

$$\begin{aligned} \theta_i &= d(v-d) + i(i-v-1), \quad 0 \le i \le d, \\ \theta_i^* &= (v-1)\left(1 - \frac{v}{d(v-d)} i\right), \quad 0 \le i \le d, \end{aligned}$$

$$\lambda_i = -\frac{v(v-1)}{d(v-d)}(i+1)(i-d)(i-v+d), \quad 0 \le i \le d-1,$$
$$\hat{\lambda}_i = -\frac{v(v-1)}{d(v-d)}(i+1)^2(i-d), \quad 0 \le i \le d-1,$$
$$b_i^* = \frac{v(v-1)}{d(v-d)} \frac{(d-i)(v+1-i)(v-d-i)}{(v-2i)(v+1-2i)}, \quad 0 \le i \le d-1,$$
$$c_i^* = \frac{v(v-1)}{d(v-d)} \frac{i(d+1-i)(v+1-d-i)}{(v+2-2i)(v+1-2i)}, \quad 1 \le i \le d,$$
$$k_i = \binom{d}{i}\binom{v-d}{i}, \quad 0 \le i \le d,$$
$$m_i = \frac{v+1-2i}{v+1-i}\binom{v}{i} = \binom{v}{i} - \binom{v}{i-1}, \quad 0 \le i \le d.$$

(a) If $v = 2d$, $J(v,d)$ is imprimitive and antipodal.
(b) If $v = 2d+1$, $J(v,d)$ has the second P-polynomial ordering $R_0, R_d, R_1, R_{d-1}, R_2, R_{d-2}, \ldots$. The distance-regular graph derived on R_d is the odd graph O_{d+1}.

(ii) The q-Johnson scheme $J_q(v,d)$ $(1 \le d \le \frac{v}{2})$
(This is also called the Grassmann scheme.)
Let V be the v-dimensional vector space over the finite field $\mathbb{F}_q$, and let $X = \binom{V}{d}_q$ be the set of d-dimensional subspaces of V. Assume $1 \le d \le \frac{v}{2}$. Define the relation R_i on X as follows:

$$(x,y) \in R_i \iff \dim(x \cap y) = d - i.$$

The full projective collineation group $P\Gamma L(v,q)$ acts transitively on each R_i. The ordering $R_0, R_1, \ldots, R_d$ gives a P-polynomial ordering of the association scheme $J_q(v,d) = (X, \{R_i\}_{0\le i\le d})$. The parameters $b_i = p^i_{1,i+1}$, $c_i = p^i_{1,i-1}$ of the distance-regular graph determined by R_1 are given as follows:

$$b_i = q^{2i+1}(q^{d-i}-1)(q^{v-d-i}-1)/(q-1)^2, \quad 0 \le i \le d-1,$$
$$c_i = (q^i-1)^2/(q-1)^2, \quad 1 \le i \le d.$$

The q-Johnson scheme $J_q(v,d)$ is a Q-polynomial scheme with the following AW-parameters:

$$\Lambda_{\mathrm{I}} = (r_1, r_2, s, s^*; h, h^*, \theta_0, \theta_0^* \mid q, d),$$
$$r_1 = q^{-v+d-1}, \quad r_2 = 0, \quad s = q^{-v-2}, \quad s^* = 0,$$
$$h = q^{v+1}/(q-1)^2, \quad h^* = q(q^v-1)(q^{v-1}-1)/(q-1)(q^{v-d}-1)(q^d-1),$$
$$\theta_0 = q(q^{v-d}-1)(q^d-1)/(q-1)^2, \quad \theta_0^* = q(q^{v-1}-1)/(q-1).$$

The other parameters are as follows:

$$\theta_i = \frac{1}{(q-1)^2}(q(q^{v-d}-1)(q^d-1) - (q^i-1)(q^{v+1-i}-1)), \quad 0 \le i \le d,$$

$$\theta_i^* = h^*\left(\frac{(q^{v-d}-1)(q^d-1)}{q^v-1} - 1 + q^{-i}\right), \quad 0 \le i \le d,$$
$$\lambda_i = hh^* q^{-v-1}(1-q^{i+1})(q^{d-i}-1)(q^{v-d-i}-1), \quad 0 \le i \le d-1,$$
$$\hat{\lambda}_i = hh^* q^{-i-v-2}(q^{i+1}-1)^2(q^{d-i}-1), \quad 0 \le i \le d-1,$$
$$b_i^* = h^* \frac{(q^{d-i}-1)(q^{v+1-i}-1)(q^{v-d-i}-1)}{q^i(q^{v-2i}-1)(q^{v+1-2i}-1)}, \quad 0 \le i \le d-1,$$
$$c_i^* = h^* \frac{(q^i-1)(q^{v-d+1-i}-1)(q^{d+1-i}-1)}{q^i(q^{v+2-2i}-1)(q^{v+1-2i}-1)}, \quad 1 \le i \le d,$$
$$k_i = q^{i^2}\binom{d}{i}_q \binom{v-d}{i}_q, \quad 0 \le i \le d,$$
$$m_i = q^i \frac{q^{v+1-2i}-1}{q^{v+1-i}-1}\binom{v}{i}_q = \binom{v}{i}_q - \binom{v}{i-1}_q, \quad 0 \le i \le d.$$

The q-Johnson scheme $J_q(v,d)$ is a q-analogue of the Johnson scheme $J(v,d)$, which is also called the Grassmann scheme. Compared with the example listed in (iii) below, it is called Type $A_{v-1}(q)$.

There exists a P- and Q-polynomial scheme which has the same parameters as $J_q(2d+1,d)$ but is not isomorphic to $J_q(2d+1,d)$. It is called the twisted Grassmann scheme (Section 6.4.2 (d)).

(iii) Dual polar schemes

Let V be a vector space over the finite field $\mathbb{F}_q$ equipped with a non-degenerate form. In the table below, we list the dimension $\dim(V)$ of V, the non-degenerate form on V, and the notation of the group which is determined by the form.

Name	$\dim(V)$	Form	e	Group
$B_d(q)$	$2d+1$	Quadratic form	0	$P\Gamma O(2d+1,q)$
$C_d(q)$	$2d$	Symplectic form	0	$P\Gamma Sp(2d,q)$
$D_d(q)$	$2d$	Quadratic form Witt index d	-1	$P\Gamma O^+(2d,q)$
${}^2D_{d+1}(q)$	$2d+2$	Quadratic form Witt index d	1	$P\Gamma O^-(2d+2,q)$
${}^2A_{2d}(r)$	$2d+1$	Hermitian form	$\frac{1}{2}$	$P\Gamma U(2d+1,r)$
${}^2A_{2d-1}(r)$	$2d$	Hermitian form	$-\frac{1}{2}$	$P\Gamma U(2d,r)$

In the table above, for the case of the Hermitian form, we require that the field has an automorphism of order 2, and hence we set $q = r^2$, where $r = p^f$, and p is a prime. Note that for any case of the table above, the dimension of a *maximal totally isotropic subspace* is d.

For each case of the above table, let X be the set of maximal totally isotropic subspaces of V. When the characteristic of the base field is 2, we need to take special care of the quadratic form. A bilinear form (u, v) is said to be symmetric if $(u, v) = (v, u)$ $(u, v \in V)$, and alternative if $(u, u) = 0$ $(u \in V)$. A quadratic form $Q : V \longrightarrow \mathbb{F}_q$ associated with a symmetric bilinear form $(\ ,\ )$ is defined by $Q(\lambda v) = \lambda^2 Q(v)$ and $Q(u + v) = Q(u) + Q(v) + (u, v)$, $(\lambda \in \mathbb{F}_q, u, v, \in V)$, and if the characteristic is not 2, we have $Q(v) = \frac{1}{2}(v, v)$ and it exists uniquely. If the characteristic is 2, it is not unique. Conversely, a symmetric bilinear form $(\ ,\ )$ associated with a quadratic form Q is defined by

$$(u, v) = Q(u + v) - Q(u) - Q(v), \tag{6.284}$$

and it is uniquely determined by Q regardless of the characteristic. If the characteristic is 2, a quadratic form is defined to be non-degenerate as follows. Let

$$\operatorname{rad}(V) = \{v \in V \mid (u, v) = 0, \text{ for all } u \in V\}, \tag{6.285}$$

$$V(Q) = \{v \in \operatorname{rad}(V) \mid Q(v) = 0\}. \tag{6.286}$$

Then the quadratic form is said to be non-degenerate if $V(Q) = 0$. (If a quadratic form is non-degenerate in this sense, the dimension of $\operatorname{rad}(V)$ is 0 or 1 depending on the parity of $\dim(V)$, so for the case of $D_d(q)$ or ${}^2D_{d+1}(q)$, we may also define it to be non-degenerate if $\operatorname{rad}(V) = 0$.) We also define a subspace U of V to be *totally isotropic* if $Q(u) = 0$ for all $u \in U$. In particular, by (6.284), if U is totally isotropic, $(u, v) = 0$ for any $u, v \in U$. Now we define the association scheme on X. Define the relation R_i $(0 \le i \le d)$ on X by

$$(x, y) \in R_i \iff \dim(x \cap y) = d - i.$$

Then the group listed on the table acts transitively on each R_i. The ordering $R_0, R_1, \ldots, R_d$ gives the P-polynomial ordering of the association scheme $\mathfrak{X} = (X, \{R_i\}_{0 \le i \le d})$. The parameters $b_i = p^i_{1,i+1}$, $c_i = p^i_{1,i-1}$ of the distance-regular graph determined by R_1 are

$$\begin{aligned} b_i &= q^{i+e+1}(q^{d-i} - 1)/(q - 1), \quad 0 \le i \le d - 1, \\ c_i &= (q^i - 1)/(q - 1), \quad 1 \le i \le d. \end{aligned}$$

The association scheme $\mathfrak{X}$ is a Q-polynomial scheme with the following AW-parameters:

$$\begin{aligned} &\Lambda_{\mathrm{I}} = (r_1, r_2, s, s^*; h, h^*, \theta_0, \theta_0^* \mid q, d), \\ &r_1 = r_2 = 0, \quad s = -q^{-d-e-2}, \quad s^* = 0, \\ &h = \frac{q^{d+e+1}}{q-1}, \quad h^* = \frac{q(q^{d+e} + 1)(q^{d+e-1} + 1)}{(q-1)(q^e + 1)}, \\ &\theta_0 = \frac{q^{e+1}(q^d - 1)}{q - 1}, \quad \theta_0^* = \frac{q^{e+1}(q^d - 1)(q^{d+e-1} + 1)}{(q - 1)(q^e + 1)}. \end{aligned}$$

The other parameters are the following:

$$\begin{aligned}
\theta_i &= \theta_0 - \frac{1}{q-1}(q^i-1)(q^{d+e+1-i}+1), \quad 0 \le i \le d,\\
\theta_i^* &= \theta_0^* + h^*(q^{-i}-1), \quad 0 \le i \le d,\\
\lambda_i &= -hh^* q^{-i-1-d}(q^{i+1}-1)(q^{d-i}-1), \quad 0 \le i \le d-1,\\
\hat{\lambda}_i &= hh^* q^{-i-e-2-d}(q^{i+1}-1)(q^{d-i}-1), \quad 0 \le i \le d-1,\\
b_i^* &= h^* \frac{q^e(q^{d-i}-1)(q^{d+e+1-i}+1)}{q^{2i}(q^{d+e-2i}+1)(q^{d+e+1-2i}+1)}, \quad 0 \le i \le d-1,\\
c_i^* &= h^* \frac{q^{e+1}(q^i-1)(q^{i-e-1}+1)}{q^{2i}(q^{d+e+2-2i}+1)(q^{d+e+1-2i}+1)}, \quad 1 \le i \le d,\\
k_i &= q^{i(2e+i+1)/2}\binom{d}{i}_q, \quad 0 \le i \le d,\\
m_i &= q^i\binom{d}{i}_q \frac{q^{d+e+1-2i}+1}{q^{d+e+1-i}+1}\prod_{j=0}^{i-1}\frac{q^{d+e-j}+1}{q^{j-e}+1}, \quad 0 \le i \le d.
\end{aligned}$$

Note that:

(a) The dual polar scheme $D_d(q)$ is imprimitive (*bipartite*).

(b) The dual polar scheme ${}^2A_{2d-1}(q)$ has the second Q-polynomial ordering $E_0, E_d, E_1, E_{d-1}, E_2, E_{d-2}, \ldots$.

(d) The dual polar schemes $B_d(q)$ and $C_d(q)$ have the same parameters but are not isomorphic. Both $B_d(q)$ and $C_d(q)$ have the extended bipartite double (Section 6.4.2 (c)) and the fusion scheme ($R_1 \cup R_2$ determines the distance-regular graph [Section 6.4.2 (c)]). The extended bipartite double of $B_d(q)$ is isomorphic to $D_{d+1}(q)$, and the fusion scheme of $B_d(q)$ is isomorphic to $\frac{1}{2}D_{d+1}(q)$. On the other hand, the extended bipartite double of $C_d(q)$ has the same parameters as $D_{d+1}(q)$, but if the characteristic is not 2, it is not isomorphic to $D_{d+1}(q)$. This P- and Q-polynomial scheme is called the *Hemmeter scheme*, denoted by $\mathrm{Hem}_{d+1}(q)$ (Section 6.4.2 (c)). Moreover, the fusion scheme of $C_d(q)$ has the same parameters as $\frac{1}{2}D_{d+1}(q)$, but if the characteristic is not 2, it is not isomorphic to $\frac{1}{2}D_{d+1}(q)$. This P- and Q-polynomial scheme is called the *Ustimenko scheme*, denoted by $\mathrm{Ust}_{[\frac{d+1}{2}]}(q)$ (Section 6.4.2 (c)). The Hemmeter scheme $\mathrm{Hem}_{d+1}(q)$ is bipartite and $\frac{1}{2}\mathrm{Hem}_{d+1}(q)$ is isomorphic to $\mathrm{Ust}_{[\frac{d+1}{2}]}(q)$.

The second family

(o) *n*-gons

We regard $X = \{0, 1, 2, \ldots, n-1\}$ as the abelian group $\mathbb{Z}/n\mathbb{Z}$, and define the relation R_i on X by

$$(x, y) \in R_i \Longleftrightarrow x - y = \pm i \pmod{n}.$$

The dihedral group of order $2n$ acts transitively on each R_i. Let $d = [\frac{n}{2}]$. The association scheme $\mathfrak{X} = (X, \{R_i\}_{0 \le i \le d})$ is a P-polynomial scheme and $R_0, R_1, \ldots, R_d$ gives a

P-polynomial ordering. The parameters $b_i = p^i_{1,i+1}$, $c_i = p^i_{1,i-1}$ of the distance-regular graph determined by R_1 are

$$b_0 = 2, \quad b_i = 1\ (1 \le i \le d-1), \quad c_i = 1\ (1 \le i \le d-1),$$

$$c_d = \begin{cases} 2, & \text{if } n \text{ is even,} \\ 1, & \text{if } n \text{ is odd.} \end{cases}$$

The association scheme $\mathfrak{X}$ is a Q-polynomial scheme with the following AW-parameters:

$$\begin{aligned} \Lambda_{\mathrm{I}} &= (r_1, r_2, s, s^*; h, h^*, \theta_0, \theta_0^* \mid q, d), \\ q &= e^{2\pi k\sqrt{-1}/n}, \quad (n,k) = 1 \quad (q \text{ is a primitive } n\text{-th root of } 1), \\ \{r_1, r_2, q^{-d-1}\} &= \{q^{-\frac{1}{2}}, -q^{-\frac{1}{2}}, -q^{-1}\}, \quad s = s^* = q^{-1}, \\ h &= h^* = 1, \quad \theta_0 = \theta_0^* = 2. \end{aligned}$$

The other parameters are the following:

$$\theta_i = \theta_i^* = q^i + q^{-i}, \quad 0 \le i \le d,$$

$$\lambda_i = q^{-2i-1}(1+q^i)(1-q^{i+1})(1-q^{2i+1}), \quad 0 \le i \le d-1,$$

$$\hat{\lambda}_i = \begin{cases} -q^{-2i-1}(1+q^i)(1-q^{i+1})(1-q^{2i+1}), & 0 \le i \le d-1, \\ & \text{if } n \text{ is even,} \\ q^{-2i-1+d}(1-q^{2i+2})(1-q^{2i+1}), & 0 \le i \le d-1, \\ & \text{if } n \text{ is odd,} \end{cases}$$

$$b_i^* = b_i\ (0 \le i \le d-1), \quad c_i^* = c_i\ (1 \le i \le d),$$

$$k_i = m_i = 2\ (1 \le i \le d-1), \quad k_0 = m_0 = 1, \quad k_d = m_d = \begin{cases} 1, & \text{if } n \text{ is even,} \\ 2, & \text{if } n \text{ is odd.} \end{cases}$$

(a) The association scheme $\mathfrak{X}$ is imprimitive if and only if n is a prime.
(b) For ℓ with $(n,\ell) = 1$, $\mathfrak{X}$ has the second P-polynomial ordering $R_0, R_\ell, R_{2\ell}, R_{3\ell}, \ldots$, where the indices are taken modulo n. Let $E_0, E_1, E_2, \ldots, E_d$ be a Q-polynomial ordering of $\mathfrak{X}$. For k with $(n,k) = 1$, $\mathfrak{X}$ has the second Q-polynomial ordering $E_0, E_k, E_{2k}, E_{3k}, \ldots$, where the indices are taken modulo n.

(i) The Hamming scheme $H(d,q)$
Let F be a q-element set and $X = F^d$. For $x \in X$, we write $x = (x_1, \ldots, x_d)$ $(x_j \in F)$. Define the relation R_i on X by

$$(x,y) \in R_i \Longleftrightarrow |\{j \mid x_j \ne y_j\}| = i$$

(Chapter 2, Section 2.10.2). The wreath product $S_q \wr S_d$ of the symmetric group S_q by the symmetric group S_d acts transitively on each relation R_i (Example 2.6). The ordering

$R_0, R_1, \ldots, R_d$ gives a P-polynomial ordering of $H(d,q)$. The parameters $b_i = p^i_{1,i+1}$, $c_i = p^i_{1,i-1}$ of the distance-regular graph determined by R_1 are

$$b_i = (q-1)(d-i), \quad 0 \le i \le d-1,$$
$$c_i = i, \quad 1 \le i \le d.$$

The Hamming scheme $H(d,q)$ is a Q-polynomial scheme with the following AW-parameters:

$$\Lambda_{\mathrm{IIC}} = (r, -, -; h, h^*, \theta_0, \theta_0^* \mid d),$$
$$r = q(q-1), \quad h = h^* = -q, \quad \theta_0 = \theta_0^* = (q-1)d.$$

The other parameters are the following:

$$\theta_i = \theta_i^* = (q-1)d - qi, \quad 0 \le i \le d,$$
$$\lambda_i = q(q-1)(i+1)(i-d), \quad 0 \le i \le d-1,$$
$$\hat{\lambda}_i = -q(i+1)(i-d), \quad 0 \le i \le d-1,$$
$$b_i^* = b_i \ (0 \le i \le d-1), \quad c_i^* = c_i \ (1 \le i \le d),$$
$$k_i = m_i = (q-1)^i \binom{d}{i}, \quad 0 \le i \le d.$$

(a) The Hamming scheme $H(d,2)$ is bipartite and antipodal. If d is odd, $\frac{1}{2}H(d,2)$ is isomorphic to $\overline{H(d,2)}$. If d is even, $\frac{1}{2}H(d,2)$ is antipodal and $\overline{H(d,2)}$ is bipartite. Moreover, $\overline{\frac{1}{2}H(d,2)}$ is isomorphic to $\frac{1}{2}\overline{H(d,2)}$ (Section 6.4.2 (a)).

(b) If d is even, $H(d,2)$ has the second P-polynomial ordering

$$R_0, R_{d-1}, R_2, R_{d-3}, R_4, R_{d-5}, \ldots, R_1, R_d$$

and the second Q-polynomial ordering

$$E_0, E_{d-1}, E_2, E_{d-3}, E_4, E_{d-5}, \ldots, E_1, E_d.$$

If d is odd, by setting $d = 2d' + 1$, $\frac{1}{2}H(d,2) \cong \overline{H(d,2)}$ has the second P-polynomial ordering

$$R_0, R_{d'}, R_1, R_{d'-1}, R_2, R_{d'-2}, \ldots$$

and the second Q-polynomial ordering

$$E_0, E_2, E_4, \ldots, E_3, E_1$$

(Section 6.4.2 (b)).

(d) There exists a P- and Q-polynomial scheme, called the Doob scheme, which has the same parameters as $H(d,4)$ but is not isomorphic to $H(d,4)$ (Section 6.4.2 (d)).

Remark 6.86. In $H(n,2)$, fix $x_0 \in X$ and set

$$Y = \Gamma_d(x_0) = \{x \in X \mid (x_0, x) \in R_d\}, \quad 1 \le d \le \frac{n}{2}.$$

The association scheme $\mathcal{Y} = (Y, \{R_{2i}\}_{0\le i\le d})$ is isomorphic to the Johnson scheme $J(n,d)$.

(ii) The bilinear forms scheme $\mathrm{Bil}_{d\times n}(q)$ $(d \le n)$

Let X be the set of $d \times n$-matrices over the finite field $\mathbb{F}_q$. Assume $d \le n$. For $0 \le i \le d$, define the relation R_i on X by

$$(x, y) \in R_i \iff \operatorname{rank}(x - y) = i.$$

The set X forms an abelian group with respect to matrix sum. The abelian group X acts naturally on X itself by addition, and the action preserves the relation R_i. Let $H = \mathrm{GL}(d,q) \times \mathrm{GL}(n,q)$ be the direct product of general linear groups $\mathrm{GL}(d,q)$ and $\mathrm{GL}(n,q)$. Then H acts on X as follows:

$$x^{(a,b)} = a^{-1}xb, \quad x \in X,\ (a,b) \in H,$$

and the action preserves the relation R_i. Combining the above actions on X yields the semidirect product $G = HX \rhd X$ of X by H: For the group G, the conjugate of $x \in X$ by $(a,b) \in H$ is given by $(a,b)^{-1}x(a,b) = x^{(a,b)} = a^{-1}xb$. The group G acts transitively on each relation R_i. The ordering $R_0, R_1, \dots, R_d$ gives a P-polynomial ordering of $\mathfrak{X} = (X, \{R_i\}_{0\le i\le d})$. The parameters $b_i = p^i_{1,i+1}$, $c_i = p^i_{1,i-1}$ of the distance-regular graph determined by R_1 are

$$\begin{aligned} b_i &= q^{2i}(q^{d-i} - 1)(q^{n-i} - 1)/(q - 1), \quad 0 \le i \le d - 1,\\ c_i &= q^{i-1}(q^i - 1)/(q - 1), \quad 1 \le i \le d. \end{aligned}$$

The association scheme $\mathfrak{X}$ is a Q-polynomial scheme with the following AW-parameters:

$$\begin{aligned} &\Lambda_{\mathrm{I}} = (r_1, r_2, s, s^*; h, h^*, \theta_0, \theta_0^* \mid q, d),\\ &r_1 = q^{-n-1}, \quad r_2 = 0, \quad s = s^* = 0,\\ &h = h^* = q^{d+n}/(q-1), \quad \theta_0 = \theta_0^* = (q^d - 1)(q^n - 1)/(q - 1). \end{aligned}$$

The other parameters are the following:

$$\begin{aligned} &\theta_i = \theta_i^* = (q^{d+n-i} - q^d - q^n + 1)/(q - 1), \quad 0 \le i \le d,\\ &\lambda_i = q^{d+n-1}(1 - q^{i+1})(q^{d-i} - 1)(q^{n-i} - 1)/(q - 1)^2, \quad 0 \le i \le d - 1,\\ &\hat{\lambda}_i = q^{d+n-1}(q^{i+1} - 1)(q^{d-i} - 1)/(q - 1)^2, \quad 0 \le i \le d - 1,\\ &b_i^* = b_i\ (0 \le i \le d - 1), \quad c_i^* = c_i\ (1 \le i \le d),\\ &k_i = m_i = q^{i(i-1)/2}\binom{d}{i}_q \binom{n}{i}_q (q - 1)(q^2 - 1)\cdots(q^i - 1), \quad 0 \le i \le d. \end{aligned}$$

The P- and Q-polynomial scheme $\mathfrak{X}$ is called the *bilinear forms scheme* denoted by $\mathrm{Bil}_{d\times n}(q)$. In the following, we state the relation between $\mathrm{Bil}_{d\times n}(q)$ and the q-Johnson scheme $J_q(d+n,d)$ ([60, page 311] and [113, page 280]). As we have seen before, in $\mathrm{Bil}_{d\times n}(q)$, the group $G = HX$ acts transitively on each R_i as the automorphism group, where $H = \mathrm{GL}(d,q) \times \mathrm{GL}(n,q)$. Therefore, $\mathrm{Bil}_{d\times n}(q)$ is identical to the association scheme determined by the transitive action of the group G on X (Example 2.3). Let 0 be the zero matrix in X. The stabilizer G_0 of 0 coincides with H: $H = G_0$. Therefore, the action of G on X is isomorphic to the action of G on $H\backslash G$. Namely, the pair of G and the subgroup H of G determines the association scheme $\mathrm{Bil}_{d\times n}(q)$.

On the other hand, the group $G = HX$ is isomorphic to the *maximal parabolic subgroup*

$$P_J = \left\{ \begin{pmatrix} a & x \\ 0 & b \end{pmatrix} \mid x \in X, (a,b) \in H \right\}$$

of the general linear group $\mathrm{GL}(d+n,q)$ by the natural correspondence $x \mapsto \left(\begin{smallmatrix} 1 & x \\ 0 & 1 \end{smallmatrix}\right)$, $(a,b) \mapsto \left(\begin{smallmatrix} a & 0 \\ 0 & b \end{smallmatrix}\right)$. Namely, let $P_J = L_J U_J \rhd U_J$ be the *Levi decomposition* of P_J where

$$L_J = \left\{ \begin{pmatrix} a & 0 \\ 0 & b \end{pmatrix} \mid (a,b) \in H \right\}, \quad U_J = \left\{ \begin{pmatrix} 1 & x \\ 0 & 1 \end{pmatrix} \mid x \in X \right\}.$$

Then we have the following isomorphism:

$$G \cong P_J, \quad H \cong L_J, \quad X \cong U_J.$$

This means that $\mathrm{Bil}_{d\times n}(q)$ is determined by the maximal parabolic subgroup P_J and the Levi subgroup L_J. Then we can interpret the homogeneous space $L_J\backslash P_J$ on which P_J acts in the framework of the q-Johnson scheme $J_q(d+n,d)$ as follows.

Let

$$V = \mathbb{F}_q^{d+n} = \{v = (v_1,\dots,v_{d+n}) \mid v_1,\dots,v_{d+n} \in \mathbb{F}_q\}$$

be the $(d+n)$-dimensional vector space over the finite field $\mathbb{F}_q$. Choose the n-dimensional subspace

$$x_0 = \{v \in V \mid v_1 = \cdots = v_d = 0, v_{d+1},\dots,v_{d+n} \in \mathbb{F}_q\}$$

of V and define the family Y of d-dimensional subspaces of V as follows:

$$Y = \left\{ y \in \binom{V}{d}_q \mid y \cap x_0 = \{\mathbf{0}\} \right\}.$$

Moreover, define the relation S_i on Y by

$$(y,y') \in S_i \iff \dim(y \cap y') = d - i.$$

Then $\mathfrak{Y} = (Y, \{S_i\}_{0 \le i \le d})$ becomes an association scheme isomorphic to $\mathrm{Bil}_{d \times n}(q)$. Note that $\mathfrak{Y}$ is the restriction of the q-Johnson scheme $J_q(d+n, d)$ on Y. We call Y the *attenuated space*.

We can check that $\mathfrak{Y}$ is isomorphic to $\mathrm{Bil}_{d \times n}(q)$ by considering the group isomorphism $G \cong P_J$ as follows. First, the general linear group $\mathrm{GL}(d+n, q)$ acts naturally on V from the right, and also acts transitively on the set $\binom{V}{n}_q$ of n-dimensional subspaces of V. The stabilizer of x_0 is equal to the maximal parabolic subgroup P_J: $\mathrm{GL}(d+n, q)_{x_0} = P_J$. Since P_J fixes x_0, it acts on Y. Moreover, the action is transitive on Y and transitive on each relation S_i. Namely, the association scheme determined by the action of the group P_J on Y coincides with $\mathfrak{Y}$. On the other hand, if we choose an element

$$y_0 = \{v \in V \mid v_1, \dots, v_d \in \mathbb{F}_q, v_{d+1} = \dots = v_{d+n} = 0\}$$

of Y, the stabilizer of y_0 coincides with L_J: $(P_J)_{y_0} = L_J$. Namely, the action of P_J on Y is isomorphic to the action of P_J on $L_J \backslash P_J$. By the isomorphism $G \cong P_J$, we obtain the isomorphism

$$(G, X) \cong (G, H\backslash G) \cong (P_J, L_J \backslash P_J) \cong (P_J, Y)$$

of permutation groups, and hence $\mathrm{Bil}_{d \times n}(q)$ is isomorphic to $\mathfrak{Y}$ as association schemes.

The correspondence of X and Y in this isomorphism is described specifically as follows. Since in the isomorphism $(G, H\backslash G) \cong (P_J, L_J\backslash P_J)$, H is the stabilizer of $\mathbf{0} \in X$ and L_J is the stabilizer of $y_0 \in Y$, in the isomorphism $(G, X) \cong (P_J, Y)$, the element $x = \mathbf{0} + x$ of X corresponds to the element $y_x = y_0\left(\begin{smallmatrix} 1 & x \\ 0 & 1 \end{smallmatrix}\right)$ of Y: $Y = \{y_x \mid x \in X\}$. Now we note that $y_0 = \{(u, \mathbf{0}) \mid u \in \mathbb{F}_q{}^d\}$. Since we have

$$y_x = \{(u, ux) \mid u \in \mathbb{F}_q{}^d\},$$

we have

$$\dim(y_x \cap y_{x'}) = \dim(\{u \in \mathbb{F}_q{}^d \mid ux = ux'\}),$$

and thus we have

$$d - \dim(y_x \cap y_{x'}) = \mathrm{rank}(x - x').$$

Namely, for the q-Johnson scheme $J_q(d+n, d)$, if we restrict the relation on $\binom{V}{d}_q$ to Y, we obtain the association scheme isomorphic to $\mathrm{Bil}_{d \times n}(q)$.

(iii) Affine schemes (or classical forms schemes)

By X and the relations defined in (iii-1), (iii-2), (iii-3) below, we obtain three types of association schemes $\mathfrak{X} = (X, \{R_i\}_{0 \le i \le d})$. We denote $\mathrm{Alt}_n(r)$, $\mathrm{Her}_d(r)$, and $\mathrm{Quad}_n(r)$ and call them the *alternating bilinear forms scheme*, the *Hermitian forms scheme*, and the *quadratic forms scheme*, respectively. The association schemes $\mathrm{Bil}_{d+n}(r)$ defined in (ii) and $\mathrm{Alt}_n(r)$, $\mathrm{Her}_d(r)$, $\mathrm{Quad}_n(r)$ defined in (iii) are called the *affine schemes* or the *classical forms schemes.*

(iii-1) The alternating bilinear forms scheme $\mathrm{Alt}_n(r)$
Let V be the n-dimensional vector space over the finite field $\mathbb{F}_r$, and let X be the set of alternating forms on V. (If the characteristic of the base field is not 2, X can be regarded as the set of skew symmetric matrices of size n). Let $d = [\frac{n}{2}]$ and define the relation R_i on X by

$$(x, y) \in R_i \Longleftrightarrow \mathrm{rank}(x - y) = 2i.$$

(Note that the rank of an alternating bilinear form is always even.)

(iii-2) Hermitian forms scheme $\mathrm{Her}_d(r)$
Let V be the d-dimensional vector space over the finite field $\mathbb{F}_{r^2}$ and let X be the set of Hermitian forms on V. (X can be regarded as the set of Hermitian matrices of size d.) Define the relation R_i on X by

$$(x, y) \in R_i \Longleftrightarrow \mathrm{rank}(x - y) = i.$$

(iii-3) Quadratic forms scheme $\mathrm{Quad}_n(r)$
Let V be the n-dimensional vector space over the finite field $\mathbb{F}_r$, and let X be the set of quadratic forms on V. (If the characteristic of the base field is not 2, X can be regarded as the set of symmetric matrices of size n.) Let $d = [\frac{n+1}{2}]$ and define the relation R_i on X by

$$(x, y) \in R_i \Longleftrightarrow \mathrm{rank}(x - y) = 2i - 1, 2i,$$

where we define $\mathrm{rank}(x) = \dim(V/V(x))$ by using the notation in (6.286). (If the characteristic of the base field is not 2, it coincides with the usual rank, that is, $\mathrm{rank}(x) = \dim(V/\mathrm{rad}(x))$.)

We give the table of affine schemes $\mathfrak{X} = (X, \{R_i\}_{0 \le i \le d})$ with parameters which will be appear later. In any case, the diameter is d.

Name	dim(V)	Field	q	ε	e	Form
$\mathrm{Bil}_{d \times n}(r)$	$d \times n\ (d \le n)$	$\mathbb{F}_r$	r	1	$n - d$	Bilinear
$\mathrm{Alt}_{2d}(r)$	$2d$	$\mathbb{F}_r$	r^2	1	$-\frac{1}{2}$	Alternating bilinear
$\mathrm{Alt}_{2d+1}(r)$	$2d + 1$	$\mathbb{F}_r$	r^2	1	$\frac{1}{2}$	Alternating bilinear
$\mathrm{Her}_d(r)$	d	$\mathbb{F}_{r^2}$ $(r > 0)$	$-r$	-1	0	Hermitian
$\mathrm{Quad}_{2d-1}(r)$	$2d - 1$	$\mathbb{F}_r$	r^2	1	$-\frac{1}{2}$	Quadratic
$\mathrm{Quad}_{2d}(r)$	$2d$	$\mathbb{F}_r$	r^2	1	$\frac{1}{2}$	Quadratic

Let $\mathfrak{X} = (X, \{R_i\}_{0\le i\le d})$ be the association scheme of diameter d with parameters q, ε, e in the above table. The ordering $R_0, R_1, \ldots, R_d$ gives a P-polynomial ordering of $\mathfrak{X}$. The parameters $b_i = p^i_{1,i+1}$, $c_i = p^i_{1,i-1}$ of the distance-regular graph determined by R_1 are

$$\begin{aligned} b_i &= \varepsilon q^{2i}(q^{d-i}-1)(q^{d+e-i}-\varepsilon)/(q-1), \quad 0 \le i \le d-1, \\ c_i &= q^{i-1}(q^i-1)/(q-1), \quad 1 \le i \le d. \end{aligned}$$

The association scheme $\mathfrak{X}$ is a Q-polynomial scheme with the following AW-parameters:

$$\begin{aligned} &\Lambda_{\mathrm{I}} = (r_1, r_2, s, s^*; h, h^*, \theta_0, \theta_0^* \mid q, d), \\ &r_1 = \varepsilon q^{-d-e-1}, \quad r_2 = 0, \quad s = s^* = 0, \\ &h = h^* = \varepsilon q^{2d+e}/(q-1), \quad \theta_0 = \theta_0^* = \varepsilon(q^d-1)(q^{d+e}-\varepsilon)/(q-1). \end{aligned}$$

The other parameters are the following:

$$\begin{aligned} &\theta_i = \theta_i^* = (\varepsilon q^{d+e}(q^{d-i}-1) - q^d + 1)/(q-1), \quad 0 \le i \le d, \\ &\lambda_i = q^{2d+e-1}(1-q^{i+1})(q^{d-i}-1)(q^{d+e-i}-\varepsilon)/(q-1)^2, \quad 0 \le i \le d-1, \\ &\hat{\lambda}_i = \varepsilon q^{2d+e-1}(q^{i+1}-1)(q^{d-i}-1)/(q-1)^2, \quad 0 \le i \le d-1, \\ &b_i^* = b_i \ (0 \le i \le d-1), \quad c_i^* = c_i \ (1 \le i \le d), \\ &k_i = m_i = \varepsilon^i q^{i(i-1)/2} \binom{d}{i}_q \prod_{j=0}^{i-1} (q^{d+e-j} - \varepsilon), \quad 0 \le i \le d. \end{aligned}$$

We state the relation between affine schemes $\mathrm{Alt}_n(r)$, $\mathrm{Her}_n(r)$, $\mathrm{Quad}_n(r)$ and dual polar schemes $D_n(r)$, ${}^2A_{2n-1}(r)$, $C_n(r)$ ([60, page 311] and [113, pages 286–287]). In what follows, we denote the $2n$-dimensional vector space over the finite field $\mathbb{F}$ by

$$\mathbb{F}^{2n} = \{v = (v_1, v_2) \mid v_1, v_2 \in \mathbb{F}^n\},$$

and fix n-dimensional subspaces of V as follows:

$$\begin{aligned} x_0 &= \{v = (0, v_2) \in V \mid v_2 \in \mathbb{F}^n\}, \\ y_0 &= \{v = (v_1, 0) \in V \mid v_1 \in \mathbb{F}^n\}. \end{aligned}$$

(iii-1) The association schemes $\mathrm{Alt}_n(r)$ and $D_n(r)$ are related as follows. Let $|\mathbb{F}| = r$ and assume that *the characteristic of the field* $\mathbb{F}$ *is not* 2. Define a non-degenerate quadratic form on V by

$$\langle u, v\rangle = u \begin{pmatrix} 0 & 1 \\ 1 & 0 \end{pmatrix} {}^t v,$$

where 1 denotes the identity matrix of size n. Let $O^+(2n,r)$ be the orthogonal group determined by the above quadratic form. Namely,

$$O^+(2n,r) = \left\{ \begin{pmatrix} a & b \\ c & d \end{pmatrix} \in \mathrm{GL}(2n,r) \mid \begin{pmatrix} a & b \\ c & d \end{pmatrix} \begin{pmatrix} 0 & 1 \\ 1 & 0 \end{pmatrix} {}^t\!\begin{pmatrix} a & b \\ c & d \end{pmatrix} = \begin{pmatrix} 0 & 1 \\ 1 & 0 \end{pmatrix} \right\}.$$

The orthogonal group $O^+(2n,r)$ acts on V as an isometry, and also acts transitively on the set of maximal totally isotropic subspaces $\binom{V}{n}'_r$ of V. The association scheme determined by the action of $O^+(2n,r)$ on $\binom{V}{n}'_r$ is $D_n(r)$. For an element $x_0 = \{(0,v_2) \mid v_2 \in \mathbb{F}^n\}$ of $\binom{V}{n}'_r$, let

$$\Gamma_n(x_0) = \left\{ y \in \binom{V}{n}'_r \mid x_0 \cap y = \{0\} \right\}.$$

Then the stabilizer $O^+(2n,r)_{x_0}$ of x_0 in $O^+(2n,r)$ acts transitively on $\Gamma_n(x_0)$. If we let $P = O^+(2n,r)_{x_0}$, P is a maximal parabolic subgroup of $O^+(2n,r)$ and has the Levi decomposition $P = LU \rhd U$. Here we have

$$L = \left\{ \begin{pmatrix} a & 0 \\ 0 & {}^t a^{-1} \end{pmatrix} \mid a \in \mathrm{GL}(n,r) \right\}, \quad U = \left\{ \begin{pmatrix} 1 & x \\ 0 & 1 \end{pmatrix} \mid x \in X \right\},$$

where X is the set of skew symmetric matrices of size n. Since the Levi subgroup L coincides with the stabilizer of $y_0 = \{(v_1,0) \mid v_1 \in \mathbb{F}^n\} \in \Gamma_n(x_0)$ in P, the action of P on $\Gamma_n(x_0)$ is isomorphic to the action of P on $L\backslash P$: $(P,\Gamma_n(x_0)) \cong (P, L\backslash P)$.

It can be checked that the association scheme determined by the action of P on $\Gamma_n(x_0)$ is the affine scheme $\mathrm{Alt}_n(r)$ as follows. First note that the set of alternating forms on $\mathbb{F}^n$ is identified with the set X of skew symmetric matrices of size n. With respect to the matrix sum, X becomes an abelian group, and the general linear group $H = \mathrm{GL}(n,r)$ acts on X by

$$x^a = a^{-1} x\, {}^t a^{-1}$$

as an automorphism group of the abelian group. Thus the semidirect product $G = HX \rhd X$ of X by H is determined (the conjugate of $x \in X$ by $a \in H$ is x^a), and G acts transitively on X by

$$y^{ax} = a^{-1} y\, {}^t a^{-1} + x \quad (y \in X,\ a \in H,\ x \in X).$$

The association scheme determined by this action is the affine scheme $\mathrm{Alt}_n(r)$. If we let 0 be the zero matrix of X, since the stabilizer of 0 in G is identical to H, the action of G on X is isomorphic to the action of G on $H\backslash G$: $(G,X) \cong (G,H\backslash G)$. Namely, the pair of the group G and the subgroup H determines $\mathrm{Alt}_n(r)$. On the other hand, since the natural isomorphisms $H \cong L$ $(a \mapsto \left(\begin{smallmatrix} a & 0 \\ 0 & {}^t a^{-1} \end{smallmatrix}\right))$, $X \cong U$ $(x \mapsto \left(\begin{smallmatrix} 1 & x \\ 0 & 1 \end{smallmatrix}\right))$ are extended to the isomorphism $G \cong P$ $(ax \mapsto \left(\begin{smallmatrix} a & 0 \\ 0 & {}^t a^{-1} \end{smallmatrix}\right)\left(\begin{smallmatrix} 1 & x \\ 0 & 1 \end{smallmatrix}\right))$, we obtain the isomorphism as permutation groups as follows:

$$(G, X) \cong (G, H\backslash G) \cong (P, L\backslash P) \cong (P, \Gamma_n(x_0)).$$

Namely, the association scheme determined by the action of P on $\Gamma_n(x_0)$ is isomorphic to $\mathrm{Alt}_n(r)$.

Similarly to the case of (ii), the correspondence between X and $\Gamma_n(x_0)$ in the above isomorphism is described explicitly as follows. As for the isomorphism $(G, H\backslash G) \cong (P, L\backslash P)$, H is the stabilizer of 0 $(\in X)$ and L is the stabilizer of y_0 $(\in \Gamma_n(x_0))$. Hence as for the isomorphism $(G, X) \cong (P, \Gamma_n(x_0))$, an element $x = 0 + x = 0^x$ of X corresponds to the element

$$y_x = y_0 \begin{pmatrix} 1 & x \\ 0 & 1 \end{pmatrix}$$

of $\Gamma_n(x_0)$: $\Gamma_n(x_0) = \{y_x \mid x \in X\}$. Note that $y_0 = \{(v, 0) \mid v \in \mathbb{F}^n\}$. Since we have

$$y_x = \{(v, vx) \mid v \in \mathbb{F}^n\},$$

we have

$$\dim(y_x \cap y_{x'}) = \dim(\{v \in \mathbb{F}^n \mid vx = vx'\}),$$

and hence we obtain

$$n - \dim(y_x \cap y_{x'}) = \mathrm{rank}(x - x').$$

Namely, restricting the relations $R_0, R_1, \dots, R_n$ of $D_n(r)$ on $\binom{V}{n}'_r$ to $\Gamma_n(x_0)$ yields the relations $R_0, R_2, R_4 \dots, R_{2d}$ $(d = [\frac{n}{2}])$, and the association scheme on $\Gamma_n(x_0)$ with these relations is isomorphic to $\mathrm{Alt}_n(r)$.

(iii-2) The association schemes $\mathrm{Her}_n(r)$ and ${}^2A_{2n-1}(r)$ are related as follows.
Let $|\mathbb{F}| = r^2$, and denote the automorphism of the field $\mathbb{F}$ of order 2 by $\lambda \mapsto \bar{\lambda}$. For a vector v and a matrix a over $\mathbb{F}$, we denote the vector and the matrix obtained by applying the above automorphism by $\bar{v}$, $\bar{a}$, respectively. Fix an element $\xi \neq 0$ of $\mathbb{F}$ such that $\bar{\xi} = -\xi$, and define a non-degenerate Hermitian form on $V = \mathbb{F}^{2n}$ by

$$\langle u, v \rangle = \xi u \begin{pmatrix} 0 & 1 \\ -1 & 0 \end{pmatrix} {}^t\bar{v},$$

where 1 denotes the identity matrix of size n. The unitary group

$$U(2n, r) = \left\{ \begin{pmatrix} a & b \\ c & d \end{pmatrix} \in \mathrm{GL}(2n, r^2) \mid \begin{pmatrix} a & b \\ c & d \end{pmatrix} \begin{pmatrix} 0 & 1 \\ -1 & 0 \end{pmatrix} {}^t\overline{\begin{pmatrix} a & b \\ c & d \end{pmatrix}} = \begin{pmatrix} 0 & 1 \\ -1 & 0 \end{pmatrix} \right\}$$

determined by this Hermitian form acts on V as an isometry and also acts transitively on the set $\binom{V}{n}'_{r^2}$ of maximal totally isotropic subspaces of V. The association scheme determined by the action of $U(2n,r)$ on $\binom{V}{n}'_{r^2}$ is ${}^2A_{2n-1}(r)$. For an element $x_0 = \{(0,v) \mid v \in \mathbb{F}^n\}$ of $\binom{V}{n}'_{r^2}$, let

$$\Gamma_n(x_0) = \left\{y \in \binom{V}{n}'_{r^2} \mid x_0 \cap y = \{0\}\right\}.$$

Then the stabilizer $U(2n,r)_{x_0}$ of x_0 in $U(2n,r)$ acts transitively on $\Gamma_n(x_0)$. If we let $P = U(2n,r)_{x_0}$, P is a maximal parabolic subgroup of $U(2n,r)$ and has the Levi decomposition $P = LU \rhd U$, where

$$L = \left\{\begin{pmatrix} a & 0 \\ 0 & {}^t\overline{a}^{-1} \end{pmatrix} \mid a \in \mathrm{GL}\left(n, r^2\right)\right\}, \quad U = \left\{\begin{pmatrix} 1 & x \\ 0 & 1 \end{pmatrix} \mid x \in X\right\}.$$

Here X is the set of Hermitian matrices of size n. Since the Levi subgroup L coincides with the stabilizer of $y_0 = \{(v,0) \mid v \in \mathbb{F}^n\} \in \Gamma_n(x_0)$ in P, the action of P on $\Gamma_n(x_0)$ is isomorphic to the action of P on $L\backslash P$: $(P, \Gamma_n(x_0)) \cong (P, L\backslash P)$.

Let $H = \mathrm{GL}(n, r^2)$. Let H act on X as an automorphism of the group X with respect to the matrix sum as follows:

$$x^a = a^{-1}x{}^t\overline{a}^{-1}.$$

Let $G = HX \rhd X$ be the semidirect product of X by H. Then by the mappings $a \mapsto \left(\begin{smallmatrix} a & 0 \\ 0 & {}^t\overline{a}^{-1} \end{smallmatrix}\right)$ and $x \mapsto \left(\begin{smallmatrix} 1 & x \\ 0 & 1 \end{smallmatrix}\right)$, the semidirect product G is isomorphic to P: $G \cong P$, $H \cong L$, $X \cong U$. The association scheme determined by the transitive action

$$y^{ax} = a^{-1}y{}^t\overline{a}^{-1} + x \quad (y \in X,\ a \in H,\ x \in X)$$

of the group G on X is $\mathrm{Her}_n(r)$. Let 0 be the zero matrix of X. Since the stabilizer of 0 in G equals H, we obtain the isomorphism $(G,X) \cong (G, H\backslash G)$ as permutation groups. Since by the isomorphism $G \cong P$ of groups, we have the isomorphism

$$(G,X) \cong (G, H\backslash G) \cong (P, L\backslash P) \cong (P, \Gamma_n(x_0))$$

as permutation groups, the association scheme determined by the action of P on $\Gamma_n(x_0)$ is isomorphic to $\mathrm{Her}_n(r)$.

Similarly to the case of (iii-1), the correspondence between X and $\Gamma_n(x_0)$ in this isomorphism is described explicitly as follows. For $x \in X$, let

$$y_x = y_0\begin{pmatrix} 1 & x \\ 0 & 1 \end{pmatrix} = \{(v, vx) \mid v \in \mathbb{F}^n\}.$$

Then we have

$$\Gamma_n(x_0) = \{y_x \mid x \in X\},$$

where X is the set of Hermitian matrices of size n. Since we have

$$n - \dim(y_x \cap y_{x'}) = \operatorname{rank}(x - x'),$$

restricting the relations on $\binom{V}{n}'_{r^2}$ of the association scheme ${}^2A_{2n-1}(r)$ to $\Gamma_n(x_0)$ yields the association scheme isomorphic to $\mathrm{Her}_n(r)$.

(iii-3) The association schemes $\mathrm{Quad}_n(r)$ and $C_n(r)$ are related as follows.
Let $|\mathbb{F}| = r$ and assume that *the characteristic of the field* $\mathbb{F}$ *is not* 2. Define a non-degenerate alternating bilinear form on $V = F^n$ by

$$\langle u, v\rangle = u\begin{pmatrix} 0 & 1 \\ -1 & 0 \end{pmatrix}{}^t v,$$

where 1 is the identity matrix of size n. The symplectic group

$$S_p(2n,r) = \left\{\begin{pmatrix} a & b \\ c & d \end{pmatrix} \in \mathrm{GL}(2n,r) \mid \begin{pmatrix} a & b \\ c & d \end{pmatrix}\begin{pmatrix} 0 & 1 \\ -1 & 0 \end{pmatrix}{}^t\begin{pmatrix} a & b \\ c & d \end{pmatrix} = \begin{pmatrix} 0 & 1 \\ -1 & 0 \end{pmatrix}\right\}$$

determined by this alternating bilinear form acts on V as an isometry, and also acts transitively on the set $\binom{V}{n}'_r$ of maximal totally isotropic subspaces of V. The association scheme determined by the action of $S_p(2n,r)$ on $\binom{V}{n}'_r$ is $C_n(r)$. For $x_0 = \{(0,v) \mid v \in \mathbb{F}^n\} \in \binom{V}{n}'_r$, let

$$\Gamma_n(x_0) = \left\{y \in \binom{V}{n}'_r \mid x_0 \cap y = \{0\}\right\}.$$

Then the stabilizer $S_p(2n,r)_{x_0}$ of x_0 in $S_p(2n,r)$ acts transitively on $\Gamma_n(x_0)$. Let $P = S_p(2n,r)_{x_0}$. Then P is a maximal parabolic subgroup of $S_p(2n,r)$, and has the Levi decomposition $P = LU \rhd U$, where

$$L = \left\{\begin{pmatrix} a & 0 \\ 0 & {}^t a^{-1} \end{pmatrix} \mid a \in \mathrm{GL}(n,r)\right\}, \quad U = \left\{\begin{pmatrix} 1 & x \\ 0 & 1 \end{pmatrix} \mid x \in X\right\}.$$

Here, X is the set of symmetric matrices of size n. Since the Levi subgroup L coincides with the stabilizer of $y_0 = \{(v,0) \mid v \in \mathbb{F}^n\} \in \Gamma_n(x_0)$ in P, the action of P on $\Gamma_n(x_0)$ is isomorphic to the action of P on $L\backslash P$: $(P, \Gamma_n(x_0)) \cong (P, L\backslash P)$.

Let $H = \mathrm{GL}(n,r)$. Let H act on X as an a belian group with respect to the matrix sum as follows:

$$x^a = a^{-1}x{}^t a^{-1}.$$

Let $G = HX \rhd X$ be the semidirect product of X by H. By the mappings $a \mapsto \begin{pmatrix} a & 0 \\ 0 & {}^t a^{-1} \end{pmatrix}$ and $x \mapsto \begin{pmatrix} 1 & x \\ 0 & 1 \end{pmatrix}$, the semidirect product G is isomorphic to P: $G \cong P$, $H \cong L$, $X \cong U$.

Consider the transitive action of G on X:

$$y^{ax} = a^{-1}y^t a^{-1} + x \quad (y \in X, a \in H, x \in X).$$

Let 0 be the zero matrix of X. Since the stabilizer of 0 in G coincides with H, we get the isomorphism $(G, X) \cong (G, H\backslash G)$ as permutation groups. By the isomorphism $G \cong P$ as groups, we obtain the isomorphism

$$(G, X) \cong (G, H\backslash G) \cong (P, L\backslash P) \cong (P, \Gamma_n(x_0))$$

as permutation groups. Namely, the action of P on $\Gamma_n(x_0)$ is isomorphic to the action of G in X.

Similarly to the cases of (iii-1), (iii-2), the correspondence between X and $\Gamma_n(x_0)$ in this isomorphism is described explicitly as follows. For $x \in X$, let

$$y_x = y_0 \begin{pmatrix} 1 & x \\ 0 & 1 \end{pmatrix} = \{(v, vx) \mid v \in \mathbb{F}^n\}.$$

Then we have

$$\Gamma_n(x_0) = \{y_x \mid x \in X\},$$

where X is the set of symmetric matrices of size n. Similarly to the cases of (iii-1), (iii-2), we have

$$n - \dim(y_x \cap y_{x'}) = \operatorname{rank}(x - x').$$

Unlike the cases of (iii-1), (iii-2), in the action of G on X, the stabilizer $H = \mathrm{GL}(n, r)$ of the zero matrix 0 is not transitive on

$$X_i = \{x \in X \mid \operatorname{rank}(x) = i\},$$

and X_i is divided into two H-orbits. (If i is odd, by the scalar multiple of a non-square in $\mathbb{F}$, these two H-orbits are merged. If i is even, they are divided into essentially different classes of symmetric matrices.) Due to this, the association scheme determined by the action of G on X (hence the action of P on $\Gamma_n(x_0)$) is not necessarily a P-polynomial scheme. By merging the relations of rank $2i - 1$ and $2i$ (four H-orbits) to make one relation, it becomes a P-polynomial scheme and obtain $\mathrm{Quad}_n(r)$ [177]. Therefore, restricting the relation on $\binom{V}{n}_r$ of the association scheme $C_n(r)$ to $\Gamma_n(x_0)$ and merging the relations R_{2i-1}, R_{2i} yields the association scheme isomorphic to $\mathrm{Quad}_n(r)$.

(d) The association schemes $\mathrm{Quad}_n(r)$ and $\mathrm{Alt}_{n+1}(r)$ have the same parameters but are not isomorphic. The automorphism group of $\mathrm{Quad}_n(r)$ does not act transitively on R_i, which is divided into 3 orbits. On the other hand, the automorphism group of $\mathrm{Alt}_{n+1}(r)$ acts transitively on each relation R_i.

6.4.2 P- and Q-polynomial schemes derived from the core part

In this section, among known P- and Q-polynomial schemes $\mathfrak{X} = (X, \{R_i\}_{0\le i\le d})$, we give the list of those derived from the "core" P- and Q-polynomial schemes in Section 6.4.1. Depending on the way they are derived from the "core" P- and Q-polynomial schemes, the list is divided into the following 4 cases:

(a) derived from imprimitive schemes;
(b) derived from the second P- or Q-polynomial orderings;
(c) derived as the extended bipartite double or a fusion scheme;
(d) derived as a "cospectral scheme," which has the same parameters as a "core" P- and Q-polynomial scheme $\mathfrak{X}$ but is not isomorphic to $\mathfrak{X}$.

(a) The case derived from imprimitive schemes

Let $\mathfrak{X} = (X, \{R_i\}_{0\le i\le d})$ be a P- and Q-polynomial scheme. If $\mathfrak{X}$ is an imprimitive association scheme, it is bipartite or antipodal as a P-polynomial scheme, and it is dual bipartite or dual antipodal as a Q-polynomial scheme (Section 6.1.3). If $\mathfrak{X}$ is bipartite as a P-polynomial scheme, then it is dual antipodal as a Q-polynomial scheme, and the bipartite half $\frac{1}{2}\mathfrak{X}$ also becomes P- and Q-polynomial scheme. Moreover, if $\mathfrak{X}$ is antipodal as a P-polynomial scheme, then it is dual bipartite as a Q-polynomial scheme, and the antipodal quotient scheme $\overline{\mathfrak{X}}$ also becomes a P- and Q-polynomial scheme.

Note that $\mathfrak{X}$ is bipartite if and only if $a_i = 0$ $(0 \le i \le d)$, and $\mathfrak{X}$ is antipodal if and only if $b_i = c_{d-i}$ $(0 \le i \le d,\ i \ne [\frac{d}{2}])$, where $a_i = p^i_{1,i}$, $b_i = p^i_{1,i+1}$, $c_i = p^i_{1,i-1}$ are the intersection numbers of $\mathfrak{X}$. Then the intersection numbers of the P-polynomial scheme $\frac{1}{2}\mathfrak{X}$ are $b'_i = b_{2i}b_{2i+1}/c_2$ $(0 \le i \le [\frac{d}{2}] - 1)$, $c'_i = c_{2i-1}c_{2i}/c_2$ $(1 \le i \le [\frac{d}{2}])$. Moreover, the intersection numbers of the P-polynomial scheme $\overline{\mathfrak{X}}$ are $\tilde{b}_i = b_i$ $(0 \le i \le [\frac{d}{2}] - 1)$, $\tilde{c}_i = c_i$ $(1 \le i \le [\frac{d}{2}] - 1)$, and

$$\tilde{c}_{[\frac{d}{2}]} = \begin{cases} (1 + k_d)c_{[\frac{d}{2}]}, & \text{if } d \text{ is even}, \\ c_{[\frac{d}{2}]}, & \text{if } d \text{ is odd}, \end{cases}$$

where $k_d = b_{[\frac{d}{2}]}/c_{[\frac{d}{2}]}$ if d is even.

Note that a Q-polynomial scheme $\mathfrak{X}$ is dual bipartite if and only if $a^*_i = 0$ $(0 \le i \le d)$, and $\mathfrak{X}$ is dual antipodal if and only if $b^*_i = c^*_{d-i}$ $(0 \le i \le d-1,\ i \ne [\frac{d}{2}])$, where $a^*_i = q^i_{1,i}$, $b^*_i = q^i_{1,i+1}$, $c^*_i = q^i_{1,i-1}$ are the dual intersection numbers of $\mathfrak{X}$. Then the dual intersection numbers of the Q-polynomial scheme $\frac{1}{2}\mathfrak{X}$ are $\tilde{b}^*_i = b^*_i$ $(0 \le i \le [\frac{d}{2}] - 1)$, $\tilde{c}^*_i = c^*_i$ $(1 \le i \le [\frac{d}{2}] - 1)$, and

$$\tilde{c}^*_{[\frac{d}{2}]} = \begin{cases} (1 + m_d)c^*_{[\frac{d}{2}]}, & \text{if } d \text{ is even}, \\ c^*_{[\frac{d}{2}]}, & \text{if } d \text{ is odd}, \end{cases}$$

where $m_d = b^*_{[\frac{d}{2}]}/c^*_{[\frac{d}{2}]}$ if d is even. Moreover, the dual intersection numbers of the Q-polynomial scheme $\overline{\mathfrak{X}}$ are $\hat{b}^*_i = b^*_{2i}b^*_{2i+1}/c^*_i$ $(0 \le i \le [\frac{d}{2}]-1)$, $\hat{c}^*_i = c^*_{2i-1}c^*_{2i}$ $(1 \le i \le [\frac{d}{2}])$.

(i) $\overline{J(2d,d)}$: The antipodal quotient of the Johnson scheme $J(2d,d)$
Since the intersection numbers $b_i = p^i_{1,i+1}$, $c_i = p^i_{1,i-1}$ of the Johnson scheme $J(2d,d)$ are $b_i = (d-i)^2$ $(0 \le i \le d-1)$, $c_i = i^2$ $(1 \le i \le d)$, we have $b_i = c_{d-i}$ $(0 \le i \le d-1)$, which implies $J(2d,d)$ is antipodal. Moreover, since the dual intersection numbers $b^*_i = q^i_{1,i+1}$, $c^*_i = q^i_{1,i-1}$ of $J(2d,d)$ are $b^*_i = (2d-1)(2d+1-i)(d-i)/d(2d+1-2i)$ $(0 \le i \le d-1)$, $c^*_i = (2d-1)i(d+1-i)/d(2d+1-2i)$ $(1 \le i \le d)$, we have $a^*_i = 0$ $(0 \le i \le d)$, which implies $J(2d,d)$ is dual bipartite. We rewrite the intersection numbers of the antipodal quotient $\overline{J(2d,d)}$ of $J(2d,d)$ as $b_i = p^i_{1,i+1}$, $c_i = p^i_{1,i-1}$. Then they are given by

$$b_i = (d-i)^2, \quad 0 \le i \le [\frac{d}{2}] - 1,$$
$$c_i = i^2, \quad 1 \le i \le [\frac{d}{2}] - 1,$$
$$c_{[\frac{d}{2}]} = \begin{cases} 2[\frac{d}{2}]^2, & \text{if } d \text{ is even,} \\ [\frac{d}{2}]^2, & \text{if } d \text{ is odd.} \end{cases}$$

The antipodal quotient $\overline{J(2d,d)}$ is also a Q-polynomial scheme with the following AW-parameters:

$$\Lambda_{\mathrm{II}} = \left(r_1, r_2, s, s^*; h, h^*, \theta_0, \theta^*_0 \mid \left[\frac{d}{2}\right]\right),$$
$$r_1 = -d-1, \quad r_2 = \begin{cases} -\frac{d}{2} - \frac{1}{2}, & \text{if } d \text{ is even,} \\ -\frac{d}{2} - 1, & \text{if } d \text{ is odd,} \end{cases}$$
$$s = -d - \frac{3}{2}, \quad s^* = -d-1,$$
$$h = 4, \quad h^* = \frac{2(2d-1)(2d-3)}{d(d-1)}, \quad \theta_0 = d^2, \quad \theta^*_0 = d(2d-3).$$

The other parameters are the following:

$$\theta_i = d^2 + 2i(2i-2d-1), \quad 0 \le i \le \left[\frac{d}{2}\right],$$
$$\theta^*_i = d(2d-3) + \frac{2(2d-1)(2d-3)}{d(d-1)} i(i-d), \quad 0 \le i \le \left[\frac{d}{2}\right],$$
$$\lambda_i = hh^*(i+1)\left(i - \frac{d-1}{2}\right)\left(i - \frac{d}{2}\right)(i-d), \quad 0 \le i \le \left[\frac{d}{2}\right] - 1,$$
$$\hat{\lambda}_i = hh^*(i+1)^2\left(i - \left[\frac{d}{2}\right]\right)\left(i + \frac{1}{2} - \left[\frac{d}{2}\right]\right), \quad 0 \le i \le \left[\frac{d}{2}\right] - 1,$$
$$b^*_i = \frac{h^*}{2} \frac{(2d+1-2i)(d-2i)(d-i)(d-2i-1)}{(2d+1-4i)(2d-1-4i)}, \quad 0 \le i \le \left[\frac{d}{2}\right] - 1,$$

$$c_i^* = \frac{h^*}{2}\frac{i(2i-1)(d+2-2i)(d+1-2i)}{(2d+3-4i)(2d+1-4i)}, \quad 0 \le i \le \left[\frac{d}{2}\right],$$

$$k_i = \binom{d}{i}^2, \quad 0 \le i \le \left[\frac{d}{2}\right] - 1,$$

$$k_{[\frac{d}{2}]} = \begin{cases} \frac{1}{2}\binom{d}{d/2}^2, & \text{if } d \text{ is even}, \\ \binom{d}{\frac{d-1}{2}}^2, & \text{if } d \text{ is odd}, \end{cases}$$

$$m_i = \frac{2d+1-4i}{2d+1-2i}\binom{2d}{2i}, \quad 0 \le i \le \left[\frac{d}{2}\right].$$

(ii) $\frac{1}{2}D_d(q)$: The bipartite half of the dual polar scheme $D_d(q)$

Since the intersection numbers $b_i = p^i_{1,i+1}$, $c_i = p^i_{1,i-1}$ of the dual polar scheme $D_d(q)$ are $b_i = q^i(q^{d-i} - 1)/(q-1)$ $(0 \le i \le d-1)$, $c_i = (q^i - 1)/(q-1)$ $(1 \le i \le d)$, we have $a_i = 0$ $(0 \le i \le d)$, which implies $D_d(q)$ is bipartite. Moreover, since the dual intersection numbers $b_i^* = q^i_{1,i+1}$, $c_i^* = q^i_{1,i-1}$ of $D_d(q)$ are

$$b_i^* = h^*(q^{d-i}-1)(q^{d-i}+1)/q^{2i+1}(q^{d-1-2i}+1)(q^{d-2i}+1), \quad 0 \le i \le d-1,$$
$$c_i^* = h^*(q^i-1)(q^i+1)/q^{2i}(q^{d+1-2i}+1)(q^{d-2i}+1), \quad 1 \le i \le d,$$

we have $b_i^* = c_{d-i}^*$, which implies $\frac{1}{2}D_d(q)$ is dual antipodal.

We rewrite the intersection numbers of the bipartite half $\frac{1}{2}D_d(q)$ of $D_d(q)$ as $b_i = p^i_{1,i+1}$, $c_i = p^i_{1,i-1}$. Then they are given by

$$b_i = \frac{q^{4i+1}(q^{d-2i}-1)(q^{d-2i-1}-1)}{(q-1)(q^2-1)}, \quad 0 \le i \le \left[\frac{d}{2}\right] - 1,$$
$$c_i = \frac{(q^{2i-1}-1)(q^{2i}-1)}{(q-1)(q^2-1)}, \quad 1 \le i \le \left[\frac{d}{2}\right].$$

The bipartite half $\frac{1}{2}D_d(q)$ is also a Q-polynomial scheme with the following AW-parameters:

$$\Lambda_I = \left(r_1, r_2, s, s^*; h, h^*, \theta_0, \theta_0^* \mid q^2, \left[\frac{d}{2}\right]\right),$$

$$r_1 = \begin{cases} q^{-d-1} & \text{if } d \text{ is even}, \\ q^{-d-2} & \text{if } d \text{ odd}, \end{cases} \quad r_2 = 0, \quad s = q^{-2(d+1)}, \quad s^* = 0,$$

$$h = \frac{q^{2d}}{(q-1)(q^2-1)}, \quad h^* = \frac{q^2(q^{d-1}+1)(q^{d-2}+1)}{q^2-1},$$

$$\theta_0 = \frac{q(q^d-1)(q^{d-1}-1)}{(q-1)(q^2-1)}, \quad \theta_0^* = \frac{q(q^{d-2}+1)(q^d+1)}{q^2-1}.$$

The other parameters are the following:

$$\theta_i = \theta_0 - \frac{(q^{2i}-1)(q^{2d-2i}-1)}{(q-1)(q^2-1)}, \quad 0 \le i \le \left[\frac{d}{2}\right],$$

$$\theta_i^* = \theta_0^* - h^*(1 - q^{-2i}), \quad 0 \le i \le \left[\frac{d}{2}\right],$$

$$\lambda_i = -hh^* q^{-2d-1}(q^{2i+2} - 1)(q^{d-2i} - 1)(q^{d-2i-1} - 1), \quad 0 \le i \le \left[\frac{d}{2}\right] - 1,$$

$$\hat{\lambda}_i = hh^* q^{-2d-2-2i}(q^{2i+2} - 1)(q^{2i+1} - 1)(q^{2[\frac{d}{2}]-2i} - 1), \quad 0 \le i \le \left[\frac{d}{2}\right] - 1,$$

$$b_i^* = h^* \frac{q^{2(d-i)} - 1}{q^{2i+1}(q^{d-1-2i} + 1)(q^{d-2i} + 1)}, \quad 0 \le i \le \left[\frac{d}{2}\right] - 1,$$

$$c_i^* = h^* \frac{q^{2i} - 1}{q^{2i}(q^{d+1-2i} + 1)(q^{d-2i} + 1)}, \quad 1 \le i \le \left[\frac{d}{2}\right] - 1,$$

$$c^*_{[\frac{d}{2}]} = \begin{cases} \frac{(q^d-1)(q^{d-1}+1)(q^{d-2}+1)}{q^{d-2}(q+1)(q^2-1)}, & \text{if } d \text{ is even}, \\ \frac{(q^{d-1}-1)(q^{d-1}+1)(q^{d-2}+1)}{q^{d-3}(q+1)(q^4-1)}, & \text{if } d \text{ is odd}, \end{cases}$$

$$k_i = q^{i(2i-1)} \binom{d}{2i}_q, \quad 0 \le i \le \left[\frac{d}{2}\right],$$

$$m_i = q^i \binom{d}{i}_q \frac{q^{d-2i} + 1}{q^{d-i} + 1} \prod_{j=0}^{i-1} \frac{q^{d-1-j} + 1}{q^{j+1} + 1}, \quad 0 \le i \le \left[\frac{d}{2}\right] - 1,$$

$$m_{[\frac{d}{2}]} = \begin{cases} q^{\frac{d}{2}} \binom{d}{\frac{d}{2}}_q \frac{1}{q^{\frac{d}{2}+1}} \prod_{j=0}^{\frac{d}{2}-1} \frac{q^{d-1-j}+1}{q^{j+1}+1}, & \text{if } d \text{ is even}, \\ q^{\frac{d-1}{2}} \binom{d}{\frac{d-1}{2}}_q \frac{q+1}{q^{\frac{d+1}{2}}+1} \prod_{j=0}^{\frac{d-1}{2}-1} \frac{q^{d-1-j}+1}{q^{j+1}+1}, & \text{if } d \text{ is odd}. \end{cases}$$

(iii) $\frac{1}{2}H(d,2)$ and $\overline{H(d,2)}$: the bipartite half and the antipodal quotient of the Hamming scheme $H(d,2)$

For the intersection numbers $b_i = p^i_{1,i+1}$, $c_i = p^i_{1,i-1}$ and the dual intersection numbers $b_i^* = q^i_{1,i+1}$, $c_i^* = q^i_{1,i-1}$ of the Hamming scheme $H(d,2)$, we have $b_i = b_i^* = d - i$ $(0 \le i \le d-1)$, $c_i = c_i^* = i$ $(1 \le i \le d)$. Hence $H(d,2)$ is bipartite and dual antipodal, and is also antipodal and dual bipartite.

We rewrite the intersection numbers of the bipartite half $\frac{1}{2}H(d,2)$ of $H(d,2)$ as $b_i = p^i_{1,i+1}$, $c_i = p^i_{1,i-1}$. Then they are given by

$$b_i = \frac{1}{2}(d-2i)(d-2i-1), \quad 0 \le i \le \left[\frac{d}{2}\right] - 1,$$

$$c_i = i(2i-1), \quad 1 \le i \le \left[\frac{d}{2}\right].$$

The bipartite half $\frac{1}{2}H(d,2)$ is also a Q-polynomial scheme with the following AW-parameters:

$$\Lambda_{\text{IIA}} = \left(r_1, -, s, -; h, h^*, \theta_0, \theta_0^* \mid \left[\frac{d}{2}\right]\right),$$

$$r_1 = \begin{cases} -\frac{d}{2} - \frac{1}{2}, & \text{if } d \text{ is even,} \\ -\frac{d}{2} - 1, & \text{if } d \text{ is odd,} \end{cases} \qquad s = -d - 1,$$

$$h = 2, \quad h^* = -4, \quad \theta_0 = \frac{1}{2}d(d-1), \quad \theta_0^* = d.$$

The other parameters are the following:

$$\theta_i = \frac{1}{2}d(d-1) - 2i(d-i), \quad 0 \le i \le \left[\frac{d}{2}\right], \quad \theta_i^* = d - 4i, \quad 0 \le i \le \left[\frac{d}{2}\right],$$

$$\lambda_i = hh^*(i+1)\left(i - \frac{d}{2}\right)\left(i - \frac{d-1}{2}\right), \quad 0 \le i \le \left[\frac{d}{2}\right] - 1,$$

$$\hat{\lambda}_i = hh^*(i+1)\left(i + \frac{1}{2}\right)\left(i - \left[\frac{d}{2}\right]\right), \quad 0 \le i \le \left[\frac{d}{2}\right] - 1,$$

$$b_i^* = d - i, \quad 0 \le i \le \left[\frac{d}{2}\right] - 1,$$

$$c_i^* = i, \quad 1 \le i \le \left[\frac{d}{2}\right] - 1, \quad c^*_{[\frac{d}{2}]} = \begin{cases} d, & \text{if } d \text{ is even,} \\ \frac{d-1}{2}, & \text{if } d \text{ is odd,} \end{cases}$$

$$k_i = \binom{d}{2i}, \quad 0 \le i \le \left[\frac{d}{2}\right],$$

$$m_i = \binom{d}{i}, \quad 0 \le i \le \left[\frac{d}{2}\right] - 1, \quad m_{[\frac{d}{2}]} = \begin{cases} \frac{1}{2}\binom{d}{\frac{d}{2}}, & \text{if } d \text{ is even,} \\ \binom{d}{\frac{d-1}{2}}, & \text{if } d \text{ is odd.} \end{cases}$$

Next, we rewrite the intersection numbers of the antipodal quotient $\overline{H(d,2)}$ of $H(d,2)$ as $b_i = p^i_{1,i+1}$, $c_i = p^i_{1,i-1}$. Then they are given by

$$b_i = d - i, \quad 0 \le i \le \left[\frac{d}{2}\right] - 1,$$

$$c_i = i, \quad 1 \le i \le \left[\frac{d}{2}\right] - 1, \quad c_{[\frac{d}{2}]} = \begin{cases} d, & \text{if } d \text{ is even,} \\ \frac{d-1}{2}, & \text{if } d \text{ is odd.} \end{cases}$$

The antipodal quotient $\overline{H(d,2)}$ is also a Q-polynomial scheme with the following AW-parameters:

$$\Lambda_{\text{IIB}} = \left(r_1, -, -, s^*; h, h^*, \theta_0, \theta_0^* \mid \left[\frac{d}{2}\right]\right),$$

$$r_1 = \begin{cases} -\frac{d}{2} - \frac{1}{2}, & \text{if } d \text{ is even,} \\ -\frac{d}{2} - 1, & \text{if } d \text{ is odd,} \end{cases} \qquad s^* = -d - 1,$$

$$h = -4, \quad h^* = 2, \quad \theta_0 = d, \quad \theta_0^* = \frac{1}{2}d(d-1).$$

The other parameters are the following:

$$\theta_i = d - 4i, \quad 0 \le i \le \left[\frac{d}{2}\right],$$

$$
\begin{aligned}
\theta_i^* &= \frac{1}{2}d(d-1) - 2i(d-i), \quad 0 \le i \le \left[\frac{d}{2}\right], \\
\lambda_i &= hh^*(i+1)\left(i - \frac{d}{2}\right)\left(i - \frac{d-1}{2}\right), \quad 0 \le i \le \left[\frac{d}{2}\right] - 1, \\
\hat{\lambda}_i &= hh^*(i+1)\left(i - \left[\frac{d}{2}\right]\right)\left(i - \left[\frac{d}{2}\right] + \frac{1}{2}\right), \quad 0 \le i \le \left[\frac{d}{2}\right] - 1, \\
b_i^* &= \frac{1}{2}(d-2i)(d-2i-1), \quad 0 \le i \le \left[\frac{d}{2}\right] - 1, \\
c_i^* &= i(2i-1), \quad 1 \le i \le \left[\frac{d}{2}\right], \\
k_i &= \binom{d}{i}, \quad 0 \le i \le \left[\frac{d}{2}\right] - 1, \quad k_{[\frac{d}{2}]} = \begin{cases} \frac{1}{2}\binom{d}{\frac{d}{2}}, & \text{if } d \text{ is even}, \\ \binom{d}{\frac{d-1}{2}}, & \text{if } d \text{ is odd}, \end{cases} \\
m_i &= \binom{d}{2i}, \quad 0 \le i \le \left[\frac{d}{2}\right].
\end{aligned}
$$

Note that the structure of $\frac{1}{2}H(d,2)$ as a P-polynomial scheme is the same as the structure of $\overline{H(d,2)}$ as a Q-polynomial scheme, and the structure of $\frac{1}{2}H(d,2)$ as a Q-polynomial scheme is the same as the structure of $\overline{H(d,2)}$ as a P-polynomial scheme.

If d is odd, $\frac{1}{2}H(d,2)$ and $\overline{H(d,2)}$ are isomorphic as association schemes. They have two P-polynomial orderings and two Q-polynomial orderings: By the isomorphism, the natural P-polynomial ordering of one corresponds to the second P-polynomial ordering of the other and the natural Q-polynomial ordering of one corresponds to the second Q-polynomial ordering of the other (Section 6.4.2 (b)).

Both $\frac{1}{2}H(d,2)$ and $\overline{H(d,2)}$ are imprimitive if and only if d is even. In this case, $\frac{1}{2}H(d,2)$ is antipodal as a P-polynomial scheme (and dual bipartite as a Q-polynomial scheme), and $\overline{H(d,2)}$ is bipartite as a P-polynomial scheme (and dual antipodal as a Q-polynomial scheme). This fact is observed from the intersection numbers b_i, c_i (and dual intersection numbers b_i^*, c_i^*).

Here we rewrite the even diameter d as $2d$. Then the following holds:

$$\overline{\frac{1}{2}H(2d,2)} \cong \frac{1}{2}\overline{H(2d,2)}.$$

Namely, the antipodal quotient of $\frac{1}{2}H(2d,2)$ and the bipartite half of $\overline{H(2d,2)}$ are isomorphic. This fact can be easily checked if we note that in the bipartite half of $H(2d,2)$, the antipodal relation $R_0 \cup R_{2d}$ is closed. We rewrite the intersection numbers of $\overline{\frac{1}{2}H(2d,2)} \cong \frac{1}{2}\overline{H(2d,2)}$ as $b_i = p^i_{1,i+1}$, $c_i = p^i_{1,i-1}$. Then they are given by

$$
\begin{aligned}
b_i &= (d-i)(2d-2i-1), \quad 0 \le i \le \left[\frac{d}{2}\right] - 1, \\
c_i &= i(2i-1), \quad 1 \le i \le \left[\frac{d}{2}\right] - 1,
\end{aligned}
$$

$$c_{[\frac{d}{2}]} = \begin{cases} d(d-1), & \text{if } d \text{ is even,} \\ \frac{(d-1)(d-2)}{2}, & \text{if } d \text{ is odd.} \end{cases}$$

The association scheme $\overline{\frac{1}{2}H(2d,2)} \cong \frac{1}{2}\overline{H(2d,2)}$ is a Q-polynomial scheme with the following AW-parameters:

$$\Lambda_{\text{II}} = \left(r_1, r_2, s, s^*; h, h^*, \theta_0, \theta_0^* \mid \left[\frac{d}{2}\right]\right),$$

$$r_1 = \begin{cases} -\frac{d}{2} - \frac{1}{2}, & \text{if } d \text{ is even,} \\ -\frac{d}{2} - 1, & \text{if } d \text{ is odd,} \end{cases} \qquad r_2 = -d - \frac{1}{2},$$

$$s = s^* = -d - 1, \quad h = h^* = 8, \quad \theta_0 = \theta_0^* = d(2d-1).$$

The other parameters are the following:

$$\theta_i = \theta_i^* = d(2d-1) - 8i(d-i), \quad 0 \le i \le \left[\frac{d}{2}\right],$$

$$\lambda_i = 8(i+1)(2i-d)(2i-d+1)(2i-2d+1), \quad 0 \le i \le \left[\frac{d}{2}\right] - 1,$$

$$\hat{\lambda}_i = 8(i+1)(2i+1)\left(2i - 2\left[\frac{d}{2}\right]\right)\left(2i - 2\left[\frac{d}{2}\right] + 1\right), \quad 0 \le i \le \left[\frac{d}{2}\right] - 1,$$

$$b_i^* = b_i, \quad 0 \le i \le \left[\frac{d}{2}\right] - 1,$$

$$c_i^* = c_i, \quad 1 \le i \le \left[\frac{d}{2}\right],$$

$$k_i = m_i = \binom{2d}{2i}, \quad 0 \le i \le \left[\frac{d}{2}\right] - 1,$$

$$k_{[\frac{d}{2}]} = m_{[\frac{d}{2}]} = \begin{cases} \frac{1}{2}\binom{2d}{d}, & \text{if } d \text{ is even,} \\ \binom{2d}{d-1}, & \text{if } d \text{ is odd.} \end{cases}$$

(b) The case derived from the second P- or Q-polynomial orderings

In this section, if a symmetric association scheme $\mathfrak{X} = (X, \{R_i\}_{0 \le i \le d})$ has two P-polynomial orderings, we consider the two P-polynomial schemes are distinct. For Q-polynomial schemes which have two Q-polynomial orderings, we consider similarly.

Let $\mathfrak{X} = (X, \{R_i\}_{0 \le i \le d})$ be a P- and Q-polynomial scheme. Let $R_0, R_1, \ldots, R_d$ be the P-polynomial ordering and A_1 the adjacency matrix of R_1. Note that the numbering of primitive idempotents $\{E_i\}_{0 \le i \le d}$ is determined by $A_1 = \sum_{i=0}^d \theta_i E_i$, where $\{\theta_i\}_{0 \le i \le d}$ are the eigenvalues of A_1. The intersection numbers $b_i = p_{1,i+1}^i$ $(0 \le i \le d-1)$, $c_i = p_{1,i-1}^i$ $(1 \le i \le d)$ are determined by the P-polynomial ordering $R_0, R_1, \ldots, R_d$ and they are expressed by AW-parameters Λ_Z, where Z denotes the type which is one of I, IA, II, IIA, IIB, IIC, and III. The AW-parameters Λ_Z determine the ordering of the eigenvalues $\theta_0, \theta_1, \ldots, \theta_d$ of the adjacency matrix A_1, and hence they determine the Q-polynomial ordering $E_0, E_1, \ldots, E_d$. Therefore, if a P- and Q-polynomial scheme has two Q-polynomial orderings, the intersection numbers b_i $(0 \le i \le d-1)$, c_i $(1 \le i \le d)$, which are determined by a fixed P-polynomial ordering, have two distinct expressions by AW-parameters.

Similarly, if a P- and Q-polynomial scheme $\mathfrak{X} = (X, \{R_i\}_{0 \le i \le d})$ has two P-polynomial orderings, the dual intersection numbers b_i^* $(0 \le i \le d-1)$, c_i^* $(1 \le i \le d)$, which are determined by a fixed Q-polynomial ordering of the primitive idempotents $E_0, E_1, \ldots, E_d$, have two distinct expressions by AW-parameters.

(i) The Johnson scheme $J(2d+1, d)$

As was seen before, the Johnson scheme $J(v, d)$ $(1 \le d \le \frac{v}{2})$ is a P- and Q-polynomial scheme with AW-parameters of Type IIA: $\Lambda_{\mathrm{IIA}} = (r_1, -, s, -; h, h^*, \theta_0, \theta_0^* \mid d)$, $r_1 = -v + d - 1$, $s = -v - 2$; $h = 1$, $h^* = -v(v-1)/d(v-d)$; $\theta_0 = d(v-d)$, $\theta_0^* = v - 1$. If $v = 2d + 1$, the dual intersection numbers $b_i^* = q^i_{1,i+1}$, $c_i^* = q^i_{1,i-1}$ are

$$\begin{aligned} b_i^* &= \frac{2d+1}{d+1}\frac{(d-i)(2d+2-i)}{2d+1-2i}, \quad 0 \le i \le d-1, \\ c_i^* &= \frac{2d+1}{d+1}\frac{i(d+2-i)}{2d+3-2i}, \quad 1 \le i \le d, \end{aligned}$$

and these dual intersection numbers of $J(2d+1, d)$ have another expression by AW-parameters:

$$\begin{aligned} &\Lambda_{\mathrm{III}} = (r_1, r_2, s, s^*; h, h^*, \theta_0, \theta_0^* \mid d), \\ &r_1 = -2d - 3, \quad r_2 = -d - 1, \quad s = -2d - 3, \quad s^* = -2d - 2, \\ &h = -\frac{1}{2}, \quad h^* = -\frac{2d+1}{2(d+1)}, \quad \theta_0 = d + 1, \quad \theta_0^* = 2d. \end{aligned}$$

The other parameters are the following:

$$\begin{aligned} &\theta_i = (-1)^i(d+1-i), \quad 0 \le i \le d, \\ &\begin{cases} \theta_{2i}^* = 2d - 2\frac{2d+1}{d+1}i, & 0 \le i \le \frac{d}{2}, \\ \theta_{2i+1}^* = 2d - 2\frac{2d+1}{d+1}(d-i), & 0 \le i \le \frac{d-1}{2}, \end{cases} \end{aligned} \tag{6.287}$$

$$\begin{aligned} &\lambda_i = -\frac{2d+1}{d+1}(i-d)(i+1)^{\varepsilon(i+1)}(i-2d-2)^{\varepsilon(i)}, \quad 0 \le i \le d-1, \\ &\hat{\lambda}_i = (-1)^d\frac{2d+1}{d+1}(i-d)(i+1)^{\varepsilon(i+1)}(i+2)^{\varepsilon(i)}, \quad 0 \le i \le d-1 \\ &(\varepsilon(j) = 1 \text{ if } j \text{ is even and } \varepsilon(j) = 0 \text{ if } j \text{ is odd}), \end{aligned} \tag{6.288}$$

$$\begin{aligned} &b_i = d + 1 - \left[\frac{i+1}{2}\right], \quad 0 \le i \le d-1, \\ &c_i = \left[\frac{i+1}{2}\right], \quad 1 \le i \le d, \end{aligned}$$

$$k_i = \binom{d}{i}\binom{d+1}{[\frac{i+1}{2}]}, \quad 0 \le i \le d,$$
$$m_i = \frac{2(d+1-i)}{2d+2-i}\binom{2d+1}{i}, \quad 0 \le i \le d.$$

The Q-polynomial ordering $E_0, E_1, \dots, E_d$ of the Johnson scheme $J(2d+1, d)$ is fixed, and the dual intersection numbers b_i^* $(0 \le i \le d-1)$, c_i^* $(1 \le i \le d)$ determined by this Q-polynomial ordering have two expressions by AW-parameters of Type IIA and of Type III. Therefore, the dual eigenvalues $\{Q_1(i)\}_{0\le i\le d}$ of the primitive idempotent E_1 have the expression $\theta_i^*((\mathrm{IIA}))$ $(0 \le i \le d)$ determined by AW-parameters Λ_{IIA} and the expression $\theta_i^*(\mathrm{III})$ $(0 \le i \le d)$ determined by AW-parameters Λ_{III}. Since we have $\theta_i^*(\mathrm{IIA}) = 2d - 2\frac{2d+1}{d(d+1)}i$ $(0 \le i \le d)$, by (6.287) the following hold:

$$\theta_{2i}^*(\mathrm{III}) = \theta_i^*(\mathrm{IIA}), \quad 0 \le i \le \frac{d}{2},$$
$$\theta_{2i+1}^*(\mathrm{III}) = \theta_{d-i}^*(\mathrm{IIA}), \quad 0 \le i \le \frac{d-1}{2}.$$

This implies that if we let $R_0, R_1, \dots, R_d$ be the P-polynomial ordering determined by the expression by AW-parameters of Type IIA, then the P-polynomial ordering determined by the expression by AW-parameters of Type III is

$$R_0, R_d, R_1, R_{d-1}, \dots.$$

The distance-regular graph determined by the relation R_1 is the Johnson graph $J(2d + 1, d)$, and the distance-regular graph determined by the relation R_d is the odd graph O_d. If we start from the P-polynomial ordering of the odd graph, then the P-polynomial ordering of the Johnson graph $J(2d+1, d)$ becomes

$$R_0, R_2, R_4, \dots, R_3, R_1,$$

and this is also checked by the fact that the odd graph O_d is almost bipartite, i. e., $a_i = 0$ $(0 \le i \le d-1, a_d \ne 0)$.

(ii) The dual polar scheme $^2A_{2d-1}(r)$

As was seen before, the dual polar scheme is a P- and Q-polynomial scheme with AW-parameters of Type I: $\Lambda_{\mathrm{I}} = (r_1, r_2, s, s^*; h, h^*, \theta_0, \theta_0^* \mid d)$, $r_1 = r_2 = s^* = 0$, $s = -r^{-2d-3}$, $h = r^{2d+1}/(r^2-1)$, $h^* = r^3(r^{2d-1}+1)(r^{2d-3}+1)/(r^2-1)(r+1)$, $\theta_0 = r(r^{2d}-1)/(r^2-1)$, $\theta_0^* = r^2(r^{2d}-1)(r^{2d-3}+1)/(r^2-1)(r+1)$. The intersection numbers $b_i = p_{1,i+1}^i$, $c_i = p_{1,i-1}^i$ are

$$b_i = \frac{r^{2d+1}}{r^2-1}(1 - r^{2(i-d)}), \quad 0 \le i \le d-1,$$
$$c_i = \frac{1}{r^2-1}(r^{2i}-1), \quad 1 \le i \le d,$$

and these intersection numbers of ${}^2A_{2d-1}(r)$ have another expression by AW-parameters:

$$\Lambda_{\mathrm{I}} = (r_1, r_2, s, s^*; h, h^*, \theta_0, \theta_0^* \mid -r, d),$$
$$r_1 = (-1)^d r^{-d-1}, \quad r_2 = 0, \quad s = r^{-2d-2}, \quad s^* = 0,$$
$$h = \frac{r^{2d+1}}{r^2-1}, \quad h^* = \frac{r(r^{2d-1}+1)}{r+1}, \quad \theta_0 = \frac{r(r^{2d}-1)}{r^2-1}, \quad \theta_0^* = h^*.$$

The other parameters are the following:

$$\begin{cases} \theta_{2i} = \theta_0 - \frac{1}{r^2-1}(r^{2i}-1)(r^{2d+1-2i}+1), & 0 \le i \le \frac{d}{2}, \\ \theta_{2i+1} = \theta_0 - \frac{1}{r^2-1}(r^{2d-2i}-1)(r^{2i+1}+1), & 0 \le i \le \frac{d-1}{2}, \end{cases} \tag{6.289}$$

$$\theta_i^* = \theta_0^* - h^*(1-(-r)^i), \quad 0 \le i \le d,$$
$$\lambda_i = -hh^* r^{-2i-1}(1-r^{2i-2d})(1-(-r)^{i+1}), \quad 0 \le i \le d-1,$$
$$\hat{\lambda}_i = (-1)^d hh^* r^{-d-2i-2}(1-r^{2i+2})(1-(-r)^{i-d}), \quad 0 \le i \le d-1,$$

$$\begin{cases} b_{2i}^* = h^* \frac{1+r^{2i-2d-1}}{1+r^{4i-2d-1}}, & 0 \le i \le \frac{d-1}{2}, \\ b_{2i+1}^* = h^* \frac{1-r^{2i-2d}}{1+r^{4i-2d+1}}, & 0 \le i \le \frac{d-2}{2}, \end{cases}$$

$$\begin{cases} c_{2i}^* = -h^* r^{2i-2d-1} \frac{1-r^{2i}}{1+r^{4i-2d-1}}, & 1 \le i \le \frac{d}{2}, \\ c_{2i+1}^* = h^* r^{2i-2d} \frac{1+r^{2i+1}}{1+r^{4i-2d+1}}, & 1 \le i \le \frac{d-1}{2}, \end{cases}$$

$$k_i = r^{i^2} \binom{d}{i}_{r^2}, \quad 0 \le i \le d,$$

$$\begin{cases} m_{2i} = r^{2i} \binom{d}{i}_{r^2} \frac{r^{2d+1-4i}+1}{r^{2d+1-2i}+1} \prod_{j=0}^{i-1} \frac{r^{2d-1-2j}+1}{r^{2j-1}+1}, & 0 \le i \le \frac{d}{2}, \\ m_{2i+1} = r^{2d-2i} \binom{d}{i}_{r^2} \frac{r^{4i+1-2d}+1}{r^{2i+1}+1} \prod_{j=0}^{d-i-1} \frac{r^{2d-1-2j}+1}{r^{2j-1}+1}, & 0 \le i \le \frac{d-1}{2}. \end{cases}$$

The P-polynomial ordering $R_0, R_1, \ldots, R_d$ of the dual polar scheme ${}^2A_{2d-1}(r)$ is fixed, and the intersection numbers b_i $(0 \le i \le d-1)$, c_i $(1 \le i \le d)$ determined by this P-polynomial ordering have two distinct expressions by AW-parameters of Type I. Therefore, the eigenvalues $\{P_1(i)\}_{0 \le i \le d}$ of the adjacency matrix A_1 of the relation R_1 have the expression $\theta_i(\text{I-1})$ $(0 \le i \le d)$ determined by the first AW-parameters $\Lambda_{\text{I-1}}$ and the expression $\theta_i(\text{I-2})$ $(0 \le i \le d)$ determined by the second AW-parameters $\Lambda_{\text{I-2}}$. Since we have

$$\theta_i(\text{I-1}) = \frac{r(r^{2d}-1)}{r^2-1} - \frac{1}{r^2-1}(r^{2i}-1)(r^{2d+1-2i}+1), \quad 0 \le i \le d,$$

by (6.289) the following hold:

$$\theta_{2i}(\text{I-2}) = \theta_i(\text{I-1}), \quad 0 \le i \le \frac{d}{2},$$
$$\theta_{2i+1}(\text{I-2}) = \theta_{d-i}(\text{I-1}), \quad 0 \le i \le \frac{d-1}{2}.$$

This implies that if we let $E_0, E_1, \dots, E_d$ be the Q-polynomial ordering determined by the first AW-parameters $\Lambda_{\text{I-1}}$, then the Q-polynomial ordering determined by the second AW-parameters $\Lambda_{\text{I-2}}$ becomes

$$E_0, E_d, E_1, E_{d-1}, \dots.$$

If we start from the second Q-polynomial ordering, the first Q-polynomial ordering becomes

$$E_0, E_2, E_4, \dots, E_3, E_1,$$

and this is also checked by the fact that the second Q-polynomial ordering is almost dual bipartite, i. e., $a_i^* = 0$ $(0 \le i \le d-1, a_d^* \ne 0)$.

(iii) The Hamming scheme $H(2d, 2)$

Let $R_0, R_1, \dots, R_{2d}$ be the natural P-polynomial ordering of the binary Hamming scheme $H(2d, 2)$ of diameter $2d$. Namely, R_i is the relation determined by the Hamming distance i (Example 2.6 in Chapter 2). The intersection numbers $b_i = 2d - i$ $(0 \le i \le 2d-1)$, $c_i = i$ $(1 \le i \le 2d)$ determined by this P-polynomial ordering have the following AW-parameters:

$$\begin{aligned} \Lambda_{\text{IIC}} &= (r, -, -, -; h, h^*, \theta_0, \theta_0^* \mid 2d), \\ r &= 2, \quad h = h^* = -2, \quad \theta_0 = \theta_0^* = 2d, \end{aligned} \tag{6.290}$$

and $H(2d, 2)$ becomes a Q-polynomial scheme. Let $E_0, E_1, \dots, E_{2d}$ be the Q-polynomial ordering determined by the AW-parameters Λ_{IIC}. The other parameters are the following:

$$\begin{aligned} &\theta_i = \theta_i^* = 2(d-i), \quad 0 \le i \le 2d, \\ &\lambda_i = -\hat{\lambda}_i = 2(i+1)(i-2d), \quad 0 \le i \le 2d-1, \\ &b_i^* = b_i = 2d - i, \quad 0 \le i \le 2d-1, \quad c_i^* = c_i = i, \quad 1 \le i \le 2d, \\ &k_i = m_i = \binom{2d}{i}, \quad 0 \le i \le 2d. \end{aligned}$$

It can be easily seen that $b_i = c_{2d-i}$ $(0 \le i \le 2d-1)$, $a_i = 0$ $(0 \le i \le 2d)$. Hence $H(2d, 2)$ has the second P-polynomial ordering as follows:

$$R_0, R_{2d-1}, R_2, R_{2d-3}, R_4, R_{2d-5}, \dots\dots, R_5, R_{2d-4}, R_3, R_{2d-2}, R_1, R_{2d}$$

(Section 6.1.3). Moreover, since $b_i^* = c_{2d-i}^*$ $(0 \le i \le 2d-1)$, $a_i^* = 0$ $(0 \le i \le 2d)$, $H(2d, 2)$ has the second Q-polynomial ordering

$$E_0, E_{2d-1}, E_2, E_{2d-3}, E_4, E_{2d-5}, \dots\dots, E_5, E_{2d-4}, E_3, E_{2d-2}, E_1, E_{2d}.$$

Therefore there are four combinations of P-polynomial orderings and Q-polynomial orderings, and corresponding to them, four P- and Q-polynomial schemes arise from $H(2d, 2)$ (see Table 6.1).

Firstly, we fix the first Q-polynomial ordering. Then the dual intersection numbers $b_i^* = 2d - i$ $(0 \leq i \leq 2d-1)$, $c_i^* = i$ $(1 \leq i \leq 2d)$ have the following AW-parameters:

$$\begin{aligned} &\Lambda_{\mathrm{III}} = (r_1, r_2, s, s^*; h, h^*, \theta_0, \theta_0^* \mid 2d), \\ &r_1 = r_2 = -d - \frac{1}{2}, \quad s = s^* = -2d - 1, \\ &h = h^* = -1, \quad \theta_0 = \theta_0^* = 2d. \end{aligned} \tag{6.291}$$

The other parameters are the following:

$$\begin{aligned} &\begin{cases} \theta_{2i} = \theta_{2i}^* = 2(d - 2i), & 0 \leq i \leq d, \\ \theta_{2i+1} = \theta_{2i+1}^* = 2(2i + 1 - d), & 0 \leq i \leq d - 1, \end{cases} \\ &\lambda_i = -\hat{\lambda}_i = \begin{cases} -4(i - d + \frac{1}{2})(i - 2d), & 0 \leq i \leq 2d - 1 \quad (\text{if } i \text{ is even}), \\ -4(i - d + \frac{1}{2})(i + 1), & 0 \leq i \leq 2d - 1 \quad (\text{if } i \text{ is odd}), \end{cases} \\ &b_i = b_i^* = 2d - i, \quad 0 \leq i \leq 2d - 1, \quad c_i = c_i^* = i, \quad 1 \leq i \leq 2d, \\ &k_i = m_i = \binom{2d}{i}, \quad 0 \leq i \leq 2d. \end{aligned}$$

Let $\theta_i^*(\mathrm{IIC})$ be the expression of the dual eigenvalues of E_1 by AW-parameters Λ_{IIC}, and let $\theta_i^*(\mathrm{III})$ be the expression of them by AW-parameters Λ_{III}. Then we have

$$\begin{aligned} \theta_{2i}^*(\mathrm{III}) &= \theta_{2i}^*(\mathrm{IIC}), \quad 0 \leq i \leq d, \\ \theta_{2i+1}^*(\mathrm{III}) &= \theta_{2d-2i-1}^*(\mathrm{IIC}), \quad 0 \leq i \leq d - 1. \end{aligned} \tag{6.292}$$

So the second P-polynomial ordering arises from the ordering of $\theta_i^*(\mathrm{III})$.

Next, we fix the first P-polynomial ordering. Then the intersection numbers $b_i = 2d - i$ $(0 \leq i \leq 2d - 1)$, $c_i = i$ $(1 \leq i \leq 2d)$ have the same AW-parameters as (6.291), i. e., Λ_{III}. Let $\theta_i(\mathrm{IIC})$ be the expression of the eigenvalues of the adjacency matrix of R_1 by AW-parameters Λ_{IIC}, and let $\theta_i(\mathrm{III})$ that by AW-parameters Λ_{III}. Then we have

$$\begin{aligned} \theta_{2i}(\mathrm{III}) &= \theta_{2i}(\mathrm{IIC}), \quad 0 \leq i \leq d, \\ \theta_{2i+1}(\mathrm{III}) &= \theta_{2d-2i-1}(\mathrm{IIC}), \quad 0 \leq i \leq d - 1. \end{aligned} \tag{6.293}$$

So the second Q-polynomial ordering arises from the ordering of $\theta_i(\mathrm{III})$.

Consider the case a P- and Q-polynomial scheme $H(2d, 2)$ is defined by the second P-polynomial ordering and the first Q-polynomial ordering. Now we fix the second P-polynomial ordering. Then the eigenvalues of the adjacency matrix A_{2d-1} of R_{2d-1} have the expression (6.291) by AW-parameters of Type III Λ_{III}, and we denote it by $\theta_i(\mathrm{III})$ $(0 \leq i \leq 2d)$. Note that $\theta_i(\mathrm{III})$ corresponds to the first Q-polynomial ordering. By (6.291), the intersection numbers by the second P-polynomial ordering are $b_i = 2d - i$

$(0 \le i \le 2d-1)$, $c_i = i$ $(1 \le i \le 2d)$, and hence have the same expression as (6.290) by AW-parameters of Type IIC Λ_{IIC}. Let $\theta_i(\mathrm{IIC})$ be the expression of the eigenvalues of A_{2d-1} by (6.290). Then (6.293) holds and hence $\theta_i(\mathrm{IIC})$ corresponds to the second Q-polynomial ordering.

As for Λ_{IIC}, we have $h = h^*$, $\theta_0 = \theta_0^*$, and as for Λ_{III}, we have $s = s^*$, $h = h^*$, $\theta_0 = \theta_0^*$.

Table 6.1: AW-parameters of $H(2d, 2)$.

$Q \setminus P$	$R_0, R_1, \ldots$	$R_0, R_{2d-1}, R_2, \ldots, R_1, R_{2d}$
$E_0, E_1, \ldots$	Λ_{IIC}, (6.290)	Λ_{III}, (6.291)
$E_0, E_{2d-1}, E_2, \ldots, E_1, E_{2d}$	Λ_{III}, (6.291)	Λ_{IIC}, (6.290)

(iv) $\frac{1}{2}H(2d+1,2)$: The bipartite half of the Hamming scheme $H(2d+1,2)$

The bipartite half $\frac{1}{2}H(2d+1,2)$ of the Hamming scheme $H(2d+1,2)$ have AW-parameters of Type IIA:

$$\begin{aligned} \Lambda_{\mathrm{IIA}} &= (r_1, -, s, -; h, h^*, \theta_0, \theta_0^* \mid d), \\ r_1 &= -d - \frac{3}{2}, \quad s = -2d-2, \\ h &= 2, \quad h^* = -4, \quad \theta_0 = d(2d+1), \quad \theta_0^* = 2d+1, \end{aligned} \tag{6.294}$$

and it becomes a P- and Q-polynomial scheme. Let $R_0, R_1, \ldots, R_d$ be the P-polynomial ordering and let $E_0, E_1, \ldots, E_d$ be the Q-polynomial ordering determined by the AW-parameters Λ_{IIA}. Then the intersection numbers and the dual intersection numbers are

$$\begin{aligned} b_i &= (d-i)(2d+1-2i), \quad 0 \le i \le d-1, \quad c_i = i(2i-1), \quad 1 \le i \le d, \\ b_i^* &= 2d+1-i, \quad 0 \le i \le d-1, \quad c_i^* = i, \quad 1 \le i \le d. \end{aligned}$$

The other parameters are the following:

$$\begin{aligned} \theta_i &= d(2d+1) - 2i(2d+1-i), \quad 0 \le i \le d, \\ \theta_i^* &= 2d+1-4i, \quad 0 \le i \le d, \\ \lambda_i &= hh^*(i+1)\left(i-d-\frac{1}{2}\right)(i-d), \quad 0 \le i \le d, \\ \hat{\lambda}_i &= hh^*(i+1)\left(i+\frac{1}{2}\right)(i-d), \quad 0 \le i \le d, \\ k_i &= \binom{2d+1}{2i}, \quad 0 \le i \le d, \quad m_i = \binom{2d+1}{i}, \quad 0 \le i \le d. \end{aligned}$$

First we fix the Q-polynomial ordering $E_0, E_1, \ldots, E_d$. Then the dual intersection numbers $b_i^* = 2d + 1 - i$ $(0 \le i \le d-1)$, $c_i^* = i$ $(1 \le i \le d)$ have the following AW-parameters:

$$\begin{aligned} \Lambda_{\mathrm{III}} &= (r_1, r_2, s, s^*; h, h^*, \theta_0, \theta_0^* \mid d), \\ r_1 &= -2d - 2, \quad r_2 = -d - 1, \quad s = s^* = -2d - 2, \\ h &= h^* = -1, \quad \theta_0 = \theta_0^* = 2d + 1. \end{aligned} \tag{6.295}$$

The other parameters are the following:

$$\begin{cases} \theta_{2i} = \theta_{2i}^* = 2d + 1 - 4i, & 0 \le i \le \frac{d}{2}, \\ \theta_{2i+1} = \theta_{2i+1}^* = 2d + 1 - 4(d - i), & 0 \le i \le \frac{d-1}{2}, \end{cases}$$

$$\lambda_i = \begin{cases} -4(i - d)(i - 2d - 1), & 0 \le i \le d - 1 \quad (\text{if } i \text{ is even}), \\ -4(i + 1)(i - d), & 0 \le i \le d - 1 \quad (\text{if } i \text{ is odd}), \end{cases}$$

$$\hat{\lambda}_i = (-1)^d 4(i + 1)(i - d), \quad 0 \le i \le d - 1,$$

$$b_i = b_i^* = 2d + 1 - i, \quad 0 \le i \le d - 1, \quad c_i = c_i^* = i, \quad 1 \le i \le d,$$

$$k_i = m_i = \binom{2d + 1}{i}, \quad 0 \le i \le d.$$

Let θ_i^* (IIA) be the expression of the dual eigenvalues of E_1 by AW-parameters Λ_{IIA}, and let θ_i^* (III) be that by AW-parameters Λ_{III}. Then we have

$$\begin{aligned} \theta_{2i}^*(\mathrm{III}) &= \theta_i^*(\mathrm{IIA}), \quad 0 \le i \le \frac{d}{2}, \\ \theta_{2i+1}^*(\mathrm{III}) &= \theta_{d-i}^*(\mathrm{IIA}), \quad 0 \le i \le \frac{d-1}{2}. \end{aligned} \tag{6.296}$$

So from the ordering of θ^* (III), the second P-polynomial ordering

$$R_0, R_d, R_1, R_{d-1}, R_2, R_{d-2}, \ldots$$

arises. This is also checked by the fact that the second P-polynomial ordering is almost bipartite, i. e., $a_i = 0$ $(0 \le i \le d - 1)$, $a_d \ne 0$.

Next, we fix the P-polynomial ordering $R_0, R_1, \ldots, R_d$. Then the intersection numbers $b_i = (d - i)(2d + 1 - 2i)$ $(0 \le i \le d - 1)$, $c_i = i(2i - 1)$ $(1 \le i \le d)$ have the following AW-parameters:

$$\begin{aligned} \Lambda_{\mathrm{II}} &= (r_1, r_2, s, s^*; h, h^*, \theta_0, \theta_0^* \mid d), \\ r_1 &= -\frac{d}{2} - \frac{3}{4}, \quad r_2 = -\frac{d}{2} - \frac{5}{4}, \quad s = s^* = -d - \frac{3}{2}, \\ h &= h^* = 8, \quad \theta_0 = \theta_0^* = d(2d + 1). \end{aligned} \tag{6.297}$$

The other parameters are the following:

$$\theta_i = \theta_i^* = d(2d + 1) + 8i\left(i - d - \frac{1}{2}\right), \quad 0 \le i \le d,$$

$$\lambda_i = hh^*(i + 1)(i - d)\left(i - \frac{d}{2} + \frac{1}{4}\right)\left(i - \frac{d}{2} - \frac{1}{4}\right), \quad 0 \le i \le d - 1,$$

$$\hat{\lambda}_i = hh^*(i+1)(i-d)\left(i-\frac{d}{2}+\frac{1}{4}\right)\left(i-\frac{d}{2}+\frac{3}{4}\right), \quad 0 \le i \le d-1,$$
$$b_i^* = b_i = (d-i)(2d+1-2i), \quad 0 \le i \le d-1, \quad c_i^* = c_i = i(2i-1), \quad 1 \le i \le d,$$
$$k_i = m_i = \binom{2d+1}{2i}, \quad 0 \le i \le d.$$

Let $\theta_i(\mathrm{IIA})$ be the expression of the eigenvalues of the adjacency matrix A_1 of R_1 by AW-parameters Λ_{IIA}, and let $\theta_i(\mathrm{II})$ be the expression of them by AW-parameters Λ_{II}. Since we have

$$\begin{cases} \theta_i(\mathrm{II}) = \theta_{2i}(\mathrm{IIA}), & 0 \le i \le \frac{d}{2}, \\ \theta_{d-i}(\mathrm{II}) = \theta_{2i+1}(\mathrm{IIA}), & 0 \le i \le \frac{d-1}{2}, \end{cases} \tag{6.298}$$

from the ordering of $\theta_i(\mathrm{II})$, the second Q-polynomial ordering

$$E_0, E_2, E_4, E_6, \ldots\ldots, E_5, E_3, E_1$$

arises. This is also checked by the fact that the first Q-polynomial ordering is almost dual bipartite, i. e., $a_i^* = 0$ $(0 \le i \le d-1)$, $a_d^* \neq 0$.

As was seen before, the second P-polynomial ordering and the first Q-polynomial ordering are associated with the expression (6.295) for $\frac{1}{2}H(2d+1,2)$ by AW-parameters of Type III. Here we fix the second P-polynomial ordering and consider the intersection numbers $b_i = 2d+1-i$ $(0 \le i \le d-1)$, $c_i = i$ $(1 \le i \le d)$. Then these intersection numbers have the following expression by AW-parameters:

$$\begin{aligned} \Lambda_{\mathrm{IIB}} &= (r_1, -, -, s^*; h, h^*, \theta_0, \theta_0^* \mid d), \\ r_1 &= -d - \frac{3}{2}, \quad s^* = -2d-2, \\ h &= -4, \quad h^* = 2, \quad \theta_0 = 2d+1, \quad \theta_0^* = d(2d+1). \end{aligned} \tag{6.299}$$

The other parameters are the following:

$$\theta_i = 2d+1-4i, \quad 0 \le i \le d, \quad \theta_i^* = d(2d+1) + 2i(i-2d-1), \quad 0 \le i \le d,$$
$$\lambda_i = hh^*(i+1)(i-d)\left(i-d-\frac{1}{2}\right), \quad 0 \le i \le d-1,$$
$$\hat{\lambda}_i = -hh^*(i+1)(i-d)\left(i-d+\frac{1}{2}\right), \quad 0 \le i \le d-1,$$
$$b_i = 2d+1-i, \quad 0 \le i \le d-1, \quad c_i = i, \quad 1 \le i \le d,$$
$$b_i^* = (d-i)(2d+1-2i), \quad 0 \le i \le d-1, \quad c_i^* = i(2i-1), \quad 1 \le i \le d,$$
$$k_i = \binom{2d+1}{i}, \quad 0 \le i \le d, \quad m_i = \binom{2d+1}{2i}, \quad 0 \le i \le d.$$

Let $\theta_i(\mathrm{III})$ be the expression of the eigenvalues of the adjacency matrix A_d of R_d by AW-parameters Λ_{III}, and let $\theta_i(\mathrm{IIB})$ be the expression of them by AW-parameters Λ_{IIB}.

Since we have

$$\begin{aligned}\theta_i(\mathrm{IIB}) &= \theta_{2i}(\mathrm{III}), \quad 0 \le i \le \frac{d}{2},\\ \theta_{d-i}(\mathrm{IIB}) &= \theta_{2i+1}(\mathrm{III}), \quad 0 \le i \le \frac{d-1}{2},\end{aligned} \tag{6.300}$$

from the ordering of $\theta_i(\mathrm{IIB})$, the second Q-polynomial ordering

$$E_0, E_2, E_4, E_6, \ldots\ldots, E_5, E_3, E_1$$

arises. This is also checked by the fact that the first Q-polynomial ordering is almost dual bipartite, i. e., $a_i^* = 0$ $(0 \le i \le d-1)$, $a_d^* \ne 0$. Note also that (6.300) is the dual of (6.296). Moreover, (6.299) is obtained from (6.297) as follows. We fix the second Q-polynomial ordering associated with the expression (6.297) for $\frac{1}{2}H(2d+1,2)$ by AW-parameters of Type II, and consider the dual intersection numbers $b_i^* = (d-i)(2d+1-2i)$ $(0 \le i \le d-1)$, $c_i^* = i(2i-1)$ $(1 \le i \le d)$. Then these dual intersection numbers have another expression (6.299) by AW-parameters of Type IIB. The expression $\theta_i^*(\mathrm{II})$ of the dual eigenvalues of the primitive idempotent E_2 by Λ_{II} and the expression $\theta_i^*(\mathrm{IIB})$ by Λ_{IIB} have the following relation:

$$\begin{aligned}\theta_{2i}^*(\mathrm{IIB}) &= \theta_i^*(\mathrm{II}), \quad 0 \le i \le \frac{d}{2},\\ \theta_{2i+1}^*(\mathrm{IIB}) &= \theta_{d-i}^*(\mathrm{II}), \quad 0 \le i \le \frac{d-1}{2}.\end{aligned} \tag{6.301}$$

So the second P-polynomial ordering

$$R_0, R_d, R_1, R_{d-1}, R_2.R_{d-2}\ldots$$

arises. Note that (6.301) is the dual of (6.298).

In the expression for $\frac{1}{2}H(2d+1,2)$ by AW-parameters, if we fix r_1 and replace s by s^*, h by h^*, and θ_0 by θ_0^*, then Λ_{IIA} is exchanged by Λ_{IIB}. On the other hand, in Λ_{II}, Λ_{III}, note that $s = s^*$, $h = h^*$, $\theta_0 = \theta_0^*$ hold.

We summarize the argument on the combination of P-polynomial orderings and Q-polynomial orderings in Table 6.2.

Table 6.2: AW-parameters of $\frac{1}{2}H(2d+1,2)$.

$Q \setminus P$	$R_0, R_1, \ldots$	$R_0, R_d, R_1, R_{d-1}, R_2, R_{d-2}\ldots$
$E_0, E_1, \ldots$	Λ_{IIA}, (6.294)	Λ_{III}, (6.295)
$E_0, E_2, E_4, \ldots, E_3, E_1$	Λ_{II}, (6.297)	Λ_{IIB}, (6.299)

(v) $\overline{H(2d+1,2)}$: The antipodal quotient of the Hamming scheme $H(2d+1,2)$

The antipodal quotient $\overline{H(2d+1,2)}$ of the Hamming scheme $H(2d+1,2)$ has AW-parameters of Type IIB:

$$\Lambda_{\text{IIB}} = (r_1, -, -, s^*; h, h^*, \theta_0, \theta_0^* \mid d),$$
$$r_1 = -d - \frac{3}{2}, \quad s^* = -2d-2,$$
$$h = -4, \quad h^* = 2, \quad \theta_0 = 2d+1, \quad \theta_0^* = d(2d+1).$$

Note that AW parameters Λ_{IIB} for $\overline{H(2d+1,2)}$ coincide with AW-parameters Λ_{IIB} for $\frac{1}{2}H(2d+1,2)$ in (6.299). Let $R_0, R_1, \ldots, R_d$ be the P-polynomial ordering of $\overline{H(2d+1,2)}$ determined by the AW-parameters Λ_{IIB}, and let $E_0, E_1, \ldots, E_d$ be the Q-polynomial ordering. This is equivalent to rewriting the second P-polynomial ordering of $\frac{1}{2}H(2d+1,2)$ as $R_0, R_1, \ldots, R_d$ and rewriting the second Q-polynomial ordering as $E_0, E_1, \ldots, E_d$. Therefore, $\overline{H(2d+1,2)}$ has the second P-polynomial ordering

$$R_0, R_2, R_4, R_6, \ldots, R_5, R_3, R_1$$

and the second Q-polynomial ordering

$$E_0, E_d, E_1, E_{d-1}, E_2, E_{d-2}, \ldots.$$

The AW-parameters for the P- and Q-polynomial schemes $\overline{H(2d+1,2)}$ arising from these combinations of P-polynomial orderings and Q-polynomial orderings are listed in Table 6.3.

Table 6.3: AW-parameters of $\overline{H(2d+1,2)}$.

$Q \setminus P$	$R_0, R_2, R_4, \ldots, R_3, R_1$	$R_0, R_1, \ldots$
$E_0, E_d, E_1, E_{d-1}, E_2, E_{d-2}, \ldots$	Λ_{IIA}, (6.294)	Λ_{III}, (6.295)
$E_0, E_1, \ldots$	Λ_{II}, (6.297)	Λ_{IIB}, (6.299)

In fact, $\frac{1}{2}H(2d+1,2)$ and $\overline{H(2d+1,2)}$ are isomorphic as association schemes. This can be verified as follows. If we denote $H(2d+1,2)$ by $\mathfrak{X} = (X, \{R_i\}_{0\le i\le 2d+1})$, then $\mathfrak{X}$ is bipartite and two bipartite halves X_1 and X_2 of X arise: $X = X_1 \cup X_2$, $X_1 \cap X_2 = \emptyset$. For $x \in X$, the antipodal point x' ($(x,x') \in R_{2d+1}$) is uniquely determined, and the quotient $\overline{X}$ of X by the equivalence relation $R_0 \cup R_{2d+1}$ arises. Since the antipodal point x' of $x \in X_1$ belongs to X_2, there is a one-to-one correspondence between the bipartite half X_1 and the quotient $\overline{X}$. By this correspondence, the association schemes $\frac{1}{2}\mathfrak{X}$ and $\overline{\mathfrak{X}}$ are isomorphic.

(c) The Hemmeter scheme $\mathrm{Hem}_d(q)$ and the Ustimenko scheme $\mathrm{Ust}_{[\frac{d}{2}]}(q)$

Let $\Gamma = (X, R)$ be a distance-regular graph of diameter d and let $\mathfrak{X} = (X, \{R_i\}_{0\le i\le d})$ be the P-polynomial scheme determined by Γ. Namely, we define $(x,y) \in R_i$ if the distance

$\partial(x, y)$ of $x, y \in X$ in Γ is i. Let $b_i = p^i_{1,i+1}$ $(0 \le i \le d-1)$, $c_i = p^i_{1,i-1}$ $(1 \le i \le d)$ be the intersection numbers of Γ.

The graph $\Gamma^{(1,2)} = (X, R_1 \cup R_2)$ is called the *distance* 1-*or*-2 graph determined by Γ. Namely, the adjacency relation of $x, y \in X$ in $\Gamma^{(1,2)}$ is defined by $\partial(x, y) = 1, 2$. Then $\Gamma^{(1,2)}$ is a distance-regular graph if and only if

$$b_{i-1} + c_{i+1} - a_i = b_0 + c_2 - a_1 \quad (2 \le i \le d-1)$$

([113, page 150]).

Let A be the adjacency matrix of a distance-regular graph $\Gamma = (X, R)$ and define

$$\tilde{A} = \begin{pmatrix} 0 & A+I \\ A+I & 0 \end{pmatrix},$$

where I is the identity matrix. We write $\tilde{\Gamma} = (\tilde{X}, \tilde{R})$ to denote the graph which has $\tilde{A}$ as its adjacency matrix and call it the *extended bipartite double* of Γ. The graph $\tilde{\Gamma}$ is a bipartite graph of diameter $d+1$, and a bipartite half $\frac{1}{2}\tilde{\Gamma}$ of $\tilde{\Gamma}$ is isomorphic to the distance 1-or-2 graph $\Gamma^{(1,2)}$ of Γ. The graph $\tilde{\Gamma}$ is a distance-regular graph if and only if

$$b_i + c_{i+1} = b_0 + 1, \quad (1 \le i \le d-1)$$

([113, page 26]). In this case, the intersection numbers $\tilde{b}_i$, $\tilde{c}_i$ of $\tilde{\Gamma}$ are

$$\tilde{b}_{i+1} = b_i \ (0 \le i \le d-1), \quad \tilde{b}_0 = b_0 + 1,$$
$$\tilde{c}_i = c_i \ (1 \le i \le d), \quad \tilde{c}_{d+1} = b_0 + 1.$$

The dual polar scheme $B_d(q)$ has the same intersection numbers as $C_d(q)$, which are

$$b_i = \frac{q^{i+1}(q^{d-i} - 1)}{q - 1} \quad (0 \le i \le d-1),$$
$$c_i = \frac{q^i - 1}{q - 1} \quad (1 \le i \le d).$$

Therefore both $B_d(q)$ and $C_d(q)$ have the extended bipartite doubles, whose intersection numbers are

$$\tilde{b}_i = \frac{q^i(q^{d+1-i} - 1)}{q - 1} \quad (0 \le i \le d),$$
$$\tilde{c}_i = \frac{q^i - 1}{q - 1} \quad (1 \le i \le d+1).$$

They coincide with the intersection numbers of the dual polar scheme $D_{d+1}(q)$. In fact, the extended bipartite double of $B_d(q)$ is isomorphic to $D_{d+1}(q)$. However, if q is odd, the extended bipartite double of $C_d(q)$ is not isomorphic to $D_{d+1}(q)$ (note that if q is even, $B_d(q)$ and $C_d(q)$ are isomorphic) [115, 214]. This graph is called the *Hemmeter graph* and the corresponding P-polynomial scheme is called the *Hemmeter scheme*, which is denoted by $\mathrm{Hem}_{d+1}(q)$. The Hemmeter scheme $\mathrm{Hem}_{d+1}(q)$ is bipartite and has the same intersection numbers as $D_{d+1}(q)$, and thus it is a Q-polynomial scheme.

The bipartite half $\frac{1}{2}\mathrm{Hem}_{d+1}(q)$ of $\mathrm{Hem}_{d+1}(q)$ has the same intersection numbers as $\frac{1}{2}D_{d+1}(q)$ but is not isomorphic to it when q is odd. This graph is called the *Ustimenko graph* and the corresponding P-polynomial scheme is called the *Ustimenko scheme*, which is denoted by $\mathrm{Ust}_{[\frac{d+1}{2}]}(q)$. Since $\mathrm{Ust}_{[\frac{d+1}{2}]}(q)$ has the same intersection numbers as $\frac{1}{2}D_{d+1}(q)$, it is a Q-polynomial scheme. Since the bipartite half $\frac{1}{2}\tilde{\Gamma}$ of the extended bipartite double $\tilde{\Gamma}$ is isomorphic to the distance 1-or-2 graph $\Gamma^{(1,2)}$, $\frac{1}{2}D_{d+1}(q)$ is isomorphic to the distance 1-or-2 graph of $B_d(q)$ and $\mathrm{Ust}_{[\frac{d+1}{2}]}(q)$ is isomorphic to the distance 1-or-2 graph (a fusion scheme) of $C_d(q)$. In fact, $\mathrm{Ust}_{[\frac{d+1}{2}]}(q)$ was found as a fusion scheme of $C_d(q)$ [491, 263, 115, 130].

For a distance-regular graph $\Gamma = (X, R)$ of diameter n, fix $x_0 \in X$ and set

$$X_n = X_n(x_0) = \{y \in X \mid \partial(x_0, y) = n\}.$$

The graph whose R is restricted on X_n is denoted by Γ_n:

$$\Gamma_n = \Gamma_n(x_0) = (X_n, R|_{X_n \times X_n}).$$

Let R_2 be the distance-2 relation on X, and denote the graph with vertex set X_n and the adjacency relation R_2 by $\Gamma_n^{(2)}$:

$$\Gamma_n^{(2)} = \Gamma_n^{(2)}(x_0) = (X_n, R_2|_{X_n \times X_n}).$$

Namely, we define the adjacency relation for $x, y \in X_n$ in $\Gamma_n^{(2)}$ by $\partial(x, y) = 2$.

As we have seen in Section 6.4.1, the second family (iii), if Γ is the dual polar scheme $D_{d+1}(q)$, then $\Gamma_{d+1}^{(2)}$ is isomorphic to the affine scheme $\mathrm{Alt}_{d+1}(q)$. On the other hand, if Γ is the Hemmeter scheme $\mathrm{Hem}_{d+1}(q)$, $\Gamma_{d+1}^{(2)}$ is isomorphic to the affine scheme $\mathrm{Quad}_d(q)$ [130]. Note that $\mathrm{Alt}_{d+1}(q)$ and $\mathrm{Quad}_d(q)$ have the same intersection numbers but are not isomorphic. Moreover, if Γ is $\frac{1}{2}D_{d+1}(q)$, then $\Gamma_{[\frac{d+1}{2}]}$ is isomorphic to $\mathrm{Alt}_{d+1}(q)$, and if Γ is $\mathrm{Ust}_{[\frac{d+1}{2}]}(q)$, then $\Gamma_{[\frac{d+1}{2}]}$ is isomorphic to $\mathrm{Quad}_d(q)$.

The above facts are visualized as the following figure.

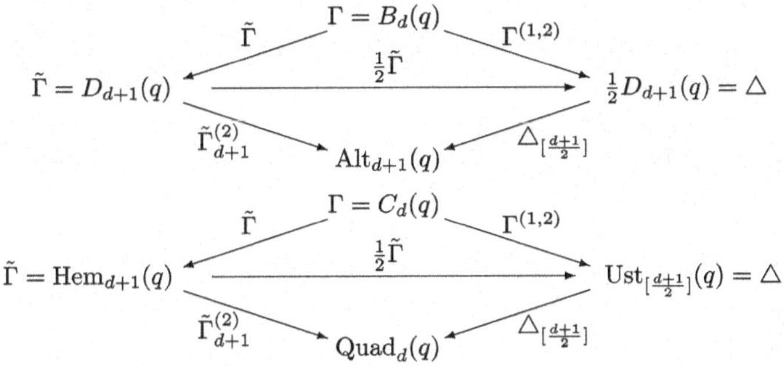

Remark 6.87. If a distance-regular graph Γ of diameter n is bipartite and Q-polynomial, $\Gamma_n^{(2)}$ is always distance-regular [128].

Remark 6.88. Neither $\mathrm{Hem}_{d+1}(q)$ nor $\mathrm{Ust}_{[\frac{d+1}{2}]}(q)$ is distance-transitive. Note that $\mathrm{Quad}_d(q)$ is not distance-transitive, either.

(d) Twisted Grassmann schemes and Doob schemes

We have considered dual polar schemes $B_d(q)$ and $C_d(q)$, affine schemes $\mathrm{Alt}_n(q)$ and $\mathrm{Quad}_{n-1}(q)$, $D_d(q)$ and $\mathrm{Hem}_d(q)$, and $\frac{1}{2}D_d(q)$ and $\mathrm{Ust}_{[\frac{d}{2}]}(q)$ as examples of pairs of P- and Q-polynomial schemes such that the intersection numbers are the same but they are not isomorphic. As other such examples, we have the q-Johnson scheme $J_q(2d+1,d)$ (or Grassmann scheme) and the twisted q-Johnson scheme (or twisted Grassmann scheme), and the Hamming scheme $H(d,4)$ and the Doob scheme.

Let V be the $(2d+1)$-dimensional vector space over the finite field $\mathbb{F}_q$ $(d \geq 2)$. Fix a $2d$-dimensional subspace H of V. Let X_1 be the set of $(d+1)$-dimensional subspaces of V not contained in H, let X_2 be the set of $(d-1)$-dimensional subspaces of V contained in H, and set $X = X_1 \cup X_2$:

$$X_1 = \{U \subset V \mid \dim(U) = d+1,\ U \not\subset H\},$$
$$X_2 = \{U \subset H \mid \dim(U) = d-1\}.$$

For $U, U' \in X$, we define U and U' are adjacent and write $U \sim U'$ if they satisfy one of the following:

$$\begin{cases} U, U' \in X_1 \quad \text{and} \quad \dim(U \cap U') = d, \\ U \in X_1, \quad U' \in X_2, \quad \text{and} \quad U' \subset U, \\ U, U' \in X_2 \quad \text{and} \quad \dim(U \cap U') = d-2. \end{cases}$$

Let Γ be the graph with vertex set X and the above adjacency relation $\sim$. Then Γ becomes a distance-regular graph and has the same intersection numbers as the q-Johnson graph $J_q(2d+1,d)$, but it is not isomorphic to $J_q(2d+1,d)$ (van Dam and Koolen [154]). This distance-regular graph Γ is called the *twisted q-Johnson graph* (or *twisted Grassmann graph*), and the corresponding P-polynomial scheme is called the *twisted q-Johnson scheme* (or *twisted Grassmann scheme*), which is denoted by ${}^2J_q(2d+1,d)$. The twisted q-Johnson scheme ${}^2J_q(2d+1,d)$ has the same intersection numbers as $J_q(2d+1,d)$ and so it is Q-polynomial.

The automorphism group of ${}^2J_q(2d+1,d)$ is the stabilizer of H in $P\Gamma L(2d+1,q)$ [187]. It is known that ${}^2J_q(2d+1,d)$ is not distance-transitive and the automorphism group does not act transitively on X.

Remark 6.89.

(1) Other than ${}^2J_q(2d+1,d)$, known examples of distance-regular graphs of large diameter which are not distance-transitive are $\mathrm{Quad}_{n-1}(q)$, $\mathrm{Hem}_d(q)$, $\mathrm{Ust}_{[\frac{d}{2}]}(q)$, and

the Doob graphs, which will be introduced below, but their automorphism groups act vertex-transitively. In this sense, ${}^2J_q(2d+1, d)$ is a special example.

(2) A. Munemasa proved that applying the Godsil–Mckay switching [199] to $J_q(2d+1, d)$ yields ${}^2J_q(2d+1, d)$.

Let $\Gamma^{(i)}$ be the graph with vertex set X_i $(1 \le i \le r)$. We define the adjacency relation $\sim$ between $x = (x_1, \dots, x_r)$, $y = (y_1, \dots, y_r)$ in $X = X_1 \times \cdots \times X_r$ as follows: we define $x \sim y$ if there exists i such that $x_j = y_j$ for $j \neq i$, and x_i and y_i are adjacent in $\Gamma^{(i)}$. The graph with vertex set X and the above adjacency relation $\sim$ is called the direct product of $\Gamma^{(i)}$ $(1 \le i \le r)$ and is denoted by $\Gamma^{(1)} \times \cdots \times \Gamma^{(r)}$.

Let Γ be the Shrikhande graph (Chapter 2, Example 2.10). The Shrikhande graph Γ is a distance-regular graph with the same intersection numbers as the Hamming graph $H(2,4)$, but is not isomorphic to $H(2,4)$. Let Γ^m be the direct product of m copies of Γ. Then Γ^m is a distance-regular graph with the same intersection numbers as the Hamming graph $H(2m,4)$ but is not isomorphic to $H(2m,4)$.

Moreover, the direct product $\Gamma^m \times H(n,4)$ of Γ^m and the Hamming graph $H(n,4)$ is a distance-regular graph with the same intersection numbers as $H(n+2m,4)$, but is not isomorphic to $H(n+2m,4)$. We call $\Gamma^m \times H(n,4)$ $(m \ge 1,\ n \ge 0)$ the *Doob graph*, and call the corresponding P-polynomial scheme the *Doob scheme*. The Doob scheme is also a Q-polynomial scheme. The automorphism group of the Doob scheme is vertex-transitive, but is not distance-transitive. Note that $H(1,4)$ is the complete graph K_4 on 4 vertices, and $H(n,4)$ is the direct product K_4^n of n copies of K_4. Since the neighbor of a vertex of the Doob scheme $\Gamma^m \times H(n,4)$ is a union of m hexagons and n triangles, if $(m,n) \neq (m',n')$, then $\Gamma^m \times H(n,4)$ and $\Gamma^{m'} \times H(n',4)$ are not isomorphic.

6.4.3 Towards the classification of P- and Q-polynomial schemes

This section is about the classification of P- and Q-polynomial schemes: how the classification problem of them arose and developed, what the present status is, and how it, in our personal view, could finish.

The starting point was the studies of distance-transitive graphs by D. G. Higman and N. L. Biggs [215, 91, 97]. Distance-transitive graphs are the terminology adopted by Biggs, whereas Higman called them "permutation groups with maximal diameter." These were studies on distance-regular graphs, in view of the fact that they did not make essential use of group actions. In fact, they soon dropped the assumption of group actions from the frameworks in which they developed their subsequent work. In the framework of distance-regular graphs, Biggs formulated the Lloyd theorem for perfect codes in the Hamming scheme [93]. Around the same time, at least by 1971, Ray-Chaudhuri defined tight designs in the Johnson scheme and gave the theorem of Lloyd type for them; they were published much later in 1975 [514, 400]. It was Delsarte who formulated these theorems of Lloyd type systematically in the framework of association schemes (Delsarte theory) [159]. To do so, he introduced the concept of P-polynomial schemes in place of distance-regular graphs and defined Q-polynomial schemes as the dual concept of them: they serve as the underlying spaces for coding theory and design theory, respectively.

Bannai, who was at Ohio State University and gave lectures on Delsarte theory in 1979, came to realize that P-polynomial schemes are the finite analogue of compact 2-point homogeneous spaces and Q-polynomial schemes are that of compact symmetric spaces of rank 1. From this point of view, since compact symmetric spaces are classified by Élie Cartan and since it is known by the theorem of Hsien Chung Wang that compact 2-point homogeneous spaces are symmetric spaces of rank 1, and vice versa [503], it is possible to guess how to collect examples of P-/Q-polynomial schemes as their finite analogue. Thus Bannai made a list of P- and Q-polynomial schemes; he also referred to the spherical functions on homogeneous spaces of finite Chevalley groups by certain maximal parabolic subgroups, which had been in a timely manner calculated by D. Stanton [435]. Based on this list, Bannai proposed the classification of P- and Q-polynomial schemes with the following conjecture:

(1) P- and Q-polynomial schemes with sufficiently large diameter are either in the list or "relatives" of those in the list.
(2) Primitive P-polynomial schemes with sufficiently large diameter are Q-polynomial. Conversely, primitive Q-polynomial schemes with sufficiently large diameter are P-polynomial.

There may be various ways to interpret the meaning of the "relatives": for example, they could be those that were derived from the list by means of (a), (b), (c), (d) in the previous section. In what follows, we understand them in a narrow sense: the "relatives" are those that have the same parameters as in the list.

Since then four new series of P- and Q-polynomial schemes have been found:
affine schemes $\mathrm{Quad}_n(r)$ in Section 6.4.1 (Egawa [177]),
Hemmeter schemes $\mathrm{Hem}_{d+1}(q)$ in Section 6.4.2 (c) [115],
Ustimenko schemes $\mathrm{Ust}_{[\frac{d+1}{2}]}(q)$ in Section 6.4.2 (c) [491, 263], and
twisted Grassmann schemes ${}^2J_q(2d+1, d)$ in Section 6.4.2 (d) [154].

Conjecture (1) holds for each of the newly found series. No counter-examples have been found to Conjecture (2). For Conjecture (2), we cannot drop the assumption that the association scheme is primitive: there are examples with arbitrarily large diameter of imprimitive P-polynomial schemes that are not Q-polynomial [60, Remark (5) on page 313], and of imprimitive Q-polynomial schemes that are not P-polynomial [354]. At present, it seems hopeless to prove Conjecture (2), if we think of its implication: Conjecture (2) reduces the classification of distance-regular graphs to Conjecture (1). Conjecture (1) was also a dream when it was put forward, but since then there have been a considerable number of breakthroughs that enable us to consider the classification of P- and Q-polynomial schemes as a realistic target. Details will be explained in the following, where the classification means the classification of P- and Q-polynomial schemes *with sufficiently large diameter.*

The classification problem of P- and Q-polynomial schemes is divided into two parts:

(A) to show that P- and Q-polynomial schemes with sufficiently large diameter have the same parameters as in Bannai's list and

(B) to characterize P- and Q-polynomial schemes in Bannai's list with sufficiently large diameter by the parameters.

First, breakthroughs were made by Y. Egawa and D. Leonard, who attended the lectures of Bannai at Ohio State University. Concerning (B), Egawa completed the characterization of the Hamming schemes $H(d, q)$ by the parameters [176]. He also added a new series of P- and Q-polynomial schemes to Bannai's list by constructing the affine scheme $\mathrm{Quad}_n(r)$ [177], which had been conjectured in Bannai's lectures to exist. Concerning (A), Leonard showed that P- and Q-polynomial schemes must have AW-parameters [310, 311] (for more details see [60, Sections 3.8, 3,9 of Chapter 3]).

Then, based on Bannai's lectures (1978–1982) at Ohio State University, the book [60] was published. The theme of the book was the classification of P- and Q-polynomial schemes, though the lectures of Bannai were mainly aimed at Delsarte theory: the part of underlying spaces for Delsarte theory was first published as Part I. *Algebraic Combinatorics*, the title of the book, has been widely accepted since then as the name of the field, in which the classification of P- and Q-polynomial schemes is one of the central problems. All the important things written in [60] about P- and Q-polynomial schemes are contained in this book, except for the theorem below.

Theorem 6.90 ([60] page 358, Theorem 7.11). *For a P- and Q-polynomial scheme which is not an n-gon and has diameter large enough, all the entries of the first eigenmatrix are rational integers.*

In [60], Theorem 6.90 is shown on the condition that the diameter $d \geq 34$. In [170], the condition is improved to be $d \geq 5$.

Let $\mathfrak{X}$ be a P- and Q-polynomial scheme with diameter d. Suppose that $d \geq 5$ and $\mathfrak{X}$ is not an n-gon. Then the eigenvalues θ_i $(0 \leq i \leq d)$ of the intersection matrix of $\mathfrak{X}$ are rational integers by Theorem 6.90. Let T be the Terwilliger algebra of $\mathfrak{X}$. Apply Theorem 6.44 to the principal T-module of $\mathfrak{X}$. Then we find that

$$\beta = \frac{\theta_{i+1} - \theta_i + \theta_{i-1} - \theta_{i-2}}{\theta_i - \theta_{i-1}}, \quad 2 \leq i \leq d-1,$$

is a rational number which is a constant not depending on i. Set

$$\beta = q + q^{-1}.$$

Then q is the parameter that determines the type of AW-parameters. Suppose that $q \neq \pm 1$ and q is a primitive ℓ-th root of unity. Since β is rational, q has degree at most 2 over $\mathbb{Q}$ and so we have $\ell = 3, 4, 6$. If $\ell = 3$, then we have $\beta + 1 = 0$, which contradicts (6.124). If $\ell = 4, 6$, then $d \geq 6$ cannot occur because of the list of eigenvalues θ_i $(0 \leq i \leq d)$ that appears after Remark 6.46. Therefore we conclude that if a P- and Q-polynomial scheme with $d \geq 6$ is not an n-gon, then the parameter q, which determines the type of AW-parameters, cannot be a root of unity in Type I and Type IA. This fact has an important meaning when we treat irreducible representations of the Terwilliger algebras later on.

Then in 1982, Terwilliger joined, though not for long, the Bannai school at Ohio State University, and the classification project of P- and Q-polynomial schemes was accelerated to reach a peak. The major results during this period include: concerning Problem (B), the characterization of the Johnson scheme $J(v, d)$ by the parameters (Terwilliger [460], Neumaier [376]); concerning Problem (A), the classification by Terwilliger of P- and Q-polynomial schemes of Type II [462], of Type IIA [465], of Type IIB [466], and of Type III [463]. Exactly speaking, the classification of Type II was completed later in the 1990s by the auxiliary work of Bussemaker and Neumaier [117], and Metsch [343]. Although the Johnson scheme $J(v, d)$ had been almost characterized by the parameters by Moon [353] and the classification of Type IIC had been completed by Egawa [176], a new method by means of root systems was devised and gave another proof to these results, finishing the classification of Type IIA as well. P- and Q-polynomial schemes of Type IA do not exist (unpublished work of Terwilliger, also refer to [155]). It is during this period that the Ustimenko scheme [491, 263] and the Hemmeter scheme [214, 115] were found. The first stage of the classification could be considered to have come to an end by the beginning of the 1990s with Bannai's moving to Kyushu University in 1989; Terwilliger moved to University of Wisconsin-Madison in 1985. Readers are referred to [61]. Terwilliger himself seemed to believe he could finish the classification within several years, but the remaining case, Type I, was tough and the classification project proposed by [60] was stuck in the framework of Bose–Mesner algebras.

Among other things that happened around 1990, two things are particularly worth mentioning here. One thing is the outflow of Russian mathematicians to the West, caused by Gorbachev's perestroika and glasnost, which were followed by the coup of August 1991 and the collapse of the Soviet Union. In the field of algebraic combinatorics, the Bannai school started active communication with Russian mathematicians such as I. A. Faradzhev, M. H. Klin, A. A. Ivanov, M. E. Muzichuk, and S. V. Shpectrov, and came to know that in the Soviet Union there were studies of cellular rings that could be traced back to Schur; the cellular ring is a concept very close to the association scheme. Readers are referred to [183], which was first published as an appendix of the Russian translation (1987) of [60]; at that time, they had no choice but to publish it as an appendix of a translation. Another thing worth mentioning is the publication of [113]. There are detailed descriptions, particularly from a geometric point of view, of known distance-regular graphs in this book, which are indispensable when we deal with Problem (B). This book was conceived early in the 1980s, based on [136], which was inspired by Bannai's lecture at Oberwalfach in 1980, but was published in 1989.

The second stage of the classification started with the creation of Terwilliger algebras [468, 469, 470], which Terwilliger himself called subconstituent algebras. When an association scheme $\mathfrak{X} = (X, \{R_i\}_{0\leq i\leq d})$ comes from an action of a group G, i. e., when G acts transitively on each R_i, the Bose–Mesner algebra of $\mathfrak{X}$ coincides with the centralizer algebra $\mathrm{Hom}_G(V, V)$ of G, where V is the standard module $V = \bigoplus_{x\in X} \mathbb{C}x$ of $\mathfrak{X}$. On the other hand, the Terwilliger algebra T of $\mathfrak{X}$ contains $\mathrm{Hom}_G(V, V)$ and is contained in the centralizer algebra $\mathrm{Hom}_H(V, V)$ of the stabilizer H in G of the base point x: $\mathrm{Hom}_G(V, V) \subseteq T \subseteq \mathrm{Hom}_H(V, V)$, with $H = G_x$, $T = T(x)$. In most cases, T coincides or almost coincides with $\mathrm{Hom}_H(V, V)$. It is natural to come up with the present definition of the Terwilliger algebra if we seek something like $\mathrm{Hom}_H(V, V)$ in general for an association scheme that does not necessarily come from an action of a group. Nevertheless it would also be natural to ask if there could be another definition for it, even in the case of a P- and Q-polynomial scheme. For example, the Terwilliger algebra T of the bilinear forms scheme $\mathrm{Bil}_{d\times n}(q)$ is properly contained in $\mathrm{Hom}_H(V, V)$ and it may cause a serious problem of how big the gap is between T and $\mathrm{Hom}_H(V, V)$.

The hidden aim of [468, 469, 470] is to introduce the concept of L-systems and apply it to the classification of P- and Q-polynomial schemes: L-systems, a concept in the framework of representation theory, turn the Leonard theorem into a theorem of representations, in view of the fact that L-systems are equivalent to dual systems of orthogonal polynomials and the Leonard theorem is the characterization of Askey–Wilson polynomials with finite support as dual systems of orthogonal polynomials. For that aim, Terwilliger introduced the Terwilliger algebras T and realized L-systems in the standard T-modules of P- and Q-polynomial schemes. In fact, he wrote a paper in 1987 on L-systems and dual systems of orthogonal polynomials, but it was never published. Perhaps, it contained Theorem 6.25 of Section 6.1.4 or something like that. If all the irreducible T-modules were thin (Remark 6.30), which Terwilliger might have expected, the classification would have been completed at this point. For known examples of P- and Q-polynomial schemes, thin irreducible T-modules are described in detail in [470]. All known P- and Q-polynomial schemes are thin except for the Doob scheme [448], affine schemes, and the twisted Grassmann scheme ${}^2J_q(2d + 1, d)$: each of the exceptions has non-thin irreducible T-modules of endpoint 1. The twisted Grassmann scheme ${}^2J_q(2d + 1, d)$ was found after [470], and the Terwilliger algebra of ${}^2J_q(2d + 1, d)$ depends on the base point, so it can be non-thin when we choose a right base point [20].

With the observation of non-thin irreducible T-modules emerging for some examples of P- and Q-polynomial schemes, the classification entered a new phase, in which the most important target was to describe all the irreducible T-modules, including non-thin ones, for all the known examples of P- and Q-polynomial schemes, thus completing the work of [470]. Progress, however, has not yet been made for that target. This is because we could not build up representation theory of non-thin irreducible T-modules until recently. It has now been established in the form of the classification of TD-pairs [256, 257, 258, 260, 253]: the case where q is a root of unity, including $q = \pm 1$, has not yet been published. If the second stage of the classification is some 15 years up to this point, we must admit that it was a period of stagnation, for most of the important tools or concepts for the classification, such as the balanced set condition (Chapter 2, Definition 2.100, Corollary 2.103, and Theorem 2.107) [464, 467, 472], the concepts of TD-pairs and TD-relations [256], not to mention Terwilliger algebras [468, 469, 470], had been obtained substantially in the 1980s by Terwilliger. Essential breakthroughs in this period are as few as the weight space decomposition, which led to the classification of TD-pairs by the series of work that started with [256]. The weight space decomposition was found by Terwilliger in the middle of the 1990s and he himself called it the split decomposition.

Nonetheless we can see steady efforts to push ahead with the classification during this period. As for Problem (A), the Terwilliger school (G. Dickie, B. Curtin, J. Caughman, M. MacLean, M. Lang, etc.) classified P- and Q-polynomial schemes that are almost dual bipartite [170], dual bipartite [171], or almost bipartite [131, 304]. As a byproduct of the classification of dual bipartite ones and almost dual bipartite ones, P- and Q-polynomial schemes with second Q-polynomial orderings were classified [170]; all of them are thin. Bipartite P- and Q-polynomial schemes were also deeply studied [126, 127, 129], though they have not yet been classified completely. Multiplicity formulas of irreducible T-modules [127] were also produced by the Terwilliger school; they are important as pioneering work. On the other hand, Terwilliger started the study of irreducible T-modules with endpoint 1 during this period ([476, 477, 196, 269, 324], see also [325]), but many of the results are in his unpublished lecture notes. For example, the Terwilliger polynomials [190] were born during this period but were hidden in his unpublished lecture notes for long. The detailed study of non-thin irreducible T-modules with endpoint 1 for P- and Q-polynomial schemes with classical parameters [231] was carried out during this period, but it owed the key part to his unpublished lecture notes. Also during this period, irreducible T-modules of the Doob scheme were completely described by Tanabe [448]. This was the first work that determined irreducible T-modules for a non-thin P- and Q-polynomial scheme. Moreover, looking back on it, it is a pioneering piece of work in which representations of the Onsager algebra appear.

As for Problem (B), the affine scheme $\mathrm{Her}_d(r)$ was characterized by the parameters [265, 471]. The previous characterization by Ivanov and Shpectrov [266] assumed some geometric local condition when $r \geq 3$, and the local condition was dropped in [471], making good use of the balanced set condition. The q-Johnson scheme $J_q(v, d)$ $(3 \leq d \leq \frac{v}{2})$ was characterized by the parameters by Metsch [341] except for the following:[1]

$$q \geq 4, \quad v = 2d, \quad 2d+1,$$
$$q = 3, \quad v = 2d, \quad 2d+1, \quad 2d+2,$$
$$q = 2, \quad v = 2d, \quad 2d+1, \quad 2d+2, \quad 2d+3.$$

Note that when $d = 2$, it is impossible to characterize $J_q(v, 2)$ by the parameters, and that when $v = 2d+1$, the twisted Grassmann scheme ${}^2J_q(2d+1, d)$ has the same parameters as $J_q(2d+1, d)$. The bilinear forms scheme $\mathrm{Bil}_{d\times n}(q)$ $(d \leq n)$ was characterized by the parameters also by Metsch [342] except for the following:[2]

$$q \geq 3, \quad n = d, \quad d+1, \quad d+2,$$
$$q = 2, \quad n = d, \quad d+1, \quad d+2, \quad d+3.$$

Note that assuming some geometric local condition, T. Huang characterized $\mathrm{Bil}_{d\times n}(q)$ by the parameters in the middle of the 1980s [242]. Assuming some geometric local condition, Munemasa and Shpectrov [358, 357] characterized the affine schemes $\mathrm{Alt}_{2d}(r)$, $\mathrm{Alt}_{2d+1}(r)$ by the parameters.

The present status of the classification of P- and Q-polynomial schemes is as follows, assuming diameters are sufficiently large. As for Problem (A), the classification has been completed except for Type I. For Type I, dual bipartite, dual almost bipartite, or almost bipartite ones are classified. Bipartite ones are almost classified: they are classified up to the 1-parameter family that includes the dual polar scheme $D_d(q)$ and so the Hemmeter scheme $\mathrm{Hem}_d(q)$ as well. As a result, P- and Q-polynomial schemes with second Q-polynomial orderings or second P-polynomial orderings[3] are classified. As for Problem (B), only the dual polar schemes (hence $\mathrm{Hem}_d(q)$, $\frac{1}{2}D_d(q)$, $\mathrm{Ust}_{[\frac{d}{2}]}(q)$ as well) and the affine schemes are left open to the characterization by the parameters. Among them, the dual polar scheme ${}^2A_{2d-1}(r)$ [264] and the affine scheme $\mathrm{Her}_d(r)$ are already characterized by their parameters. For other dual polar schemes, some geometric conditions were assumed for their characterization by the parameters ([113, 155, 120]). For the q-Johnson scheme $J_q(v, d)$ and the bilinear forms scheme $\mathrm{Bil}_{d\times n}(q)$, few but serious cases are left open to the characterization by the parameters (see the results above by Metsch).

The following observations are made about Problem (A) in the list of known P- and Q- polynomial schemes:

---

1 Recently, $J_q(2d, d)$ was characterized by the parameters [191, 193].

2 Recently, $Bil_{d\times d}(2)$ was characterized by the parameters [191, 192].

3 In the Japanese version, it is written that P- and Q-polynomial schemes with second P-polynomial orderings are not yet classified. It is pointed out by Jack Koolen that they can be classified in principle, because almost bipartite P- and Q-polynomial schemes are classified and bipartite ones are classified up to the 1-parameter family.

(i) In the core part of the list, the AW-parameter s^* vanishes: $s^* = 0$. It is only in the part derived from the core that $s^* \neq 0$ occurs.
(ii) In the core part of the list, known P- and Q-polynomial schemes of Type I have the AW-parameter q that satisfies $q = p^f > 1$ for some prime p, except for the affine scheme $\mathrm{Her}_d(r)$.

Note that $q < -1$ holds for $\mathrm{Her}_d(r)$ and for this reason $\mathrm{Her}_d(r)$ becomes kite-free, which is the key property used for the characterization of $\mathrm{Her}_d(r)$ by the parameters. Further, the following holds for known P- and Q- polynomial schemes with the unique exception of the twisted Grassmann scheme ${}^2J(2d+1, d)$:
(iii) Non-thin irreducible T-modules appear only when $s = s^* = 0$.

It is difficult to explain in the framework of Bose–Mesner algebras why known P- and Q- polynomial schemes have such properties. We hope that it will be explained in the framework of Terwilliger algebras.

A TD-pair arises from an irreducible T-module and it is expressed as a tensor product of L-pairs [260]. An irreducible T-module with endpoint 0 is nothing but the principal T-module, which exists uniquely, and it contains exactly the same information as the Bose–Mesner algebra. On the other hand, an irreducible T-module with endpoint 1 produces a TD-pair which has a very special structure, i. e., it is expressed as a tensor product of L-pairs in a very special shape. This structure of the tensor product determines the eigenvalues of the first subconstituent $\Gamma_1(x_0)$; in many cases, it determines geometric structure of $\Gamma_1(x_0)$ further. We expect that the same will hold for irreducible T-modules with endpoint 2. Suppose so. Then the structure of $\Gamma_1(x_0) \cup \Gamma_2(x_0)$, the first and second subconstituents, will be extended to the whole space by the balanced set condition and the TD-relations, determining the global structure. This global structure depends on the base point x_0 and must be in harmony when we vary x_0, perhaps leading to properties (i), (ii), (iii) above. Also if we get multiplicity formulas of irreducible T-modules in the standard module in general, we will be able to eliminate many possibilities of the parameters. Anyway, it is a top priority to find the decomposition of the standard T-module into a sum of irreducible T-modules for all the known P- and Q-polynomial schemes, thus completing the list of [470], which only treats the thin cases and does not calculate the multiplicities of irreducible T-modules in the standard T-module. For the decomposition of the standard T-module, readers are referred to [478].

Problem (A) mainly concerns non-existence and Problem (B) uniqueness, but the same methods are often used when we treat them; if there is progress in one of them, the method can be applied to the other. The characterization by the parameters (Problem (B)) proceeds as follows in general: first (1) to determine the geometric structure of the i-th subconstituent $\Gamma_i(x_0)$ $(i = 1, 2)$, and then (2) to extend the local structure of (1) to the whole space. In the case of [266] which characterizes $\mathrm{Her}_d(r)$ by the parameters, the local structure is assumed for $i = 1$ for the first step (1), while for the second step (2), the farthest subconstituent $\Gamma_d(x_0)$ of ${}^2A_{2d-1}(r)$, which is isomorphic to $\mathrm{Her}_d(r)$, is extended to the whole space, which turns out to have the same parameters as ${}^2A_{2d-1}(r)$, thus reducing the Problem to the characterization of ${}^2A_{2d-1}(r)$ by the parameters that has been already solved in [264]. Although T-modules of ${}^2A_{2d-1}(r)$ are hidden behind the argument of [266], the second step can be understood more easily in the framework of T-modules. If some method of T-modules is developed for Problem (A), it can be used for Problem (B), and vice versa.

Finally, we give an outline of the classification of TD-pairs in the case where q is not a root of unity. We hope it will be of help to those who read [256, 257, 258, 259, 260], in which full details are found. Section 6.2 covers [256]. Expository articles are [252, 253, 254].

Let $A, A^* \in \mathrm{End}(V)$ be a TD-pair of Type I. Denote by $\mathcal{A} = \langle z, z^* \rangle$ the associative $\mathbb{C}$-algebra generated by z, z^* subject to the following relations (TD):

$$\text{(TD)} \quad \begin{cases} [z,[z,[z,z^*]_q]_{q^{-1}}] = -\varepsilon(q^2-q^{-2})^2[z,z^*], \\ [z^*,[z^*,[z^*,z]_q]_{q^{-1}}] = -\varepsilon^*(q^2-q^{-2})^2[z^*,z], \end{cases}$$

where $\varepsilon, \varepsilon^* \in \{1, 0\}$, $[X,Y] = XY - YX$, $[X,Y]_q = qXY - q^{-1}YX$. $\mathcal{A}$ becomes an infinite-dimensional algebra. $\mathcal{A}$ is called the TD-algebra of Type I. Affine transformations $\lambda A + \mu$, $\lambda^* A^* + \mu^*$ of A, A^* are also a TD-pair, where $\lambda \neq 0$, $\lambda^* \neq 0$, $\lambda, \lambda^*, \mu, \mu^* \in \mathbb{C}$, and for appropriate $\lambda, \lambda^*, \mu, \mu^*$, $\lambda A + \mu$ and $\lambda^* A^* + \mu^*$ satisfy the relations (TD). So from the beginning we may assume A, A^* satisfy the relations (TD). Then

$$\rho : \mathcal{A} \longrightarrow \mathrm{End}(V), \quad (z, z^* \longmapsto A, A^*) \tag{6.302}$$

becomes a finite-dimensional irreducible representation of $\mathcal{A}$ (Theorem 6.44). Here we have chosen q^2 in place of the main AW-parameter q: $\beta = q^2 + q^{-2}$. We may assume that q is not a root of unity (Theorem 6.90). Conversely, if a finite dimensional irreducible representation ρ of $\mathcal{A}$ satisfies

$$A = \rho(z), \quad A^* = \rho(z^*) \quad \text{are diagonalizable,} \tag{6.303}$$

then A, A^* become a TD-pair (Theorem 6.47). Therefore, the classification of TD-pairs of Type I is reduced to that of finite-dimensional irreducible representations of $\mathcal{A}$ that satisfy (6.303). In the case of $\varepsilon = \varepsilon^* = 0$, $\mathcal{A}$ is isomorphic to the positive part of the quantum affine algebra $U_q(\hat{\mathfrak{sl}}_2)$ and (TD) is the q-Serre relations. In the case of $\varepsilon = \varepsilon^* = 1$, $\mathcal{A}$ is called the *q-Onsager algebra.*[4] To summarize, our goal is to make a q-analogue of the representation theory of the Onsager algebra for $\mathcal{A}$.

We choose a TD-system $(A, A^*; \{V_i\}_{i=0}^d, \{V_i^*\}_{i=0}^d)$ for the TD-pair A, A^* and fix it. With orderings of the eigenspaces fixed, we regard the TD-pair A, A^* as a TD-system. Let

4 In the Japanese version, it is written that finite-dimensional irreducible representations of the q-Onsager algebra always satisfy (6.303), but it is a false statement.

$V = \bigoplus_{i=0}^{d} U_i$ be the weight space decomposition. First, we construct a basis of $\mathcal{A}$ as a linear space, which we call the zigzag basis. As a corollary to the construction of the zigzag basis, we can prove the shape conjecture: $\dim U_0 = 1$ holds in the weight space decomposition $V = \bigoplus_{i=0}^{d} U_i$ of the TD-system A, A^*. Let R, L be the raising and lowering maps, respectively, associated with the TD-system A, A^* and the weight space decomposition $V = \bigoplus_{i=0}^{d} U_i$. Define the sequence $\{\sigma_i\}_{i=0}^{d}$ by

$$L^i R^i|_{U_0} = \sigma_i \in \mathbb{C} \quad (i = 0, 1, \ldots, d). \tag{6.304}$$

We remark that $\sigma_d \neq 0$ holds. Then the isomorphism class of the TD-system A, A^* is determined by the eigenvalues $\{\theta_i\}_{i=0}^{d}$, $\{\theta_i^*\}_{i=0}^{d}$ of A, A^* and the sequence $\{\sigma_i\}_{i=0}^{d}$. This is obtained as a corollary to the shape conjecture (see also [255]).

The next stage is to embed $\mathcal{A}$ into $U_q(L(\mathfrak{sl}_2))$ and construct finite-dimensional irreducible representations of $\mathcal{A}$ through the embedding, where $U_q(L(\mathfrak{sl}_2))$ is the q-analogue of the universal enveloping algebra $U(L(\mathfrak{sl}_2))$ of the loop algebra. The embedding is constructed in two steps: to construct an embedding ι_t of $\mathcal{A}$ into the augmented TD-algebra $\mathcal{T}$ and to construct an embedding φ_s of $\mathcal{T}$ into $U_q(L(\mathfrak{sl}_2))$, where $t, s \in \mathbb{C}^\times$ are parameters for the embeddings, and finally to embed $\mathcal{A}$ into $U_q(L(\mathfrak{sl}_2))$ by $\varphi_s \circ \iota_t$. For a finite-dimensional irreducible representation ρ of $U_q(L(\mathfrak{sl}_2))$, we ask when $\rho \circ \varphi_s \circ \iota_t$ is irreducible as a representation of $\mathcal{A}$, how we can distinguish the isomorphism class of $\rho \circ \varphi_s \circ \iota_t$ as a representation of $\mathcal{A}$ if it is irreducible, and whether all the irreducible representations of $\mathcal{A}$ can be obtained in this manner as $\rho \circ \varphi_s \circ \iota_t$. These questions are answered in terms of the Drinfeld polynomial $P_V(\lambda)$:

$$P_V(\lambda) = \sum_{i=0}^{d} (-1)^i \frac{\sigma_i}{(q - q^{-1})^{2i}([i]!)^2} \prod_{j=i+1}^{d} (\lambda - \varepsilon s^{-2} q^{2(d-j)} - \varepsilon^* s^2 q^{-2(d-j)}), \tag{6.305}$$

where $[i] = \frac{q^i - q^{-i}}{q - q^{-1}}$, $[i]! = [i][i-1] \cdots [1]$, s is the parameter for the embedding φ_s, and σ_i is the eigenvalue of $L^i R^i$ on U_0 defined by (6.304).

The irreducibility of $\rho \circ \varphi_s \circ \iota_t$ as a representation of $\mathcal{A}$ is judged by s, t, and the structure of zeros of $P_V(\lambda)$. The isomorphism class of $\rho \circ \varphi_s \circ \iota_t$ as a representation of $\mathcal{A}$ is determined by $P_V(\lambda)$ and the eigenvalues $\{\theta_i\}_{i=0}^{d}$, $\{\theta_i^*\}_{i=0}^{d}$ of A, A^*, if it is irreducible, where

$$\begin{aligned} \theta_i &= stq^{2i-d} + \varepsilon s^{-1} t^{-1} q^{-2i+d}, \quad 0 \leq i \leq d, \\ \theta_i^* &= \varepsilon^* st^{-1} q^{2i-d} + s^{-1} tq^{-2i+d}, \quad 0 \leq i \leq d. \end{aligned} \tag{6.306}$$

An important and interesting fact is that $P_V(\lambda)$ is independent of the eigenvalues of A, A^*, though the parameter s appears in the definition (6.305) of $P_V(\lambda)$: in fact, s disappears from the right-hand side of (6.305) eventually. Finally, it is shown that all the finite-dimensional irreducible representations of $\mathcal{A}$ are obtained in this manner as $\rho \circ \varphi_s \circ \iota_t$. This means that a TD-pair is a tensor product of L-pairs, because a finite-dimensional irreducible module of $U_q(L(\mathfrak{sl}_2))$ is isomorphic to a tensor product of evaluation modules.

The classification of TD-pairs is summarized in [253], including the cases of Type I with q a root of unity, of Type II and of Type III, but the details of the last three cases are not yet published. In the case of Type II or of Type III, the classification of TD-pairs is also reduced to the classification of finite-dimensional irreducible representations of the TD-algebra that satisfy (6.303). The construction of representations, however, is through the limits of those of Type I, since we have not yet found algebras such as $U_q(L(\mathfrak{sl}_2))$ into which the TD-algebra of Type II or Type III can be embedded; the Onsager algebra, which is the most degenerated TD-algebra of Type II, can be embedded in $U(L(\mathfrak{sl}_2))$.

The q-Onsager algebra, which was found during the course of the classification of P- and Q-polynomial schemes, finds applications nowadays in the field of statistical mechanics such as the *XXZ* model [78, 79, 80, 81, 82, 83, 84]. If the TD-algebra of Type II or of Type III can be embedded into an algebra that has a coproduct, the algebra will be bigger[5] than $U(L(\mathfrak{sl}_2))$ or $U(L(\mathfrak{sl}_2))$ at $q = -1$ [487, 488, 489] and will have further applications in various fields of research. It was in [153] that distance-regular graphs were first related to a quantum group, notably in the study of spin models, which belongs to statistical mechanics again.

L-pairs correspond to AW-polynomials with finite support, so most probably TD-pairs, which are tensor products of L-pairs, will be corresponding to some sort of orthogonal polynomials, but so far we cannot pin them down. It is important to find them – while writing this part, there seems to have been progress: some kind of multivariable AW-polynomials have been found to be related to TD-pairs [85]. It is often observed that a P- and Q-polynomial scheme contains another P- and Q-polynomial scheme as a subset. Structure of the inclusion hierarchy seems to have important meaning in the classification of P- and Q-polynomial schemes. See [473, 450, 452, 453, 454], [114]. The attempt to define L-pairs of rank 2 is also challenging and important [244].

The classification problem of P- and Q-polynomial schemes has been producing new concepts, deepening the interactions with other areas of research, each time it meets difficulty. This means that P- and Q-polynomial schemes promise fertile land of mathematics. It is not easy to foresee how and when the classification of P- and Q-polynomial schemes finishes, but our mathematics will certainly be deepened each time we meet another difficulty in this problem.

5 $U(L(\mathfrak{sl}_2))$ may be big enough for the TD-algebra of Type II to be embedded. We found some strong evidence for it after the publication of the Japanese version.

Bibliography

[1] K. Abdukhalikov, E. Bannai and S. Suda, *Association schemes related to universally optimal configurations, Kerdock codes and extremal Euclidean line-sets*, J. Comb. Theory, Ser. A 116 (2009), 434–448.

[2] N. N. Andreev, *An extremal property of the icosahedron*, East J. Approx. 2 (1996), 459–462.

[3] N. N. Andreev, *Location of points on a sphere with minimal energy (Russian)*, Tr. Mat. Inst. Steklova 219 (1997), 27–31. Translation in Proc. Steklov Inst. Math. 219 (1997), 20–24.

[4] J. Arias de Reyna, *A generalized mean-value theorem*, Monatshefte Math. 106 (1988), no. 2, 95–97.

[5] K. Anstreicher, *The thirteen spheres: a new proof*, Discrete Comput. Geom. 31 (2004), 613–625.

[6] E. F. Assmus and H. F. Mattson, Jr., *New 5-designs*, J. Comb. Theory 6 (1969), 122–151.

[7] M. Atiyah and P. Sutcliffe, *Polyhedra in physics, chemistry and geometry*, Milan J. Math. 71 (2003), 33–58.

[8] C. Bachoc, *On harmonic weight enumerators of binary codes*, Des. Codes Cryptogr. 18 (1999), 11–28.

[9] C. Bachoc, E. Bannai and R. Coulangeon, *Codes and designs in Grassmannian spaces*, Discrete Math. 277 (2004), no. 1–3, 15–28.

[10] C. Bachoc, R. Coulangeon and G. Nebe, *Designs in Grassmannian spaces and lattices*, J. Algebraic Comb. 16 (2002), no. 1, 5–19.

[11] C. Bachoc and M. Ehler, *Tight p-fusion frames*, Appl. Comput. Harmon. Anal. 35 (2013), no. 1, 1–15.

[12] C. Bachoc and G. Nebe, *Extremal lattices of minimum* 8 *related to the Mathieu group* $M22$, J. Reine Angew. Math. 494 (1998), 155–171.

[13] C. Bachoc and F. Vallentin, *New upper bounds for kissing numbers from semidefinite programming*, J. Am. Math. Soc. 21 (2008), no. 3, 909–924.

[14] B. Bajnok, *Construction of designs on the 2-sphere*, Eur. J. Comb. 12 (1991), 377–382.

[15] B. Bajnok, *Construction of spherical t-designs*, Geom. Dedic., 43 (1992), 167–179.

[16] B. Bajnok, *On Euclidean designs*, Adv. Geom. 6 (2006), no. 3, 423–438.

[17] B. Bajnok, *Orbits of the hyperoctahedral group as Euclidean designs*, J. Algebraic Comb. 25 (2007), no. 4, 375–397.

[18] B. Ballinger, G. Blekherman, H. Cohn, N. Giansiracusa, E. Kelly and A. Schürmann, *Experimental study of energy-minimizing point configurations on spheres*, Exp. Math. 18 (2009), no. 3, 257–283.

[19] S. Bang, A. Dubickas, J. H. Koolen and V. Moulton, *There are only finitely many distance-regular graphs of fixed valency greater than two*, Adv. Math. 269 (2015), 1–55.

[20] S. Bang, T. Fujisaki and J. H. Koolen, *The spectra of the local graphs of the twisted Grassmann graphs*, Eur. J. Comb. 30 (2009), no. 3, 638–654.

[21] Ei. Bannai, *On perfect codes in the Hamming schemes $H(n,q)$ with q arbitary*, J. Comb. Theory, Ser. A 23 (1977), 52–67.

[22] Ei. Bannai, *On tight designs*, Q. J. Math. 28 (1977), 433–448.

[23] Ei. Bannai, *Codes in bipartite distance-regular graphs*, J. Lond. Math. Soc. 16 (1977), 197–202.

[24] Ei. Bannai, *Algebraic combinatorics*, Sūgaku 31 (1979), 126–143 (Japanese).

[25] Ei. Bannai, *Spherical t-designs which are orbits of finite groups*, J. Math. Soc. Jpn. 36 (1984), 341–354.

[26] Ei. Bannai, *Spherical t-designs and group representations*, Contemp. Math. 34 (1984), 85–107.

https://doi.org/10.1515/9783110630251-007

[27] Ei. Bannai, *Character tables of commutative association schemes*, in: *Finite Geometries Buildings, and Related Topics* (ed. by W. M. Kantor et al.), Oxford Sci. Publ., Oxford Univ. Press, New York, 1990, pp. 105–128.

[28] Ei. Bannai, *Subschemes of some association schemes*, J. Algebra, 144 (1991), 167–188.

[29] Ei. Bannai, *Algebraic combinatorics: recent topics on association schemes*, Sūgaku Expo. 7 (1994), 181–207.

[30] Ei. Bannai, *Combinatorics as pure mathematics*, Sūgaku Expo. 26 (2013), 225–246.

[31] Ei. Bannai and Et. Bannai, *How many P-polynomial structures can an association scheme have?*, Eur. J. Comb. 1 (1980), 289–298.

[32] Ei. Bannai and Et. Bannai, *Algebraic Combinatorics on Spheres*, Springer-Verlag Tokyo, 1999 (Japanese).

[33] Ei. Bannai and Et. Bannai, *Tight Gaussian 4-designs*, J. Algebraic Comb. 22 (2005), 39–63.

[34] Ei. Bannai and Et. Bannai, *A note on the spherical embeddings of strongly regular graphs*, Eur. J. Comb. 26 (2005), 1177–1179.

[35] Ei. Bannai and Et. Bannai, *On Euclidean tight 4-designs*, J. Math. Soc. Jpn. 58 (2006), 775–804.

[36] Ei. Bannai and Et. Bannai, *On antipodal spherical t-designs of degree s with* $t \geq 2s - 3$ (Special volume honoring the 75th birthday of Prof. D. K. Ray-Chaudhuri), J. Comb. Inf. Syst. Sci. 34 (2009), no. 1–4, 33–50.

[37] Ei. Bannai and Et. Bannai, *Spherical designs and Euclidean designs*, in: *Recent Developments in Algebra and Related Areas (Beijing, 2007)*, Adv. Lect. Math. 8, Higher Education Press/International Press, Beijing/Boston, 2009, pp. 1–37.

[38] Ei. Bannai and Et. Bannai, *A survey on spherical designs and algebraic combinatorics on spheres*, Eur. J. Comb. 30 (2009), 1392–1425.

[39] Ei. Bannai and Et. Bannai, *Euclidean designs and coherent configurations*, Contemp. Math. 531, (2010), 59–93.

[40] Ei. Bannai and Et. Bannai, *Tight 9-designs on two concentric spheres*, J. Math. Soc. Jpn. 64 (2011), no. 4, 1359–1376.

[41] Ei. Bannai and Et. Bannai, *Remarks on the concepts of t-designs*, in: *Proceedings of AGC2010*, J. Appl. Math. Comput. 40 (2012), no. 1–2, 195–207.

[42] Ei. Bannai and Et. Bannai, *Tight t-designs on two concentric spheres*, Mosc. J. Comb. Number Theory 4 (2014), 52–77.

[43] Ei. Bannai, Et. Bannai and H. Bannai, *Uniqueness of certain association schemes*, Eur. J. Comb. 29 (2008), no. 6, 1379–1395.

[44] Ei. Bannai, Et. Bannai and H. Bannai, *On the existence of tight relative 2-designs on binary Hamming association schemes*, Discrete Math. 314 (2014), 17–37.

[45] Ei. Bannai, Et. Bannai, M. Hirao and M. Sawa, *Cubature formulas in numerical analysis and Euclidean tight designs*, Eur. J. Comb. 31 (2010), 423–441.

[46] Ei. Bannai, Et. Bannai, M. Hirao and M. Sawa, *On the existence of minimum cubature formulas for Gaussian measure on* $\mathbb{R}^2$ *of degree t supported by* $[\frac{t}{4}] + 1$ *circles*, J. Algebraic Comb. 35 (2012), 109–119.

[47] Ei. Bannai, Et. Bannai and F. Jaeger, *On spin models, modular invariance, and duality*, J. Algebraic Comb. 6 (1997), 203–228.

[48] Ei. Bannai, Et. Bannai and J. Shigezumi, *A new Euclidean tight 6-design*, Ann. Comb. 16 (2012), no. 4, 651–659.

[49] Ei. Bannai, Et. Bannai, S. Suda and H. Tanaka, *On relative t-designs in polynomial association schemes*, Electron. J. Comb. 22 (2015), P4.47.

[50] Ei. Bannai, Et. Bannai and D. Suprijanto, *On the strong non-rigidity of certain tight Euclidean designs*, Eur. J. Comb. 28 (2007), 1662–1680.

[51] EI. BANNAI, ET. BANNAI, H. TANAKA and Y. ZHU, *Design theory from the viewpoint of algebraic combinatorics*, Graphs Comb. 33 (2017), 1–41.

[52] EI. BANNAI, ET. BANNAI, H. TANAKA and Y. ZHU, *Tight relative t-designs on two shells in hypercubes, and Hahn and Hermite polynomials*, arXiv:2006.02054v1.

[53] EI. BANNAI, ET. BANNAI and Y. ZHU, *Relative t-designs in binary Hamming association scheme* $H(n,2)$, Des. Codes Cryptogr. 84 (2017), 23–53.

[54] EI. BANNAI, A. BLOKHUIS, P. DELSARTE and J. J. SEIDEL, *An addition theorem for hyperbolic spaces*, J. Comb. Theory, Ser. A 36 (1984), 332–341.

[55] EI. BANNAI and R. M. DAMERELL, *Tight spherical designs I*, J. Math. Soc. Jpn. 31 (1979), no. 1, 199–207.

[56] EI. BANNAI and R. M. DAMERELL, *Tight spherical designs II*, J. Lond. Math. Soc. (2) 21 (1980), no. 1, 13–30.

[57] EI. BANNAI, S. HAO and S. Y. SONG, *Character tables of the association schemes of finite orthogonal groups acting on the nonisotropic points*, J. Comb. Theory, Ser. A 54 (1990), 164–200.

[58] EI. BANNAI and S. G. HOGGAR, *Tight t-designs and squarefree integers*, Eur. J. Comb. 10 (1989), 113–135.

[59] EI. BANNAI and T. ITO, *On finite Moore graphs*, J. Fac. Sci., Univ. Tokyo, Sect. 1A, Math. 20 (1973), 191–208.

[60] EI. BANNAI and T. ITO, *Algebraic Combinatorics I: Association Schemes*, Benjamin/Cummings, Menlo Park, California, 1984.

[61] EI. BANNAI and T. ITO, *Current research on algebraic combinatorics: supplements to our book, algebraic combinatorics. I*, Graphs Comb. 2 (1986), 287–308.

[62] EI. BANNAI and T. ITO, *On distance-regular graphs with fixed valency. III*, J. Algebra 107 (1987), no. 1, 43–52.

[63] EI. BANNAI and T. ITO, *On distance-regular graphs with fixed valency*, Graphs Comb. 3 (1987), no. 2, 95–109.

[64] EI. BANNAI and T. ITO, *On distance-regular graphs with fixed valency. II*, Graphs Comb. 4 (1988), no. 3, 219–228.

[65] EI. BANNAI and T. ITO, *On distance-regular graphs with fixed valency. IV.*, Eur. J. Comb. 10 (1989), no. 2, 137–148.

[66] EI. BANNAI, M. KOIKE, M. SHINOHARA and M. TAGAMI, *Spherical designs attached to extremal lattices and the modulo p property of Fourier coefficients of extremal modular forms*, Mosc. Math. J. 6 (2006), 225–264.

[67] EI. BANNAI and T. MIEZAKI, *Toy models for D. H. Lehmer's conjecture*, J. Math. Soc. Jpn. 62 (2010), no. 3, 687–705.

[68] EI. BANNAI and T. MIEZAKI, *On a property of 2-dimensional integral Euclidean lattices*, J. Number Theory 132 (2012), 371–378.

[69] EI. BANNAI and T. MIEZAKI, *Toy models for D. H. Lehmer's conjecture, II*, in: *Proceedings of Florida Conference, Quadratic and Higher Degree Forms* (ed. by K. ALLADI ET AL.), Dev. Math. 31, 2013, pp. 1–27.

[70] EI. BANNAI, T. MIEZAKI and V. A. YUDIN, *An elementary approach to toy models for D. H. Lehmer's conjecture*, Izv. Ross. Akad. Nauk Ser. Mat. 75 (2011), no. 6, 3–16 (Russian). Translation in Izv. Math.

[71] EI. BANNAI, A. MUNEMASA and B. VENKOV, *The nonexistence of certain tight spherical designs*, Algebra Anal. 16 (2004), 1–23. MR 2090848 (2005e:05022). Translation in St. Petersburg Math. J. 16 (2005), 609–625.

[72] EI. BANNAI and N. J. A. SLOANE, *Uniqueness of certain spherical codes*, Can. J. Math. 33 (1981), no. 2, 437–449.

[73] Ei. Bannai, S. Y. Song, S. Hao and H. Z. Wei, *Character tables of certain association schemes coming from finite unitary and symplectic groups*, J. Algebra 144 (1991), 189–213.

[74] Et. Bannai, *On antipodal Euclidean tight* $(2e + 1)$*-designs*, J. Algebraic Comb. 24 (2006), 391–414.

[75] Et. Bannai, *New examples of Euclidean tight 4-designs*, Eur. J. Comb. 30 (2009), 655–667.

[76] A. Barg and O Musin, *Bounds on sets with few distances*, J. Comb. Theory, Ser. A 118 (2011), 1465–1474.

[77] A. Barg and W.-H. Yu, *New bounds for spherical two-distance sets*, Exp. Math. 22 (2013), no. 2, 187–194.

[78] P. Baseilhac, *An integrable structure related with tridiagonal algebras*, Nucl. Phys. B 705 (2005), no. 3, 605–619.

[79] P. Baseilhac, *Deformed Dolan-Grady relations in quantum integrable models*, Nucl. Phys. B 709 (2005), no. 3, 491–521.

[80] P. Baseilhac, *The q-deformed analogue of the Onsager algebra: beyond the Bethe ansatz approach*, Nucl. Phys. B 754 (2006), no. 3, 309–328.

[81] P. Baseilhac, *A family of tridiagonal pairs and related symmetric functions*, J. Phys. A 39 (2006), 11773.

[82] P. Baseilhac and K. Koizumi, *A new (in)finite-dimensional algebra for quantum integrable models*, Nucl. Phys. B 720 (2005), no. 3, 325–347.

[83] P. Baseilhac and K. Koizumi, *A deformed analogue of Onsager's symmetry in the XXZ open spin chain*, J. Stat. Mech. Theory Exp. (2005), no. 10, P10005.

[84] P. Baseilhac and K. Koizumi, *Exact spectrum of the XXZ open spin chain from the q-Onsager algebra representation theory*, J. Stat. Mech. (2007), P09006.

[85] P. Baseilhac and X. Martin, *A bispectral q-hypergeometric basis for a class of quantum integrable models*, J. Math. Phys. 59 (2018), 011704.

[86] S. N. Bernstein, *On quadrature formulas with positive coefficients (Russian)*, Izv. Akad. Nauk. SSSR, Ser. Mat. 1 (1937), no. 4, 479–503 (Reprinted in Collected Works 2, Izdat. Akad. Nauk SSSR, Moskow (1954), 205–2227). See also announcements in C. R. Acad. Sci. Paris 204 (1937), 1294–1296; 1526–1526.

[87] S. N. Bernstein, *Sur les formulas de quadrature de cotes et Tchebycheff*, Dokl. Akad. Nauk SSSR 14 (1937), 323–327.

[88] M. R. Best, *A contribution to the nonexistence of perfect codes*, (academisch proefschrift), Amsterdam, Math-Centrum, (1982). MR 0726757 (Mathematisch Centrum, Amsterdam (1983)).

[89] K. Betsumiya, M. Harada and A. Munemasa, *A complete classification of doubly even self-dual codes of length* 40, Electron. J. Comb. 19 (2012), P18. http://www.math.is.tohoku.ac.jp/~munemasa/research/codes/sd2.htm.

[90] A. Bezdek and W. Kuperberg, *Packing Euclidian space with congruent cylinders and with congruent ellipsoids*, in: *Applied Geometry and Discrete Mathematics: The Victor Klee Festschrift* (ed. by P. Gritzmann, B. Sturmfels), DIMACS Ser. Discrete Math. Theoret. Comput. Sci. 4, 1991, pp. 71–80.

[91] N. L. Biggs, *Intersection matrices for linear graphs*, in: *Combinatorial Mathematics and Its Applications (Proc. Oxford, 7–10 July 1969)* (ed. by D. J. A. Welsh), Acad. Press, London, 1971, pp. 15–23.

[92] N. L. Biggs, *Finite Groups of Automorphisms*, London Math. Soc. Lecture Note Ser. 6, Cambridge University Press, 1971.

[93] N. L. Biggs, *Perfect codes in graphs*, J. Comb. Theory, Ser. B 15 (1973), 289–296.

[94] N. L. Biggs, *Algebraic Graph Theory*, Cambridge Univ. Press, Cambridge, 1974.

[95] N. L. Biggs, *Automorphic graphs and the Krein condition*, Geom. Dedic. 5 (1976), 117–127.

[96] N. L. Biggs, A. G. Boshier and J. Shawe-Taylor, *Cubic distance-regular graphs*, J. Lond. Math. Soc. (2) 33 (1986), no. 3, 385–394.

[97] N. L. Biggs and D. H. Smith, *On trivalent graphs*, Bull. Lond. Math. Soc. 3 (1971), 155–158.

[98] A. V. Bondarenko and M. S. Viazovska, *New asymptotic estimates for spherical designs*, J. Approx. Theory 152 (2008), 101–106.

[99] A. V. Bondarenko and M. S. Viazovska, *Spherical designs via Brouwer fixed point theorem*, Discrete Math. 24 (2010), 207–217.

[100] A. Bondarenko, D. Radchenko and M. Viazovska, *Optimal asymptotic bounds for spherical designs*, Ann. Math. (2) 178 (2013), no. 2, 443–452.

[101] A. Bondarenko, D. Radchenko and M. Viazovska, *Well separated spherical designs*, Constr. Approx. 41 (2015), no. 1, 93–112.

[102] R. C. Bose, S. S. Shrikhande and E. T. Parker, *Further results on the construction of mutually orthogonal Latin squares and the falsity of Euler's conjecture*, Can. J. Math. 12 (1960), 189–203.

[103] K. Böröczky, *The Newton-Gregory problem revisited*, in: *Discrete Geometry* (ed. by A. Bezdek), Dekker, 2003, pp. 103–110.

[104] S. Boumova, P. Boyvalenkov, H. Kulina and M. Stoyanova, *Polynomial techniques for investigation of spherical designs*, Des. Codes Cryptogr. 51 (2009), 289–300.

[105] P. Boyvalenkov, *Extremal polynomials for obtaining bounds for spherical codes and designs*, Discrete Comput. Geom. 14 (1995), 167–188.

[106] P. Boyvalenkov, D. Danev and P. Kazakov, *Indices of spherical codes*, DIMACS Ser. Discret. Math. Comput. Sci. 58 (2001), 47–57.

[107] P. Boyvalenkov, D. Danev and I. Landgev, *On maximal spherical codes. II*, J. Comb. Des. 7 (5) (1999), 316–326.

[108] P. Boyvalenkov, D. Danev and S. Nikova, *Nonexistence of certain spherical designs of odd strengths and cardinalities*, Discrete Comput. Geom. 21 (1) (1999), 143–156.

[109] P. Boyvalenkov and S. Nikova, *On lower bounds on the size of designs in compact symmetric spaces of rank* 1, Arch. Math. (Basel) 68 (1) (1997), 81–88.

[110] M. Braun, A. Kerber and R. Laue, *Systematic construction of q-analogs of t-(v,k,λ)-designs*, Des. Codes Cryptogr. 34 (2005), 55–70.

[111] A. Bremner, *A Diophantine equation arising from tight 4-designs*, Osaka J. Math. 16 (1979), 353–356.

[112] M. Broué and M. Enguehard, *Polynômes des poids de certains codes et fonctions thêta de certains réseaux*, Ann. Sci. Éc. Norm. Supér. (4) 5 (1972), 157–181 (French).

[113] A. E. Brouwer, A. M. Cohen and A. Neumaier, *Distance-Regular Graphs*, Ergebnisse der Mathematik und ihrer Grenzgebiete (3) (Results in Mathematics and Related Areas (3)) 18, Springer-Verlag, Berlin, 1989.

[114] A. E. Brouwer, C. D. Godsil, J. H. Koolen and W. J. Martin, *Width and dual width of subsets in polynomial association schemes*, J. Comb. Theory, Ser. A 102 (2003), 255–271.

[115] A. Brouwer and J. Hemmeter, *A new family of distance-regular graphs and the $\{0,1,2\}$-cliques in dual polar graphs*, Eur. J. Comb. 13 (1992), no. 2, 71–79.

[116] H. Bruck, H. J. Ryser, *The nonexistence of certain finite projective planes*, Can. J. Math. 1 (1949), 88–93.

[117] F. C. Bussemaker and A. Neumaier, *Exceptional graphs with smallest eigenvalue -2 and related problems*, Math. Comput. 59 (1992), 583–608.

[118] A. R. Calderbank, R. H. Hardin, E. M. Rains, P. W. Shor and N. J. A. Sloane, *A group-theoretic framework for the construction of packings in Grassmannian spaces*, J. Algebraic Comb. 9 (1999), 129–140.

[119] P. J. CAMERON and J. H. VAN LINT, *Graph Theory, Coding Theory and Block Designs*, London Math. Soc. Lecture Note Ser. 19, Cambridge Univ. Press, 1975.

[120] P. J. CAMERON, *Dual polar spaces*, Geom. Dedic. 12 (1982), 75–85.

[121] P. J. CAMERON, *There are only finitely many finite distance-transitive graphs of given valency greater than two*, Combinatorica 2 (1982), no. 1, 9–13.

[122] P. J. CAMERON, C. E. PRAEGER, J. SAXL and G. M. SEITZ, *On the Sims conjecture and distance transitive graphs*, Bull. Lond. Math. Soc. 15 (1983), no. 5, 499–506.

[123] P. J. CAMERON and C. E. PRAEGER, *Block-transitive t-designs, II. Large t*, in: *Finite Geometry and Combinatorics (Deinze, 1992)*, London Math. Soc. Lecture Note Ser. 191, Cambridge Univ. Press, Cambridge, 1993, pp. 103–119.

[124] E. CARTAN, *Oeuvres complètes, Prtie I.Groupes de Lie*, Gauthier-Villars, Paris, 1952 (French).

[125] B. CASSELMAN, *The difficulties of kissing in three dimensions*, Not. Am. Math. Soc. 51 (2004), 884–885.

[126] J. S. CAUGHMAN, IV, *Spectra of bipartite P- and Q-polynomial association schemes*, Graphs Comb. 14 (1998), no. 4, 321–343.

[127] J. S. CAUGHMAN, IV, *The Terwilliger algebras of bipartite P- and Q-polynomial schemes*, Discrete Math. 196 (1999), no. 1–3, 65–95.

[128] J. S. CAUGHMAN, IV, *The last subconstituent of a bipartite Q-polynomial distance-regular graph*, Eur. J. Comb. 24 (2003), 459–470.

[129] J. S. CAUGHMAN, IV, *Bipartite Q-polynomial distance-regular graphs*, Graphs Comb. 20 (2004), no. 1, 47–57.

[130] J. S. CAUGHMAN, IV, E. J. HART and J. MA, *The last subconstituent of the Hemmeter graph*, Discrete Math. 308 (2008), 3056–3060.

[131] J. S. CAUGHMAN, IV, M. S. MACLEAN and P. M. TERWILLIGER, *The Terwilliger algebra of an almost-bipartite P- and Q-polynomial association scheme*, Discrete Math. 292 (2005), no. 1–3, 17–44.

[132] D. R. CERZO and H. SUZUKI, *Non-existence of imprimitive Q-polynomial schemes of exceptional type with $d = 4$*, Eur. J. Comb. 30 (2009), 674–681.

[133] X. CHEN, A. FROMMER and B. LANG, *Computational existence proofs for spherical t-designs*, Numer. Math. 117 (2011), 289–305.

[134] X. CHEN and R. S. WOMERSLEY, *Existence of solutions to systems of underdetermined equations and spherical designs*, SIAM J. Numer. Anal. 44 (2006), no. 6, 2326–2341.

[135] L. CHIHARA, *On the zeros of the Askey-Wilson polynomials, with applications to coding theory*, SIAM J. Math. Anal. 18 (1987), 191–207.

[136] A. M. COHEN, *A synopsis of known distance-regular graphs with large diameters*, Mathematisch Center Report ZW 168/81, Amsterdam, 1981.

[137] H. COHN, *Order and disorder in energy minimization*, in: *Proceedings of the International Congress of Mathematicians. IV*, Hindustan Book Agency, New Delhi, 2010, pp. 2416–2443.

[138] H. COHN, J. H. CONWAY, N. ELKIES and A. KUMAR, *The D_4 root system is not universally optimal*, Exp. Math. 16 (2006), 313–320.

[139] H. COHN and N. ELKIES, *New upper bounds on sphere packings I*, Ann. Math. 157 (2003), 689–714.

[140] H. COHN and A. KUMAR, *Universally optimal distribution of points on spheres*, J. Am. Math. Soc. 20 (2007), 99–148.

[141] H. COHN and A. KUMAR, *Algorithmic design of self-assembling structures*, Proc. Natl. Acad. Sci. USA 106 (2009), 9570–9575.

[142] H. COHN and A. KUMAR, *Optimality and uniqueness of the Leech lattice among lattices*, Ann. Math. 170 (2009), 1003–1050.

[143] H. Cohn, A. Kumar, S. D. Miller, D. Radchenko and M. Viazovska, *The sphere packing problem in dimension* 24, Ann. Math. 185 (2017), 1017–1033.

[144] H. Cohn and J. Woo, *Three-point bounds for energy minimization*, J. Am. Math. Soc. 25 (2012), 929–958.

[145] J. H. Conway and V. Pless, *On the enumeration of self-dual codes*, J. Comb. Theory, Ser. A 28 (1980), 26–53.

[146] J. H. Conway, V. Pless and N. J. A. Sloane, *The binary self-dual codes of length up to 32: a revised enumeration*, J. Comb. Theory, Ser. A 60 (1992), 183–195.

[147] J. H. Conway and N. J. A. Sloane, *Sphere Packings, Lattices and Groups*, Grundlehren Der Mathematischen Wissenschaften, Springer-Verlag, 1998.

[148] J. H. Conway and D. A. Smith, *On Quaternions and Octonions: Their Geometry, Arithmetic, and Symmetry*, A K Peters, Ltd., Natick, MA, 2003.

[149] R. Cools and H. J. Schmid, *A new lower bound for the number of nodes in cubature formulae of degree* $4n + 1$ *for some circularly symmetric integrals*, in: *Numerical Integration, IV (Oberwolfach, 1992)*, Internat. Ser. Numer. Math. 112, Birkhäuser, Basel, 1993, pp. 57–66.

[150] R. Coulangeon and A. Schürmann, *Energy minimization, periodic sets and spherical designs*, Int. Math. Res. Not. (2012), no. 4, 829–848.

[151] J. Crann, R. Pereira, D. W. Kribs, *Spherical designs and anticoherent spin states*, J. Phys. A, Math. Theor. 43 (2010), 255307.

[152] Z. Cui, J. Xia and Z. Xiang, *Rational designs*, Adv. Math. 352 (2019), 541–571.

[153] B. Curtin and K. Nomura, *Distance-regular graphs related to the quantum enveloping algebra of* $sl(2)$, J. Algebraic Comb. 12 (2000), no. 1, 25–36.

[154] E. R. Van Dam and J. H. Koolen, *A new family of distance-regular graphs with unbounded diameter*, Invent. Math. 162 (2005), 189–193.

[155] E. R. van Dam, J. H. Koolen and H. Tanaka, *Distance-regular graphs*, Electron. J. Comb. DS22 (2016), 1–156.

[156] M. Damerell, *On Moore graphs*, Proc. Camb. Philol. Soc. 74 (1973), 227–236.

[157] L. Danzer, *Finite point-sets on* S^2 *with minimum distance as large as possible*, Discrete Math. 60 (1986), 3–66.

[158] P. Delsarte, *Four fundamental parameters of codes and their combinatorial significance*, Inf. Control 23 (1973), 407–438.

[159] P. Delsarte, *An algebraic approach to the association schemes of the coding theory*, Thesis, Universite Catholique de Louvain, Philips Res. Repts Suppl. 10, 1973.

[160] P. Delsarte, *Association schemes and t-designs in regular semilattices*, J. Comb. Theory, Ser. A 20 (1976), 230–243.

[161] P. Delsarte, *Pairs of vectors in the space of an association scheme*, Philips Res. Rep. 32 (1977), 373–411.

[162] P. Delsarte, J. M. Goethals and J. J. Seidel, *Bounds for Systems of Lines and Jacobi Polynomials*, Philips Research Reports 30, 1975.

[163] P. Delsarte, J. M. Goethals and J. J. Seidel, *Spherical codes and designs*, Geom. Dedic. 6 (1977), no. 3, 363–388.

[164] P. Delsarte and J. J. Seidel, *Fisher type inequalities for Euclidean t-designs*, Linear Algebra Appl. 114–115 (1989), 213–230.

[165] P. de la Harpe and C. Pache, *Cubature formulas, geometrical designs, reproducing kernels, and Markov operators*, in: *Infinite Groups: Geometric, Combinatorial and Dynamical Aspects*, Progr. Math. 248, Birkhauser, Basel, 2005, pp. 219–267.

[166] P. de la Harpe, C. Pache and B. Venkov, *Construction of spherical cubature formulas using lattices*, Algebra Anal. 18(1) (2006), 162–186. Translation in St. Petersburg Math. J. 18 (2007), no. 1, 119–139.

[167] P. Dembowski, *Finite Geometries*, Ergebnisse der Mathematik und ihrer Grenzgebiete 44, Springer-Verlag, Berlin-New York, 1968.

[168] M. Detour-Sikirić, A. Schürmann and F. Vallentin, Personal communication.

[169] M. Deza and P. Frankl, *Bounds on the maximum number of vectors with given scalar products*, Proc. Am. Math. Soc. 95 (1985), 323–329.

[170] G. A. Dickie, *Q-polynomial structures for association schemes and distance-regular graphs*, Ph.D. Thesis, University of Wisconsin-Mdison, 1995.

[171] G. A. Dickie and P. M. Terwilliger, *Dual bipartite Q-polynomial distance-regular graphs*, Eur. J. Comb. 17 (1996), no. 7, 613–623.

[172] G. A. Dickie and P. M. Terwilliger, *A note on thin P-polynomial and dual-thin Q-polynomial symmetric association schemes*, J. Algebraic Comb. 7 (1998), no. 1, 5–15.

[173] P. Dukes and J. Short-Gershman, *Nonexistence results for tight block designs*, J. Algebraic Comb. 38 (2013), 103–119.

[174] C. F. Dunkl, *Orthogonal functions on some permutation groups*, in: *Relations Between Combinatorics and Other Parts of Mathematics (Proc. Sympos. Pure Math., Ohio State Univ., Columbus, Ohio, 1978)*, Proc. Sympos. Pure Math., XXXIV, Amer. Math. Soc., Providence, R.I., 1979, pp. 129–147.

[175] W. Ebeling, *Lattices and Codes*, Vieweg Wiesbaden, 1994.

[176] Y. Egawa, *Characterization of $H(n, q)$ by the parameters*, J. Comb. Theory, Ser. A 31 (1981), 108–125.

[177] Y. Egawa, *Association schemes of quadratic forms*, J. Comb. Theory, Ser. A 38 (1985), 1–14.

[178] S. Elliahou, *Enumerative combinatorics and coding theory*, Enseign. Math. (2) 40 (1994), no. 1–2, 171–185.

[179] H. Enomoto, N. Ito, R. Noda, *Tight 4-designs*, Osaka J. Math. 16 (1979), 39–43.

[180] T. Ericson and V. Zinoviev, *Codes on Euclidean Spheres*, North-Holland Mathematical Studies, Elsevier Science & Technol., 2001.

[181] T. Etzion, *On perfect codes in the Johnson scheme*, DIMACS Discret. Math. Theor. Comput. Sci. 56 (2001), 125–130.

[182] I. A. Faradjev, A. A. Ivanov, A. V. Ivanov, *Distance-transitive graphs of valency 5, 6 and 7*, Eur. J. Comb. 7 (1986), no. 4, 303–319.

[183] I. A. Faradžev, N. H. Klin and A. A. Ivanov, *Galois correspondence between permutation groups and cellular rings (association schemes)*, Graphs Comb. 6 (1990), 303–332.

[184] A. Fazeli, S. Lovett and A. Vardy, *Nontrivial t-designs over finite fields exist for all t*, J. Comb. Theory, Ser. A 127 (2014), 14–160.

[185] W. Feit and G. Higman, *The non-existence of certain generalized polygons*, J. Algebra 1 (1964), 114–131.

[186] R. Feng, Y. Wang, C. Ma and J. Ma, *Eigenvalues of association schemes of quadratic forms*, Discrete Math. 308 (2008), no. 14, 3023–3047.

[187] T. Fujisaki, J. Koolen M. Tagami, *Some properties of the twisted Grassmann graphs*, Innov. Incid. Geom. 3 (2006), 81–87.

[188] P. Gaborit, *Construction of new extremal unimodular lattices*, Eur. J. Comb. 25 (2004), no. 4, 549–564.

[189] G. Gasper and M. Rahman, *Basic Hypergeometric Series. With a Foreword by Richard Askey*, Encyclopedia Math. Appl. 35, Cambridge University Press, Cambridge, 1990.

[190] A. L. Gavrilyuk and J. H. Koolen, *The Terwilliger polynomial of a Q-polynomial distance-regular graph and its application to the pseudo-partition graphs*, Linear Algebra Appl. 466 (2015), 117–140.

[191] A. L. Gavrilyuk and J. H. Koolen, *On some recent progress in the classification of (P and Q)-polynomial association schemes*, Arab. J. Math. (2019). https://doi.org/10.1007/s40065-019-00273-x.

[192] A. L. Gavrilyuk and J. H. Koolen, *A characterization of the graphs of $(d \times d)$-bilinear forms over $\mathbb{F}_2$*, Combinatorica 39(2) (2019), 289–321.

[193] A. L. Gavrilyuk and J. H. Koolen, *On a characterization of the Grassmann graphs*, arXiv:1806.02652.

[194] D. Gijswijt, A. Schrijver and H. Tanaka, *New upper bounds for nonbinary codes based on the Terwilliger algebra and semidefinite programming*, J. Comb. Theory, Ser. A 113 (2006), 1719–1731.

[195] A. Glazyrin and W.-H. Yu, *Upper bounds for s-distance sets and equiangular lines*, Adv. Math. 330 (2018), 810–833.

[196] J. T. Go and P. M. Terwilliger, *Tight distance-regular graphs and the subconstituent algebra*, Eur. J. Comb. 23 (2002), no. 7, 793–816.

[197] C. D. Godsil, *Algebraic Combinatorics*, Chapman & Hall, 1993.

[198] C. D. Godsil, *Polynomial spaces*, Discrete Math. 73 (1988), 71–88.

[199] C. D. Godsil and B. McKay, *Constructing cospectral graphs*, Aequ. Math. 25 (1982), no. 2–3, 257–268.

[200] C. D. Godsil and G. Royle, *Algebraic Graph Theory*, Grad. Texts in Math. 207, Springer, 2001.

[201] J. M. Goethals and J. J. Seidel, *Cubature formulae, polytopes and spherical designs*, in: *The Geometric Vein: The Coxeter Festshrift* (ed. by C. Davis, B. Grünbaum, F. A. Sherk), Springer-Verlag, 1981, pp. 203–218.

[202] D. H. Gottlieb, *A certain class of incidence matrices*, Proc. Am. Math. Soc. 17 (1966), 1233–1237.

[203] M. Gräf and D. Potts, *On the computation of spherical designs by a new optimization approach based on fast spherical Fourier transforms*, Numer. Math. 119 (2011), 699–724.

[204] J. E. Graver and W. B. Jurkat, *The module structure of integral designs*, J. Comb. Theory, Ser. A 15 (1973), 75–90.

[205] R. L. Griess, Jr., *Rank 72 high minimum norm lattices*, J. Number Theory 130 (2010), no. 7, 1512–1519.

[206] R. Guralnick, P. Tiep, *Decompositions of small tensor powers and Larsen's conjecture*, Represent. Theory 9 (2005), 138–208.

[207] T. C. Hales, *A proof of the Kepler conjecture*, Ann. Math. 162 (2005), 1063–1185.

[208] M. Hall, Jr., *Hadamard matrices of order* 16, J. P. L. Research Summary No. 36-10, 1 (1961), 21–26.

[209] M. Hall, Jr., *Hadamard matrices of order* 20, J. P. L. Technical Report 32–761, 1965.

[210] M. Hall, Jr., *Combinatorial Theory*, Blaisdell Publishing Co. Ginn and Co., Waltham, Mass.-Toronto, Ont.-London, (1967). Second edition. Wiley-Interscience Series in Discrete Mathematics. A Wiley-Interscience Publication. John Wiley & Sons, Inc., New York, 1986.

[211] M. Harada, *An extremal doubly even self-dual code of length* 112, Electron. J. Comb. 15 (2008), N33.

[212] R. H. Hardin and N. J. A. Sloane, *McLaren's improved snub cube and other new spherical designs in three dimensions*, Discrete Comput. Geom. 15 (1996), 429–441.

[213] S. Helgason, *Differential Geometry and Symmetric Spaces*, Academic Press, New York, 1962.

[214] J. Hemmeter, *Distance-regular graphs and halved graphs*, Eur. J. Comb. 7 (1986), 119–129.

[215] D. G. Higman, *Intersection matrices for finite permutation groups*, J. Algebra 6 (1967), 22–42.

[216] D. G. Higman, *Coherent configurations. I*, Rend. Semin. Mat. Univ. Padova 44 (1970), 1–25.

[217] D. G. Higman, *Coherent configurations. I. Ordinary representation theory*, Geom. Dedic. 4 (1975), 1–32.

[218] D. G. HIGMAN, *Coherent configurations. II. Weights*, Geom. Dedic. 5 (1976), 413–424.

[219] D. G. HIGMAN, *Coherent algebras*, Linear Algebra Appl. 93 (1987), 209–239.

[220] D. G. HIGMAN, *Strongly regular designs and coherent configurations of type* (3 2 3), Eur. J. Comb. 9 (1988), 411–422.

[221] A. HIRAKI, *An improvement of the Boshier-Nomura bound*, J. Comb. Theory, Ser. B 61 (1994), no. 1, 1–4.

[222] A. HIRAKI, *A constant bound on the number of columns* $(1, k-2, 1)$ *in the intersection array of a distance-regular graph*, Graphs Comb. 12 (1996), no. 1, 23–37.

[223] A. HIRAKI, *Distance-regular subgraphs in a distance-regular graph. VI*, J. Comb. 19 (1998), no. 8, 953–965.

[224] A. HIRAKI, *Strongly closed subgraphs in a regular thick near polygon*, J. Comb. 20 (1999), no. 8, 789–796.

[225] A. HIRAKI, *Retracing argument for distance-regular graphs*, J. Comb. Theory, Ser. B 79 (2000), no. 2, 211–220.

[226] A. HIRAKI, *A distance-regular graph with strongly closed subgraphs*, J. Algebraic Comb. 14 (2001), no. 2, 127–131.

[227] A. HIRAKI, *A characterization of the doubled Grassmann graphs, the doubled odd graphs, and the odd graphs by strongly closed subgraphs*, Eur. J. Comb. 24 (2003), no. 2, 161–171.

[228] A. HIRAKI, *A characterization of the Hamming graph by strongly closed subgraphs*, Eur. J. Comb. 29 (2008), no. 7, 1603–1616.

[229] M. HIRAO and M. SAWA, *On minimal cubature formulae of odd degrees for circularly symmetric integrals*, Adv. Geom. 12 (2012), no. 3, 483–500.

[230] M. HIRAO, M. SAWA and Y. ZHOU, *Some remarks on Euclidean tight designs*, J. Comb. Theory, Ser. A 118 (2011), no. 2, 634–640.

[231] S. HOBART and T. ITO, *The structure of nonthin irreducible T-modules of endpoint 1: ladder bases and classical parameters*, J. Algebraic Comb. 7 (1998), no. 1, 53–75.

[232] A. J. HOFFMAN and R. R. SINGLETON, *On Moore graphs with diameters 2 and 3*, IBM J. Res. Dev. 4 (1960), 497–504.

[233] S. G. HOGGAR, *t-Designs in projective spaces*, Eur. J. Comb. 3 (1982), 233–254.

[234] S. G. HOGGAR, *Tight 4 and 5-designs in projective spaces*, Graphs Comb. 5 (1989), no 1, 87–94.

[235] S. G. HOGGAR, *t-Designs in Delsarte spaces*, in: *Coding Theory and Design Theory, Part II. IMA Volumes in Mathematics and Its Applications* 21 (ed. by RAY-CHAUDHURI), Springer-Verlag, 1990, pp. 144–165.

[236] G. HÖHN, *Conformal designs based on vertex operator algebras*, Adv. Math. 217 (5) (2008), 2301–2335.

[237] Y. HONG, *On the nonexistence of unknown perfect 6- and 8-codes in Hamming schemes $H(n, q)$ with q arbitrary*, Osaka J. Math. 21 (1984), no. 3, 687–700.

[238] Y. HONG, *On the nonexistence of nontrivial perfect e-codes and tight 2e-designs in Hamming schemes $H(n, q)$ with $e \geq 3$ and $q \geq 3$*, Graphs Comb. 2 (1986), no. 2, 145–164.

[239] N. HORIGUCHI, T. MIEZAKI and NAKASORA *On the support designs of extremal binary doubly even self-dual codes*, Des. Codes Cryptogr. 72 (2014), no. 3, 529–537. *Erratum to: On the support designs of extremal binary doubly even self-dual codes*, Des. Codes Cryptogr. 73 (2014), 1027–1028.

[240] W.-Y. HSIANG, *Least Action Principle of Crystal Formation of Dense Packing Type and Kepler's Conjecture*, World Scientific, 2001.

[241] L.-K. HUA and Z.-X. WAN, *Tien Hsing Shun (Classical Groups)*, Shanghai Science Technology Press, Shanghai, 1963.

[242] T. Y. HUANG, *A characterization of the association schemes of bilinear forms*, Eur. J. Comb. 8 (1987), no. 2, 159–173.

[243] D. R. HUGHES and F. C. PIPER, *Projective Planes*, Grad. Texts in Math. 6, Springer-Verlag, New York-Berlin, 1973.

[244] P. ILIEV and P. TERWILLIGER, *The Rahman polynomials and the Lie algebra* $\mathfrak{sl}_3(C)$, Trans. Am. Math. Soc. 364 (2012), no. 8, 4225–4238.

[245] N. ITO, *On tight 4-designs*, Osaka J. Math. 12 (1975), 493–522.

[246] N. ITO, *Corrections and supplements to "On tight 4-designs"*, Osaka J. Math. 15 (1978), 693–697.

[247] N. ITO, *On Hadamard groups*, J. Algebra 168 (1994), 981–987.

[248] N. ITO, *On Hadamard groups III*, Kyushu J. Math. 51 (1997), 369–379.

[249] N. ITO, J. S. LEON and J. Q. LONGYEAR, *Classification of* 3-(24, 12, 5) *designs and 24-dimensional Hadamard matrices*, J. Comb. Theory, Ser. A 27 (1979), 289–306.

[250] T. ITO, *Bipartite distance regular graphs of valency three*, Linear Algebra Appl. 46 (1982), 195–213.

[251] T. ITO, *Designs in a coset geometry: Delsarte theory revisited*, Eur. J. Comb. 25 (2004), no. 2, 229–238.

[252] T. ITO, *TD pairs and q-Onsager algebras*, Sūgaku 65 (2013), no. 1, 69–92 (Japanese).

[253] T. ITO, *The classification of TD-pairs*, RIMS Kokyuroku 1926 (2014), 146–164 (Japanese).

[254] T. ITO, *TD-pairs and the q-Onsager algebra*, Sūgaku Expo. 32 (2019), 205–232.

[255] T. ITO, K. NOMURA and P. TERWILLIGER, *A classification of sharp tridiagonal pairs*, Linear Algebra Appl. 435 (2011), no. 8, 1857–1884.

[256] T. ITO, K. TANABE and P. TERWILLIGER, *Some algebra related to P- and Q-polynomial association schemes*, in: *Codes and Association Schemes (Piscataway, NJ, 1999)*, DIMACS Ser. Discrete Math. Theoret. Comput. Sci. 56 (2001), pp. 167–192.

[257] T. ITO and P. TERWILLIGER, *The shape of a tridiagonal pair*, J. Pure Appl. Algebra 188 (2004), no. 1–3, 145–160.

[258] T. ITO and P. TERWILLIGER, *Two non-nilpotent linear transformations that satisfy the cubic q-Serre relations*, J. Algebra Appl. 6 (2007), no. 3, 477–503.

[259] T. ITO and P. TERWILLIGER, *Tridiagonal pairs of q-Racah type*, J. Algebra 322 (2009), no. 1, 68–93.

[260] T. ITO and P. TERWILLIGER, *The augmented tridiagonal algebra*, Kyushu J. Math. 64 (2010), no. 1, 81–144.

[261] Y. IWAKATA, *Minimal subschemes of the group association schemes of Mathieu groups*, Graphs Comb. 6 (1990), no. 3, 239–244.

[262] A. A. IVANOV, *Bounding the diameter of a distance-regular graph*, Sov. Math. Dokl. 28 (1983), 149–152.

[263] A. A. IVANOV, M. E. MUZICHUK, V. A. USTIMENKO, *On a new family of (P and Q)-polynomial schemes*, Eur. J. Comb. 10 (1989), 337–345.

[264] A. A. IVANOV and S. V. SHPECTOROV, *The association schemes of dual polar space of type* ${}^2A_{2d-1}(p^f)$ *are characterized by their parameters if* $d \geq 3$, Linear Algebra Appl. 114/115 (1989), 133–135.

[265] A. A. IVANOV and S. V. SHPECTOROV, *Characterization of the association schemes of Hermitian forms over* $GF(2^2)$, Geom. Dedic. 30 (1989), no. 1, 23–33.

[266] A. A. IVANOV and S. V. SHPECTOROV, *A characterization of the association schemes of Hermitian forms*, J. Math. Soc. Jpn. 43 (1991), no. 1, 25–48.

[267] G. J. JANUSZ, *Overlap and covering polynomials with applications to designs and self-dual codes*, SIAM J. Discrete Math. 13 (2000), no. 2, 154–178.

[268] P. Jenkins and J. Rouse, *Bounds for coefficients of cusp forms and extremal lattices*, Bull. Lond. Math. Soc. 43 (2011), 927–938.

[269] A. Jurišić, J. Koolen and P. M. Terwilliger, *Tight distance-regular graphs*, J. Algebraic Comb. 12 (2000), no. 2, 163–197.

[270] G. A. Kabatiansky and V. I. Levenshtein, *On bounds for packings on a sphere and in space*, Probl. Inf. Transm. 14 (1978), 3–25.

[271] W. M. Kantor, *k-Homogeneous groups*, Math. Z. 124 (1972), 261–265.

[272] W. M. Kantor, *On incidence matrices of finite projective and affine spaces*, Math. Z. 124, (1972), 315–318.

[273] P. Keevash, *Counting designs*, arXiv:1504.02909v1.

[274] H. Kharaghani and B. Tayfeh-Rezaie, *A Hadamrd matrix of order* 428, J. Comb. Des. 13 (2005), 435–440.

[275] H. Kharaghani and B. Tayfeh-Rezaie, *Hadamard matrices of order* 32, J. Comb. Des. 21 (2013), no. 5, 212–221.

[276] G. B. Khosrovshahi and B. Tayfeh-Rezaie, *A new proof of a classical theorem in design theory*, J. Comb. Theory, Ser. A 93 (2001), no. 2, 391–396,.

[277] H. Kimura, *New Hadamard matrix of order* 24, Graphs Comb. 5 (1989), no. 3, 235–242.

[278] H. Kimura, *Classification of Hadamard matrices of order* 28, Discrete Math. 133 (1994), no. 1–3, 171–180.

[279] O. D. King, *The mass of extremal doubly-even self-dual codes of length* 40, IEEE Trans. Inf. Theory 47 (2001), no. 6, 2558–2560.

[280] O. D. King, *A mass formula for unimodular lattices with no roots*, Math. Comput. 72 (2003), no. 242, 839–863.

[281] M. Kitazume, T. Kondo and I. Miyamoto, *Even lattices and doubly even codes*, J. Math. Soc. Jpn. 43 (1991), no. 1, 67–87.

[282] H. Koch, *On self-dual, doubly even codes of length* 32, J. Comb. Theory, Ser. A 51 (1989), 63–76.

[283] A. V. Kolushov and V. A. Yudin, *On the Korkin-Zolotarev construction*, Diskretn. Mat. 6 (1994), 155–157 (Russian). Translation in Discrete Math. Appl. 4 (1994), 143–146.

[284] A. V. Kolushov and V. A. Yudin, *Extremal dispositions of points on the sphere*, Anal. Math. 23 (1997), 25–34.

[285] T. Koornwinder, *The addition formula for Jacobi polynomials and spherical harmonics*, SIAM J. Appl. Math. 25 (1973), 236–246.

[286] T. H. Koornwinder, *A note on the absolute bound for systems of lines*, Indag. Math. 38 (1976), 152–153.

[287] J. Korevaar and J. L. H. Meyers, *Spherical Faraday cage for the case of equal point charges and Chebyshev-type quadrature on the sphere*, Integral Transforms Spec. Funct. 1 (1993), 105–117.

[288] M. Kotani and T. Sunada, *Spectral geometry of crystal lattices*, Contemp. Math. 338 (2003), 271–305.

[289] N. O. Kottelina and A. B. Pevnyĭ, *Extremal properties of spherical semidesigns*, Algebra Anal., 22 (2010), no. 5, 131–139; St. Petersburg Math. J. 22 (2011), 795–801.

[290] A. B. Kuijlaars, *The minimal number of nodes in Chebyshev type quadrature formulas*, Indag. Math. 4 (1993), 339–362.

[291] G. Kuperberg, *Special moments*, Adv. Appl. Math. 34 (2005), no. 4, 853–870.

[292] G. Kuperberg, L. Shachar and R. Peled, *Probabilistic existence of rigid combinatorial structures*, in: *Proceedings of the 2012 ACM Symposium on Theory of Computing*, ACM, New York, 2012, 1091–1105.

[293] G. Kuperberg, S. Lovett and R. Peled, *Probabilistic existence of regular combinatorial structures*, Geom. Funct. Anal. 27 (2017), 919–972.

[294] H. Kurihara, *(1) On spherical designs obtained from standard realization of association schemes, (2) Character tables of m-flat association schemes*, Master Thesis, Kyushu University, 2009.

[295] H. Kurihara, *Character tables of m-flat association schemes*, Adv. Geom. 11 (2011), no. 2, 293–301.

[296] H. Kurihara, *Character tables of association schemes based on attenuated spaces*, Ann. Comb. 17 (2013), 525–541.

[297] H. Kurihara and H. Nozaki, *A characterization of Q-polynomial association schemes*, J. Comb. Theory, Ser. A 119 (2012), no. 1, 57–62.

[298] H. Kurihara and H. Nozaki, *A spectral equivalent condition of the P-polynomial property for association schemes*, Electron. J. Comb. 21 (2014), no. 3, Paper 3.1.

[299] Hannu Laakso, *Nonexistence of nontrivial perfect codes in the case* $q = p_1^a p_2^b p_3^c$, Ph.D. Thesis, University of Turku, Finland, 1979.

[300] C. W. H. Lam, *The search for a finite projective plane of order* 10, Am. Math. Mon. 98 (1991), no. 4, 305–318.

[301] C. W. H. Lam, G. Kolesova and L. H. Thiel, *A Computer Search for a Finite Projective Plane of Order* 9, Discrete Math. 92 (1991), no 1–3, 187–195.

[302] C. W. H. Lam, L. H. Thiel and S. Swiercz, *The nonexistence of finite projective planes of order* 10, Can. J. Math. 41 (1989), no. 6, 1117–1123.

[303] E. S. Lander, *Symmetric Designs: An Algebraic Approach*, London Math. Soc. Lecture Note Ser. 74, Cambridge University Press, Cambridge, 1983.

[304] M. S. Lang and P. M. Terwilliger, *Almost-bipartite distance-regular graphs with the Q-polynomial property*, Eur. J. Comb. 28 (2007), no. 1, 258–265.

[305] D. G. Larman, C. A. Rogers and J. J. Seidel, *On two-distance sets in Euclidean space*, Bull. Lond. Math. Soc. 9 (1977), 261–267.

[306] R. Laue, S. S. Magliveras and A. Wassermann, *New large sets of t-designs*, J. Comb. Des. 9 (2001), no. 1, 40–59.

[307] J. Leech, *The problem of the thirteen spheres*, Math. Gaz. 41 (1956), 22–23.

[308] J. Leech, *Equilibrium of sets of particles on a sphere*, Math. Gaz. 41 (1957), 81–90.

[309] D. H. Lehmer, *The vanishing of Ramanujan's function* $\tau(n)$, Duke Math. J. 14 (1947), 429–433.

[310] D. A. Leonard, *Orthogonal polynomials, duality and association schemes*, SIAM J. Math. Anal. 13(4) (1982), 656–663.

[311] D. A. Leonard, *Parameters of association schemes that are both P- and Q-polynomial*, J. Comb. Theory, Ser. A 36 (1984), no. 3, 355–363.

[312] V. I. Levenshtein, *On bounds for packings in n-dimensional Euclidean space*, Dokl. Akad. Nauk SSR 245 (1979), 1299–1303; Sov. Math. Dokl. 20 (1979), 417–421.

[313] V. I. Levenshtein, *Designs as maximum codes in polynomial metric spaces. Interactions between algebra and combinatorics*, Acta Appl. Math. 29 (12) (1992), 1–82.

[314] V. I. Levenshtein, *On designs in compact metric spaces and a universal bound on their size*, in: *Discrete Metric Spaces (Villeurbanne, 1996)*, Discrete Math. 192 (13) (1998), 252–271.

[315] V. I. Levenshtein, *Universal bounds for codes and designs*, in: *Handbook of Coding Theory, vol. I, II*, North-Holland, Amsterdam, 1998, pp. 499–648.

[316] Z. Li, Ei. Bannai and Et. Bannai, *Tight relative 2- and 4-designs on binary Hamming association schemes*, Graphs Comb. 30 (2014), 203–227.

[317] J. H. van Lint, *Coding Theory*, Lecture Notes in Math. 201, Springer-Verlag, 1971.

[318] J. van Lint, *Recent Results on Perfect Codes and Related Topics in Combinatorics I*, Math. Centre Tracts 55, Amsterdam (1974), 158–178.

[319] J. H. van Lint and R. M. Wilson, *A Course in Combinatorics*, Cambridge University Press, Cambridge, 1992. (Second edition in 2001).

[320] D. Livingstone and A. Wagner, *Transitivity of finite permutation groups on unordered sets*, Math. Z. 90 (1965), 393–403.

[321] Y. I. Lyubich and L. N. Vaserstein, *Isometric embeddings between classical Banach spaces, cubature formulas, and spherical designs*, Geom. Dedic. 47 (1993), 327–362.

[322] Y. I. Lyubich, *Lower bounds for projective designs, cubature formulas and related isometric embeddings*, Eur. J. Comb. 30 (2009), no. 4, 841–852.

[323] J. Ma and K. Wang, *Nonexistence of exceptional 5-class association schemes with two Q-polynomial structures*, Linear Algebra Appl. 440 (2014), 278–285.

[324] M. S. MacLean and P. Terwilliger, *Taut distance-regular graphs and the subconstituent algebra*, Discrete Math. 306 (2006), no. 15, 1694–1721.

[325] M. S. MacLean and P. Terwilliger, *The subconstituent algebra of a bipartite distance-regular graph; thin modules with endpoint two*, Discrete Math. 308 (2008), no. 7, 1230–1259.

[326] F. J. MacWiliams and N. J. A Sloane, *The Theory of Error-Correcting Codes*, North-Holland Mathematical Library 16, North-Holland Publishing Co., Amsterdam-New York-Oxford, 1977.

[327] H. Maehara, *Isoperimetric theorem for spherical polygons and the problem of 13 spheres*, Ryukyu Math. J. 14 (2001), 41–57.

[328] H. Maehara, *The problem of thirteen spheres—a proof for undergraduates*, Eur. J. Comb. 28 (2007), 1770–1778.

[329] H. Maehara, *On the number of concyclic points in planar lattices*, Bull. Center Educ. Res. Dev. Tokai Univ., 5 (2009), 3–16 (Japanese).

[330] S. S. Magliveras, *Large sets of t-designs from groups*, Math. Slovaca 59 (2009), no 1, 1–20.

[331] A. A. Makhnev, *On the nonexistence of strongly regular graphs with the parameters* (486, 165, 36, 66), Ukr. Mat. ž. 54 (2002), 941–949 (Russian). Translation in Ukr. Math. J. 54 (2002), 1137–1146.

[332] C. L. Mallows, A. M. Odlyzko and N. J. A. Sloane, *Upper bounds for modular forms, lattices, and codes*, J. Algebra 36 (1975), 68–76.

[333] Y. I. Manin, *What is the maximum number of points on a curve over* F_2? J. Fac. Sci., Univ. Tokyo, Sect. 1A, Math. 28 (1981), no. 3, 715–720.

[334] Y. I. Manin, *A computability challenge: asymptotic bounds for error-correcting codes*, in: *Computation, Physics and Beyond*, Lecture Notes in Comput. Sci. 7160, Springer, Heidelberg, 2012, pp. 174–182.

[335] W. J. Martin, M. Muzychuk and J. Williford, *Imprimitive cometric association schemes: constructions and analysis*, J. Algebraic Comb. 25 (2007), 399–415.

[336] W. J. Martin and H. Tanaka, *Commutative association schemes*, Eur. J. Comb. 30 (2009), no. 6, 1497–1525.

[337] W. J. Martin and J. Williford, *There are finitely many Q-polynomial association schemes with given first multiplicity at least three*, Eur. J. Comb. 30 (2009), 698–704.

[338] A. Matsuo, *Norton's trace formulae for the Griess algebra of a vertex operator algebra with larger symmetry*, Commun. Math. Phys. 224 (2001), 565–591.

[339] R. J. McEliece, E. R. Rodemich, H. Rumsey Jr. and L. R. Welch, *New upper bounds on the rate of a code via the Delsarte-MacWilliams inequalities*, IEEE Trans. Inf. Theory 23 (1977), 157–166.

[340] J. McKay, *A setting for the Leech lattice*, in: *Finite Groups '72, Proceedings of the the Gainesville Conference on Finite Groups*, North-Holland Mathematics Studies 7, 1973, pp. 117–118.

[341] K. Metsch, *A characterization of Grassmann graphs*, Eur. J. Comb. 16 (1995), 639–644.

[342] K. Metsch, *On a characterization of bilinear forms graphs*, Eur. J. Comb. 20 (1999), 293–306.

[343] K. Metsch, *On the characterization of the folded halved cubes by their intersection arrays*, Des. Codes Cryptogr. 29 (2003), 215–225.

[344] B. Meyer, *Extreme lattices and vexillar designs*, J. Algebra 322 (2009), 4368–4381.

[345] H. N. Mhaskar, F. J. Narcowich and J. D. Ward, *Spherical Marcinkiewicz-Zygmund inequalities and positive quadrature*, Math. Comput. 70 (2001), 1113–1130.

[346] T. Miezaki, *On spherical designs constructed from finite subgroups of the orthogonal group of degree 4*, in: *Proceedings, The 27th Algebraic Combinatorics Symposium*, 2010 (Japanese).

[347] T. Miezaki, *Conformal designs and D. H. Lehmer's conjecture*, J. Algebra 374 (2013), 59–65.

[348] T. Miezaki and M. Tagami, *On Euclidean designs and the potential energy*, Electron. J. Comb. 19 (2012), no. 1, Paper 2.

[349] T. Miezaki and H. Nakasora, *An upper bound of the value of t of the support t-designs of extremal binary doubly even self-dual codes*, Des. Codes Cryptogr. 79 (2016), 37–46.

[350] Y. Miura, *Codes snd designs in complex grassmannian spaces*, Master Thesis, Kyushu University, 2004.

[351] T. Molien, *Über die Invarianten der linear Substitutionsgruppe*, Sitz.ber. K. Preuss. Akad. Wiss. (1897), 1152–1156.

[352] H. M. Möller, *Lower bounds for the number of nodes in cubature formulae*, in: *Numerische Integration (Tagung, Math. Forschungsinst., Oberwolfach, 1978)*, Internat. Ser. Numer. Math. 45, Birkhäuser, Basel-Boston, Mass., 1979, pp. 221–230.

[353] A. Moon, *The graphs $G(n, k)$ of the Johnson schemes are unique for $n \geq 20$*, J. Comb. Theory, Ser. B 37 (1984), 173–188.

[354] G. E. Moorhouse and J. Williford, *Double covers of symplectic dual polar graphs*, Discrete Math. 339 (2016), no. 2, 571–588.

[355] A. Munemasa, *An analogue of t-designs in the association schemes of alternating bilinear forms*, Graphs Comb. 2 (1986), no. 3, 259–267.

[356] A. Munemasa, *The Geometry of Orthogonal Groups over Finite Fields*, Lecture Note in Mathematics 3, Sophia University, Tokyo, Japan, 1996.

[357] A. Munemasa, D. V. Pasechnik and S. V. Shpectorov, *A local characterization of the graphs of alternating forms and the graphs of quadratic forms over $GF(2)$*, in: *Finite Geometry and Combinatorics* (ed. by F. De Clerck et al.), Cambridge University Press, 1993, pp. 303–317.

[358] A. Munemasa and S. V. Shpectorov, *A local characterization of the graph of alternating forms*, in: *Finite Geometry and Combinatorics* (ed. by F. De Clerck et al.), Cambridge University Press, 1993, pp. 289–302.

[359] O. R. Musin, *The kissing problem in three dimensions*, Discrete Comput. Geom. 35 (2006), 375–384.

[360] O. R. Musin, *The kissing number in four dimensions*, Ann. Math. (2) 168 (2008), 1–32.

[361] O. R. Musin, *Spherical two-distance sets*, J. Comb. Theory, Ser. A 116 (2009), 988–995.

[362] O. R. Musin and H. Nozaki, *Bounds on three- and higher-distance sets*, Eur. J. Comb. 32 (2011), no. 8, 1182–1190.

[363] O. Musin and A. Tarasov, *The strong thirteen spheres problem*, Discrete Comput. Geom. 48 (2012), no. 1, 128–141.

[364] O. Musin and A. Tarasov, *The Tammes problem for $N = 14$*, Exp. Math. 24 (2015), no. 4, 460–468.

[365] M. E. Muzichuk, *V-lings of the Permutation Groups with Invariant Metric*, Ph.D. Thesis, Kiev State University, 1987.

[366] H. Nagao, *Groups and Designs*, Iwanami shoten, 1974 (Japanese).

[367] G. Nebe, *Some cyclo-quaternionic lattices*, J. Algebra 199 (1998), 472–498.

[368] G. Nebe, *An even unimodular 72-dimensional lattice of minimum* 8, J. Reine Angew. Math. 673 (2012), 237–247.

[369] G. NEBE, *Boris Venkov's theory of lattices and spherical designs*, in *Diophantine Methods, Lattices, and Arithmetic Theory of Quadratic Forms*, Contemp. Math. 587, 2013, pp. 1–19.

[370] G. NEBE, *A fourth extremal even unimodular lattice of dimension* 48, Discrete Math. 331 (2014), 133–136.

[371] http://www.math.rwth-aachen.de/~Gabriele.Nebe/papers/aut5.pdf.

[372] http://www.math.rwth-aachen.de/~Gabriele.Nebe/papers/nebeOW.pdf.

[373] G. NEBE, E. M. RAINS and N. J. A. SLOANE, *Self-Dual Codes and Invariant Theory*, Springer-Verlag, New York, 2006.

[374] G. NEBE and B. VENKOV, *On tight spherical designs*, Algebra Anal. 24 (2012), no. 3, 163–171. Translation in St. Petersburg Math. J. 24 (2013), no. 3, 485–491.

[375] A. NEUMAIER, *Combinatorial configurations in terms of distances* (mimeographed notes), Memorandum 81-09, Eindvohen Univ. of Technology, 1981.

[376] A. NEUMAIER, *Characterization of a class of distance regular graphs*, J. Reine Angew. Math. 357 (1985), 182–192.

[377] A. NEUMAIER and J. J. SEIDEL, *Discrete measures for spherical designs, eutactic stars and lattices*, Proc. K. Ned. Akad. Wet., Ser. A, Indag. Math. 50 (1988), 321–334.

[378] S. NIKOVA, V. NIKOV, *Improvement of the Delsarte bound for τ-designs when it is not the best bound possible*, Des. Codes Cryptogr. 28 (2) (2003), 201–222.

[379] R. NODA, *On orthogonal arrays of strength* 4 *achieving Rao's bound*, J. Lond. Math. Soc. (2) 19 (1979), no. 3, 385–390.

[380] R. NODA, *On orthogonal arrays of strength* 3 *and* 5 *achieving Rao's bound*, Graphs Comb. 2 (1986), no. 3, 277–282.

[381] R. NODA, *Some bounds for the number of blocks*, Eur. J. Comb. 22 (2001), 91–94.

[382] R. NODA, *Some bounds for the number of blocks II*, Eur. J. Comb. 22 (2001), 95–100.

[383] K. NOMURA and P. TERWILLIGER, *The structure of a tridiagonal pair*, Linear Algebra Appl. 429 (2008), 1647–1662.

[384] K. NOMURA and P. TERWILLIGER, *On the shape of a tridiagonal pair*, Linear Algebra Appl. 432 (2010), 615–636.

[385] H. NOZAKI, *Inside s-inner product sets and Euclidean designs*, Combinatorica 31(6) (2011), 725–737.

[386] H. NOZAKI, *A generalization of Larman-Rogers-Seidel's theorem*, Discrete Math. 311 (2011), 792–799.

[387] H. NOZAKI and M. SAWA, *Remarks on Hilbert identities, isometric embeddings, and invariant cubature*, Algebra Anal. 25 (2013), no. 4, 139–181. Translation in St. Petersburg Math. J. 25 (2014), no. 4, 615–646.

[388] H. NOZAKI and M. SHINOHARA, *On a generalization of distance sets*, J. Comb. Theory, Ser. A 117 (2010), 810–826.

[389] A. M. ODLYZKO and N. J. A. SLOANE, *New bounds on the of numbers of unit spheres that can touch a unit sphere in n dimensions*, J. Comb. Theory, Ser. A 26 (1979), 210–214.

[390] T. OKUDA, *Relation between spherical designs through a Hopf map*, arXiv:1506.08414.

[391] M. OZEKI, *On the notion of Jacobi polynomials for codes*, Math. Proc. Camb. Philos. Soc. 121 (1997), no. 1, 15–30.

[392] C. PACHE, *Shells of selfdual lattices viewed as spherical designs*, Int. J. Algebra Comput. 15 (5–6) (2005), 1085–1127.

[393] A. A. PASCASIO, *On the multiplicities of the primitive idempotents of a Q-polynomial distance regular graph*, Eur. J. Comb. 23 (2002), no 8, 1073–1078.

[394] C. PETERSON, *On tight 6-designs*, Osaka J. Math. 14 (1977), 417–435.

[395] A. JA. PETRENJUK, *On Fisher's inequality for tactical configurations (Russian)*, Mat. Zametki 4 (1968), 417–425.

[396] F. Pfender and G. M. Ziegler, *Kissing numbers, sphere packings, and some unexpected proofs*, Not. Am. Math. Soc. 51 (2004), 873–883.

[397] V. S. Pless and W. C. Huffman, *Handbook of Coding Theory*, North Holland, 1998.

[398] P. Rabau and B. Bajnok, *Bela Bounds for the number of nodes in Chebyshev type quadrature formulas*, J. Approx. Theory 67 (1991), no. 2, 199–214.

[399] D. K. Ray-Chaudhuri and R. M. Wilson, *Solution of Kirkman's schoolgirl problem*, in: *Combinatorics: Proc. Sympos. Pure Math., Vol. XIX, Univ. California, Los Angeles, Calif., 1968*, Amer. Math. Soc., Providence, R.I., 1971, pp. 187–203.

[400] D. K. Ray-Chaudhuri and R. M. Wilson, *On t-designs*, Osaka J. Math. 12 (1975), 737–744.

[401] J. M. Renes, R. Blume-Kohout, A. J. Scott and C. M. Caves, *Symmetric informationally complete quantum measurements*, J. Math. Phys. 45 (2004), 2171–2180.

[402] H. F. H. Reuvers, *Some Non-existence Theorems for Perfect Codes Over Arbitrary Alphabets*, Ph.D. Thesis, Technische Universiteit Eindhoven, 1977.

[403] B. Reznick, *Sums of even powers of real linear forms*, Mem. Am. Math. Soc. 96 (1992) no. 463.

[404] R. M. Robinson, *Arrangement of 24 circles on a sphere*, Math. Ann. 144 (1961), 17–48.

[405] C. Roos, *A note on the existence of perfect constant weight codes*, Discrete Math. 47 (1983), 121–123.

[406] A. Roy, *Bounds for codes and designs in complex subspaces*, J. Algebraic Comb. 31 (2010), 1–32.

[407] A. Roy and A. J. Scott, *Weighted complex projective 2-designs from bases: optimal state determination by orthogonal measurements*, J. Math. Phys. 48 (2007), 072110.

[408] A. Roy and S. Suda, *Complex spherical designs and codes*, J. Comb. Des. 22 (2014), no. 3, 105–148.

[409] B. Runge, *Codes and Siegel modular forms*, Discrete Math. 148 (1996), no. 1–3, 175–204.

[410] E. B. Saff and A. B. J. Kuijlaars, *Distributing many points on a sphere*, Math. Intell. 19 (1997), 5–11.

[411] B. E. Sagan and J. S. Caughman, *The multiplicities of a dual-thin Q-polynomial association schemes*, Electron. J. Comb. 8 (2001), no. 1, Note 4.

[412] M. Sawa, M. Hirao and S. Kageyama, *Euclidean Design Theory*, Springer Briefs in Statics, JSS Research Series in Statics, 2019.

[413] M. Sawa and Y. Xu, *On positive cubature rules on the simplex and isometric embeddings*, Math. Comput. 83 (2014), no. 287, 1251–1277.

[414] A. Schrijver, *New code upper bounds from the Terwilliger algebra and semidefinite programming*, IEEE Trans. Inf. Theory 51 (2005), 2859–2866.

[415] A. Schürmann, *Dense ellipsoid packings*, Discrete Math. 247 (2002), no. 1–3, 243–249.

[416] K. Schütte and B. L. van der Waerden, *Das Problem der dreizehn Zugeln*, Math. Ann. 125 (1953), 325–334.

[417] L. L. Scott, *Some properties of character products*, J. Algebra 45 (1977), no. 2, 259–265.

[418] A. J. Scott, *Tight informationally complete quantum measurements*, J. Phys. A, Math. Gen. 39 (2006), 13507–13530.

[419] A. J. Scott and M. Grassl, *Symmetric informationally complete positive-operator-valued measures: a new computer study*, J. Math. Phys. 51 (2010), 042203.

[420] J. P. Serre, *A Course in Arithmetic*, Graduate Texts in Mathematics, 7, Springer-Verlag, New York-Heidelberg, 1973. Translated from the French.

[421] J. P. Serre, *Sur la lacunarite des puissances de η*, Glasg. Math. J. 27 (1985), 203–221 (French).

[422] P. D. Seymour and T. Zaslavsky, *Averaging sets: a generalization of mean values and spherical designs*, Adv. Math. 52 (3) (1984), 213–240.

[423] H. SHEN and H. Z. WEI, *The construction of PBIB designs from 2-dimensional nonisotropic subspaces in symplectic geometry* (Chinese). An English summary appears in Chin. Ann. Math., Ser. B 6 (1985), no. 4, 494; Chin. Ann. Math., Ser. A 6 (1985), no. 5, 587–594.

[424] G. C. SHEPHARD and J. A. TODD, *Finite unitary reflection groups*, Can. J. Math. 6 (1954), 274–304.

[425] J. SHIGEZUMI, *A construction of spherical designs from finite graphs with the theory of crystal lattice* (2008). arXiv:0804.2956

[426] O. SHIMABUKURO, *On the nonexistence of perfect codes in* $J(2w + p^2, w)$, Ars Comb. 75 (2005), 129–134.

[427] P. W. SHOR and N. J. A. SLOANE, *A family of optimal packings in Grassmannian manifolds*, J. Algebraic Comb. 7 (1998), no. 2, 157–163.

[428] V. M. SIDELNIKOV, *Spherical 7-designs in* 2^n*-dimensional Euclidean space*, J. Algebraic Comb. 10 (1999), 279–288.

[429] V. M. SIDELNIKOV, *Orbital spherical 11-designs in which the initial point is a root of an invariant polynomial*, Algebra Anal. 11 (1999), no. 4, 183–203 (Russian). Translation in St. Petersburg Math. J. 11 (2000), no. 4, 673–686.

[430] S. SMALE, *Mathematical problems for the next century*, Math. Intell. 20 (1998), 7–15.

[431] D. H. SMITH, *Primitive and imprimitive graphs*, Q. J. Math. (2) 22 (1971), 551–557.

[432] D. H. SMITH, *An improved version of Lloyd's theorem*, Discrete Math. 15 (1976), 175–184.

[433] S. Y. SONG and H. TANAKA, *Group-case commutative association schemes and their character tables*, in: *Proceedings, Algebraic Combinatorics, An International Conference in Honor of Eiichi Bannai's 60th Birthday, Sendai, Japan*, 2006, pp. 204–213.

[434] R. P. STANLEY, *Invariants of finite groups and their applications to combinatorics*, Bull., New Ser., Am. Math. Soc. 1 (1979), no. 3, 475–511.

[435] D. STANTON, *Some q-Krawtchouk polynomials on Chevalley groups*, Am. J. Math. 102 (1980), no. 4, 625–662.

[436] D. STANTON, *Three addition theorems for some q-Krawtchouk polynomials*, Geom. Dedic. 10 (1981), no. 1–4, 403–425.

[437] D. STANTON, *Orthogonal polynomials and Chevalley groups*, in: *Special Functions: Group Theoretical Aspects and Applications*, Math. Appl. 18, Dordrecht, 1984, pp. 87–128.

[438] D. STANTON, *t-Designs in classical association schemes*, Graphs Comb. 2 (1986), no. 3, 283–286.

[439] D. STEHLÉ and M. WATKINS, *On the extremality of an 80-dimensional lattice*, in: *ANTS2010: Algorithmic Number Theory* (ed. by G. HANROT, F. MORAIN and E. THOMÉ), Lecture Notes in Computer Science 6197, Springer, Berlin, Heidelberg, 2010, pp. 340–356.

[440] R. J. STROEKER, *On the Diophantine equation* $(2y^2 - 3)^2 = x^2(3x^2 - 2)$ *in connection with the existence of nontrivial tight 4-designs*, Proc. K. Ned. Akad. Wet., Ser. A, Indag. Math. 43 (1981), no. 3, 353–358.

[441] S. SUDA, *Coherent configurations and triply regular association schemes obtained from spherical designs*, J. Comb. Theory, Ser. A 117 (2010), no. 8, 1178–1194.

[442] T. SUNADA, *Why do Diamonds Look so Beautiful?* Springer-Verlag Tokyo, 2006 (Japanese).

[443] T. SUNADA, *Crystals that nature might miss creating*, Not. Am. Math. Soc. 55 (2008), 208–215.

[444] H. SUZUKI, *Association schemes with multiple Q-polynomial structures*, J. Algebraic Comb. 7 (1998), 181–196.

[445] G. SZEGÖ, *Orthogonal Polynomials*, Amer. Math. Soc. Colloq. Publ. 23 (2003) (first edition 1939).

[446] T. TAKAGI, *Sugaku Shokei (Small Landscape of Mathematics)* Iwanami shoten, 1943 (Japanese).

[447] P. M. L. TAMMES, *On the origin of number and arrangement of places of exit on the surface of pollen grains*, Recl. Trav. Bot. Néerl. 27 (1930), 1–84.

[448] K. Tanabe, *The irreducible modules of the Terwilliger algebras of Doob schemes*, J. Algebraic Comb. 6 (1997), no. 2, 173–195.

[449] K. Tanabe, *A new proof of the Assmus-Mattson theorem for non-binary codes*, Des. Codes Cryptogr. 22 (2001), 149–155.

[450] H. Tanaka, *Classification of subsets with minimal width and dual width in Grassmann, bilinear forms and dual polar graphs*, J. Comb. Theory, Ser. A 113 (2006), no. 5, 903–910.

[451] H. Tanaka, *New proofs of the Assmus-Mattson theorem based on the Terwilliger algebra*, Eur. J. Comb. 30 (2009), 736–746.

[452] H. Tanaka, *Vertex subsets with minimal width and dual width in Q-polynomial distance-regular graphs*, Electron. J. Comb. 18, (2011), no. 1, P167.

[453] H. Tanaka, *The Erdős-Ko-Rado theorem for twisted Grassmann graphs*, Combinatorica 32 (2012), no. 6, 735–740.

[454] H. Tanaka, *The Erdős-Ko-Rado basis for a Leonard system*, Contrib. Discrete Math. 8 (2013), no. 2, 41–59.

[455] H. Tanaka and R. Tanaka, *Nonexistence of exceptional imprimitive Q-polynomial association schemes with six classes*, Eur. J. Comb. 32 (2011), 155–161.

[456] G. Tarry, *Lé probléme des 36 Officiers*, C. R. Assem. Franc. Pour Avacement Sci. 2 (1901), 170–203.

[457] L. Teirlinck, *Nontrivial t-designs without repeated blocks exist for all t*, Discrete Math. 65 (1987), no. 3, 301–311.

[458] P. Terwilliger, *The diameter of bipartite distance-regular graphs*, J. Comb. Theory, Ser. B 32 (1982), no. 2, 182–188.

[459] P. Terwilliger, *Distance-regular graphs with girth 3 or 4. I*, J. Comb. Theory, Ser. B 39 (1985), no. 3, 265–281.

[460] P. Terwilliger, *The Johnson graph $J(d,r)$ is unique if $(d,r) \neq (2,8)$*, Discrete Math. 58 (1986), no. 2, 175–189.

[461] P. Terwilliger, *A new feasibility condition for distance-regular graphs*, Discrete Math. 61 (1986), no. 2–3, 311–315.

[462] P. Terwilliger, *A class of distance-regular graphs that are Q-polynomial*, J. Comb. Theory, Ser. B 40 (1986), no. 2, 213–223.

[463] P. Terwilliger, *P and Q polynomial schemes with $q = -1$*, J. Comb. Theory, Ser. B 42 (1987), no. 1, 64–67.

[464] P. Terwilliger, *A characterization of P- and Q-polynomial association schemes*, J. Comb. Theory, Ser. A 45 (1987), no. 1, 8–26.

[465] P. Terwilliger, *Root systems and the Johnson and Hamming graphs*, Eur. J. Comb. 8 (1987), no. 1, 73–102.

[466] P. Terwilliger, *The classification of distance-regular graphs of type IIB*, Combinatorica 8 (1988), no. 1, 125–132.

[467] P. Terwilliger, *Balanced sets and Q-polynomial association schemes*, Graphs Comb. 4 (1988), no. 1, 87–94.

[468] P. Terwilliger, *The subconstituent algebra of an association scheme I*, J. Algebraic Comb. 1 (1992), 363–388.

[469] P. Terwilliger, *The subconstituent algebra of an association scheme II*, J. Algebraic Comb. 2 (1993), 73–103.

[470] P. Terwilliger, *The subconstituent algebra of an association scheme III*, J. Algebraic Comb. 2 (1993), 177–210.

[471] P. Terwilliger, *Kite-free distance-regular graphs*, Eur. J. Comb. 16 (1995), no. 4, 405–414.

[472] P. Terwilliger, *A new inequality for distance-regular graphs*, Discrete Math. 137 (1995), no. 1–3, 319–332.

[473] P. Terwilliger, *Quantum matroid*, in: *Progress in Algebraic Combinatorics*, Adv. Stud. Pure Math. 24, Math. Soc. Japan, Tokyo, 1996, pp. 323–441.

[474] P. Terwilliger, *Two relations that generalize the q-Serre relations and the Dolan-Grady relations*, in: *Physics and Combinatorics 1999*, Proceedings of the International Workshop, Nagoya, 1999 (ed. by A. N. Kirilov et al.), World Scientific, River Edge, NJ, 2001, pp. 377–398.

[475] P. Terwilliger, *Two linear transformations each tridiagonal with respect to an eigenbasis of the other*, Linear Algebra Appl. 330 (2001), no. 1–3, 149–203.

[476] P. Terwilliger, *The subconstituent algebra of a distance-regular graph; thin modules with endpoint one*, in: *Special Issue on Algebraic Graph Theory (Edinburgh, 2001)*, Linear Algebra Appl. 356 (2002), 157–187.

[477] P. Terwilliger, *An inequality involving the local eigenvalues of a distance-regular graph*, J. Algebraic Comb. 19 (2004), no. 2, 143–172.

[478] P. Terwilliger, *The displacement and split decompositions for a Q-polynomial distance-regular graph*, Graphs Comb. 21 (2005), no. 2, 263–276.

[479] P. Terwilliger, *An algebraic approach to the Askey scheme of orthogonal polynomials*, in: *Orthogonal Polynomials and Special Functions*, Lecture Notes in Math. 1883, Springer, Berlin, 2006, pp. 255–330.

[480] P. Terwilliger and R. Vidunas, *Leonard pairs and the Askey-Wilson relations*, J. Algebra Appl. 3 (2004), no. 4, 411–426.

[481] P. H. Tiep, *Finite groups admitting grassmannian 4-designs*, J. Algebra 306 (1) (2006), 227–243.

[482] A. Tietäväinen, *On the nonexistence of perfect codes over finite fields*, SIAM J. Appl. Math. 24 (1973), no. 1, 88–96.

[483] A. Tietäväinen, *Nonexistence of nontrivial perfect codes in case $q = p_1^r p_2^s, e \geq 3$*, Discrete Math. 17 (1977), 199–205.

[484] A. Tietäväinen and A. Perko, *There are no un-known perfect binary codes*, Ann. Univ. Turku. Ser. A 148 (1971), 3–10.

[485] M. A. Tsfasman, S. G. Vlădut, *Algebraic-Geometric Codes* Mathematics and Its Applications (Soviet Series) 58, Kluwer Academic Publishers Group, Dordrecht, 1991 (Translated from the Russian by the authors).

[486] M. A. Tsfasman, S. G. Vlădut and Th. Zink, *Modular curves, Shimura curves, and Goppa codes, better than Varshamov-Gilbert bound*, Math. Nachr. 109 (1982), 21–28.

[487] S. Tsujimoto, L. Vinet and A. Zhedanov, *Jordan algebras and orthogonal polynomials*, J. Math. Phys. 52 (2011), 103512.

[488] S. Tsujimoto, L. Vinet and A. Zhedanov, *From slq(2) to a parabosonic Hopf algebra*, SIGMA 7 (2011), 093.

[489] S. Tsujimoto, L. Vinet and A. Zhedanov, *Dunkl shift operators and Bannai-Ito polynomials*, Adv. Math. 229 (2012), 2123–2158.

[490] W. T. Tutte, *A family of cubical graphs*, Proc. Camb. Philos. Soc. 43 (1947), 459–474.

[491] V. A. Ustimenko, On some properties of the geometries of the Chevalley groups and their generalizations, in *Studies in Algebraic Theory of Combinatorial Objects, Moscow*, 1986 (in Russian).

[492] O. Veblen and J. W. Young, *Projective Geometry, vol. 1 and vol. 2*, Blaisdell Publishing Co. Ginn and Co., New York-Toronto-London, 1965.

[493] B. B. Venkov, *Even unimodular extremal lattices*, Tr. Mat. Inst. Steklova 165 (1984), 43–48.

[494] B. Venkov, *Réseaux et designs sphériques, Réseaux euclidiens, designs sphériques et formes modulaires*, Monogr. Enseign. Math. 37, Enseignement Math., Geneva (2001), 10–86.

[495] P. Verlinden and R. Cools, *On cubature formulae of degree $4k + 1$ attaining Möller's lower bound for integrals with circular symmetry*, Numer. Math. 61 (1992), 395–407.

[496] M. Viazovska, *The sphere packing problem in dimension* 8, Ann. Math. 185 (2017), no. 3 991–1015.

[497] N. Victoir, *Asymmetric cubature formulae with few points in high dimension for symmetric measures*, SIAM J. Numer. Anal. 42 (2004), no. 1, 209–227.

[498] L. Vretare, *Formulas for elementary spherical functions and generalized Jacobi polynomials*, SIAM J. Math. Anal. 15 (1984), no. 4, 805–833.

[499] G. Wagner, *On averaging sets*, Monatshefte Math. 111 (1991), no. 1, 69–78.

[500] S. F. D. Waldron, *An Introduction to Finite Tight Frames*, Birkhäuser (2018).

[501] Z. -X. Wan, *Geometry of Matrices*, World Scientific, Singapore, 1996.

[502] Z. -X. Wan, *Geometry of Classical Groups over Finite Fields*, Science Press, Beijing, New York, 2002.

[503] H.-C. Wang, *Two-point homogeneous spaces*, Ann. Math. (2) 55 (1952), 177–191.

[504] Y. Wang, Y. Huo, C. Ma, *Association Schemes of Matrices*, Science Press, Beijing China (2006) (English translation by J. Ma (2010)).

[505] K. Wang, F. Li, J. Guo and J. Ma, *Association schemes coming from minimal flats in classical polar spaces*, Linear Algebra Appl. 435 (2011), no. 1, 163–174.

[506] Y. Wang and J. Ma, *Association schemes of symmetric matrices over a finite field of characteristic two*, in: *Shanghai conference issue on designs, codes, and finite geometries, part 2 (Shanghai, 1993)*, J. Stat. Plan. Inference 51 (1996), no. 3, 351–371.

[507] Y. Wang, C. Wang, C. Ma and J. Ma, *Association schemes of quadratic forms and symmetric bilinear forms*, J. Algebraic Comb. 17 (2003), no. 2, 149–161.

[508] H. N. Ward, *Quadratic residue codes and symplectic groups*, J. Algebra 29 (1974), 150–171.

[509] M. Watkins, *Another* 80*-dimensional extremal lattice*, J. Théor. Nr. Bordx. 24 (2012), no. 1, 237–255.

[510] H. Z. Wei and Y. Wang, *Enumeration theorems for finite singular unitary geometries and the construction of PBIB designs*, Acta Math. Sci. 15 (1995), no. 2, 228–240 (Chinese).

[511] H. Z. Wei and Y. X. Wang, *Suborbits of the set of m-dimensional totally isotropic subspaces under actions of pseudosymplectic groups over finite fields of characteristic* 2, Acta Math. Sin. 38 (1995), no. 5, 696–707 (Chinese).

[512] H. Wei and Y. Wang, *Suborbits of the transitive set of subspaces of type $(m, 0)$ under finite classical groups*, Algebra Colloq. 3 (1996), no. 1, 73–84.

[513] R. Weiss, *On distance-transitive graphs*, Bull. Lond. Math. Soc. 17 (1985), no. 3, 253–256.

[514] R. M. Wilson and Ray-Chaudhuri, *Generalization to Fisher's inequality to t-designs*, Not. Am. Math. Soc. 18 (1971), 805.

[515] R. M. Wilson, *An existence theory for pairwise balanced designs. I. Composition theorems and morphisms*, J. Comb. Theory, Ser. A 13 (1972), 220–245.

[516] R. M. Wilson, *An existence theory for pairwise balanced designs. II. The structure of PBD-closed sets and the existence conjectures*, J. Comb. Theory, Ser. A 13 (1972), 246–273.

[517] R. M. Wilson, *The necessary conditions for t-designs are sufficient for something*, Util. Math. 4 (1973), 207–215.

[518] R. M. Wilson, *An existence theory for pairwise balanced designs. III. Proof of the existence conjectures*, J. Comb. Theory, Ser. A 18 (1975), 71–79.

[519] J. A. Wolf, *Spaces of Constant Curvature*, 5th ed. University of California, Berkeley, CA (1984) (First edition: MacGraw-Hill Book Co., New York-London-Sydney 1967).

[520] W. K. Wootters, *Quantum measurements and finite geometry*, in: *Special issue of invited papers dedicated to Asher Peres on the occasion of his seventieth birthday*, Found. Phys. 36 (2006), no. 1, 112–126.

[521] Z. Xiang, *Fisher type inequality for regular t-wise balanced design*, J. Comb. Theory, Ser. A 119 (2012), 1523–1527.

[522] Z. Xiang, *Nonexistence of nontrivial tight 8-designs*, J. Algebraic Comb. 47 (2018), 301–318.
[523] K. Yamamoto, *Combinatorics*, Asakura shoten, 1989 (Japanese).
[524] W.-H. Yu, *New bounds for equiangular lines and spherical two-distance sets*, SIAM J. Discrete Math. 31 (2017), no. 2, 908–917.
[525] V. A. Yudin, *Minimum potential energy of a point system of charges*, Diskretn. Mat. 4 (1992), 115–121 (Russian). Translation in Discrete Math. Appl. 3 (1993), 75–81.
[526] V. A. Yudin, *Lower bounds for spherical designs*, Izv. Ross. Akad. Nauk Ser. Mat. 61 (3) (1997), 211–233. Translation in Izv. Math. 61 (1997), 673–683.
[527] S. Zhang, *On the nonexistence of extremal self-dual codes*, Discrete Appl. Math. 91 (1999), 277–286.
[528] Y. Zhu, Ei. Bannai and Et. Bannai, *Tight relative 2-designs on two shells in Johnson association schemes*, Discrete Math. 339 (2016), 957–973.

Index

De Gruyter Series in Discrete Mathematics and Applications

Volume 4
Zoran Stanić
Regular Graphs. A Spectral Approach, 2017
ISBN 978-3-11-035128-6, e-ISBN (PDF) 978-3-11-035134-7,
e-ISBN (EPUB) 978-3-11-038336-2

Volume 3
Christos Koukouvinos, Dimitros E. Simos
Combinatorial Designs. With Applications to Coding Theory and Cryptography
ISBN 978-3-11-029408-8, e-ISBN (PDF) 978-3-11-029434-7,
e-ISBN (EPUB) 978-3-11-038192-4

Volume 2
Nikos Tzanakis
Elliptic Diophantine Equations. A Concrete Approach via the Elliptic Logarithm, 2013
ISBN 978-3-11-028091-3, e-ISBN 978-3-11-028114-9

Volume 1
Petrică C. Pop
Generalized Network Design Problems. Modeling and Optimization, 2012
ISBN 978-3-11-026758-7, e-ISBN 978-3-11-026768-6